Vol. 3
Human Reflexes, Pathophysiology of Motor Systems,
Methodology of Human Reflexes

New Developments in Electromyography and Clinical Neurophysiology

Vol. 3

S. Karger · Basel · München · Paris · London · New York · Sydney

Human Reflexes
Pathophysiology of Motor Systems
Methodology of Human Reflexes

Editor: JOHN E. DESMEDT, M.D.
Professor of Neurophysiology and of Pathophysiology of the Nervous System, University of Brussels

359 figures, 15 tables

S. Karger · Basel · München · Paris · London · New York · Sydney   1973

New Developments in Electromyography
and Clinical Neurophysiology

Complete set in a box (Vol. 1–3):
XXVII + 2094 p., 869 fig., 82 tab., 4 cpl., 1973
ISBN 3–8055–1409–3

Vol. 1: New Concepts of the Motor Unit, Neuromuscular Disorders, Electromyographic
Kinesiology.
XII + 710 p., 286 fig., 38 tab., 4 cpl., 1973
ISBN 3–8055–1451–4

Vol. 2: Pathological Conduction in Nerve Fibers, Electromyography of Sphincter Muscles,
Automatic Analysis of Electromyogram with Computers.
VII + 514 p., 224 fig., 29 tab., 1973
ISBN 3–8055–1452–2

Vol. 3: Human Reflexes, Pathophysiology of Motor Systems, Methodology of Human
Reflexes.
VIII + 870 p., 359 fig., 15 tab., 1973
ISBN 3–8055–1453–0

S. Karger · Basel · München · Paris · London · New York · Sydney
Arnold-Böcklin-Strasse 25, CH-4011 Basel (Switzerland)

Contents

Clinical Criteria of Motor Disorders

Anatomical and Physiological Background

Contents

Motor Unit Discharge and Proprioceptive Input

Proprioceptive Reflexes and the H-Reflex

Servo-Control in Coordination of Movement

Reflex Effects of Muscle Vibration

Pathophysiology of Spasticity

Unloading Reflexes and the Silent Period

Blink Reflexes

Exteroceptive, Interoceptive and Long-Loop Reflexes

Methodology of the Triceps Surae Proprioceptive Reflexes

Clinical Criteria of Motor Disorders

New Developments in Electromyography and Clinical Neurophysiology, edited by J. E. Desmedt, vol. 3, pp. 1–10 (Karger, Basel 1973)

Muscular Hypertonia – The Clinical Viewpoint

R. D. ADAMS

Kennedy and Shriver Laboratories, Massachusetts General Hospital, Fernald State School and Department of Neuropathology-Neurology, Harvard Medical School, Boston, Mass.

The clinical examination of patients with disordered motor function often discloses alterations in the tonus of the musculature, either in the form of an excess (hypertonia) or diminution (hypotonia). Abnormality of the tonus has thus come to stand, along with paralysis and involuntary movements, as an important sign of nervous disease.

To specify different types and degrees of abnormality of tone clinicians have introduced a rather complex and confusing nomenclature. SIGWALD and RAVERDY [1969] remind us that MULLER in 1838 was the first to use the term 'tonus' to refer to any state of contraction of resting muscle, and that VULPIAN in 1866 widened the definition further by including all states of 'permanent muscular tension'. The words 'spasticity' and 'rigidity' have come to designate distinct forms of hypertonus, largely now on the basis of the clinical setting in which they are found. Dystonia, another clinical term, has been used with the greatest inconsistency to denominate such diverse states as fixed deformities of trunk and proximal limbs consequent upon abnormal distributions of excess tone, any enduring attitude related to heightened tone, or athetosis of trunk and proximal limb muscles. The term 'decerebrate rigidity' applies strictly to particular combinations of posture, i. e. transient extensor hypertonus and cerebral deficit seen first in the laboratory animal, and paratonia to improper relaxation usually in obvious relation to cerebral disease. 'Tonic innervation' and intention spasm are terms which specify disorders of action in which volitional movement initiates and is followed by an excess of muscular tone.

From the clinician's viewpoint the investigation of hypertonus involves the study of muscular tonus during all states of rest and activity, more particularly, during relaxation, during the maintenance of certain postures and attitudes and during and following voluntary movement. It becomes evident, when approached in this manner, that some forms of hypertonus are observed to occur only under fairly precise conditions of rest and activity whereas others are more pervasive and present at all times in the waking organism, and occasionally even in sleep.

The traditional way of studying hypertonia clinically is to examine in detail a series of patients who exhibit one of the standard syndromes of which hypertonus is a regular part. The reason for organizing the discussion along these lines is that to write of such conditions as spasticity or rigidity as abstract disturbances of the physiology of the nervous system and to relate them to phenomena observed in the 'lesioned' experimental animal presupposes a degree of knowledge of clinical physiology which the present writer does not possess.

A. Hypertonus in the Patient with Nonparalytic Cerebral Disease: Paratonia or Gegenhalten

Probably this term comprehends the least and most subtle forms of hypertonia. In mildest degree it is represented by an uneven opposition of the musculature of the patient's limb to passive manipulation. Then it verges on the normal, for individuals are known to vary in their capacity to relax upon command. Notably, the patient with paratonia usually shows no prevailing attitude or posture. When lying quietly in bed the muscles are visibly and palpably slack; the patient seems only inadvertently to interfere with the examiner's attempts to flex and extend his limbs. Contact and holding seem to excite the resistance. When more exaggerated, as in disease states and ageing, the resistance assumes a more plastic character. Paucity of movement, reflected in both action and speech, are then frequent accompaniments, and often inconsistent grasp reactions to suitable palmar stimuli and even 'magnet' and groping reactions are added. The irregular resistance to passive flexion and extension of limbs is not attended by consistent enhancement of stretch (myotatic) reflexes. Volitional actions when made are appropriately quick, facile and strong but often there is a curious impersistence of commanded action. Extemes of this state verge on rigid and other nonparalytic forms of akinetic mutism and parkinsonian rigidity, often with a loss of ability to maintain upright stance and to walk. The latter, called cerebral paraplegia in flexion, is the final stage. Exceptionally, avoidance movements are evoked by contact; excesses of tone, outlasting the stimulus, occur in relationship to such activities.

Viewed physiologically an accompanying hypo- or akinesia is present, often so subtle as to be overlooked or dismissed as the consequence of a reduced level of alertness and attention. In Jacksonian terms the motor system is decompensated with the slightest degree of release of certain automatic actions which are conditioned by contact and the passive movement itself. However, it may be that the deficit is an inability to relax or command, an active process impaired by cerebral disease. DENNY-BROWN [1950] conceives of the state as one in which a cortical automatism becomes manifest as a consequence of a lesion in one part of the cerebrum. For example the frontal lesion liberates the tactile and proprioceptive mechanisms of the sensorimotor regions; a parietal lesion releases high-level avoidance reactions to contact. Seemingly compulsive activities evoked by visual stimulation have a similar basis in removal of temporo-occipital control of volitional movement.

Although not usually classified with the other more classic disorders of tone, nevertheless the paratonias, of which there are obviously more than one type and mechanism, are amongst the most frequent symptoms of disordered motor function and the ones most difficult to couch in physiological terms.

B. The Hemiplegic Patient with a Cerebral Lesion: Spasticity

Here the hypertonus is invariably combined with a profound weakness of extensors of upper extremity and flexors of the lower. The hypertonus predominates in the opposite muscle groups. Thus one of the principal attributes of the hypertonus called spasticity is its predilection for certain groups of muscles. Another feature is its easy activation and enhancement by passive movement, and a manifest overactivity of spinal reflexes and automatisms. Hyperactivity of phasic, as well as myotatic (stretch) reflexes and exaggerated reflexes of defense of which the Babinski is a part, are so constant as to serve as clinical signs, though the latter is by no means invariable.

Spasticity which has been more thoroughly studied in the laboratory than any other form of hypertonus is interpretable in terms of segmental tonic and phasic reflexes, giving rise to lengthening and shortening reactions, etc. In man persistence of the posture of flexion of the arm and extension of the leg, the predilection pattern of hemiplegia, has been termed hemiplegic dystonia by DENNY-BROWN; French neurologists have referred to this fixed state as 'contracture', a term which has another meaning in physiology and in English neurology.

Spastic limbs when inactive reveal no increase in turgor and are normally relaxed. In purest form the spasticity is evoked only during passive and active movements. For its elicitation a certain speed and angle of movement are required and it is under these precise conditions that the first few mm of passive movement are noted to encounter no resistance (a free interval); then follows a quick relatively unyielding catch which gradually melts away (the clasp-knife phenomenon) as the passive movement continues. Such changes are obviously based on an abnormal facility of spinal reflexes. An interesting feature of the spasticity is that in acute disease it does not appear immediately but only some weeks after a state of flaccid paralysis. This latter and the associated hyporeflexia, i.e. 'spinal shock' rarely precedes spasticity in cerebral hemiplegia. Exceptionally the spasticity and paralysis develop together, or the paralyzed limbs remain flaccid.

Not all hemiplegias due to 'capsular' or cerebral white matter lesions present a hypertonus that corresponds exactly to the type called spastic. Not infrequently the paralyzed limbs, while maintaining the hemiplegic posture of flexed arm and extended leg, exhibit a more plastic type of rigidity, as described below. Or, the arm flexors and leg extensors may be spastic while the antagonist muscles are rigid. When rigidity is present there is usually a fair amount of volitional movement. The tendon reflexes are hyperactive and the flexor reflexes also but to a lesser degree. Cerebral spastic hemiplegias may remain in a stable state for months or years, then dominating all other types of movement disorder such as parkinsonian tremor, chorea or cerebellar ataxia. The position of the body, its contact with a surface as in the right or left lateral decubitus position, suspension of the body with face down, and turning or tilting of the head are reported to modify the tone and posture of the affected limbs but these effects are inconsistent and difficult to study. Spastic hemiplegia may, during partial recovery, evolve into a hemichorea, hemiathetosis or hemitremor. As this happens stretch reflexes become less conspicuously overactive, flexor reflexes are more difficult to evoke and finally disappear, and the so-called hemiplegic posture is variably altered. With more complete recovery of volitional power the spastic hypertonia disappears, a fact which emphasizes the close relationship of paralysis and spasticity in the human.

Clinical comparison in man of a hemiplegia resulting from a lesion in the cerebral white matter or internal capsule with one located in the cerebral peduncle or basis pontis has been relatively uninformative. In spite of current theory from which differences would be anticipated, the clinical neurologist cannot localize the level of the lesion from the type of posture and the pattern of tonus and paralysis alone. Even a lesion as low as the pontomedullary junction has resulted in a typical spastic hemiplegia.

C. Spinal Hemiplegia, Quadriplegia, and Paraplegia: Spasticity, Involuntary Spasms

The opportunity to study hypertonus due to spinal cord lesions in man most often is afforded by cervical trauma, tumors and vascular lesions. In acute lesions, unlike capsular hemiplegia, the spinal reflexes of the leg are abolished – the state of 'pyramidal spinal shock' supervenes. Some weeks later, or from the onset in gradually developing spinal cord diseases, the paralysis of voluntary movement of legs is accompanied by hypertonus which appears first and is maximal in the adductor and extensor muscles of the thighs. When it is fully developed, the relaxed, unstimulated leg muscles are soft and electromyographically silent, though at times a summation of exteroceptive and proprioceptive stimuli evokes what seem to be spontaneous spasms. Passive movements of flexion of knee and abduction encounter a reflex hypertonus similar in pattern to that of the hemiplegic patient; it is manifestly based on the 'uncomplicated increase of stretch reflexes' which are believed to characterize all spastic states. With the most severe transverse lesions of the spinal cord which usually lead to paraplegia in flexion, the flexor reflexes appear to have overcome in large measure the stretch reflexes. Indeed the latter may be difficult to elicit. With flexor spasms of the legs a more widespread discharge occurs which empties the bladder and elicits vasomotor, pilomotor and sweating reactions. These tend to disappear with time and other reflex increases in tone are observed: biphasic responses of flexion then extension or the contralateral extension of the opposite leg or even abortive stepping movements. A series of pin-pricks is the most effective stimulus in eliciting spasms. Thus, the resistance of the legs to passive extension in the state of paraplegia in flexion is clearly dependent on some other type of reflex activity resulting from exteroceptive as well as proprioceptive stimuli. As in cerebral hemiplegias, with the recovery of voluntary power spasticity disappears. For the reasons given, the clinician cannot accept the theory that the reflex hypertonus in spinal man is reducible simply to overactivity of phasic myotatic reflexes. In the less severe spinal cord lesions, extensor postures, extensor spasms and hyperactivity of the phasic myotatic reflexes are preponderant and flexor reflexes, though usually elicitable, are less active. The legs tend to remain in extension and to adduct when the body is supine and in walking. When the patient is placed in the lateral decubitus position the legs may flex at knee and hip, a change thought to be the expression of body righting reflexes. Turning and tilting of the head to evoke cervical and labyrinthine righting reflexes do not alter the hypertonus and posture. Another notable feature of the spinal spastic paralyses is that the hypertonus varies considerably from time to time. Once a widespread spasm is induced by stimulation or by the patient's own effort to move, it may persist long after the stimulation or volitional effort ceases. Such hypertonus retains few of the characteristics of the clasp-knife type of spasticity.

When volitional movement is analyzed in greater detail in both cerebral hemiplegia and spinal paraplegia it is evident that it is slow and fatigable, and the patient's capacity for discrete movement of a group of muscles may be lost even though the same muscles may still be engaged in more general willed activity. Attempts by the patient to move one part of a hand or foot often result in the activation of other more proximal muscles not needed in the movement, a phenomenon seen to a lesser degree in normal individuals. Also, volitional movement of the healthy limb, when only one side is affected, may evoke imitative movements in the spastic one. These pecularities of residual volitional activity have attracted much attention, especially in French neurology, having been described by PIERRE MARIE and CHARLES FOIX [FOIX, 1922] as synkinesias.

According to Jacksonian-Sherringtonian theory the hypertonia is interpreted as a manifestation of excessively facile (released) labyrinthine, contact neck righting and body contact reflexes. Indeed, the effects of such conditions as body position, head and neck movement or position do modify the tone in some instances, but in my experience this is exceptional. The typical 'clasp-knife phenomenon' of catch, maximum resistance and yielding are not difficult to demonstrate in capsular hemiplegias and spinal quadriplegias or paraplegias. But it is the spinal lesion that leads to the most extreme forms of spasticity and at times in spinal cord disease it so vastly exceeds paresis of voluntary movement as to persuade the clinician that these two states depend on closely related but separate mechanisms.

D. Parkinsonian Syndrome: Rigidity

This is more definitely a clinical condition for which perfect laboratory models have not been found. Its characteristics are a constant, even resistance to passive movement (without the clasp-knife phenomenon) disposed in all muscle groups and persisting even during inactivity. The patient's own action seems not to be necessary to evoke the hypertonia, and abolition of it by efforts to relax are relatively unsuccessful. Not only is resting tonus increased but it persists as well during the maintenance of all postures and support. When the limb has been passively flexed the shortened muscle is at once hypertonic in its new state of contraction, a phenomenon upon which much emphasis was placed by FOIX and THEVENARD [1923] and FOERSTER [1921].

The play of volitional and automatic movement is simultaneously reduced and the limbs appear less readily engaged in action than are the normal ones. This latter quality, called hypokinesia, is most evident when only one side of the body is affected and all natural activities are carried out with the normal side. Such movements as are made are slow and may be impeded by tremor but at no time is the impairment of voluntary movement in such obvious relationship to the hypertonus as in the 'spastic' states. Indeed, stereotaxic surgery may abolish rigidity while leaving the hypokinesia; and also paradoxical kinesia, wherein excellent volitional movement may be made briefly by an almost completely inert patient, is another indication of this same feature. Other relatively automatic or even reflex activities such as blinking are also diminished. Attitudinal flexion dystonia of the entire body probably reflects a slightly greater tone in all flexor than in extensor muscles and it imparts to the parkinsonian patient his characteristic stance.

In contrast to spasticity the tendon reflexes are not consistently altered. Nor is there obvious release of the other spinal reflexes and automatisms, such as the reflexes of defense, though abnormalities of flexor reflexes such as reduced habituation, etc. may be demonstrated in the laboratory [SHAHANI and YOUNG, 1971]. The resistance to passive movement is usually immediate, without free interval, and is fairly even except when interrupted by coexistent tremor which rhythmically interrupts the rigidity – the Negri sign. It is the combination of slowness and difficulty in initiating movement, tremor during attitudes of repose and plastic rigidity that constitutes the myostatic syndrome of Parkinson. This characterizes the disease known as paralysis agitans but appears also as a sequel to encephalitis lethargica and certain other degenerative diseases such as Hallevorden-Spatz disease, Wilson's disease, etc.

Persistent rigidity of muscle from central causes is plainly based on enhanced activity of spinal and brain-stem motoneurons but the mechanisms have eluded physiologists. Theories that explain either rigidity or spasticity on the basis of overactivity of the γ motoneurons are open to criticism, as pointed out by LANDAU [1967]. The detailed anatomy of the lesions has also remained largely unsettled because of imprecise neuropathologic techniques.

E. Decerebrate Patients

Our concepts of this condition, which was produced for the first time in the physiology laboratory by transecting the brain stem between the superior colliculi and lower pons, have been transferred rather unsuccessfully to the bedside. Unlike spastic paralysis it is observed to come on immediately after the receipt of injury (no intervening spinal shock), and is associated with release of lower brain-stem reticulospinal and vestibular mechanisms.

The situation in man remains confusing. Decerebrate rigidity, meaning extension and pronation of arms, extension of legs on one or both sides and dorsiflexion of neck, is observable in man. Most often the patient is at the same time stuporous or comatose, but this need not be so; we have seen patients fully conscious with certain degrees of it. Unlike the rigidity described under C and D above, the phasic myostatic reflexes are hyperactive and typical spasticity may be present. Often painful stimulation of face, neck, or arms is required for its evocation. The hyperextension and pronation of limbs are manifestly an expression of excessive segmental reflex activity.

Persistent decerebrate rigidity is most likely to appear with respiratory embarrassment, as was pointed out by LANDAU [1967], and has a grave prognosis, whereas fragmentary states of evoked decerebrate posturing do not.

F. Patients with Hypertonia in States of Dystonia and Athetosis (Intention Spasms)

Here the intermittency of hypertonus is even more noteworthy than in spinal diseases. When relaxed, the muscles may be soft or are maintained in contraction depending on the nature of the neurological syndrome. In some cases of congenital athetosis there is constant hypertonus at rest and during the maintenance of posture with the same shortening reactions and difficulties of movement as in parkinsonian rigidity; tendon reflexes may be exaggerated or remain at an average level. In other cases of so-called birth injury or in early cases of dystonia musculorum deformans the hypertonus occurs unpredictably during an automatic movement sequence such as walking, or during and following a willed movement. In the latter instance a synkinesia occurs in cooperating muscles or in all the muscles in a limb, agonist and antagonist alike; and it persists for many seconds after the movement should have ceased. Naturally, the intended willed movement is hampered. Such 'tension states' or 'intention spasms' probably reflect the same physiological disorder as underlies some of the synkinesias mentioned above. Persistence of a particular willed movement such as grasping may occur, even after its purpose has been accomplished, a phenomenon that has been called 'tonic innervation', to separate it from reflex grasping which is different. This form of hypertonus may merge with involuntary spasms such as are seen in spasmodic torticollis or retrocollis.

G. Hypertonia of Other Types

The above states do not exhaust the list of muscular hypertonias seen in the clinic. Tetany, tetanus, myotonia, myokymia, physiological contracture, spinal hypertonia and the 'stiff-man syndrome' should be included, if only for the sake of completeness. They involve a number of rather different lower motoneuron, peripheral nerve and muscular mechanisms and will not be discussed here.

In conclusion, it may be restated that the human organism in conditions of disease exhibits a wide variety of hypertonias each of which reveals something of the complex organization of the motor system. Although much clarification of the physiological basis of these conditions has already been attained from laboratory study, many fundamental problems remain unsolved. Current explanations as to the relationship of paralysis to spasticity in so-called pyramidal tract disorders, the relationship of hypokinesia to rigidity in extrapyramidal states, the relationship of tremor to rigidity and to hypertonia or hypotonia in basal ganglia and cerebellar diseases do not satisfy the clinician any more than they do the physiologist. Full descriptions of the functional abnormalities with detailed anatomical study are still needed. Although we have grouped the hypertonias roughly according to location of disease (e. g., cerebral vs. spinal) or according to disease (e. g., paralysis agitans) a more exact statement of the anatomy of the lesions in specific physiological systems cannot be made.

Animal models of hypertonias will probably never provide complete explanations of disturbed human motility where deficits of volitional movement are of central importance. The motor organization of the animal is too different from that of man, and his volitional activity and motor skills cannot be compared. It is certain that a number of hypertonic states resulting from disease of the CNS are occurring in man that have never been seen in animals. Their study may reveal new physiological derangements hitherto unknown in animal physiology.

Author's address: Dr. RAYMOND D. ADAMS, Kennedy and Shriver Laboratories, Massachusetts General Hospital, *Boston, MA 02114* (USA)

New Developments in Electromyography and Clinical Neurophysiology,
edited by J. E. Desmedt, vol. 3, pp. 11–12 (Karger, Basel 1973)

Clinical Features of the Pyramidal Syndrome

P. CASTAIGNE

Clinique des Maladies du Système Nerveux, Hôpital de la Salpêtrière, Paris

The pyramidal syndrome is a clinical entity, the diagnosis of which is usually straightforward, but which includes a variety of signs as diverse as the flaccid paraplegia of spinal shock, the spastic hemiplegia secondary to an old capsular lesion or the flexor spasms associated with certain chronic spinal lesions. Pathological studies have shown that the various clinical features correspond to different anatomical lesions. Reflex studies should lead to a better understanding of the pathophysiological mechanisms involved and they will be made both easier and more effective by taking into account the wide range of topographic and etiologic varieties of the pyramidal syndrome. It is thus necessary to propose a classification based on the site, nature and mode of evolution of the lesion.

The topographic distribution of the pyramidal syndrome and of its clinical signs depends on the site of the lesion. Thus, exaggeration of withdrawal reflexes, leading sometimes to flexor spasms is most marked in spinal lesions. Spastic increase in tone is often slight or even absent with lesions of the brain stem. Synkinesia is particularly frequent in patients with lesions in the cerebral hemisphere.

If the actual site of the lesion can only be established by anatomical study, clinical examination suggests the topographic localization in the majority of cases with a precision depending largely on other signs accompanying the pyramidal syndrome, be they sensory, cerebellar, radicular or cranial nerve signs, or disturbances of higher function. The precise localization may be difficult on clinical grounds in certain patients when several discrete lesions are present (e. g., in vertebro-basilar insufficiency). On the other hand, it is almost always possible to state, even without anatomical evidence, whether the lesion is situated at the level of the cortex, the internal capsule, the brain stem or the spinal cord.

Such a topographic classification may appear over-simple, but it is the only one which is always possible and which provides useful criteria for grouping the physiological results obtained in different patients.

While the extent of the anatomical lesion is often difficult to outline precisely by clinical examination, the mode of evolution is an important parameter depending on whether the syndrome appeared suddenly (trauma, vascular accident) or developed progressively. However, even for lesions identical in site and pathology, the exact chronology of the clinical modifications of reflexes varies greatly from one patient to another. This is why it appears difficult to group patients in terms of the evolution or duration of the lesion. I would rather suggest the grouping of patients with the same clinical features even though their lesions would vary in type and duration. It is thus necessary to supplement the topographic classification by the following clinical criteria:

1. The voluntary strength of the muscles studied by reflexological techniques distinguishing, for example, 3 grades: (a) normal or almost normal force; (b) definite weakness, and (c) no voluntary movement.

2. The state of proprioceptive reflexes, distinguishing 4 categories: (a) absent tendon reflexes; (b) present tendon reflexes; (c) elicitation of a brief stretch reflex or of a prolonged stretch reflex on a passive movement, and (d) clonus. Spasticity implies the existence of a stretch reflex on passive mobilization of the limb.

3. The state of exteroceptive reflexes, distinguishing 3 categories: (a) isolated inversion of the plantar cutaneous reflex (Babinski); (b) exaggeration of the withdrawal reflex ('triple retrait') elicited by stimuli which are less painful than those required in a normal subject, and (c) withdrawal reflex occurring with the slightest stimulus (flexor spasms).

Patients with multiple sclerosis and heredo-degenerative disorders of the central nervous system present special problems. The interpretation of reflex studies in such patients is extremely difficult, in view of the dissemination of lesions which can affect other systems (sensory or cerebellar) without necessarily producing clinical signs of such involvement.

Author's address: Prof. P. CASTAIGNE, Clinique des Maladies du Système Nerveux, Hôpital de la Salpêtrière, 47, Boulevard de l'Hôpital, *75-Paris 13*e (France)

New Developments in Electromyography and Clinical Neurophysiology,
edited by J. E. Desmedt, vol. 3, pp. 13–14 (Karger, Basel 1973)

Some Comments on Spasticity and Rigidity

P. NATHAN

The National Hospital, London

Spasticity

Spasticity is a condition in which the stretch reflexes that are normally latent become obvious. The tendon reflexes have a lowered threshold to tap, the response of the tapped muscle is increased, and usually muscles besides the tapped one respond; tonic stretch reflexes are affected in the same way.

Clonus may be induced in spasticity. It occurs in no other state but spasticity.

When spasticity is severe, cutaneous stimuli, pressure or tension on a limb, or a part of a limb or the abdominal wall cause massive spreading and lasting muscular responses. Even when no stimuli are applied, the musculature is usually found not to be relaxed as it is in the normal.

Tone

Tone is a term used in clinical neurology meaning resistance to stretch. Clinicians recognise hypertonia and hypotonia; between these 2 pathological states is normal tone. Hypertonia is due to lowering of threshold and increase in the response in tonic stretch reflexes together with certain viscoelastic properties of the muscles; normal tone is due to the visco-elastic properties, often with the addition of normal stretch reflexes, and hypotonia is due only to the visco-elastic properties of the muscles.

Rigidity

This term is used to mean at least 3 different states: intense spasticity, in which the muscles seldom relax, decerebrate rigidity, parkinsonian rigidity. The term should not be used to mean intense spasticity because there is no need of a different word to describe spasticity when it is most severe, and because it may suggest that the cause is not that of spasticity. Whether or not the term 'rigidity' is good when used in the combination of 'decerebrate rigidity', it is bound to stay in the form in which it was introduced by Sherrington and is now time-honoured. It differs from severe spasticity not only in its mechanism but in the fact that the co-contraction of antagonists rarely relaxes and that the activity of muscles is far more influenced by phases of respiration, very noticeable in Cheyne-Stokes respiration, and by strong and lasting movements of the head, neck and of the upper limbs and, less so, of the lower limbs.

Apart from decerebrate rigidity, the term 'rigidity' should be kept for the state of the musculature characteristic of parkinsonism. This state seems impossible or very difficult to define, so that the definition does not include aspects of severe spasticity. But in the practice of neurology there is usually little difficulty in telling one from the other. It is usually said that in parkinsonian rigidity, both the muscle being passively stretched and its antagonist are active throughout the stretching and, again, throughout the relaxation induced by the examiner. This is so, but it is also so for severe spasticity. Nevertheless, the feel of the 2 conditions is different. This difference may be due to the fact that in severe spasticity the resistance tends to increase throughout a passive movement, whereas it feels as if it remains constant in parkinsonian rigidity.

Author's address: Dr. Peter Nathan, The National Hospital, Queen Square, *London W.C.1* (England)

New Developments in Electromyography and Clinical Neurophysiology,
edited by J. E. Desmedt, vol. 3, pp. 15–19 (Karger, Basel 1973)

Clinical Changes in Muscle Tone

G. Noël

Department of Neurology, Hôpital Civil, Charleroi

The term 'muscle tone' introduced by MUELLER in 1838 is not very precise and has been defined in various ways. Indeed the uncertainties of the clinical study contrasts with the finesse of electrophysiological analysis of reflexes. Furthermore, the older reports of BABINSKI, DEJERINE, SHERRINGTON, WALSHE and others have only too often been ignored.

Tone is the state of contraction of muscles maintaining a certain posture, and tending to oppose passive movements of parts of a limb.

Hypotonia is characterized by a visible change in the normal outline of a muscle, by the posture of the limb, by muscle softness on palpation, by hyperextensibility on passive movement and by a characteristic 'ballottement' of the hand or foot which occurs when the observer shakes the limb. Hypotonia may be generalized, as in cerebellar syndromes or limited, for example to the legs immediately after a spinal transection, or to the limbs on one side immediately after a cerebro-vascular accident. Hypotonia is also found in those diseases in which there is marked loss of postural sense such as tabes and in lesions of the lower motor neurone.

Hypertonia is characterized by an augmented resistance to passive movement. It must be distinguished from false hypertonias, the origin of which are not to be found in the central nervous system. One example is the disturbance of the normal equilibrium between flexors and extensors which occurs in acute poliomyelitis when the motor neurons which have escaped infection produce excessive contraction, thus stretching the paralyzed antagonistic muscles. Another example is the spasm of a muscle related to irritation in a neighbouring organ. Examples include abdominal rigidity in peritonitis, spasm of the shoulder muscles in angina pectoris or in capsulitis of the shoulder joint and spasm of the paravertebral muscles in radicular lesions.

The character of hypertonia varies according to the site and stage of evolution of the lesion.

1. In the pyramidal syndrome due to cortical and/or sub-cortical lesions there is spasticity in flexion of the upper limb and in extension of the lower limb. Apart from the paralysis of the muscles of the upper limb which is most marked distally, the stretch reflexes are exaggerated. In the upper limb, the increased stretch reflexes result in a characteristic posture with the fingers flexed into the palm (by both deep and superficial flexors), the thumb adducted, the hand flexed and pronated, the forearm flexed and the shoulder internally rotated. Passive movement produces a resistance to stretch which partially gives way after several seconds. Eventually complete extension can be obtained unless this is prevented by a secondary fibrous fixation. The exaggeration of the myotatic reflex also produces brisk tendon reflexes, sometimes poly-kinetic. The area from which these reflexes can be elicited is enlarged [reflex irradiation; see LANCE et al., this volume]. Clonus of the wrist may be obtained by sudden passive extension and abduction of the hand. The small finger reflexes may be exaggerated. These are the Hoffmann reflex produced by forced flexion of the terminal finger phalanges followed by sudden release, and the Trömmer reflex produced by percussion of the palmar aspect of the ter-minal finger phalanges. Both stimuli result in flexion of the fingers and adduc-tion of the thumb. The reflexes are present in many normal subjects and have no diagnostic value except when they differ on the affected and normal side.

In the lower limb, a stretch reflex can often be found in the adductor magnus muscle of normal subjects and also in the dorsiflexors of the toes which are seen to contract under the skin when the subject is gently pushed back-wards. In pyramidal hypertonia the stretch reflexes are exaggerated most markedly in the adductors, quadriceps and triceps surae. This results in a posture of adduction and extension of the thigh and plantar flexion of the foot. The knee and ankle tendon reflexes are exaggerated and the area from which they are elicited is enlarged. They may be polykinetic with sustained patellar and ankle clonus. The clinical picture is completed by an extensor plantar response (Babinski's sign) or sometimes just an absent plantar response. It is interesting that, on walking, the patient tends to place the leg outwards and thus seems able to overcome the adductor stretch reflex.

When the lesion is of recent onset, the hemiplegia is flaccid and reflexes absent. The first reflex to appear is the Babinski response, followed by the tonic stretch reflexes which rapidly increase in intensity. If the lesion resolves, the signs disappear in the reverse order. If it persists, hypertonia remains in the same groups of muscles, even if the paralysis partially resolves. The spasticity may be increased by the repeated passive movements of untimely physio-therapy.

2. Decerebrate rigidity, well-known in intensive care units, is a posture with adduction of the upper limbs to the mid-line, slight flexion of the forearms with pronation and internal rotation. The lower limbs, neck and spine are all extended. The hypertonia may be episodic but it is continuous in the more serious cases. Passive movements are resisted by the hypertonia whatever the direction in which the limb is moved. The posture of decerebrate rigidity is changed by movements of the head. One example is the reflex of Magnus- de Kleyn, in which rotation of the head produces extension of the upper limb towards which the face is turned, and a flexion of the upper limb on the contra-lateral side.

3. The rigidity associated with choreic or athetotic movements varies greatly from patient to patient. The dystonia is most marked in the axial musculature. Asymmetrical hypertonia in these muscles results in postures of torsion [RONDOT, 1968]. Dystonia may only be present in certain situations – for example, on standing, walking or during voluntary movements against the direction of pull of the increased tone, and in certain automatic activities. In this type of patient, contraction of a number of muscles can be provoked by auditory, visual, tactile and kinaesthetic stimuli and by mental arithmetic. Passive movements are resisted by increased tone which appears to be an exaggeration of postural reflexes and which is most marked in one or another of different groups of muscles. Passive movement or voluntary contraction can produce distant reactions in limbs which are not, at that moment, being manipulated by the examiner.

4. In Parkinson's disease, motor expressivity is reduced. Examination of passive movements shows 2 phenomena. The first is an exaggeration of the tonic myotatic stretch reflex, with the peculiarity that the passive movement is resisted by the cogwheel phenomenon; this increase in tone is exaggerated by simultaneous contralateral effort. The second is an exaggeration of the postural reflex which tends to fix a segment of a limb in the position in which it is placed [FOIX and THEVENARD, 1923]. This can be seen in the biceps of a normal subject. When the forearm is passively flexed the tendon of the biceps muscle can be felt to harden under the palpating finger at the end of the move-ment. This reflex, which is hardly detectable in normals, is exaggerated in Parkinson's disease. In this disease the increase in the stretch reflex is present in all groups of muscles and not only in those concerned with support against gravity. These complex reactions can be related to the plasticity of parkin-sonian hypertonia.

5. Bulbar lesions only rarely cause hypertonia. It seems likely that a patient could hardly survive both a lesion extensive enough to cause bulbar symptoms and a complete lesion of pyramidal and vestibular spinal fibres.

6. Spinal lesions produce different signs depending on whether they are complete or incomplete.

(a) Complete transection (spinal man) is seen more frequently in dorsal than in cervical lesions. After the phase of spinal shock which lasts several weeks in man and during which the lower limbs are flaccid and hypotonic, the cutaneous and tendon reflexes return and then automatic activities appear. At this stage the paraplegia is flaccid at rest. Nevertheless, the tendon reflexes are brisk, polykinetic and show irradiation, but this is less marked than in patients with incomplete transection or with hemiplegia. The plantar responses are extensor and the superficial abdominal reflexes are absent. Passive movements are not strongly resisted but elicit definite but weak stretch reflexes in the adductors of the thighs, the hamstrings and in quadriceps.

Stimulation, particularly painful stimulation of the sole of the foot, or passive forced flexion of the toes produces a flexor withdrawal response of the ipsilateral limb and, sometimes if one persists, a contralateral response as well. This reflex [BABINSKI, 1898] is most easily induced by painful stimulation of the skin and deep tissues of the foot and by forced flexion at the metatarsophalangeal joint [MARIE and FOIX, 1912]. However, the area from which the reflex can be induced may be very much larger and may include even the abdomen. In certain patients the whole limb flexes – dorsiflexion of the toes and ankle, flexion of the knee and hip. If the stimulus is prolonged the contralateral limb may also flex. If the threshold for this reflex is low, stimulation may produce simultaneous flexion of both lower limbs and incontinence of urine and faeces – the mass reflex. As time goes on the flexor spasms become more and more frequent if nothing is done to prevent them, and the threshold for their elicitation becomes lower and lower. It may happen that the spinal patient tries to induce these spasms by mechanical stimulation of the lower limbs as he sees hoped-for movements in his paralyzed limbs which results in a further lowering of threshold leading eventually to paraplegia in flexion which may be secondarily maintained by fibrous fixation. Patients should, of course, be instructed against this practice and should receive appropriate drug treatment, e.g. with Diazepam. When flexor spasms are correctly treated they become less and less troublesome and, in the chronic phase, the spinal man then shows only moderate spasticity and vestigial spasms.

One may also see a crossed extension reflex of the contralateral limb with the flexor withdrawal in the stimulated limb, and an ipsilateral extension reflex induced by passive dorsiflexion of the foot. These reflexes are relatively rare in cases of complete transection but are more frequent in paraplegics with partial lesions.

(b) Incomplete paraplegia. In cases of incomplete section, even if there is no voluntary movement, spasticity is much greater. The lower limbs are extended and adducted. The stretch reflexes are much brisker, being easily produced by passive movement. The tendon reflexes are polykinetic. Sustained patellar and ankle clonus occurs. Babinski and Rossolimo signs are easily elicited. This clinical picture occurs soon after a traumatic or acute ischaemic lesion. In progressive lesions, in contrast, voluntary power diminishes only gradually and spasticity may play the dominant role in the clinical picture (multiple sclerosis, amyotrophic lateral sclerosis, etc.). When the patient is standing and trying to walk, he may be hampered by strong contraction of the adductors of the hip and extensors of the knee and of the ankle. Hypertonia may be so extreme that the tendon reflexes do not appear to be exaggerated on clinical testing.

Conclusion. Neurophysiological investigation of spasticity would be more useful if the cases investigated form homogeneous groups. I suggest the following divisions:

1. Cortical or juxta-cortical vascular lesions. Tumours or traumatic lesions can have distant effects which introduce other variables.

2. Extra-pyramidal and parkinsonian rigidity which form a very different patho-physiological group from the above.

3. Complete spinal lesions (spinal man) at similar times of their clinical course, thus distinguishing: (a) spinal shock; (b) return of reflexes and of the flexor withdrawal response, and (c) the chronic phase. It is emphasized that the patients in the latter group vary according to the efficiency of the treatment received.

4. Incomplete spinal lesions include heterogeneous who cannot be dealt with as a single group.

5. In cases of 'brain palsy', multiple sclerosis and degenerative disorders (e.g. Friedreich's ataxia), the lesions are scattered and numerous and the patients are difficult to compare with each other.

Acknowledgement

The author is indebted to Dr. J. E. Desmedt for suggestions in the preparation of this paper.

Author's address: Dr. G. Noël, 2, rue Ferrer, *B-6000 Charleroi* (Belgium)

Anatomical and Physiological Background

New Developments in Electromyography and Clinical Neurophysiology,
edited by J. E. Desmedt, vol. 3, pp. 20–37 (Karger, Basel 1973)

Cyto- and Neuropil Architecture of the Spinal Cord[1]

J. Szentágothai and M. Réthelyi

Department of Anatomy, Semmelweis University Medical School, Budapest

The increasing demands of neurophysiology for structural data on spinal cord architectonics prompted an upsurge in histological studies after a lull of several decades between the classical period in the late 19th century and the early fifties of this century. When methods for tracing axonal and synaptic degeneration became available, identification of the termination sites of various pathways began gradually to occupy the focus of interest. Due to technical difficulties, such attempts were at first rather sporadic [Schimert, 1938, 1939; Szentágothai-Schimert, 1941] but after the development of the specific suppression silver stains by Nauta and Gygax [1954] the studies on fiber terminations in the spinal cord and elsewhere in the CNS became one of the most important lines of investigation in neuroanatomy. The more exact definition of the various specific sites in the grey matter in a new cytoarchitectonic description of the spinal cord by Rexed [1952, 1954] was soon universally accepted and, in spite of certain shortcomings to be discussed presently, will probably remain the practical basis for defining locations reliably in the future.

Understandably, the importance of neuropil architectonics was not sufficiently appreciated in the early degeneration studies. Although already Ramón y Cajal [1909] correctly evaluated on the basis of Golgi studies the crucial significance for synaptic connectivity of the mutual interrelationships of axonal and dendritic arborizations, these ideas re-emerged only much later and rather hesitantly in the investigations using axonal and terminal degeneration [Szentágothai and Albert, 1955; Sterling and Kuypers, 1967a, b; Szentágothai, 1964a]. However, it was not until the Golgi technique quite recently reoccupied its central and primary position in neuroanatomical

1   This article is dedicated to the memory of L. Laruelle (1876–1960) the eminent Belgian neurologist, who was one of the first to understand the importance of spatial orientation of the neuropil in various parts of the spinal grey matter.

studies that the systematic analysis of neuropil architecture again received due attention in the study of the spinal cord [SZENTÁGOTHAI, 1964a, b; SCHEIBEL and SCHEIBEL, 1966, 1968, 1969]. LARUELLE [1937], in the thirties, looms as a lonely figure who correctly appreciated the importance of studying and understanding neuropil architectonics in the spinal cord, and who in making a plea for the study of the spinal grey matter in longitudinal sections, anticipated a technical development that might appear to us quite obvious now, but which still took many years to come into effect.

The object of this paper is to summarize, very briefly, some of the main principles of spinal cord cyto- and neuropil architectonics, that have emerged recently mainly from the SCHEIBELS' and our own studies. Our considerations will remain here on the level of the light microscope; however, we shall try to show later how highly significant this type of knowledge will probably become for synaptological studies using the electron microscope. Although most of these studies have been made on the spinal cord of the cat and the dog, some Golgi material of the human spinal cord available in this department, and a complete longitudinal section series of the entire human spinal cord prepared by LUDWIG and KLINGLER in 1939 (neuro-fibrillar stain) in the Anatomy Department of Basle University [unpublished work prompted by the studies of LARUELLE, 1937], indicate that the structural principles discussed hold true also in the human cord.

Cytoarchitectonics and Dendrite Arborization Patterns

The main parameters determining the cytoarchitectural patterns of any grey matter are: size, density, shape and axis orientation (if any) of neuron perikarya. Intracellular structural details, contributing also significantly to the cytoarchitectonic picture, will be neglected here entirely, since they are of no relevance to the questions to be discussed. Dendritic arborization patterns are mainly indirectly related to the cellular patterns (apart from determining the axis orientation of cells having one or two dendritic poles) by contributing secondarily to the formation of laminated, columnar or various other arrangements of cells. Certainly neither of the two patterns can fully explain the other, hence in order to understand neuronal, not to speak of synaptic, arrangements both patterns have always to be compared. Such a comparison can be summarized for the spinal grey matter as follows.

Lamina I of REXED [1954] contains relatively large cells of irregular dendritic arborizations spreading strictly tangentially in the thin layer [SZENTÁ-

GOTHAI, 1964b; SCHEIBEL and SCHEIBEL, 1968]. Usually only terminal branches of these dendrites enter the neighboring lamina II (and also the white matter).

Lamina II contains the typical small substantia gelatinosa neurons. The dense dendritic tree of these cells is flattened in the sagittal plane to a cranio-caudally elongated sheet [SCHEIBEL and SCHEIBEL, 1968; RÉTHELYI and SZENTÁGOTHAI, 1969]. The dendritic arborization is extremely dense and the thinner branches bear numerous spine-like appendages.

Lamina III. The change from lamina II is gradual. Neurons become some-what larger but their dendritic arborizations are similarly arranged in sagit-tally oriented sheets. In most cells the dendritic sheet extends into lamina II, i.e. in a dorsal direction. In the deeper parts of this lamina many neurons be-come fusiform with cranio-caudal axes of the dendritic trees. They lose the sagittal orientation of the arborization, but many of the dendrite branches still tend to ascend in dorsal direction.

Lamina IV is characterized by medium size and large neurons having a number of dorsally oriented dendrites that penetrate all the way through the gelatinosa complex (lamina II and partly lamina III). These cells have been clearly described already by RAMÓN Y CAJAL [1909] and more detailed recent descriptions [SZENTÁGOTHAI, 1964b; SCHEIBEL and SCHEIBEL, 1968] agree in the main points: the numerous dendritic appendages (spines) and the spread in a transverse direction of the dorsally oriented dendrites; i.e., they are not confined to a single sagittal sheet of tissue space but may enter several parallel (or slightly radially diverging) neuropil spaces of laminae II and III. These neurons have been termed recently antenna-type neurons [RÉTHELYI and SZENTÁGOTHAI, 1972]. SCHEIBEL and SCHEIBEL [1968] emphasize additionally the occurrence of shorter and smoother horizontal dendrites.

Up until *lamina V*, the horizontal (or slightly curved) layering of the dorsal horn grey matter, in the sense described by REXED [1952, 1954] is ob-vious enough, and the arrangement of the dendritic and especially of the axonal neuropil in sagittally (or taking into account the slight curvature of the cell layers, somewhat radially) oriented longitudinal sheets does not detract from this lamination. However, even in the Nissl picture the REXED layering becomes obscured and rather arbitrary from lamina V through to lamina VIII.

Fig. 1. Sections through the intermediate region of the spinal cord in the kitten ori-ented transversally (A), sagittally (B), and horizontally (C). Directions are indicated in all photographs. Scale 50 μm. Rapid Golgi stain. It becomes clear from the three photographs that both dendrites and most parts of the neuronal neuropil are arborizing in the transversal plane of the spinal cord.

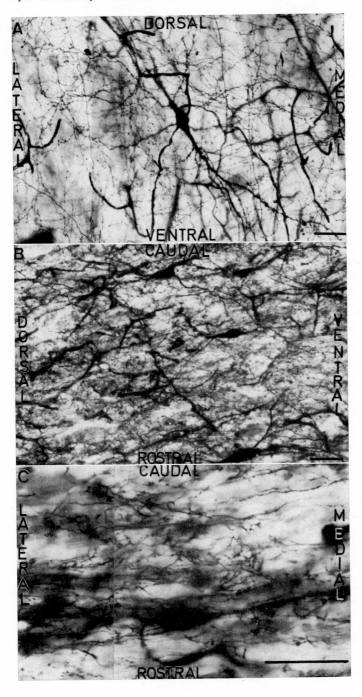

As shown in figure 1, the dendritic arborizations of intermediate region neurons are confined to transversally oriented flat disc-shaped spaces [SCHEIBEL and SCHEIBEL, 1968], being thus parallel to the main orientation pattern of the axonal neuropil (fig. 4–5). Another orientation pattern of the dendrites in the intermediate region has escaped detection until recently, because it is inconspicuous in the newborn or very young animals generally used for Golgi staining. Application of the perfusion Golgi-Kopsch technique to adult cats [RÉTHELYI, 1972] has revealed the strange arrangement of the dendrites along the surface of a longitudinal (cranio-caudal) cylinder (fig. 2). This cylinder is equally conspicuous in the cervical, in the thoracic as well as in the lower lumbar region of the spinal cord, and occupies mainly the intermediate region; however, it protrudes both into the dorsal and particularly into the ventral horn. This strange arrangement gives some *a posteriori* structural justification for an architectural concept proposed some time ago on the basis of synaptic connectivity [SZENTÁGOTHAI, 1967a]. The concept envisaged the spinal cord grey matter as a 'double-barrel shaped central core' corresponding to the intermediate region in the widest sense with the dorsal horn and the motor nuclei attached as 'dorsal and ventral appendages' [see some further discussion of this concept by RÉTHELYI and SZENTÁGOTHAI, 1972].

This cylinder protrudes well into the middle of lamina V and reaches even lamina IV; however, two nuclei at the base of the dorsal horn – the known medial (internal) and lateral (external) nuclei of the base of the dorsal horn [RAMÓN Y CAJAL, 1909] – are largely outside. More exactly speaking, most parts of the dendritic tree of the cells in those two nuclei are outside the cylinder, but several authors have recognized earlier that most of these cells have a group of horizontally oriented dendrites [ZHUKOVA, 1968; SCHEIBEL and SCHEIBEL, 1968]. Actually these dendrites are not really horizontal but fit into the dorsal surface of the cylinder.

We do not want to over-emphasize the cylindrical shape resulting from the overall dendritic arrangement. It might come about quite accidentally by some expansive postnatal growth process occurring in the core of the cylinder. Since the cylindrical arrangement is lacking in the neonatal state or in very

---

Fig. 2. Transverse section of the spinal cord at segment L₅ of an adult cat; perfusion Golgi-Kopsch stain. While the horizontal (slightly curved) lamination of the dorsal horn proper is apparent, the dendritic arborizations of the intermediate region result in a clear circular pattern, protruding both into the base of the dorsal and into the center of the ventral horn. The motor nuclei are clearly outside the circle. As this circular pattern is maintained consequently throughout the entire length of the cord this corresponds to a double-barrel shaped cylindrical organization of the central grey core. Scale 500 μm.

young animals, this would be the simplest explanation, not needing the assumption of extensive rearrangement of the dendritic arborizations already present at birth. As shown convincingly in recent studies by SCHEIBEL and SCHEIBEL [1971], axonal arborizations, particularly of the initial collateral systems, show quite spectacular growth and elaboration in pattern during postnatal development of kittens. As quite a substantial part of this growth occurs in the intermediate region, one might easily envisage a considerable increase of tissue mass caused by invasion and further branching of these collateral axon branches.

It may be rewarding to try some quantitative approach on the electron microscope level, to assess the postnatal growth and elaboration of neuropil (and synaptic) architecture in the central core of the intermediate region (the cylinder shown in figure 2) as compared to other regions of the grey matter. In fact, the massive Golgi staining of the central part of the intermediate region was observed by the early neurohistologists. This is generally a nuisance as it renders tracing of axonal elements through this region virtually impossible (fig. 5). It can be recognized, however, from a great many drawings of the classical Golgi studies, mainly those of RAMÓN Y CAJAL [1909] that this tangle is due to two factors: (1) numerous collaterals and branchings issued by the axons running through, and (2) a crossing of the long collaterals traversing the entire grey matter. This crossing occurs in the center of the cylinder and some speculations have been made recently on its significance [RÉTHELYI and SZENTÁGOTHAI, 1972].

The possible functional significance of this will be discussed in the concluding section. But even at this stage of our considerations it becomes obvious that the subdivision of the intermediate region of the spinal grey matter into the more or less horizontal laminae V, VI, VII and VIII, although of certain practical value for defining various localizations, does not reflect any real architectural principle of the spinal cord (fig. 3), whereas the lamination of the dorsal horn (laminae I–IV) certainly does. Additionally there are two specific territories in the intermediate region (in the widest sense) that are completely different. One is Clarke's column with a spectacular longitudinal organization of both dendrites and axonal arborizations [LARUELLE and REUMONT, 1938; SZENTÁGOTHAI and ALBERT, 1955; RÉTHELYI, 1968, 1970]. Clarke's column is situated within the territory of the intermediate cylindrical core of the spinal grey matter. The other peculiar region, attached to the cylinder at its lateral border, is the sympathetic intermediolateral nucleus or column [LARUELLE, 1936]. This column has an even more rigid longitudinal (cranio-caudal) orientation of both dendritic and axonal arborizations [RÉTHELYI, 1972]. Both nuclei are virtually close in the sense of MANNEN's 'noyeau fermé' [1960],

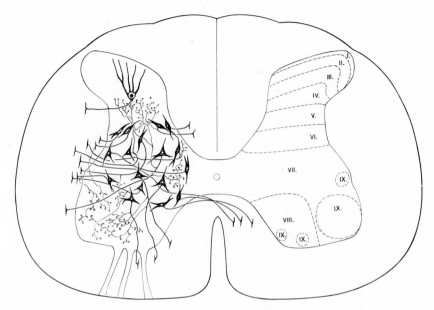

Fig. 3. Comparison between Rexed's cytoarchitectonic lamination in L_5 of the cat spinal cord (right side) and the cylindric arrangement yielded by the general orientation of the dendrites in the intermediate region (left side). A single large antenna-type neuron of Rexed's lamina IV is indicated. The initial courses of axons and some of their initial collaterals are also indicated. Most of these initial collateral arborizations are also transversally oriented.

i. e. dendrites or the afferent axons do not invade the neighboring territories significantly, nor is their territory significantly invaded by dendrites and axons of the neighboring grey matter.

In the *ventral horn proper* (lamina IX of REXED) two radically different types of dendritic arborizations can be observed. The larger cells, particularly (but by no means only) motoneurons, have spherical or often cranio-caudally elongated dendritic trees [STERLING and KUYPERS, 1967b; SCHEIBEL and SCHEIBEL, 1969]. Conversely, a group of medium size cells, easily recognized on the basis of their axonal arborizations as interneurons, have disc-shaped dendritic arborizations oriented strictly in the transversal plane (fig. 4).

Gross Axonal Neuropil Arrangement

Appreciation of the significance of the general orientation of terminal axon ramifications emerged gradually. RAMÓN Y CAJAL [1909] emphasized the

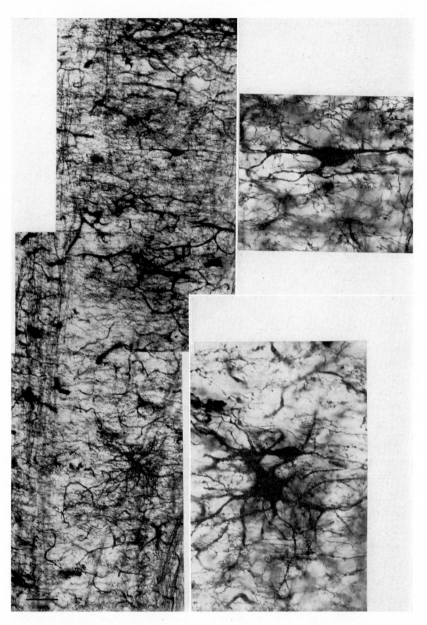

Fig. 4. Paramedian sagittal section through the ventral horn of a kitten (ventral is to the right). Rapid Golgi stain, scale 100 μm. Some details are enlarged on the right, the upper figure of which shows the clearly transversally oriented dendritic tree of an interneuron. The lower shows a motoneuron with spherical dendritic arborization (the cranio-caudal elongation of the dendritic arborization develops later) and a neighboring interneuron with transversal dendritic arborization. The axonal neuropil is clearly oriented transversally.

Fig. 5. Paramedian sagittal section through the dorsal horn and the intermediate region of a kitten (the section passes medially from the main motor nucleus); scale 100 μm. Sagittally oriented brick-shaped neuropil lobuli of the substantia gelatinosa complex are recognizable on the top of the photomontage (arrow). The dark region in the middle corresponds to the region where most of the crossing of collaterals occurs (see text). The large neurons with spherical or even cranio-caudally elongated dendritic trees (at bottom) are probably large funicular interneurons giving rise to crossed ascending tracts. The transversal orientation of the axonal neuropil can be recognized at this very low power.

longitudinal orientation of the axonal neuropil in the dorsal horn. The functional consequences of the parallel (cranio-caudal) orientation of terminal axon branches and dendrites by giving opportunity to climbing type (repeated) synaptic contacts were understood first in Clarke's column [SZENTÁGOTHAI and ALBERT, 1955] and the importance of a longitudinally oriented plexus of primary afferents in lamina III of the dorsal horn was stressed by STERLING and KUYPERS [1967a]. It was only after the recent large-scale revival of Golgi studies that this side of neuropil architecture was fully realized [SCHEIBEL and SCHEIBEL, 1968, 1969].

The neuropil of lamina I is entirely tangentially organized. Apart from the collateral bundles penetrating through this layer all preterminal and terminal axon branches tend to arborize within this thin sheet, but with dominantly longitudinal orientation. Thus, both dendritic and axonal arborizations are restricted to a relatively thin space with the geometrical consequence of a relatively higher probability of multiple contacts between the branches of any dendrite and axon.

The *substantia gelatinosa complex* (lamina II and partly lamina III) consists of sagittally (or slightly radially) oriented flat sheets of axonal neuropil, brought about by the ascending arborizations of large cutaneous afferents [SCHEIBEL and SCHEIBEL, 1968; RÉTHELYI and SZENTÁGOTHAI, 1969] ramifying in narrow sagittally arranged brick-shaped spaces. These correspond to the large flame-shaped ascending arborizations described first by RAMÓN Y CAJAL [1909] and later termed lobuli [SZENTÁGOTHAI, 1964b] before their sagittal organization was recognized. Also the ramifications of other afferents, mainly arising from Lissauer's tract, terminate within similar sagittally oriented spaces of the gelatinosa complex [SZENTÁGOTHAI, 1964b, and see particularly detailed discussion by SCHEIBEL and SCHEIBEL, 1968]. Lamina II contains additionally a very dense longitudinal plexus of delicate preterminal axons arising mainly from the small substantia gelatinosa cells [SZENTÁGOTHAI, 1964b]. It is not quite clear whether the longitudinal plexus of primary afferents described by STERLING and KUYPERS [1967a] in lamina III and partly IV corresponds only to the preterminal part of the ascending large cutaneous afferents, directed mainly to lamina II, or whether there is also a genuine terminal neuropil of longitudinal orientation, but not necessarily sagittally arranged in lamina III and IV. From the occurrence, in the central part of lamina III, of nerve cells having extremely elongated cranio-caudal dendritic arborizations, not invading lamina II, one might assume that the latter is the more probable.

The organization of both axonal and dendritic neuropil of the dorsal horn in relatively narrow sagittal (radiate) sheets has some aspects of somatotopy of rather intricate nature.

The terminal branches of primary afferents are shifted medially during their ascent within the dorsal horn. This was shown first to apply for Clarke's column [SZENTÁGOTHAI, 1961a] but according to recent investigations [IMAI and KUSAMA, 1969] it seems to be a general rule throughout the dorsal horn. Another somatotopic principle between the dermatome and the substantia gelatinosa described earlier [SZENTÁGOTHAI and KISS, 1949] is that dorsal parts of the dermatome are projected on the lateral, and ventral parts on the medial portion of the substantia gelatinosa. At first sight these two projection principles seem to be conflicting and mutually exclusive. With some additional speculation, however, the two may be reconciled [RÉTHELYI and SZENTÁGOTHAI, 1972] which, if verified, might be of considerable help for functional interpretations.

As shown convincingly by SCHEIBEL and SCHEIBEL [1968, 1969] the neuropil of the intermediate region and the ventral horn, i.e. from lamina IV onwards, is consequently organized in transversally oriented sheets or discs. Relatively few axon arborizations, mainly short axons or initial collaterals of interneurons, depart from the transversal plane of arborization. The transversal orientation of the axonal neuropil in the motor nucleus (fig. 4), and through a complete sagittal section of the grey matter (fig. 5) can be recognized even in low-power photomicrographs. There are, however, two regions of the intermediate grey matter with strictly longitudinally (cranio-caudally) oriented axonal neuropil. Clarke's column [LARUELLE and REUMONT, 1938; SZENTÁGOTHAI and ALBERT, 1955; SZENTÁGOTHAI, 1961a, b; RÉTHELYI, 1968, 1970], and the sympathetic intermediolateral nucleus [LARUELLE, 1936; RÉTHELYI, 1972].

The general arrangements of dendritic and axonal arborizations are summarized diagrammatically in figure 6. Apart from the characteristics discussed earlier, some general features of spinal cord synaptic connectivity are included in this diagram that have been analyzed in detail before [SZENTÁGOTHAI, 1964a], and that will be summarized in the next section.

The Consequences for Synaptic Connectivity
of Spatial Neuropil Arrangement

It is quite obvious that synaptic connectivity strongly depends on the mutual interrelations between the principal orientation of axonal and dendritic arborizations [RAMÓN-MOLINER, 1962]. The two extreme cases of this would be dendritic and axonal arborizations systematically arranged (1) in parallel, and (2) conversely at right angles. As a third intermediate case one would have to consider that of (3) both axons and dendrites arborizing two-dimensionally (i.e., in relatively flat sheets of tissue space), with the additional condition that the flat spaces or arborizations are arranged in parallel (i.e., dendrites and axons share arborization spaces entirely or with some alternation or shifted overlap). From the foregoing it became apparent that the most common arrangement of neuropil in the spinal grey matter is case 3.

It is obvious for purely geometrical reasons, that the probability of having synaptic contacts between any two of the individual axonal or dendritic arborizations of the two sets entering each others' spaces is statistically the highest in case 1 and lowest in case 2, while in case 3 the probability is intermediate. These probabilities can be calculated [UTTLEY, 1955] if the assumption is made that, apart from the general orientation, intertwinement of axons and dendrite branches is random. We know, of course, that the interrelations between axonal and dendritic arborizations are by no means really random but subject to highly complex specific selections and attractions. However, this does not render the above reasonings irrelevant; specific selections between certain axonal and dendritic elements would only elevate (by unknown factors) the probability of contacts within each of these categories. Also specificity of selection between various neuron elements for establishing synapses has certain limits, and randomness inside certain categories remains an important aspect of neuronal connectivity.

Fig. 6. Highly schematized stereodiagram indicating the main geometrical principles of neuropil orientation in the spinal grey matter. This diagram is a further development and combination of ideas proposed by SZENTÁGOTHAI [1967a] and by SCHEIBEL and SCHEIBEL [1968]. Lamina I of Rexed is a thin layer of tangentially oriented dendrite and axonal arborizations. Connexions mainly with Lissauer's tract (Li. T.) some of which are small primary afferents. Sagittally (slightly radially) oriented brick-shaped spaces of neuropil are shared by the arborizations of large cutaneous afferents (LCA) and dendritic arborization of small substantia gelatinosa neurons (in laminae II and partly III). Large antenna-type neurons of lamina IV send dendrites through the entire depth of the substantia gelatinosa. Here they have 'crossing-over' type synapses with vertical axonal neuropil arising from substantia gelatinosa cells. The large antenna neurons give rise to fibers of the spino-cervical tract (SCT). The cylindrically organized central core of the grey matter is shown as a horizontally laminated structure built of parallel discs, to which both dendritic arborizations to the interneurons and axonal ramifications are confined. Clarke's column (Cl. C.), where present, is incorporated into the central core as a continuous longitudinally oriented cylinder. The sympathetic intermediolateral nucleus (ILN) is a similar longitudinally oriented cell and neuropil mass attached to the central core on the lateral side. The motor horn proper (lamina IX) has also a horizontally oriented axonal neuropil; however, the more longitudinally oriented motoneurons penetrate through several successive neuropil layers. Interneurons are generally confined with their dendritic arborizations to one or a few neighbouring horizontal neuropil discs. Neuronal connectivity is treated only very cursorily in this diagram.

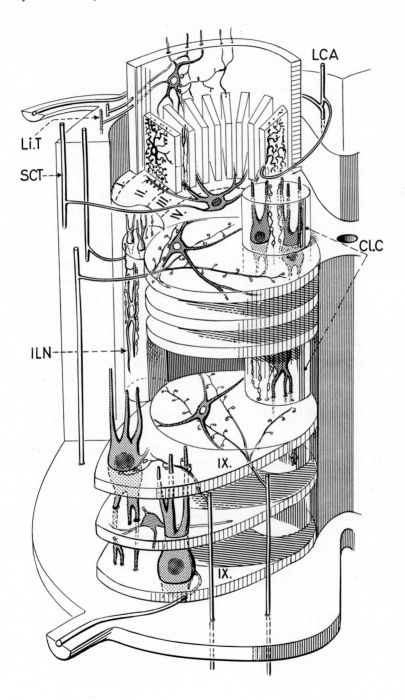

Case 1, or systematically parallel arrangement between terminal axon ramifications and dendrite arborizations, occurs in two places: in Clarke's column and in the sympathetic intermediolateral nucleus. The occurrence of climbing type axo-dendritic attachments [SZENTÁGOTHAI and ALBERT, 1955], with numerous repeated synaptic specializations [RÉTHELYI, 1970] in Clarke's column fits well with the powerful transmission characteristics of the main monosynaptic relay from I_A primary afferents and Clarke neurons [LLOYD and McINTYRE, 1950; JANSEN, NICOLAYSEN and RUDJORD, 1966; KUNO and MIYAHARA, 1968]. Systematic parallel arrangement between axon terminals and dendrites, and climbing type repeated contacts have been shown recently to prevail also in the sympathetic lateral column [RÉTHELYI, 1972]. Although this had been suggested by STERLING and KUYPERS [1967b] on the basis of degeneration findings for the synaptic relay between primary I_A afferents and motoneurons, this is not so according to Golgi pictures [SCHEIBEL and SCHEIBEL, 1969; see also fig. 4], and also the electron microscope degeneration studies of CONRADI [1969] would not support such a view. More detailed Golgi and electron microscope studies would be needed to find out whether the cells with longitudinally oriented dendrites [RÉTHELYI and SZENTÁGOTHAI, 1972] in the so-called center of the dorsal horn (corresponding mainly to the middle part of lamina III and partly IV) have a similar synaptic arrangement. From the Golgi pictures so far available, and also from degeneration findings of STERLING and KUYPERS [1967a], the synaptic arrangement would appear to be similar to that in Clarke's column and in the intermediolateral nucleus, but no final conclusion is possible until specific electron microscope studies support this notion.

Case 3 is the most general arrangement in the spinal grey matter between axonal and dendritic neuropils. But the same spatial principle is applied in two rather different forms (a) in the substantia gelatinosa complex, and (b) in the intermediate region and the ventral horn:

(a) The sagittal sheets, brought about mainly by the dense axonal arborizations of large cutaneous afferents, overlap completely with the sagittally oriented dense dendrite arborizations of the typical substantia gelatinosa cells. Both arborizations are so dense and irregular that one cannot speak of any real parallel arrangement between axons and dendrites. It is rather a dense irregular tangle in which both arborizations became intertwined. The dendrites are supplied by numerous rather complex spine-like appendages so that there is ample opportunity for very numerous contacts between any substantia gelatinosa cell and the axonal arborization sheet to which it belongs. According to the elegant explanation of SCHEIBEL and SCHEIBEL [1968] the same would apply to the dendrites of the large (antenna-type) neurons of lamina IV, which pierce the entire depth of the substantia gelatinosa complex. Our own explanation – which is also conjectural – differs from the reasoning of the SCHEIBELS in that it assumes a 'crossing-over' (case 2: see below) synaptic arrangement between the longitudinal axonal plexus arising from the axons of the substantia gelatinosa neurons [SZENTÁGOTHAI, 1964b; RÉTHELYI and SZENTÁGOTHAI, 1969].

(b) The disc-shaped dendritic arborizations of the majority of the intermediate region (fig. 1) and of the ventral horn interneurons (fig. 4) are arranged in parallel with the transversally oriented preterminal and terminal axon aborizations of the neuropil. Since the dendritic arborizations are of stellate character, and the preterminal axon branches show characteristically straight courses in the intermediate region [SZENTÁGOTHAI, 1964a] the relations between the two elements could be described with some approximation as the geometric relations between two sets of straight lines (fig. 6) lying in the same plane but running at random angles. Obviously some dendrite and axon branches will be approximately parallel, while others will cross each other at right angles, with all intermediates between. The straight axon branches do not, in general, establish synapses directly but have short terminal side-branches with bulbous synaptic endings. These terminal side-branches are issued about every 10–20 μm in random directions all over the course of the preterminal axons in the grey matter [SZENTÁGOTHAI, 1964a]. Hence, each preterminal axon can be envisaged as surrounded by a cylindrical space within which it can potentially establish synapses with any dendrite or cell body that comes within the radius of this imaginary cylinder. The geometrical and mathematical implications of this have been considered already by UTTLEY [1955]. (The situation is entirely analogous [only reverse] to the spiny dendrites, when one has to imagine a cylinder of a radius corresponding to the average length of the spines (+ thickness of the dendrite) within which synapses can be established with any axon branch entering or touching the space of the cylinder.)

Parallel arrangement between any part of the dendrites and axons is obviously the exception, hence the probability of numerous synapses between a certain axon and dendrite in the 'rope-ladder' fashion is rather low. However, there is a specific region at the base of the dorsal horn where the preterminal collaterals enter from the lateral side and penetrate the base of the dorsal horn in fan-shaped fashion. Here many of the preterminal branches may run quite consistently parallel with the similar fan-shaped dendritic arborization of the cells of the lateral basal nucleus of CAJAL [SCHEIBEL and SCHEIBEL, 1966].

From this, one would expect a quite powerful kind of synaptic action of corticospinal tract fibers upon the interneurons (for the relay of corticospinal impulses towards motoneurons), which is indeed the case [Vasilenko, Zadorozhny and Kostyuk,1967].

The dendritic trees of some of the interneurons (fig. 4) and particularly of the motoneurons (fig. 5) being either spherical in shape or even cranio-caudally elongated [Sterling and Kuypers, 1967b; Scheibel and Scheibel, 1969] do not correspond to the transversal layering of the neuropil. Consequently they may have synaptic relations to a whole set of neighboring parallel axonal neuropil layers, and thus the synaptic arrangement corresponds more to the second category, to be considered next.

Case 2 does not occur in the spinal grey matter in the pure form of 'crossing-over' synapses [Hámori and Szentágothai, 1964] as found, for example, in the molecular layer of the cerebellar cortex. A somewhat similar arrangement has, however, been described in the substantia gelatinosa between the intrinsic longitudinal plexus of very delicate terminal axons (given rise to mainly by the small substantia gelatinosa nerve cells) and the penetrating spiny dendrites of the large antenna type cells of lamina IV [Szentágothai, 1964b; Réthelyi and Szentágothai, 1969]. There is some uncertainty in this relation as it is difficult both in the Golgi picture and also under the electron microscope to separate from these the possible synaptic relations between the antenna cell dendrites and the sagittally oriented dense arborizations of the cutaneous afferents. The Scheibels [1968] appear to favour the view that the antenna-type dendrites are synaptically engaged mainly with the cutaneous afferents, while our own electron microscope observations of chronically isolated dorsal horn preparations [Réthelyi and Szentágothai, 1969] would be more in favor of the first explanation, i.e. of synapses between the longitudinal axon plexus of local origin and the antenna dendrites. Also the overall connectivity of the substantia gelatinosa neuron system points in this direction [Szentágothai, 1964b]. The two propositions are by no means mutually exclusive; both types of synaptic relations may be present simultaneously.

For the motoneurons Scheibel and Scheibel [1968] have developed similar ideas on a synaptic arrangement of parallel presynaptic domains (i.e., the horizontal neuropil layers) and more or less longitudinally oriented dendrites penetrating a number of successive neuropil layers. Thus any dendrite (or body) has several synaptic sites with each of these presynaptic domains, some of which are at the cell body itself, others proximal, and still others at remote parts of the dendrites. This very elegant reasoning and its illustrations have considerable implications in the relative effectiveness of postsynaptic potentials as lucidly discussed by RALL [1962, 1964]. The general arrangement between axonal neuropil and motoneurons is undoubtedly as described by the SCHEIBELS; however, it is not quite clear whether, and how far, this general arrangement holds true for the I_A primary afferents. Individual primary I_A afferents tend to establish relatively numerous synaptic contacts around the cell body of one or two specific motoneurons in the newborn [SZENTÁGOTHAI, 1967b]. This does not mean, of course, that this relation is permanent. There are many examples in the CNS where the early synaptic relations established with the cell body are transient and the terminal axons creep along the dendrites at various distances and lose the early contacts with the cell body. This is the case, for example, with the climbing fibers in the cerebellum [RAMÓN Y CAJAL, 1911]. A similar change has to occur also with the I_A collaterals which according to the degeneration findings of CONRADI [1969] can be found mainly on the proximal parts of motoneuron dendrites. But at any rate it seems unlikely that a significant portion of the synapses on the remote parts of the dendrites are I_A afferents. This may be the case also in the very large collaterals of the antero-lateral funiculus, reaching in the newborn cat only one single motoneuron, and establishing 20 synaptic knobs or more on the soma and the proximal dendrites [SZENTÁGOTHAI, 1967b]. It is conceivable that during later development these contacts are shifted to more peripheral parts of the dendrites but less likely that they make other contacts with the dendrites of other cells in the neighborhood. This is mentioned only to indicate that mutual interrelations of dendritic and axonal arborizations, although of crucial importance, do not necessarily and fully explain synaptic architecture and connectivity.

Conversely, synaptic architecture cannot be understood, and even the best degeneration results or electron microscope observations on synapses cannot be correctly interpreted, if the spatial relations of dendrite and axonal arborizations are not sufficiently well-known. The same synapse seen in an electron micrograph has an entirely different functional significance if it is made by an axon crossing a dendrite at right angles, and hence is a single contact, or if the axons and dendrites are running parallel with the consequence that numerous contacts of the same kind exist between the two. The predominant orientation of dendrites and axons is also of great importance for any electron microscopical approach in order to study the structure in the most appropriate planes of sectioning.

Author's address: Dr. J. SZENTÁGOTHAI, 1st Department of Anatomy, Semmelweis University Medical School, Tüzoltó-utca 58, *Budapest IX* (Hungary)

New Developments in Electromyography and Clinical Neurophysiology,
edited by J. E. Desmedt, vol. 3, pp. 38–68 (Karger, Basel 1973)

The Anatomical Organization of the Descending Pathways and their Contributions to Motor Control Especially in Primates

H. G. J. M. KUYPERS

Department of Anatomy, Rotterdam Medical Faculty, Rotterdam

The present essay deals with the organization of the descending pathways from the cerebral cortex and the brain stem to the spinal cord with special emphasis on the rhesus monkey. The connections of these pathways have been studied in a series of anatomical experiments and their contributions to motor control investigated in a series of functional studies based on the anatomical data. The structural and functional findings of these studies all seem to fit together and point to a certain plan of organization of the central motor system.

When considering the connections of the descending pathways to the spinal cord, it should be realized that their motor capacities are not so much determined by the location of their cells of origin, for example in the precentral gyrus or the brain stem reticular formation, as by an entirely different set of factors, i.e. the motor capacities of the interneurons and motoneurons on which these pathways terminate. As a consequence, in order to lead to a functionally meaningful concept the anatomical characterization of the various descending pathways should be based on their termination patterns rather than on the location of their cells of origin [KUYPERS, 1963, 1964].

The descending pathways from the cerebral cortex and the brain stem converge on neurons in the spinal gray matter which consists mainly of (a) the nucleus proprius of the dorsal horn capped by the substantia gelatinosa; (b) the motoneuronal cell groups of the ventral horn, and (c) the intermediate zone (REXED's [1954] laminae VI–VIII in the cat), which contains the majority of the spinal interneurons (fig. 1).

The fibers of the descending pathways from the brain stem terminate mainly in the spinal intermediate zone. This holds true for the cat [NYBERG-HANSEN, 1966; PETRAS, 1967] and the rhesus monkey [KUYPERS, FLEMING and

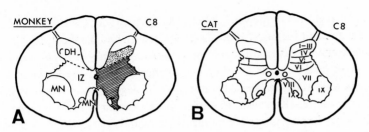

Fig. 1[1]. *A* shows the nucleus proprius (stippled) of the spinal dorsal horn, the intermediate zone (hatched) and the motoneuronal cell groups of the ventral horn (blank) in C8 of the rhesus monkey. *B* shows the location of Rexed's laminae in C8 of the cat.

FARINHOLT, 1962] and presumably also applies in man. Only a few of these brain-stem fibers are distributed in the motoneuronal cell groups. However, their possible termination on motoneuronal dendrites outside these cell groups is anatomically difficult to determine[2].

The fibers from the cerebral cortex also terminate in the intermediate zone, but in addition terminate in the nucleus proprius of the dorsal horn (lamina IV and the medial parts of lamina V) [CHAMBERS and LIU, 1957, 1965; KUYPERS, 1960; NYBERG-HANSEN, 1966] and in some species also in moto-

1   *Key to abbreviations used in all figures:* AS, arcuate sulcus. BC, (decussation of) superior cerebellar peduncle. CN, cochlear nuclei. CP, cerebral peduncle. CS, central sulcus. DC, dorsal column nuclei. DH, dorsal horn of spinal gray matter. F, fastigial nucleus. IA, nucleus interpositus anterior. IC, inferior colliculus. IP, nucleus interpositus posterior. IPS, interparietal sulcus. IS, interstitial nucleus of CAJAL. IZ, intermediate zone of spinal gray matter. L, lateral (dendate) nucleus. LF, lateral fissure. LL, nucleus of lateral lemniscus. LRF, lateral reticular formation. MV, motor nucleus of Vth nerve. ML, medial lemniscus. MN, motoneuronal cell groups of spinal gray matter. MRF, medial reticular formation. PV, principal nucleus of Vth nerve. PD, precentral dimple. PN, pontine nuclei. PS, principal sulcus. PT, pretectum. PY, pyramidal tract. RB, inferior cerebellar peduncle. RN, red nucleus. RS, rubrospinal tract. SV, spinal trigeminal complex. SC, superior colliculus. SN, substantia nigra. SO, superior olive. VC, vestibular complex. III, oculomotor nuclei. V, fifth nerve. VII, facial nucleus. VIII, eighth nerve.
2   Many of the motoneuronal dendrites in the spinal cord are oriented longitudinally and remain within the confines of the motoneuronal cell group [SCHEIBEL and SCHEIBEL, 1966, 1970, 1971; STERLING and KUYPERS, 1967]. However, some of the radially oriented dendrites pass into the intermediate zone and the funiculi where they may receive synaptic terminals from the long descending pathways. The same applies to the radial dendrites of the hypoglossal neurons [CAJAL, 1952]. These dendrites extend into the lateral bulbar reticular formation where they may receive synaptic terminals from the fibers of the descending pathways which terminate in this area.

neuronal cell groups of the ventral horn [KUYPERS, 1960; CHAMBERS and LIU, 1965; PETRAS and LEHMAN, 1966; JANE, CAMPBELL and YASHON, 1965].

The terminal distribution of the *descending brain-stem pathways* was investigated in the rhesus monkey by placing lesions in the lower medulla oblongata and studying the distribution of the degenerating elements in the spinal gray matter [KUYPERS, FLEMING and FARINHOLT, 1962] by means of selective silver impregnation techniques [NAUTA and GYGAX, 1954; ALBRECHT and FERNSTRÖM, 1959; FINK and HEIMER, 1967]. The findings in this study (fig. 2) indicate that the fiber bundles which traverse the paramedian, the ventromedial and the ventrolateral portions of the medullary cross-section distribute fibers throughout the spinal cord including the sacral segments. However, fiber bundles which traverse the area of the lateral reticular formation adjoining the trigeminal nuclei contain mainly short descending fibers which do not travel beyond the cervical cord.

The terminal distribution of the long descending brain-stem fibers was found to be arranged as follows. The fiber bundles which traverse the paramedian and ventromedial portion of the medullary cross-section, descend into the ventral and ventrolateral funiculi and distribute their fibers preferentially to the ventromedial part of the intermediate zone, to some extent bilaterally. In contrast, the fiber bundles which traverse the ventrolateral portion of the medullary cross-section descend into the dorsolateral funiculus and distribute their elements preferentially to the dorsolateral portion of the intermediate zone. The two groups of brain-stem fibers which terminate in the ventro-medial and dorsolateral parts of the spinal intermediate zone will be labeled the *ventromedial* and the *lateral brain-stem pathway*, respectively.

In cat [NYBERG-HANSEN, 1966; PETRAS, 1967] and rhesus monkey [KUYPERS, FLEMING and FARINHOLT, 1962; KUYPERS, 1964] these brain-stem pathways encompass several tracts which originate in different brain-stem cell groups (fig. 3). The ventromedial brain-stem pathway to each half of the spinal cord contains the descending fibers from (a) the ipsilateral lateral vestibular nucleus; (b) the ipsilateral medial reticular formation of the pons and the ipsilateral and contralateral medial reticular formation of the medulla oblongata; (c) the ipsilateral interstitial nucleus of CAJAL; (d) the deep layers of the contralateral superior colliculus, and (e) the ipsilateral and contralateral medial vestibular nuclei.

The fibers from the mesencephalic medial reticular formation do not reach the spinal cord [TORVIK and BRODAL, 1957; NYBERG-HANSEN, 1965]; however, many of them terminate in the medial bulbar reticular formation [KUYPERS, 1964]. For this reason the medial mesencephalic reticular for-

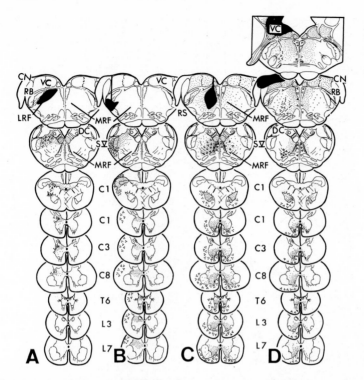

Fig. 2. Schematic diagram of the distribution of fiber degeneration following lesions of different parts of the medullary decussation in the monkey [KUYPERS *et al.*, 1962]. Note that the fibers which traverse the lateral reticular formation (A) are distributed mainly to the dorsolateral part of the spinal intermediate zone of the cervical cord, while those which traverse the area ventral to the trigeminal complex (B) are distributed to the same area as in A but throughout the cord. Note also that the fibers which traverse the medial reticular formation (C) and the fibers from the vestibular complex (D) are distributed mainly to the ventromedial parts of the intermediate zone. (See footnote to figure 1 for key.)

mation will be regarded as associated with the ventromedial brain-stem pathway.

The component tracts of the ventromedial brain-stem pathway share a common termination area in the ventromedial part of the spinal intermediate zone. However, some differences occur in their terminal distribution in this area. For example, the fibers of the medial vestibulospinal and the interstitiospinal tracts terminate most medially [NYBERG-HANSEN, 1966; PETRAS, 1967]. The termination areas of the tectospinal and the medullary reticulospinal tracts in the cat extend further laterally in the ventromedial part of the inter-

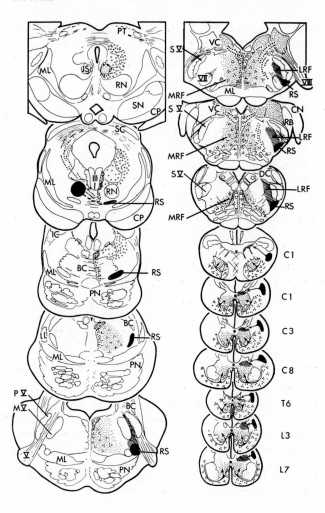

Fig. 3. Diagram of the course of the ventromedial and the lateral brain-stem pathways and their termination in the brain stem and the spinal intermediate zone. The lateral brain-stem pathway (solid black) originates mainly in the contralateral magnocellular red nucleus and terminates in the lateral bulbar reticular formation and the dorsolateral part of the spinal intermediate zone (shaded). The ventromedial brain-stem pathway (coarse and fine stippling) originates mainly in the interstitial nucleus of CAJAL, superior colliculus, vestibular complex and bulbar medial reticular formation, and terminates in the ventromedial parts of the spinal intermediate zone bilaterally. Open circles: medial mesencephalic reticular formation. (See footnote to figure 1 for key.)

mediate zone [NYBERG-HANSEN, 1966; PETRAS, 1967]. Further, the tectospinal and medial vestibulospinal tracts in the cat do not distribute throughout the spinal cord but terminate mainly at cervical levels [RASMUSSEN, 1932; NYBERG-HANSEN, 1966][3].

The lateral brain-stem pathway which traverses the area ventral to the lateral reticular formation and the trigeminal nuclei occupies this position throughout the lower brain stem. It contains primarily fibers from the contralateral red nucleus, which cross in the mesencephalon (COLLIER and BUZZARD, 1901; METTLER, 1944; CARPENTER and PINES, 1956; KUYPERS, 1964; NYBERG-HANSEN, 1966]. This pathway also contains fibers which are distributed from the mesencephalon contralaterally to the bulbar lateral reticular formation [KUYPERS, FLEMING and FARINHOLT, 1962; KUYPERS, 1964; MARTIN and DOM, 1970]. The red nucleus in many species consists of a rostral parvicellular part and a caudal magnocellular part. In the cat the rubrospinal fibers are derived from cells in both parts [POMPEIANO and BRODAL, 1957]. In the rhesus monkey these fibers seem to be derived almost exclusively from the magnocellular part [KUYPERS and LAWRENCE, 1967]. The same may apply to the chimpanzee and man [STERN, 1938]. In these species the red nucleus is very large. However, in comparison with the rhesus monkey it contains only a limited number of magnocellular elements and only a limited number of large rubrospinal fibers descend into the spinal cord [SIE PEK GIOK, 1956; SCHOEN, 1969 b]. This phylogenetic decline in the number of magnocellular elements in the red nucleus and of large rubrospinal fibers in the cord, suggests that the lateral brain-stem pathway may become less prominent in the highest primates. This argument would only be valid if the lateral brain-stem pathway contains mainly fibers from the magnocellular red nucleus. In the cat, rubrospinal fibers are accompanied by few fibers from the pontine reticular formation [BUSCH, 1964]. However, their spinal termination has not been completely clarified.

The brain-stem cell groups which give rise to the component tracts of the ventromedial brain-stem pathway are extensively interconnected and some of them also distribute fibers to eye muscle nuclei [LORENTE DE NÓ, 1933; SZENTÁGOTHAI, 1943; BRODAL and POMPEIANO, 1957; POMPEIANO and WALBERG, 1957; NAUTA and KUYPERS, 1958; SCHEIBEL and SCHEIBEL, 1958; PEARCE,

---

3   The tectospinal fibers in the cat terminate especially laterally in the ventral parts of the intermediate zone in the upper cervical segments but terminate centrally in the low cervical segments [NYBERG-HANSEN, 1966]. This difference in the distribution may merely reflect the lateral expansion of the intermediate zone in low cervical as compared to upper cervical segments [cf. RUSTIONI, KUYPERS and HOLSTEGE, 1971].

1958; ALTMAN and CARPENTER, 1961; BRODAL, POMPEIANO and WALBERG, 1962; MCMASTERS, WEISS and CARPENTER, 1966; LADPLI and BRODAL, 1968; CARPENTER, HARBISON and PETER, 1970]. However, despite the abundance of interconnections between these brain-stem cell groups, their efferent projections consistently avoid the magnocellular red nucleus and the lateral bulbar reticular formation which are related to the lateral brain-stem pathway.

The differential relationship between the descending pathways and the motoneurons of various muscles may be clarified by determining the fiber connections of the cells in the different parts of the *spinal intermediate zone* on which the two brain-stem pathways terminate.

Since the days of CAJAL it has been reported that all cells in the intermediate zone, in addition to their local collaterals, send their main fibers into the funiculi [CAJAL, 1952; TESTA, 1964; SCHEIBEL and SCHEIBEL, 1966; MATSUSHITA, 1970]. These fibers and their collaterals re-enter the gray matter at different levels and terminate mainly in the intermediate zone and the motoneuronal cell groups. In order to further elucidate the organization of these propriospinal connections, the differential distribution of the propriospinal fibers which travel in different parts of the ventral and lateral funiculi was studied [STERLING and KUYPERS, 1968; RUSTIONI, KUYPERS and HOLSTEGE, 1971]. This was done in the cat by making small funicular lesions and comparing the resulting fiber degeneration in the intermediate zone and the motoneuronal cell groups. The cat was chosen as experimental animal because few of its supraspinal fibers terminate in the spinal motoneuronal cell groups [NYBERG-HANSEN, 1969]. As a consequence, the funicular projections to these cell groups are almost exclusively of spinal origin.

The findings in these studies indicate the existence of the following arrangement. In the brachial and the lumbosacral cord (fig. 4), the fibers in the ventral funiculus are distributed mainly to the ventromedial part of the intermediate zone, to some extent bilaterally, and to motoneurons in the ventral part of the ipsilateral ventral horn. The fibers in the dorsal and the intermediate portions of the lateral funiculus show a contrasting pattern and are distributed mainly to the dorsal and lateral part of the ipsilateral intermediate zone and to motoneurons in the dorsal and dorsolateral part of the ipsilateral ventral horn. These findings suggest that the propriospinal connections follow the same organizational principles as those governing the descending brain-stem pathways.

The bulk of the propriospinal fibers were found to be short and to terminate in nearby segments. However, the fibers in the ventral funiculus which are distributed to the ventromedial portion of the intermediate zone travel

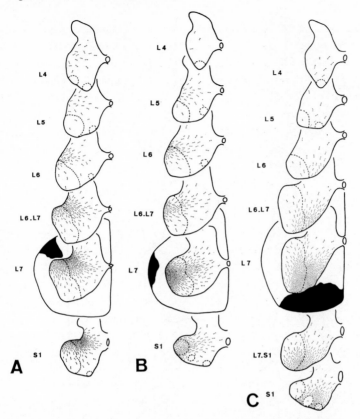

Fig.4. Semidiagrammatic representation of the differential distribution of the degenerating elements in the intermediate zone and the lateral motoneuronal cell group of the lumbosacral cord following lesions of the ventral and lateral funiculi. [After RUSTIONI *et al.*, 1971; reprinted with permission from Brain Research.] Following lesions of the dorsal (A) and intermediate (B) portions of the lateral funiculus, degeneration occurs mainly in the dorsal and lateral parts of the intermediate zone and of the lateral motoneuronal cell group. Following a ventral funicular lesion (C), degeneration occurs mainly in the ventromedial parts of the intermediate zone and of the motoneuronal cell groups.

over much longer distances (fig. 5), some of them interconnecting the enlargements [LLOYD, 1942; WILLIS and WILLIS, 1966; GIOVANELLI and KUYPERS, 1969; RUSTIONI, KUYPERS and HOLSTEGE, 1971].

In order to clarify the differential distribution of the propriospinal fibers in the motoneuronal cell groups their somatotopic organization will be reviewed [MARINESCO, 1904; REED, 1940; SPRAGUE, 1948; ROMANES, 1951; SHARRARD, 1955; STERLING and KUYPERS, 1967]. The motoneurons which

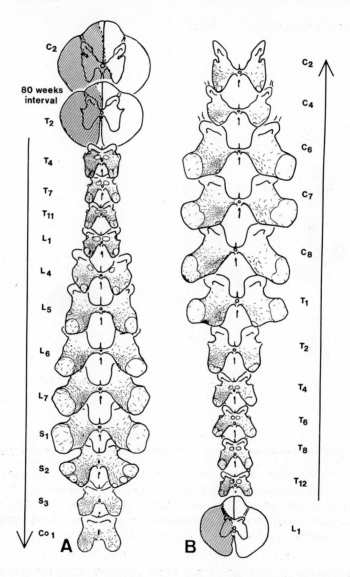

Fig.5. Semidiagrammatic representation of the distribution of the degenerating descending (A) and ascending (B) long propriospinal fibers in the spinal gray matter of the cat following hemisection of the spinal cord at T_2 and L_1. In order to obtain the degeneration of propriospinal fibers only, the supraspinal descending pathways were interrupted by a hemisection at C_2, 80 weeks prior to the hemisection at T_2. [From GIOVANELLI and KUYPERS, 1969; reprinted with permission from Brain Research.]

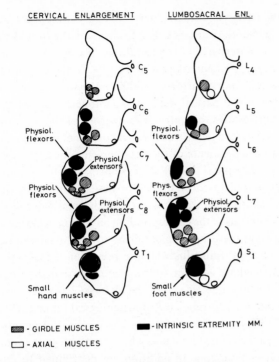

CERVICAL ENLARGEMENT   LUMBOSACRAL ENL.

Fig.6. Diagram of the motoneuronal cell column in the cervical and lumbosacral enlargements of the cat. Note the differential location of the cell columns which innervate axial, girdle and intrinsic extremity muscles.

innervate the striated muscles of body and limbs are clustered into a medial and a lateral cell group (fig. 6) which in the thoracic segments tend to be fused to one nucleus. The medial motoneuronal cell group is located in the ventro-medial portion of the ventral horn and innervates the muscles along the vertebral column. The lateral motoneuronal cell group is located in the lateral part of the ventral horn and varies in size at different levels. In the upper cervical and lumbar segments it is small and its neurons innervate the ventral and lateral neck muscles (e.g. scaleni, longus capitis and colli, levator scapulae, rhomboids) and the lumbar muscles acting on the hip joint (e.g. iliacus and psoas), respectively. Proceeding caudally into both enlargements, the lateral motoneuronal cell group expands dorsolaterally by the addition of new sub-groups. Thus, in the lateral motoneuronal cell group of the enlargements a ventral and a dorsal part may be distinguished. The motoneurons in the ventral part innervate girdle and proximal limb muscles. In the brachial cord they innervate the muscles which act on the shoulder joint (e.g. deltoid, pectorales, latissimus dorsi) and some of the muscles which bridge both the shoulder and the elbow joint. Their counterparts in the lumbosacral cord innervate the muscles which act on the hip joint and those which bridge both the hip and the knee joint (e.g. glutei, rectus femoris, sartorius, adductores, hamstring muscles). The motoneurons in the dorsal part of the lateral motoneuronal cell group innervate the muscles intrinsic to the limbs. The neurons of the most distal intrinsic limb muscles are located in the most dorsal portion of the dorsal part especially in the caudalmost segments of the enlargements. In addition, the motoneurons which innervate physiological flexors intrinsic to the limbs tend to be located in both enlargements close to the dorsolateral funiculus. In the brachial cord they are located mainly dorsal and in the lumbosacral cord dorsolateral to the motoneurons of the corresponding physiological extensors.

These data suggest that the propriospinal fibers in the dorsal and intermediate portions of the lateral funiculus which terminate in the dorsal and lateral parts of the intermediate zone in addition are distributed to motoneurons of the distal extremity muscles and of the intrinsic extremity flexors. This is in general agreement with some electrophysiological findings [BERNHARD and REXED, 1945; KOSTYUK, VASILENKO and LANG, 1971]. In contrast, the propriospinal fibers in the ventral funiculus which terminate in the ventromedial part of the intermediate zone in addition are distributed to motoneurons of axial and proximal limb muscles. In keeping with this pattern, it has been observed that the long ascending propriospinal fibers to the ventromedial part of the cervical intermediate zone are accompanied by lumbar and low thoracic fibers which are distributed to motoneurons of shoulder muscles [STERLING and KUYPERS, 1967, 1968; GIOVANELLI and KUYPERS, 1969]. This is in agreement with recent electrophysiological findings [MILLER, 1970].

The location of the cells of origin of the different propriospinal fibers is somewhat uncertain. However, a contrast appears to exist in that some cells in the most lateral part of the intermediate zone send their fibers into the lateral funiculus while cells in the most ventromedial part of the intermediate zone send them into the ventral funiculus [CAJAL, 1952; SZENTÁGOTHAI, 1964; STERLING and KUYPERS, 1968; MATSUSHITA, 1969, 1970]. These cells in the two extremes of the intermediate zone must be mainly under the influence of the lateral and the ventromedial brain-stem pathway, respectively. This, in turn, suggests that the lateral brain-stem pathway differs from the ventromedial one in that the former, by way of its connections with propriospinal elements, influences especially distal extremity muscles and the intrinsic extremity flexors, while the latter influences especially axial and proximal limb muscles. However, in view of the local connections between the cells in the intermediate zone, the anatomical findings do not imply that the two brain-stem pathways exclusively influence motoneurons of these respective groups of muscles. Instead, the findings only suggest that when these pathways are compared they would show this bias.

The differential relationship between the descending brain-stem pathways and the spinal motoneurons as suggested by the anatomical findings had also been observed physiologically. Many studies demonstrated that the rubro-spinal tract facilitates mainly flexor motoneurons by way of interneurons in the dorsolateral part of the intermediate zone [POMPEIANO, 1957; SASAKI, NAMIKAWA and HASHIRAMOTO, 1960; THULIN, 1963; HONGO, JANKOWSKA and LUNDBERG, 1969; KOSTYUK and PILYAVSKY, 1969]. In contrast, the lateral vestibulospinal tract has been found to facilitate extensor motoneurons and to inhibit their flexor counterparts [FULTON, LIDDELL and RIOCH, 1931; SPRAGUE and CHAMBERS, 1954; BRODAL, POMPEIANO and WALBERG, 1962; SASAKI, TANATA and MORI, 1962; WILSON and YOSHIDA, 1968, 1969 b; GRILL-NER, HONGO and LUND, 1970]. The difference between the two brain-stem pathways in respect to their relation to distal *versus* axial and proximal limb muscles has seldom been mentioned in the literature. However, SHERRINGTON [1906] had observed that decerebrate rigidity affects especially axial and proxi-mal limb muscles and that this rigidity may be abolished by transection of the upper cervical ventral funiculus, in agreement with the anatomical findings. In addition, it has been shown that the vestibulospinal components of the ventromedial brain-stem pathway preferentially influence proximal reflex movements [McCOUCH, LIU and CHAMBERS, 1966]. In contrast, pathways to the dorsolateral part of the intermediate zone have been shown to influence distal reflex movements and to facilitate preferentially motoneurons of distal extremity muscles [BROOKHART, 1952; CORAZZA, FADIGA and PARMEGGIANI, 1963; ENGBERG, 1964; PRESTON, SHENDE and UEMARA, 1967; HONGO, JAN-KOWSKA and LUNDBERG, 1969; ASANUMA, STONEY and THOMPSON, 1971]. The arrangement suggested by the anatomical data is also consistent with recent electrophysiological findings concerning the distribution of monosynaptic connections from the descending brain-stem pathways to the motoneurons [SHAPOVALOV, KURCHAVAYI, KARANYAN and REPINA, 1971; SHAPOVALOV, this volume][4]. The connections from the vestibular complex and the bulbar

4   The anatomical findings [KUYPERS, FLEMING and FARINHOLT, 1962; NYBERG-HANSEN, 1969] suggest that these monosynaptic connections are either limited in number or are established mainly with radial motoneuronal dendrites which extend into the funiculi and the intermediate zone.

reticular formation in the rhesus monkey were found to be distributed mainly to motoneurons of proximal limb muscles, in agreement with anatomical findings in monkey [KUYPERS, FLEMING and FARINHOLT, 1962] and with some physiological findings in cat [WILSON and YOSHIDA, 1969a, b]. On the other hand, the connections from the red nucleus were found to be distributed mainly to motoneurons of distal extremity muscles [SHAPOVALOV, KURCHAVAYI, KARANYAN and REPINA, 1971; SHAPOVALOV, this volume].

The descending brain-stem pathways to the spinal cord are paralleled by the corticospinal pathway. Many of these cortical fibers also terminate in the intermediate zone where their termination area overlaps with that of the descending brain-stem pathways. However, the cortical fibers in addition terminate in the somatosensory nuclei (nuclei cuneatus and gracilis and the nucleus proprius of the dorsal horn) [WALBERG, 1957; KUYPERS, 1958a, b, 1960; CHAMBERS and LIU, 1957, 1965; NYBERG-HANSEN, 1966; SCHOEN, 1969 a] and in some species also in the motoneuronal cell groups [KUYPERS, 1960; CHAMBERS and LIU, 1965; PETRAS and LEHMAN, 1966; JANE, CAMPBELL and YASHON, 1969]. The fibers to the somatosensory nuclei especially modulate sensory transmission while those to the intermediate zone mainly influence the motoneurons [ANDERSEN, ECCLES and SEARS, 1964; MARCHIAFAVA and POMPEIANO, 1964; MORRISON and POMPEIANO, 1965; FETZ, 1968].

The termination area of the cortical fibers in the *intermediate zone* is rather restricted in opossum and goat [BAUTISTA and MATZKE, 1965; HAARTSEN and VERHAART, 1967]. It is larger in cat [SZENTÁGOTHAI-SCHIMERT, 1941; CHAMBERS and LIU, 1957; NYBERG-HANSEN, 1966] and is very extensive in higher primates, i.e. rhesus monkey, chimpanzee and man [KUYPERS, 1960, 1964; SCHOEN, 1964; CHAMBERS and LIU, 1965; KUYPERS and BRINKMAN, 1970]. In the cat, the cortical fibers from each hemisphere are distributed to the dorsolateral part of the intermediate zone contralaterally (fig. 7). In contrast, in higher primates these cortical fibers are distributed bilaterally, i.e. to the dorsolateral part of the intermediate zone contralaterally and to its ventromedial parts bilaterally. However, the cells along the medial border of the ventral horn ($\pm$ lamina VIII) receive only a limited number of fibers from the contralateral hemisphere. Cortical fibers from each hemisphere are also distributed to contralateral motoneuronal cell groups, especially those which innervate distal extremity muscles [BERNHARD and BOHM, 1954; PHILLIPS and PORTER, 1964; PHILLIPS, 1969; KUYPERS and BRINKMAN, 1970]. These fibers are also present in some other animals but are most numerous in chimpanzee and man.

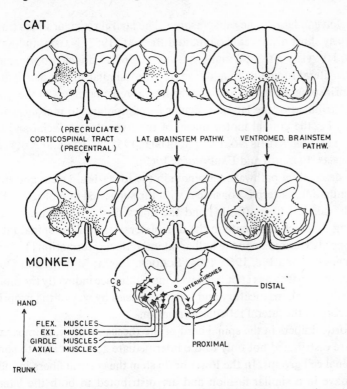

Fig. 7. Diagram of the distribution of cortical and brain-stem pathways in the spinal intermediate zone and motoneuronal cell groups in cat and rhesus monkey. Note the differences between the distributions of the cortical fibers in cat and monkey.

The discussion so far has dealt only with the termination of the descending pathways in the spinal cord, arbitrarily separating it from the bulbus, i.e. the lower brain stem. However, these two parts of the neuraxis are closely related and several cell groups may be traced from one structure into the other. For example, the spinal trigeminal complex represents the uninterrupted rostral continuation of the spinal dorsal horn, while the bulbar reticular formation represents the uninterrupted continuation of the spinal intermediate zone [OLSZEWSKI and BAXTER, 1954]. However, owing to the unfolding of the neural tube in the lower brain stem the bulbar homologues of the dorsolateral and the ventromedial parts of the spinal intermediate zone come to lay laterally and medially in the lower brain stem. In view of this, the lateral reticular fibers would represent the bulbar homologue of the propriospinal fibers in the dorsal parts of the lateral funiculus. This is supported by the fact that the two groups of fibers behave in largely the same fashion. The lateral reticular fibers travel over relatively short distances [TORVIK and BRODAL, 1957; NAUTA and KUYPERS, 1958; KUYPERS, FLEMING and FARINHOLT, 1962] as do the propriospinal fibers in the lateral funiculus.In addition, in a way similar to the distribution of the propriospinal fibers in the cord, the lateral reticular fibers in the brain stem are distributed to the lateral reticular formation and to motoneuronal cell groups, i.e. the cranial motor nuclei of the hypoglossal, facial and trigeminal nerves [LORENTE DE NÓ, 1933; SCHEIBEL, 1955; NAUTA and KUYPERS, 1958; KUYPERS, FLEMING and FARINHOLT, 1962].

Some descending pathways terminate in the lower brain stem and their bulbar termination pattern is consistent with the suggested homologues. For example the cortical fibers (fig. 14) and the fibers of the lateral brain-stem pathway (fig. 3) which terminate in the dorsolateral part of the spinal intermediate zone also terminate in the bulbar lateral reticular formation [TORVIK, 1956; KUYPERS, 1958 a, b, c, 1960, 1964; SCHOEN, 1969 a; MARTIN and DOM, 1970]. Thus, these pathways at bulbar levels may influence indirectly the motoneurons of the facial, masticatory and tongue muscles by way of the propriobulbar neurons in the lateral reticular formation.

The cortical fibers in the spinal cord of the rhesus monkey, chimpanzee and man are distributed not only to the intermediate zone but also to some motoneuronal cell groups. In the lower brain stem the cortical fibers in these species behave in a similar fashion and are distributed to both the lateral bulbar reticular formation and to the cranial motor nuclei of the hypoglossal, facial and trigeminal nerves [KUYPERS, 1958 b, c, 1960; SCHOEN, 1969 a]. However, in the lower brain stem some of the fibers of the lateral brain-stem pathway also terminate directly on bulbar motoneurons, i.e. in the dorsal and lateral part of the facial nucleus [KUYPERS, 1964; SCHOEN, 1969 b; MARTIN and DOM, 1970].

It has been pointed out already that the motor capacities of the individual pathways are mainly determined by the motor capacities of the interneurons and motoneurons on which they terminate [KUYPERS, 1963]. When this idea is applied to the entire population of descending pathways it suggests that cortical and brain-stem pathways which terminate on the same interneurons should subserve a similar function and way be grouped together.

On this basis in the *cat*, two major groups of pathways may be distinguished which influence motoneurons by way of the interneurons in the different parts of the intermediate zone (fig. 7). One group would consist of the ventromedial brain-stem pathway to the ventromedial parts of the intermediate zone. The other would be composed of the corticospinal pathway and the lateral brain-stem pathway to its dorsolateral part. These two groups of pathways would differ in that the former would influence especially motoneurons of axial and proximal limb muscles and of external eye muscles, while the latter would influence especially motoneurons of distal extremity muscles and of the intrinsic extremity flexors as well as motoneurons of face, tongue and chewing muscles.

The propriospinal fibers in the *rhesus monkey* are in all likelihood distributed in the same way as in the cat. In that case the same two groups of pathways may be distinguished in this animal (fig. 7). However, in the rhesus monkey, the pathways to the ventromedial part of the spinal intermediate zone would also contain corticospinal elements. Further, in the rhesus monkey, chimpanzee and man a third pathway may be distinguished which consists mainly of corticospinal fibers and leads directly to the motoneuronal cell groups of distal extremity muscles, face muscles and tongue muscles. On the basis of this termination these fibers may mobilize selectively small individual groups of motoneurons of distal extremity muscles and of facial and tongue muscles.

The anatomical data, when viewed in this perspective, imply a functional organization in which body movements and distal extremity movements are differentially controlled. The importance of these functional implications made it seem worthwhile to try to test their validity. First this was done in the cat by comparing the motor defects resulting from the interruption of one or other of the two groups of descending pathways which terminate in the different parts of the intermediate zone [KUYPERS, 1963, 1964].

In one group of cats the ventromedial brain-stem pathway was interrupted by a large lesion in the core of the medulla oblongata. In another group both pyramidal tracts were interrupted in combination with one lateral brain-stem pathway. Consistent with the anatomical findings, these two groups of animals showed two contrasting motor defects. The animals of the first group did not right themselves for several days. In addition, they showed unsteadiness of head and trunk, displayed exaggerated flexion reflexes and stood in a crouched position. However, the animals were able to use their extremities effectively in clawing, and visual and tactile placing reactions could be elicited almost as readily as in the normal animal. In addition, the animals walked bars very well by gripping them with their 'hands' and 'feet', provided their body was properly supported. The second group of animals displayed a different type of motor defect affecting mainly distal extremity movements. The animals had little trouble with righting and did not show unsteadiness of head and trunk. However, they suffered from some extremity weakness, which was

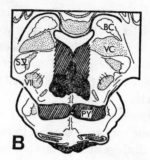

A   B

Fig.8. Semidiagrammatic representation of brain-stem sections of animals with *A* bilateral pyramidotomy, and *B* bilateral pyramidotomy plus a lesion of the ventromedial brain-stem pathways [drawn after LAWRENCE and KUYPERS, 1968a, plate VI, fig.2; 1968b, plate VIII, fig.5, lesion solid black, gliosis shaded]. (See footnote to figure 1 for key.)

most obvious in the wrist. Visual placing and contact placing of the extremities occurred but the extremity with both pathways interrupted was placed mainly by a proximal extremity movement with little wrist and elbow flexion. This paucity of distal extremity movements was obvious even when the animal was fighting in anger. These findings seem to support the functional implications of the anatomical findings.

The functional capacities of the descending pathways were also studied in the rhesus monkey [LAWRENCE and KUYPERS, 1968a, b]. However, in this animal separate interruption of the two groups of pathways to the respective dorsolateral and ventromedial parts of the intermediate zone is difficult to achieve since both groups contain pyramidal elements. For this reason a different approach was followed. First in all animals both pyramidal tracts were interrupted. After recovery from the pyramidal lesions either the ventromedial or the lateral brain-stem pathway was interrupted and the resulting motor defects compared. Immediately after the bilateral pyramidotomy (fig. 8 A) all animals sat up, walked and climbed. After recovery they were able to extend fully either arm and pick up morsels of food by closing all fingers together. However, relatively independent finger movements as observed in the normal animal never returned (fig. 10 A, B). This was demonstrated by testing the animals on a modified Klüver board [LAWRENCE and KUYPERS, 1968a, b].

In some of the monkeys with bilateral pyramidotomy the ventromedial brain-stem pathway was interrupted bilaterally, immediately caudal to the abducens nucleus (fig. 8 B). As in the cat, these animals could not right themselves for several weeks and in addition showed a flexion bias of trunk and

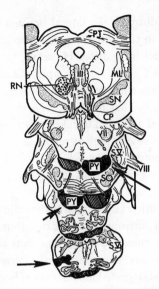

Fig. 9. Semidiagrammatic representation of brain-stem sections of an animal with a bilateral pyramidotomy (solid black) and a unilateral lesion of the *left* lateral brain-stem pathway (solid black). Note cell loss in the crossed red nucleus. Pyramidal gliosis, shaded. [Drawn after LAWRENCE and KUYPERS, 1968b, plate IV, fig. 3.] (See footnote to figure 1 for key.)

limbs. When they finally righted themselves they tended to slump forward and frequently fell over (fig. 11 D), apparently due to an impairment of correcting movements, particularly those involving the body axis and proximal part of the limbs. The few animals which could ultimately walk had difficulty in directing their course of progression. Despite these severe defects, 1 week after the operation the animals could pick up pieces of food with their hands in much the same manner as after recovery from bilateral pyramidotomy (fig. 10C). However, when approached with food their general behaviour was different from before. After pyramidotomy, when sitting in the examining chair, the animals immediately looked at the approaching food, turned towards it and eagerly reached for it with an outstretched arm and an open hand (fig. 10B). After the additional interruption of the ventromedial brain-stem pathways they did not turn towards the food and did not reach for it from the shoulder. Instead, they followed the food mainly with their eyes and picked it up only when it was close enough by moving primarily their elbows and hands (fig. 10C).

In other monkeys with bilateral pyramidotomy, the lateral brain-stem pathway was interrupted (fig.9). In contrast to the findings in the previous group these animals seldom showed any defect in righting. However, some motor defect occurred in the ipsilateral arm and hand. For example, the animals had difficulty in flexing the arm when it was extended and in closing the hand. When offered food soon after the operation they moved their arm with the hand and fingers extended towards the food and tried to rake it into their mouth with the extended hand. After further recovery they were able to close the hand but only as a part of a total arm movement (fig. 11, E1, E2). Yet they remained unable to execute independent closing movements of the hand on an extended arm.

In respect to the rhesus monkey these and other findings led to the following conclusions. The ventromedial brain-stem pathway, which is distributed bilaterally to the ventromedial parts of the intermediate zone, represents the basic system by which the brain governs movements. This control is concerned in particular with movements of the head, with integrated movements of body and limbs and with total limb movements. This control is also concerned with the maintenance of erect posture and with directing the course of progression. The lateral brain-stem pathway which terminates in the dorsolateral part of the intermediate zone, adds to this basic control the capacity for independent use of the extremities, especially of the hands. The corticospinal pathway which terminates in both the dorsolateral and the ventromedial parts of the intermediate zone exerts an additional control over both types of movements and strongly amplifies the capacity for independent use of the extremities especially their distal parts. In addition, the corticospinal pathway characteristically provides the capacity for a high degree of fractionation of movements as exemplified by relatively independent finger movements.

These functional findings are consistent with previous observations in freely moving animals. In respect to the ventromedial brain-stem pathway electrical stimulation of the superior colliculus and of the area of the posterior commissure, the interstitial nucleus of Cajal and the medial vestibular nuclei was found to elicit raising and lowering of the head as well as turning and rotatory movements of head and trunk, frequently in combination with similarly directed eye movements. Unilateral electrical stimulation of the medial reticular formation in the mesencephalon and the lower brain stem elicits turning movements of the head and the trunk with flexion of the ipsilateral and extension of the contralateral extremities [INGRAM, RANSON, HANNETT, ZEISS and TERWILLIGER, 1932; BURGI and MONNIER, 1943; APTER, 1946; HESS, BURGI and BUCHER, 1946; MONNIER, 1946; HASSLER and HESS,

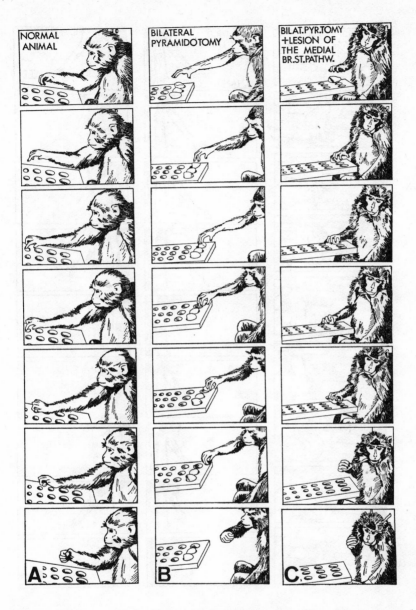

Fig. 10. Drawings of successive photographic frames from films of LAWRENCE and KUYPERS, 1968a, b and BRINKMAN *et al.*, 1970, showing 3 monkeys taking food morsels from a modified Klüver board. *A* Normal animal; note relatively independent finger movements. *B* Animal with bilateral pyramidotomy (cf. fig. 8 A). Note lack of relatively independent finger movements. *C* Animal with bilateral pyramidotomy combined with a lesion of the ventromedial brain-stem pathway (cf. fig. 8 B). Note similarity of hand movements in *B* and *C*. Note also that arm movements in *C* occur mainly at the elbow.

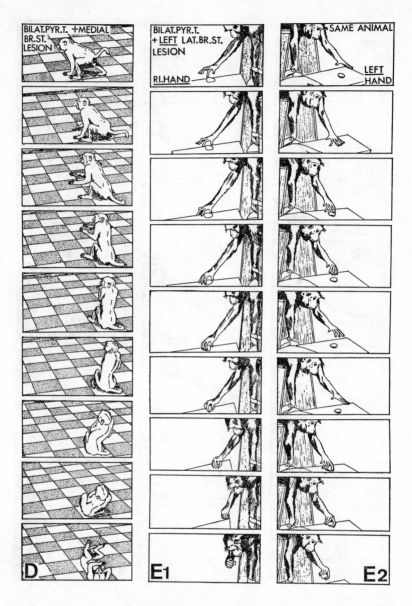

Fig. 11. Drawing of successive photographic frames from film of LAWRENCE and KUYPERS [1968 b] of an animal (D) with a bilateral pyramidotomy combined with a lesion of the ventromedial brain-stem pathway (cf. fig. 8 B). Sitting down, the animal topples over. E_1 and E_2 show an animal with a bilateral pyramidotomy and a lesion of the *left* lateral brain-stem pathway (cf. fig. 9) picking up food morsels. *Right* hand *(E_1)* picks up food morsel in the same fashion as in figure 10B (bilateral pyramidotomy). *Left* hand *(E_2)* shows limited action. The food is caught with a sweeping movement of the arm.

CORTICOSPINAL FIBERS

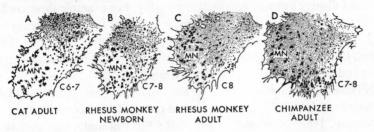

CAT ADULT   RHESUS MONKEY   RHESUS MONKEY   CHIMPANZEE
           NEWBORN   ADULT   ADULT

Fig. 12. Semidiagrammatic representation of the distribution of the cortical fibers in the intermediate zone and the motoneuronal cell groups (MN) of the cervical cord in cat (A), newborn rhesus monkey (B), adult rhesus monkey (C) and chimpanzee (D). Note especially the differences in the distributions in the motoneuronal cell groups. [Drawn after fig. 14, KUYPERS, 1963.]

1954; SPRAGUE and CHAMBERS, 1954; MONTANDON and MONNIER, 1964; HESS, 1968; MABUCHI, 1970]. Further stimulation of this area at the isthmus elicits walking movements. Both turning and walking movements may be elicited independent of the corticospinal and rubrospinal tracts [HINSEY, RANSON and DIXON, 1930; SHIK, ORLOVSKII and SEVERIN, 1968].

In respect to the lateral brain-stem pathway lesions of the area of the rubrospinal tract in the monkey have been reported to produce hypokinesia of the ipsilateral extremities [ORIOLI and METTLER, 1956, 1957]. Lesions of the area of the red nucleus in the cat have been found to interfere with conditioned flexor responses of the contralateral forepaw [SMITH, 1970].

In respect to the corticospinal pathway, our findings are in keeping with many of the findings of TOWER [1940, 1949]. In general, they are also in agreement with those of BUCY and others in the rhesus monkey and in man [BUCY, KEPLINGER and SIQUEIRA, 1964; MASPES and PAGNI, 1964; WALKER and RICHTER, 1966; BUCY, LADPLI and EHRLICH, 1966]. However, the marked recovery which occurred in some of their cases may in part be due to remaining corticospinal fibers which were spared by the lesions [cf. LAWRENCE and KUYPERS, 1968a, b]. Finally, the more recent findings of BECK and CHAMBERS [1970] further demonstrated the impaired capacity of the pyramidotomized monkey to execute relatively independent finger movements.

This capacity depends in all likelihood on the direct cortical projections to the motoneurons of the distal extremity muscles. This is supported by the following observations. Direct cortico-motoneuronal connections (fig. 12) occur in those animals which possess the capacity for individual finger movements and display unusual manipulatory agility. For example these connections are present in increasing numbers in the slow loris *(Nycticebus coucang)*, raccoon, rhesus monkey, chimpanzee and man (KUYPERS, 1960; SCHOEN, 1964; CHAMBERS and LIU, 1965; JANE, CAMPBELL and YASHON, 1965; PETRAS and LEHMAN, 1966; BUXTON and GOODMAN, 1967]. However, they are largely lacking in the opossum, tree shrew *(Tupaia glis)*, rat, goat, cat and dog [CHAMBERS and LIU, 1957; BAUTISTA and MATZKE, 1965; GOODMAN, JARRARD and NELSON, 1966; NYBERG-HANSEN, 1966; BUXTON and GOODMAN, 1967; HAARTSEN and VERHAART, 1967; SCHRIVER and NOBACK, 1967; JANE, CAMPBELL and YASHON, 1969]. Secondly, in newborn rhesus monkeys, the direct cortico-motoneuronal connections are not established as yet but develop during the first 6 months of postnatal life [KUYPERS, 1962; cf. also BUXTON and GOODMAN, 1967; DOMINIK and WIESENDANGER, 1971]. Consistent with the anatomical findings it has been demonstrated that the rhesus monkey during the first 2 months of its postnatal life does not possess the capacity for relatively independent finger movements and that its subsequent development depends exclusively on the pyramidal tract [LAWRENCE and HOPKINS, 1970, 1972].

The cortical projections to the spinal and bulbar interneurons and motoneurons and to the cell groups of the descending brain-stem pathways in the rhesus monkey are derived especially from the frontal lobe [KUYPERS and LAWRENCE, 1967]. The projections to nuclei of descending brain-stem pathways are derived from a large area which encompasses the precentral gyrus and the rostrally adjoining 'premotor' area. The fibers from different parts of this frontal area are distributed to different brain-stem cell groups. (The cortical projections to other brain-stem cell groups will not be dealt with. They have been described in the original paper [KUYPERS and LAWRENCE, 1967] but do not seem to be immediately relevant to the present subject.)

The fibers from the premotor area, including the frontal eye field, avoid cranial motor nuclei and are distributed to cell groups of the ventromedial brain-stem pathway, i.e. the deep layers of the superior colliculus and the pretectum, the nucleus of the posterior commissure and the medial reticular formation of the mesencephalon and lower brain stem (fig. 13). The fibers to the lower brain stem travel by way of the pyramidal tract and pes lemnisci. The descending pathways from these cell groups are particularly concerned with the control of integrated movements of head, body and limbs [LAWRENCE and KUYPERS, 1968 a, b]. The distribution of the premotor projections to these cell groups suggests that the premotor area subserves a similar function. This

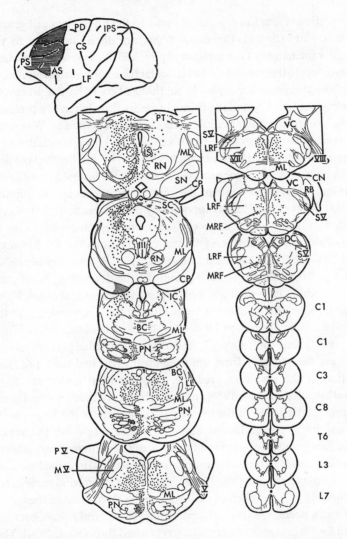

Fig. 13. Diagram of the distribution of the fiber degeneration following lesions of the premotor area rostral to the precentral gyrus. Note that the fibers are distributed mainly to cell groups of the ventromedial brain-stem pathway. (See footnote to figure 1 for key.)

is in keeping with the observation that electrical stimulation of this area in the freely moving animal elicits turning movements of the eyes, the head and neck and the body as well as other total body movements [LILLY, 1958; HASSLER, 1960; WAGMAN, 1964].

The projections from the precentral gyrus to the brain-stem cell groups terminate differently (fig. 14). Precentral fibers are distributed also to the medial reticular formation of the mesencephalon and of the lower brain stem. However, the fibers to the pontine medial reticular formation are less numerous than those from the premotor areas. In addition fibers from the precentral gyrus of the cat, the monkey and the chimpanzee are distributed in a topically organized fashion to the magnocellular red nucleus [RINVIK and WALBERG, 1963; KUYPERS, 1966; KUYPERS and LAWRENCE, 1967] from which the main components of the lateral brain-stem pathway arise. This implies that the precentral gyrus by means of its connections to the lateral brain-stem pathway may govern individual movements of the contralateral fore- and hindlimbs, especially their distal parts. This is supported by the observations that electrical stimulation of these cortical areas elicits arm and leg movements even in the absence of the pyramidal tracts [WOOLSEY, SETTLAGE, MEYER, SPENCER, HAMUY and TRAVIS, 1950; LEWIS and BRINDLEY, 1965; WOOLSEY et al., 1972].

The above projections to the cell groups of the descending brain-stem pathways lead indirectly to the bulbar and spinal interneuronal areas. However, the frontal lobe also distributes fibers directly to these areas, i.e. to the bulbar lateral reticular formation bilaterally, to the dorsolateral part of the spinal intermediate zone contralaterally and to its ventromedial part bilaterally. These fibers are derived only from the precentral gyrus and from the most caudal portion of the premotor area [i.e. the precentral motor cortex of WOOLSEY and his collaborators, 1950] and travel by way of the pyramidal tract. The face representation area of the motor cortex constitutes the main source of the cortical fibers to the lateral bulbar reticular formation [KUYPERS, 1958b]. The hand and foot representation areas distribute fibers almost exclusively to the dorsolateral part of the contralateral spinal intermediate zone in the corresponding enlargements. The fibers to the ventromedial parts of the spinal intermediate zone are derived mainly from the rostral part of the motor cortex and from a portion of its caudal part along the central sulcus, between the hand and foot representation areas [KUYPERS and BRINKMAN, 1970]. This seems to be in keeping with the findings of WOOLSEY and his collaborators [1950] who found that electrical stimulation of the latter two areas in the rostral and caudal parts of the motor cortex elicited body and proximal limb movements.

In summary, the area rostral to the precentral gyrus in the rhesus monkey projects to the ventromedial part of the intermediate zone indirectly by way of the ventromedial brain-stem pathway and subserves mainly the control of integrated movements of eyes, head, body and limbs. When proceeding to-

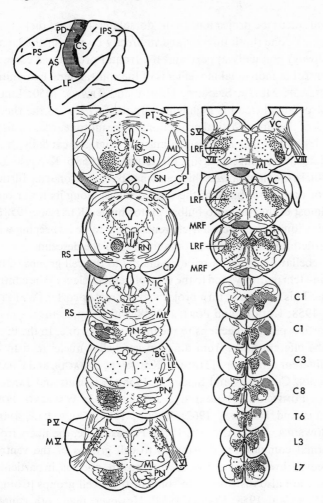

Fig. 14. Schematic diagram of the distribution of the degenerated fibers in the brainstem and spinal cord following a lesion of the precentral gyrus. Note their distribution to the magnocellular red nucleus, lateral reticular formation, spinal intermediate zone and bulbar and spinal motoneuronal cell groups. Distribution in the dorsolateral mesencephalic tegmentum has been omitted. The precentral origin of the bulk of the cortical fibers to bulbar and spinal motoneuronal cell groups is indicated in solid black. (See footnote to figure 1 for key.)

wards the central sulcus the projections to the dorsolateral part of the contralateral intermediate zone (both directly and indirectly by way of the lateral
brain-stem pathway) gradually appear and the frontal projections come to
subserve the control of individual movements of the contralateral extremities
[WOOLSEY, SETTLAGE, MEYER, SPENCER, HAMUY and TRAVIS, 1950; LILLY,
1958]. This rostro-caudal trend is highlighted by the fact that in the rhesus
monkey the bulk of the direct cortical projections to the motoneurons of distal
extremity muscles and of face and tongue muscles originate caudally, from
specific areas along the central sulcus (fig. 14) [KUYPERS, 1958b; KUYPERS and
BRINKMAN, 1970]. In the chimpanzee, this trend may be carried one step further
since portions of the rostral bank of the central sulcus along its lower one-
third project almost exclusively into bulbar motor nuclei [KUYPERS, 1958b].

The organization of the cerebellar efferents seems to be in keeping with
the proposed structural and functional organization of the descending pathways. These cerebellar fibers are distributed mainly to the cell groups of the
descending brain-stem pathways and to the ventrolateral nucleus of the contralateral thalamus. This nucleus in turn projects to the motor cortex [WALKER,
1938; MACCHI, 1958; KRÜGER and PORTER, 1958] which represents the main
source of the corticospinal projections to the intermediate zone. In the brain
stem (fig. 15), the efferent fibers from different cerebellar nuclei tend to be
distributed to different cell groups [THOMAS, KAUFMAN, SPRAGUE and CHAM-
BERS, 1956; COHEN, CHAMBERS and SPRAGUE, 1958; WALBERG and JANSEN,
1961; WALBERG, POMPEIANO, BRODAL and JANSEN, 1962; WALBERG, POM-
PEIANO, WESTRUM and HANSSEN, 1962; VOOGD, 1964; COURVILLE, 1966;
ANGAUT and BOWSHER, 1970; ANGAUT, 1970]. For example, the fibers from
the fastigial-vermal complex are distributed to cell groups of the ventromedial brain-stem pathway. Fibers from the interpositus nucleus, in particular
its posterior part, are also distributed to some of these cell groups [COHEN,
CHAMBERS and SPRAGUE, 1958; VOOGD, 1964]. However, many other interpositus fibers which are especially derived from the anterior part of the nucleus
are distributed to the contralateral magnocellular red nucleus [VOOGD, 1964;
COURVILLE, 1966] from which a major portion of the lateral brain-stem pathway arises [POMPEIANO and BRODAL, 1957; KUYPERS and LAWRENCE, 1967].

The influence exerted from the cerebellar nuclei on the contralateral
motor cortex by way of the ventrolateral nucleus of the thalamus as observed
in the cat seems to conform to this general pattern. For example, the fastigial
nuclei influence mainly the rostral part of the motor cortex [RISPAL-PADEL and
MASSION, 1970; RISPAL-PADEL, LATREILLE and VANUXEM, 1971; MASSION and
PADEL-RISPAL, 1972] which, in the rhesus monkey, projects mainly to cell

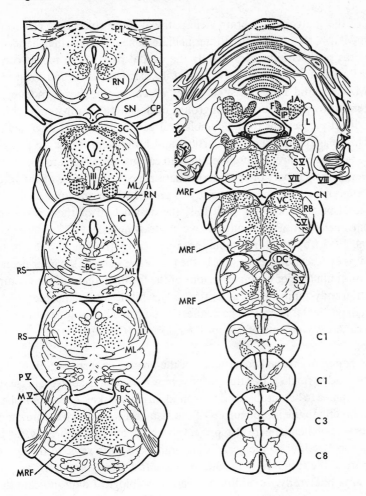

Fig. 15. Diagram of the differential distribution of cerebellar fibers from the vermal-fastigial complex and the interpositus nuclei to cell groups of the descending brain-stem pathways. Vermal-fastigial fibers are distributed to cell groups of the ventromedial brain-stem pathway (stippled). Interpositus fibers, mainly from the n. interpositus posterior are also distributed to these cell groups. Many other interpositus fibers, mainly from the n. interpositus anterior are distributed to the magnocellular red nucleus (crosses) which contributes to the lateral brain-stem pathway. [Based on data of THOMAS *et al.*, 1956; COHEN *et al.*, 1958; VOOGD, 1964; COURVILLE, 1966; ANGAUT, 1970; ANGAUT and BOWSHER, 1970.] (See footnote to figure 1 for key.)

groups of the ventromedial brain-stem pathway and projects directly to the ventromedial parts of the spinal intermediate zone. On the other hand, the interpositus nucleus influences mainly the caudal part of the motor cortex [RISPAL-PADEL and MASSION, 1970; RISPAL-PADEL, LATREILLE and VANUXEM, 1971; MASSION and PADEL-RISPAL, 1972] which, in the rhesus monkey, is characterized by projections to cell groups of the lateral brain-stem pathway and by direct projections to the dorsolateral part of the spinal intermediate zone. This suggests that both at cortical and brain-stem levels the fastigial efferents tend to be directed preferentially to cell groups which are connected to the ventromedial part of the intermediate zone, while many of the interpositus efferents tend to be directed to cell groups which are connected to its dorsolateral part. In keeping with this, the destruction of these respective cerebellar nuclei in the cat results in two contrasting defects [CHAMBERS and SPRAGUE, 1955a, b] which resemble those caused by the interruption of the two groups of descending pathways to the respective parts of the intermediate zone in this animal. For example, lesions of the fastigial nuclei affect posture, locomotion and equilibrium of the entire body, while lesions of the interpositus nucleus, in particular its rostral part, affect placing and hopping reactions and the individual movements of the ipsilateral limbs [CHAMBERS and SPRAGUE, 1955a, b].

The apparent internal consistency of the present concept encouraged us to test its validity further. For this purpose motor control in split-brain rhesus monkeys seemed to be suitable testing ground. In the monkey, each hemisphere projects directly to the motoneurons of the distal muscles of the contralateral extremities. Each hemisphere also projects both directly and indirectly, i.e. by way of its connections to the descending brain-stem pathways, to the dorsolateral part of the spinal intermediate zone contralaterally and to its ventromedial parts bilaterally. In addition, electrophysiological data indicate that the projections by way of the descending brain-stem pathways to the dorsolateral and ventromedial parts of the intermediate zone include some projections to motoneurons of distal and proximal extremity muscles, respectively [SHAPOVALOV, KURCHAVAYI, KARANYAN and REPINA, 1971]. This would imply that contralaterally each hemisphere has full control over the movements of arm, hand and fingers, while ipsilaterally it has full control only over proximal arm movements and complex arm-hand movements and lacks such control over individual movements of the hand and the fingers. Some preliminary findings in split-brain monkeys seem to support these assumptions.

The optic chiasma, and the telencephalic, diencephalic and dorsal mesencephalic commissures were transected in 8 rhesus monkeys. Their motor performance was studied while they were picking up pieces of food from a forceps with one eye covered. Under these circumstances the movements of the arm and hand ipsilateral or contralateral to the open eye did not differ strikingly. However, the agility of the hand and finger movements ipsilateral to the open eye seemed to depend on contact between the hand and the food morsels. Such contact presumably recruits the motor guidance of the non-seeing hemisphere by way of the somatosensory pathways. In order to avoid this guidance of the ipsilateral hand a new test board was designed [BRINKMAN, KUYPERS and LAWRENCE, 1970]. It contains small food wells and is constructed in such a fashion that it emphasizes visual contrast between the food morsels and the background while it minimizes their tactual contrast. When the animals with one eye covered were presented with a food morsel in this test board, the hand contralateral to the open eye picked the food morsel out of the food wells by means of delicate relatively independent hand and finger movements. However, the extremity ipsilateral to the open eye behaved quite differently. The arm and hand which are presumably guided by way of the ventromedial part of the intermediate zone were brought to the proper place on the board but the hand and fingers did not pick up the food morsels. Instead, they began to explore the area tactually as if blind, lacking visual guidance. This suggests that each half of the brain has only limited control over the individual movements of the ipsilateral hand and fingers. This supports our original idea, provided the movements of the ipsilateral arm are not dependent on the descending frontal projections from the nonseeing hemisphere.

In conclusion, our anatomical findings concerning the connections of the descending pathways and of the propriospinal pathways as well as the behavioral findings following interruption of the various pathways and following commissural disconnections all seem to fit together. These structural and functional findings point to a plan of organization of the motor system by which integrated movements of body and limbs and independent movements of the limbs—especially their distal parts—are differentially controlled.

Some observations in human patients [SPERRY, GAZZANIGA and BOGEN, 1968] have much in common with our findings in the rhesus monkey. For example, in patients with callosal section, good ipsilateral control was found for responses of axial and proximal limb muscles and in respect to total body movements. However, ipsilateral control of distal limb movements was found to be impaired. These findings have much in common with those of GESCHWIND [1970]. They suggest that the plan of organization of the descending pathways based upon the findings in the rhesus monkey may also apply in man. This additional perspective should form a stimulus to further sharpen our concept of the organization of the neuronal systems by which the brain governs movements.

Acknowledgement

The author thanks Miss J. BRINKMAN for her help in the preparation of the manuscript. The author also thanks Dr. D. G. LAWRENCE and Dr. G. GORDON for reading the manuscript.

Author's address: Prof. H. G. J. M. KUYPERS, Department of Anatomy, Rotterdam Medical Faculty, *Rotterdam* (The Netherlands)

New Developments in Electromyography and Clinical Neurophysiology,
edited by J. E. Desmedt, vol. 3, pp. 69–94 (Karger, Basel 1973)

On the Central Nervous System Control of Fast and Slow Twitch Motor Units

R. E. BURKE

Laboratory of Neural Control, NINDS, National Institutes of Health, Bethesda, Md.

Following RANVIER's classic description [1874] of the histological and physiological characteristics of slow twitch *red* and fast twitch *pale* muscles, GRÜTZNER [1884] noted that many mammalian skeletal muscles appeared to contain mixtures of muscle fibers with different histological appearance. He suggested that such muscles consisted of subsets of muscle fibers, some with the properties of red muscle and others with the properties of pale. Early histological work has been abundantly confirmed and greatly extended in detailed histochemical and ultrastructural studies [e. g., see ROMANUL, 1964; PADYKULA and GAUTHIER, 1967; GAUTHIER, 1969; YELLIN and GUTH, 1970; BARNARD, EDGERTON, FURUKAWA and PETER, 1971]. From the standpoint of morphology and histochemistry, many limb muscles in mammals, particularly those of the grossly pale variety, are quite heterogeneous with respect to the types of muscle fibers in them.

Looking at the organization of muscles from the viewpoint of the physiologist, SHERRINGTON [1925] introduced the idea of the motor unit, which includes an α motoneuron plus the bundle of muscle fibers innervated by that cell. For convenience, the muscle fibers belonging to one motor unit will be called the *muscle unit* [cf. BURKE, 1967a]. Physiological studies have shown that histochemically heterogeneous muscles contain motor units with muscle unit properties varying over a wide range. On the basis of muscle unit contraction time, more or less distinguishable groups of fast twitch and slow twitch motor units have been described in trunk [ANDERSEN and SEARS, 1964; STEG, 1964], limb [GORDON and PHILLIPS, 1953; ECCLES, ECCLES and LUNDBERG, 1958; WUERKER, MCPHEDRAN and HENNEMAN, 1965; OLSON and

SWETT, 1966, 1971; BURKE, 1967a, 1968b; CLOSE, 1967] and digit muscles [APPELBERG and EMONET-DENAND, 1967; BESSOU, EMONET-DENAND and LAPORTE, 1963], as well as in extraocular muscles [HESS and PILAR, 1963; BACH-Y-RITA and ITO, 1966]. Recently, it has become possible to correlate directly physiological properties with histochemical appearance in one and the same muscle unit [EDSTRÖM and KUGELBERG, 1968] and such a physio-logical-histochemical correlation has been shown to be quite precise for the population of motor units in the cat gastrocnemius [BURKE, LEVINE, ZAJAC, TSAIRIS and ENGEL, 1971 and this series, volume 1].

Before leaving this discussion of motor unit populations in various muscles, it should be pointed out that such populations are in some muscles much more homogeneous than, for example, in the cat gastrocnemius. The soleus of the cat is quite homogenous, both in histochemistry and in the distribution of physiological properties of its motor units [HENNEMAN and OLSON, 1965; MCPHEDRAN, WUERKER and HENNEMAN, 1965; BURKE, 1967a; BURKE, LEVINE, SALCMAN and TSAIRIS, unpublished observations]. Similarly, the forearm muscles of the baboon [ECCLES, PHILLIPS and WU, 1968] appear to contain motor units with a much more restricted range of properties than seen in large limb muscles with both postural and 'phasic' functions. The remainder of this essay will be confined to an examination of some of the evidence available which bears on the question of how motor units of different types are controlled by the central nervous system (CNS), taking the motor unit pool of the cat gastrocnemius as a model system for study.

Control of Fast and Slow Twitch Muscle

Investigations into the mechanisms by which the CNS exercises differ-ential control over muscles of different types were among the earliest chapters written in the development of neurophysiology. The situation as of the early 30s was summarized as follows by SHERRINGTON and his colleagues:

'Early investigators postulated a dual mechanism for muscular contraction, slowly contracting muscles for tonic sustained contraction and rapidly contracting muscles for phasic contractions. Recent investigations [DENNY-BROWN, 1929] have shown that there is a certain degree of truth in this hypothesis, for the slower muscles are, in fact, those in which the stretch reflex is most easily provoked and... the rapidly contracting muscles are those which take the earliest and greatest part in reflexes which involve rapid contraction... But both types of muscle are potentially involved in all reflexes and the difference between them is only relative, expressed in terms of threshold of spinal excitation.' [CREED, DENNY-BROWN, ECCLES, LIDDELL and SHERRINGTON, 1932, pp. 59–60.]

It seems fair to say that the above formulation is still as useful today as it was in 1932. However, several problems should be noted. The above authors were really referring to collections of motor units rather than strictly to *muscles* in this quote, since the concepts of *reflexes* and *thresholds* actually deal with CNS elements, both motoneurons and interneurons, and with afferent neurons as well. Secondly, as noted in the Introduction to this essay, most rapidly contracting limb muscles (and certainly the gastrocnemius used in much of both early and more recent reflex work) are mixed muscles containing motor units with widely varying contraction speeds and tension outputs. The fact that many of the grossly fast twitch muscles used for reflex studies actually contain a sizeable proportion of units with slow twitch properties complicates interpretation of these investigations. The same problem affects interpretation of studies in which the properties of motoneurons of fast muscles (primarily of gastrocnemius in the cat) have been compared with cells innervating the much more purely slow soleus muscle. For these reasons, we have for several years been investigating a number of factors, important to the problem of central control, in motoneurons innervating muscle units with defined properties. Thus, in essence, we are examining the inter-relation between the electrophysiology of motoneurons and the mechanical properties of the muscle units innervated by the same cells. The results have enabled greater precision in the definition of factors important to the question of control of different types of motor units, even within the motor pool of a single muscle.

A Comment on Methods

Our approach to the study of the central control of different types of motor units has utilized intracellular recording and stimulation techniques, as first applied to motor units studies by DEVENANDAN, ECCLES and WESTERMAN [1965]. The intracellular approach has the great advantage that properties intrinsic to motoneurons can be studied with conventional techniques [BURKE, 1967a], the patterns of synaptic inputs to the same cells can be recorded during natural or electrical stimulation of afferents [BURKE, 1968a, b; BURKE, JANKOWSKA and TEN BRUGGENCATE, 1970], and finally, the mechanical properties of the muscle units innervated by the same cells can be recorded during intracellular stimulation of the motoneurons with transmembrane currents passed through the micro-electrode [DEVENANDAN, ECCLES and WESTERMAN, 1965; BURKE, 1967a]. The ability to stimulate single motoneurons intracellularly ensures activation of one and only one muscle unit, so that this approach appears useful also for studies of the mechanical properties of muscle units alone [BURKE, RUDOMIN and ZAJAC, 1970].

A disadvantage of the intracellular technique of sampling motor units is the introduction of possible bias in the sample population, since small motoneurons are presumably more difficult to penetrate successfully than are the larger cells. It is difficult to assess the relative importance of this factor, but it may be that the proportion of motor units with small motoneurons is actually greater than the observed numbers in our gastrocnemius studies.

Classification of Motor Units

In order to facilitate discussion it is convenient to have a system for classification of motor units into groups. It appears that the most useful classification criteria relate to the properties of the muscle unit portions of motor units. Recent work has shown that the population of motor units present in the gastrocnemius muscle of cats can be divided into three clearly distinguishable groups on the basis of two physiological parameters of mechanical muscle unit responses. The same study has given direct evidence that the muscle fibers in units of each group share the same histochemical profile, and that the three physiological groups correspond to the three basic histochemical types of muscle fibers found in many normal mammalian muscles [BURKE, LEVINE, ZAJAC, TSAIRIS and ENGEL, 1971]. In the above classification, two groups of rapidly contracting units have been found, called types FF and FR, which together comprise roughly 70–75% of the population of gastrocnemius units. FF and FR motor units differ primarily in their resistance to fatigue during repetitive stimulation. The remaining 25–30% of gastrocnemius units are slowly contracting and very fatigue resistant, and are termed type S.

In an earlier series of papers, I proposed that twitch contraction time alone might be used to separate two groups of motor units in cat gastrocnemius, type F (rapidly contracting) and type S (slowly contracting) [BURKE, 1967a, 1968b]. There is evidence which suggests that the type F group of these earlier results was made up primarily if not exclusively of both FF and FR units of the newer classification scheme, while the type S groups in both classification systems are essentially the same. Most of the evidence bearing on the question of central nervous system control of the different types of motor units was obtained using only twitch contraction time as a criterion of unit classification and the remainder of this paper will discuss gastrocnemius motor units as either type F or type S on this basis. Further work is underway to clarify the relative position of type FF and FR units in relation to the central control factors to be discussed.

I will not deal in any detail with the physiological properties of the muscle units of F and S motor units, except to mention that, in general, type F units produce considerably greater twitch and tetanus tension than type S units [BURKE, 1967a].

Intrinsic Properties of Motoneurons of F and S Motor Units

With regard to the CNS control of motor units, the first set of factors to be considered relates to characteristics of the motoneuron portion which are intrinsic to the cell. These factors can be considered as related to specific cell membrane properties and to factors of cell size and geometry. Both membrane and geometric factors then enter into a further consideration of what can be termed the electrotonic architecture of α-motoneurons.

To deal first with motoneuron size, it has long been held that the anatomic size of a motoneuron should be directly related to the diameter (and therefore to the conduction velocity) of the axon arising from it [GRANIT, HENATSCH and STEG, 1956; HENNEMAN, SOMJEN and CARPENTER, 1965]. This correlation has now been directly demonstrated by BARRETT and CRILL [1971] in dye-marked motoneurons. There has been considerable indirect evidence to support the idea that the input resistance of motoneurons, measured as the equivalent electrical resistance to current passed from an intracellular micro-electrode through the cell membrane, must also be related to motoneuron size, although in this case the relationship is an inverse one [KERNELL, 1966a]. Both BARRETT and CRILL [1971] and LUX, SCHUBERT and KREUTZBERG [1970], measuring cell input resistance and motoneuron dimensions in marked cells, have demonstrated this inverse relationship directly. Furthermore, both studies provide evidence that the factor of dendritic to somatic conductance ratio, which is a key factor in the relation between input resistance and cell membrane area [RALL, 1959], is *not* a function of cell size. Thus, motoneuron input resistance can be related quite specifically to cell membrane area.

Motoneurons innervating grossly fast muscles such as gastrocnemius have rapidly conducting axons as compared to cells innervating slow muscles [HAY, 1901; ECCLES, ECCLES and LUNDBERG, 1958; KUNO, 1959] and this relation was confirmed for single motor units by WUERKER, MCPHEDRAN and HENNEMAN [1965]. KERNELL [1966a] showed that motoneuron input resistances and axonal conduction velocities were closely correlated with one an-

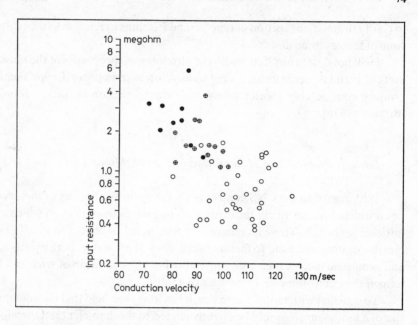

Fig.1. Semilogarithmic plot of the relation between axonal conduction velocity (abscissa, linear scale) and motoneuron input resistance (ordinate, logarithmic scale) in a sample of motoneurons from cat gastrocnemius and soleus muscles. Motor unit twitch type is denoted by the symbols: type F units by open circles, type S units in gastrocnemius by circles containing crosses, and type S units in soleus by solid circles. (The soleus muscle contains only type S units in the cat.) Note the inverse relationship between conduction velocity and input resistance, and the apparently continuous scatter of points. Data and figure slightly modified from Burke [1967a].

other and, as shown in figure 1, there is a further correlation of these physiological indices of motoneuron size with the twitch type of the motoneuron to which the cells belong. Thus, there is a great deal of evidence that the motoneurons innervating slowly contracting muscle units in large limb muscles are smaller and have less membrane surface area than the cells which innervate faster contracting muscle units. Note that there is no clear dividing line between F and S motoneurons, however (fig. 1), so that motoneurons of F and S motor units cannot be reliably distinguished from one another on the basis of indices of cell size alone [cf. Burke, 1967a].

Turning to properties of motoneuron membrane, it has been recognized for some time that the duration of post-spike hyperpolarization is generally longer in cells innervating grossly slow muscles than in fast muscle motoneurons [Eccles, Eccles and Lundberg, 1958; Kuno, 1959]. This holds true also for motoneurons of F and S motor units within the gastrocnemius population [BURKE, 1967a]. The post-spike hyperpolarization is an important factor controlling the rate of firing of motoneurons and KERNELL [1965] has shown a significant relation between after-hyperpolarization duration and the rate at which motoneurons respond to steady transmembrane stimulation, finding that cells with longer after-hyperpolarization exhibit lower maximal firing rates and less variation in firing intervals than cells with shorter afterpotentials. As discussed by KERNELL [1966b], the correlation between afterhyperpolarization duration and firing rate provides an appropriate matching to the mechanical properties of the muscle units innervated. The same sort of matching is also apparent in the maximal firing rates of F and S motoneurons responding to purely synaptic drive in the decerebrate preparation, type F cells tending to fire significantly more rapidly than S [BURKE, 1968b; table I].

Another membrane property importantly related to motoneuron firing in response to synaptic or transmembrane current drive is the phenomenon of accommodation. Several studies have shown that motoneurons with relatively short duration after-hyperpolarizations (and thus presumably innervating fast twitch muscles) tend to show greater degrees of accommodation to transmembrane current ramps than cells with longer duration after-potentials [SASAKI and OTANI, 1961; USHIYAMA, KOIZUMI and BROOKS, 1966]. A correlation between motor unit twitch type and the trend toward accommodation of the innervating motoneurons has been directly demonstrated [BURKE and NELSON, 1971] and, as expected, cells innervating type F units showed significantly greater tendency to exhibit accommodation to transmembrane current ramps than cells of type S motor units. However, this tendency did not absolutely differentiate F from S units; in particular, given the criteria for accommodation used, about half of type F motoneurons showed no accommodation. The overall pattern of accommodative behavior in F and S motor units displays another example of appropriate matching between motoneuron and muscle unit characteristics, in that the muscle units of many type F motor units evidence rapid fatigue during prolonged activity [BURKE, LEVINE, ZAJAC, TSAIRIS and ENGEL, 1971] and the presence of accommodative responses in their motoneurons is one mechanism tending to limit the duration of repetitive activity.

A number of recent studies have indicated that the membrane time constant of α-motoneurons can vary over a 4- to 5-fold range, with mean value about 5 msec [BURKE, 1968a; NELSON and LUX, 1970; BURKE and TEN BRUGGENCATE, 1971; JACK, MILLER, PORTER and REDMAN, 1971]. In motoneurons of F and S motor units, the data suggest that the time constant of type F cells tends to be, on average, about 20% shorter (mean about 5.6 msec) than that of type S motoneurons [mean about 6.7 msec; BURKE and TEN BRUGGENCATE, 1971]. Because of the wide range in observed time constant values in different cells, which may well be due in part to variations in ongoing synaptic activity [cf. LUX, 1967], it is difficult to interpret the 20-percent difference in mean time constants of F and S neurons very rigidly in terms of specific membrane resistance. However, it is of some interest that in the direct measurement studies of BARRETT and CRILL [1971], the mean specific membrane resistivity for small motoneurons (with input resistance values greater than 1.5 MΩ) was about 20% greater than the mean specific resistance for larger cells (with input resistance values less than 1.5 MΩ).

A final point regarding the intrinsic properties of F and S motoneurons concerns what may be called the electrotonic architecture of the cells. Consideration of the effect which distal dendritic synapses may have on the firing of motoneurons involves the notion of dendritic electrotonic length, which affects the attenuation and distortion of synaptic potentials located at some distance from the cell soma [RALL, 1967, 1970]. Electrophysiological measurements of the electrotonic length of the dendritic tree in unidentified motoneurons has indicated that the average length is about 1.2–1.5 electrotonic length constants [NELSON and LUX, 1970; JACK, MILLER, PORTER and REDMAN, 1971]. This figure has been confirmed in combined electrophysiological and anatomical studies in marked motoneurons [LUX, SCHUBERT and KREUTZBERG, 1970; BARRETT and CRILL, 1971]. Examined in motoneurons of F and S motor units varying over a wide size range (judged by input resistance values), the mean electrotonic length in both large F and small S cells is about the same, that is, about 1.5 length constants in both sets of cells [BURKE and TEN BRUGGENCATE, 1971]. These data indicate that distal dendritic synaptic terminals can play a significant part in controlling motoneuron firing behavior in both type F and type S motor units [cf. BURKE, 1967b]. The invariance in dendritic mean electrotonic length, together with the other properties intrinsic to the cells, are important factors in the interpretation of observations relating to synaptic organization in F and S motoneurons [cf. BURKE and TEN BRUGGENCATE, 1971].

Organization of Synaptic Input to Motoneurons of F and S Motor Units

The results of a number of studies [including, e.g., ECCLES, ECCLES and LUNDBERG, 1957a, b; ECCLES and LUNDBERG, 1958a, 1959a; ECCLES, ECCLES,

IGGO and ITO, 1961; HOLMQVIST and LUNDBERG, 1961; SASAKI and TANAKA, 1964; HONGO, JANKOWSKA and LUNDBERG, 1969] have suggested that there may be significant differences in synaptic input to motoneurons innervating flexor versus extensor muscles, and in cells innervating grossly fast *versus* slow muscles, even when a close synergistic relation exists between the muscles, as between gastrocnemius and soleus.

However, it is now clear that the gastrocnemius is a heterogeneous muscle and since it may have some differences in function from its synergist soleus, simply on the basis of somewhat different physical orientation, the extent to which the gastrocnemius-soleus system can be used to explore details of synaptic organization is limited. Rather, what is needed is an examination of the organization of synaptic input to motoneurons within the same motor unit pool, for example gastrocnemius alone. Information is now available regarding the patterns of synaptic input from a number of sources to type F and type S motoneurons of gastrocnemius, mainly the medial head [BURKE, 1968a,b; BURKE, JANKOWSKA and TEN BRUGGENCATE, 1970] and these data will be reviewed briefly here. For purposes of exposition, the data can be viewed as illustrating two contrasting patterns of synaptic organization: (1) input which is qualitatively the same to both F and S motoneurons but which shows differences of a *quantitative* nature correlated with motor unit type, and (2) input which shows evidence of *qualitative* differences in organization related to motor unit type.

Quantitative Differences

Muscle spindle primary afferents (group I_A afferents) make monosynaptic connection with α-motoneurons innervating motor units in the muscle of afferent origin (homonymous connection) as well as with other motoneurons innervating certain synergist muscles [ECCLES, ECCLES and LUNDBERG, 1957a]. Within the motor unit pool of the cat gastrocnemius muscle, simultaneous electrical stimulation of all of the homonymous group I_A afferents produce larger amplitude monosynaptic excitatory postsynaptic potentials (EPSP) in motoneurons of type S units than in cells of type F [BURKE, 1968a,b]. As shown in figure 2, the amplitude of such composite homonymous group I_A EPSP is significantly correlated with the input resistance (and therefore with the size) of motoneurons although, as is evident in the graph, a good deal of scatter is present in this correlation. The data support the long-held opinion [cf. DENNY-BROWN, 1929] that motoneurons innervating slowly con-

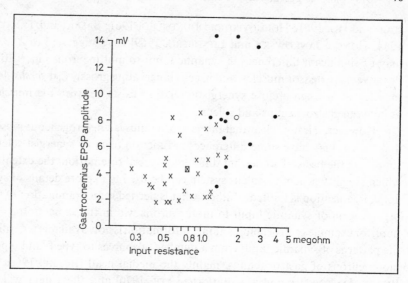

Fig.2. Semilogarithmic plot of the relation between the amplitude of homonymous group I_A EPSP produced by synchronous I_A volleys (ordinate; linear scale), and cell input resistance (abscissa; logarithmic scale) in a sample of gastrocnemius motor units. Type F units designated by crosses and type S by solid circles. The sample population consisted entirely of gastrocnemius motor units. The two-dimensional mean for the scatter of type F points is shown by the shaded rectangle containing a cross; the similar mean for the S scatter is denoted by the shaded circle. Data from BURKE [1968a].

tracting muscle are more powerfully excited by muscle spindle afferents than are the cells of rapidly contracting muscle and extend this conclusion to the population of F and S motor units residing within a single mixed muscle.

In addition to the amplitude correlation discussed above, the shape of composite homonymous group I_A EPSP recorded in F or S motoneurons is, in general, also correlated with twitch type and with EPSP amplitude. As parameters of shape, we have studied the EPSP time to peak (rise time) and the half-width (or duration at half amplitude) of the EPSP wave [inset in fig.3; RALL, BURKE, SMITH, NELSON and FRANK, 1967; BURKE, 1968a, b]. As shown in the graph in figure 3, both of these 'shape indices' tend to be shorter in duration for EPSP recorded from type F motoneurons (crosses) than for PSP in type S cells (solid circles). This difference in EPSP shapes is only partially explained by the statistical difference in membrane time constant between the F and S groups of motoneurons [see above references, and BURKE, 1968a]. In addition, it appears necessary to postulate that there is greater relative

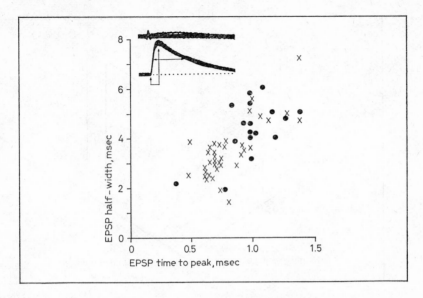

Fig.3. Linear plot of the relation between the time to peak (abscissa) and the half-width (ordinate) of composite homonymous group I_A EPSP in a sample population of F and S gastrocnemius motoneurons. Each point represents the EPSP in a different cell; type F cells are denoted by crosses and type S by solid circles. The inset shows a typical monosynaptic group I_A EPSP in intracellular recording (lower trace), together with the incoming afferent volley recorded at the dorsal root entry to the spinal cord (upper trace). The points for measurement of the 'shape indices' are indicated on the EPSP outline. Data from BURKE [1968a].

dendritic 'weighting' in the distribution of group I_A afferent terminals impinging on type S motoneurons as compared with type F cells.

It seems clear from a number of studies of group I_A termination patterns on motoneurons that I_A synapses are widely distributed to somatic and dendritic membrane regions [BURKE, 1967b; RALL, BURKE, SMITH, NELSON and FRANK, 1967; confirmed by JACK, MILLER, PORTER and REDMAN, 1970, 1971; PORTER and HORE, 1969; KUNO and MIYAHARA, 1969; MENDELL and HENNEMAN, 1971]. The available evidence suggests that there is a subtle but nonetheless significant shift in the distribution of I_A terminals, favoring more dendritic dominance in type S cells [BURKE, 1968a]. This conclusion has recently received some support from results of JACK and coworkers, using a different method of approach [JACK, MILLER, PORTER and REDMAN, 1970]. More distal dendritic location of group I_A terminals favors production of steady depolariz-

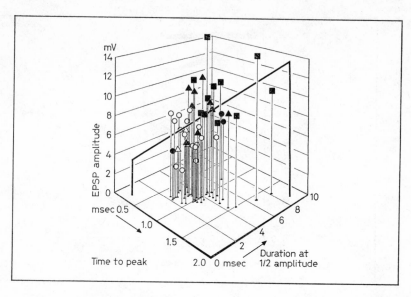

Fig.4. Linear three-dimensional graph on which are plotted 5 variables. The three axes of the graph concern the shape indices (horizontal two-dimensional plane) and the amplitude (vertical dimension) of composite homonymous group I_A EPSP recorded in a population of medial gastrocnemius motor units in decerebrate preparations. Characteristics of the muscle unit portions of the units studied is denoted by the shape of the symbols used: type F units with relatively large tension output by circles, type F units with relatively small tension output by triangles, and type S units by squares. The open symbols denote units which *did not* respond to muscle stretch with sustained firing; solid symbols denote units which *did* respond with sustained firing to similar medial gastrocnemius stretch. Figure from BURKE [1968b, Fig.9].

ing potential shifts in the motoneuron initial segment during sustained spindle afferent firing because of the electrotonic slowing and smoothing of synaptic potentials generated at some distance from the cell soma [cf. RALL, 1967]. Thus, both the amplitude and the shape of group I_A EPSP, as well as many of the intrinsic properties of the motoneurons, are in concert with the hypothesis that type S motor units are particularly suited to sustained firing in response to sustained input from muscle stretch receptors.

An obvious inference from the above is that type S motor units should participate vigorously in stretch reflexes under conditions of high activity in the muscle spindle loop. The classic decerebrate state is one of high spindle loop activity [ELDRED, GRANIT and MERTON, 1953], accounting for much of the sensitivity of extensor muscle stretch reflexes in such preparations [CREED,

DENNY-BROWN, ECCLES, LIDDELL and SHERRINGTON, 1932]. In the motor unit pool of the medial gastrocnemius muscle, type S motor units are very powerfully driven by muscle stretch and often are observed to be firing even without stretch [BURKE, 1968 b]. In addition, about 40% of the type F units studied could be made to fire during stretch in decerebrate preparations, although F units seldom showed spontaneous activity without stretch. In general, type F units which could be made to fire in decerebrate animals were those which produced relatively small tension outputs, while most of the F units with large tension output did not fire *at all* under the same conditions [BURKE, 1968 b; fig. 4].

In decerebrate preparations, the presence or absence of sustained repetitive firing of motor units in response to muscle stretch is significantly correlated with the amplitude and shape of homonymous group I_A EPSP, as well as with motor unit twitch type. These interrelations are illustrated in the multidimensional graph in figure 4. The scatter of data points shows clearly that larger I_A EPSP, in both F and S motoneurons, tend to be longer in duration than the smaller PSP, which are found mainly in type F cells. Further, the units with larger, longer I_A EPSP tended to show repetitive firing during gastrocnemius stretch (solid symbols) while units with smaller, shorter I_A EPSP did not (open symbols). These results provide quite direct evidence that quantitative and spatial distribution of group I_A synapses to gastrocnemius motoneurons are important factors governing the response of the motor unit population to muscle spindle afferent activation. However, the fact that many type S units were active without stretch, and the complimentary observation that many type F units could not be made to fire at all under the same conditions, suggest that other synaptic systems may also play a part in the pattern of motor unit recruitment in the decerebrate state.

The Question of Synaptic Density

It is of some interest to examine possible explanations for the correlation between the amplitudes of composite group I_A EPSP and the characteristics of the motoneurons in which the EPSP were recorded. It has been suggested that the direct correlation between composite I_A EPSP amplitude and motoneuron input resistance (fig. 2) is best explained by postulating an increase in the density of group I_A synapses per unit of cell receptive area as one progresses through the α-motoneuron size spectrum from large to small cells [BURKE, 1968 a]. Since this hypothesis has recently been challenged [DAVIS, 1971;

Mendell and Henneman, 1971], it seems appropriate to re-examine the evidence available.

The problem is most easily dealt with by considering a model system with a set of spherical neurons with varying diameter, d, but equal specific membrane resistivity, R_m. The input resistance, r_m, of such cells is

$$r_m = \frac{R_m}{\pi d^2},$$

(1)

where πd^2 is the membrane surface area. Now consider a unit patch of surface membrane which receives a certain number, D, of excitatory synapses from a given afferent system. If each of these synapses produces the same conductance change, g_e, and the conductance of each synapse is independent of the others, then the total conductance change, G_e, produced in the unit patch of membrane during simultaneous activation of all of the synapses is

$$G_e = g_e \cdot D.$$

(2)

Since D represents the number of synapses per unit membrane area, it is equivalent to the density of synapses. The density of synaptic current, I_m, for the unit patch of membrane during simultaneous activation of the synapses is

$$I_m = (V_m - E_e) G_e,$$

(3)

where V_m is the actual membrane potential and E_e is the equilibrium potential for excitatory conductances, and the difference between these two potentials is the driving force for EPSP.

Now consider the voltage change produced in a spherical neuron by simultaneous activation of all of the synapses from the given system. For simplicity, consider the steady-state voltage change, V_{ss}, which would be produced by indefinite prolongation of the excitatory conductance changes:

$$V_{ss} = I \cdot r_m.$$

(4)

Voltage perturbations produced by shorter-lasting conductance changes would be directly proportional to V_{ss}, and use of the latter simplifies the model system, assuming that the membrane time constants and conductance change transients are the same in all cells of the model system. The total synaptic current, I, produced by simultaneous activation of all of the synapses under discussion is

$$I = I_m (\pi d^2)$$

(5)

Using equations 1, 2, 3 and 5, the voltage change produced by the synaptic system under examination (equation 4) reduces to

$$V_{ss} = (V_m - E_e)(g_e \cdot D)(R_m) \tag{6}$$

for all of the cells in the model system. Thus, the amplitude of synaptic potentials produced by simultaneous activation of a given set of synapses is *independent* of cell size, but depends directly on the density of synapses per unit of surface membrane. This treatment implies that the proportion of membrane area subsynaptic to the active terminals (and therefore not contributing to the load resistance, r_m) is small relative to the total membrane area.

It should be clear that the situation in α-motoneurons with dendritic trees is considerably more complex than in a set of hypothetical spherical neurons. However, with a number of constraints derived from experimental data, plus one so-far unsupported assumption, the case of group I_A synaptic input to α-motoneurons approximates the spherical case sufficiently closely that the results of the model discussed above can be applied to motoneurons. This application has been discussed at some length elsewhere [BURKE, 1968a] and all of the experimental evidence available at present supports the case. Of particular importance is recent evidence that the electrotonic architecture of α-motoneurons is more or less constant over the size range of these cells [KERNELL, 1966a; LUX, SCHUBERT and KREUTZBERG, 1970; BARRETT and CRILL, 1971; BURKE and TEN BRUGGENCATE, 1971]. Although there is some evidence which suggests that there may be a trend toward somewhat higher specific membrane resistivity in small motoneurons (see above references), this is insufficient in itself to explain the very large range in I_A composite EPSP amplitudes observed (e.g., fig. 2). With regard to factors of group I_A synaptic distribution over the receptive membrane of motoneurons, the wide spatial distribution of these endings on motoneurons and the suggested trend toward some dendritic dominance in small cells are both completely in accord with the hypothesis that I_A synaptic density plays a controlling role in determining composite group I_A EPSP amplitude [cf. BURKE, 1968a]. The only factor in the argument so far unsupported by experimental data is the assumption that group I_A terminals produce the same conductance change (g_e, in the model formulation) everywhere, without regard to postsynaptic cell size. However, it seems quite reasonable to make this assumption in the absence of contradictory information. Thus, the available evidence suggests that the positive correlation between composite I_A EPSP amplitude and motoneuron input resistance is due in very large measure to an increase in density of the group I_A synaptic terminals through the size spectrum of α-motoneurons. The same conclusion has been reached using a somewhat different model formulation kindly communicated to me by Dr. R. S. ZUCKER of Stanford University [ZUCKER, personal communication].

There is another line of evidence which points to the same conclusion. Recent studies suggest that most, if not all, of the group I_A afferent fibers from a given muscle project to its own α-motoneurons, both large and small [MENDELL and HENNEMAN, 1968, 1971; KUNO and MIYAHARA, 1969]. The data of KUNO and MIYAHARA [1969] suggest that, on the average, group I_A fibers liberate the same number of transmitter quanta (i.e., generate the same total conductance change) irrespective of the size of the postsynaptic cell (see especially their figure 4). In accord with this conclusion, single group I_A afferent fibers tend to produce larger unitary EPSP in small motoneurons than in large cells [BURKE, 1968a, fig. 9; KUNO and MIYAHARA, 1969, fig. 6; MENDELL and HENNEMAN, 1971, fig. 13]. If, as has been suggested [KUNO, 1964], a single group I_A terminal liberates on average one quantum of transmitter, the above evidence suggests that the *total number* of group I_A boutons ending on each cell within a given group of α-motoneurons (such as medial gastrocnemius) may be essentially constant without regard to the size of the postsynaptic cell. Thus, since total membrane surface must be smaller in smaller motoneurons, this line of reasoning suggests that the density of group I_A terminals must increase as smaller cells are examined.

Another synaptic system which appears to exhibit a pattern of synaptic organization similar to group I_A excitation is the inhibitory input to gastrocnemius motoneurons arising from group I_A afferents from antagonist muscles (tibialis anterior, primarily) and operating through an interposed interneuron [disynaptic group I_A inhibition; cf. ECCLES and LUNDBERG, 1958b]. It has been found that, in general, the disynaptic I_A IPSP in gastrocnemius motoneurons is larger in type S cells than in type F, with and without facilitation of transmission through the inhibitory I_A interneuron by rubrospinal stimulation [BURKE, JANKOWSKA and TEN BRUGGENCATE, 1970, esp. fig. 12]. The locus of synaptic termination of I_A inhibitory interneurons appears to be on and quite close to the cell soma [BURKE, FEDINA and LUNDBERG, 1971], although possible systematic variations in this factor with motoneuron size have not been specifically tested. Nevertheless, it seems reasonable, using a line of reasoning such as outlined above, to postulate that the density of inhibitory synapses (in terms of total cell membrane area) from this I_A inhibitory system increases through the size spectrum from large to small motoneurons.

The synaptic organization patterns exhibited by both the group I_A excitatory and the I_A disynaptic inhibitory systems are basically similar. Input from both systems to gastrocnemius motoneurons is distributed such that the effect of activation of the given system is greater in small motoneurons, which in turn innervate slowly contracting muscle units. However, it should be noted that there is a fraction of fast twitch motor units which receive spindle afferent input in almost the same amount as do the majority of type S units (fig. 2 and 4), and which are repetitively active in decerebrate preparations (fig. 4). Such units occupy a middle ground in the motor unit spectrum between S units and the large tension type F units. These latter appear to be less tightly bound into spindle afferent loops and do not exhibit activity, as a rule, in decerebrate animals.

If, as appears likely, the synaptic densities of both group I_A excitatory and I_A interneuron inhibitory synapses are higher on smaller cells than on larger motoneurons, it can be argued that, if the *total* synaptic density is the same on large and small cells, one will 'run out of room' for all the synaptic systems impinging on the smaller motoneurons. One solution to this paradox would be apparent if the *total synaptic density from all systems* also increased on smaller cells. There is some evidence suggesting that this may be the case. GELFAN and RAPISARDA [1964] assessed the relation between the total density of synaptic knobs (stained with Rasmussen's method) and the cross-sectional area of ventral horn cell profiles in thin (8–10 μm) spinal cord sections. Their data (especially their figure 2) shows that the highest observed synaptic densities were found on the cell profiles of smallest cross-sectional area, while lower synaptic densities were observed on profiles of both large and small area. It is reasonable to assume that large motoneurons were represented in this sample by both large and small area profiles, since some of the cells undoubtedly were cut tangentially. Thus the shape of the scatter in the relation between synaptic density and cell profile area observed by GELFAN and RAPISARDA [1964, fig. 2] is compatible with the hypothesis that total synaptic density may well be greater on the smaller motoneurons than on the larger ones. The small increase in membrane time constant [BURKE and TEN BRUGGENCATE, 1971], and in calculated specific membrane resistance [BARRETT and CRILL, 1971], already referred to are also compatible with this hypothesis [cf. BURKE, 1968a].

An assumption implicit in the above discussion, and indeed in most analyses of motor control, is that the motoneurons of a given muscle nucleus all receive qualitatively similar synaptic inputs [cf. HENNEMAN, SOMJEN and CARPENTER, 1965]. The differences in recruitment of various motor units within a muscle nucleus would then be explained on the basis of quantitative considerations such as those discussed in the preceding section. However, there is now evidence that there may be *qualitative* as well as quantitative differences in the synaptic input to motoneurons within the gastrocnemius nucleus. This may represent a possible further mitigation of the problem of total synaptic density, since large motoneurons may receive some inputs *not* shared by small cells, and vice versa.

Qualitative Differences

The possibility that closely synergistic slow and fast muscles receive qualitatively different sorts of synaptic inputs is implicit in an old observation of SHERRINGTON and coworkers. In decerebrate preparations, stimulation of the common peroneal nerve sometimes caused enhancement of stretch induced tension in the fast gastrocnemius and simultaneous suppression of activity in the slow soleus [CREED, DENNY-BROWN, ECCLES, LIDDELL and SHERRINGTON, 1932, pp. 73–77]. Using gastrocnemius and soleus cells as prototypic fast and slow muscle motoneurons, respectively, a number of studies have demonstrated apparent qualitative differences in the organization of segmental afferent and descending input to these two groups of motoneurons [PRESTON and WHITLOCK, 1963; BOSEMARK, 1966; HONGO, JANKOWSKA and LUNDBERG, 1969; see also ROSENBERG, 1970]. In one study [SASAKI and TANAKA, 1964], the duration of post-spike hyperpolarization was used as an index of 'phasic' or 'tonic' nature of the motoneurons studied and a variety of synaptic effects was observed during supraspinal stimulation which were correlated with this index of motoneuron identity. Finally, the effect of spinal cord temperature changes is different in gastrocnemius and soleus muscles; the excitability of some gastrocnemius motoneurons apparently increases with cord cooling while soleus cell excitability appears to decrease [STELTER, SPAAN and KLUSS-MANN, 1969].

Recently, further evidence has been obtained which suggests qualitative differences in the organization of two polysynaptic systems impinging on type F and type S motoneurons within the gastrocnemius nucleus alone [BURKE, JANKOWSKA and TEN BRUGGENCATE, 1970]. Electrical stimulation of low to medium threshold afferents in some skin nerves produces, in some type F cells, predominantly excitatory polysynaptic PSP while in some type S motoneurons in the same animal, predominantly IPSP are recorded (fig. 5, sural records). A large number of cells, including both types F and S, exhibit a mixture of excitatory and inhibitory potentials without predominance of either, so that PSP recorded could be described only as 'mixed' (fig. 5, sural, center set).

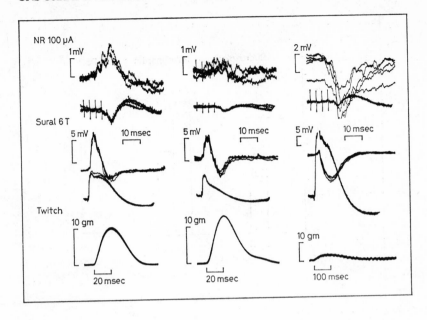

Fig. 5. Intracellular records of PSP, and mechanical records of muscle unit twitches, obtained from three different medial gastrocnemius motor units in the same animal. Top row of records (NR) shows intracellular potential (upper traces) and potential at dorsal root entry zone (lower traces) following 4 volleys at 100 μA delivered through a tungsten micro-electrode to the contralateral red nucleus. The middle row of records (sural) shows the polysynaptic PSP (upper traces) and cord dorsum potentials (lower traces) produced by single pulse stimulation of the ipsilateral sural nerve at about 6 times threshold strength. (Gain in cord dorsum records varies, as did the placement of the electrode, accounting for the variation in shape and amplitude of the recorded potential in the different records.) The bottom set of records shows mechanical twitch responses of the three motor units studied, recorded following intracellular stimulation of the motoneurons from which the PSP records had been obtained. Note that the PSP from both red nucleus and sural stimulation in the unit shown in the left column were predominantly excitatory; the twitch response had a short time to peak and the unit was classed as type F. The PSP from red nucleus and sural stimulation in the type F unit in the middle column were mixtures of excitatory and inhibitory components without a clear predominance of one or the other. The right column shows that both red nucleus and sural stimulation produced in this unit predominantly inhibitory potentials; the twitch response of the muscle unit innervated by this cell exhibited a long contraction time and the unit was classed as type S. Data from experiments of BURKE, JANKOWSKA and TEN BRUGGENCATE [1970]; chloralose anesthesia.

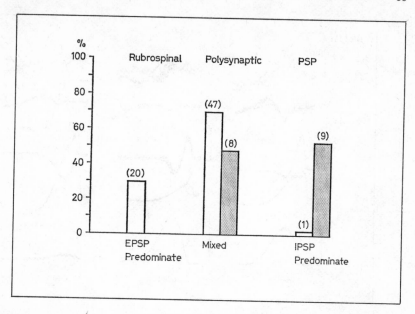

Fig.6. Histogram of the predominant deflection in polysynaptic PSP, evoked by short train stimulation of the contralateral red nucleus, recorded in a total sample population of 85 gastrocnemius motor units. Data expressed as percentage of either the type F sample (open bars) or the type S sample (shaded bars) exhibiting PSP classed according to predominating deflection (as in fig. 5). Absolute number of units in each category is indicated by the number in brackets above each bar. Note the asymmetric distribution with predominating excitation in about 30% of type F units and no type S, and at the other extreme predominating inhibition in over 50% of type S units. The latter PSP category was observed in only one type F unit. Data taken from experiments of BURKE, JANKOWSKA and TEN BRUGGENCATE [1970].

A similar pattern has been observed with the polysynaptic PSP produced in F and S motoneurons by repetitive stimulation in the contralateral red nucleus [BURKE, JANKOWSKA and TEN BRUGGENCATE, 1970]. As illustrated in the NR records in figure 5, some type F cells exhibit predominantly EPSP to this stimulation while some type S motoneurons show predominantly IPSP. As with skin nerve stimulation, there is a rather large group of units with mixed rubrospinal PSP impossible to classify with the crude criteria available. Using the rough categories of predominating deflection, the PSP produced by rubrospinal stimulation were correlated with motor unit twitch type in a sample of gastrocnemius motor units, and figure 6 shows a histogram of the distribution of the qualitative PSP types. Despite the large number of mixed

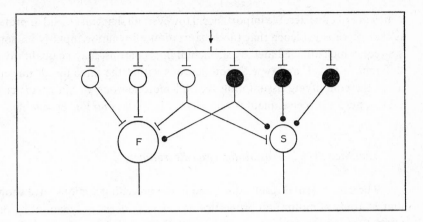

Fig. 7. Schematic diagram of possible connections of a set of last order interneurons (upper row) with type F and type S motoneurons of gastrocnemius pool, which are meant to represent the extremes in the system distribution. Interneurons with excitatory synaptic effects indicated by open circles; interneurons with inhibitory effect shown by solid circles.

responses, the distribution is clearly skewed toward predominating excitation for type F units and toward predominating inhibition for type S.

The assessment of polysynaptic PSP produced by electrical stimulation of nerves or central nuclei is difficult and a number of technical problems limit the extent to which interpretation can be carried. Nevertheless, the observations appear to indicate that there are some qualitative differences in the distribution of last-order interneurons mediating polysynaptic activity from skin afferents and rubrospinal axons to gastrocnemius motoneurons which are related to the twitch type of the target motor units. The scheme illustrated in figure 7 may explain the observations, assuming that the F and S cells shown represent the opposite ends of a spectrum, with many intermediate units with different combinations of connection with the set of last-order interneurons pictured. Activation of the set of interneurons would produce predominantly (but not exclusively) excitatory effects in the schematic F cell and predominantly (but again not exclusively) inhibitory potentials in the S cell. Different mixtures of excitation and inhibition, of course, would result depending on the relative number and strength of connections from the excitatory and inhibitory interneurons between the two extremes. The data available at present do not permit any assessment of the synaptic potentials from skin and rubrospinal stimulation in spatial or quantitative terms, as was done with group I_A input; there may well be significant quantitative factors in the system which are of considerable importance. However, all that can be said at present is that there is evidence that the organization of synaptic input from some sources, which are mediated by segmental interneurons, may be qualitatively different to type F and type S motoneurons within the same motor nucleus. Thus the simplifying assumption that all motoneurons within a particular muscle nucleus receive qualitatively identical inputs is no longer tenable.

The 'Size Principle' of Motoneuron Recruitment

The experimental results discussed in the preceding sections have shown that a variety of motor unit properties, as well as patterns of synaptic organization, form apparently meaningful correlations with the mechanical characteristics of the units' muscle unit portions. The motor units of cat gastrocnemius can be quite clearly classified according to muscle unit characteristics [cf. BURKE et al., 1971 and this series, volume 1]. No such clear categories appear on consideration of motoneuron intrinsic properties or synaptic patterns, or any combination of these. Nevertheless, it is obvious that the entire range of observations available may also be ordered on the basis of the size spectrum of α-motoneurons.

It has been known for some time that under some conditions smaller motoneurons are recruited into movements before larger cells become active. This has been attributed to a difference in activation *threshold*, dependent on factors associated with motoneuron size. In this context, the word threshold is not used in a strict sense connoting voltage or current threshold as understood from intracellular recording, but rather as a loose and almost qualitative term. In decerebrate preparations, GRANIT and coworkers [GRANIT, HENATSCH and STEG, 1956; GRANIT, PHILLIPS, SKOGLUND and STEG, 1957] suggested that small α-motoneurons respond to synaptic drive with *tonic* sustained firing while larger cells tend to give only unsustained *phasic* responses. They argued that the difference between tonic and phasic motoneurons represents a fundamental difference between the two groups, probably due to a combination of motoneuron intrinsic properties and synaptic input characteristics. Since then, HENATSCH, SCHULTE and BUSCH [1959], and KERNELL [1965, 1966a], have demonstrated that the response patterns of motoneurons may be less clearly separable into tonic and phasic categories than originally proposed. Nevertheless, there is evidence that some type F motoneurons may be incapable of prolonged repetitive firing even when driven by strong transmembrane currents [MISHELEVICH, 1969], an observation very likely related to the degree of accommodation present in some F cells [BURKE and NELSON, 1971].

Henneman and his colleagues have discussed a somewhat different view of the motor pool [HENNEMAN, 1957; HENNEMAN, SOMJEN and CARPENTER, 1965; SOMJEN, CARPENTER and HENNEMAN, 1965]. They studied the recruitment order of motoneurons of different sizes (judged by the amplitude of the motor axon spike in ventral root filament recordings) in response to a variety of stimulation paradigms in decerebrate preparations. They found that smaller motoneurons usually responded earlier (with lower 'threshold' in the broad sense) than larger cells, and concluded that recruitment order in motoneurons is a predictable function of cell size. In this hypothesis, the motoneuron pool of a given muscle is viewed as a continuous spectrum with cell size as the critical variable, and subclasses of cells such as *tonic* and *phasic* are not recognized. This hypothesis has been advanced as a 'size principle' governing motoneuron activity during a wide variety of movements [HENNEMAN, SOMJEN and CARPENTER, 1965].

As should be evident from the material discussed in preceding parts of this essay, the notion that the size of a motoneuron, *in and of itself*, can govern activity patterns is a serious oversimplification. The evidence presently available indicates that multiple factors, including other intrinsic motoneuron properties in addition to synaptic organization, must be considered in explanation of experimental observations. Nevertheless, many of these factors are clearly *correlated with* motoneuron size, and thus the 'size principle' can serve as a convenient shorthand in dealing with a large amount of related information.

Most animal experiments dealing with the question of motoneuron recruitment patterns have been done with decerebrate cats. Although interpretations differ somewhat, the observations obtained in such studies are for the most part in basic agreement. Smaller motoneurons (as determined on the basis of physiological criteria) tend to be recruited by stretch reflexes or by other types of stimulation earlier and more powerfully than larger cells. Studies which have utilized ventral root recording techniques [GRANIT, HENATSCH and STEG, 1956; GRANIT, PHILLIPS, SKOGLUND and STEG, 1957; HENNEMAN, SOMJEN and CARPENTER, 1965; SOMJEN, CARPENTER and HENNEMAN, 1965] have clearly demonstrated this point within the set of motoneurons which become active at some point during stimulation. Intracellular study of decerebrate preparations [BURKE, 1968b] has confirmed the size-excitability relation deduced from ventral root experiments and has also shown that there is a significant fraction of the gastrocnemius motor pool, consisting entirely of type F units, which may not be activated by muscle stretch, even in decerebrate preparations with well-developed rigidity. The latter study, as noted above, also demonstrated directly the correlation between motoneuron recruitment and the organization of group I_A synaptic input (fig. 4). However, it seems proper to question the extent to which generalizations can be made on the basis of results obtained from decerebrate animals. The decerebrate state produces marked effects not only among motoneurons but also among segmental reflex pathways [ECCLES and LUNDBERG, 1959b; KUNO and PERL, 1960; CARPENTER, ENGBERG, FUNKENSTEIN and LUNDBERG, 1963] and it seems possible that synaptic systems other than group I_A may play a part in determining recruitment patterns [cf. BURKE, 1968b]. Indeed, even in the decerebrate state, SHERRINGTON and coworkers [CREED, DENNY-BROWN, ECCLES, LIDDELL and SHERRINGTON, 1932] have provided some evidence that it may be possible to reverse the expected order of motoneuron recruitment.

There is much less information concerning the order of recruitment of various motor unit types under conditions other than the decerebrate state in animals. Most of the available evidence stems from EMG studies in humans. It has been known for some time that, in the human, motor units giving small EMG spikes tend to appear in records taken during slowly increasing tension before units with larger amplitude potentials [SMITH, 1934; DENNY-BROWN and PENNYBACKER, 1938]. Interpretation of such results in terms of the size principle is complicated by lack of detailed information about motor unit architecture in the muscles studied [cf. OLSON, CARPENTER and HENNEMAN, 1968], and factors of technique play an important part in such observations.

Recently, it has been shown that, in human leg muscles, the same order of recruitment of motor units was observed during slow voluntary movements as during flexion or stretch reflex activation [ASHWORTH, GRIMBY and KUGELBERG, 1967]. Using a technique by which motor unit mechanical responses could be recorded in human muscle, BUCHTHAL and SCHMALBRUCH [1970] have shown that the more slowly contracting units in human soleus were the most responsive during the H-reflex. Both of these sets of observations are what might be predicted from the size principle. However, such stereotyped recruitment order is not always the rule. GRIMBY and HANNERZ [1968] have demonstrated that recruitment order may be altered depending on the speed of movement onset, on the resistance to the movement, and on the presence or absence of co-contraction of other muscles in the limb [see also WAGMAN, PIERCE and BURGER, 1965]. Even more marked variations in recruitment order have been observed in human subjects with complete spinal transsections during phasic flexion reflexes [GRIMBY and HANNERZ, 1970]. BUCH-

THAL and SCHMALBRUCH [1970] have reported results which suggest that, while relatively small tension motor units are recruited early into slow voluntary movements of human biceps, units with considerably greater tension output may be active during spontaneous fasciculations.

Conclusions

During many sorts of movements which are governed primarily by spinal reflex mechanisms or by descending systems playing upon spinal reflex mechanisms, including the γ loop, the motor units which respond most readily are probably those with smaller motoneurons innervating slowly contracting muscle units (the type S units in the cat). LUNDBERG [1969] has discussed evidence suggesting that normal stepping may be such a movement. It may well be that a proportion of rapidly contracting units also respond almost as readily to such activation and it can be supposed that these correspond to the cat type FR units. In the cat, both S and FR units are resistant to fatigue [BURKE et al., 1971 and this series, volume 1] and such an organization thus makes teleological 'sense'. In addition, there is now evidence that at least some voluntary movements in man are accompanied by activation of the γ loop (α–γ co-activation) [VALLBO, 1970a, b and this volume].

Taking together the available evidence, it appears likely that the same set of motor units is active during both 'tonic' or 'postural' movements and during at least some voluntary or 'phasic' movements. During the performance of such movements, the order of motor unit recruitment probably follows the spectrum of motoneuron size as described by the 'size principle'.

However, the recruitment order from small to large motoneurons may not necessarily be immutable. As noted above, variations in recruitment order have been demonstrated. The fact that there are qualitative differences in the organization of some synaptic systems impinging on gastrocnemius motor units in the cat suggests that the CNS may have available alternative pathways which can activate large motoneurons with large, fast twitch muscle units, while perhaps simultaneously suppressing activity in the other units of the same muscle, which are more tightly bound into reflex loops. Such a 'switching mechanism' may well have biological advantage to the organism when placed in circumstances requiring very rapid movement with large tension output, such as may be encountered under 'fight or flight' situations. The organization of control systems in the motor unit pool within single mixed muscles appears to be rather complex, but this complexity seems appropriate when it is recognized that mixed limb muscles may be called upon to provide a very wide range of output states during the life of the animal.

Acknowledgement

I wish to thank Dr. ELZBIETA JANKOWSKA and Dr. GERRIT TEN BRUGGENCATE for their permission to reproduce data from our collaborative studies.

Author's address: Dr. R. E. BURKE, Laboratory of Neural Control, NINDS, National Institutes of Health, *Bethesda, MD 20014* (USA)

New Developments in Electromyography and Clinical Neurophysiology,
edited by J.E. Desmedt, vol. 3, pp. 95–125 (Karger, Basel 1973)

The Advances of the Last Decade
of Animal Experimentation upon Muscle Spindles

The Background to Current Human Work

P.B.C. MATTHEWS

University Laboratory of Physiology, Oxford

The last 10 years have seen notable progress in our understanding of muscle receptors and particularly of the muscle spindle, that end-organ which RUFFINI [1898] long ago recognized to be so intricate that 'apart from the organs of special sense (eye, ear, etc.) the body possesses no terminal organ that can compare with these in richness of nerve-fibres and nerve endings'. The present article sets out to review briefly some of the newer findings on mammals and to indicate which of them have now received general acceptance among neurophysiologists and which of them still remain controversial. Fortunately, all the indications suggest that the findings which have been made chiefly on the cat can be directly transposed to man. A much fuller discussion of the whole field may be found in two recent monographs [GRANIT, 1970; MATTHEWS, 1972] and in an earlier review [MATTHEWS, 1964]; these also provide a wealth of additional references to the matters presently discussed.

The Golgi Tendon Organ

Even for this simple receptor our views have recently had to be modified. Since B.H.C. MATTHEWS' work of 1933, this has been correctly seen as a stretch receptor which is arranged in series with the extrafusal muscle fibres and so is excited by any increase of muscular tension, irrespective of whether the increase is produced by a passive pulling upon a muscle or by its active contraction. In the early work, which was chiefly upon the soleus muscle, the threshold of the receptor to passive stretch of the muscle was often found to be rather high. For the ensuing 30-odd years it therefore came to be widely believed that tendon organs existed simply in order to provide the afferent path of a reflexly operated safety stop, by means of which excessive tensions in a muscle were counteracted by its tendon organs reflexly inhibiting its motoneurones and thereby abolishing contraction and bringing the tension down to a safe level. Such a restrictive view of the function of tendon organs was never very convincing and it has now been abandoned following the demonstration that the amount of active muscular tension required to excite tendon organs is invariably quite small in relation to the potential strength of a muscle. This was first shown by JANSEN and RUDJORD [1964] using the soleus muscle. They found that the threshold tension required to excite a tendon organ was generally much lower when the tension was developed by contraction than when it was developed by passive stretch. This is partly because the tension in a passive stretch is borne by fascial strands bridging the muscle as well as by the tendon proper, and partly because the tendon organ responds more readily to the dynamic stimulus of a twitch than it does to slowly applied pulls. When passive stretches are applied rapidly the difference in thresholds is almost non-existent [STUART, GOSLOW, MOSHER and REINKING, 1970]. In m. tibialis anterior the tendon organs tend to have generally a lower threshold than those in soleus, and many respond to slow passive stretches restricted to the physiological range; moreover, they manifest no gross difference between their thresholds to actively and to passively generated tensions [ALNAES, 1967]. However, relatively little passive tension is normally produced by stretching a muscle over its physiologically permitted range of lengths and so for most practical purposes tendon organs may be looked upon as 'contraction receptors'.

A more dramatic indication of the sensitivity of tendon organs to the right kind of stimulus is provided by HOUK and HENNEMAN's [1967] demonstration that they can be excited by the contraction of a single motor unit, activated by stimulating a single motor fibre in a ventral root filament. Individual tendon organs were found to be excited by from 4 to 15 different motor units. Earlier histological observations had shown that a similarly small number of muscle fibres pull directly upon individual tendon organs. Thus it appears that the absolute threshold of the tendon organs is low enough for them to be regularly excited by the tetanic contraction of individual muscle fibres, provided that these latter are appropriately placed to pull directly upon the particular tendinous fascicle upon which the tendon organ lies. Other muscle fibres, however, are largely ineffective showing that individual tendon organs are sampling the contraction rather than taking a widespread average. Thus it is now clear that some tendon organs must be excited during even quite weak muscle contractions. Hence tendon organs may reasonably be presumed to be playing some continuous part in the central regulation of muscle contraction, rather than simply being reserved for use in emergencies.

Muscle Spindles

Quite surprisingly in the late 1950s, after a full century of microscopic observation of the spindle, further patient and thorough histological study unaccompanied by any particular new technical advance yielded the important discovery that mammalian intrafusal muscle fibres are of two distinct kinds, and not just one as had hitherto been supposed. The transformation in our thinking introduced by this finding is still not complete as the detailed microphysiological study of the two kinds of fibre is still in progress. But enough has been achieved to allow a rational if somewhat speculative picture of the significance of the two kinds of fibre. Yet many textbook statements on the role of the spindle in the body still limit themselves to discussing a simplified model in which the spindle sends information to the CNS by only a single type of afferent fibre—the I_A fibres from the primary endings—and with the nervous system controlling the spindle by only a single motor system—the γ efferents. In fact, there are secondary afferent endings within the spindle as well as primary endings, each kind of ending having its own particular relationship with the two kinds of intrafusal fibre; in addition, there are two functionally distinct kinds of γ motor fibre, the static and dynamic fusimotor fibres. Any discussion of function which ignores this established complexity seems doomed from the start to be seriously incomplete.

Classical Histology

For the first half of the present century the accepted structure of the spindle of both man and animals was that described by SHERRINGTON and by RUFFINI late in the 1890s and which is illustrated in figure 1. This may be

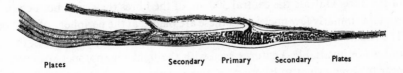

Plates   Secondary   Primary   Secondary   Plates

Fig. 1. The muscle spindle as seen by RUFFINI in 1898.

termed the classical picture and it fortunately remains true as far as it goes. The spindle consists of a bundle of up to 10 specialized small striated muscle fibres around and upon which are placed a variety of nerve terminals. The whole structure is several mm long and about 100 μm wide at its central equator. The equatorial part of the intrafusal bundle is surrounded by a capsule which encloses a certain amount of free fluid; this probably has the effect of insulating the nerve terminals from local mechanical disturbances. RUFFINI sub-divided the nerve terminals on the basis of their morphology into primary endings, secondary endings and plate endings; these names remain applicable. Degeneration experiments have since established that the primary and secondary endings are afferent, and that the plate endings are motor. The primary ending, of which there is only one in any spindle, lies on the central equatorial regions of the intrafusal fibres where they lose their striations and so would appear to be non-contractile. The secondary endings, of which there may be 0–5 within any particular spindle, lie on either side of the primary ending on regions of intrafusal fibre which are moderately well striated; when several secondary endings are present they are supplied by separate afferent fibres. The plate endings mostly lie towards the pole of the spindle and there may be several of them on each individual intrafusal muscle fibre, apparently often arising from different motor fibres, though individual motor fibres branch to supply several different plate endings within a particular spindle.

Delimitation of Nuclear Bag and Nuclear Chain Fibres

All mammals so far studied, including man, have now been found to possess two distinct kinds of intrafusal muscle fibre. These are called the nuclear bag and the nuclear chain fibres after the arrangement of nuclei found in their equatorial regions; they were originally described independently by BOYD [1962] in the cat and by COOPER and DANIEL [1963] in man. The differences between the bag and chain fibres are illustrated diagrammatically in figure 2. At the spindle equator the bag fibres have a cluster of nuclei lying 2 or 3 abreast, while those of the chain fibres lie in single file down the centre of the fibre. Outside the central 300 μm of the fibres they soon both come to contain similarly few nuclei, and these are placed at the periphery of the fibre. The bag fibres are commonly longer and thicker than the chain fibres, though this is not invariable. Features seen with the light microscope, however, appear to some extent to be epi-phenomena only distantly related to function, for they may show appreciable variation for spindles from different muscles

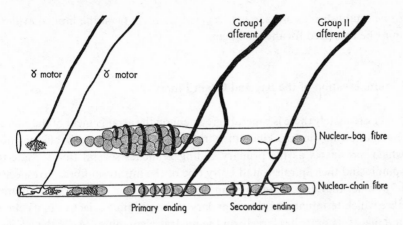

Fig. 2. Simplified diagram of the central region (about 1 mm) of the muscle spindle as currently recognized. [From MATTHEWS, 1964.]

of the same animal and also between species, without apparently affecting the overall behaviour of the spindle. Most notably, the intrafusal fibres of the rabbit all appear to be of the bag type when they are examined with classical histological techniques and for a time this confused the issue. Electron-microscopic examination, however, has now confirmed the existence of 2 kinds of intrafusal fibre in all species so far examined, including in the rabbit [see, for example, ADAL, 1969; CORVAJA, MARINOZZI and POMPEIANO, 1969; CORVAJA and POMPEIANO, 1970]. The bag fibres have a regularly arranged array of myofilaments, contain rather little sarcoplasm with few mitochondria, and lack an M line in the middle of the sarcomere. The chain fibres have their myofilaments less regularly arrayed, and have prominent mitochondria and an appreciable amount of sarcoplasm between the myofilaments; in addition, they possess a well developed M line. The difference in the band structure of the two kinds of fibre suggests that the chain fibres contract the more rapidly, for M lines are found in the frog twitch fibres but not in the frog slow fibres. Various histochemical differences have also been described between the bag and chain fibres. In particular, the chain fibres are richer in both mitochondrial and myofibrillary ATPase. The morphological duality of the bag and chain intrafusal fibres may thus now be considered to be firmly established and appears to be of a sufficient degree to betoken a functional duality. There have been various attempts to subdivide the intrafusal fibres still further on the basis of various histochemical tests. These further subdivisions have as yet to be established as specific entities, and perforce from the limited evidence must be neglected for the time being.

Innervation of the Bag and Chain Fibres

Fortunately there is general agreement on the arrangement of the afferent innervation of the two kinds of intrafusal fibre. The large I_A afferent fibre which terminates as the primary ending branches several times inside the spindle and then spirals round every one of the intrafusal fibres, irrespective of whether they are of bag or chain type. The medium sized group II afferent fibre which terminates as the secondary ending is more selective in its choice and places its branches largely on the nuclear chain fibres. As in the classical picture the primary ending lies equatorially and any secondary endings lie juxta-equatorially. In RUFFINI's view, the two kinds of ending could be differentiated by their fine structure. He saw the primary endings as exclusively *annulo-spiral* and the secondary endings as *flower-spray* or *flower-wreath* in their appearance. These differences are often apparent in the cat, but are not universal throughout the mammalia and are not particularly well marked in man. Under the electron microscope the two kinds of nerve terminal appear just the same except for their size. Both are closely apposed to the surface of the intrafusal muscle fibres, with the muscular and neural membranes separated by only 10–20 nm without any intervening basement membrane. The closeness of contact is often increased by either kind of nerve terminal lying in a groove on the surface of the muscle fibre, sometimes with lips of sarcoplasm partly overlying the sensory terminal.

The arrangement of the motor terminals unfortunately still continues to be a matter for controversy. It is generally agreed that the plate endings of the classical picture provide only a part of the motor innervation and that there is also a more diffuse type of motor ending which somewhat resembles the *en grappe* terminations of the slow fibres of the frog. These diffuse intrafusal endings are now termed the *trail endings* [BARKER, STACEY and ADAL, 1970]; an earlier but now discarded name for them was the γ_2 *network*. The trail ending differs from the plate ending in having no discrete point of termination as viewed with the light microscope. Instead it wanders some 100 μm along the length of an intrafusal fibre, apparently making synaptic contact more or less continuously along its line of physical contact with the underlying intrafusal fibre. Under the electron microscope there is a post-junctional folding of the intrafusal membrane beneath the plate endings which is just like that seen at motor endplates on extrafusal muscle fibres. In contrast, such folding is completely absent underneath the trail endings even though synaptic contact is indicated by the presence of the usual synaptic vesicles in the fine trail terminals.

The plate terminals have also become more complicated than classically recognized, for they have been subdivided into two kinds named p_1 and p_2 plates [BARKER, STACEY and ADAL, 1970]. p_1 plates very closely resemble the motor endplates on the extrafusal muscle fibres. p_2 plates are about twice as long as p_1 plates and are associated with rather less infolding of the underlying membrane and fewer sole plate nuclei. Very occasionally it has been possible to show that a p_1 plate is derived from a branch of a motor fibre which also supplies motor plates to extrafusal muscle fibres. Whether all p_1 plates are so derived is an open question. The relative numbers of p_1 and p_2 plates vary greatly from muscle to muscle. The p_2 plates and the trail endings are both thought to be derived from γ motor fibres.

A given fusimotor fibre is believed to have all its terminations of the same morphological kind. All 3 kinds of ending have been seen on both kinds of intrafusal fibre; but, in the main, plate endings occur on bag fibres and trail endings on chain fibres. The chief present uncertainty is as to the extent of the cross-innervation of the two kinds of intrafusal fibre by individual fusimotor fibres. After a spirited controversy lasting some years [see for example the contributions by BARKER and by BOYD in GRANIT, 1966] the problem seemed to be no nearer resolution by the continued application of purely histological techniques. The deadlock arose largely because of the extraordinary difficulty of following a particular single fine motor fibre for any distance inside the spindle without getting it mixed up with other fine fibres. Fortunately the difficulty has very recently been circumvented by the elegant combination of electrophysiological and histological techniques and the matter may soon be expected to be cleared up [BARKER, EMONET-DÉNAND, LAPORTE, PROSKE and STACEY, 1970]. In an initial electrophysiological experiment, performed under sterile conditions, a fine ventral root filament is detected which contains only one of the 20 or so γ fibres to the tenuissimus muscle of the cat. This filament is then preserved intact while all the remaining portions of ventral root supplying the muscle are severed. An appropriate interval is then left for the unwanted fibres to degenerate. This leaves a preparation in which the spindles of m. tenuissimus receive their motor innervation solely from the single fusimotor fibre which had originally been isolated in continuity. So far, only fibres terminating in trails have been successfully prepared, which incidentally have all been static fusimotor fibres identified electrophysiologically.

These trail fibres have proved on occasion to send branches to both kinds of intrafusal fibres, but the extent of such cross-innervation has yet to be put on a quantitative basis. This observation of cross-innervation is an awkward finding, since the bulk of the physiological evidence suggests that the two kinds of intrafusal fibre are largely innervated by separate fusimotor fibres. It still seems likely, however, that the bag and chain fibres receive a sufficiently distinct motor innervation for their contraction to be the basis of the distinctive *static* and *dynamic* fusimotor effects (see later), and to provide the peripheral machinery to allow the CNS to control each kind of intrafusal fibre largely independently of the other.

The Functional Distinctiveness of the Primary and Secondary Endings

The fact that the primary and secondary endings differ in their anatomical arrangement has long been taken to indicate that they have different functions and send different sorts of messages to the CNS. That this is indeed so has now been amply demonstrated by experiment. The initial problem in electrophysiological experimentation was to develop a way of recognizing whether any particular spindle afferent fibre which is being recorded from comes from a primary or from a secondary ending. RUFFINI's classical histological work showed that at the spindle the primary ending is supplied by a larger afferent fibre than is the secondary ending, and it has long been assumed that the same is true when the fibres are running in the main nerve trunk. Work on the cat has now verified this suggestion by showing that the spindle receptors connected to large fibres have quite different properties from those connected to the smaller fibres. In the cat, the dividing line between the afferents from the two kinds of ending has long been placed at 72 m/sec (corresponding to a diameter of 12 μm), and this has turned out to be a reasonably appropriate figure for this species in spite of the arbitrariness of the original reasons for choosing this point. In man, the fastest fibres conduct at only about 80 m/sec, as compared with 120 m/sec in the cat, so the dividing line must inevitably be somewhat lower. Moreover, it should be noted that the present evidence is against the existence of a razor-sharp dividing line at any particular position, but rather for the diameters of the afferents from the two kinds of receptor overlapping by a certain amount. In addition, it has yet to be proven that all receptors with afferents conducting near the dividing line have functional properties which place them firmly into one or other of two functionally classified groups. Nonetheless, the method of measuring conduction velocity has enabled a large number of important differences to be established between the responsiveness of primary and secondary endings to a variety of stimuli.

For large amplitude stretches the behaviour of the 2 kinds of ending may be summarized to a first approximation by saying that the primary ending is very much more sensitive to the dynamic components of any stimulus than is the secondary ending, but that they are about equally sensitive to changes in muscle length *per se*. These differences probably arise as a consequence of the mechanical properties of the underlying regions of intrafusal fibre rather than from any differences in the transducing properties of the nerve terminals themselves. In contrast, when very small stretches are used, less than 100 μm, the relative behaviour of the two kinds of ending is rather different. The primary ending then shows no particularly greater degree of velocity sensitivity than does the secondary ending, but the primary ending is now appreciably the more sensitive to any absolute change of length, provided that this lasts no longer than a few seconds. The similarities in velocity sensitivity under these conditions probably reflect similarities in the dynamic transducing properties of the two kinds of nerve terminal, while the dissimilarities in static sensitivity probably once again depend upon intrafusal mechanics. These differences in behaviour will next be described in more detail as they have an undoubted functional significance for the role of the spindle in the body. Discussion on their origin may be found elsewhere [MATTHEWS, 1972]; the matter is controversial and currently under active investigation by various workers. Comparison of the behaviour of the two kinds of ending is complicated by the fact that the responsiveness of both of them may be modified by fusimotor control. The description which follows applies both to denervated spindles and to those seen in the decerebrate cat which maintains a steady intrafusal contraction, probably of both kinds of intrafusal fibre; this may be hoped to make the spindles behave more typically of those in the awake animal than do spindles whose intrafusal fibres are lying flaccid.

Figure 3 shows the contrasting responses of the two kinds of ending to a large rapidly applied stretch, both in the presence and absence of steady fusimotor activity. The difference in dynamic sensitivity and approximate equivalence of static sensitivity of the kinds of ending is readily apparent. Figure 4 shows similar responses by means of a direct display of the frequency of firing of the endings. A convenient way of quantifying the velocity sensitivity of an ending is to measure the so-called dynamic index, as illustrated in figure 4. This index is simply the amount by which the frequency of firing of an ending decreases in the first 0.5 sec after completing the dynamic phase of a ramp stretch at constant velocity. (The increase in frequency at the beginning

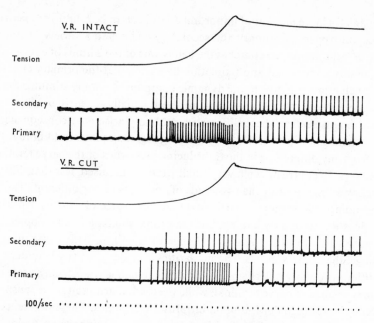

Fig.3. The contrasting responses of spindle primary and secondary endings to a rapidly applied stretch (approximately 14 mm at 70 mm/sec). The stretches were applied first in the presence of a steady intrafusal contraction which was maintained by the 'spontaneous' fusimotor discharge of the decerebrate cat (V.R. intact), and second to the same endings after cutting the ventral roots to abolish the intrafusal contraction (V.R. cut). [From MATTHEWS, 1972.]

of a stretch does not provide as satisfactory an index since quite often an ending is initially silent.) The dynamic index increases monotonically with the velocity of stretching, as required of a measure of velocity sensitivity, but the relation is not linear. By and large, the dynamic indices of primary endings lie well above those of secondary endings, but the difference appears to be particularly clear-cut for endings in the soleus muscle which was the one used to introduce the measure. Another point of difference between the endings is seen at the very beginning of a ramp stretch when the primary ending, but not the secondary ending, often gives an initial burst of impulses at higher frequency than that seen as the dynamic phase of stretching continues. On attaining the final length the frequency of firing of the primary may briefly dip below the subsequent semi-equilibrium level. These may be interpreted as responses to the stimulus of acceleration, over and above responses to the stimulus of

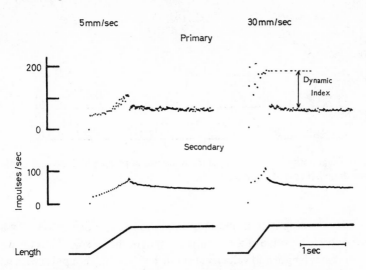

Fig. 4. Comparison of the responses of primary and secondary endings by means of a direct record of their frequency of firing. Each action potential contributes one spot to the display. The deflection of the spot above zero gives the instantaneous frequency of firing of the ending determined from the preceding interspike interval. [From BROWN, CROWE and MATTHEWS, 1965.]

velocity; but as they appear to depend partly upon a mechanism akin to static friction between the myofilaments their functional significance is still debatable.

The functional distinctiveness of primary and secondary endings has been particularly well shown by experiments in which it has been possible to study the responses of both kinds of ending from the same muscle spindle. A way of doing this was introduced by BESSOU and LAPORTE [1962] using the m. tenuissimus of the cat. This peculiar muscle runs the full length of the thigh but contains only about 15 muscle spindles. As a result, particularly at the bottom of the muscle, the spindles are so spread out that they do not overlap with each other in the longitudinal direction. This means that if a pair of afferent fibres can be studied which originate from a primary and a secondary ending in the same region of the muscle then the two endings are assured to be lying on the same spindle. Figure 5 summarizes in diagrammatic form how the two kinds of afferent ending respond to various stimuli. These responses may all be covered by the generalization that the primary ending responds to the combined stimuli of the instantaneous values of the length of the muscle and the velocity at which the length is changing (and possibly also acceleration), while

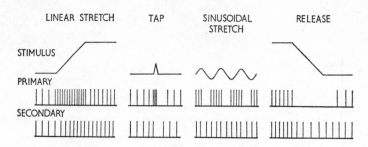

Fig.5. Diagrammatic comparison of the responses of primary and secondary endings to various stimuli. [From MATTHEWS, 1964.]

the secondary ending is to a first approximation insensitive to velocity and responds only to the value of the absolute length of the muscle. Thus the ramp stretch combines length and velocity stimuli. The brief tap is a pulse with large velocity and acceleration components but only a small displacement component. A sinusoidal stretch of 1 mm or so amplitude has an appreciable velocity component but a small length component. A release combines a negative value of velocity with a positive but declining value of length. (It should be noted, however, that in the presence of appreciable static fusimotor activity the primary ending fires on release.)

In line with the greater concern of the secondary ending with the measurement of muscle length is the recent finding that when the length of a muscle is held constant the secondary ending discharges appreciably more regularly than does the primary ending [MATTHEWS and STEIN, 1969b]. The variability of the intervals between successive spikes discharged by the secondary ending is about a quarter of that found for the primary ending under similar circumstances. The greater regularity of the secondary ending probably allows it to give the more accurate signal of absolute length. The greater irregularity of the primary ending may perhaps help it to signal rapid phasic changes the more effectively, for they would not be so liable to get mixed up with the rhythm of discharge of the afferent fibre itself.

Vibration Sensitivity

Another interesting difference between primary and secondary endings which is proving of the greatest importance for reflex studies lies in their sensitivity to vibration at frequencies of 100 c/sec and more. In the soleus of

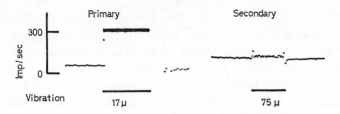

Fig.6. The different effects of longitudinally applied muscle vibration on a primary and on a secondary ending. The vibration was at 300 c/sec; its peak-to-peak amplitude is shown below. [From MATTHEWS, 1967.]

the cat the primary ending can be driven by vibration to discharge an impulse on every cycle when the peak-to-peak amplitude of movement is below 30 µm, even though the ending is deprived of fusimotor support [BROWN, ENGBERG and MATTHEWS, 1967]. In contrast, under similar conditions the secondary ending is barely excited at all by much larger amplitudes of movement, as illustrated in figure 6. Fusimotor activity lowers the threshold of the primary ending so that in favourable circumstances it may be driven to fire at several hundred impulses/sec by peak-to-peak amplitudes of only 5 µm. With fusimotor support the secondary ending also becomes slightly more sensitive, but it still shows very little increase in its mean frequency of firing when the muscle is vibrated with amplitudes of 100–200 µm. Vibration thus provides a convenient means of selectively activating the primary ending. It is thus coming into widespread use for central studies, particularly in man.

One or two provisos are, however, unfortunately necessary. To begin with, when the vibrated muscle responds with a reflex contraction, as it commonly does, then the Golgi tendon organs become more responsive to vibration. They may then be somewhat excited by amplitudes of around 100 µm to which they were previously quite oblivious. Moreover, contraction of the main muscle also makes the spindles less responsive to vibration applied to the tendon. This is probably because a contracting muscle becomes stiffer and so more of the vibratory deformation is taken up in the tendon and correspondingly less applied to the spindles. Next, a selective activation of the primary ending without the secondary ending is best shown when the vibration is applied to the tendon rather than directly to the surface of the muscle itself. BIANCONI and VAN DER MEULEN [1963] applied vibration directly over the spindle and then found rather little difference between the responsiveness of the two kinds of ending. However, the secondary endings could only be excited by careful placing of the vibrator directly over the spindle, thus it seems likely that even with direct muscular application vibration will excite many more primary than secondary endings. Finally, it must be remembered that vibration is far from being a localized form of stimulation and tends to spread far beyond the vibrated region itself. Thus it may well excite spindle primaries in muscles outside the ones being intentionally studied, and probably also excites low threshold cutaneous mechanoreceptors and Pacinian corpuscles over a wide area. In the experimental animal most of these difficulties can be avoided by widespread denervation and so on; but in man there is no escape.

Sensitivity to Small Amplitude Sinusoidal Stretching

The response of spindle endings to sinusoidal stretching is of particular interest both because this form of input has been favoured by engineers for analyzing inanimate control systems and also because the movements of tremors may be closely sinusoidal. Large amplitudes of sinusoidal movement produce the intermittent firing diagrammatically illustrated in figure 5. These are very hard to analyze in a quantitative manner for it is not clear which facets of the response are best measured. Moreover, such responses are 'non-linear' in the engineering sense so that doubling the size of the stimulus fails to produce a doubling in the size of the response. Thus knowledge of the response of an ending to any one stimulus does not allow prediction of its response to another stimulus. However, if the amplitude of stretching is sufficiently reduced, namely to below about 0.1 mm for the 5-cm long soleus of the cat, then the responses of both primary and secondary endings are found to be linearly related to the stimulus. This then permits the determination of a unique frequency response curve for an ending which is independent of the precise amplitude of stretching. For these reasons MATTHEWS and STEIN [1969a] and POPPELE and BOWMAN [1970] have recently studied the response of primary and secondary endings to small sinusoidal stretches. This has led to some unexpected findings.

Figure 7 portrays the response of the two kinds of ending to a wide range of frequencies of stretching applied in the decerebrate cat with continuous background fusimotor activity. The response at each frequency was determined by averaging the firing evoked from the ending by a large number of cycles of stretching. The 'sensitivity' of an ending to the stretching is assessed by dividing the amplitude of the sinusoidal modulation of its firing frequency by the amplitude of the stretching. This gives a value at each frequency which is the

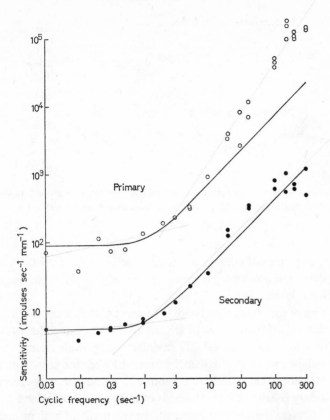

Fig. 7. Comparison of the sensitivity to sinusoidal stretching of a primary and of a secondary ending. The amplitude of the stretching was made sufficiently small to fall within the linear range of the endings. The endings were under the continuous influence of the spontaneous fusimotor activity of the decerebrate cat. [From MATTHEWS and STEIN, 1969a.]

change in firing, in impulses per sec, per mm of applied extension (i.e. impulses sec^{-1}mm^{-1}). For both kinds of ending, the graph of sensitivity against frequency is usually approximately flat from 0.1 to 1 c/sec. It then begins to increase along the curve which describes the behaviour of a system which is sensitive to the sum of the length and velocity components of the stretching. The 'corner frequency' at which the curve begins to rise provides a measure of the relative sensitivities of an ending to length and to velocity; the lower the corner frequency the more sensitive is the ending to velocity. Somewhat unexpectedly, the principal corner frequency turns out to be approximately

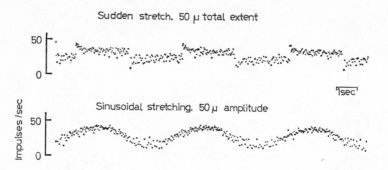

Fig. 8. The high sensitivity of the primary ending to small stretches. The stretching was applied in the absence of fusimotor activity. [From MATTHEWS and STEIN, 1969a.]

the same for the two kinds of ending, about 1.5 c/sec, in spite of the fact that the primary ending is so much the more sensitive to velocity when the amplitude of stretching is large.

The responses of the two kinds of ending to small stretches are, however, far from identical. At all frequencies of stretching the absolute value of the sensitivity of the primary ending is usually an order of magnitude greater than that of the secondary ending. At frequencies around 1 c/sec the sensitivity of the primary in decerebrate cats is usually around 100 impulses/sec/mm whereas that of the secondary is below 10. On the basis of earlier results obtained with large stretches, of several millimeters in extent, the sensitivity of both primary and secondary endings to stretching at below 1 c/sec would have been expected to be around 5 impulses/sec/mm. Thus the secondary ending is seen to be behaving very much as expected from what was previously known, whereas the primary ending gives a paradoxically large response when the stimulus amplitude is restricted. This high sensitivity is probably simply a consequence of intrafusal mechanics and due to the existence of a certain amount of cross-bridge formation between the myofilaments of even non-contracting intrafusal fibres, but this will not be discussed here. What matters for the moment is that the high-sensitivity effect would appear to be of physiological significance for the normal functioning of the reflex control mechanisms. Although the small linear range is only a fraction of the physiological range of movement of the muscle, yet the sensitivity of the primary ending within the linear range is sufficiently high for small movements to produce an appreciable modulation in the frequency of firing of the primary. This is illustrated in figure 8. Moreover, on changing the length of a muscle its spindles can 'reset' themselves so that the high sensitivity soon transfers itself to the new length and the endings avoid becoming saturated. On the other hand, the importance of the rather different type of response evoked from the primary ending by large stretches should not now be minimized. The fact that the ending is then behaving non-linearly in the engineering sense does not mean that its responses become physiologically unimportant. Large movements of muscles are as normal as postural fixation, and the spindle primary responds to them in a systematic manner with variations in their extent and velocity of application. The CNS can be expected to be able to make good use of all this information and to know the properties of its peripheral receptors well enough to be able to decode their messages, even though at first sight the messages appear to us to be unduly complicated and to be derived from an instrument with a variable calibration.

Fusimotor Control of the Spindle

The Gamma and the Beta Motor Fibres

In most mammals, the motor fibres to voluntary muscle fall into two groups differing both in diameter and in function. First, there are the large α motor fibres (above 12 μm diameter in the cat) which supply the ordinary muscle fibres and cause contraction of the main muscle. Second, there are smaller motor fibres (2–8 μm in diameter) which produce no direct effect on the extrafusal muscle fibres but which instead elicit an intrafusal contraction. These γ motor fibres are thus specifically fusimotor. This was first shown by LEKSELL [1945] using a pressure block to inactivate the large fibres and then confirmed by KUFFLER, HUNT and QUILLIAM [1951] by stimulating isolated motor fibres. The separation of fusimotor from ordinary motor control is not, however, found in many lower animals. For example, in the frog the motor supply to the spindle is entirely derived from branches of the ordinary motor fibres. Such an arrangement of a shared fusimotor-skeletomotor innervation has now been found to occur on occasion in the mammal also. Mixed motor fibres in the mammal are currently called beta (β) fibres in order to fit in with the pre-existing α–γ classification. There is no real logic behind this nomenclature, however, since β fibres may fall into either the normal α or γ ranges of velocity, though not usually at the extremes of either. The extent of β innervation in the mammal is still problematical, but it appears unlikely to be at all common in large muscles which receive an abundant motor supply. So far it has only been established in lumbrical muscles, and possibly also in tail muscles; these little muscles are supplied by a rather limited total number of nerve fibres and the possession of β fibres might appear to allow a muscle to economize yet further in this respect. The experimental difficulties of investigating large muscles are currently so formidable that all opinions exist about the extent, if any, of their mixed innervation.

The proof of the existence of β fibres has been twofold. On the one hand, in a very few cases individual motor fibres have been observed histologically to branch and to supply both intrafusal and extrafusal muscle fibres [ADAL and BARKER, 1965]. On the other hand, electrophysiological study of single motor fibres has shown that some of them may elicit both an extrafusal contraction and specific effects on the spindle [BESSOU, EMONET-DÉNAND and LAPORTE, 1965]. The specificity of the excitatory effects on the spindle was shown by demonstrating that they did not run parallel with the extrafusal tension changes when the frequency of stimulation was varied or the muscle curarized. In the cat muscles so far studied the β fibres appear to be additional to specific fusimotor fibres of γ diameter of both static and dynamic kinds. However, in the rabbit lumbrical the situation has been suspected sometimes to approach that in the frog, with an absence of specific fusimotor fibres and a consequent complete reliance of the spindles upon β fibres [EMONET-DÉNAND, JANKOWSKA and LAPORTE, 1970]. In the rabbit, β fibres have been found capable of exerting either a static or a dynamic action; in the cat, only dynamic actions have as yet been observed. Pending the more exact determination of the frequency of occurrence of β fibres and of the power of their action, attention may reasonably be concentrated on the control of the spindle by specific fusimotor fibres. These appear to be the more usual and the more advanced method of fusimotor control in the mammal.

Static and Dynamic Fusimotor Fibres

Ten years ago, with the histological demonstration of the two kinds of intrafusal muscle fibres it became natural to speculate on the nature of the distinctive effects which each kind of fibre might be presumed to produce. BOYD [1962] originally described the two kinds of intrafusal fibre as receiving a completely separate motor supply; although this has proved to be something of a simplification it can still reasonably be taken to express the essence of the matter. The suggestion originally had the tremendous advantage that by its very dogmatism it simplified and so encouraged speculative thinking about

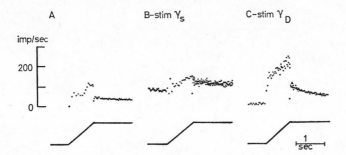

Fig.9. The contrasting effects of static and dynamic fusimotor stimulation on a primary ending. *A* Response to stretching in the absence of fusimotor stimulation. *B* Response to similar stretch applied during repetitive stimulation of a single static (γs) fusimotor fibre. *C* Response during stimulation of a single dynamic (γD) fusimotor fibre. [From BROWN and MATTHEWS, 1966.]

the teleological purpose of the body possessing two kinds of intrafusal fibre. JANSEN and MATTHEWS [1962] suggested that the separateness of the bag and chain fibres might provide a way for the CNS to control more or less independently of each the two facets of the response of the spindle primary ending to large stretches, namely its response to length and its response to velocity. In favour of this idea they found that in the decerebrate cat the length component and the velocity component of the responses of a primary ending to a ramp stretch sometimes varied independently of each other with centrally induced changes in fusimotor activity. This suggestion was followed up by investigating the effect of stimulating single fusimotor fibres on the responses of primary endings to stretches of various velocities. It then duly turned out that fusimotor fibres of γ conduction velocity could indeed be divided into two functionally distinct kinds [MATTHEWS, 1962]. The validity and usefulness of the sub-division has come to be generally accepted, in spite of the fact that there has been continuing uncertainty about what the functional classification corresponds to in histological terms. The two kinds of efferents were named static fusimotor fibres and dynamic fusimotor fibres, on the basis of their effects on the response of the primary ending to ramp stretches. Figure 9 illustrates the rationale for this. Stimulation of either kind of fusimotor fibre excites the primary ending when the muscle is at a constant length, and the difference between their action can be seen only at the beginning and end of the dynamic phase of stretching. The dynamic fibre then causes the normal velocity response of the primary ending to be augmented, whereas the static

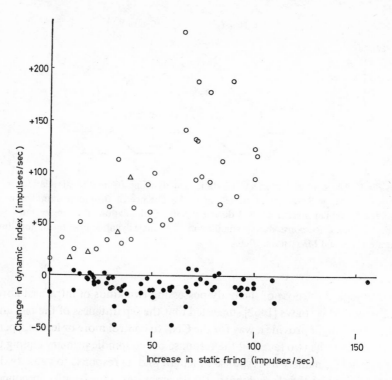

Fig. 10. Scatter diagram relating the effect of stimulating single fusimotor fibres on the dynamic index of a number of primary endings to their effect in exciting the same endings with the muscle at a constant length. Each point shows the effect of a particular motor fibre upon a particular primary ending. Open symbols, effect of dynamic fibres; closed symbols, effect of static fibres. [From BROWN, CROWE and MATTHEWS, 1965.]

fibre causes the velocity response to be decreased in spite of producing a considerable excitatory action on the length response of the ending.

The effect of fusimotor stimulation on the velocity responsiveness of the primary may be crudely assessed by measuring the dynamic index, as used in the differentiation of primary and secondary endings. As illustrated in figure 10, a population of dynamic fibres increases the dynamic index of primary endings *pari passu* with their action in exciting the same endings with the muscle at a constant length. In contrast, static fibres decrease the dynamic index or leave it unchanged, irrespective of the extent of their action in exciting the ending under static conditions. Increasing the frequency of stimulation of individual fusimotor fibres produces stronger effects of the same kind but does not change their type of action [CROWE and MATTHEWS, 1964 a]. Figure 11 also shows for a population of fusimotor fibres, which was believed to be randomly selected, that the static-dynamic classification is a natural one; there is no obvious sign of intermediate fibres, at any rate for those fibres which have a strong action. In retrospect, with the histological demonstration of a degree of motor cross-innervation between bag and chain fibres, there is a lurking suspicion that the apparently clear cut static-dynamic dichotomy arises partly because any weak intrafusal contractions are swamped in their action by strong contraction of other fibres. Thus if a fusimotor fibre were to have a mixed static-dynamic action the weaker effect might fail to declare itself. Such things remain for detailed study.

A specially important finding supporting the physiological validity of the static-dynamic classification is that when several primary endings can be studied which are all excited by any one particular fusimotor fibre, then they are all found to be influenced in the same way, whether static or dynamic. As there is only one primary ending in any particular muscle spindle this provides the most cogent argument available that the classification is physiologically significant, for it proves that a given fusimotor fibre has a static or a dynamic action by virtue of its own right, rather than of some chance relation that it enters into with the intrafusal fibres of one particular spindle. This functional specificity is not, however, matched by any gross anatomical difference between the two kinds of motor fibre where they run in the main nerve trunk, for in this region the conduction velocities of the two kinds of fibre overlap to a very high degree. However, for some muscles the majority of slow γ fibres have a static action. It may be noted that static fibres appear to be 2–3 times as common as dynamic fibres. Moreover, quite often individual static fibres are more powerful in producing excitation under static conditions than are individual dynamic fibres. Thus the combined effects of all static fibres to any particular spindle on its length response must be several times that produced by all its dynamic fibres.

A good many other differences have now been established between the actions of the two kinds of fusimotor fibre. Perhaps the most important of these for thinking about their physiological role in the body is that the secondary ending has been found to be controlled exclusively by the static fibres. Stimulation of static fibres has regularly been found to have an action on the secondary ending rather like their action on the primary ending, namely to excite an increased afferent discharge while leaving the ending largely unresponsive to dynamic stimuli. Stimulation of dynamic fibres fails to excite the secondary ending even when the particular dynamic fibre stimulated can

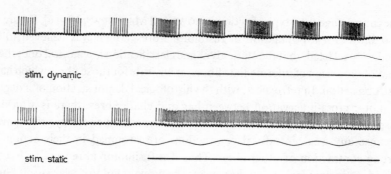

stim. dynamic

stim. static

Fig. 11. The effect of fusimotor stimulation on the response of a primary ending to sinusoidal stretching of medium amplitude. [1 mm peak-to-peak at 3 c/sec. From CROWE and MATTHEWS, 1964 b.]

be guaranteed to be causing an intrafusal contraction in the very spindle in which the secondary ending studied lies. Such a categorical statement is possible because of the experiments of APPELBERG, BESSOU and LAPORTE [1966] on a primary and a secondary ending lying in the same spindle using the tenuissimus preparation. Individual static fibres excited both kinds of afferent ending, but dynamic fibres excited only the primary ending.

Another important finding is that static fusimotor stimulation is much more effective than dynamic stimulation at counteracting the silence that the primary ending normally shows on release of a pre-existing stretch. In particular, LENNERSTRAND and THODEN [1968] have shown that increasing the frequency of static fusimotor stimulation while allowing a muscle to shorten can still make the firing of the primary ending increase above its pre-existing level. An increasing frequency of dynamic stimulation, however, failed to do so under their conditions. This argues that only the static fibres can be playing the part of conveying the command signal in any postulated servo system producing movement by way of the muscle spindle.

The response of the primary ending to sinusoidal stretching is also altered in an interesting way on stimulating the two kinds of fusimotor fibre, as shown in figure 11. During dynamic stimulation and using a medium amplitude of stretching, the primary ending fires much more intensively than normal on the stretching phase of the sine wave but still falls silent on release. During static fusimotor stimulation the mean discharge frequency is considerably increased, but the discharge is then only slightly modulated by the rhythmical mechanical stimulus. The different effects of the two kinds of fusimotor fibre remain prominent when the amplitude of stretching is sufficiently reduced for

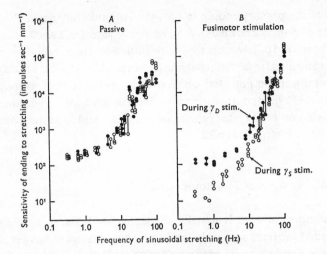

Fig. 12. The effect of fusimotor stimulation on the response of spindle primary endings to sinusoidal stretching. Two different primary endings (○, ●) were studied in two different preparations. Their behaviour in the absence of fusimotor stimulation was very similar (A). One ending was then studied during repetitive stimulation of a single static fusimotor fibre (γ_S) and the other ending during stimulation of a single dynamic (γ_D) fusimotor fibre (B). The two types of response were typical for the two types of fusimotor fibre. [From GOODWIN and MATTHEWS, 1972.]

the primary ending to be behaving linearly; their behaviour can then be described quantitatively by a frequency response curve. As shown in figure 12 at frequencies of stretching in the physiologically important range below 20 c/sec the sensitivity of the ending to stretching is much higher during dynamic stimulation than it is during static stimulation. The convergence of the two curves at higher frequencies is of considerable interest from the point of view of intrafusal mechanisms, but that need not detain us here. It is, however, related to the fact that the sensitivity of the primary ending to being driven by high frequency vibration is enhanced on stimulation of either kind of fusimotor fibre, and no differences have been found between the action of static and dynamic fibres in this respect.

Two further differences between static and dynamic action have been described which are chiefly of interest from the point of view of intrafusal mechanisms. The first is that the static fibres appear to produce their action via an intrafusal system which contracts more 'quickly' than does that activated by the dynamic fibres. This has been shown by analyzing the time course of the response of the primary ending to fusimotor stimulation, particularly by using the type of display known as a *frequencygram* [BESSOU, LAPORTE and PAGÈS, 1968]. The second difference is that some static fibres have been shown, by intracellular recording from intrafusal muscle fibres, to be capable of eliciting an all-or-nothing action potential which overshoots the resting potential [BESSOU and PAGÈS, 1969]. In contrast, dynamic fibres would appear to produce all their actions via non-propagated junctional potentials and to be incapable of eliciting a full-sized intrafusal spike.

Intrafusal Mechanisms of Fusimotor Action

The understanding of what is actually happening inside the spindle has, however, proceeded far from smoothly, and universal agreement has yet to be reached. This is simply because of the lack of suitably direct evidence. On their discovery, the suggestion was made that the dynamic fusimotor fibres achieved their particular action by virtue of supplying the nuclear bag intrafusal muscle fibres and the static fibres achieved a different action by virtue of supplying the nuclear chain fibres [JANSEN and MATTHEWS, 1962; MATTHEWS, 1962; CROWE and MATTHEWS, 1964b]. On its inception this hypothesis ran far ahead of the available evidence, and in certain respects the arguments originally advanced in its favour have not proved entirely correct. Even so, this original suggestion continues to lead the field and is widely accepted as the most likely state of affairs; nor has any comprehensive alternative to it been put forward. The hypothesis has two separate components. First, that the two kinds of fusimotor fibre achieve their separate actions by each supplying one or other of the two different kinds of intrafusal muscle fibre. Second, that the correspondence between static and dynamic action, and bag and chain intrafusal fibres falls out one particular way round. BOYD [1971] recently found that stimulation of individual fusimotor fibres usually produces a detectable contraction of only one or other kind of intrafusal fibre, and not of both. He did this by watching individual spindles under the microscope while stimulating their nerve of supply with graded shocks. Thus the first component of the hypothesis receives some direct experimental support, quite independent of the original histological evidence. But on about 10% of occasions BOYD observed that a single fusimotor fibre produced a contraction of both kinds of intrafusal fibre, and the histological evidence for a degree of motor cross-innervation appears to be inescapable. It would be surprising if such cross-innervation did not turn out to have some physiological significance. But on present evidence it may reasonably be held to be insufficiently widespread to prevent the static and dynamic actions being a result of individual fusimotor fibres supplying predominantly one or other kind of intrafusal fibre.

The evidence that the contraction of bag fibres is responsible for the dynamic effect and that of the chain fibres for the static effect depends mainly upon a correspondence between their respective speeds of action. As already noted the static fibres produce a much more rapid action on the time course of primary firing than do the dynamic fibres. Direct microscopy of isolated spindles shows that the chain fibres contract more rapidly than do the bag fibres, irrespective of whether they are activated by direct electrical stimulation or via their nerves [BOYD, 1966; SMITH, 1966]. Incidentally, it may be noted that this now irrefutable observation was initially surprising in view of the fact that the chain fibres mainly receive a diffuse motor innervation whereas the bag fibres receive mainly discrete plate motor endings; in the frog, the fast twitch fibres receive plates and the slow tonic fibres receive diffuse endings. Further evidence for which type of fusimotor fibre supplies which type of intrafusal fibre is provided by the exclusive excitation of the secondary ending by static fibres. As already noted the secondary ending lies predominantly, if not entirely, upon the nuclear chain fibres. This suggests that contraction of the chain fibres alone should excite the secondary ending, while contraction of the bag fibres should not usually do so since there are usually no secondary terminals upon them. Thus the failure of the dynamic fibres to excite the secondary ending fits in with the idea that it is these which supply the bag fibres. Likewise the ability of the static fibres to excite secondary endings.

An earlier objection to the present scheme was based on certain findings in the rabbit. For a time it was believed that all the intrafusal fibres of this species were of the nuclear bag type. Yet functionally separable static and dynamic fusimotor fibres proved to be regularly isolatable in this species also, and so it appeared that the two functionally distinct kinds of fusimotor fibre could not achieve their specific actions by virtue of supplying different kinds of intrafusal fibre. This argument collapsed with the demonstration that two kinds of intrafusal fibre could be demonstrated in the rabbit on using the appropriate histochemical and electron-microscopic techniques. There is certainly a great deal more to be done on isolated spindles to unravel the full complexity of their internal operations, but for the time being it seems reasonable to continue to accept the working hypothesis of a correspondence between dynamic fusimotors and bag intrafusals, and between static fusimotors and chain intrafusals.

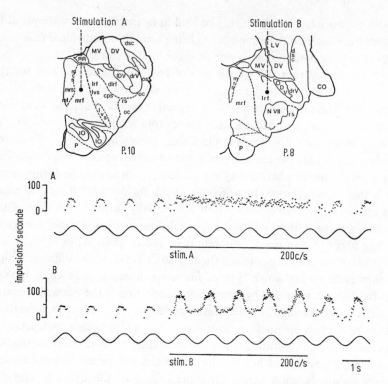

Fig. 13. Typical static *(A)* and dynamic *(B)* actions on a primary ending elicited by stimulating at two separate points in the reticular formation. [From VEDEL and MOUILLAC-BAUDEVIN, 1969.]

Possible Functional Roles for the Two Kinds of Fusimotor Fibre

There is now plenty of evidence to show that the CNS can exert an independent control over the static and dynamic fusimotor fibres. This has been demonstrated by stimulating a variety of sites within the CNS while recording from primary and/or secondary endings and observing the appropriate change in their behaviour [see, for example JANSEN and MATTHEWS, 1962; APPELBERG and EMONET-DÉNAND, 1965; VEDEL and MOUILLAC-BAUDEVIN, 1969]. Relatively pure actions of either static or dynamic type have then been obtained on many occasions. Figure 13 illustrates striking examples on stimulating in the midbrain. But it is still premature to attempt to attach meaning to the findings from the point of view of which central region produces which action, and to use this to argue about the function of various higher structures. A more

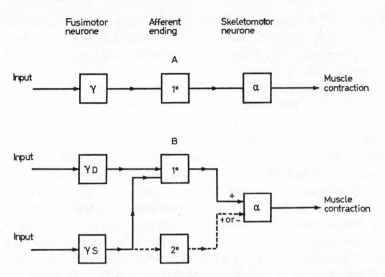

Fig. 14. Diagram showing possible routes for the activation of muscle via the fusimotor fibres and the spindle afferents. *A* Classical scheme. *B* Modified scheme incorporating some of the newer findings. [From BROWN and MATTHEWS, 1966.]

immediate need is to achieve understanding of the kind of use to which the CNS, taken as a whole, puts the two kinds of fusimotor fibre. When there was thought to be only a single system of fusimotor fibres they were widely believed to provide an indirect or γ route for producing activation of the main muscle, functioning rather in the way that would lead an engineer to call the system a *follow-up length servo* [MERTON, 1953; GRANIT, 1955; HAMMOND, MERTON and SUTTON, 1956]. The classical scheme for this is illustrated at the top of figure 14. In this mode of operation the higher centres were suggested to command a muscle contraction by exciting in the first place the γ-moto-neurones rather than the α-motoneurones as would normally be expected. Such γ excitation would activate the primary spindle endings with consequent monosynaptic excitation of the α-motoneurones and contraction of the main muscle. Thus the γ discharge was looked upon as a 'command signal' for eliciting muscle contraction. This apparently roundabout way of doing things was seen as having the advantage of continuously utilizing the feedback properties of the stretch reflex loop so that a movement would be able to continue on its way irrespective of any external obstacle or fatigue of the muscle. Unfortunately, we cannot continue to think in terms of such a simple scheme for it ignores the secondary ending and also the two types of fusimotor fibres.

The lower half of figure 14 shows the currently more realistic scheme, though even this is still a gross simplification of the real state of affairs; it omits, among other things, the Golgi tendon organ and any controlling reactions mediated by higher centres.

The main question about the new scheme is whether only one or other of the two types of fusimotor fibres provides the input command to the postulated servo. The answer appears to be that the static fibres are chiefly if not wholly responsible. This can be said because the static fibres alone appear to be able to elicit an intrafusal contraction which is fast enough to keep up with the extrafusal muscle fibres. This is essential to make the spindle fire faster during shortening of the muscle than before and thus enable it to continue to provide a reflex excitation of the α-motoneurones to keep the contraction going. The dynamic fibres appear to produce an intrafusal contraction which is just too slow to be able to do this. A number of experiments now leave no doubt that the spindle primary afferents usually do indeed fire faster during physiologically evoked contractions, and thus seem fitted to play a part in reflexly helping the contractions along.

Casting the static fusimotor fibres in the role of the pathway for the command signal for muscle shortening immediately raises a new problem, namely that the static fibres automatically activate the secondary spindle endings along with the spindle primary endings. Since the primary and secondary endings are approximately equally numerous any static fusimotor activity will lead to an approximately equal inflow of either kind of spindle impulses to the spinal cord, where both may be presumed to produce reflex actions. The conventional view of the spinal action of the group II fibres from the secondary endings is that they produce a non-specific flexor reflex involving excitation of flexor motoneurones and inhibition of extensor motoneurones, whatever the muscle of origin of the afferent fibres themselves. If so, transmission of a command signal along the static fusimotor fibres of an extensor muscle will produce a mixture of inhibitory and excitatory reflex effects on the α-motoneurones of the muscle; the balance of these effects will depend upon their relative strengths. If the excitation and the inhibition were equally powerful the net effect of the static command would be zero. Flexor motoneurones, however, would be expected to behave quite differently from extensor ones and to be reflexly excited by static fusimotor activity increasing both primary and secondary firing. It would be surprising if the autogenetic reflexes initiating and controlling movement were to be so differently organized for flexor and for extensor muscles. One way out of the difficulty is to suggest that the interneurones, via which the reflex effects of the secondary endings are mediated, are normally kept out of action by being tonically inhibited by supraspinal centres. This is an unsatisfying view for it seems quite unnecessary to waste the rather accurate information which the secondary endings purvey about the length of a muscle. Such information might be expected to be of use for immediate reflex control as well as for the instruction of higher centres. An alternative way out of the quandary is to question the conventional view of the reflex action of the spindle secondaries and to suggest that under some conditions the interneuronal networks via which they act can be biassed by supraspinal action so that they produce an autogenetic excitation of extensors rather than an inhibition. Recent indirect experiments suggest that the rigidity of the decerebrate cat is partly mediated by the spindle secondary endings contributing to the tonic stretch of extensor muscles rather than opposing it [MATTHEWS, 1969; WESTBURY, 1971], though the matter is still controversial.

What then is the function of the dynamic fibres? The original suggestion was that by increasing the sensitivity of the primary ending to velocity they served to damp the stretch reflex arc and hence reduce its tendency to go into spontaneous oscillation or tremor. This remains a possible idea, but it should be emphasized that it has yet to be tested in any quantitative manner; only thus can the truth or otherwise of the idea be established. The more recent finding that dynamic fusimotor activity maintains a high sensitivity of the primary ending to small amplitude stimuli seems just as important as the dynamic fusimotor action in augmenting the velocity response of the primary to large stretches. Both actions, however, would have the valuable effect of sensitizing the stretch reflex and thus increasing the efficacy of postural holding reactions. It may be noted that the stretch reflex of the decerebrate cat may show a larger sensitivity to small stretches than it does to large ones, presumably as a result of such behaviour on the part of the primary ending. Thus in general terms the dynamic fibres may be seen as controlling the parameters of response of the primary ending to external stimuli, and thereby controlling the sensitivity of the stretch reflex loop.

Central Co-Activation of α and γ Motor Fibres

In spite of the popularity of the *follow-up servo* hypothesis there has never been any very cogent evidence that fusimotor fibres normally function on their own, and that movements are actually solely produced by the CNS using the static fibres to inject a command signal into the stretch reflex loop. Virtually all recordings of motor discharges during centrally initiated muscle contractions show a co-activation of α and γ motor fibres, as has long been recognized [GRANIT, 1955]. Quite often the increased activity of the α fibres precedes that of the γ fibres so there is no question of the α fibres having been initially activated via the γ pathway. Several descending pathways would appear to be organized so as to excite the two kinds of motor fibre in parallel, though in varying degree. Co-activation of α and γ fibres gives what may be crudely termed a 'servo assistance' of movement rather than a follow-up servo. One essential difference between the two modes of operation lies in the gain required of the feedback loop. For a follow-up system to work successfully the gain of the spindle feedback must be rather high in order to ensure that the same movement occurs whatever the load into which the muscle is working. In contrast, in a servo-assisted system the higher centres have to size up the situation and send the appropriate commands to the α-motoneurones to deal with the expected load. Even quite a moderate gain in the spindle feedback would then help to deal with an unexpected disturbance, and it becomes irrelevant that the increase in spindle firing during a movement is insufficient to elicit reflexly the whole of the observed α firing. Experiments on both animals and man suggest that the gain of the stretch reflex is relatively low for a follow-up system to be effective [MATTHEWS, 1966, 1967; TARDIEU, TABARY and TARDIEU, 1968]. Co-activation, of course, also avoids the loss of time involved in trying to initiate a movement solely via the γ route.

It may next be asked why the body should have bothered to improve upon the shared skeleto-fusimotor innervation found in lower animals if the α and γ motoneurones are still going to be co-activated. One possible answer is that it may be advantageous to keep the spindles firing at much the same rate throughout the course of a movement as long as the movement proceeds according to plan; the spindles would then continue to be at the ready to signal any deviations. As noted earlier [MATTHEWS, 1964], 'this would be achieved if the relative amounts of α and γ activity were adjusted to be appropriate for the velocity of shortening "expected" under any particular set of conditions. Then if the shortening proceeded faster than "intended" by the higher centres it would be slowed by servo action, and if shortening were hindered by some unexpected load it would be speeded up by servo action'. The independence of the fusimotor pathway from the normal pathway provides the essential degree of freedom which would allow the fusimotor discharge to be appropriate primarily to the trajectory of the planned movement. The direct motor discharge would, in addition, have to be appropriate to the external load. However, all such suggestions are clearly highly speculative until more can be found out about the behaviour of spindles during a variety of voluntary movements and also about the accompanying phasic variations in the sensitivity of the reflex centres. For this sort of thing, human recording appears the method *par excellence* and puts the animal experimenter in the shade. Perhaps the laboratory physiologist has his next major contribution to make in unravelling the use made of the spindle afferent discharges by the higher centres, rather than by just the spinal cord.

Author's address: Dr. P.B.C. MATTHEWS, University Laboratory of Physiology, *Oxford, OX 1 3 PT* (England)

New Developments in Electromyography and Clinical Neurophysiology,
edited by J.E. Desmedt, vol. 3, pp. 126–135 (Karger, Basel 1973)

Cortical Control of Dynamic
and Static ϒ Motoneurone Activity

J. P. VEDEL

CNRS, Institut de Neurophysiologie et de Psychophysiologie, Marseille

The γ motoneurones are subjected to excitatory and inhibitory effects,
independent of any reflex activity, by numerous supraspinal structures and in
particular by stimulation of the sensorimotor cortex and of the pyramidal
tract [GRANIT and KAADA, 1952; MORTIMER and AKERT, 1961; KATO, TAKA-
MURA and FUJIMORI, 1964; LAURSEN and WIESENDANGER, 1966; KOEZE,
PHILLIPS and SHERIDAN, 1968].

In 1962, MATTHEWS made the distinction between dynamic and static
γ motoneurones excercising different influences on the activity of the primary
afferent endings of the muscle spindles [see MATTHEWS, this volume]. It thus
becomes necessary to reconsider the centrifugal control of fusimotricity,
taking account not only of both types of γ fibre, but also of the functions of
the antagonist muscles whose proprioceptors are supplied by them.

With this intention we did a comparative study of the changes in dynamic
sensitivity and static discharge of the primary intrafusal endings of an extensor
muscle (soleus) and a flexor muscle (tibialis anterior) following repetitive
stimulation of the sensorimotor cortical areas and of the pyramidal tract,
under different degrees of anaesthesia. We investigated the pathways involved
in the diffusion of cortical effects and, taking the same line as CROWE and
MATTHEWS [MATTHEWS, 1962; CROWE and MATTHEWS, 1964a,b], we attempted to specify the dynamic or static nature of the fusimotor system concerned in each of the supraspinal actions observed.

Methods

The experimental subjects were 48 adult cats, under fluothane anaesthesia [VEDEL and
MOUILLAC-BAUDEVIN, 1970].

With the animal in the prone position, the head was fixed in a Horsley-Clarke type of restraining apparatus. For cortical stimulation bipolar electrodes made of two platinum balls were used. Pyramidal stimulation was performed with concentric electrodes inserted stereotactically through the cerebellum and medulla while recording the pyramidal response to cortical stimulation.

The choice of stimulation intensity was found to be crucial in order to differentiate activation of the α motoneurones from that of the γ motoneurones. As for many supraspinal structures, the excitation threshold for α motoneurones from the sensorimotor cortex and the pyramidal tract is slightly higher than that of γ motoneurones [GRANIT and KAADA, 1952]. By selecting the minimal stimulation intensity necessary to cause a regular change in intrafusal discharges, we were able to eliminate interfering muscle contraction.

Classical techniques were used to denervate the hind limb and to dissect the dorsal spinal roots L7, L6, S1 in order to record unit proprioceptive activity from soleus and tibialis anterior. The unit activity recorded was identified as originating intrafusally by the increase in frequency in discharge during muscle stretch and the cessation of discharge during a contraction. Calculation of conduction velocity then indicated whether the active fibre was group I_A or group II and, therefore, whether the intrafusal ending was primary or secondary. The higher dynamic sensitivity of the primary intrafusal endings also differentiated their discharges from those of secondary endings.

The action potentials triggered a frequency meter displaying on a CRO the I_A fibre instantaneous frequency [BESSOU, LAPORTE and PAGES, 1968].

The dynamic sensitivity of the primary endings, indicated by their response to a constant velocity muscle stretch, was expressed quantitatively by the dynamic index, i.e. the difference between the frequency of impulses measured at the end of the phasic muscle stretch and 0.5 sec later [CROWE and MATTHEWS, 1964a].

The soleus and tibialis anterior tendons were isolated and connected to an electromechanical device which performed constant velocity stretches between 0.5 and 6 mm/sec. Tension in the two muscles was recorded with strain gauge devices to check for absence of muscle contractions during central stimulation. The stretch velocity was recorded with a linear transducer.

Results

Previous animal work [SHIMAZU, HONGO and KUBOTA, 1962; VEDEL and MOUILLAC-BAUDEVIN, 1969] has shown that the fusimotor effects caused by central structures often depend on the degree of anaesthesia and we therefore varyied this factor experimentally.

Soleus Muscle (Extensor)

Effects of Cortical Stimulation
1. Inhibitory effect. Under light anaesthesia EMG activity persists in the muscles of the neck and shoulders, and repetitive bipolar stimulation of the

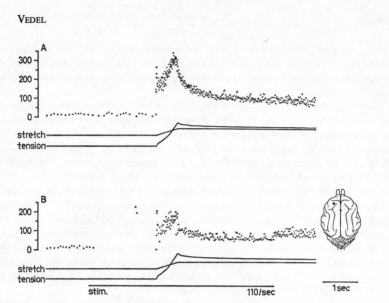

Fig. 1. Inhibitory effect. Decrease of the static discharge and of the dynamic sensitivity of a I_A afferent from soleus following stimulation of the contralateral sensorimotor cortex, under light anaesthesia. *A* Static discharge and response of the ending to muscle stretch at a constant velocity without stimulation. *B* Effects of cortical stimulation at 110 c/sec, 0.3 msec, 5 V. Amplitude of stretch: 5 mm. Conduction velocity of the afferent fibre: 97 m/sec. In all figures the position of the stimulating cathode is shown on the diagram of the cortex. The frequency of firing of the I_A spindle afferent is plotted as spikes/sec in the ordinate (frequencygram). The lines below indicate the stretching of the muscle, the variations of tension and the time of electrical stimulation of cortex or pyramid. Horizontal calibration, 1 sec.

contralateral sensorimotor cortex provokes a decrease and often complete cessation, of variable duration, of the spontaneous discharge of primary afferent endings. It also causes a reduction of the response of these endings to constant velocity phasic muscle stretching. This reduction in dynamic sensitivity is seen by the decrease in mean and maximal frequency of the intrafusal discharge during lengthening of the muscle, the dynamic index being either decreased or unchanged (fig. 1).

2. Excitatory effects. Under deeper anaesthesia, and contrary to previous observations, stimulation of the contralateral sensorimotor cortex causes reinforcement of the static discharge of I_A endings of the soleus. Depending on the nature of the increase in spindle discharge and the changes in dynamic

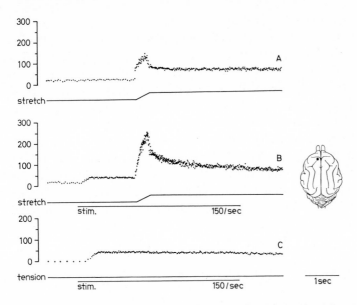

Fig. 2. Dynamic facilitating effect. Increase in static discharge and in dynamic sensitivity of a I_A afferent from soleus following stimulation of the contralateral sensorimotor cortex. *A* Response of the ending to a muscle stretch at constant velocity without stimulation. *B* Response to the same stretch during cortical stimulation. *C* Effect of stimulation on the static discharge of the ending (constant muscle tension). Stimulation at 150 c/sec, 0.5 msec, 3 V. Amplitude of stretch: 3 mm. Conduction velocity of the afferent fibre 93 m/sec.

sensitivity which accompany it, two types of excitatory effect may be differentiated, one dynamic and the other static.

(a) A dynamic effect was seen in 61 primary endings of the soleus. Cortical stimulation causes moderate and regular reinforcement of the static discharge when the muscle is kept under constant tension. Simultaneously, a large increase in dynamic sensitivity is seen by the increase of mean and maximal frequencies and of the dynamic index of the response of the sensory endings to phasic muscle stretch. This effect is shown in figure 2.

(b) A static excitatory effect was obtained very irregularly in 7 of the 61 endings in the soleus which had shown the dynamic excitatory effect. This also involves an increase of the static discharge which is often greater than in the previous case and is accompanied by a clear increase in variability of the interval between successive nerve impulses. At the same time, stimulation of the contralateral sensorimotor cortex causes a decrease in dynamic sensitivity.

Effects of Pyramidal Stimulation

Under the same conditions of anaesthesia, constant muscle tension or phasic muscle stretch, repetitive stimulation of the contralateral medullary pyramid evokes changes in the activity of I_A spindle endings in the soleus similar to those obtained with cortical stimulation, with exactly the same temporal evolution but a greater intensity. Also the same effects are obtained with stimulation voltages much lower than those used at the cortex.

In summary, under light anaesthesia, stimulation of the sensorimotor cortex and of the pyramidal tract reduces the static discharge of I_A spindle endings of the soleus and decreases their dynamic sensitivity. Under deeper anaesthesia, the same stimulation reinforces the static discharge of these endings; on the whole, this increases response to phasic muscle stretch and diminishes it only in a few cases.

Tibialis Anterior Muscle (Flexor)

Effects of Cortical Stimulation

Unlike the effects seen in the soleus, stimulation of the sensorimotor cortex, whatever the level of anaesthesia, causes only facilitation of the static discharge in I_A afferents from tibialis anterior. As for soleus, the results provide differentiation between dynamic and static types of facilitating effect.

(a) Dynamic facilitation was seen in 31 primary endings. Cortical stimulation caused moderate and regular increase in static discharge of the endings and at the same time greatly reinforced their dynamic sensitivity. The general characteristics of this facilitation were identical to those seen in the soleus.

(b) Static facilitation was found irregularly in 8 of 31 primary endings which had shown the dynamic facilitating effect. It was similar to that seen in the soleus. It comprises a large increase in static discharge and a decrease of the dynamic sensitivity of the primary endings (fig. 3).

Effects of Pyramidal Stimulation

As for the soleus muscle all the fusimotor effects seen with cortical stimulation were reproduced with pyramidal stimulation. To summarize, stimulation of the contralateral sensorimotor cortex and pyramidal tract reinforces the static discharge of the primary spindle endings in tibialis anterior whatever the level of anaesthesia. This facilitation is usually accompanied by an increase but occasionally a decrease in the response of the endings to phasic muscle stretch.

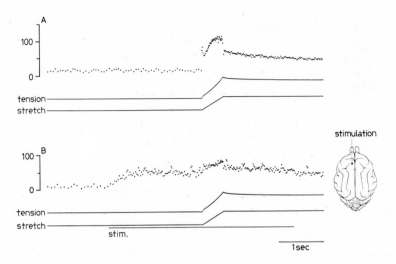

Fig. 3. Static facilitating effect. Increase in the static discharge and decrease of dynamic sensitivity of a I_A afferent from tibialis anterior following stimulation of the contralateral sensorimotor cortex. *A* Static discharge and response of the ending to a muscle stretch at constant velocity without stimulation. *B* Effect of cortical stimulation at 90 c/sec, 0.3 msec, 6 V. Amplitude of stretch: 4 mm. Conduction velocity of afferent fibre: 101 m/sec.

Identification of the Efferent Pathway Responsible
for Transmitting the Fusimotor Effects Arising from the Cortex

Since there is cortical control of the reactivity of muscle spindles, the problem arises of identifying the efferent pathway responsible for corticofugal transmission to the gamma motoneurones. The possibilities that this might be rubrospinal or pyramidal were investigated.

The red nucleus ipsilateral to the cortex stimulated was completely destroyed by electrolytic lesions. These lesions never modified, in our experiments, either the sequence of appearance or the characteristics of the effects studied.

The total or partial interruption of the pyramidal tract suppressed all modifications in the activity of the I_A afferents of extensor and flexor muscles previously obtained from the contralateral sensorimotor cortex. After electrocoagulation, cortical stimuli at the same or a higher voltage than that used in the control situation, and repeated at regular intervals for 2–3 h, remained ineffective whatever the level of anaesthesia (fig. 4D).

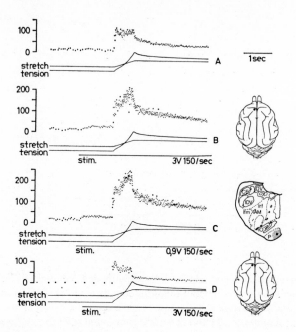

Fig.4. Dynamic facilitating effect on the discharge of a I_A afferent from soleus by stimulation of the sensorimotor cortex and of the pyramidal tract. *A* Static discharge and response of the ending to a constant velocity muscle stretch without stimulation. *B* Effect of cortical stimulation. *C* Effect of pyramidal stimulation. *D* Effect of cortical stimulation after electrocoagulation of the pyramidal tract. Stimulation at 150 c/sec and 0.5 msec, at 3 V in *B* and *D*, at 0.9 V in *C*. Amplitude of stretch: 5 mm. Conduction velocity of the afferent fibre: 107 m/sec. On the diagram of a cross-section of the medulla the shaded area around the point of stimulation shows the extent of the electrolytic lesion.

Discussion

These results confirm the existence of functional relations between the sensorimotor cortex and both the pyramidal tract and the motricity of the muscle spindles, which had already been demonstrated [GRANIT and KAADA, 1952; MORTIMER and AKERT, 1961; KATO, TAKAMURA and FUJIMORI, 1964; FIDONE and PRESTON, 1969; KOEZE, PHILLIPS and SHERIDAN, 1968; YOKOTA and VOORHOEVE, 1969].

The combination of effects observed suggests that under our experimental conditions the fusimotor effects obtained with stimulation of the sensorimotor cortex and of the pyramidal tract result from the involvement, at different levels, of the same efferent mechanism. They appear to demonstrate the existence of a contralateral cortical control of the activity of the γ moto-neurones via the pyramidal tract.

The arguments for this interpretation are: (1) the localization on the frontal cortex of stimulation points initiating fusimotor effects in the regions from which many pyramidal fibres leave; (2) the suppression of all the corti-cally originating fusimotor effects by electrolytic lesion of the medullary pyramid, and (3) the similarity of cortical and pyramidal effects observed.

We shall now discuss the resemblance between effects of stimulating the cortex or pyramid and the actions of the pyramidal system. If we compare our studies of static discharge and dynamic sensitivity of the primary spindle endings with the effects obtained by MATTHEWS [1962] and CROWE and MATTHEWS [1964a,b] from stimulating fusimotor fibres directly, we may define the role of the pyramidal system in the control of activity of both dy-namic and static γ motoneurones.

The inhibitory effect which is exerted on the primary spindle discharges of the soleus and not the tibialis anterior may be obtained independently of any muscle contraction. It appears, therefore, to be purely fusimotor in origin and probably reflects a decrease in activity of γ motoneurones of the type ob-served by FIDONE and PRESTON [1969] in the nerves of extensor muscles when stimulating the pyramidal system.

The mechanisms responsible for the decrease in spindle activity may be discussed for both the central level and the peripheral level. Centrally, it may be assumed that the stimulation of the pyramidal system blocks supraspinally originating impulses which maintain the activity of fusimotor neurones and thus assure a certain degree of sensitivity of the intrafusal endings. Peripher-ally, knowing the functional properties of dynamic and static fusimotor fibres and accepting that the two types of fusimotricity are active spontane-ously we may suggest that the strong decrease in the static discharge of the primary spindle endings, and the more moderate reduction in their responses to muscle stretch, manifest a depression of activity in the static γ moto-neurones.

The dynamic facilitating effect seen very regularly in the primary ending discharge of soleus and tibialis anterior is characterized by a moderate in-crease in the static discharge of the endings and a very large reinforcement of their responses to phasic muscle stretch, comparable to those seen by CROWE and MATTHEWS [1964a, b] when stimulating dynamic fusimotor fibres directly.

The static facilitating effect which appears less frequently and very ir-regularly is also seen as an increase in the static discharge of I_A endings but this is greater than that seen in the previous case and is accompanied by a decrease in dynamic sensitivity. These changes are comparable to those obtained by direct stimulation of static fusimotor fibres [Matthews, 1962; Crowe and Matthews, 1964a,b; Bessou, Laporte and Pages, 1968].

These facts suggest that in our experiments stimulation of the pyramidal system under light anaesthesia caused a decrease in the activity of the static γ motoneurones in the extensor muscles, and under deeper anaesthesia facilitation of the activity of the dynamic γ motoneurones and less commonly of their static γ motoneurones. At the same time whatever the level of anaesthesia, the same stimulation reinforces the activity of the dynamic γ motoneurones of the flexor muscles and less regularly of their static γ motoneurones.

As other authors have proposed [Fidone and Preston, 1969; Yokota and Voorhoeve, 1969], it appears that the sensorimotor cortex is able to control the activity of both types of γ motoneurone. It seems also that there may be a difference in corticofugal fusimotor effects depending on whether the γ motoneurones involved supply flexor or extensor muscles, and in the same muscle group whether they are dynamic or static. This suggests that the cortex controls all the sensory properties of the muscle spindles which allows it to control the spinal and supraspinal functions in which the proprioceptive afferents participate.

Current theories on the role of muscle spindles in movement suggest various hypotheses as to the significance of cortical control of fusimotricity. According to Kuffler and Hunt [1952], it is possible that the effect of the cortex on the γ motoneurones protects the muscle spindles from being unloaded by contraction and thus assures continuity of proprioceptive information regardless of the state of the muscle. According to Merton [1953] the muscle spindles act as length detectors in a servo mechanism which by reflex pathways maintains the muscle at a given length, and this would mean, as Matthews has suggested [1964] that the action of fusimotricity is to allow the system to adapt itself to different conditions. In this way the cortex could determine the length of the muscle to be maintained via the servo mechanism proposed by Merton.

Apart from its effect on spinal reflex activity which depends on muscle proprioceptors, the cortical control of fusimotor neurones appears to be a mechanism for modulating supraspinal sensory projections [Phillips, 1969]. It is becoming more and more apparent that the spindle afferents are relayed to supraspinal nervous structures and in particular to certain cortical areas [Amassian and Berlin, 1958; Oscarsson and Rosen, 1963; Albe Fessard and Liebeskind, 1966; Landgren and Silfvenius, 1969; Phillips, Powell and Wiesendanger, 1970].

The sensorimotor cortex thus appears to constitute with the mɔuscl spindles a functional loop since it can control the activity of the proprioceptors from which at the same time it receives information. Thus the sensorimotor cortex would control the activity of the mechanism for constant estimation of muscle state (the cortical projections of the muscle spindles) and that the information transmitted by this mechanism would allow the cortex to readjust its effect on α motoneurones.

A system of regulation via the cortex would obviously be slower than via the spinal cord since the ascending projections have several synapses but it would permit the necessary analysis of proprioceptive information for reorganisation of motor control.

Summary

1. Effects of repetitive stimulation of the sensorimotor cortex and the pyramidal tract were studied on the static discharge and dynamic sensitivity of contralateral flexor (tibialis anterior) and extensor (soleus) muscle spindle primary endings in cats anaesthetized with Halothane (Fluothane).

2. Under light anaesthesia, cortical and pyramidal stimulation decreases the static discharge and the dynamic sensitivity of the soleus spindle primary endings (depressant effect).

3. Under moderate anaesthesia, cortical and pyramidal stimulations mainly induce a powerful increase of the dynamic sensitivity and a slight increase of the static discharge in the soleus muscle spindle primary endings (facilitatory dynamic effect). Less often, the same stimulations increase the static discharge and decrease the dynamic sensitivity of the soleus primary endings (facilitatory static effect).

4. Whatever the level of anaesthesia, cortical and pyramidal stimulations induce only facilitatory effects, mainly dynamic and sometimes static on the tibialis anterior spindle primary endings.

5. Cortical fusimotor effects are suppressed by electrocoagulation of the ipsilateral pyramidal tract.

6. Cortical and pyramidal fusimotor effects were compared with the effects of dynamic and static γ fibre stimulation.

Author's address: Dr. J. P. VEDEL, CNRS, Institut de Neurophysiologie et de Psychophysiologie, 31, chemin Joseph Aiguier, *F-13 Marseilles 9e* (France)

New Developments in Electromyography and Clinical Neurophysiology,
edited by J.E. Desmedt, vol. 3, pp. 136–144 (Karger, Basel 1973)

Pyramidal Apparatus for Control
of the Baboon's Hand

C. G. Phillips

University Laboratory of Physiology, Oxford

The hands of primates have remained morphologically primitive; their
claws have actually degenerated into flattened nails [Le Gros Clark, 1962];
but their finger-pads, which are firmly supported by the flattened nails, are
richly innervated and richly and extensively represented in the cerebral cortex
[Mountcastle and Powell, 1959]. In the course of evolution, enlarging
brains have enormously increased their powers of exploratory tactile discrimination. Kuypers' article [this volume] has described the connections whereby
the pyramidal tract (PT) can control the input from the hand at the level of
the second-order neurones in the dorsal horn of the spinal cord and the dorsal
column nuclei, and the importance of such control can scarcely be exaggerated.
But electrophysiological experiments on this control have been performed so
far mainly on the cat [for a review, see Gordon, 1968].

Exploratory tactile discrimination by the primates has been associated
with a corresponding increase in their powers of manipulation and in the
repertory of actual and possible movements that enlarging brains can impress
on morphologically primitive hands [Phillips, 1971]. These motor refinements
have developed *pari passu* with enlargement of the cortico-motoneuronal (CM)
component of the PT, which Kuypers [this volume] has also described, and
which was demonstrated electrophysiologically by Bernhard and his colleagues in Stockholm [Bernhard and Bohm, 1954; Preston and Whitlock,
1960, 1961]. This evolutionary trend can be followed in the series of living
primates [Phillips, 1971]. The hands of Prosimii show stereotyped prehensive
patterns, and reach and grasp with a whole-arm movement culminating in
extension-divergence followed by flexion-convergence of all 5 digits: they do
not explore surfaces with their touch-pads [Bishop, 1964]. Their PT have a
negligible CM component [Campbell, Yashon and Jane, 1966]. The hands

of Anthropoidea (especially those of the Old World monkeys, apes and man) show an increasing wealth of prehensive patterns, in which their hands adapt their shapes to the shapes of the surfaces they are about to grasp; and a clear differentiation of *power* and *precision grips* [NAPIER, 1956], the former involving the hand as a whole, the latter involving the independent activity of thumb and index finger which enables (for example) any baboon to extract the sting from a scorpion [SCHULTZ, 1961]. Cutting the PT destroys the precision grip and the independent activity of the index finger in adult monkeys [TOWER, 1940; LAWRENCE and KUYPERS, 1968]; the same operation prevents their development in infant monkeys [LAWRENCE and HOPKINS, 1970]. Use of the hands for prehension then resembles in general that of the Prosimii, but there is also difficulty in relaxing the fingers' hold on a morsel of food when it is brought to the mouth. The normal reactions of the hands to contact are permanently impaired [DENNY-BROWN, 1966]. Their use in walking and climbing is fully recovered soon after the operation.

The α-motoneurones which innervate the muscles involved in the baboon's precision grip are preferentially accessible to electrical stimulation of the motor cortex. A brief, high-frequency burst of impulses [cf. PHILLIPS, 1969] discharged by a population of PT cells centred on the area outlined in figure 1, elicits a brief motor response ('twitch') that is confined to those parts. The same stimulus, applied elsewhere to the cortex exposed in figure 1, and exciting corticofugal discharges of similar size, frequency and duration, has no overt motor effect. The response is not obtainable from monkeys after chronic section of the corresponding PT, even if only 30% of the cross-section of the tract is destroyed: the control response from the opposite hemisphere is normal [FELIX and WIESENDANGER, 1971]. The characteristic preferential accessibility therefore depends on the integrity of the PT.

Figure 1A shows the pure excitatory postsynaptic potential (EPSP) evoked at monosynaptic latency in a motoneurone of the radial nerve by a single small volley in the fastest PT axons from the area shown on the photograph [LANDGREN, PHILLIPS and PORTER, 1962]. Figure 1 B shows the EPSP evoked in the same motoneurone by a volley in the lowest-threshold (group I_A) muscle afferent fibres. Figure 2 shows that the PT and I_A synapses applied to the same motoneurones have different transmitting properties: those of the PT exhibit marked presynaptic facilitation when activated repetitively [PHILLIPS and PORTER, 1964; PORTER, 1970]; those of the I_A input do not. Figure 3A shows that the α-motoneurones of the intrinsic hand muscles, and also those of the extensors of the digits which must open the hand to adapt its shape to what it is about to grasp, receive the largest quantities of CM excit-

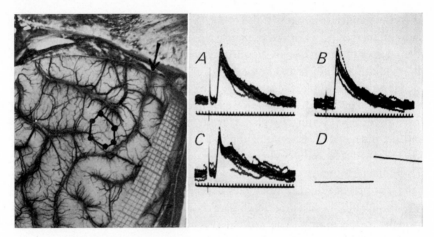

Fig.1. Left: rolandic cortex of baboon, showing area from which flick abduction of index was evoked by single, unifocal, rectangular, surface-cathodal current pulses, strength 1.4 mA, duration 4.5 msec. No other movement was evoked from this area by these pulses, and no movement from any other part of the exposed cortex. Scale in mm. Arrow marks central (rolandic) fissure. *A–C* Intracellular records from α-motoneurone of radial nerve of baboon. *A* EPSP evoked by unifocal surface-anodal pulses, strength 1.25 mA, applied to cortical area corresponding to that outlined in figure 1. Time, 1,000c/sec in *A*, *B* and *C*. *B* Group I$_A$ monosynaptic EPSP. *C* Same pulses applied to cortex at a point 3 mm distant from point stimulated in *A*: note smaller EPSP and inhibitory erosion of its decaying phase. *D* Calibrating 1.5 mV rectangular voltage step to amplifier input [LANDGREN, PHILLIPS and PORTER, 1962].

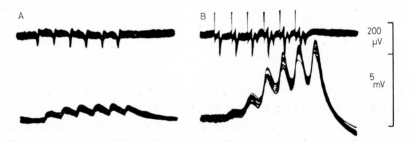

Fig.2. Intracellular records from α-motoneurone of median nerve of baboon. *A* Group I$_A$ volleys repeated at 200 c/sec (upper record); monosynaptic EPSP evoked by them in motoneurone (lower record). *B* Corticospinal volleys repeated at 200 c/sec (upper record); monosynaptic EPSP evoked in motoneurone (lower record). Strength of cortical pulses adjusted so that first CM EPSP was similar in size to first group I$_A$ EPSP [PHILLIPS and PORTER, 1964].

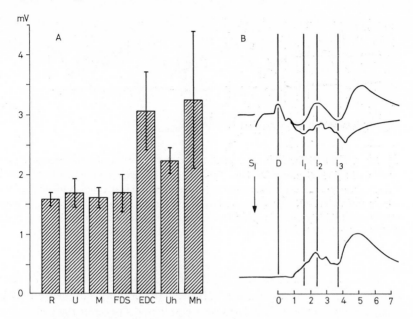

Fig.3. A Mean amplitudes of maximal CM EPSP in α-motoneurones classified by antidromic invasion from the following nerves: whole radial (R); whole ulnar (U); whole median including 4 from palmaris longus (M); flexor digitorum sublimis (FDS); extensor digitorum communis (EDC); ulnar to intrinsic hand muscles (Uh); median to intrinsic hand muscles (Mh) [CLOUGH, KERNELL and PHILLIPS, 1968]. *B* Tracings of intracellular records from γ-motoneurone of radial nerve (top tracing) and extracellular field potential recorded immediately outside the cell membrane (middle tracing). The difference between these curves (bottom tracing) gives the active response (EPSP) of the cell membrane to single unifocal surface-anodal cortical pulses, strength 11 mA, duration 0.1 msec, which discharged D and I volleys whose earliest impulses arrived in spinal segment at times marked by vertical lines. Arrow marks instant of stimulation. Time scale: msec. Vertical scale: mV depolarization. Conduction velocity of γ-axon 22.5 m/sec [CLOUGH, PHILLIPS and SHERIDAN, 1971].

atory drive. This fact, together with the fact of presynaptic CM facilitation, explains the preferential accessibility of the 'precision' parts to those cortical stimuli which evoke brief high-frequency bursts of corticofugal impulses.

It has long been appreciated that the cortex can inhibit as well as excite movement. Figure 1 C shows that the volley discharged by a stimulus to a point 3 mm distant from the 'best point' for excitation evoked a smaller EPSP, and also an inhibitory postsynaptic potential (IPSP) at disynaptic latency [PRESTON and WHITLOCK, 1960, 1961], clearly visible as an erosion of the down-stroke of the EPSP.

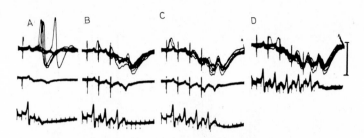

Fig. 4. Intracellular records of predominating IPSP from γ-motoneurone of radial nerve (conduction velocity 25.5 m/sec). Spike potential deteriorating from *A* to *D*. Each set of records comprises intracellular records (upper); extracellular controls (middle, in *A*, *B* and *C*); and arrival of corticospinal volleys in spinal segment, with time in msec (lower). *A* Responses to single cortical pulses; *B* to 3 pulses; *C* to 4 pulses; *D* to 5 pulses at 385 c/sec. Vertical scale: 4 mV hyperpolarization [CLOUGH, PHILLIPS and SHERIDAN, 1971].

Figures 1 and 2 show that the PT and I_A inputs converge on the membranes of the same α-motoneurones. If the γ- (fusimotor) motoneurones were also accessible to the PT, the apparatus would make possible the servo assistance of movement by coactivation of muscle spindles and α-motoneurones [see articles by MARSDEN, MATTHEWS and VALLBO, this volume].

CLOUGH, PHILLIPS and SHERIDAN [1971] found that many of the γ-motoneurones innervating the forearm and hand could not be fired by a brief burst of PT volleys. Of 19 fusimotor neurones recorded intracellularly, 6 showed EPSP related monosynaptically to the volley in the fastest PT axons (fig. 3 B), and 2 showed EPSP at latencies which were either polysynaptic from the fastest PT axons, or monosynaptic from the more slowly conducting PT axons which form the bulk of the cortico-spinal tract. The sample is small because fusimotor neurones are difficult to impale with intracellular micro-electrodes. In the more numerous extracellular experiments it was not possible to discriminate between monosynaptic firing from fast or slow PT axons and polysynaptic firing [CLOUGH, PHILLIPS and SHERIDAN, 1971]. GRIGG and PRESTON [1971] recorded from single fusimotor axons of flexor and extensor muscles of the baboon's ankle, and detected the earliest modulation of their 'spontaneous' firing by PT volleys in large numbers of trials. In favourable conditions, facilitation at monosynaptic latency could be detected in 9 of 17 fusimotors tested by this method. But fusimotors which were not firing spontaneously could not be fired at monosynaptic latency by PT volleys.

CLOUGH, PHILLIPS and SHERIDAN [1971] saw IPSP in 4 of 19 intracellularly recorded fusimotors of distal forelimb muscles (fig. 4), and GRIGG and PRESTON

[1971] found that 10 of 27 fusimotors of ankle muscles were inhibited by PT volleys. The central latency indicates an interneuronal link, as in the case of α-motoneurones. KOEZE, PHILLIPS and SHERIDAN [1968] found that the discharges of 2 of 53 spindle primary endings of m. extensor digitorum communis were slowed by cortical stimulation. Cortical inhibition of fusimotors would be important in the learning of manual skills, in which the need to relax 'unwanted' muscles is a matter of everyday experience.

KOEZE, PHILLIPS and SHERIDAN [1968] stimulated the cortex with trains of pulses at a frequency of pulses at a frequency of 50–100 c/sec and train duration 3 sec. The effects that they observed on spindle primary afferents need not, therefore, have been exclusively due to the PT, but could have been due to cortico-rubro-spinal and cortico-reticulo-spinal projections issuing from the same area of cortex (outlined in fig. 1). But the coactivation of α-motoneurones and spindle primaries was always well synchronized, and comparable to that seen in VALLBO's human studies [see his article, this volume]: prolonged stirring up of fusimotor activity, as seen in response to stimulation of the cat's reticular formation [GRANIT and KAADA, 1952], was not observed. The actions on fusimotor and on α-motoneurones could be dissociated by varying the level of anaesthesia, showing that the cortex would be able to control them independently (fig. 5).

Servo assistance of movement could be effected by PT and I_A impulses reinforcing one another at the α-motoneurone membranes. The absolute size of the I_A EPSP is small [CLOUGH, KERNELL and PHILLIPS, 1968] compared to their size in the hind-limb motoneurones of the cat [ECCLES, ECCLES and LUNDBERG, 1957], so the effect is likely to be marginal, but could be significant if the α-motoneurones were already firing steadily [GRANIT, 1970]. LUNDBERG [1959] pointed out that where the central distribution of I_A actions is widespread, it would be unlikely that the γ-loop, if activated by itself, could subserve more precise movements. Some of the I_A feedback from the muscles of the baboon's hand is distributed to motoneurones of muscles other than the muscle of origin; but CLOUGH, KERNELL and PHILLIPS [1968] found that the largest I_A EPSP are found in those α-motoneurones which also receive the largest CM inputs. 'In cortically-driven movements of the hand, any spindle feedback directed to motoneurones other than those selected by the CM projection could well remain subthreshold, and need not blur the precision of the movements. The output of the CM-selected motoneurones could be governed automatically by marginal changes in a cortically-sustained spindle feedback, reinforcing or withdrawing support from the CM input in response to changes in peripheral load' [PHILLIPS, 1969].

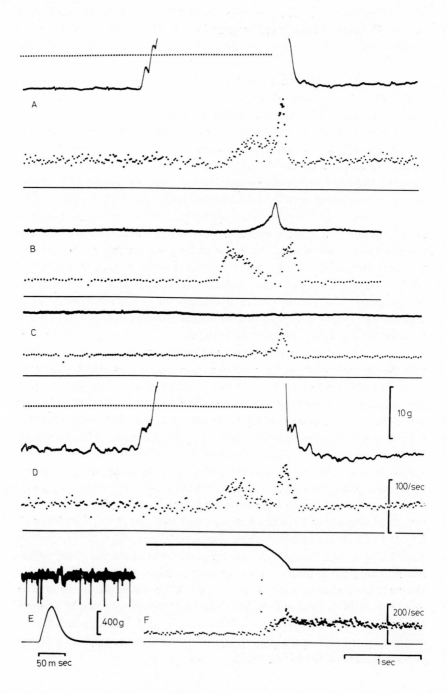

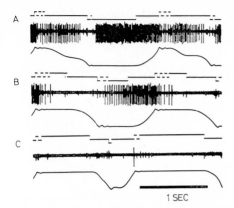

Fig.6. Discharges of monkey's PT neurone (conduction velocity 80 m/sec) recorded during conditioned flexion of wrist. *A* Flexion resisted by weight: neurone discharged more strongly during flexion, and continued to discharge during extension, when paying-out of weight was controlled by diminishing flexor force. *B* Movement unresisted; neurone discharged during flexion (upward movement of lower trace). *C* Flexion aided by weight, and controlled by *extensor* force: PT neurone discharged one impulse only [EVARTS, 1967].

A second possibility [PHILLIPS, 1969] is that the spindle feedback drives the PT cells which would then drive the α-motoneurones by the direct and powerful CM pathway. Figure 6 shows that the discharges of a monkey's PT cell are increased when a rewarded flexion of the wrist is resisted by a load. More force is then demanded to earn the reward [EVARTS, 1967]. But although cells in the motor cortex, including PT cells, are accelerated by muscle afferents

---

Fig.5. Dissociation of α and spindle responses to cortical stimulation with unifocal 0.1 msec surface anodal pulses at 52 c/sec for 3.3 sec, by varying level of anaesthesia. Dorsal roots cut and muscle (extensor digitorum communis) held at minimum physiological length. Conduction velocity of spindle axon 65 m/sec. *A* Light anaesthesia. Myogram (upper record) shows tremulous background discharge of α-motoneurones. Record of spindle frequency shows high level of background discharge. Cortical stimulation at strength 1.4 mA fires α-motoneurones before it accelerates the spindle. *B* Deeper anaesthesia. Background α discharge abolished. Background spindle discharge reduced and its firing less irregular, due to reduced fusimotor activity. Cortical stimulation (1.8 mA) accelerates spindle as in *A*; α response reduced and delayed. *C* Deeper anaesthesia. α response abolished, spindle response reduced. *D* Anaesthetic lightened again. Background activity and response to cortical stimulation as in *A*. *E* Muscle spindle unloaded by maximal motor nerve twitch. *F* Response of spindle to stretching muscle from minimal to maximal physiological length at 24 mm/sec [KOEZE, PHILLIPS and SHERIDAN, 1968].

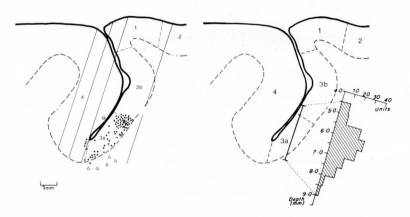

Fig.7. Composite diagrams of parasagittal sections through arm area of precentral gyrus (area 4) and postcentral gyrus (areas 3a, 3b, 1, 2) of baboons. *Left:* Dots show histologically verified recording sites from which potential waves and single-neurone responses were evoked by muscle afferent volleys in deep radial nerve or deep palmar branch of ulnar nerve. Open circle indicates only response ever recorded from area 4 to this input. Oblique lines show representative micro-electrode tracks from which no responses were recorded. Triangles show sites of recording from fibres in white matter in response to a cutaneous input (superficial radial nerve) of impulses presumably *en route* for areas 3b and 1. *Right:* Distribution in depth of 138 single neurones which responded to muscle afferent volleys in deep radial and deep ulnar nerves [PHILLIPS, POWELL and WIESENDANGER, 1971].

[ALBE-FESSARD and LIEBESKIND, 1966; WIESENDANGER, 1972], there is no oligosynaptic pathway from spindle primary afferents to PT cells. There is, however, a dense oligosynaptic pathway to area 3a in the depths of the postcentral gyrus (fig. 7) [PHILLIPS, POWELL and WIESENDANGER, 1971]. The increased activity of the PT cell seen in figure 6 would not, therefore, have been due to a crude oligosynaptic excitatory action of the I_A input; an input from the secondary spindle endings is not, however, excluded. The alternative hypothesis [PHILLIPS, 1969] that the PT cell acceleration is in response to an error signal computed elsewhere, e.g. in the cerebellum, awaits investigation.

Author's address: Dr. C.G. PHILLIPS, University Laboratory of Physiology, *Oxford* (England)

New Developments in Electromyography and Clinical Neurophysiology,
edited by J.E. Desmedt, vol. 3, pp. 145–158 (Karger, Basel 1973)

Extrapyramidal Control of Primate Motoneurons

A. I. SHAPOVALOV

Laboratory of Physiology of the Nerve Cell, Sechenov Institute of Evolutionary
Physiology and Biochemistry, Leningrad

The outstanding feature of the organization of the supraspinal projections in primates and man is the powerful and progressive development of a
corticospinal system which, in contrast to other mammals, establishes direct
monosynaptic articulations with segmental motoneurons. The direct connections between corticopyramidal cells and α-motoneurons are thought to provide an important mechanism for the fine control of skilled voluntary movements especially of distal muscles of extremities. The phylogenetically older
motor outflow (the brain stem-spinal or extrapyramidal system) is assumed
to be concerned with the execution of the rather crude reflex contractions of
posture and locomotion governed through diffuse multisynaptic projections.
However, it has become increasingly evident that the pyramidal tract (PT)
is not the sole pathway which is important for execution of voluntary motor
performances. Clinical observations of PUTNAM [1940] have shown that transection of the lateral columns only slightly affects voluntary movements in
man. BUCY and co-workers [BUCY, 1957; BUCY and KEPLINGER, 1961; BUCY,
KEPLINGER and SIQUERA, 1964; BUCY, LADPLI and EHRLICH, 1966] found in
man and monkeys that following interruption of the PT in the cerebral
peduncle there was considerable retention of useful finger movements and
also the ability to walk, climb and jump. The monkeys with bilateral PT interruption at the medullary level were shown to be capable of directing a wide
range of activity of body and limbs [LAWRENCE and KUYPERS, 1965]. In man,
the leg area of the motor cortex can be removed completely, monitoring with
stimulation, without impairing the ability to walk [PENFIELD, 1964]. Thus,
the results of spinal, medullary, peduncular and even cortical lesions have
shown that the lateral corticospinal system is not a prerequisite for volitional

or discrete movements and that there are other descending pathways which can serve a similar function.

In recent studies on several subprimate species the existence of rather simple fast monosynaptic and disynaptic connections between reticulo-, vestibulo- and rubrospinal fibers and spinal α-motoneurons have been established by modern electrophysiological methods [SHAPOVALOV, 1966, 1969; SHAPOVALOV, KURCHAVYI and STROGANOVA, 1966; SHAPOVALOV, GRANTYN and KURCHAVYI, 1967; LUND and POMPEIANO, 1968; HONGO, JANKOWSKA and LUNDBERG, 1969; WILSON and YOSHIDA, 1969]. From anatomical studies in monkeys it is known that corresponding brain stem-spinal pathways descend to sacral levels of the cord [KUYPERS, FLEMING and FARINHOLT, 1962] and that their interruption may induce severe impairment of motor performances [LAWRENCE and KUYPERS, 1968]. However, until very recently [SHAPOVALOV, KARAMAYAN, KURCHAVYI and REPINA, 1971; SHAPOVALOV, KURCHAVYI, KARAMAYAN and REPINA, 1971; SHAPOVALOV, TAMAROVA, KARAMJAN and KURCHAVYI, 1971; SHAPOVALOV, 1972] synaptic effects mediated by the extrapyramidal pathways in the primates had not been analyzed. Taking into account the existence of the monosynaptic cortico-motoneuronal projections in the monkey, it would be of particular importance to know whether primates possess the direct brain stem-motoneuronal linkage similar to those found in subprimate forms.

The present survey is based on investigations made on 38 rhesus monkeys. The results were obtained by local electrical stimulation (pulses of 0.05 to 0.2 msec in duration) of different brain-stem structures and intracellular recording from hind limb motoneurons. In some experiments intracellular microelectrodes were filled with a solution of Procion yellow dye for staining the neurons recorded for the localization of their somas and dendrites within the ventral gray matter. Special precautions were taken to exclude simultaneous transmission of impulses via the collaterals of the pyramidal fibers. For this purpose: (1) the bulbar pyramids were transected in the course of experiments in 12 animals, and in 11 animals acute lesions of the spinal cord were accomplished. Histological checks of stimulating points and of pyramidal and cord sections were regularly performed; (2) 5 monkeys were subjected to unilateral or bilateral lesions in the precentral gyrus (upper two-thirds) including the adjoining areas. The paresis of the hind and the forelimb whose corresponding motor cortex area had been removed was evidenced by imperfection of willed movement. After an appropriate postoperative survival time (2–3 weeks) the animals were used in acute experiments, and (3) an averaging computer was employed to obtain accurate summation of the synaptic actions from motor cortex and brain-stem structures to demonstrate the absence of any occlusion and, thus, to show that corresponding monosynaptic effects are mediated by separate descending pathways.

Distribution of Stimulating Sites in the Brain Stem

Single-shock stimulation of the contralateral red nucleus and the nuclei of the medial reticular formation of the lower pons and medulla produced clear tract volleys recorded from the dorsal surface in the lumbar segments and excitatory or inhibitory postsynaptic potentials (EPSP or IPSP) in the impaled hind limb motoneurons. Rubral shocks frequently evoked both early and late monosynaptic EPSP with latencies of 2.7–3.5 and 3.5–4.5 msec, respectively, and corresponding descending volleys. Systematic mapping of the sites of stimulation (fig. 1) reveals that the points with low threshold for evoking early (I) components of rubro-motoneuronal excitation are concentrated in the caudal magnocellular part of the red nucleus (RNm), whereas points with low threshold for evoking late (II) components are concentrated in the rostral, parvocellular part of the red nucleus (RNp) (fig. 1A). This seems to imply that the source of the early rubrospinal effects is in the RNm. As the bulk of rubro-spinal fibers apparently comes from the RNm [KUYPERS, FLEMING and FARIN-HOLT, 1962] it is not clear whether the late effect is due to trans-synaptic excitation of rubrospinal neurons or to direct activation of some other contingent of rubrospinal axons originating from RNp. When the stimulus strength at a given point is raised above that required to evoke one component, it usually also evokes the other, presumably due to current spread. The current intensity difference for evoking the early and late responses from caudal and rostral parts is clearly seen in figure 1A.

The stimulating points from which reticulospinal monosynaptic actions were elicited were scattered throughout the pontomedullary reticular formation, especially in the region of n. reticularis pontis caudalis (RP), n. reticularis gigantocellularis (RGC) and fasciculus medialis longitudinalis (MLF) (fig. 1). [ITO, UDO and MANO, 1970] demonstrated in the cat that the reticulospinal cells are dispersed over the pontomedullary reticular formation in both rostro-caudal and ventrodorsal directions. Corresponding electrophysiological studies have not been made in the monkey, but on account of the anatomical similarities of the reticular nuclei and reticulospinal organization [KUYPERS, FLEMING and FARINHOLT, 1962; NYBERG-HANSEN, 1965] in the cat and monkey it appears likely that functional conditions are the same in the two species.

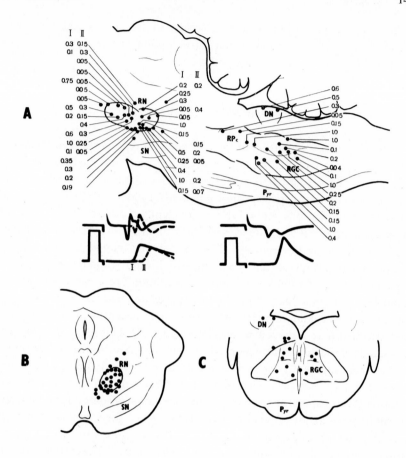

Fig. 1. Correlation of stimulating electrode position and responses recorded in the lumbar cord. *A*, *B* and *C*: Diagrams summarizing the locations of stimulating points in different animals grouped in the regions of the red nucleus and pontomedullary reticular formation. The maps of the brain stem of the rhesus monkey were adopted from the atlas of OERTEL [1969]. The drawings show schematic representation of the sagittal (*A*) and transverse (*B*, *C*) plane of the brain stem. Figures at each site give the intensity of stimulation (in mA) required to produce a detectable monosynaptic EPSP and a tract volley. The Roman numerals indicate early (I) and late (II) rubrospinal responses. Examples of descending responses are illustrated below, the superimposed traces to the left show early (solid line) and late (broken line) rubrospinal responses elicited in the same motoneuron by shocks applied to the RNm and RNp, respectively. The upper traces are records of the cord dorsum potentials, the lower traces are intracellular records. A square wave calibration pulse 1 mV in height and 1 msec in duration precedes the EPSP.

Control experiments made on monkeys with chronic destruction of the motor cortex or after acute pyramidotomy gave similar results, hence there is no evidence that responses evoked from the red nucleus and reticular formation can be due to stimulation of collaterals of the pyramidal fibers.

Descending Volleys Recorded from the Cord Dorsum Surface

Conduction velocity of descending volleys was calculated from the latency of the positive peak of the tract potential and the distance between the stimulating and recording points. The conduction velocity of the early rubrospinal volley was found to be 69–92.2, mean 82.8 ± 1.7 m/sec. The conduction difference between early rubrospinal and direct corticospinal volleys (57–76, mean 67.9 ± 2.1 m/sec) is statistically significant ($p < 0.001$). The rate of conduction of the reticulospinal and vestibulospinal volleys was 61–88.5 m/sec (mean 78 ± 2.05 m/sec). Thus the conduction velocities observed in these cases indicate that the fibers mediating brain stem-spinal effects belong to the fast myelinated fiber group.

The characteristic property of the rubrospinal (as also pyramidal) volley is a prominent negative deflection which was absent in the case of bulbospinal

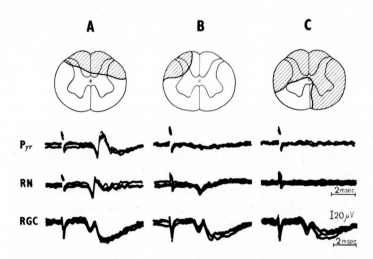

Fig. 2. Spinal location of the descending pathways with monosynaptic effects on lumbar motoneurons. A, B and C are surface potentials from the L_6 dorsal root entry zone and drawings of transverse section of the spinal cord showing acute lesions (hatched area) made before recording.

potentials. This suggests that fibers of the rubrospinal tract are located more dorsally than those of bulbospinal pathways. Acute sections of the spinal cord have shown (fig. 2) that all brain stem-spinal volleys remain practically unaffected after a lesion involving the dorsal funiculi. However, after damage to the dorsal part of the lateral funiculus the negative deflection of the rubrospinal volley was lost. Destruction of the whole lateral funiculus completely abolished the rubrospinal response, suggesting that the relevant fibers are located in the lateral funiculus of the ipsilateral lumbar cord. After a lesion involving dorsal and lateral funiculi and contralateral ventral funiculi, the reticulospinal volley could still be recorded suggesting that the relevant fibers are located mainly in the ventral funiculi of the ipsilateral cord.

Monosynaptic Brain Stem-Motoneuronal Actions

The short-latency EPSP and IPSP were regularly evoked in lumbar motoneurons by rubrospinal, reticulospinal and vestibulospinal impulses. The interval between the peak of the early positive deflection of the corresponding tract volley and the onset of the EPSP (but not the IPSP) in a large population of cells was less than 1.0 msec (usually 0.5–0.8 msec) indicating monosynaptic linkage.

In contrast to subprimates [Shapovalov, 1972], in the rhesus monkey the reticulospinal and vestibulospinal monosynaptic input was found preferentially in motoneurons innervating proximal muscles of extremities and sometimes also in DP motoneurons. Examples are shown in figure 3, which also illustrates the lack of cortico-motoneuronal input in cells monosynaptically activated by bulbospinal volley. The motoneurons of distal muscles (flexor digitorum longus [FDL] and peroneus profundus [DP], extensor digitorum longus [EDL], plantaris), in addition to direct pyramidal projections, tend also to receive monosynaptic excitatory input from the red nucleus (fig. 3 C). The staining of impaled motoneurons with Procion yellow dye shows (fig. 4) that cells innervating distal muscles of the hind limb are located in the dorsolateral motor pool, whereas motoneurons of proximal muscles are located in the ventromedial motor nuclei. It may be seen also in figure 4 that dendrites of these motoneurons spread into the zones where, according to degeneration studies of Kuypers, Fleming and Farinholt [1962] the terminations of rubrospinal and reticulospinal axons respectively are concentrated. Thus the best explanation for the demonstrated monosynaptic connections to spinal α-motoneurons is direct fiber connection to the dendrites.

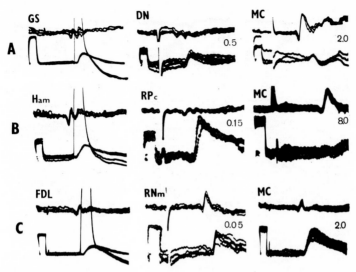

Fig. 3. Monosynaptic actions evoked from the Deiters' nucleus (*A*) reticular formation of the caudal pons (*B*) and the large-celled red nucleus (*C*) in three different hind-limb motoneurons (GS, Ham. and FDL). Effects of cortical stimulation (MC) are also shown. The strength of stimulating current (in mA) applied to the supraspinal structures is noted beside each record. A calibration pulse 1 mV for supraspinal effects and 5 mV for nerve stimulation, 1 msec, precedes the intracellular responses. For abbreviations see text.

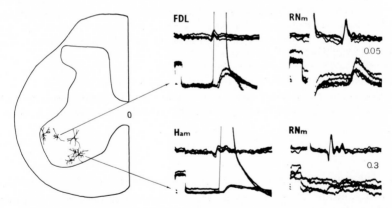

Fig. 4. Drawings of different Procion yellow stained motoneurons and examples of responses elicited in them by rubral and nerve shocks. The strength of stimulating current (in mA) is noted beside records, a calibration pulse 5 mV for nerve, and 1 mV for rubral stimulation, 1 msec. The figure also illustrates how the spread of the dendrites of Ham. motoneurons (three lower cells) invade the sites of termination of reticulospinal axons, and dendrites of EDL and FDL motoneurons (three upper cells) may contribute to the formation of rubro- motoneuronal linkage with terminations of rubrospinal axons.

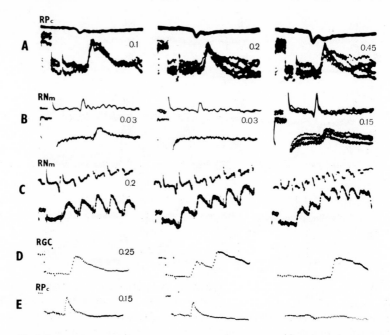

Fig. 5. Properties of monosynaptic EPSP evoked from different supraspinal structures in pyramidotomized monkeys. *A* Reticulo-motoneuronal EPSP elicited by different current intensities. *B* Single traces of rubro-motoneuronal EPSP evoked by shocks decreased to the threshold value (0.03 mA) and occurring in an all-or-none manner. Final record: superimposed traces of EPSP evoked by shocks incremented to 0.15 mA. *C* Effects of repetitive stimulation of the RNm at different frequencies. *D* Averaged record of reticulo-motoneuronal EPSP evoked by single (first trace), paired shocks (middle trace), and the result of subtraction of the single response from the double by the computer (third trace). *E* Averaged records of reticulo-motoneuronal EPSP. Left, no polarizing current; center, a hyperpolarizing current at intensity 1.10^{-8} A is passed; right, difference between two other records. The strength of stimulating current (mA) is noted beside records. Calibration pulses 1 mV, 1 msec.

The maximal amplitude of the rubro-motoneuronal EPSP ranged from 0.2 to 1.7, mean 0.59 $\pm$ 0.03 mV and that of the reticulo-motoneuronal EPSP was in the range 0.15–2.1, mean 0.64 $\pm$ 0.03 mV. For most of the motoneurons recorded from the averaged shapes and amplitudes of rubro-motoneuronal and cortico-motoneuronal EPSP produced in the same cell were quite different. In general, the cortico-motoneuronal EPSP was larger and slower to rise and decay. This general finding held when a whole population of impaled motoneurons was examined, the difference of corresponding values being of high

statistical significance (p < 0.001) [SHAPOVALOV *et al.*, 1971]. The mean amplitude of reticulo-motoneuronal EPSP was also significantly smaller than that
of the cortico-motoneuronal EPSP. However, in some individual motoneurons
they can fall into the same range of values. In a few DP motoneurons in which
both cortico- and reticulo-motoneuronal EPSP were recorded, the former had
especially large amplitude as compared with the latter (mean 1.09 ± 0.4 and
0.55 ± 0.05 mV, respectively).

Figure 5 illustrates the characteristic properties of brain stem-motoneuronal EPSP recorded in the pyramidotomized monkeys. The monosynaptic
EPSP were usually saturated at small increase of stimulating current and once
the maximal size had been attained the stimulus intensity could be very much
increased without any further change in the height of the EPSP (fig. 5A). It
was possible to reduce the intensity of stimulating shocks just to the threshold
level so that the resulting EPSP occurred in an all-or-none manner and the
increase of stimulus intensity did not alter the height of the recorded EPSP but
restored the regularity of its generation (fig. 5B). The growth of monosynaptic
EPSP as a function of applied pulse strength and all-or-none behavior of the
unitary EPSP suggests that there may be a strictly limited number of cells connected with a particular α-motoneuron. The monosynaptic EPSP evoked by
repetitive stimulation could follow the stimulating rates up to 500–800/sec and
did not sum appreciably until the stimulus interval became less than 3–5 msec
(fig. 5C). They reveal, in contrast to cortico-motoneuronal EPSP, only moderate frequency potentiation (fig. 5D). The hyperpolarizing current does not produce demonstrable increase of the amplitude of monosynaptic EPSP and
sometimes even slightly decreases the height of the EPSP (fig. 5E), suggesting,
in accordance with anatomical data, the dendritic location of the synaptic
input.

Polysynaptic Brain-Stem Spinal Influences

It seems clear on the basis of available anatomical evidence [KUYPERS,
FLEMING and FARINHOLT, 1962] that a good deal of descending motor effect
from the brain stem-spinal pathways may be exerted on α-motoneurons by
way of internuncials located within the spinal cord. In fact, the delay between
the arrival of the tract volley and the start of postsynaptic response was in
large populations of cells more than 1.1–2.2 msec suggesting at least disynaptic
linkage. Usually the second positive component of the cord dorsum potential
preceded the beginning of disynaptic EPSP or IPSP by 0.2–0.5 msec (fig. 6A).

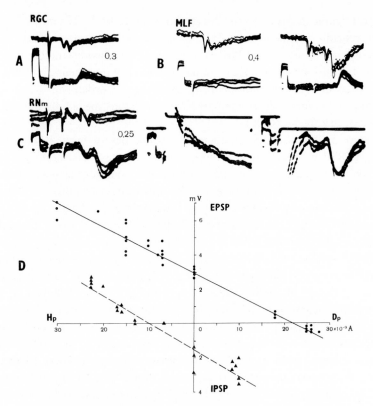

Fig. 6. Disynaptic EPSP (*A, B*) and IPSP (*C*). *D* Effect of injected current on disynaptic EPSP (upper curve) and IPSP (lower curve). Calibration pulses 1 mV, 1 msec.

It is very likely that this later component of surface potential reflects the activation of proportional elements included in the disynaptic pathway from supraspinal structures to motoneurons [SHAPOVALOV, 1969]. Responses with disynaptic delay could be evoked by single shock (fig. 6A), but in most cells detectable disynaptic effects could be elicited only by paired shocks (fig. 6B) or by a short train of stimulating pulses. In contrast to monosynaptic EPSP, disynaptic EPSP and IPSP showed progressive growth of amplitude when the stimuli were repeated (fig. 6C). The marked increase of the relayed component of the surface potential paralleled the increase of disynaptic responses. Polarizing currents passed through an intracellular micro-electrode readily influence the disynaptic IPSP and EPSP (fig. 6) and identify the accompanying membrane impedance change. These results suggest that the loci of corre-

A   **B**

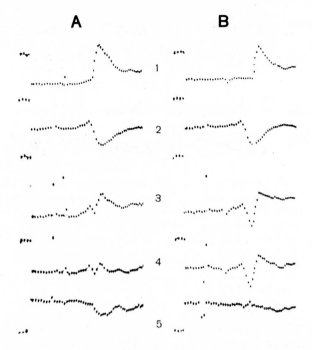

Fig. 7. Interaction of monosynaptic reticulospinal EPSP and disynaptic rubrospinal IPSP evoked at different intervals (*A* and *B*) in the same motoneuron. Averaged records of PSP evoked by reticulospinal (1) and rubrospinal (2) volley first separately and then together (4), their algebraic sum (3), and the difference between the algebraic and motoneuronal sum (5). Calibration pulse 1 mV (doubled by the computer in 3), 1 msec.

sponding synapses are relatively near the recording site, presumably the cell body. The averaged records of figure 7 further show that with superposition of the disynaptic rubrospinal IPSP on the monosynaptic reticulospinal EPSP in a hamstring motoneuron at various intervals, the EPSP is greatly reduced when superimposed on an onset of IPSP. When an EPSP is superimposed later on in an IPSP, there is simple summation of the two potentials. This behavior is explained by the fact that the postsynaptic permeability change during single disynaptic IPSP is at a significant intensity for less than 1–2 msec.

Rubrospinal IPSP were usually recorded from gastrocnemius soleus (GS), hamstring (Ham.) and quadriceps (Q) motoneurons, whereas reticulospinal IPSP were found in FDL, plantaris, GS and DP motoneurons. Disynaptic and polysynaptic EPSP were elicited by rubrospinal and reticulospinal impulses in DP, GS, plantaris, Q and Ham. motoneurons.

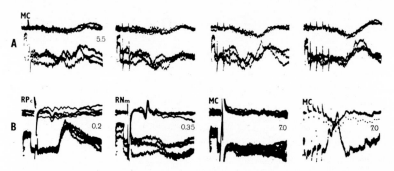

Fig. 8. Supraspinal effect evoked in two different motoneurons (*A* and *B*) of pyramidotomized monkeys. *A* Superimposition of effects evoked by 1–4 stimuli applied to motor cortex (MC). *B* Effects produced by reticulospinal and rubrospinal volley, single and repetitive stimulation of MC. The strength of stimulating current (mA) is noted beside records. Calibration pulses 1 mV, 1 msec. Time mark 1 msec for final record in *B*.

Cortico-Brain Stem-Spinal Effects

Descending extrapyramidal pathways must also carry the cortically evoked activity. Degeneration studies in the monkey [KUYPERS, 1958; KUYPERS and LAWRENCE, 1967] revealed that cortical motor areas have heavy projections to the medial reticular formation of the pontine and medullary tegmentum which give off reticulospinal fibers and fibers to the red nucleus. Stimulation of the motor cortex in the baboon has shown that section of the pyramidal tract merely reduced the number of distinct movements obtainable from the cortex and raised their threshold [LEWIS and BRINDLEY, 1965]. In the rhesus monkey stimulation of the motor cortex with a lesion in its pyramidal projection was effective in eliciting discharges in the distal muscles, but the latency of discharges and the threshold current were considerably increased as compared with control animals [FELIX and WIESENDANGER, 1971].

In monkeys with pyramidal section there was a complete abolition of direct pyramidal volley and a dramatic decrease of cortically induced synaptic actions especially in motoneurons of distal muscles of extremities. Cortical stimulation was effective only with repetitive stimuli (fig. 8) and the latency of postsynaptic response was frequently not less than 8–10 msec. The PSP were usually evoked by strong currents: above 5–6 mA, although in some cells weaker intensity of stimulation (2–3 mA) was effective. In most cases the pattern of synaptic activation of motoneurons tended to be diffuse, i.e. not

time-locked with the applied shocks. However, the pattern of cortical activation of motoneurons which as a rule do not receive direct pyramidal input
(Q, Ham., GS) may sometimes not differ appreciably in control and pyramidotomized animals. It was possible to identify EPSP time-locked to stimuli
in the train applied to the motor cortex in some cells receiving monosynaptic
reticulo-motoneuronal input (fig. 8 B). The latencies of the earliest responses
measured from the effective stimuli suggest disynaptic linkage. Thus, direct
reticulo-motoneuronal projection may serve as a relay for cortically induced
influences upon segmental motoneurons.

After pyramidotomy, the cortex may not only facilitate but also inhibit
lumbar motoneurons or may induce mixed inhibitory-excitatory effects.

Discussion and Conclusions

The main result of the electroanatomical investigation of the synaptic
organization of the extrapyramidal system in the monkey is the finding that
brain stem-spinal fibers establish monosynaptic contacts with segmental α-
motoneurons. These data indicate that primate extrapyramidal pathways
have a mechanism of action similar to that previously described only for the
pyramidal tract. There are several important similarities between direct non-
pyramidal and pyramidal projections to lumbar α-motoneurons: (1) both are
excitatory; (2) both are established by fast-conducting fibers running in the
descending tracts; (3) monosynaptic EPSP evoked by a single volley are subthreshold for the initiation of the spike, and (4) in the case of rubro-motoneuronal projection both are found in motoneurons presumably innervating
the distal muscles of extremities. However, there are certain differences between cortico-motoneuronal and brain stem-motoneuronal actions: (1) the
latter have smaller amplitude and faster time course and apparently are induced by a smaller number of descending axons [SHAPOVALOV *et al.*, 1971];
(2) the operating characteristics (frequency potentiation) of the two projections are different, and (3) the reticulo-motoneuronal and cortico-motoneuronal inputs is found mainly in functionally different motor nuclei. Thus,
the difference between the pyramidal and nonpyramidal projections may
depend not upon the complexity of corresponding connections with spinal
α-motoneurons, but on the number and properties of involved fibers and
synapses and, possibly, on the intrinsic properties and synaptic organization
of the cerebrofugal cells giving off descending axons. As the cortico-motoneuronal connections apparently outnumber the brain stem-motoneuronal

connections the former may allow a finer subdivision and fractionation of motor control.

A remarkable feature of organization of supraspinal control in the monkey is the finding that two direct descending systems, corticospinal and rubrospinal, have access to the same α-motoneurons. These data are in agreement with ablation experiments: in the absence of the pyramidal tracts, interruption of the rubrospinal pathways produces severe impairment of independent distal extremity movements. Similar lesions made prior to pyramidotomy evoked only a negligible effect [LAWRENCE and KUYPERS, 1968]. Apparently efficient supraspinal control requires the cooperation of pyramidal and extrapyramidal influences.

As the descending pyramidal pathway increases in bulk and importance in the phylogenetic scale of mammals, the relative role of nonpyramidal connections may decrease. However, the latter continue to play an essential role and apparently are significant in man. This is clearly demonstrated by the results of pyramidal lesion in man mentioned above. Some other observations on human subjects also testify to the probably functional importance of the brain stem-spinal pathways. Anatomical studies of human embryos have shown that reticulospinal fibers running in the ventral funiculi establish direct contacts with cervical motoneurons [SHULEIKINA and GLADKOVITCH, 1965]. The general opinion that the rubrospinal tract is rudimentary and functionally insignificant in man and subhuman primates is at variance with some clinical and anatomical observations [cf. BRODAL, 1969]. It may be supposed, therefore, that in order to understand supraspinal control, attention must not be focused exclusively on the cortico-pyramidal tract and that further studies of the extrapyramidal system are needed.

Author's address: Dr. A. I. SHAPOVALOV, Laboratory of Physiology of the Nerve Cell, Sechenov Institute of Academy of Sciences USSR, Thorez Pr. 52, *Leningrad K-223* (USSR)

New Developments in Electromyography and Clinical Neurophysiology,
edited by J. E. Desmedt, vol. 3, pp. 159–174 (Karger, Basel 1973)

Some Aspects of Pyramidal Tract Functions in Primates

M. Wiesendanger

Institute for Brain Research, University of Zurich, Zurich

The pyramidal tract (PT), composed of those fibres running longitudinally in the pyramids on the ventral surface of the brain-stem, has long
been regarded as the essential pathway mediating the cerebral control of
'willed' movements. This is not surprising since the decussation at the
caudal end of the medulla oblongata was recognized by simple dissection
in the early days of neuroanatomy and could explain the fact that a paralysis of voluntary movements occurred on the side contralateral to a cerebral
lesion. The course of the PT was clearly established from cases of capsular
lesions when myelin staining methods were introduced [see, for instance,
von Monakow, 1897]. It is also not surprising, therefore, that the clinical
symptomatology following lesions in the internal capsule or in the peri-
rolandic cortex has been attributed to damage to the PT.

There is no need to debate this terminology, but my point is to demonstrate that animal experiments including morphological, electrophysiological and behavioural techniques have revealed a complexity of the
pyramidal pathway that makes one reluctant to consider the PT as a unit.
Although many textbooks still describe *the* function of pyramidal fibres,
they probably mean mainly only one component of the PT, namely the
fast-conducting direct *corticomotoneuronal system* [Bernhard and Bohm,
1954]. This subsystem of the PT is phylogenetically new and has also a late
ontogenetic development. It is only present in primates, and within primate
evolution it has developed *pari passu* with the increasingly sophisticated
cerebral control of the hand [Phillips, 1971]. Cortico-motoneuronal
control develops in parallel with other functional systems such as the
pontine nuclei, the inferior olive, the cerebellar cortex and dentate nuclei,
all of which contribute to evolutionary increase in motor skill.

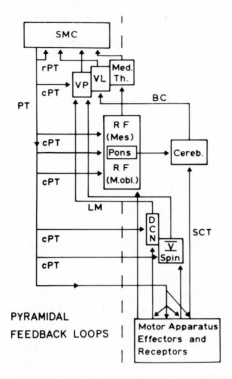

Fig. 1. Diagram of pyramidal feedback connections. The pyramidal tract fibres give off recurrent collaterals (rPT), collaterals (cPT) to thalamic nuclei (VP, VL, Med. Th.), the reticular formation (RF) of the mesencephalon (Mes) and of the lower brain-stem (M. obl.), the pontine nuclei (Pons), the dorsal column nuclei (DCN), and the spinal trigeminal nucleus (VSpin). The pyramidal fibres influence the motor units and muscle spindles (*via* the gamma motoneurones) as well as spinal neurones projecting to higher centres, especially to the cerebellum (SCT, Cereb.). All recipients of the pyramidal fibre endings have again connections with the sensory motor cortex (SMC). BC brachium conjunctivum, LM lemniscus medialis [WIESENDANGER, 1969].

According to LASSEK [1954] the human pyramid has about 1 million fibres. Most of the fibres have a small diameter and only about 10% are larger than 5 μm. Since the fibres connecting the motor cortex monosynaptically with the motoneurones are rapidly conducting, it can be assumed that they constitute not more than 10% of the total pyramidal fibre population. The remainder of the tract is composed mainly of small fibres (<5 μm) whose functions are not established. The corticospinal tract of rabbits which has no monosynaptic motor connections, reaches only to thoracic levels.

In all species, a large proportion of pyramidal fibres terminate in sensory relay stations (dorsal column nuclei, trigeminal nuclei), in the medial reticular formation of the brain-stem, in the lateral reticular nucleus, in the inferior olive and in the dorsal horn of the spinal cord. Figure 1 illustrates some feedback loops of which the PT with its collaterals forms the efferent limb. Much recent electrophysiological work has been devoted to analyses of the effects of corticofugal volleys on sensory transmission [for review, see WIESENDANGER, 1969]. From these studies emerges the general concept of centrifugal control of sensory transmission, thought to be important in the continuous shaping of perceptual processes. Although some of the sensory control mechanisms, such as presynaptic and post-synaptic inhibition and facilitation, are fairly well understood in neurophysiological terms, the significance of these mechanisms in accounting for behaviour has scarcely been broached.

There have been two main approaches to studying the PT. The classical procedure, first described by MAGENDIE [1839], is to section the PT at the bulbar level. Essentially one is then looking for the remaining motor capacity. Although it may be dangerous to draw conclusions about the function of the system on the basis of the *deficits* following its removal, this approach still has its place in brain research and may be helpful in giving us indications about some of the pyramidal functions. The other traditional method consists of studying the pyramidal system in isolation. This is effected by making a brain-stem transection destroying all connections from the cerebrum and cerebellum to the spinal cord with exception of the PT (fig. 2A). This so-called 'pyramidal preparation', introduced by LLOYD [1941], can be used to study effects of electrical stimulation of the cortex or the PT at a level rostral to the transection on spinal neurones. The disadvantage is that these animals are deeply comatose and the spinal cord is deprived of most of its tonic supraspinal influences which normally contribute so importantly to the general excitatory state of the 'final common path' [LAURSEN and WIESENDANGER, 1966].

In the following section I wish to demonstrate that we are confronted with puzzling cases from human neuropathology which underline the difficulty of drawing conclusions about normal functions. An additional factor is that clinical descriptions vary depending upon whether one is mainly interested in the *remaining capacity* (and this is of course what the physician is usually interested in) or in the sometimes very subtle *deficits*.

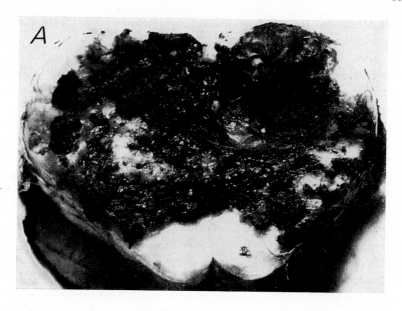

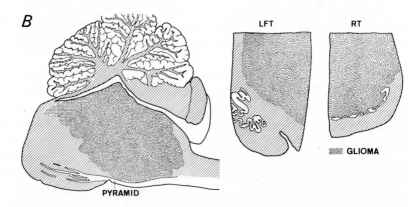

Fig. 2. A Pyramidal preparation. The lower brain-stem of a cat was transected with a spatula which spared both bulbar pyramids. The effects of electrical stimulation rostral to the transection were mediated exclusively by pyramidal tract fibres to the spinal cord [LAURSEN and WIESENDANGER, 1966]. *B* Pyramidal isolation in man. A malignant glioma occupied most of the structures dorsal to the bulbar pyramids. This patient had no paralysis of his extremities.

*What Can Human Neuropathology Teach us About the Normal
Functioning of the Pyramidal Tract?*

A review of the literature on this subject is somewhat deceptive. If one
is taking into account the fact that the bulbar pyramid is the only location
where the pyramidal fibre system is uncontaminated with other fibre
systems, cases of 'pure' pyramidal lesions are lacking. Lesions of the bulbar
pyramids of vascular origin have been reported by DAVISON [1937], BROWN
and FANG [1956], MEYER and HERNDON [1962] and FISHER and CURRY
[1965]. In general, these cases first exhibited a flaccid paralysis which, after
some weeks, changed into a spastic paresis with enhanced tendon reflexes
and the Babinski sign. These cases could be assumed to define the 'pyramidal syndrome'. However, none of the lesions described was actually
confined to the bulbar pyramid. As an example, the lesion shown in MEYER
and HERNDON's paper [1962], which is one of the most discrete, extends
far into the medial lemniscus and the medial reticular formation even
reaching the floor of the 4th ventricle. LASSEK [1954] has reviewed the
literature on the clinico-pathological correlation between PT degeneration
and the degree of paresis. It was concluded that the amount of degeneration
was not paralleled by the degree of paresis. In particular, there were many
cases of severe hemiparesis which showed no degeneration of the bulbar
pyramid. Transection of the middle third of the cerebral peduncle (where
pyramidal fibres are concentrated) was used with therapeutic benefit in
cases of hemiballism [WALKER, 1949; BUCY, KEPLINGER and SIQUEIRA,
1964]. This lesion probably comes nearest to pyramidotomy of animal
experiments (although one should expect a more pronounced deficit when
the PT is severed at a mesencephalic level because of the many collaterals
there which are also put out of function). It was emphasized by BUCY and
co-workers that the patients not only lost their involuntary movements, but
surprisingly rapidly regained a good control of voluntary movements. One
case has come to autopsy which revealed a subtotal degeneration of pyramidal fibres at the medullary level. According to the author's statement,
this patient regained 'an almost normal volitional control' of the contralateral extremities, although more refined motor testing might have
revealed deficits of skills requiring independent finger movements. These
cases, nevertheless, stand in sharp contradiction to the familiar 'pyramidal
syndrome'.

Finally, a case of a brain-stem tumour will be briefly reported; this
came to autopsy and in its final stages was virtually identical to the 'py-

ramidal preparation' in animal experiments[1]. A 60-year-old patient had a short illness of 2.5 months duration, starting with vertigo, vomiting and loss of weight. An initial neurological examination revealed a horizontal nystagmus, muscle twitches in the face and a discrete palsy of the right facial nerve. Soon the neurological symptomatology became more complex: both eyes became immobile, the pupillary reaction was diminished on the right side, there was a total paresis of the 7th nerve on the right, difficulty in swallowing and a diminished sensibility in the right trigeminal distribution. The tendon reflexes were not elicitable, abnormal 'pyramidal reflexes' were lacking. Two months after the onset of symptoms, there were signs of involvement of most of the cranial nerves on the right side (Garcin syndrome). There were, however, no changes in muscle tone; voluntary motility and the force of movement in the extremities remained unaltered and the tendon reflexes were bilaterally weak. Plantar reflexes remained normal. Diadochokinesis was only slightly disturbed and the finger-to-nose test was described as being insecure. Thereafter, the general state of the patient deteriorated and signs of a cardiopulmonary insufficiency developed rapidly so that operative exploration of the brain-stem region could not be undertaken. The day before his death the patient suffered respiratory troubles and a slight, *transient* left-sided hemiparesis. That same evening the patient was still conscious and able to move his limbs symmetrically. Death occurred from respiratory failure.

In summary, this patient developed rapidly increasing signs of brain-stem dysfunction which affected practically all cranial nerves on the right side but left the motility of his extremities bilaterally intact. Since the patient was in a poor condition generally, it may have been that there was some proximal motor deficit although he could lift his limbs voluntarily without difficulty. There was no evidence of an impairment of the motility of his hands, fingers and toes. The stretch reflexes were not increased and there was no Babinski sign. As shown in figure 2B, the right pyramid is represented by a flattened band and the inferior olive is hardly visible, the left pyramid is of normal configuration and also the olive seems to be intact. Structures dorsal to the pyramids in the medulla were replaced by a large malignant glioma which disrupted all connections between cerebrum and spinal cord except for the PT. From this case, one must recognize that

1   I am grateful to Prof. G. YASARGIL, Department of Neurosurgery, University of Zurich, for having sent me the brain-stem of this interesting case for histological investigation.

a great deal of supraspinal motor control can be mediated by the PT, at least in terms of careful clinical observation.

In summary, then, it is evident from both types of lesions, those affecting mainly the PT and those destroying most of the descending and ascending tracts in the brain-stem except the PT, that it is difficult to draw precise conclusions about the *specific* functions of the PT in humans. In the following paragraphs I shall briefly summarize experimental results obtained in sub-human primates with pyramidal lesions.

Effect of Pyramidal Lesions in Sub-Human Primates

Early reports on pyramidotomized monkeys [SCHÄFER, 1900; SCHÜLLER, 1906; ROTHMANN, 1907] were reviewed by MARSHALL [1936]. Noteworthy in those early experiments is the remark by ROTHMANN [1907] that the PT can be compensated for by non-pyramidal systems. No specific function was attributed to the PT except a certain contribution to the speed of movements. All subsequent investigators have also been struck by the fact that monkeys with pyramidal lesions have a rapid and a remarkable recovery of their motor capacity. TOWER [1940] first pointed out the crucial deficits, namely those affecting the skilled movements of the hand. It is only recently that LAWRENCE and KUYPERS [1968] have systematically re-investigated this subject and carefully analyzed a large series of monkeys with bilateral and unilateral lesions of the PT. As mentioned by KUYPERS [this volume], the most constant and permanent deficit of pyramidotomized monkeys is the lack of independent finger movements as evident from cinematography of animals working on the 'foodboard'. The fine control of distal motor units is most likely to depend on an intact cortico-motoneuronal system which has been investigated so intensively by PHILLIPS and his co-workers [PHILLIPS, 1969].

Effect on Conditioned Hand Movements

In our own experiments [HEPP-REYMOND and WIESENDANGER, 1972] we have looked for specific changes of a conditioned hand movement occurring after lesions confined to one bulbar pyramid. Figure 3 shows the movement which was performed many hundreds of times before and after placement of the lesion. It required opposition of thumb and index finger

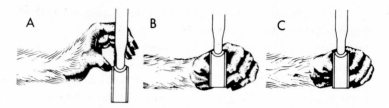

Fig. 3. Conditioned finger grip of monkeys. *A, B, C* Successive stages of the conditioned movement were drawn from cinematographic recordings. The manipulandum contains a force transducer. On reaching a preset threshold of force the animal is rewarded by a small amount of fruit juice. This movement was studied over several months, before and after transection of the pyramidal tract.

(or another finger). The manipulandum contained a microtransducer whose output indicated the strength of the finger grip. In each session, 4 different pre-set force thresholds (100, 200, 400 and 700 g), alternating every 5 min in random fashion, had to be reached in order to release the alimentary reinforcement (fruit juice). Four such monkeys have been subjected to unilateral pyramidotomy. Later investigation of their lesions by means of histological serial sections revealed that one animal had a 60-percent lesion, one an 80-percent lesion, one a 90-percent lesion and one a 100-percent lesion.

After pyramidotomy the contralateral hand of all animals was too weak to perform the motor task. Intensive retraining however resulted in a gradual improvement in performance and, in the course of the post-operative period, lasting 6 months, even the animal with the total lesion succeeded in activating the manipulandum at the highest force threshold. The movement itself was performed more slowly, i.e. the building up of force to reach threshold was significantly prolonged, a fact that is best seen for the higher thresholds, and concomitantly the EMG summation time was increased as shown in the example of the monkey with a 90-percent lesion (fig. 4 A, B).

One animal whose lesion has been verified histologically (it was a 75-percent pyramidal lesion) has been studied for reaction times both before and after pyramidotomy. Post-operatively, the response time for essentially the same motor task increased by about 30-percent. Since, however, an equivalent increase of the EMG latencies was not observed, it was concluded that the *execution* of the motor task rather than its initiation was delayed. The results of both types of experiments are taken to indicate that the pyramidal system plays an important role in the control of force. This is in accordance with another type of results reported by EVARTS [1968]: the intensity of spike discharges of a given PT cell, related to a conditioned hand movement, was found to depend on the force rather than the displacement required of the movement. In summary, it was possible to establish small motor deficits which probably would escape ordinary clinical observation. Although the residual deficits do not mirror the full capacity of the pyramidal system since some compensatory mechanisms have come into play, such chronic, quantitative, behavioural observations seem to be useful to analyze both the immediate severe deficits produced by pyramidotomy and the remarkable processes of recovery.

BECK and CHAMBERS [1970] compared the strength of the two arms in monkeys with chronic unilateral pyramidotomy. The affected arm was reported as being weaker; furthermore, a thumb-forefinger grasp (pulling a rod which had to be held between two opposing fingers against variable weights) was poorly performed on the affected side. Although the results point in the same direction as ours, one has to keep in mind that BECK and CHAMBERS were comparing the two sides only postoperatively. Secondly, these authors evaluated the grasp test during a few sessions whereas several months of daily training gave our monkeys a chance to reach their optimal proficiency after pyramidotomy.

In electrophysiological experiments on the same and some other monkeys with chronic, unilateral lesions of the pyramid, it has been shown [FELIX and WIESENDANGER, 1971] that when the corresponding motor cortex was stimulated repetitively for several seconds (50/sec, 1 msec), distal motor units could be activated despite a complete interruption of the pyramidal projection. The summation time was longer and the current intensity for threshold effects higher than on stimulation of the opposite side. This kind of stimulation involves a pronounced intracortical as well as spinal summation. If lesions in the pyramids were made as small as 30%, single long (3 msec) anodal shocks failed to activate the distal motor units. On the normal side such stimuli evoked short latency synchronized motor responses. These experiments as well as similar ones performed on baboons [LEWIS and BRINDLEY, 1965] have demonstrated that the motor cortex still has access to the distal motor units if deprived of its pyramidal proection. The need for prolonged temporal summation to discharge the motor units may reflect the behavioural deficit described above.

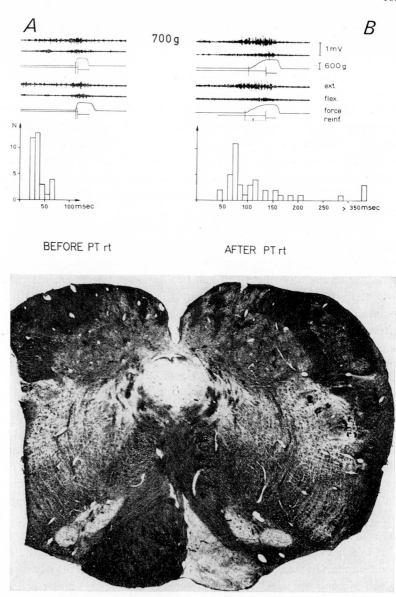

A

700 g

B

I 1 mV
I 600 g
ext.
flex.
force
reinf.

N
10
5
0
50   100 msec

50   100   150   200   250   > 350 msec

BEFORE PT rt   AFTER PT rt

Fig. 4. A Recordings from a monkey before and after right pyramidotomy at the highest force threshold (700 g); left hand. Representative recordings of the EMG activity of the extensor (ext.) and flexor (flex.) muscles of the wrist and fingers and of the pressure exerted on the transducer (force). The downward deflection of the lowest trace indicates the moment when the force threshold was reached and the reinforcement (reinf.) delivered. The

Tonic Effects of Pyramidal Lesions
on Reflexes and Posture

Besides the effects seen on phasic actions of distal motor units, it has
already been emphasized by Tower [1940] that pyramidal lesions also
affect muscle tone. The author did not find any spasticity, but rather a
diminished muscle tone. Denny-Brown [1966] reviewing Tower's work
and comparing it with his own results, maintained that some mild signs of
spasticity were present in pyramidotomized monkeys. Thus, Denny-Brown
[1966] observed overflexion of limbs and a tendency for adduction. Passive
stretching resulted in a 'mild, soft resistance' and the tendon reflexes were
'larger' on the affected side. The spasticity was said to be more evident in
the distal muscles. For instance, the fingers were 'curled in semiflexion'.
EMG recordings revealed enhanced stretch evoked activity on the affected
side.

In agreement with Tower [1940] we have consistently observed a
reduction of muscle tone tested by passive stretching of the limbs. As a
standard procedure the EMG response to plantar flexion of the foot has
been recorded from m. tibialis anterior. Although the reflex activity may
vary from moment to moment, dependent for instance on the emotional
state of the animal, it has been observed regularly that the stretch reflexes
are weaker or absent on the affected side. This is shown in the examples of
figures 5 and 6. The results are in agreement with those reported by
Gilman [this volume] on pyramidotomized monkeys. He found that the
stretch sensitivity of muscle spindle endings were markedly reduced. It was
concluded that this was due to a lack of tonic fusimotor activity. Such a
loss of tonic γ bias has been observed in cats after unilateral chronic
pyramidal lesions [Wiesendanger and Tarnecki, 1966].

The affected limbs of monkeys with pyramidal lesions were often in an
extended position, but this was not due to an enhanced extensor tonus, but
rather to a diminished tonic activity in the flexor muscles as seen from the
EMG recordings (fig. 6A).

---

force development time is denoted by t. The t values are represented in the histograms
below for the pre- and post-operative period. *B* Degeneration of the right bulbar pyramid
of the monkey performing the conditioned finger movement of figure 5*A*. Subtotal degeneration of pyramidal fibres (about 90%) caudal to the lesion (Weil's technique). Some
dorsolateral and lateral fibres were spared. Serial sections of the rostral brain-stem revealed
no damage to other structures [Hepp-Reymond and Wiesendanger, 1971].

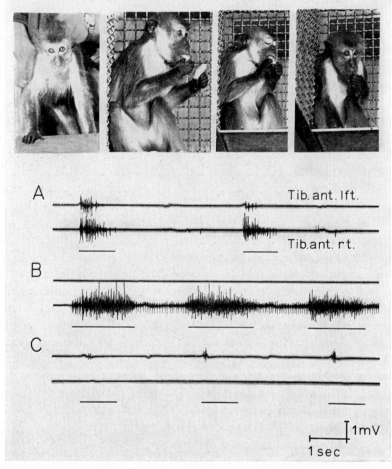

5A

Fig.5. A Monkey with a partial lesion (60%) of the right pyramid. Clinical observation revealed hardly any deficit in the manual skill. The placing reaction was, however, lacking on the left side. Below are EMG tracings from the left and right m. tibialis anterior (Tib. ant.). *A* Activity evoked by simultaneous lifting of both feet ('Stützreaktion'). *B* Activity evoked by slow plantar flexion of the right foot (indicated by bar). *C* The same manoeuvre was effected on the left side. Note difference in the evoked EMG activity. *B* Monkey with a complete lesion of the left pyramid. It was already noted on clinical observations that the right hand was used less than the left. Below are EMG tracings from right and left m. tibialis anterior. On the upper two traces stretch reflexes evoked by passive plantar flexion of the foot on the left side (indicated by bar); on the lower two traces stretching of the right muscles resulted in acceleration of a single unit discharge.

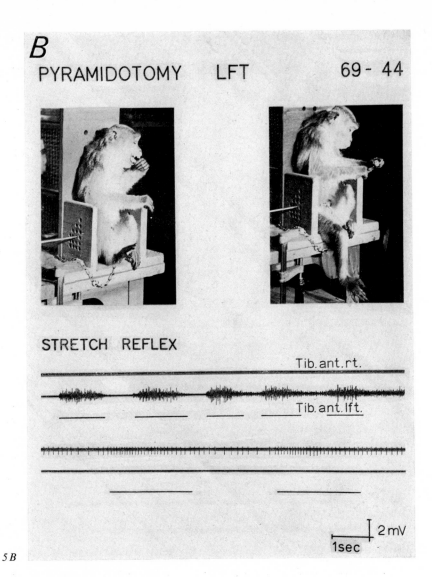

5B

One of the most constant and persistent signs of pyramidotomy is a defect in the placing reaction (fig. 5, 6). In the first post-operative weeks, the legs are often not placed even if visual feedback is permitted. In later stages, the only deficit may be a delay in contact placing.

Sometimes bizarre postures of the limbs have been observed (fig. 6). In our experience this always occurred in association with some damage to neighbouring structures of the PT, especially to the medial lemniscus.

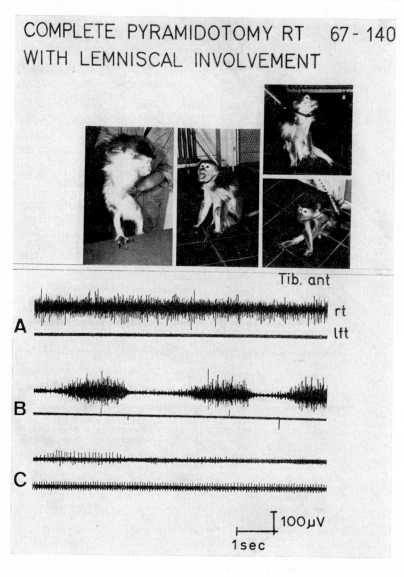

Fig.6. Monkey with a complete lesion of the right pyramid and some damage to the medial lemniscus. Note bizarre postures of the left extremities and lacking placing reaction. Below are EMG tracings of the right and left m. tibialis anterior. *A* Animal was sitting in a chair with its left foot hanging down in an extended position. Note lack of flexor tonus in left muscle. *B* Stretch-evoked activity in the right muscle. *C* Lack of stretch-evoked activity in the muscle on plantar flexion of left foot.

Summary and Conclusion

1. It is evident from both morphological and electrophysiological investigations that the pyramidal system does not represent a homogenous system. One cannot escape the conclusion that various subsystems with different fibre types, different cortical origin and different projections probably also subserve different functions.

2. The clinical observations of patients with pyramidal lesions or with lesions involving the brain-stem except the PT are conflicting. The difficulty lies chiefly in the fact that neuropathologically 'pure' pyramidal lesions are still lacking and that quantitative motor tasks have not been used. A remarkable case is presented of a patient with a tumour in the pons and the medulla which resulted in a pyramidal isolation. This patient had no difficulty with voluntary movement of his extremities, which fact emphasizes the extraordinary capability of the PT in man.

3. The subsystem which has been investigated most intensively is the direct cortico-motoneuronal projection. It is a phylogenetically new system, existing in primates only and also has a late ontogenetic development. Experimental evidence has accumulated that this system is chiefly concerned with the control of distal motor units. On the behavioural level the cortico-motoneuronal system is probably responsible for digital skills, including the capacity to move the fingers independently. It is suggested that inhibition of many muscles in prevention of mass movements is an important aspect of PT function. Single unit recordings of PT neurones in awake monkeys performing a conditioned hand movement have shown further that these cells are phasically activated prior to the movement. The intensity of spike discharges is correlated with the amount of *force* with which the movement was performed. This is in keeping with behavioural experiments showing that pyramidotomized monkeys have a greater difficulty in achieving a prescribed force threshold. The possible role of proprioceptive afferents in the control of force has been discussed by PHILLIPS in this volume.

4. It is not established whether pyramidal fibres which have an effect on neurones of sensory systems are the same as those which influence the motor apparatus. Some preliminary results of the study of corticofugal effects on neurones of the dorsal column nuclei in monkeys [HARRIS, JABBUR, MORSE and TOWE, 1965] reveal the involvement of rapidly conducting fibres, but slow fibres may also be engaged in other central sensory control mechanisms [GORDON and MILLER, 1969].

5. Monkeys deprived of their PT still have a cortical motor control of distal motor units. Electrical stimulation of the motor cortex, however, reveals the difference between the non-pyramidal projection system and the motor control system of intact animals: thresholds are higher and the summation time (at the cortical and probably also at the segmental level) prolonged. Single anodal pulses of 3 msec duration fail to activate distal motor units in monkeys with pyramidal lesions.

Monkeys with unilateral pyramidal lesions undergo an impressive recovery of their remaining motor capacity as judged from clinical observations. Refined behavioural tasks have to be employed to reveal lasting deficits in a manual task. The main result is a slower contraction of a finger grip.

6. Some tonic changes and a deficit in postural reactions following chronic section of the PT in monkeys have also been observed. The EMG responses to passive stretch are reduced as compared with the normal side or even completely lacking. These findings are consistent with results on single muscle spindle endings in monkeys with pyramidal lesions. Another conspicuous sign of a pyramidal lesion is a lack in the initial stage and a delay in the chronic stages of placing responses. Often the animals are seen with their limbs hanging down loosely in an extended position. More awkward positions are usually due to lesions encroaching upon the medial lemniscus or reticular systems. It is likely that these tonic pyramidal effects are mediated by fibres other than the rapidly conducting cortico-motoneuronal fibres which typically are phasically active. Whether smaller, slowly conducting pyramidal fibres may be instrumental in these mechanisms remains to be investigated.

Acknowledgements

I owe much to the collaborators named in this paper.

This work was supported by grants from the Swiss National Research Foundation (No. 3.415.70 and 3.133.69). Additional grants for instrumentation were obtained by the 'Stiftung für wissenschaftliche Forschung an der Universität Zürich' and the 'Jubiläumsspende für die Universität Zürich'. The technical help of Mrs. GYARMATI, Mrs. RÜFENACHT, Mrs. HADVARY, Mr. KÄGI, Mr. FREI, and Mr. FÄH is gratefully acknowledged.

Author's address: Prof. M. WIESENDANGER, Institut für Hirnforschung der Universität Zürich, August-Forel-Strasse 1, *CH-8008 Zürich* (Switzerland)

New Developments in Electromyography and Clinical Neurophysiology,
edited by J. E. Desmedt, vol. 3, pp. 175–193 (Karger, Basel 1973)

Significance of Muscle Receptor Control Systems in the Pathophysiology of Experimental Postural Abnormalities

S. GILMAN

Department of Neurology, College of Physicians and Surgeons of Columbia University, New York, N.Y.

One of the great challenges in clinical neurology is to understand more completely the mechanisms underlying abnormalities of posture and movement. We have little information about the basic pathophysiology of progressive crippling disorders such as Parkinson's disease, dystonia musculorum deformans, Wilson's disease, Huntington's chorea, and Hallervorden-Spatz disease. Even the mechanisms of spasticity resulting from vascular or neoplastic disorders of the nervous system are not fully understood. In humans, these disorders can be studied only be indirect means, but in animals far greater insight into mechanisms can be obtained when abnormalities of posture and movement replicating human disorders can be produced. It has proved possible to simulate certain human disorders, such as cerebellar hypotonia and hemiplegic hypertonia, by using ablation techniques in monkeys. Recent studies, which will be reviewed in this communication, have indicated that abnormalities of muscle receptor control systems in the spinal cord lie at the core of certain of these postural abnormalities. Accordingly, the current concepts of muscle receptors and the cells governing their behavior will be reviewed briefly before presenting studies bearing directly on the pathophysiology of hypotonia and hypertonia [see also MATTHEWS, this volume].

Spinal Motor Control of the Responses of Muscle Spindle Primary and Secondary Afferents
The intrafusal fibers of mammalian striated muscle contain specialized types of muscle fiber and nerve ending. There are two types of intrafusal muscle fiber, nuclear bag and nuclear chain [WALKER, 1958; SWETT and

ELDRED, 1960; BOYD, 1962; BARKER, 1962; ELDRED, BRIDGMAN, SWETT, and ELDRED, 1962]. Spindle primary afferent nerve fibers arise from spirals in the equatorial region of both bag and chain fibers, then form the large diameter (12–20 μ) group I_A afferents [BOYD, 1962; COOPER and DANIEL, 1956; BARKER, 1962]. Spindle secondary afferent nerve fibers arise chiefly from nuclear chain muscle fibers, though a small number arise from nuclear bag fibers [BARKER and IP, 1961; BOYD, 1962]. The secondary afferents, located on the myotube or polar region of the muscle fiber, form nerve fibers of group II (4–12 μ). In the absence of motor innervation there are striking differences in the responses to muscle stretch of primary and secondary afferent endings [COOPER, 1959, 1961; HARVEY and MATTHEWS, 1961; BESSOU and LAPORTE, 1962]. Primary afferents respond to both the dynamic and the static phases of muscle extension whereas secondary afferents respond chiefly to the static phase. Accordingly, the primaries signal rate of change of muscle length as well as length itself whereas the secondaries record only muscle length [MATTHEWS, 1964; and this volume]. The motor fibers innervating the muscle spindles terminate either as endplates or as diffuse ('trail') endings [BOYD, 1962; BARKER and COPE, 1962]. Both types of ending are terminations of γ efferent fibers and are now called 'γ plates' (γ-1 ending of BOYD) or 'γ trails' (γ-2), respectively [GRANIT, 1970]. These motor fibers have been termed 'fusimotor' fibers, a term coined by HUNT and PAINTAL [1958] to signify intrafusal innervation regardless of the diameter of the innervating fiber, whether of α, β, or γ range. There is disagreement as to the distribution of the two types of γ ending on the two types of intrafusal muscle fiber. BOYD and DAVEY [1966] state that, with few exceptions, γ plates innervate selectively the nuclear bag fibers and γ trails innervate the nuclear chain fibers. However, BARKER and IP [1964] and ADAL and BARKER [1965a, b] find both types of nerve ending on both types of intrafusal muscle.

The net effect of fusimotor stimulation is mechanical deformation of the spindle endings, thereby initiating or enhancing spindle afferent discharge. Two types of fusimotor fiber have been identified by examining the effect of stimulating individual fibers in ventral roots upon the responses of spindle endings to dynamic or static muscle extension [MATTHEWS, 1962]. Dynamic fusimotor fibers enhance the responses of the spindle primary endings during dynamic muscle extension, producing little effect when the muscle remains static in length. Static fusimotor fibers raise the firing level of spindle primary endings when the muscle is kept at a fixed length but fail to increase the response to dynamic extension of the muscle [MATTHEWS, 1962, 1964; BESSOU and LAPORTE, 1966; BROWN and MATTHEWS, 1966; JANSEN, 1966].

Muscle spindle secondary endings are influenced by stimulation of static fusimotor fibers, but not by stimulation of dynamic fusimotor fibers, except for the few secondary endings that show some sensitivity to the rate of change of muscle length (dynamic sensitivity) [CROWE and MATTHEWS, 1964; APPELBERG, BESSOU and LAPORTE, 1965, 1966]. Attempts to relate the two physiological types of fusimotor fibers to the two types of fusimotor endings (plate and trail) on spindles have not been entirely satisfactory. Thus, no definite correlation has been established between axonal conduction velocity and the static or dynamic effect of stimulating fusimotor fibers, though the lowest velocity fibers tend to be associated with the static type of effect [CROWE and MATTHEWS, 1964; BROWN, CROWE, and MATTHEWS, 1965; BESSOU, LAPORTE, and PAGÈS, 1966]. However, it has been suggested that dynamic fusimotor fibers engage the terminals on the nuclear bag intrafusal fibers and that static fusimotor fibers activate the terminals on the nuclear chain intrafusal fibers [GRANIT, 1970]. There is evidence, reviewed recently [GRANIT, 1970], that α-motoneurons may provide innervation for some intrafusal muscle fibers.

Effects of Muscle Receptor Discharge on α-Motoneuron Responses

Impulses initiated in primary endings course through I_A afferents, uniformly exciting postsynaptically the motoneurons innervating the muscle of origin of the endings as well as the synergistic muscles [LLOYD, 1946; LAPORTE and LLOYD, 1952; ECCLES, ECCLES, and LUNDBERG, 1957a]. These impulses inhibit motoneurons of related antagonistic muscles across a disynaptic arc [BROCK, COOMBS, and ECCLES, 1952; ECCLES, FATT, and LANDGREN, 1956]. Since a large proportion of spindle primary afferents in a hindlimb extensor muscle of the intact, lightly anesthetized cat [GILMAN and McDONALD, 1967a] or monkey [GILMAN, 1969] discharge even in the absence of applied stretch (i.e with the muscle at its shortest natural length), there is likely to be strong tonic facilitation of α-motoneurons by spindle primary afferents when the fusimotor γ loop is intact. In addition, activation of spindle endings across the γ loop is capable of activating α-motoneurons [GRANIT, KELLERTH, and SZUMSKI, 1966].

The impulses from muscle spindle secondary endings course through group II afferent fibers, exciting motoneurons of flexor muscles and inhibiting those of extensor muscles, regardless of the muscle of origin of the secondary endings [LLOYD, 1946; BROCK, ECCLES, and RALL, 1951; LAPORTE and BESSOU, 1959]. Golgi tendon organs, which have no direct efferent control from the CNS, respond with lower thresholds to twitch than to passive stretch in an extensor muscle [JANSEN and RUDJORD, 1964], but with equal sensitivity to both in a flexor muscle [ALNAES, 1967]. In either situation, it seems likely that the Golgi tendon organ serves the function of a tension recorder [GRANIT, 1970]. The central effects of activating Golgi tendon organs are inhibition of the motoneurons projecting to the muscle containing the receptor as well as the motoneurons of synergistic muscles [LAPORTE and LLOYD, 1952; ECCLES, ECCLES, and LUNDBERG, 1957b,c; GRANIT, 1970].

Two types of α-motoneurons were identified by GRANIT, HENATSCH, and STEG [1956] and labeled 'tonic' and 'phasic' on the basis of their responses to muscle extension after post-tetanic potentiation. The tonic motoneurons responded with sustained discharge to maintained muscle stretch after potentiation whereas the phasic motoneurons responded with only a short-duration discharge. Furthermore, the tonic motoneurons were associated with small amplitude spikes, suggesting that the conducting axons were of small diameter, whereas the phasic motoneurons were associated with large-amplitude spikes. Later it was found that α-motoneurons with axons of low conduction velocity (and therefore small diameter) had after-potentials of longer duration than those with axons of high conduction-velocity [ECCLES, ECCLES, and LUNDBERG, 1958; KUNO, 1959; KERNELL, 1965; BURKE, 1967]. Next, it was found that total membrane resistance was greater in small than in large motoneurons [KERNELL, 1966; BURKE, 1967]. In addition, slowly conducting α motor axons innervate slowly contracting muscle fibers whereas rapidly conducting axons innervate rapidly contracting fibers [STEG, 1962, 1964]. It is now thought that the large diameter, rapidly conducting axons belong to large diameter, phasic α motoneurons which innervate pale, rapidly contracting muscle fibers, whereas the small diameter, slowly conducting axons belong to small-diameter, tonic α-motoneurons which innervate red, slowly contracting muscle fibers [BURKE, 1968a, b and this volume]. The tonic α-motoneurons appear to be influenced synaptically by spindle afferent fibers to a greater degree on the average than the phasic motoneurons [ECCLES, ECCLES, and LUNDBERG, 1957b; GRANIT, 1970].

In recent studies, based upon combined intracellular stimulation of various α-motoneurons, measurements of the resulting muscle tension, and histochemical staining of the stimulated muscle fibers, three types of motor units have been described: type FF (fast contracting, fast fatigue); FR (fast contracting, fatigue resistant); and S (slowly contracting) [BURKE, LEVINE, ZAJAC, TSAIRIS, and ENGEL, 1971]. The histochemical profiles of these three different types of motor units show distinct characteristics. Muscle fibers of FF units are large in diameter and sparsely supplied with capillaries; fibers of

FR units are variable in diameter and liberally supplied with capillaries; fibers of S units are small in diameter and liberally supplied with capillaries. FF and FR unit fibers are rich in glycogen whereas those of S fibers are poor in glycogen. Thus the FF fibers probably depend primarily on anaerobic glycolysis for energy whereas the S fibers probably utilize aerobic energy pathways. Fibers of FR units evidently have both aerobic and anaerobic metabolic properties.

The Pathophysiology of Cerebellar Disorders

The nature of the symptoms resulting from cerebellar disease in humans has been discussed in the classical papers of GORDON HOLMES [1922]. The effects of cerebellar lesions in animals have been analyzed by a number of investigators [cf. DOW and MORUZZI, 1958]. Cerebellar lesions in the primate result in clinical abnormalities closely akin to those observed in humans. Accordingly, study of the effects of cerebellar ablations in the monkey are pertinent to the understanding of the mechanisms of cerebellar dysfunction in the human. The effects of complete cerebellar ablation in the monkey have been reported previously by several investigators, among them LUCIANI [1891], CARREA and METTLER [1947], and DENNY-BROWN [1966]. I have studied 13 monkeys made completely decerebellate, 8 in collaboration with Dr. D. DENNY-BROWN [DENNY-BROWN and GILMAN, 1965; DENNY-BROWN, 1966] and 5 subsequently [GILMAN, 1969]. Histological examination has shown no damage to brain stem in these preparations.

The most remarkable abnormality following complete cerebellar ablation was a severe disturbance of stance and gait. For the first several weeks after operation, a typical animal could not stand or walk. Usually the animal would lie on the floor in the prone position with the head erect and all limbs closely flexed under the body. Any active movement of the trunk or limbs initiated a coarse, rhythmical tremor of the trunk and head. Movement about the cage was performed by ataxic progression movements of the limbs, which were kept in wide abduction, the abdomen dragging along the floor, or with briefly sustained galloping movements. If stood erect on all four limbs the animal, when let go, immediately collapsed to the floor. Passive manipulation of the limbs revealed a soft, plastic resistance in the flexor muscles, particularly at proximal joints, but clearly diminished resistance (hypotonia) in the extensor muscles. The deep tendon reflexes of the ankles, knees and wrists were slow and pendular. If the animal was suspended in air in the upright or prone position from the trunk or pelvis, the limbs remained flexed close to the body. In the second week after operation, with the animal suspended in air held from the trunk, a light contact with the plantar surface of the distal pads of the toes or fingers led to a gradual extension of the limb. This response has been termed the 'magnet reaction' [RADEMAKER, 1931]. However, despite the sustained extension of the limb following the magnet reaction, the limb failed to sustain the weight of the body and, if stood on the extended limb, the animal would collapse to the floor.

In time, usually by 6–8 weeks post-operatively, the supporting reactions of the limbs recovered partially, so that progressively the animal was able to sustain the upright posture, usually only when the side of the body remained in contact with a wall. Eventually the animal was able to walk in the erect position on all four limbs, developing a coarse tremor of trunk and limbs. As the proprioceptive positive supporting reactions of the limbs increased in intensity, the hypotonia (decreased resistance to passive manipulation) of the limb extensor muscles progressively decreased. Correspondingly, the tendon reflexes became less pendular.

The effects of denervation and deafferentation procedures on the magnet and positive supporting reactions were studied in decerebellate monkeys by DENNY-BROWN and GILMAN [1965]. Denervation of the foot or hand by sectioning all nerves at the ankle or wrist abolished the magnet reaction completely. If a single cutaneous nerve in the hand or foot was preserved, contact with the innervated area of skin could still evoke the magnet reaction. Interactions between the magnet and the positive supporting reactions were studied by sectioning the dorsal roots L 3, 4 and 5 in a monkey decerebellate for 23 days. This procedure deafferented the quadriceps muscle but left intact the cutaneous innervation of the foot. The result was a depression of both the positive supporting and the magnet reactions for an interval of several weeks, leaving preserved only sustained postures of flexion at the knee. Eventually, however, a strong magnet reaction reappeared and, although sustained extension of the leg could be provoked, the knee could support only a weight of 1.5 pounds, completely inadequate for sustaining the weight of the animal.

It was concluded from these observations that the prolonged initial loss of positive supporting reactions after cerebellectomy resulted from a decrease of the responses to stretch of spindle afferents in limb muscles, particularly the antigravity extensor muscles [DENNY-BROWN and GILMAN, 1965]. This disorder was superimposed upon a series of preserved tactile reactions of opposite type: a magnet reaction which could evoke limb extension and a more general truncal contactual reaction which usually dominated, producing the predominant postures of limb flexion. In time, compensatory processes restored the spindle responses partially, so that the proprioceptive

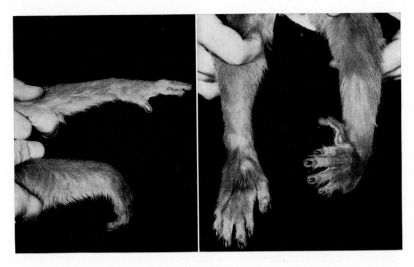

Fig. 1. Postures of the extended upper (left panel) and lower (right) limbs of a monkey 2 days after right hemicerebellectomy, showing flexion of the right wrist and plantar flexion with pronation of the right ankle under the influence of gravity.

positive supporting reactions reappeared and the hypotonia diminished. The observation that a partially compensated positive supporting reaction in the hind limb was depressed severely by deafferentiation of the knee musculature emphasized the importance of the integrity of the spindle effects on α-motoneurons in evoking the supporting reactions.

Study of the effects of hemicerebellar ablation in the monkey permitted a quantitative comparison between the two sides of the body of the impairment of the positive supporting response and the abnormal resistance to passive manipulation of the limbs [GILMAN, 1969]. In a typical animal, following recovery from anesthesia after a right hemicerebellar ablation, the animal could stand and walk but often fell to the right. When the animal sat or stood quietly, the right limbs adopted postures of close flexion. Passive manipulation of the limbs revealed diminished resistance in the extensor muscles of the right limbs. When the limbs were held extended in the horizontal position, the wrist and ankle flexed under the influence of gravity more on the right than on the left (fig. 1). The deep tendon reflexes of the ankle, knee and wrist were slow and pendular on the right, rapid and non-pendular on the left. The positive supporting reactions of the limbs were tested by standing the animal on a scale, one limb at a time, pressing firmly downward on the animal's shoulder or back, and measuring the weight required to cause collapse of the

Fig. 2. EMG recorded from the right (upper trace) and left (lower) quadriceps muscles of a monkey 2 days after right hemicerebellar ablation. Middle trace records the rate, extent, and direction of passive movements of lower limbs at the knees, downward deflection signifying flexion. Voltage and time calibration apply to both EMG tracings. Angular movement calibration applies to the middle trace, indicating extent of movement at the knees. Note that a greater degree of flexion is required to elicit EMG responses on the right than on the left.

limb. In a representative animal, the supporting reactions of the right limbs were 20% of those on the left in the first post-operative week, 50% in the fifth and 75% in the eighth. The degree of hypotonia of the right limbs was determined by clinical estimation and quantitated by EMG measurements. The latter were performed by inserting needle electrodes into the quadriceps muscles of each leg, passively manipulating the lower legs from full extension through 90° of flexion at the knees, and comparing the degree of flexion required to initiate stretch responses on the two sides. The data, which have been published in detail [GILMAN, 1969], showed that a greater degree of extension was required to initiate EMG activity on the right side compared with the left (fig. 2). This abnormality lessened in time. The recovery of the positive supporting response and the lessening of the hypotonia (measured both clinically and electromyographically) showed a similar time course, suggesting that they resulted from a common abnormality, possibly a depression of spindle afferent responses, as had been suggested earlier [VAN DER MEULEN and GILMAN, 1965; DENNY-BROWN and GILMAN, 1965]. Figure 3 shows a representative histological section from the brain stem and cerebellum of one of the hemidecerebellate monkeys described above.

Direct recordings of the responses of spindle primary afferents to muscle stretch in monkeys confirmed the assumption of a decrease of responsiveness after recent cerebellectomy [GILMAN, 1969]. The responses to static (fig. 4) or dynamic extension of muscle spindle primary afferents in medial gastrocnemius muscle were severely depressed but the responses of spindle secondary afferents were essentially unaffected. Marked similarities were noted between the effects of muscle stretch at various rates on the behavior of spindle primary afferent units in dorsal root filament recordings and of extrafusal muscle

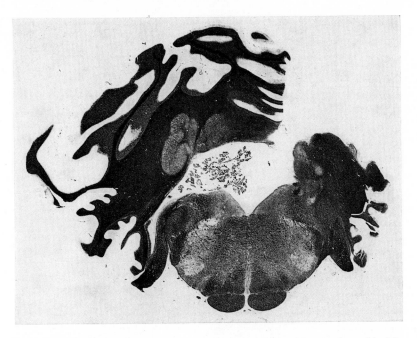

Fig. 3. Histological section showing the extent of the lesion in a monkey with right hemicerebellectomy of 99-day duration. Luxol fast blue stain. Other records from this animal appear in figures 1 and 2.

units in EMG recordings. These similarities suggested that the disorder of spindle afferent response is directly and possibly causally related to the EMG abnormalities. The EMG abnormalities detect graphically the disorder described clinically as hypotonia, a decrease of the resistance to passive manipulation of the limbs. Accordingly, the present evidence suggests strongly that a depression of spindle responsiveness underlies the clinical hypotonia that is so characteristic of the acute cerebellar lesion, particularly in higher apes and man [FULTON and DOW, 1937; HOLMES, 1922].

Cerebellectomy affects chiefly the spindle primary afferents in the upper range of axonal conduction velocities [GILMAN and McDONALD, 1967a, b], depressing the responses to either static or dynamic extension [GILMAN, 1969]. In addition, cerebellectomy decreases the initial increase of muscle tension induced by sinusoidal stretching of the ankle extensor muscles in decerebrate cats [GLASER and HIGGINS, 1966]. This latter effect could also be achieved without cerebellectomy by blocking fusimotor γ fibers with a local anesthetic agent [GLASER and HIGGINS, 1966]. Accordingly, it may now be concluded

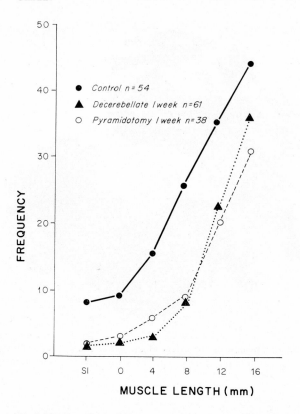

Fig.4. Graphs of mean discharge frequency of medial gastrocnemius muscle spindle primary afferents as a function of static length of gastrocnemius muscle in monkeys: control (●—●); 5–6 days after complete cerebellectomy (▲...▲) and 4–7 days after bilateral pyramidotomy (○---○). On the abscissa, Sl indicates that the muscle is slack, free of external tension; 0 denotes extension of the muscle sufficient to deflect the tracing of a myograph set at 2 g/mm; subsequent numbers indicate extension of the muscle in 4 mm increments beyond 0 length. On the ordinate, frequency is given in impulses/sec.

that cerebellar ablation leads to hypotonia and a reduction of the proprioceptive positive supporting reaction by decreasing the resting discharge of both static and dynamic fusimotor γ fibers, thus impairing the corresponding responses of the spindle primary afferents in extensor muscles. Indeed, direct recordings from fusimotor fibers in dissected peripheral nerves of cats have shown a significant decrease of resting discharge after cerebellectomy [GILMAN and EBEL, 1970]. Interestingly, however, the depressed fusimotor unit activity responds to essentially a normal extent when natural vestibular, cutaneous, or proprioceptive stimuli are applied. These findings may be related to the abnormalities of coordinated movement of the decerebellate animal, but direct evidence for this assumption is lacking. In addition, it has not been established whether the depressed spindle responses affect differentially the activity of the three categories of α motor units described above. In any case, the depressed spindle afferent responses would be expected to provide a decreased level of facilitation upon the α-motoneurons to which the spindle afferents project. The anticipated result of this withdrawal of facilitation would be a decrease of the tonic output of the corresponding α-motoneurons, leading to hypotonia.

The Pathophysiology of Pyramidal Tract Disorders

In 1940, TOWER found that an enduring *hypotonic* paresis of the limbs resulted from section of the medullary pyramidal tracts in the monkey. In addition, TOWER [1940] pointed out that one of the most conspicuous defects after pyramidotomy was a loss of skilled movements of the hands. Systematic reinvestigation of this finding by LAWRENCE and KUYPERS [1968] showed that the lack of independent finger movements is, indeed, one of the major deficiencies of the pyramid-sectioned animal. However, TOWER's finding of a hypotonic paresis contradicted the classical clinical neurological view that pyramidal lesions produce a hypertonic (spastic) paresis. Subsequent studies have provided conflicting evidence concerning the nature of the postural defects after pyramid lesions in the monkey. DENNY-BROWN [1964, 1966] reported that submaximal unilateral lesions of the pyramid resulted in an *increase* of the stretch reflexes during the second post-operative week. There was a definite, though slight spasticity in all muscles of the affected limbs, both flexors and extensors, with clonic finger and toe jerks, positive Hoffmann responses and very brisk, repetitive biceps and quadriceps tendon reflexes. However, GOLDBERGER [1969] studied the unilateral pyramid-sectioned monkey both clinically and electromyographically but found no evidence of increased myotatic reflexes. He described diminished resistance in some muscle groups of the limbs opposite the section, normal resistance in other muscle groups, and no increase of resistance in any group. WIESENDANGER [this volume] has also found only a hypotonic paresis in the animal with unilateral pyramidotomy even of subtotal extent.

In the light of the controversial nature of the reported findings, it was of interest to examine anew the clinical and EMG effects of unilateral and bilateral pyramid section in the monkey. Furthermore, it was of interest to determine the effects of pyramid section upon the responses to stretch of

Fig. 5. Posture of a monkey 5 days after section of the right medullary pyramid, showing extension of the left limbs under the influence of gravity and the natural postures of flexion of the other limbs.

muscle spindle primary afferents in the monkey. Accordingly, Dr. Luis Marco and I have studied 22 monkeys with bilateral and 2 with unilateral medullary pyramidal tract section. The results, which have recently been reported in full [Gilman and Marco, 1971; Gilman, Marco, and Ebel, 1971], will be summarized briefly here. The major clinical defects following bilateral pyramidotomy consisted of loss of contactual orienting responses of the hands and feet (tactile placing, grasping and avoiding); defective use of the fingers and toes for fine coordinated movements; hypotonia (decreased resistance to passive manipulation) of the limbs without hypertonia; and various transient disorders of posture and limb placement. In the chronic

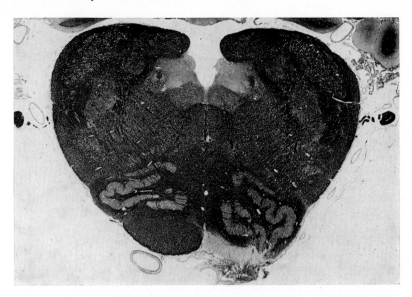

Fig.6. Histological section showing the extent of the lesion in a monkey with right pyramidotomy of 8-month duration. Luxol fast blue stain.

animal these abnormalities showed incomplete recovery. Unilateral pyramid section produced in the contralateral limbs all the defects observed after bilateral section, and in addition, more severe impairment of the affected limbs owing to defects of visually guided projected movements and certain proprioceptive and cutaneous responses. The typical posture of the animal with unilateral pyramidotomy is illustrated in figure 5, and an example of the histological verification in figure 6. Defects similar to those in the unilaterally sectioned animals were found in the limbs opposite the more extensive pyramid lesions in animals with bilateral lesions of unequal extent. EMG responses of the quadriceps muscles to passive manipulation of the legs were studied in animals with unilateral pyramidotomy. During the immediate post-operative period there was a depression of EMG responses in the limb opposite the sectioned pyramid (fig. 7) which regressed partially with time. The depression of EMG response was similar to that described in monkeys with cerebellar ablation [GILMAN, 1969], consisting of a raised threshold for the onset of stretch responses. Neither clinical nor EMG evidence of heightened myotatic responses was found after pyramidotomy.

In the above studies, the afferent responses to passive extension of medial gastrocnemius muscle spindle primary endings were studied in 3 monkeys

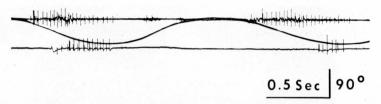

0.5 Sec | 90°

Fig. 7. EMG recorded from the right (upper trace) and left (lower) quadriceps muscles of a monkey 10 days after right medullary pyramidotomy. Middle trace records the rate, extent, and direction of passive movements of lower limbs at the knees, downward deflection signifying flexion, angular movement calibrated below. Note that a greater degree of flexion is required to elicit EMG responses on the left than on the right.

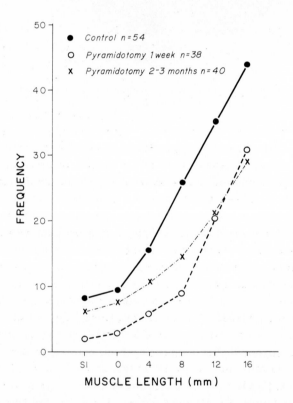

Fig. 8. Graphs of mean discharge frequency of medial gastrocnemius muscle spindle primary afferents as a function of static length of gastrocnemius muscle in monkeys: control (●——●); 4–7 days after bilateral pyramidotomy (○---○); and 2–3 months after bilateral pyramidotomy (× —·—·— ×).

with bilateral pyramidotomy of 4 to 7-day duration [GILMAN, MARCO, and EBEL, 1971]. The responses in these animals were significantly decreased relative to the responses of the unoperated control group (fig. 4). However, the responses of the pyramid-sectioned animals were not significantly different from those of a previously reported group of monkeys subjected to complete cerebellar ablation [GILMAN, 1969]. In one of the animals recorded 7 days after pyramid section, the responses of the spindle afferents were recorded before and after acute ablation of the neocerebellar cortex and dentate nuclei. The ablation did not alter the responses of the units, but subsequent denervation of the muscle spindles by ventral root section produced a significant decrease. These findings suggested that the depression of spindle afferent response by pyramidotomy had already interrupted a CNS reflex arch which involves the cerebellum. Hence, neocerebellar ablation produced no further decrement of spindle activity. However, residual facilitation from other CNS sources was present, as shown by the effects of ventral root section. Spindle afferent responses were also recorded in 4 animals with bilateral pyramidotomy of 2- to 3-month duration. These responses were greater than those of the animals with pyramidotomy of 4- to 7-day duration, but less than those of the control group (fig. 8). It was concluded from these studies that section of the medullary pyramidal tract decreases the tonic fusimotor innervation of muscle spindles in a hind limb extensor muscle, resulting in a depression of the afferent responses to passive extension of muscle spindle primaries. These depressed responses contribute to the hypotonia observed clinically and electromyographically in animals with pyramid lesions.

Facilitation of Muscle Spindle Responses from Cerebellum and Pyramidal Tract: A Cerebello-Thalamo-Corticospinal Circuit

In his discussion on the evolution of cerebellar regulation of postural tonus, BREMER [1935] concluded that the paleocerebellum tonically inhibits postural extensor mechanisms of the spinal cord through bulbospinal relays, whereas the neocerebellum tonically facilitates these mechanisms by way of the cerebral cortex. In support of this hypothesis, DOW and MORUZZI [1958] cited the similarity in primates of the symptoms of hypotonia resulting from cerebellar lesions [BOTTERELL and FULTON, 1938a, b] and those resulting from medullary pyramidotomy [TOWER, 1940] as evidence of a neocerebellar facilitation of cerebral cortical postural mechanisms. According to the current concepts of cerebellar physiology, only inhibitory or disfacilitatory effects can be derived from the outflow of the cerebellar cortex in the cat [ECCLES, ITO, and SZENTÁGOTHAI, 1967]. It has not been established whether the same principle holds for the primate and human cerebellum. It appears at present, however, that the paleocerebellar inhibition results from the inhibitory effects of paleocerebellar cortex upon the vestibular nuclei. In contrast, the tonic neocerebellar facilitation is derived not from the neocerebellar cortex, but from the neocerebellar nuclei. From this site the facilitation is mediated to neurons in the ventrolateral nucleus (VLN) of the thalamus and from there to the precentral cortex. Evidence of tonic cerebellar facilitation of VLN neurons has been presented by MASSION and CROIZE [1970]. Accordingly, it should be possible to demonstrate a depression of spindle afferent responses through lesions of the lateral (dentate) nuclei, the superior cerebellar peduncles or the VLN. In addition, spindle responses should be affected by lesions of the cortical projection area of the VLN of the thalamus.

There is considerable evidence already available that these assumptions are correct. GILMAN and McDONALD [1967a] found that interruption of the outflow of the cerebellar lateral nuclei in the cat by sectioning the superior cerebellar peduncles reduced the afferent responses of muscle spindle primaries. GILMAN, LIEBERMAN, and COPACK [unpublished] have studied the effects of lesions in VLN on the responses to progressive muscle extension of spindle primary afferents in medial gastrocnemius (MG) muscle. Lesions of thalamus were produced with a stereotaxically placed cryogenic probe. It was found that lesions of VLN depressed severely the spindle afferent responses when the probe tip temperature reached $-20°$ to $-40°C$. Placement of the probe in pulvinar, carried out for control observations, did not affect spindle afferent responses despite cooling to levels of $-60°C$, producing large lesions. It was concluded that lesions of VLN decrease the tonic activity of MG fusimotor neurons leading to a depression of the responses to stretch of MG spindle primary afferents. Consequently, the VLN has a net excitatory effect on hind limb extensor fusimotor neurons. These findings are compatible with the concept that the VLN mediates a tonic neocerebellar facilitation of the cerebral cortical mechanisms influencing segmental postural reflex responses through fusimotor neurons. It is likely that a depression of fusimotor activity from VLN lesions may be an important mechanism underlying the beneficial effects of VLN lesions in humans with limb hypertonia [COOPER, 1969].

The Pathophysiology of Experimental Hypertonia

The foregoing evidence has indicated that a depression of fusimotor activity leading to decreased spindle afferent responses to muscle stretch constitutes a fundamental mechanism underlying hypotonia of the limbs in animal models of human cerebellar, pyramidal tract and VLN lesions. We have recently undertaken an investigation of the level of activity of muscle spindle afferents in experimentally produced hypertonia. The results have been presented in brief [GILMAN, MARCO, and LIEBERMAN, 1971] and a full account is currently in preparation. We selected as an experimental model the monkey in which areas 4 and 6 were ablated in one hemisphere, since this preparation provides initially a hypotonic hemiplegia followed in time by a hypertonic hemiplegia [DENNY-BROWN, 1966]. Accordingly, in 10 monkeys, areas 4 and 6 of Brodmann were ablated in one hemisphere by subpial aspiration under pentobarbital anesthesia. Clinical examination revealed a marked hypotonic hemiparesis of the contralateral limbs for the first 3 weeks post-operatively. There was diminished resistance to passive manipulation at all joints of the opposite limbs, diminished to absent deep tendon reflexes, and absence of the oriented grasping, avoiding and traction responses. EMG recordings taken from the triceps surae muscles of both legs simultaneously revealed a raised threshold for muscle discharge in response to passive dorsiflexion of the ankles on the paretic side relative to the control side. In the third week, a slight resistance to passive manipulation of the hemiparetic limbs became detectable and then progressively increased in intensity until the fourth week, when a definite heightening of resistance to dorsiflexion of the ankle and wrist of these limbs could be detected. At this time, the deep tendon reflexes of the hemiparetic limbs became hyperactive and abnormally spreading. During the subsequent month, increasing plastic resistance to passive manipulation involved the shoulder, wrist and ankle; little or no increase of resistance appeared at the knee, and the resistance at the elbow was variable. Correspondingly, EMG recordings now revealed a lowered threshold for the onset of responses to passive stretch in the triceps surae muscles of the hemiparetic limb.

The afferent responses to stretch of 39 MG muscle spindle primaries were recorded from the hemiplegic hind limbs of 2 monkeys with cortical ablation during the first post-operative week. The responses were decreased significantly relative to those recorded from unoperated control animals, and also decreased in comparison with those recorded from animals with medullary pyramid section of comparable duration. Recordings were also taken from 64 MG spindle primaries in the hemiplegic hind limbs of 6 monkeys with cortical ablation during the period of hypertonic hemiparesis. The spindle afferent responses in these animals were essentially equal to those of control animals, and considerably greater than those of monkeys with medullary pyramidotomy of comparable duration.

Ablation of areas 4 and 6 in monkey cortex interrupts the projections of both a large portion of pyramidal tract efferents and a sizeable number of extrapyramidal neurons. Accordingly, the clinical effect is an initial hypotonic paresis which later evolves into a hypertonic paresis. Corresponding to these clinical changes, the responses to stretch of MG spindle primary afferents are depressed during the hypotonic phase and recover to control levels during the hypertonic phase. Despite clear-cut hypertonia of the ankle extensor muscles, there is no evidence of heightened spindle afferent responses. These findings indicate that the effects of cortical ablation of areas 4 and 6 in the chronic animal represent an algebraic summation of the depressant effects on fusimotor neurons of interrupting the pyramidal projections and the heightening effects of interrupting the extrapyramidal projections. The extrapyramidal effects on fusimotor neurons may contribute to the hypertonia following cortical ablation, but these effects do not appear to represent a major mechanism underlying the hypertonia. Direct effects of released supraspinal mechanisms on α-motoneurons or on interneurons articulating with α-motoneurons must be considered as the source of the hypertonia. Consequently, it would be of great interest to study the responses of the various types of α-motoneurons in animals with experimental hypertonia.

Summary

The neuroanatomy and neurophysiology of the fusimotor-γ and α-motor neuron control systems have been worked out in considerable detail. However, we have only begun to study the significance of these motor control systems in the pathophysiology of human disease processes. Evidence has been presented that hypotonia, i.e. decreased resistance to passive manipulation of the limbs, results from a depression of the tonic discharge of fusimotor neurons innervating muscle spindle afferents which, in turn, provide α-motoneurons with decreased tonic facilitatory influences. Experimental models of hypotonia in the monkey, produced by cerebellar lesions, section of the medullary pyramidal tracts, or recent ablations in the cerebral cortex, have shown a clear correlation between the hypotonia demonstrated clinically or electromyographically and a depression of spindle afferent responses demonstrated by direct recordings. In addition, evidence has been presented suggesting that interruption of a central reflex involving cerebello-thalamo-cortico-spinal structures results in a decrease of spindle afferent responses, leading to hypotonia of the limbs. However, experimental hypertonia, produced by ablation of cerebral cortical areas 4 and 6 in the monkey, does not appear to result from excessive fusimotor drive. Despite the development in the chronic animal of clear-cut hypertonia, demonstrated clinically and electromyographically, spindle afferent responses function at levels equal to those of normal control animals. Thus, the level of activity of the fusimotor system correlates well with clinical hypotonia but not with hypertonia in experimental animal models simulating human disorders.

Acknowledgements

This work has been supported by the Clinical Research Center for Parkinson's and Allied Diseases NS 05184. I am indebted to my colleagues Dr. P. COPACK, Dr. D. DENNY-BROWN, Dr. H. C. EBEL, Dr. L. A. MARCO, Dr. W. I. McDONALD, and Dr. J. S. LIEBERMAN for their participation in the studies described herein. I am also indebted to the editor of *Brain* for permission to reproduce illustrations 1–4 and 6–8.

Author's address: Dr. S. GILMAN, Department of Neurology, College of Physicians and Surgeons, Columbia University, 630 West 168th Street, *New York, NY 10032* (USA)

New Developments in Electromyography and Clinical Neurophysiology,
edited by J. E. Desmedt, vol. 3, pp. 194–208 (Karger, Basel 1973)

The Baboon *(Papio papio)* as a Model
for the Study of Spinal Reflexes

J. P. ROLL, M. BONNET and M. HUGON

Laboratoire de Neurophysiologie Comparée, Centre St.-Jérôme, Université de
Provence, Marseille

The general properties of mono- and polysynaptic reflexes in normal man
have been much studied [HOFFMANN, 1922; PAILLARD, 1955; HAGBARTH, 1960;
COQUERY, 1962; MARK, 1962; LANDAU and CLARE, 1964; TABORIKOVA and
SAX, 1969]. Analogous studies have been carried out on patients with hemiplegia, paraplegia, Parkinson's disease, and disorders of cerebellar pathways
[ANGEL and HOFFMANN, 1963; PIERROT-DESEILLIGNY, 1966; DIMITRIJEVIC and
NATHAN, 1967; RUSHWORTH, 1967; ZANDER-OLSEN and DIAMANTOPOULOS,
1967; DELWAIDE, 1970]. However, knowledge of the underlying physiological
mechanisms remains incomplete and the studies of the tonic vibration reflex
have made these deficiencies more obvious [DE GAIL, LANCE and NEILSON,
1966; HAGBARTH and EKLUND, 1966; RUSHWORTH and YOUNG, 1966; LANCE,
NEILSON and TASSINARI, 1968; MARSDEN, MEADOWS and HODGSON, 1969].

Physiological analysis carried out on the cat cannot be easily extrapdated
to man, if only because of the different postural mechanisms of this quadruped.
Corticospinal and cerebellospinal organizations are different in the two
species [PRESTON, SHENDE and UEMURA, 1967] and experiments on primates
are more promising. This work describes a primate preparation sufficiently
close to man to serve as an adequate model for the physiological study of
reflexes.

Methods

The general principle has been to apply to the baboon *(Papio papio)* techniques used
in the study of human reflexes. Techniques used in humans presuppose the co-operation of
the subjects in achieving either constant immobilization of the muscles involved in the
experiment, or specific, graded muscular activity. Animals cannot co-operate predictably

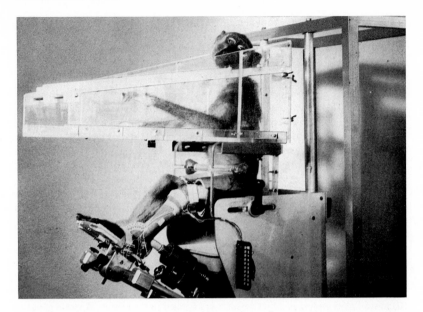

Fig. 1. System of restraint for studying reflexes in the conscious baboon. The general position of the animal is that of a sitting biped, with semi-extended legs. The forelimbs are free in a plexiglass surround. The stimulating electrodes for eliciting the H-reflex are placed at the knee. The recording electrodes are visible on the lateral part of the leg. A mechanical stimulator is placed below the tendon of the soleus.

unless they undergo a period of training. Painless restraint, however, can provide sufficient stability for studies on reflexes. The animals tolerate it well as these are only short periods of tonic activity alternating with long periods of rest. Tonic activity can be compared to spontaneous motor unit activity. Reflex responses can be studied during such spontaneous activity. Periods of rest are used for simple reflexological studies.

The stability of preparations under restraint is such that experiments can be continued for days without altering the conditions of stimulation, even in hyperbaric chambers [HUGON, CHOUTEAU, ROLL, BONNET and IMBERT, 1972]. Limited surgical operations can also be carried out, using general anaesthesia for induction, followed by local anaesthesia; for example, the popliteal fossa can be explored in this way.

A. Experiments under Normal Physiological Conditions

The animals (4–8 kg) are fixed in a restraining chair arranged in such a way that the lower limbs are restrained, but the upper limbs and head remain free (fig. 1). Setting up the animal in this way is usually carried out under intravenous 0.5 mg/kg Droperidol (Droleptan, LeBrun laboratories, Paris). The animals are then indifferent, akinetic and slightly hypertonic. These clinical effects wear off in 2–3 h. This delay is used for setting up stimulating and recording systems, and for providing further restraint.

1. Stimulation

Stimulation of the posterior tibial nerve as employed by Hoffmann is through two popliteal surface electrodes opposite two surface anodes placed on the anterior surface of thigh and leg. The two cathodes are linked through a potentiometer resistance of 2 kΩ. The two anodes are similarly linked. The slide of each potentiometer is linked to the appropriate terminal of an isolating transformer. This quadripolar system develops in the leg a stimulus field, the orientation of which can be modified by the position of the potentiometer slides without moving the electrodes. Stimulus duration was 0.5 msec and the shocks were delivered singly or in pairs at a frequency varying between 0.1 and 10/sec.

Stimulation of the sural nerve was through hypodermic needles placed close to the nerve, under the lateral malleolus. Stimulus duration was 0.5 msec and the stimulus was single, or in a train of 2–8 pulses at a frequency of 300/sec, the frequency of such trains varying between 0.1 and 1/sec.

Mechanical stimulation of the Achilles tendon was, as in man, by means of an electromagnetic hammer (Racia, Bordeaux) whose direction, distance and speed of excursion were all variable. Stimulus frequency was in most cases 0.25/sec.

The vibratory stimulus was derived from a Keydon (E2A) vibrator fitted with a 10 cm steel plate ending in a block which was applied to the tendon. A strain gauge placed on the steel plate detected the frequency and amplitude of vibrations. The absolute values of stretch imposed on the muscle were not known, but the stimulus employed was reproducible in terms of frequency, amplitude and duration.

2. Recording

The recording electrodes were either surface discs, or fine wires isolated except at their tips. Separation of these electrodes was of the order of 1 cm.

The muscles (peroneus, gastrocnemius, soleus and tibialis anterior) were identified by the usual morphological criteria. Implantation of wire electrodes was accomplished with a hypodermic needle.

3. Restraint

The stimulating and recording electrodes being maintained in place by elastic bands, the legs of the animal were fixed in the chosen position by means of plaster bandages. The knee and ankle joints were fixed, and the foot strapped to an articulated pedal. The general position of the animal is that of a 'biped' sitting with the legs half extended. Muscular responses observed were isometric contractions.

After the experiment, the animal was freed. Experiments could be repeated on a single animal at weekly intervals, without the development of any extrapyramidal or other disturbance.

B. Experiments under Surgical Conditions

General restraint of the animal, including the pelvic and shoulder girdles, but leaving free the forearms and head, is quite painless and results in a hypotonic wakeful preparation.

A short-acting general anaesthetic is given, such as Viadril G (Pfizer-Clin Labs, Paris) or thiopental sodium (Nesdonal Specia Labs, Paris). This is followed by local anaesthesia of the popliteal fossa, allowed dissection of the appropriate branches of the sciatic nerve. After the general anaesthetic had worn off, the preparation was ready for experimentation. Local anaesthesia of the edge of the wound was maintained, and the animal did not appear to suffer. At the end of the experiment, the animal was killed by deep intravenous anaesthesia.

Results

Electrical stimulation of the sciatic nerve in the popliteal fossa, and/or rapid stretch of the triceps surae, produces a monosynaptic reflex response in the soleus muscle (H-reflex and tendon reflex). Mechanical vibration (55/sec) applied to the tendon of triceps surae produces synchronous periodic EMG responses. Controlled stimulation of a cutaneous nerve such as the sural produces polysynaptic responses in the flexor muscles of the hind limb.

The characteristics of these responses are described and compared to the homologous responses observed in man.

H-reflex

A threshold electrical stimulus in the popliteal fossa evokes a response limited to the soleus muscle. A stronger stimulus evokes a direct response of shorter latency in the same muscle. These responses are homologous to the H-reflex and to the direct M response of soleus in man. Their characteristics are as follows.

1. The H-reflex (fig. 2A–F)
The H-reflex has a simple triphasic form, due to the differential recording of synchronous motor unit potentials in a volume conductor [see Hugon, this volume]. The latency, measured to the peak of the negative phase, is very stable in any one animal, and averages about 16 msec. Its duration, on average 6 msec, is constant in any given experimental situation. The potential amplitude is of the order of 1 mV.

2. The M Response (fig. 2G–I)
With limited stimulus strength, the M response has a triphasic form, the latency to the peak of the negative phase being 6 msec. It is closely comparable

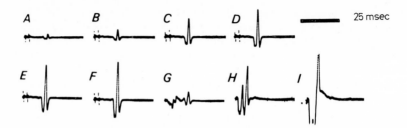

Fig. 2. H-reflex: recruitment. The effect of increasing strength of stimulus, applied in the region of the popliteal fossa on the amplitude of the H-reflex. Maximal amplitude is reached in *F. G,* Decrease in amplitude and *H* then disappearance follows, as the M response increases. In *H* and *I* the H-reflex has disappeared, and only the M response remains. This is duplicated because of the use of a double stimulus.

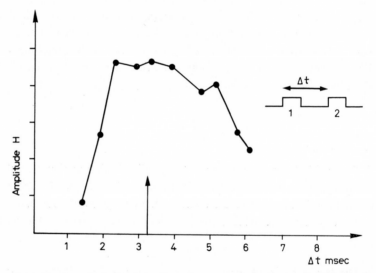

Fig. 3. H-reflex: the facilitating effect of a double stimulus with variable interstimulus delay. The H-reflex is facilitated by the use of a double stimulus if the interstimulus delay lies between 2 and 5 msec. Each point on the curve represents the mean of 15 responses. The delay of 3.2 msec (vertical arrow) has been chosen for eliciting the H-reflex by double stimulation. This curve corresponds to the phase of early facilitation of excitability (PAILLARD's phase II).

to the H-reflex evoked in the same muscle and recorded with the same system of electrodes.

When stimulus strength is increased, an M response is evoked which is stable, but more or less complex in form. The components of this response are

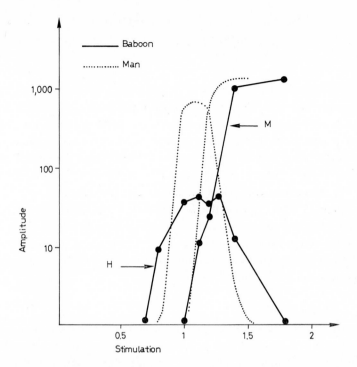

Fig. 4. H-reflex recruitment curve. *Abscissa:* stimulus strength. The threshold stimulus
for the M response has an arbitrary strength of 1. *Ordinate:* amplitude of the reflex and
direct motor response. The weak H-reflex in the baboon requires the use of a logarithmic
scale for amplitude. The maximal M values have been aligned. Note the similarity of the
different phases of recruitment between the two species. One important difference lies in the
different H maximum/M maximum ratio.

not identifiable without surgical intervention – certainly some arise from the
soleus muscle, but other components arise from nearby muscles, and the
activity is conducted in the volume of the limb to the soleus. There does not
seem to be any functional significance in these components of the M response
obtained with stronger stimuli.

3. Stimulation by Paired Shocks (fig. 3)

We have found that the baboon shows a relative hyporeflexia which only
rarely occurs in man. This hypo-excitability has been overcome by the use
of double stimuli (two shocks each of 0.5-msec duration, separated by 3.2
msec). This type of stimulation constantly evokes two M responses that are

more or less mixed together; the negative phase of the H-reflex can likewise present a composite aspect if the first stimulus was suprathreshold for the reflex. In other cases the double shock produces a single, non-bifid response.

4. Recruitment Curves (fig. 2 and 4)

Progressive increase in stimulus strength produces an H-reflex that grows, reaches a maximum, and then decreases as an M response is evoked. It is at first of simple and then more complex form. The maximal amplitude of the H-reflex is, if the muscle is as rest, at most 15% of the maximal amplitude of the M response of the soleus muscle. This maximal amplitude can be greater if the soleus muscle is active when the reflex is evoked.

Recruitment curves drawn at the time of surgical exposure (see Methods) show that stimulation of the nerve to soleus alone is sufficient to evoke an H-reflex in the soleus muscle. The influence of the nerves to the medial and lateral heads of gastrocnemius, to plantaris and the posterior tibial nerve on the recruitment curve is at present being studied.

5. Excitability Curves (fig. 5)

We have studied the amplitude of the soleus H-reflex evoked at varying intervals after a first shock which is itself suprathreshold for the H-reflex. Double stimuli were employed. They were identical, and arranged so that the first evoked a soleus response which was 50% of the maximal H-reflex. The repitition interval between pairs of shocks was at least 5 sec.

Figure 5 shows that the amplitude of the test H-reflex depends on the interval between it and the conditioning shock. The mean excitability curve (8 animals) shows that it is possible to distinguish 5 phases, following the nomenclature established by PAILLARD for man.

Phases I and II (0 to < 10 msec) are shown in figure 3. Phase II is that of early facilitation. The mechanism responsible for this effect is doubtless related to summation at motoneurone level, a phenomenon which we employ in the case of double shock stimulation.

Phase III (10 to < 80 msec) is a phase of total inhibition of the test response. It is followed by phase IV (80 to < 250 msec), a phase of restoration of the test response and of facilitation which leads to supernormal values. The return towards normal excitability takes place progressively during phase V (> 250 msec), without any marked late inhibition. Normal values are obtained at about 500 msec.

The facilitation observed with an interstimulus delay of about 120 msec corresponds to the phenomenon of facilitation obtained on repeated stimu-

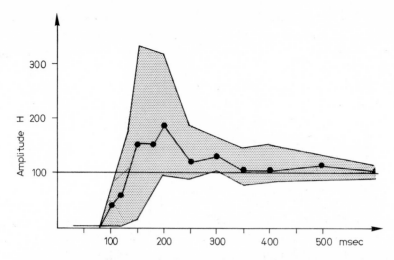

Fig. 5. H-reflex excitability curve. Effect of double stimulation on the H-reflex, recorded in triceps surae. The conditioning and test stimuli are electrical shocks which, by themselves, give identical reflex responses of about 50% maximal amplitude. *Ordinate:* amplitude of the conditioned response expressed as a proportion of the test response, which is given a value of unity. *Abscissa:* interstimulus delay. Black circles: mean excitability curve calculated for the results of 8 subjects. Shaded area: includes all experiment values for the 8 subjects.

lation with a periodicity of the same order – a frequency of 7–10/sec. This results in cumulative progressive facilitation, followed by stabilization at a high level of facilitation of monosynaptic responses. The same phenomenon has been described in man by PAILLARD [1955] under the name of 'frequency of stablization'.

Moreover, the H-reflexes do not show inhibition on repetition, as found in man and cat. This is in agreement with the absence of a period of late depression of the H-reflex, as noted above.

Curves of excitability using tendon stimulation as the conditionning stimulus have not yet been established.

6. Silent Period

A monosynaptic reflex evoked in a soleus muscle during a tonic contraction will produce a transitory inhibition of the tonic activity. There is a period of silence followed by a rebound of activity. The silent period (60 $\pm$ 20 msec) is longer if the tonic activity is weak, and the reflexogenic stimulus is strong. Moreover it should be noted that tonic activity considerably facilitates the H-reflex.

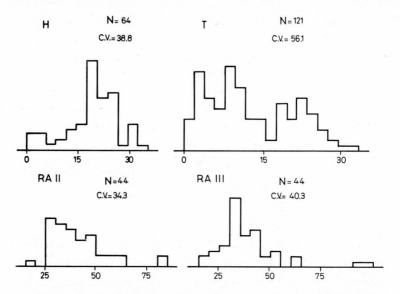

Fig. 6. Histogram of reflex response amplitude. *Abscissae:* amplitude of reflex response expressed in arbitrary units. *Ordinates;* number in each class. This figure gives an indication of the variability of reflex responses in the resting, conscious animal. The stimulus frequency is 0.1/sec. The T histograms for the three different types of reflexes (H, T, RA II, RA III) have been drawn from results obtained on different animals.

7. Variability of H-reflexes (fig. 6)

Variations in the vigilance of the animal and in the intercurrent tonic activity, introduce important variations into the results. Calm surroundings, the use of fixed electrodes and minimization of stress due to immobilization have considerably improved the stability of the reflexes so that it is now comparable to that observed in man. The M response is very stable.

Stretch Reflexes

A. Rapid Stretch

A brisk, though weak, blow to the Achilles tendor produces a tendon (T) reflex response in triceps surae, of fixed latency and duration. This response is, on the basis of analogous animal and human work, considered to be a monosynaptic reflex response due to phasic stretching of the primary spindle endings (I_A fibres).

The amplitude of this response increases with the speed of percussion which determines, other factors being equal, the speed of stretch of the spindles. Tendon percussion is less effective if the muscle is relaxed than if it is stretched. If the muscle is at rest, slightly stretched, the T-reflex is approximately equal to the H-reflex evoked in the same muscle. If the muscle is in slight tonic activity, the T response is considerably facilitated. A silent period follows the T-reflex that is little different to the silent period which follows an H-reflex. The variability of the T-reflexes does not seem to be very different from the variability found in man. It is greater than the variability of the H-reflex.

B. Slow Stretch

Slow stretch of the soleus muscle, and the constant lengthening which follows it, obtained by passive dorsiflexion of the foot, increases the discharge frequency of spontaneously active motor units.

There are two phases: (1) an initial period of considerable increase in frequency, which occurs at the same time as the movement, and which is a myotatic response to phasic stretch. The H- and T-reflexes evoked during this period are considerably increased, and (2) a later period of moderate tonic facilitation, which is a tonic myotatic response to static stretch.

A marked static stretch (30° out of a possible joint excursion of 50°) does not, in our experiments, produce an inhibition of spontaneous activity or of monosynaptic H-reflexes. (The alteration in T responses can be attributed to mechanical changes in tendon stimulation.)

Stretching the soleus muscle at a moderate speed (about 10° in 200 msec) produces a myotatic response en bouffée.

C. Vibratory Stimulus

Only one animal was examined using this stimulus. Low amplitude vibration at a frequency of 55/sec applied to the soleus muscle evoked a response made up of a succession of synchronous discharges, often triphasic, with a stable periodicity at the same frequency as the vibrator (fig. 7). This response began with the onset of vibration, and ceased with it.

Vibration inhibited the H-reflex by at least 50%, but the responses evoked after the cessation of vibration were facilitated. This post-vibratory potentiation lasted about 10 sec (fig. 8).

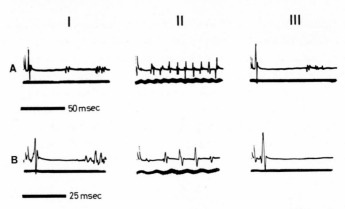

Fig. 7. Effect of vibration (55/sec), applied to the Achilles tendon, on the H-reflex. The records of line B are similar to those of line A with the exception of the time scale. *I.* H-reflexes evoked in the soleus muscle; no vibration. *II.* Vibration strongly inhibits the H-reflex. There is a succession of synchronous discharges in the soleus muscle at the same frequency as the vibration (tonic vibration reflex): *III.* When vibration stops the H-reflex appears, and it remains of higher amplitude than before for about 10 sec.

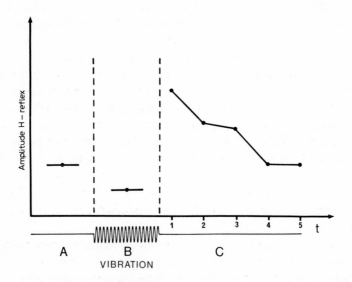

Fig. 8. Inhibition of H-reflex during vibration, and postvibratory facilitation. *Abscissa:* succession of experimental sequences *A, B,* and *C* as below. *Ordinate:* Amplitude of reflex responses expressed in arbitrary units. *A* Control. The point drawn is the mean of 20 results. *B* Period of vibration (55/sec). The point drawn is the mean of 20 responses during 10 sequences of vibration. *C* The amplitude of 5 responses following vibration. The stimulus frequency was 0.25/sec. Each point represents the mean of 10 responses.

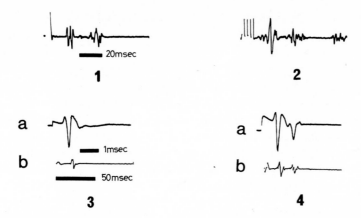

Fig.9. Polysynaptic reflexes of cutaneous origin. *1.* Early (RA II) and late (RA III) reflex responses recorded from the peronei following a sural nerve stimulus. *2.* Responses evoked during background tonic activity. Note the periods of silence following the RA II and III responses. *3.* Limited sural nerve stimulation, sufficient to excite group II fibres (neurogram *a*), evokes an RA II reflex response *(b)*. *4.* Increasing stimulus strength excites group II and group III fibres (neurogram *a*). Note the appearance of a later reflex response *(b)*.

Reflex Responses of Exteroceptive Origin

Stimulation of the sural nerve below and behind the malleolus evokes polysynaptic reflexes in the peroneal muscles. We have described two types of responses which differ in latency and in the conditions which evoke them.

In work carried out in man [HUGON, 1967] it is possible to distinguish, following a train of stimuli (1–7 shocks, duration 0.5 msec, frequency 300/sec):

1. A type RA II response, latency 25–30 msec, due to afferent activity in cutaneous fibres of low threshold (group α–β, or group II of LLOYD).

2. A type RA III response, latency 45–50 msec due to afferent activity in high threshold cutaneous fibres, some of which are nociceptive (group III of LLOYD).

RA II and RA III are polysynaptic reflexes of fluctuating latency and duration. These responses have been related to the waves of the sural neurogram recorded at operation (see Techniques). This is a direct confirmation of the interpretations of normal physiology [HUGON, 1967] and of physiopathology [COLLINS, NULSON and RANDT, 1960] (fig. 9, 3–4).

When polysynaptic reflexes are evoked in the resting peroneal muscles the RA III response develops first. When the intensity of sural nerve stimulation is increased, the earlier response appears. This seemingly paradoxical observation would be explained if strong sural stimulation can affect spinal interneurones so that a reflexogenic path can be opened which was previously closed [Hugon, 1967].

Spontaneous tonic motor activity in the muscle accessible to the efferent arc of the reflex will augment the amplitude of the RA II response, and produce a period of EMG silence. The facilitatory effect for the RA III response is much weaker and is probably masked by the silent period which follows R AII.

Extinction of reflexes by repetition of stimulation at low frequencies has been observed (0.2–0.3/sec). Extinction affects RA II very markedly. The presence of RA II expresses, in the baboon as in man, only the effect of excitatory factors outweighing inhibitory factors.

Discussion

The restrained baboon demonstrates proprioceptive reflexes (response to vibration and stretch, and T- and H-reflexes), and cutaneous exteroceptive reflexes, which can be compared to those of man to establish the validity of the model.

A. The H-Reflex

The H-reflex is closely comparable in man and baboon; the strict synchronization of motor unit potentials leads to a simple and stable EMG response, and to constancy of latency and duration. Not all the causes of the synchronization are as yet determined. The ratio between maximal H- and M response is 4–5 times smaller than the equivalent ratio for man.

The I_A afferent pathway seems to have less reflexogenic power in the baboon, and this is also indicated by the fact that a single shock is not always sufficient to evoke a reflex response. This difficulty has always been overcome by the use of paired stimuli, but this technique does introduce some physiological complications which cannot be ignored: presynaptic excitation is doubled, and excitation of α fibres, with the I_A fibres leads to doubling of the activity of recurrent circuits. Paired stimuli with short inter-stimulus duration does facilitate the H-reflex in man [Paillard, 1955]. This may be attributed to the effects of central summation of primary I_A afferents. The comparison between baboon and man can therefore be extended to include mechanisms of reflex activation of motoneurones. The similarity of the recruitment and excitability curves also confirms the homologous nature of these responses in the two species.

1. Recruitment Curve

The general shape of the recruitment curve is identical for the two species. The most characteristic point is the limitation of the maximal H-reflex compared to the maximal M response, in resting conditions. Only a small part (about 10% of the soleus motoneurone pool) can be activated by the proprioceptive reflex path. This could depend as much on anatomical factors (e. g., fibre diameter) as on physiological factors (balance between excitation and inhibition). That is one question open to experimentation which could reconcile the question as to why H-reflexes, with the exception of the soleus muscle, are so weak in man.

2. Excitability Curve

There is certainly a general similarity between excitability curves in the baboon and in man. We have already indicated the existence of an early phase of facilitation in both (phase II). Phase III is comparable in its latency and duration, although the level is less in the baboon.

This phase is followed by a marked rebound of excitation in the baboon, a phase which is less marked in man (phase IV). This sequence is identical to that observed during tonic activity, following a monosynaptic reflex: in both species there is a silent period, and then a rebound. The mechanism must be strictly comparable, although the post-inhibitory rebound is much better developed in the baboon. The study of these mechanisms is under way.

Phase V, in man, is marked by a long-lasting depression of reflexes. In the monkey, this phase is less clear-cut, either because it is truly absent, or because it is masked by the effects of a prolonged rebound. Phase IV has been interpreted as a late manifestation of presynaptic inhibition on the I_A afferent terminals [DELWAIDE, 1970]. Its weakness in the baboon is an indication of the lack of presynaptic proprioceptive inhibition in the monkey. On this hypothesis one could explain the increased myotatic reflex of the monkey, though the T reflex remains moderate.

B. Stretch Reflexes

The T reflexes are weak in the baboon when compared to the maximal M response, while the reflex response to slow stretch is well developed. This paradox is an argument in favour of a general weakness of monosynaptic reflexes, an expression of a hypoexcitability of the α-motoneurons, since the existence of dynamic myotatic responses suggest the presence of elevated afferent spindle activity. From this point of view, the general level of tendon reflexes in man is at least as high as in the baboon. We have not demonstrated that the myotatic reflex in the baboon is of spinal origin. A supraspinal factor could be involved, but not manifest in man, who can remain immobile on command in a state of deliberate inhibition.

C. Responses to Vibration

A vibrating stimulus to the tendon excites an EMG response which is, in the first approximation, similar to the tonic vibration reflex described in man. Common features include the inhibition of monosynaptic reflexes (H) the absence of response outlasting the stimulus, and a post vibratory potentiation of reflexes. The underlying mechanisms are probably the same in the monkey as in man.

D. Exteroceptive Polysynaptic Reflexes

Those reflexes obtained on stimulation of the sural nerve are in general similar to those described in man [HUGON, 1967]. Threshold stimulation produces contraction in the peronei. The magnitude and variability of these reflexes are greater in the baboon. In particular the RA II response is very sensitive to variations in the state of peroneal muscular contraction. Habituation is similar in the two species.

Summary

The conscious, restrained baboon *(Papio papio)* shows, in response to adequate stimulation, vibration, myotatic, tendon and Hoffmann reflexes, as well as polysynaptic exteroceptive reflexes, which are homologous to those obtained in normal man.

The proprioceptive reflexes are all found in the soleus muscle, and the exteroceptive reflexes first develop, in response to sural stimulation, in the peroneal muscles.

Monosynaptic reflexes efferent to the soleus muscle are relatively weak, whereas polysynaptic reflexes efferent to the peronei are relatively strong.

Proprioceptive and exteroceptive reflexes fluctuate spontaneously, principally because of altering supraspinal factors. In conclusion, the baboon is a satisfactory model for the study of reflexes in man.

Authors' address: Dr. J. P. ROLL, Dr. M. BONNET, and Prof. M. HUGON, Laboratoire de Neurophysiologie, Faculté des Sciences St-Jérôme, Traverse de la Barasse, *F-13 Marseille 13e* (France)

New Developments in Electromyography and Clinical Neurophysiology,
edited by J. E. Desmedt, vol. 3, pp. 209–224 (Karger, Basel 1973)

A Re-Evaluation of Cerebellar Function in Man

J. C. ECCLES

Department of Physiology, School of Medicine, State University of New York,
Buffalo, N.Y.

Clinical Introduction

Clinical observations on patients with cerebellar lesions and the early experimental investigations using such techniques as ablation and stimulation have led to the general conclusion that the cerebellum is essentially concerned in posture and movement [HERRICK, 1924; HOLMES, 1939; DOW and MORUZZI, 1958]. It is assumed that in the cerebellum there is integration and organization of the information flowing into it along the various neural pathways and that the consequent cerebellar output either goes down the spinal cord to the motoneurones (and so participates directly in the control of movement) or else is returned via the various relay nuclei to the cerebral cortex, there to modify the control of movement from these higher centers.

Before giving a summary of recent experimental studies on cerebellar structure and function I have thought it expedient to survey briefly the symptoms and signs of cerebellar disease that have been disclosed by refined studies on human patients who have had localized lesions of the cerebellum by penetrating gunshot wounds in the first world war. GORDON HOLMES studied these patients, and his report on disorders of movement produced by cerebellar lesions is still by far the best in the clinical literature [HOLMES, 1922, 1939]. The most striking investigations of HOLMES were on subjects that had a sharply defined lesion of the cerebellum on one side only. This gave the opportunity that he fully utilized to compare the movements on the normal side with the movements on the affected side. Another advantage of HOLMES' work was that he developed remarkable clinical recording techniques. HOLMES classified the disorders arising from cerebellar lesions under 4 main headings.

1. Firstly, he described hypotonia or diminished postural tone of muscle. As a consequence one has the flail-like movements of limbs, pendular knee jerks, etc. HOLMES summarizes this by stating that cerebellar lesions give diminution of postural tone 'which alone, or by reinforcing voluntary effort, tends to fix each part of the body in the attitude which it occupies, and they show that volitional contraction of the muscles alone is not adequate to maintain posture'. The postural tone is equally important in movement which, as HUGHLINGS JACKSON states, can be regarded as a succession of postures.

2. Secondly, HOLMES reported that asthenia and muscle fatigue were a major symptom of cerebellar lesions. He used special dynamometers to measure contractions of a wide range of limb muscles. The power on the affected side may be reduced to as low as half of the control. As I examine critically this second major subdivision of cerebellar disabilities, I am not convinced that it can effectively be distinguished from the hypotonia group of symptoms.

3. However, there can be no doubt about the great importance of the third class that HOLMES describes, namely abnormalities in the rate, regularity and force of voluntary movements. The accounts he gives of his investigations lead to most fruitful considerations in the light of our present knowledge and also provide guidance for further laboratory investigations. It is important to reproduce some of HOLMES' excellent illustrations. HOLMES developed a very simple but effective technique in studying the disorders of movement and obtaining simple photographic pictures where it was possible to compare the normal with the disordered movement.

In figure 1 the subject was asked to move his arm so that a finger pointed accurately and in smooth sequence to a series of red spots arranged in two vertical rows of 3 each. On the tip of his finger there was a small bulb giving flashes of light at 25/sec. This flashing technique ensured a photographic record of both the course and the speed of movement. On the normal side in A the subject moved his finger from point to point fairly directly and slowed up smoothly in approaching the target then accelerated to the next target and so on. A quite different record is shown in B on the side of the cerebellar lesion, where the movements were irregular in direction and showed evidence at each of the turning points of inaccuracy and indecision and tremor. This is well illustrated by the plotted measurements below. I do not know of a more effective display of the essential disorders of movement in a cerebellar lesion.

HOLMES gave many other impressive illustrations. I select a few examples. For example (fig. 2) the subject had to move his finger from spot A above his

Normal and dysmetric movements

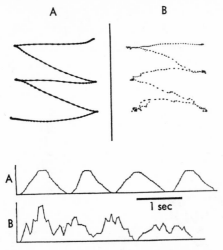

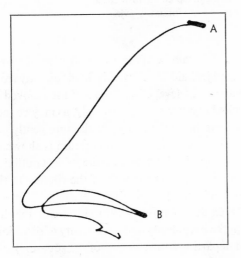

Fig. 1. A and *B* are records obtained as described in text. The range of movement was about 75 cm. The plotted curves below are calculated from the traces above to show movements on normal and abnormal sides in *A* and *B*, respectively. The range of movement is plotted vertically and there is 0.04 sec in the horizontal scale between the dots. In contrast to the normal side *A*, there is in *B* gross irregularity and failure to reach target dots in the later 3 movements [HOLMES, 1939].

Fig. 2. Movement of finger from *A* above head to nose at *B* in subject with cerebellar lesion as described in text [HOLMES, 1939].

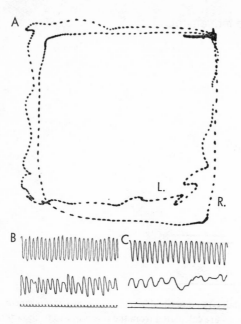

Fig. 3. In *A* the subject with a left cerebellar lesion attempts to outline by finger of outstretched arm a square on the wall of the room, there being as in figure 1 a flashing light at 25/sec on his finger tip. In *B* and *C* there are tracings of the alternating pronation and supination of the arm on the normal (upper) and on side of cerebellar lesion (lower). In *C* there is a faster rate of alternation with a corresponding further disorder of movement on the affected side. Time in seconds for *B* and *C* [HOLMES, 1939].

head to his nose at B. Here we have an example of the decomposition of movement described by HOLMES. The subject did not move his finger directly from A to B, but firstly lowered it to about the level of his nose and then moved in towards the nose in an uncontrolled manner hitting it sharply, as may be seen by the rapid rebound and the uncertain attempt to go back more gently but with disorder. Another striking example of HOLMES' illustration is shown in figure 3A where the light flash technique is also used with the subject outlining with his finger a square which is projected on the wall of the room. On the normal right side there is a reasonably good trace of a square with again the slowing up at the corners and then acceleration. With the left side the disordered movement is apparent in the irregularity and uncertainty of direction, particularly in approaching and departing from a corner of the square. B and C illustrate the disorder of movement in alternating pronation and supination (adiadochokinesis).

The type of cerebellar disability illustrated in figures 2 and 3 is often called hypermetria because of the overshooting of the movement with respect to the target. Alternatively it is called cerebellar ataxia. The essential feature is the failure of smooth control. There is delay in the initiation and cessation of movement, and tremor, particularly at the start and end of the movement. HOLMES rightly concluded that this was a disorder in the innervation of the prime movers and could not be attributed merely to hypotonia. The decomposition of movement indicates the same kind of disorder of smooth control. The patient evidently cannot carry out a movement of complex integration, but has to decompose the movement into action at one joint after another in order to bring about the desired position of the limb. HOLMES quotes from one of his intelligent observant patients who said: 'The movements of my left arm are done subconsciously, but I have to think out each movement of the right (affected) arm. I come to a dead stop in turning and have to think before I start again.' I suggest that this is one of the most penetrating statements that can be made with respect to the disability produced by a cerebellar lesion.

One of the key points of my subsequent discussion will be to try to give an account of how the cerebellum normally operates in order to avoid the intense voluntary concentration that is required to carry out movements in the absence of cerebellar control. Normally our most complex movements are carried out subconsciously and with consummate skill. We do not have any appreciation of the complexity of muscular contractions and joint movements. All that we are voluntarily conscious of is a general directive given by what we may call our voluntary command system as, for example, when we write our signature. All the finesse and skill seems naturally and automatically to flow from that. It will be my thesis that the cerebellum is concerned in all of this enormously complex organization and control of movement and that throughout life, particularly in the earlier years, we are engaged in an incessant teaching program for the cerebellum so that it can carry out all of these remarkable tasks that we set it in the whole repertoire of our skilled movements—in games, in techniques, in musical performance, in speech, dance and song.

4. The fourth category of HOLMES concerns disorders of associated and automatic movements. However, I think that these are not concerned essentially with the cerebellum but with the whole brain stem neuronal machinery.

It may be thought that the disorders of gait have been overlooked in this classification, but I agree with HOLMES that these disorders, remarkable as they are, are only special examples of the disorders of movements which come into classes 1 and 3 particularly.

An important negative finding of the study of cerebellar lesions is that the cerebellum is not in any way concerned in conscious perception, or indeed in any conscious experience [HOLMES, 1939]. PENFIELD [personal communication] has found that electrical stimulation of the cerebellum gives no sensation to the conscious patient.

In summarizing this clinical introduction, the principal disorder in human cerebellar disease undoubtedly is concerned with the fine control of movement, but clinical testing is quite inadequate to reveal the finer levels of this disability. For example, HORNABROOK [1968, and personal communication] reports that patients often come with the report of incipient Kuru, a cerebellar degenerative disease, some weeks before clinical tests can reveal the onset of the characteristic motor disabilities. They have detected a developing failure of their fine finger movement in string bag construction!

Cerebellar Anatomy and Physiology

It is, of course, well recognized that there are enormously powerful lines of communication in the circuit from the cerebral cortex, particularly the motor cortex to the cerebellum and back to the motor cortex. It will be my task to give a brief account of the anatomy and physiology of this operational circuit and to show how disorders of the circuit give rise to the principal symptoms and signs of cerebellar lesions. In addition, there are the older cerebellar pathways from the vermis to the spinal cord and from the spinal cord to the vermis. In recent years there has been intensive investigation of these simpler circuits [BRODAL, 1967; ECCLES, ITO and SZENTÁGOTHAI, 1967; POMPEIANO, 1967] and they provide a model for our attempts to understand the more complex cerebral circuits.

Localized lesions of the cerebellar cortex and of the nuclei [CHAMBERS and SPRAGUE, 1955a, b] have revealed that the vermis of both anterior and posterior lobes and the associated fastigial nuclei are concerned with posture and general movements of the animal as a whole. In contrast, the pars intermedia has been found to be much more specifically organized—to forelimb or to hindlimb and even to particular movements. Finally, the lateral hemispheres are closely inter-related with the motor cortices of the contralateral side and with the pyramidal tracts stemming therefrom. This work of CHAMBERS and SPRAGUE forms a sound basis for modern attempts to understand cerebellar function.

Figure 4A gives a simplified diagram of the input and output lines from an element of the cerebellar cortex and illustrates the integrative interactions

Diagrams of pathways to and from cerebellum

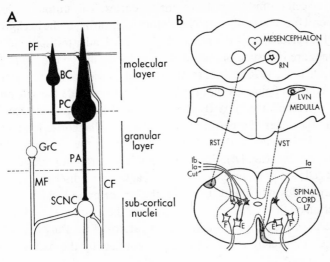

Fig. 4. A Diagram showing elements of principal neuronal connections in the cerebellum. MF is mossy fibre giving collateral to the subcortical nuclear cell, SCNC, and then making an excitatory synapse on a granule cell, GrC. The axon of GrC bifurcates in the molecular layer to form a parallel fibre, PF, that gives excitatory synapses on spines of a basket cell, BC, and a Purkyně cell, PC. BC gives an inhibitory synapse to the soma of PC, while the axon, PA, of PC gives an inhibitory synapse to SCNC. The incoming climbing fibre (CF) also gives a collateral to SCNC and then goes on to make a powerful excitatory synapse on PC. *B* Diagram showing descending pathways operated from the cerebellum (cf. fig. 5 and 6). RN is red nucleus with decussating axon of rubrospinal tract (RST) to make synapses on interneurones to flexor (F) and extensor (E) motoneurones. The interneurones are also shown acted on by primary afferents, cutaneous (Cut), and muscle (Ia and Ib). LVN is Deiters' nucleus with vestibulo-spinal tract, VST, acting on E motoneurone directly and on F motoneurone via an interneurone.

therein [cf. ECCLES, Ito and SZENTÁGOTHAI, 1967]. The climbing fibre (CF) is seen to give an axon collateral to a subcortical nuclear cell (SCNC) of the cerebellar nuclei or of Deiters' nucleus, and to go on to excite powerfully the Purkyně cell (PC). The mossy fibre (MF) also gives an excitatory collateral to the SCNC and goes on to excite the granule cell (GrC) whose axon bifurcates in a T-junction to form the parallel fibre (PF) that excites both the basket cell (BC) and the Purkyně cell, which latter inhibits the SCNC.

It can be postulated that the synaptic excitatory action on the subcortical nuclei by the axon collaterals of the climbing and mossy fibres is important in providing a background excitation which is subjected to the inhibitory action of the Purkyně cell output from the cerebellar cortex. This output achieves a spatio-temporal form by applying an inhibitory sculpturing to the background excitation of the neurones of the subcortical nuclei. Under testing conditions these neurones are in a state of incessant irregular discharge usually at 10 to 50/sec [THACH, 1968, 1970a; ECCLES, SABAH and TÁBOŘÍKOVÁ, 1972]. It will be appreciated from figure 4A that the Purkyně cells provide the sole output line from the cerebellar cortex. By the patterns of impulse discharge from Purkyně cells, the integrative performance of the cerebellar cortex is impressed on the subcortical nuclear cells, which in turn operate the descending pathways to motoneurones (fig. 4B). Furthermore, it is important to recognize that the Purkyně cells are in an incessant state of background discharge, which is modulated up or down by inputs to the cerebellar cortex and which is inhibitory in its action on the SCNCs. As a consequence these latter cells relay down their axons a negative image of the temporal pattern of the Purkyně cell discharges, which eventually reaches the motoneurones by the pathways diagrammed in figure 4B.

There has now been an intensive investigation of the mosaic patterns formed by the subsets of input from nerves and receptors of the forelimb and hindlimb to the anterior lobe of the cerebellum. In the earlier investigations the potential fields in the cerebellar cortex were used in mapping the distribution of inputs from limb nerves, but with computer averaging techniques it is now possible to obtain very reliable information on the inputs to an individual Purkyně cell, as disclosed by the modulation of its background firing frequency [ECCLES, FABER, MURPHY, SABAH and TÁBOŘÍKOVÁ, 1971a, b]. Furthermore, nerve stimulation has been replaced by precisely controlled adequate stimulation of skin or muscle receptors [ECCLES, SABAH, SCHMIDT and TÁBOŘÍKOVÁ, 1971]. For the present it is sufficient to report that there is an unexpected degree of individuality of Purkyně cells with respect to the uniqueness of afferent inputs by both mossy and climbing fibres. When testing with volleys in a wide range of afferent nerves or with many types of cutaneous stimulation, there may be similar responses in closely adjacent cells, but these small homogenous patches blend into others characterized by quite different responses. In summary it can be stated that, as defined by the responses to the range of testing inputs, Purkyně cells are found to belong to a great many subsets that have an ill-defined patchy character in their topographic relationships in the cerebellar cortex. Not only are Purkyně cells specific in their innervation pattern from the mechanoreceptors, but it has also been found that Purkyně cells of like function tend to innervate the same colonies of subcortical nuclear cells (cf. fig. 4). In general, we have developed the concept that Purkyně cells tend to be arranged in multitudes of colonies, each colony comprising cells of related integrative function. This colonial pattern is preserved in the distribution of Purkyně cell axons, which is evident by the discharge patterns of the SCNC in respect of specific cutaneous inputs [ECCLES, SABAH and TÁBOŘÍKOVÁ, 1972]. Only in the light of this concept is it possible to explain how the cerebellum carries out its vital function of control of movement.

The Cerebellum and the Dynamic Loop Control of Movement [ECCLES, 1967, 1969]

A distinction can be made between the relatively simple approach to the processing of information in a small area of the cerebellar cortex and a more comprehensive examination of the cerebellar performance as a whole. The simple approach that is usually adopted serves quite well in a limited manner, but it fails to recognize the amazing complexity of the task of the cerebellum in integrating information from an immense array of receptors of widely different modalities (cf. fig. 5). It is not sufficient for this purpose to restrict the examination to a small element of the cerebellar cortex. Rather a more global concept of cerebellar action is needed. For this purpose two diagrams will now be presented, and later there will be an account of the way in which they aid in explaining how disorders of movement occur in cerebellar lesions.

Loops to and from the Spinal Cord

Figure 5 gives a general diagram of input and output pathways to the vermis of the cerebellar cortex. No distinction is made between the different spatial distributions of the CF and MF inputs to the cerebellar cortex, which are shown diagrammatically along with the efferent pathways from the cerebellar vermis to the spinal cord. Figure 5 represents in a very simple form the operational loop in the control of posture and movement by the cerebellar vermis. It has been shown that there is a transfer of information from the Purkyně cells of the cerebellar cortex to the spinal motor centers via the pathways of figure 5, as shown in part in figure 4B.

In summarizing the postulates diagrammed in figure 5, it can be stated that there is a dynamic control of an evolving movement by feedback loops up to the cerebellar cortex. It is important to recognize that in the resting state

Dynamic loops through cerebellar vermis

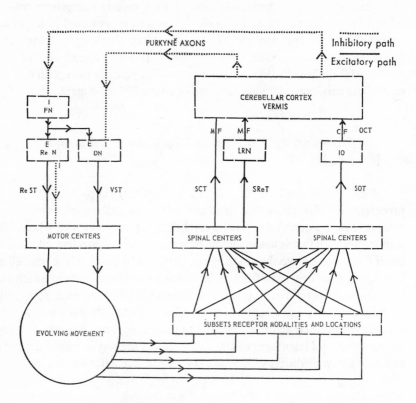

Fig. 5. Diagram showing in detail the pathways involved in the cerebello-spinal circuits, the continuous and the dotted lines showing respectively the excitatory and inhibitory neural pathways. Spinoolivary tract, SOT, to inferior olive, IO, with climbing fibres, CF, in olivocerebellar tract, OCT; MF, mossy fibres of spinocerebellar tracts, SCT; FN, nucleus fastigius; ReN, reticular nucleus with reticulo-spinal tract, ReST; DN, Deiters' nucleus with vestibulo-spinal tract, VST; LRN, lateral reticular nucleus. The MF and CF input lines are shown in figure 4A, and also the output from Purkyně axons to the subcortical nuclear cells, FN and DN. The pathway (VST) from DN to the motor centres is shown in figure 4B.

of a posture before a movement begins, there is a background discharge of impulses along all the components of these loops, and that the evolving movement merely heightens or lowers or even silences these discharges [cf. THACH, 1968, 1970a, b]. Furthermore, there is a continuous ongoing operation of these loops, not merely a staccato control with a brief burst of efferent discharges, then a pause for the return messages, then a revised efferent output, and so on. Nevertheless, consideration of the operational sequences in an imaginary isolated loop is essential for the understanding of the temporal effectiveness of the postulated dynamic loop control of movement.

Cerebro-Cerebellar Pathways

In recent years there have been intensive investigations of the neuronal connections between the motor cortex and the cerebellum, particularly by the Japanese school: ITO, TAKAHASHI, OSHIMA, TSUKAHARA and their colleagues [cf. ECCLES, ITO and SZENTÁGOTHAI, 1967, chap. 13, 14; EVARTS and THACH, 1969]. This work has been remarkable for the consistency and precision with which pathways have been defined with the fine details of synaptic connectivity. Figure 6 shows diagrammatically the principal features of the present state of our understanding. The pathways are shown in an elemental manner, with a minimum of complication by lines in parallel and with neglect of most of the neuronal machinery in the cerebral and cerebellar cortices. In order to present a model that can be easily described and understood, an economy of representation has been deemed essential. Several nuclei and their well established lines of communication have been neglected.

The axons of the two large (L.PYR.C.) and the two small (S.PYR.C.) cortical pyramidal cells form respectively the large (LPTF) and small (SPTF) pyramidal tract fibres. LPTF and SPTF both give collaterals to the immensely numerous cells in the nuclei pontis (NP) whose axons in turn form the mossy fibres, M.F., and give collaterals to the intracerebellar nuclei, interpositus, IP, and dentatus, DE (cf. fig. 4A). Meanwhile the SPTF gives branches to the inferior olivary nucleus [I.O.] whose axons give off collaterals to IP and DE and then after branching several times terminate as climbing fibres (C.F.). The cerebello-cerebral pathway is from Purkyně cells of the pars intermedia and cerebellar hemispheres that inhibit the neurones of IP and DE, respectively (cf. fig. 4A). From DE the excitatory path to the large pyramidal cell is via a synaptic relay in the ventro-lateral nucleus of the thalamus (VL THAL). From IP it is more complicated. The fibres give off collaterals to the red nucleus (RN), and so secure a pathway directly to the spinal cord via the rubrospinal tract (RST) (cf. fig. 4B). In addition there are inhibitory interneurones in the VL thalamus, and the axons from VL thalamus excite also small pyramidal cells via interneurones in the cerebral cortex.

CEREBRO-CEREBELLAR PATHWAYS

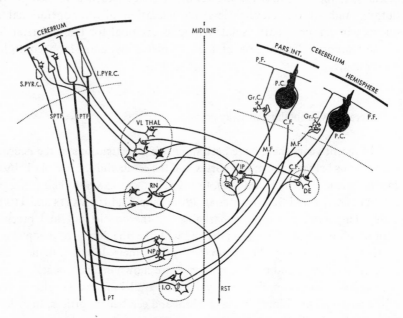

Fig.6. Diagram of cerebro-cerebellar pathways showing the reciprocal connections between cerebrum and cerebellum together with the synaptic relays in the various nuclei. All of these pathways have now been established by anatomical and electrophysiological investigations. Further description and key to abbreviations is given in the text. The pathway from the red nucleus (RN) down the RST of the opposite side to spinal motoneurones is shown in figure 4B.

In general, the MF input has a wide origin from the cerebral cortex, as shown by anatomic investigations, and is distributed widely to the cerebellar hemisphere of the opposite side. However, it should be recognized that the physiological investigations of these pathways are still at a very early stage [PROVINI, REDMAN and STRATA, 1967; EVARTS and THACH, 1969; TSUKAHARA, 1972]. The neuronal pathways of figure 6 are designed so that the cerebellum receives precise and immediate information of the discharges down the pyramidal tract (PT). As a consequence it can return appropriate corrective information to the cerebral motor areas before the PT discharge has caused any muscle contraction. We may regard the cerebellum as in this way giving to the cerebrum an ongoing comment on the pattern of its PT discharges. Considerably later there will be the feedback to the cerebrum along the spinocerebral pathways of information from skin, joints and muscle that results from the muscle contractions evoked by the PT discharge. This pathway is fairly expeditious, being via fast conducting fibres of the dorsal columns or the spino-cervico-thalamic tract, and it acts monosynaptically on PT cells as well as by more delayed and integrated cortical pathways. Nevertheless, the total loop time to the periphery and back would be perhaps 100 msec for the human forearm.

The cerebellum receives information from the evolving movement only when this causes changes in the PT discharge with a consequent input to the cerebellum via the collaterals (cf. fig. 6). Moreover, it must be appreciated that the cerebellum is continually modulating the PT discharge by virtue of its computation from the input that it receives from the PT collaterals (cf. fig. 6). Thus we can envisage that all PT discharges are provisional and are unceasingly subject to revision both by the feedback from the periphery as a consequence of the evolving movement, and more expeditiously via the cerebro-cerebellar loop. Presumably the cerebellum carries an immense store of information coded in its specific neuronal connectivities so that, in response to any pattern of PT input, computation by the integrational machinery of the cerebellum leads to an output to the cerebrum that appropriately corrects its PT discharge. Corrective information via this dynamic loop operation must be continuously provided during all movements deriving from pyramidal tract action. Figure 6 is, of course, greatly oversimplified. It may be presumed that there are many other dynamic loops forming complex interacting patterns with it and involving such structures as the caudate nucleus, the globus pallidus, the substantia nigra and the nucleus ruber parvicellularis.

Experimental Investigations on Cerebellar Control of Movement

An important investigation on cerebellar function has been performed on primates by EVARTS and THACH [1969], and specially relates to the model of figure 6. A monkey was trained to make movements of a rod to one side and the other in a repeated reciprocal manner in a cycle of 400–700 msec (fig. 7A). Figure 7B shows that the background frequency of certain pyramidal cells of the motor cortex was modulated in phase with the movements and at a timing that would be expected for those cells controlling the movement by firing down the pyramidal tract. In other monkeys the spike discharges of Purkyně cells of the contralateral hemisphere or pars intermedia were found to be also modulated in phase with the movement (fig. 7C). Evidently this is exactly in accord with the performance to be expected for the model of figure 6,

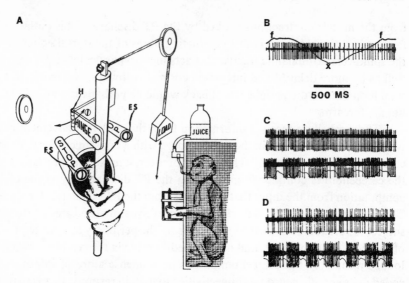

Fig. 7. A This figure shows the monkey's left hand protruding from a tube in a lucite panel attached to the front of the home cage. In order to receive a fruit juice reward, the monkey is required to grasp the vertical rod attached to a hinge and to move it back and forth from one stop to the other. The stops are labelled FS (flexor stop) and ES (extensor stop). The monkey is required to contact the flexor stop and then move the handle through the arc between the stops until the extensor stop is reached. If the period between breaking contact with the flexor stop and making contact with the extensor stop is between 400 and 700 msec, and if the previous movement in the other direction also falls within these time limits, the solenoid valve is automatically operated and a reward is delivered. *B* shows the activity of a pair of motor pyramidal cells whose discharge patterns were positively correlated during the flexor-extensor movement [EVARTS, 1967]. *C* Upper trace shows Purkyně cell background discharge with 3 climbing fibre responses indicated by dots. In lower trace the Purkyně cell discharge is positively correlated with the movement as indicated. *D* Upper trace is background discharge of a cerebellar nuclear cell and, in lower trace, this discharge is seen to be positively correlated with movement [THACH, 1968].

where there are powerful reciprocal connectivities between the motor cortex and the contralateral cerebellar cortex. Furthermore, as shown in figure 7D the cyclic modulation was also exhibited by certain cells of the cerebellar nuclei as indicated in IP and DE of figure 6. The principal discovery made by these experiments is the close locking together of cerebral motor cortex and associated cerebellar nuclei.

These continuous movements to and fro do not allow any evaluation of the timing of Purkyně cell discharge in relation to a single movement. In further studies, THACH [1970a, b] has shown that Purkyně cell discharge often precedes the first onset of the movement, and is affected in various ways, up or down or in more complex sequences, during a simple movement. It must be recognized that discharges down the pyramidal tract may initially be below threshold for eliciting movement [PHILLIPS, 1966], but could nevertheless affect Purkyně cells via the collaterals as diagrammed in figure 6.

It is implicit in the cerebellar performance as outlined above, that the cerebellum is not a fixed computing device with a connectivity solely determined by a genetically controlled development. It is postulated that at the microstructural level neuronal connections are developed in relationship to learned skills. We have to envisage that the cerebellum plays a major role in the performance of all skilled actions and hence that it can learn from experience so that its performance to any given input is conditioned by this 'remembered experience'. As yet, of course, we have no knowledge of the structural and functional changes that form the basis of this learned response. However, one can speculate that the spine synapses on the dendrites in the molecular layer (fig. 4A) are especially concerned in this and that usage gives growth of the spines and particularly the formation of the secondary spines that HÁMORI and SZENTÁGOTHAI [1964] described on Purkyně dendrites. One can, therefore, imagine that in the learning of movements and skills there is the microgrowth of such structures giving increased synaptic function, and that as a consequence the cerebellum is able to compute in an especially adapted way for each particular learned movement and thus can provide appropriate corrective information that keeps the movement on target.

Movement Disorders Resulting from Cerebellar Lesions

It is possible to provide a general explanation of these disorders on the basis of the concepts of cerebellar function outlined above. Firstly, there is an intense impulse traffic between the cerebrum and cerebellum in both directions (cf. fig. 6). Secondly the impulse coding of discharges down the pyramidal tract is signalled with precision and speed from the collaterals through the pontine and inferior olivary nuclei to the opposite cerebellar hemisphere, there to achieve a limited and restricted integration in the cerebellar cortex and to reappear in the coded discharges from the 10 million Purkyně cells in that hemisphere. Thirdly Purkyně cell discharges return to the cerebral cortex via its transmutation as a negative image because of the inhibitory action on the cells of the nucleus dentatus, whose discharges act on the ventro-lateral thalamus that in turn projects to the pyramidal cells in or near the motor cortex. It has to be appreciated that there is immense redundancy in this pathway for control of ordinary voluntary movements. As a consequence extensive cerebellar lesions may not give overt signs by the usual methods of neurological testing.

The symptoms of dysmetria or cerebellar ataxia are readily explained by the failure of the continual revisory control by the cerebellum. It can be envisaged that voluntary commands from the motor cortex give imprecise and merely provisional signals down the pyramidal tract, and that all refined control is exercised via the continuously operating cerebellar loops as diagrammed in figure 6. The tremendous development of the cerebellar hemispheres and the immensity of the reciprocal cerebro-cerebellar pathways give the potentiality for learning the most refined movement patterns in motor skills of all kinds. Rough and clumsy movements can be carried out even with complete cerebellar loss, or with a small fragment of surviving cerebellum [DOW and MORUZZI, 1958]. The neural cost for the most refined skills, as for example in musical performance or in ballet, is prodigious. For example it has been estimated that about 20 million mossy fibres project to each cerebellar hemisphere, where there are 10,000 million granule cells and 10 million Purkyně cells. In evolutionary terms it is evident that this neural cost was justified by man's pre-eminence in skills even in the paleolithic culture, with the consequence of his eventual dominance in the whole biological domain.

Essentially the cerebellum functions as a computer in its guidance of all movements. A rough analogy is to liken its control of movement to the computer control of a target-finding missile. But of course it has complexity and versatility unimaginably beyond the performance of any hard-ware computer. The essential wiring diagram shown in figure 4A is very simple. The exquisite performance derives from the integration of both excitatory and inhibitory actions at the synaptic relays in the cerebellar cortex with immense divergence and convergence adding to the dimensions of this integration. Subtlety and reliability derive from the enormous numbers of the units arranged in parallel at each stage of the circuit. The attempt to gain further understanding of all these features provides great challenges in the further scientific investigations on movement control.

Author's address: Prof. Sir JOHN ECCLES, Laboratory of Neurobiology, University of New York at Buffalo, 4234 Ridge Lea Road, *Amherst, NY 14226* (USA)

New Developments in Electromyography and Clinical Neurophysiology,
edited by J. E. Desmedt, vol. 3, pp. 225–233 (Karger, Basel 1973)

Tonic and Phasic Recruitment Order of Motor Units in Man under Normal and Pathological Conditions

L. GRIMBY and J. HANNERZ

Department of Neurology, Karolinska Hospital, Stockholm

The fact that different motor units have different properties makes it pertinent to study their order of recruitment and the mechanisms regulating the latter in individual motoneurones have been extensively studied in animal experiments [HENNEMAN, SOMJEN and CARPENTER, 1965a, b; BURKE, 1968a, b; CLOUGH, KERNELL and PHILLIPS, 1968], but not in man. Little is known of the mechanisms determining the recruitment order in voluntary activity.

If an EMG recording is made carefully with an appropriate technique, single motor units can be accurately identified by the characteristic and constant shape of their potentials [cf. ASHWORTH, GRIMBY and KUGELBERG, 1967; HANNERZ, 1972]. It is thus possible to study the functional properties of single motoneurones in man as conclusively as in animal experiments.

We have analyzed the order in which different human motor units are recruited in tonic and phasic activity under normal and pathological conditions. The m. tibialis anterior is studied mainly as the risk of displacement of the recording electrodes is small in this muscle.

Tonic Activity

In normal subjects the recruitment order of motor units in tonic voluntary activity is predominantly stable as long as there is no contraction of other muscles [GRIMBY and HANNERZ, 1968]. In patients with total interruption of the spinal cord, and consequent loss of cerebral influence on the motoneurone pool, the first motor unit recruited in tonic reflex activity is largely the same even with widely different types of stimuli [GRIMBY and HANNERZ, 1970]. In patients with partial spinal cord lesions the recruitment order is the same in

tonic voluntary and tonic reflex activity. This stability of the recruitment order
gives a rational basis for those types of physiotherapy in which reflex activity
is used to reinforce voluntary effort.

In animals, and presumably also in man, there are separate tonic and
phasic motoneurones [GRANIT, PHILLIPS, SKOGLUND and STEG, 1957; R.E.
BURKE, this volume]. The motor units first recruited in tonic activity in man
have a strong tendency to repeat and should be of the tonic type.

In animal muscles there are motor units distinguished by long or short
contraction time [DENNY-BROWN, 1929]. In human muscles the contraction
time can vary with changes in the mode of activation, indicating the existence
of slow as well as fast motor units [HOMMA and KANO, 1962; McCOMAS and
THOMAS, 1968]. The motor units first recruited in tonic activity are not prone
to discharge with intervals shorter than 50 msec. It is reasonable to assume
that the motor units fuse before this maximal frequency and that they are of
the slow type.

Motor units are uniform as regards muscle fibre type [EDSTRÖM and
KUGELBERG, 1968]. Muscle fibres, rich in oxidative enzymes, have a very high
resistance to fatigue, whilst muscle fibres, poor in oxidative and rich in gluco-
lytic enzymes, have a very low resistance to fatigue, at least in rat [KUGELBERG
and EDSTRÖM, 1968; see this series, volume 1]. The motor units first recruited
in tonic activity in man can be active for hours without signs of fatigue. It
can be assumed that they have muscle fibres rich in oxidative enzymes.

In experiments on decerebrate cats the recruitment order of motoneurones
has been shown to be determined by the input resistance which is a function of
the cell size [HENNEMAN, SOMJEN and CARPENTER, 1965a; R.E. BURKE, this
volume]. The hypothesis has been advanced that 'a particular cell receives the
same proportion of the total input from each of the systems which is afferent
to it' [HENNEMAN, SOMJEN and CARPENTER, 1965b]. Our findings are in good
agreement with this concept as long as only tonic input is concerned, but the
principle is not generally valid in man as will appear from the next section.

Phasic Activity

In phasic activity the motor units are recruited in a different order than
in tonic activity. The differences are seen both in voluntary activity in healthy
man as in figure 1 [GRIMBY and HANNERZ, 1968], and in reflex activity in
patients with spinal cord lesions so severe that the cerebral influence on the
motoneurone pool is blocked, as in figure 2 [GRIMBY and HANNERZ, 1970].

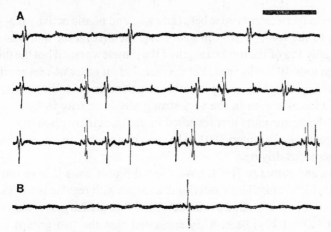

Fig. 1. Difference between tonic (A), and phasic (B) voluntary recruitment order in a normal subject. Recording in m. tibialis anterior. Time bar 50 msec.

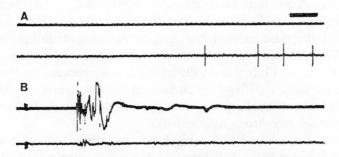

Fig. 2. Difference between the recruitment order in tonic (A) and phasic (B) nociceptive reflex in a subject with total interruption of the cervical spinal cord. Two electrodes inserted at different sites in the m. tibialis anterior a few cm apart (upper and lower traces in each recording). Stimulation in the sole of the foot. Time bar 100 msec.

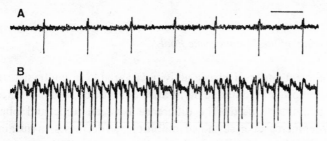

Fig. 3. Difference between the discharge patterns of motor units with low (A) and high threshold (B) in tonic voluntary activity. Time bar 100 msec.

The difference in recruitment order between tonic and phasic activity may be considerable. In the tonic activity in figure 1 A, the first motor unit was recruited when only 1 % of the total strength of the muscle was used but the third motor unit not until 10% was used. Yet the recruitment order was reversed in phasic activity (fig. 1 B). The motor unit that was number 1 in tonic activity in figure 2 A was inactive even in the very strong phasic activity in figure 2 B.

Some of the motor units first recruited in phasic activity discharge with shorter and more irregular intervals than the first motor units in tonic activity, also on simultaneous driving.

TOKIZANE and SHIMAZU [1964] have divided motor units in man into a 'tonic' group, which in m. tibialis anterior dischaiges with regular intervals at a frequency of about 7–8/sec and a 'kinetic' group that discharges regularly at a frequency of about 12–13/sec. They suggested that the two groups correspond to the tonic and phasic groups of the experimental animals.

Recordings with a more elaborate technique, permitting analysis during very strong or quick contractions, reveal motor units which do not discharge regularly even when their frequency is about 40/sec, as in figure 3 [HANNERZ, 1972]. It is likely that these extremely high frequency motor units, rather than TOKIZANE's kinetic ones, correspond to the phasic motor units in animals.

However, it is not likely that the distinction is sharp between the tonic and phasic motor units. The irregularly discharging motor units first recruited in phasic activity should be of a more phasic type than the regularly discharging motor units first recruited in tonic activity.

WARMOLTS and ENGEL [1972] have shown, in experiments on type-grouped human motor units, that those with fast twitch muscle fibres discharge at a higher and more irregular frequency than those with slow twitch fibres. It is thus likely that the change in recruitment order which we have shown to appear on changing from tonic to phasic contraction increases the relative role of fast muscle fibres and decreases that of slow muscle fibres.

'Programming' of the Motoneurone Pool

The recruitment order in phasic activity is variable in contrast to the stable recruitment order in tonic activity. Variations are obtained also when a standardized stimulus is used. The difference between tonic and phasic recruitment order increases on sustained inhibition and decreases on sustained facilitation of the motoneurone pool prior to the activation. The variation in recruitment order in a series of identically elicited phasic activities can thus be

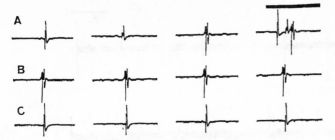

Fig. 4. Adaptability of recruitment order in phasic nociceptive spinal reflexes in a normal subject. In *A*, the subject does not expect the stimulus, and the first motor unit to be recruited varies unexpectedly from reflex to reflex. In *B*, the subject expects the stimulus and voluntarily facilitates the motoneurone pool subliminally, and so one of the motor units is always first recruited in a series of reflexes. In *C*, the subject expects the stimulus but inhibits the motoneurone pool voluntarily, and so another of the motor units is regularly first recruited. Recording in m. tibialis anterior. Stimulation in the sole of the foot by an electric shock with 10 msec duration. Latencies of reflexes less than 100 msec. Time bar 50 msec.

explained by variations in the relative importance of the two facilitating factors: (1) the long-lasting pre-existing state of facilitation that tends to activate the motor units in one 'tonic' order, and (2) the short eliciting impulse that tends to activate the motor units in another 'phasic' order.

It is functionally essential that motor units with different properties are activated in an order that is optimal for the type of movement intended. In 'spinal man' there is no stabilization of the recruitment order in a series of phasic reflexes. Normal man, however, is capable of adjusting the prefacilitation to a high or low level, thus in advance 'programming' the motoneurone pool for use of the 'tonic' or the 'phasic' recruitment order [HANNERZ and GRIMBY, 1972].

Figure 4 A shows the varying recruitment order when the motoneurone pool is not 'programmed'; figure 4 B shows one stable recruitment order when the motoneurone pool is 'programmed' by subliminal voluntary facilitation; figure 4 C shows another stable recruitment order when the motoneurone pool is 'programmed' by voluntary inhibition.

The Role of the γ Loop in the Recruitment Order

It is generally agreed that phasic voluntary activity is obtained via direct activation of the motoneurone pool but that tonic voluntary activity is dependent on support from the muscle spindles and indirect facilitation through

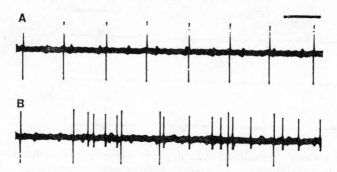

Fig. 5. Change in the recruitment pattern of tonic voluntary activity on local cooling of the muscle; *A*, before cooling, one motor unit discharging with long regular intervals; *B*, on cooling, one motor unit discharging with short irregular intervals appears without the original motor unit changing its discharge pattern. Recording in m. tibialis anterior. Time bar 100 msec.

the γ loop, though it is under debate how strong this dependence is. In cat, afferent discharges from the muscle spindles activate the motoneurones in a stable order with tonic ones before phasic ones [BURKE, 1968a, b]. It is pertinent to question to what extent the muscle spindles and the γ loop are responsible for the tonic recruitment order.

In man the discharge from the muscle spindles can be diminished by unloading the muscle [STRUPPLER, LANDAU and MEHLS, 1964]; by partial narcotization of the motor nerve with Lidocaine which may produce a roughly selective block of the γ efferents [MATTHEWS and RUSHWORTH, 1957] and possibly also acts on the muscle afferents [GASSEL and DIAMANTOPOULOS, 1964] prior to the α efferents; by local cooling of the muscle so decreasing the sensivity of the muscle spindles [ELDRED, LINDSLEY and BUCHWALD, 1960].

All three procedures can change the recruitment order in tonic voluntary activity but chiefly if motor units with nearby thresholds are studied and only temporarily. The same type of change of the recruitment order is obtained irrespective of the method used. The new motor unit that replaces the original one tends to discharge with shorter and more irregular intervals than the original motor unit (as in fig. 4) indicating that it is of a more phasic type [GRIMBY and HANNERZ, 1968].

None of the methods influences the discharge from the muscle spindles selectively, but the main effect they have in common is to decrease it. It must be concluded that the discharge from the muscle spindles, and thus the γ loop contributes to the tonic recruitment order. At present it cannot, however, be

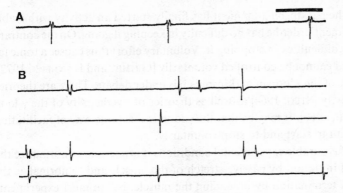

Fig. 6. Difference in recruitment pattern between gentle voluntary initiation in *A* a normal subject and *B* a parkinsonian patient with severe bradykinesia. Recording in m. tibialis anterior. Time bar 50 msec.

determined how decisive this contribution is in comparison with other factors. The temporary effect of the blocking procedures indicates that the loss of γ loop function obtained, can be compensated. In experiments on decerebrate cats, deefferentation did not influence the order of recruitment of moto neurones [HENNEMAN, SOMJEN and CARPENTER, 1965a], but in these preparations there is a strong background excitatory bias favouring tonic motor units.

Disturbances in the Recruitment Order in Parkinsonian Bradykinesia

A normal subject can voluntarily initiate an activity with a stable tonic recruitment order without difficulty. A patient with severe parkinsonian bradykinesia can make such an initiation only with considerable delay. Attempts to shorten the delay by strong voluntary effort result in bursts of irregularly alternating motor units as in phasic activity [GRIMBY and HANNERZ, 1972].

Figure 6 shows that the recruitment order in a gentle voluntary initiation is stable from the very start in a normal subject but that it takes considerable time to establish a stable recruitment order for a parkinsonian patient with severe bradykinesia. We have concluded that the latter has difficulty in initiating the way of facilitation that is normally used for tonic voluntary activity, and that he has to depend on ways of facilitation that are normally used only for phasic voluntary activity.

When the parkinsonian patient has finally started an activity with stable tonic recruitment order he has no difficulty in keeping it going. On the contrary he soon has difficulties in stopping it. Voluntary effort thus causes a tonic facilitation that cannot be controlled voluntarily [GRIMBY and HANNERZ, 1972].

The role of the γ loop in parkinsonism is under debate. There are theories of underactivity [STEG, 1964] as well as theories of overactivity of the γ loop [RUSHWORTH, 1960]. It may be that the γ loop is this tonic way of facilitation that is difficult to start and to stop voluntarily.

Under favourable experimental conditions it is possible to normalize the pathological initiation by passive stretch of the muscle and to normalize the pathological termination by unloading the muscle, but in most experiments these changes in the proprioceptive afferent activity are not enough for normalization [GRIMBY and HANNERZ, 1972]. With single nerve fibre recording it is shown that in parkinsonism the inflow from the muscle spindles at rest is more pronounced, whilst the enhancement during voluntary effort is rather smaller than in normal man [HAGBARTH, HONGELL and WALLIN, 1970]. There are thus indications that the γ loop in parkinsonism may be underactive on voluntary initiation but overactive on voluntary relaxation. At present, however, it cannot be determined what role these presumed γ loop disturbances play in comparison with other defects.

Disturbances in the Recruitment Order in Spastic Paresis

A normal subject can voluntarily activate a motor unit with low threshold in tonic activity for hours without signs of fatigue. A patient with severe spastic paresis due to a spinal cord lesion can do this for only a few minutes. Then he looses his ability to drive not only the activated but also other motor units tonically. It is thus the tonic voluntary drive of the motoneurone pool that is exhausted and not single motor units [GRIMBY, HANNERZ and RÅNLUND, 1972].

On extreme voluntary effort the spastic patient is able to activate other motor units instead of the lost tonically discharging ones. The new motor units tend to appear in bursts and to alternate irregularly without definite recruitment order. Some of them discharge with shorter and more irregular intervals than the original ones, also on simultaneous driving, as in figure 7 [GRIMBY, HANNERZ and RÅNLUND, 1972].

The fatigue thus causes a change from tonic to phasic type of recruitment. We have concluded that the way of facilitation that is normally used for tonic

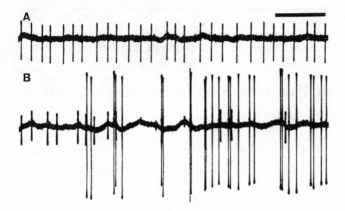

Fig. 7. Change in recruitment order in a subject with severe spastic paresis on sustained contraction; *A* before and *B* on fatigue. Recording in m. tibialis anterior. Time bar 1 sec.

voluntary activity is fatigued abnormally fast in patients with spastic paresis, and that they have to depend on ways of facilitation that are normally used for phasic voluntary activity.

The original tonic recruitment order is restored if the voluntary effort is reinforced by tonic proprioceptive as well as exteroceptive reflex activity [GRIMBY, HANNERZ and RÅNLUND, 1972]. In normal man, celocurin blockades of the intra-fusal muscle fibres result in inability to maintain voluntary discharge in at least part of the α-motoneurone fibres [FREYSCHUSS and KNUTSSON, 1971]. More experiments are needed before it is possible to decide whether the fatigue reaction in patients with spastic paresis is due to insufficiency of the proprioceptive support of voluntary contraction.

Author's address: Dr. L. GRIMBY, Neurologiska Kliniken, Karolinska sjukhuset, *S-10401 Stockholm 60* (Sweden)

New Developments in Electromyography and Clinical Neurophysiology,
edited by J. E. Desmedt, vol. 3, pp. 234–241 (Karger, Basel 1973)

Discharge Pattern of Motoneurones in Humans

A Single-Fibre EMG Study[1]

E. STÅLBERG and B. THIELE[2]

Department of Clinical Neurophysiology, Academic Hospital, Uppsala

From routine EMG investigations and experimental studies [DAS GUPTA, 1962; BASMAJIAN, 1963; TOKIZANE and SHIMAZU, 1964; CLAMANN, 1969; PETAJAN and PHILIP, 1969; FREUND and WITA, 1971] it is known that the firing frequency of voluntarily activated motoneurones is somewhat irregular, i.e. the interpotential intervals are of slightly different length, especially at long mean intervals. The purpose of this investigation is to study the nature of this innervation irregularity and to test whether it can be used for characterizing motoneurone types or whether unpredictable central and peripheral factors are so important that this is impossible. In this investigation only motoneurones activated at tonic voluntary contraction have been studied.

Material and Methods

With a single-fibre EMG technique as described by EKSTEDT and STÅLBERG [1963], EKSTEDT [1964] and STÅLBERG [1966], recording was made in voluntarily activated human muscles. Action potentials from 1 to 3 muscle fibres belonging to the same motor unit were recorded using a needle electrode with two electrode surfaces, each 25 μm in diameter. The interference from other motor units did not disturb the recording at slight and medium activity.

In 10 healthy subjects (22–37 years of age) the following muscles were investigated: m. interosseus dorsalis I, m. extensor digitorum communis, m. flexor digitorum, m. biceps brachii, m. quadriceps, m. tibialis anterior and m. longissimus dorsi. In 7 patients with

1   The investigation was supported by the Swedish Medical Research Council, Grant 14X-135.
2   Fellow of the Deutsche Forschungsgemeinschaft. Present address: Department of Clinical Neurophysiology, Klinikum Steglitz, 1 Berlin 45, Hindenburgdamm 30, FRG.

thigh amputations, the m. quadriceps femoris was studied. The subject controlled the innervation frequency both visually by means of a rate-meter and acoustically by listening to a loudspeaker. In most experiments the subjects were asked to keep a certain predicted innervation frequency as constant as possible for a period long enough to record 201 discharges. Thereafter, the innervation frequency was changed and in this way each motoneurone was tested at 3–10 different frequency levels. In most experiments the type of contraction was neither pure isometric nor pure isotonic. Pure isometric contraction was used in experiments when the influence of different muscle lengths was studied and when motoneurones with different recruitment threshold were compared.

In some experiments a vibrator was placed on the muscle under study or its antagonist, either during continuous voluntary activity or with the aim of evoking a tonic vibration reflex (TVR) in passive relaxed subjects. The vibration frequency was 150/sec.

Analysis

The mean value and the variability of the interdischarge intervals were analyzed from 200 intervals measured with a time interval counter (HP 5245 L). To eliminate the influence of slow systematic changes in the innervation rate, the variability was expressed as the mean consecutive difference (MCD) of the intervals (absolute values), calculated on a CDC 3600 computer [STÅLBERG, EKSTEDT and BROMAN, 1971]. For each motoneurone the different mean intervals were plotted as abscissa and corresponding MCD as ordinate. Values from motoneurones belonging to the same muscle were included on the same plot.

Results

After a training period (10–15 min) the subjects were able to activate a motoneurone under study at any required frequency between 8 and 20/sec with good precision. It was, however, impossible to maintain a regular innervation pattern with a frequency below a certain value. This 'limit frequency' was different for different motoneurones in the same muscle and varied even more for different muscles. Mean frequencies below this value could only be obtained by introducing compensatory pauses between trains of impulses, appearing with the limit frequency (fig. 1 and 2). Thus the plot of mean interval-MCD relations included these compensatory pauses giving the curve a steeper slope for long intervals (fig. 3 A). For a given mean interval the distribution of the interval was roughly Gaussian (fig. 2) which has been further studied statistically for m. biceps brachii by CLAMANN [1969].

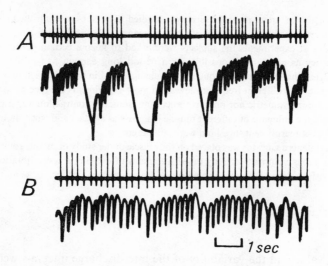

Fig. 1. Two motoneurones with an innervation frequency of 5/sec. *A* M. extensor digitorum communis. *B* M. longissimus dorsi. Upper beam: original spikes as recorded with the electrode. Lower beam: integration curve of this spike trace. The more dynamic motoneurone (*A*) cannot be activated regularly with 5/sec but discharges with periods of limit frequencies of 8/sec interrupted by compensatory pauses, while the more tonic motoneurone (*B*) has its limit frequency at about 5/sec.

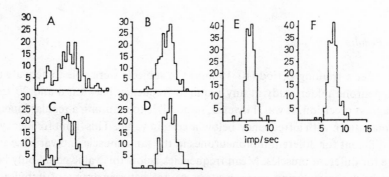

Fig. 2. Discharge frequency histograms for motoneurone from m. extensor digitorum communis (*A* and *B*) m. quadriceps femoris (*C* and *D*) and m. longissimus dorsi (*E* and *F*). In the histograms *A*, *C* and E, intended innervation frequency 5/sec, in the histograms *B*, *D*, *F*, 8/sec. Note that in *A* and *C* the histograms are bimodal corresponding to the limit frequency (about 8/sec) and compensatory pauses. The other histograms are roughly Gaussian.

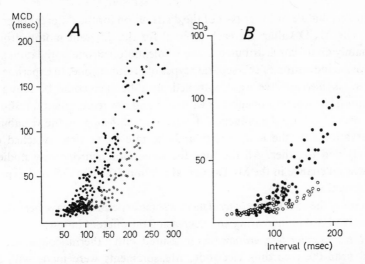

Fig. 3. *A* Interval irregularity of motoneurones from the m. extensor digitorum communis (●) and the m. longissimus dorsi (○). Abscissa: mean interval. Ordinate: interval irregularity measured as MCD. Each motoneurone is represented by 3–10 points having a characteristic local distribution in the material. No separation in two types of motoneurone occurred in each muscle, but the 2 muscles show different distributions. The limit frequencies for extensor digitorum communis motoneurones were typically 8–10/sec and for the lumbar muscle 5–7/sec causing the shift to the right for the interval-MCD curve of the latter. *B* Comparison of discharge irregularity in m. biceps brachii between material presented by TOKIZANE and SHIMAZU [1964] (○) and our material (●). The difference is a group of extreme tonic motoneurones not found in our material, which is homogenous without evidence for two separated distributions. Abscissa: mean interval. Ordinate: interval irregularity expressed as standard deviation over 9 values [according to TOKIZANE and SHIMAZU, 1964].

In the m. quadriceps femoris, m. longissimus dorsi and m. tibialis anterior the motoneurones generally had limit frequencies of about 5–7/sec. The motoneurones of m. interosseus dorsalis I, m. flexor digitorum, m. extensor digitorum communis and m. biceps brachii had limit frequencies of 8–10/sec. These values corresponded to the 'onset interval' as described by PETAJAN and PHILIP [1969] for the same muscles, except for the m. interosseus, where these authors found longer onset intervals (142 ± 39 msec). The mean interval-MCD plot was shifted to the right for muscles with lower limit frequencies in relation to those with higher limit frequencies (fig. 3 A).

The MCD values at different mean intervals for a certain motoneurone showed characteristic local distribution in relation to the total motoneurone population of that muscle. Thus one motoneurone was represented either in the upper, middle or lower part of the distribution for the different interval values. The MCD values for each interval for the different motoneurones were mainly Gaussian distributed in the whole motoneurone population and there was no indication of two separate types of motoneurone. In experiments where the isometric force was controlled, no difference could be found in distribution between motoneurones with high and low recruitment threshold.

In order to study the influence of some *central factors* on the variability of innervation rate the loudspeaker and the rate-meter were switched off separately and together. All the time the same motoneurone was studied. There was no change in the MCD-interval relation for these different experimental situations.

To study the effect of *peripheral factors* some experiments have been done which were aimed at increasing and decreasing the afferent inflow.

1. Intramuscular temperature was measured with a thermocouple needle inserted near the recording electrode. Measurements were made with an accuracy of 0.05 °C. When the muscle temperature was decreased to 27 °C and then restored to normal no change in the interval irregularity was seen in 2 long-term experiments.

2. Muscle stretching. With the wrist passively bent dorsally (extended) recording was made from wrist-flexors. The interval irregularity did not differ from the values obtained with the muscle at normal length.

3. Vibration. High frequency muscle vibration induces strong, sustained activation of muscle spindle primary afferents [GRANIT and HENATSCH, 1956]. The effect of muscle vibration was studied in two different situations. (1) The subject activated the muscle voluntarily and tried to keep the innervation frequency for a certain motoneurone constant when vibration was started. Under these conditions the vibration had no reproducible effect on the frequency irregularity when the vibration was applied either to the muscle itself or to its antagonist. (2) The subject was passive and the TVR was initiated by vibration [EKLUND and HAGBARTH, 1965] and the actual discharge frequency in different motoneurones was measured. In these cases, motoneurones were activated with a slightly more regular frequency than during voluntary activation. With a constant vibration frequency of 150 c/sec different motoneurones had different discharge frequencies. The lowest frequencies obtained in this way were somewhat lower than the so-called limit frequency for voluntary activation. No compensatory pauses were seen.

4. Lidocaine (Xylocaine) blockade. With a dilute Xylocaine solution (0.25%) on the nerve it is possible to block mainly the thin γ-fibres [MATTHEWS and RUSHWORTH, 1957]. In 7 experiments Xylocaine was injected around the peroneal nerve. The subjects were asked to keep a preferred innervation frequency as constant as possible during the whole experiment. After a few minutes the innervation frequency in the tibialis anterior became more and more irregular for one motor unit after another within the uptake area of the electrode. The interpotential intervals were not multiples of the earlier regular intervals which would have been the case if α-fibres were blocked. This, however, happened in some experiments after a while and finally no activity could be recorded.

5. *Thigh amputation.* Motoneurones innervating the m. quadriceps femoris on the operated side in 7 patients with thigh amputations were studied. All the patients functioned well with prostheses. There was no surgical fixation of the muscle to the bone and therefore it could be assumed that the afferent input from the muscles was reduced or partially lacking especially in those cases where a great mass of muscle had been removed. Many motoneurones could be activated only for a short time, too short for measurements to be done. Motoneurones which could be activated for longer times showed MCD values slightly above or in the upper part of the normal material.

Discussion

When a motoneurone is depolarized with a constant direct current it discharges with a rate proportional to the depolarization within certain limits [GRANIT, KERNELL and SHORTESS, 1963]. Even under these conditions the interspike intervals are of different length as shown in the cat motoneurone by CALVIN and STEVENS [1968]. The irregularity could be ascribed to spontaneous fluctuations in the trigger level of the motoneurone membrane, variable presynaptic inflow or short term irregularity in each depolarization curve (synaptic noise).

Fluctuation of the Trigger Level

No direct experiments have been performed to study the variability in trigger level without disturbances from synaptic noise. A comparison may perhaps be made with the motor end-plate where the fluctuation in the trigger level is assumed to be the main cause of the variability in the transmission time, called jitter [EKSTEDT, 1964; STÅLBERG, EKSTEDT and BROMAN, 1971] This variability is of the order of 10–40 μsec and thus very small in relation to the interval variability studied in this investigation. In experiments with the H-reflex in single motoneurones in man TRONTELJ [1968] reported a variability range of 950 µsec (corresponding to an MCD value of about 300–400 µsec). The variability in trigger level is certainly not responsible for the whole H-reflex jitter and can thus not explain the much greater discharge interval variability. CALVIN and STEVENS [1968] conclude from their experiments on cats that the variation in the trigger level can be responsible for the interval irregularity only to a minor degree.

Irregular Presynaptic Inflow

Slow systematic changes in the interval irregularity were seen, possibly caused by variations in motoneurone excitability, sometimes in connection with respiratory rhythm. These slow changes were eliminated by using the MCD as a measure of interval variability. Short-term variations of the presynaptic inflow from interval to interval may, however, also occur. One indication of this may be that during the strong and regular afferent input caused by vibration of passive muscle, the MCD values for the same mean interval were slightly lower than during voluntary activation. It is here assumed that the same type of motoneurones are activated in the two types of experiment, an assumption which is supported by the findings of ASHWORTH, GRIMBY and KUGELBERG [1967]. Manoeuvres which increased or decreased the afferent inflow to a moderate degree (overstretching the muscle, accessory vibration, cooling) had, however, no measurable effect upon the irregularity of motoneurone discharge. This indicates that the afferent inflow during normal circumstances is so strong that, even when slightly reduced, it depolarizes the motoneurone at maximal speed. With more pronounced decrease of the peripheral afferent input however, (Xylocaine blockade, experiments on amputees) the irregularity increased.

Synaptic Noise

The depolarization curve of the motoneurone is due to summation of EPSP and IPSP. The curve, however, shows certain irregularities, called wavelets by GRANIT. CALVIN and STEVENS [1968] found that this irregularity in the depolarization curve could fully explain the interval variability which in cats had a coefficient of variation of 5%. With a slower slope of the depolarization curve these wavelets will be of greater influence and increase the irregularity in the discharge intervals, i.e. the variability at lower innervation rate will increase. This was the finding in our experiments in which the coefficient of variation was of the order of 10%. The shape of the depolarization curve depends on the number and anatomical arrangement of the synapses, and the size of the motoneurone itself [HENNEMAN, SOMJEN and CARPENTER, 1968]. As these characteristics of the motoneurone seem to be the main cause of the innervation irregularity, this can be used as a parameter characterizing different normal motoneurones, and perhaps also to detect pathological changes of the motoneurone membrane, for example in amyotrophic lateral sclerosis.

While testing the different factors described above, it has been found that there are differences in motoneurones to the same muscle and even greater differences between motoneurones to different muscles. Two separate types of motoneurone as reported by TOKIZANE and SHIMA [1964] corresponding to tonic and phasic units have, however, not been identified in these investigations (fig. 3B). Our material fits into their so-called kinetic motor-units and in spite of the fact that we always used a tonic contraction it was not possible to find their tonic units.

Summary

Discharge pattern of single motoneurones has been studied during tonic contraction in different muscles in healthy human subjects by single fibre EMG. There was a variability in the discharge intervals, more pronounced at lower innervation frequencies. Different motoneurones in the same muscle had different mean interval irregularity curves but two separate types of motoneurone were not seen. Motoneurones of lumbar muscles and quadriceps muscles had a more regular innervation frequency than those of hand and forearm muscles.

Different tests indicated that the interval variability in healthy subjects was mainly caused by short-term irregularity in each depolarization curve (synaptic noise) and to a minor degree by spontaneous fluctuations in the trigger level of the motoneurone membrane and by variability in the presynaptic inflow. The innervation irregularity might thus be used as a parameter characterizing a motoneurone and is suspected to be changed under certain pathological conditions.

Author's address: Dr. E. STÅLBERG, Department of Clinical Neurophysiology, Academic Hospital, *S-750 14 Uppsala 14* (Sweden)

New Developments in Electromyography and Clinical Neurophysiology,
edited by J. E. Desmedt, vol. 3, pp. 242–250 (Karger, Basel 1973)

Discharge Characteristics of Single Motor Units in Normal Subjects and Patients with Supraspinal Motor Disturbances[1]

H.-J. FREUND, V. DIETZ, C. W. WITA and H. KAPP

Neurologische Klinik mit Abteilung für Neurophysiologie der Universität,
Freiburg i. Br.

Whereas the analogue features of the motor unit potentials (amplitude, duration, form) together with spontaneous activity and the grading and density of interference are commonly used in clinical EMG, the temporal order of single motor unit discharges has not been employed for diagnostic purposes. Since the motoneurone is the final common path of the motor system, disturbances of various parts of the latter should influence the pattern of impulse generation. FREUND and WITA [1971] recently described the normal motoneuronal discharge patterns in stationary isometric contraction and found them to be basically changed in patients with motor disturbances and, in particular, with *supraspinal* motor lesions.

In order to get more information on the motor units studied, their firing threshold and the conduction velocity of the motor axon were used as an estimate of the size of the corresponding motoneurones. It was found that in normal subjects the recruitment order of motor units in the hand muscles is an approximately linear function of the size of the motoneurone [FREUND, WITA and SPRUNG, 1972], thus in line with the size principle established in animal experiments [HENNEMAN, SOMJEN and CARPENTER, 1965].

Methods

Recordings were taken from the first dorsal interosseus muscle during stationary isometric pressure. Force was measured by a strain gauge and fed to a pressure dial and a tape-recorder. The subject was instructed to keep the needle of the pressure dial constantly in a

1   This research was supported by Deutsche Forschungsgemeinschaft, SFB 70.

particular position. Single unit potentials were recorded with high impedance tungsten micro-electrodes and the compound muscle potential with surface skin electrodes. Each action potential triggered a single sweep on the oscilloscope and the trigger impulse of an interface was registered on a second channel. The trigger level was adjusted so that an exact 1:1 allocation could be made. The inter-spike intervals were measured and stored in an IBM-1130 computer, where they were available for further data processing [for further details see Freund and Wita, 1971].

Results

Discharge Properties of Motor Units in Normals

The general features of the discharge properties of motor units from normal subjects are displayed for a representative motor unit in figure 1. In figure 1 A a normal frequency histogram is shown. The discharge

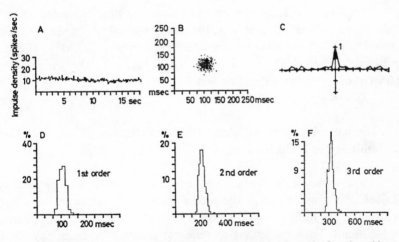

Fig. 1. Discharge properties of a single motor unit recorded from the first dorsal inter-osseus muscle of a normal subject described by: *A* frequency histogram during stationary isometric contraction. Ordinate: impulse density calculated as reciprocals of the single inter-spike intervals. Abscissa: time in msec. *B* Joint interval histogram. Ordinate and abscissa: interval length in msec. Each point represents the length of 2 adjacent intervals by the distances to the ordinate and the abscissa. *C* Autocorrelation function. Ordinate: correlation coefficient. 1 division = 0.5. Positive coefficients upward, negative coefficients downwards. Abscissa: time in msec (1 intersection = 200 msec). *D–F* Interval histograms of the 1st–3rd order inter-spike intervals. Abscissa: interval length in msec. Ordinate: frequency of occurrence of interval lengths indicated on the abscissa. Binwidth = 10 msec. Note: different scale for both co-ordinates from *D–F*.

frequency is plotted as a function of time. This display allows a fast orientation about trends and the regularity of discharge. The normal histogram shows a regular frequency profile. As shown by the interval histograms (fig. 1 D–F) of the first-, second- and third-order inter-spike intervals, the interval length varies randomly and resembles a Gaussian distribution with small variance for all three interval classes. The sequence of adjacent intervals can be displayed by the joint-interval histogram (JIH; fig. 1 B). As indicated by the typical concentric pattern, in normal subjects there is no serial dependence in the length of adjacent intervals. This can be further proved by the autocorrelation function shown in figure 1 C. There is no evidence for a rhythmic grouping as reported by LIPPOLD, REDFEARN and VUCO [1957].

The discharge pattern of normal motoneurones during stationary condition can thus be described as regular and, from the statistical viewpoint, as a stochastic point process. Because the sequence of intervals varies randomly, it is also a renewal process.

The discharge frequency of normal motor units recorded from 20 subjects varied for different units between 5 and 20 spikes/sec for various strengths of isometric contraction. The regularity of discharge increased with successively higher discharge rates of the unit. The variability coefficient of the interval duration was 10–25 %.

Discharge Properties of Motor Units in Patients
with Motor Disturbances

The changes in the discharge pattern of motor units from patients with lesions of various parts of the motor system are severe in most cases. Therefore, the differentiation between normal and non-normal patterns is not difficult. Between lesions of different motor subsystems (pyramidal, cerebellar, basal ganglia) there are characteristic differences particularly in the temporal order of adjacent inter-spike intervals, that can be seen in many cases in the joint interval histograms. Figure 2 shows the same histograms as explained in figure 1 A–F, from a patient with minor symptoms of left spastic hemiparesis. There were no other neurological deficits. The JIH shows that interval length varies much more than normally and that intervals of short duration are followed by long intervals and *vice versa*. This change was typical for the majority of pyramidal lesions.

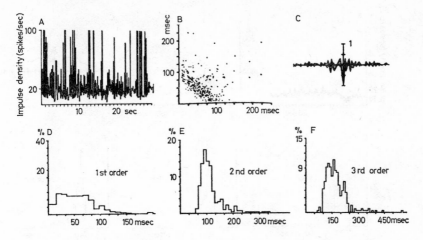

Fig. 2. Same histograms as in figure 1 for a patient with a spastic hemiparesis of his left side. Recording from the left hand.

Patients with cerebellar symptoms had less marked changes in their firing pattern. Figure 3A shows that a patient with mild cerebellar symptoms has slow trends of interval length, but the variation of adjacent interval lengths is surprisingly low. The non-sequential interval distribution (fig. 3 A–F) shows an increase in variance, but the JIH (fig. 3B) and the autocorrelation function (fig. 3 C) demonstrate that the variation in the length of adjacent intervals is extremely low and that the interval sequence is deterministic: the length of any interval depends heavily on the length of the preceding intervals. Again, the configuration seen in figure 3B is typical for many cerebellar lesions and can be seen also in cases with only minor clinical symptoms. Figure 4 shows the histograms of a patient with an akinetic parkinsonian syndrome. The JIH (fig. 4B) shows a triangle-shaped point distribution. The autocorrelation function (fig. 4C) shows the rhythmicity of higher and lower correlation coefficients that is equal to tremor frequency.

Figure 5, from a patient with polyneuropathy shows an example of histogram changes in lower motoneurone disease. As expected, the transmission down the axon fails sometimes, which is indicated by a second group of interval lengths in the JIH (fig. 5B). The interval length of this group is twice that of the first group, which is due to the failure in transmission of one spike.

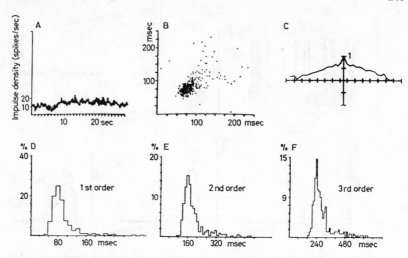

Fig. 3. Same histograms as in figure 1 for a patient with light cerebellar symptoms on both sides. Recording from the right hand.

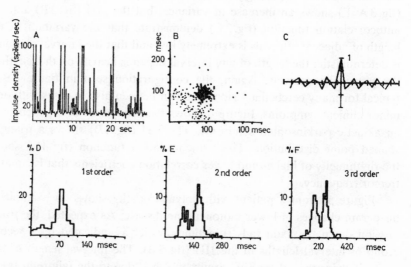

Fig. 4. Same histograms as in figure 1 for a patient with Parkinson's disease. Recording from the right hand.

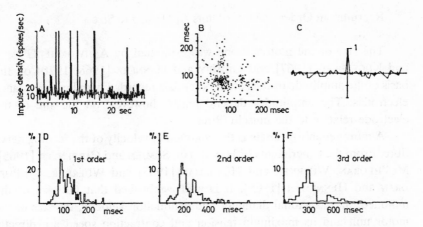

Fig. 5. Same histograms as figure 1 from a patient with a symmetric polyneuropathy. Recording from the right hand.

In figures 2–4 histograms of lesions of 3 different supraspinal motor subsystems were shown and are typical for many cases with clinically similar types of motor lesions. But these findings are not consistent for *all* cases: (1) not all units recorded from one patient are similary involved; some of them show the characteristic pattern, others do not or only weakly. Therefore, 4–6 units have to be analyzed from each patient; (2) a few patients belonging clinically to one particular group do not show typical patterns at all. One possible explanation may be that the clinical symptoms are rather coarse as an indicator for a lesion of a particular motor pathway. We are now trying to find out whether finer clinical differentations of muscle tonus, reflex excitability and paresis can provide further information about these exceptions. On the other hand, a sampling error could be involved, that is, only the normal units mentioned in (1) are included in the sample.

Because lesions of different motor subsystems may possibly involve large and small motoneurones differently, the size of the appropriate motoneurone feeding a muscle fibre should be known. Therefore, we have now determined the size of the motoneurones as well as the firing characteristics.

Recruitment Order of Motor Units According to Size

The size of the motoneurone was estimated by ASHWORTH, GRIMBY and KUGELBERG [1967] and GRIMBY and HANNERZ [1968, 1970] on the basis of the amplitude of the action potentials recorded with low impedance electrodes. The amplitudes depend partly, however, on the site of the electrode relative to the muscle fibres.

A more reliable estimate is the conduction velocity of the feeding nerve fibre. From the experiments of HENNEMAN, SOMJEN and CARPENTER [1965], McPHEDRAN, WUERKER and HENNEMAN [1965] and WUERKER, McPHE-DRAN and HENNEMAN [1965], it is well established that the size of the motoneurone, its axonal diameter and conduction velocity, the size of the motor unit and its maximum tension and contraction speed are directly correlated. For measurements on normal subjects and patients with various motor diseases, where conduction velocity may be slower in some cases, we preferred a relative measurement. The compound electrical response elicited by ulnar nerve stimulation at the wrist was recorded from the skin surface over the belly of the muscle at the optimal location where the negative component is maximal [DESMEDT, 1958]. The relative location of the peak of the single fibre action potential to the muscle action potential was measured (fig. 6A). The duration of the muscle compound electric response (negative and positive components) was normalized to 1 and the time of occurrence of the peak of the single fibre potential was expressed as a fraction of 1 (fig. 6C). The amplitude of the single fibre spike did not change with variation of stimulus strength (fig. 6A, B). The form of the single fibre spike following electrical stimulation was photographically compared with that of the voluntary activated spike. The identification of the spike forms depends critically on the use of high impedance electrodes.

The firing threshold of the motor units was determined in relation to the measured force exerted by the whole muscle during increasing isometric contraction. Figure 6C shows the firing threshold as a function of the relative conduction velocity. For these motor units randomly sampled from normal subjects, the firing threshold is an approximately linear function of conduction velocity and thus presumably of the size of the appropriate motoneurone. The size principle established for cat motoneurones seems, therefore, to apply also for the recruitment order of human motoneurones during normal voluntary innervation of the hand muscles. During station-ary innervations longer than those described in these experiments small variations or trends of the firing threshold can be seen during repetitive

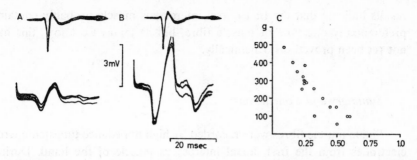

Fig. 6. A. Electric responses recorded from single muscle fibres (upper traces) from the whole muscle (surface electrodes, lower traces) following stimulation of the ulnar nerve 2 cm proximal to the wrist. *B.* Same record for higher stimulus intensity. The single fibre potential remains unchanged. *C.* Firing threshold (ordinate) of single muscle fibres as function of the conduction velocities of the feeding nerve fibres. The duration of the muscle compound electric response was normalized to 1. The position of the peak of the fibre potential within the muscle potential was measured and plotted as a fraction of 1 on the abscissa. Since the smallest values on the abscissa (left) relate to the fastest conducting nerve fibres, this plot shows increasing velocities from right to left.

threshold measurements. They may explain the 'rotation' in the recruitment of several simultaneously recorded motor units as reported by ASHWORTH, GRIMBY and KUGELBERG [1967] and GRIMBY and HANNERZ [1968, 1970; see also GRIMBY, this volume]. This rotation, however, could probably occur only between units with similar thresholds. This interpretation is not unreasonable since the recordings of GRIMBY and HANNERZ are mainly from low threshold units, because during stronger efforts interference of many other units may disturb these recordings with low impedance electrodes.

Figure 6 C presents no evidence for the existence of 2 separate groups of motor units as reported by TOKIZANE and SHIMAZU [1964]. Since the variance was calculated for each interval histogram, this parameter was compared for 60 motor units. There was no grouping of the variance according to the tonic and phasic type of the standard deviation as described by the latter authors. Also from the functional viewpoint clear differences between presumably tonic and phasic units could not be seen. All units could be activated by increasing isometric forces. For each recording position of the electrode small brisk movements were performed in order to activate phasic units, that are not recruited by the former condition. But in normal subjects no additional units could be activated in this way. These results indicate that the first dorsal interosseus muscle probably contains preferentially one type of muscle fibre, but as far as we know, this has not yet been proved histochemically.

Summary and Conclusions

Single muscle fibres were recorded by high impedance tungsten micro-electrodes from the first dorsal interosseus muscle of the hand. During stationary isometric contraction, recordings from normal subjects are characterized by a regular discharge profile, a normal distribution of the inter-spike intervals and a random interval sequence without serial dependence.

In patients with supraspinal lesions of the motor system or with lower motoneurone diseases this discharge pattern is basically changed. The common disturbance is the inability of the motoneurone to produce approximately constant intervals. Lesions of different parts of the motor system, such as pyramidal or cerebellar, show differences in the temporal order of the spike discharge. These differences are characterized by changes in the joint interval histograms. Thus, the impulse generation at the motoneurone seems to be specifically changed by disturbances of different motor centers.

The firing thresholds of the single motor units were determined and the conduction velocities of the corresponding motor axon were measured. In normal subjects, the firing threshold increases with successively higher conduction velocities. This result indicates that the size principle as established by Henneman, Somjen and Carpenter [1965] for the cat also applies to the recruitment order of motoneurones of human hand muscles during voluntary isometric contraction.

Author's address: Dr. H.J. Freund, Neurologische Universitätsklinik, Hansastrasse 9, D-78 Freiburg i. Br. (Federal Republic of Germany)

New Developments in Electromyography and Clinical Neurophysiology,
edited by J. E. Desmedt, vol. 3, pp. 251–262 (Karger, Basel 1973)

Muscle Spindle Afferent Discharge from Resting and Contracting Muscles in Normal Human Subjects

Å. B. VALLBO

Department of Physiology, University of Umeå, Umeå

The physiology of muscle spindles and the fusimotor system has been studied extensively [MATTHEWS, 1964; ELDRED, YELLIN, GADBOIS and SWEENEY, 1967; GRANIT, 1970], but there are comparatively few analyses of the activity in this system during naturally occurring motor acts [e. g., CRITCHLOW and VON EULER, 1963; DAVEY and TAYLOR, 1967; SEVERIN, ORLOVSKIJ and SHIK, 1967]. However, the findings concerning fusimotor activity in breathing, stepping and more simple reflexes all seem to agree in showing that the fusimotor system clearly takes part in these motor acts, and that it often does so with sufficient power to compensate for the unloading effect on the muscle spindle of the muscle shortening.

Direct recordings from muscle spindle afferents in waking human subjects were introduced by VALLBO and HAGBARTH [1967] who developed a method by which nerve impulses can be recorded with a flexible tungsten micro-electrode, percutaneously inserted into the nerve and freely floating in the tissues [VALLBO and HAGBARTH, 1968; VALLBO, 1972]. This made it possibile to analyze the activity in the fusimotor system and the muscle spindles during natural movements in an intact preparation without interference from anaesthesia.

In a series of investigations, the basic characteristics of the spindle afferent discharge were analysed in multifibre recordings from various muscles and in single unit recordings mainly from the wrist and finger flexor muscles [HAGBARTH and VALLBO, 1968; 1969; VALLBO, 1970a, b, 1971, 1972]. It was found that when the muscles were relaxed, there was a low and approximately constant discharge in the spindle afferents as a group, indicating that the fusimotor drive was very low or absent [HAGBARTH, HONGELL and WALLIN, 1970; VALLBO, 1970a; 1972]. To passive movements of the appropriate joints, the spindles responded with an increase of the discharge rate as the muscle was stretched and *vice versa*. There was no indication of a varying fusimotor drive to passive joint movements or skin stimulation of moderate intensity [VALLBO, 1970a; 1972]. In contrast, the spindles were strongly activated during voluntary contractions. This was true for slowly rising contractions as well as for small, fast rising muscle twitches [HAGBARTH and VALLBO, 1968; HAGBARTH, HONGELL and WALLIN, 1970; VALLBO, 1970b; 1971]. The fusimotor outflow was limited to the active muscles and it engaged the majority of the spindles within these muscles [VALLBO, 1970b]. Thus, the skeletomotor and the fusimotor outflows were directed very distinctly towards the same muscle portions, indicating that the spindles provide very detailed information concerning the state of the contracting muscles. It was further shown that the skeletomotor activity precedes the spindle afferent discharge at the onset of a contraction, implying that the initiation of the skeletomotor contraction is not dependent upon an increased spindle afferent discharge [VALLBO, 1971]. Consequently the voluntary contractions are not servo driven through the muscle spindle reflex loop with regard to the initial activity, and it seems unlikely that this would be the case for the rest of the contraction, as this would imply a complete switch from one control principle to the other during the same contraction.

It seems that these findings are all consonant with the principle of α-γ linkage [GRANIT, 1955]: firstly, as it was found that the activity in one of the two motor systems was regularly associated with activity also in the other; secondly, as it could be inferred that the excitatory drive from supraspinal structures was directed towards both types of motoneurones in voluntary contractions, and not exclusively towards the fusimotor neurones.

The present report deals with the relations between the spindle afferent discharge rate on the one hand, and the intensity of the skeletomotor activity, the muscle length and the rate of change of the muscle length on the other hand, during voluntary contractions. An analysis of these relations seems to be of some interest as any hypothesis concerning the role of the fusimotor system and the muscle spindles in motor control must be consistent with the characteristics of this signal. For instance, the hypothesis that the skeletomotor contraction is controlled exclusively through the reflex effects from the muscle spindle afferents, i.e. the follow-up length servo hypothesis [MERTON, 1951, 1953], would be supported if it was found that the spindle afferent discharge was always proportional to the intensity of the skeletomotor contraction. On the other hand, if pronounced and systematic deviations from proportionality were found between these two variables, this would rather suggest another function of the spindles. The findings to be presented, which are to some extent preliminary, do not support the follow-up length servo hypothesis but they support the notion that the spindles assist in the control of movements.

Methods

The analysis is based upon single unit recordings from muscle spindle afferents in the median nerves of normal human subjects. The discharge originated from the wrist and finger flexor muscles. All the units considered had characteristics of muscle spindle primary endings [VALLBO, 1970a]. The recording technique and the system for data collection and processing have been described in detail in other reports [VALLBO and HAGBARTH, 1968; VALLBO, 1970a; 1972].

Results

Figure 1 illustrates certain basic features of the response from a representative unit. The ending was located in the part of the m. flexor digitorum which acted on the ring finger and the figure shows the responses to passive movements of the metacarpo-phalangeal joints. During the tests illustrated in figure 1 A the muscles were all relaxed, whereas in figure 1 B the same tests were performed with the subject instructed to press his finger with a constant force against the device to which the fingers were fixed. The upper traces in each section are the instantaneous impulse frequency, the middle traces the angle at the metacarpo-phalangeal joints which were passively changed between 160° and 140°. The lower traces represent the EMG activity. It may be seen that the discharge was very low when the muscles were relaxed and that the dynamic sensitivity to muscle stretch was high. There was also a burst of impulses at the onset of the muscle stretch, suggesting that there was no fusimotor drive to the spindle [BROWN, GOODWIN and MATTHEWS, 1969]. The marked increase of the spindle discharge during the sustained voluntary contraction is obvious in figure 1 B as well as the pronounced irregularity which is characteristic of muscle spindle primary endings subjected to static fusimotor drive [STEIN and MATTHEWS, 1965; MATTHEWS and STEIN, 1969].

Preliminary observations indicated that, for many units, the discharge frequency of spindle afferents was related to the intensity of the skeletomotor contraction under isometric conditions. An example is given in figure 2A from the same unit as in figure 1. The response to an isometric voluntary flexion of the ring finger is shown with the instantaneous impulse frequency in

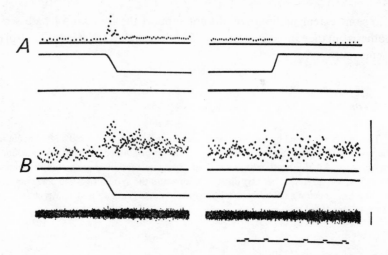

Fig. 1. Response of a muscle spindle primary ending to joint position and passive joint movements when the muscle was relaxed (*A*) and when the muscle was contracting (*B*). The spindle was located in the part of the m. flexor digitorum which acts on the ring finger. From above are shown the instantaneous impulse frequency, the angle at the metacarpo-phalangeal joints and the EMG activity. The angle at the metacarpo-phalangeal joints was passively changed between 140° (top) and 160° (bottom). The interphalangeal joints were kept in 180° extension. During the tests illustrated in *B* the subject was instructed to exert a constant pressure with his ring finger against the device to which the fingers were fixed. Same unit as in figure 2A, 3, 5 and 6. Calibrations: 0 and 100 imp/sec, 0.2 mV. Time signal: 1 sec.

the upper trace, the torque due to active contraction in the middle and the EMG activity below. It may be seen that the discharge rate was considerably higher during the contraction, and further, that this response was strictly confined in time, to the period of the skeletomotor contraction. Although the impulse frequency was quite irregular, there was a close relation between the average frequency and the torque due to active contraction. The most simple relation would be a linear one and figure 2B shows, in a plot, to what extent this was true in semi-stationary states for a representative unit. In this test series the subject performed a number of isometric contractions and he was instructed to keep the intensity constant at a number of different levels. Each point in the plot represents the impulse frequency at one of these different levels, as measured from the number of spikes during 1 sec when the contraction intensity remained within 10% of the mean, as appreciated from the torque signal. It may be seen that stronger contractions were clearly associated with higher impulse rate and *vice versa*. The plot suggests a straight line relation

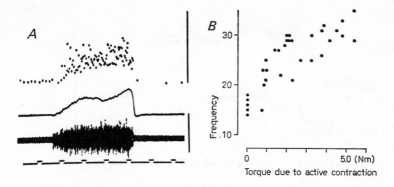

Fig. 2. A Discharge from a spindle primary ending during an isometric voluntary contraction. Same unit as in figure 1, 3, 5 and 8. From above are shown the instantaneous impulse frequency, the torque due to active flexion of the ring finger and the EMG activity of the m. flexor digitorum. Calibrations: 0 and 100 imp/sec, 0.2 mV. The peak torque was 0.13 Nm (metre-newton). Time signal: 1 sec. *B* Relation between the impulse frequency per second of a spindle primary ending and the torque due to active contraction in semi-stationary states under isometric conditions. The ending was located in the ring finger portion of the m. flexor digitorum and the torque was produced by flexions of this finger alone.

between the two variables, implying that the increase of the discharge rate above the resting rate was proportional to the torque due to active contraction. Also in non-stationary states the spindle discharge frequency was closely related to the intensity of the skeletomotor contraction as may be appreciated from figures 2A and 3, which show responses from the same spindle afferent during isometric flexions of the ring finger. The top trace in figure 3 represents the impulse frequency in an analogue form. This display seems more suited than the instantaneous impulse frequency display for a direct comparison with the analogue signal of the torque due to active contraction which is shown by the lower trace. It is obvious that the discharge rate from this spindle afferent was closely related to the contraction intensity also when this varied continuously. These findings indicate that the spindle afferent signal from a contracting muscle was broadly proportional to the intensity of the skeleto-motor output under isometric conditions. Further, as the spindle length did not change appreciably during the contractions, it may be inferred that the fusimotor drive to the spindles was largely proportional to the skeletomotor output. These conclusions are based upon the assumption that the torque due to active contraction as measured in these tests was closely related to the total skeletomotor output to the particular muscle in which the spindle was located,

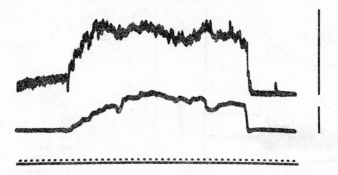

Fig.3. Discharge from a spindle primary ending during an isometric voluntary ontraction. Same unit as in figures 1, 2A, 5 and 6. The top trace represents the impulse frequency of the single afferent and the lower trace the torque due to voluntary flexion of the ring finger. The impulse frequency was not constant before the contraction started but it increased slowly. This is an after-effect of the preceding contraction which was followed by a period of lower discharge rate, as was the illustrated test contraction. Calibrations: 0 and 100 imp/sec, 0.3 Nm. Time signal: 1 sec.

and further, that the discharge rate from the single spindle studied was representative of the composite signal from the spindles as a group in the muscle. It seems that these assumptions are justified, as the same findings were obtained from the vast majority of the spindles, and from spindles located in several different muscles, and as the contractions engaged, as selectively as possible, the muscle in which the spindle was located. Considering current hypotheses concerning the functional role of the muscle spindles, the finding is consistent with the follow-up length servo hypothesis, implying that the skeletomotor output would be the result mainly of the excitatory effects of the spindle afferents. However, the finding is also consonant with the alternative hypothesis that both types of motoneurones are excited from supraspinal structures [GRANIT, 1955; MATTHEWS, 1964; PHILLIPS, 1969]. According to this view the present findings suggest that the excitatory drive from the spindle afferents varies in proportion to the excitatory drive from the supraspinal structures directly onto the skeletomotor neurones. It seems that this type of continuous adjustment of the discharge from the spindles is essential, if the spindle is to be capable of continuously exerting a control function in weak contractions as well as in strong contractions, whatever the exact nature of this control function may be.

However, there were certain deviations from the basic principle that the spindle afferent discharge rate was proportional to the skeletomotor con-

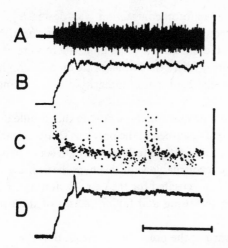

Fig. 4. Discharge from a spindle primary ending during an isometric voluntary contraction. The ending was located 5 cm proximal to the wrist and it responded to a supination of the hand. From above are shown the EMG activity (*A*), the integrated EMG activity (*B*), the instantaneous impulse frequency (*C*), and the torque due to the voluntary contraction (*D*). This signal was not calibrated as the experimental device was not adequate for the measurements of the torque at the appropriate joint. Calibrations: 0.75 mV, 0 and 100 imp/sec. Time signal: 10 sec.

traction intensity. In two respects, these deviations could be related to other elements in the motor acts. It may be seen in figure 3 that during part of the rising phase of the contraction the spindle discharge rate was higher, in relation to the torque due to active contraction at that moment, than it was later on when the contraction had reached a maximum. Another example is given in figure 4, which shows the EMG activity in figure 4A, the integrated EMG activity in figure 4B, the instantaneous impulse frequency from a single muscle spindle afferent in figure 4C, and the force due to active contraction in figure 4D, while the subject performed an isometric supination of his hand. The impulse rate was maximal initially when the contraction force was minimal, and the rate decreased more or less continuously as the contraction intensity increased. This phenomenon was seldom as pronounced as in this example but, for the majority of the spindle afferents, the discharge rate was relatively higher during the rising phase, or part thereof, than it was later on.

Obvious deviations from proportionality between the spindle discharge rate and the skeletomotor output were also seen on the falling phase of the contraction: in many afferents there was a pronounced burst of impulses during this phase, when the skeletomotor contraction was decreasing. It seems that this could well be accounted for by passive re-extension of the spindle by the elastic elements in the muscle. In some units, there were also other deviations, as shown in figure 4 where the impulse rate on occasion increased to very high values while the contraction intensity did not change appreciably.

The systematic deviations from proportionality between the spindle discharge rate and the contraction force described in the previous section clearly constitute circumstantial arguments against the follow-up length servo hypothesis. The findings would only be reconcilable with this hypothesis if mechanisms could be suggested which compensated for the extra excitatory drive from the spindle afferents during the rising and falling phases of the contractions.

The basic inference put forward in the preceding sections, that the fusimotor output was broadly proportional to the skeletomotor output, may form a basis for an analysis of the spindle afferent discharge during isotonic contractions. It was assumed that this principle holds also when a muscle is actively shortening.

In an analysis of the spindle afferent activity during isotonic contractions, it was found that most of the end organs responded with a sustained discharge during the contraction, implying that the fusimotor outflow was powerful enough to off-set the unloading effect of the muscle shortening. Thus, the basic requisite was fulfilled for the hypothesis that the spindles participate in the control of voluntary movements. This type of response was not critically dependent upon the speed of movement but the sustained discharge during shortening was present in slow movements as well as in relatively fast movements. In these tests the muscles mostly worked against a very small extra load in addition to the weight of the fingers.

An example of the events associated with an isotonic contraction is shown in figure 5, where spindle afferent impulses are shown in the top trace, the angle at the metacarpo-phalangeal joints in the middle and the EMG activity below. In the test illustrated the subject flexed his ring finger by approximately 10° from an intermediate position. The finger rested against a stop before the movement started and the flexion was opposed by an extra load corresponding to a twisting force of 0.1 Nm (metre-newton) at the metacarpo-phalangeal joints. This is equivalent to approximately 100 g weight at the finger tip, which is a rather small load in relation to the maximal capacity of the muscle. Hence, the contraction was far from maximal. It may be seen that the contraction was first isometric and the skeletomotor activity increased continuously to over-

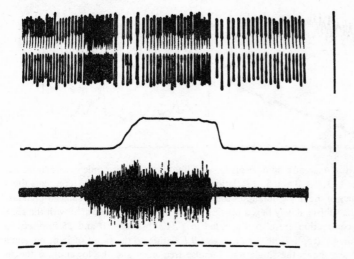

Fig. 5. Response of a spindle primary ending to an isotonic contraction. Same unit as in figures 1, 2A, 3 and 6. From the top shown the single unit impulses, the angle at the metacarpo-phalangeal joint and the EMG activity when the subject flexed his ring finger. The flexion was opposed by a load corresponding to a torque of 0.1 Nm at the metacarpo-phalangeal joint. Calibrations: 100 μV, 150° (bottom) and 140° (top), 0.2 mV. Time signal: 1 sec.

come the load before the movement started. During the movement the EMG activity was maximal and it then decreased to a lower level when the movement stopped and the finger was kept in flexion. At the onset of the contraction the spindle discharge increased, indicating that the sense organ was under a powerful fusimotor control. The impulse rate was maximal before the movement, minimal during the movement and intermediate when the finger was kept in flexion. Thus, it is obvious that the impulse rate was not proportional to the intensity of the skeletomotor activity, as, for instance, it was minimal when the EMG activity was maximal. This finding clearly militates against the follow-up lenght servo hypothesis. On the other hand, the discharge rate was dependent upon the muscle length and the rate of change of the muscle length: first, the impulse rate was higher initially when the muscle was long, compared to the rate at the shorter muscle length; second, the discharge rate was lower during the shortening than it was at the end of the movement when the muscle was short. Thus, the spindle discharge rate was a function partly of an increased fusimotor activity and partly of the muscle length and the rate of change of the muscle length during the isotonic contraction.

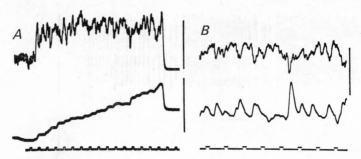

Fig.6. Response of a spindle primary ending during an isotonic contraction. Same unit as in figures 1, 2A, 3 and 5. *A* Relation to joint angle. The top trace represents the impulse frequency of the single unit and the lower trace the angle at the metacarpo-phalangeal joint when the subject slowly flexed his ring finger. The events associated with the second half of this contraction are also illustrated in *B*. Calibrations: 0 and 25 imp/sec, 155° (bottom) and 145° (top). Time signal: 1 sec. *B* Relation to the speed of joint movement. The upper trace shows the single unit impulse frequency and the lower trace the time derivative of the joint angle signal and, hence, it represents the speed of the joint movement. Calibrations: 0 and 25 imp/sec. Time signal: 1 sec.

The response during a slower movement is illustrated in figure 6A. The upper trace represents the afferent impulse rate from the same spindle as in figure 5 and the lower trace the angle at the metacarpo-phalangeal joints, when the subject flexed his ring finger. The extra load on the muscle was very small in this test, corresponding to a twisting force of less than 0.01 Nm at the metacarpo-phalangeal joints. Again the spindle discharge increased considerably when the contraction started but in this test it remained high during the whole movement. Although the impulse rate varied considerably, there was not much of a systematic change which was closely related to the muscle length. Thus, it seems that the sense organ was relatively insensitive to the muscle length. An alternative interpretation is that there was a continuous increase of the fusimotor outflow which compensated for the muscle shortening. Considering the findings from isometric contractions presented in previous sections, this is not an unreasonable assumption, as the skeleto-motor output increased continuously when the shortening progressed in this test. This was seen in a simultaneous recording of the EMG activity, not illustrated in the figure. A closer analysis of the two records suggested that the changes of the impulse frequency, as shown by the ups and downs of the upper trace, were related to the speed of joint movement. This is shown more explicitly in figure 6B where parts of the same events are illustrated on an expanded time scale, but now with the time derivative of the joint angle signal below and the impulse frequency above. Thus, the lower signal is an approximation of the speed of muscle shortening. It may be seen that the two signals go in opposite directions in most instances, implying that the faster the muscle was shortening, the lower was the discharge rate and *vice versa*. With regard to the central effects on the spinal level, this finding indicates that the excitatory drive onto the skeletomotor neurones decreases immediately when the movement progresses faster and increases when the movement progresses slower. In this way any variations in the speed of movement—which could be due to a variation in the load, a variation in the frictional resistance or an irregular skeletomotor output—would be reduced in amplitude and the result would be a smoother movement.

Thus, it seems that this finding supports the hypothesis that at least one function of the muscle spindle primary endings in voluntary contractions is to contribute to the control of the velocity of movement [MATTHEWS, 1964; PHILLIPS, 1969]. Considering, on the other hand, the follow-up length servo hypothesis, this rather predicts the opposite finding in this test, namely that the two variables would follow each other in the ups and downs in figure 6 B.

Comments

The present findings show that there are spindle afferents which may provide a signal well suited for reducing variations in the speed of movement. However, it has not been assessed whether all the group I spindle afferents may provide a signal of this character, nor under which conditions a spindle gives this type of response. It is also obvious from the present findings that the spindle primary endings provide a continuous excitatory drive to the motoneurones during voluntary contractions and movements. This excitation must certainly assist in maintaining the contraction. On the other hand, the present findings do not tell to what extent the excitatory effect from the spindle afferents is critical for the contraction. It has earlier been shown that it is not at all involved in the initiation of the skeletomotor output at the onset of a contraction [VALLBO, 1971] but it is probably not justified to extrapolate too far from the initial events with regard to the role of the muscle spindles during a lasting contraction, as the synaptic influence upon the skeletomotor neurones is different in important respects at the onset of a contraction from that prevailing during a lasting contraction: the Renshaw feed-back as well as the inhibition from Golgi tendon organs are not present initially but enter immediately after the onset of the skeletomotor output.

Acknowledgements

This investigation was supported by grants from the Swedish Medical Research Council, projects No. B 70-14 X-2075-04 B and B 71-14 X-2075-05 C, and from the Medical Research Council of the Swedish Life-Offices. The author is indebted to Mr. G. WESTLING for valuable technical assistance.

Author's address: Dr. Å.B. VALLBO, Department of Physiology, University of Umeå, S-901 87 Umeå (Sweden)

New Developments in Electromyography and Clinical Neurophysiology,
edited by J. E. Desmedt, vol. 3, pp. 263–272 (Karger, Basel 1973)

Recordings from Muscle Afferents
in Parkinsonian Rigidity[1]

B. G. WALLIN, A. HONGELL and K.-E. HAGBARTH

Department of Clinical Neurophysiology, Academic Hospital, Uppsala

Whereas the immediate cause of rigidity in Parkinson's disease is well
established (a pathological hyperactivity in the skeletomotor neurons leading
to tonic involuntary muscle contractions), the role played by the fusimotor
neurons in this syndrome still remains unclear. From animal experiments two
types of muscular hypertonus are well-known, one in which the hyperactivity
is confined to the skeletomotor neurons and another in which increased fusimotor activity leading to enhanced tonic stretch reflexes is an important feature. Most hypotheses put forward to explain parkinsonian rigidity have been
based on these two experimental models and, consequently, the rigidity has
been suggested to be either of the 'α or the γ type'. A variety of indirect experimental methods such as recordings of mechanically or electrically elicited
stretch reflexes, sometimes in combination with reinforcement manoeuvres or
fusimotor blockades, has been used to test these hypotheses in man, but both
the results and the conclusions reached have varied considerably between
different investigators. As emphasized in a recent review by DIETRICHSON
[1971] these contradictory results may partly be due to inadequately standardized test methods, and after a careful re-investigation of the problem with
special regard to the methodological difficulties, DIETRICHSON concluded [in
agreement with RUSHWORTH, 1960] that hyperactivity in static fusimotor
neurons is an essential feature in parkinsonian rigidity. However, although this
re-evaluation of previous data may remove some of the confusion in the literature, an increased understanding of the pathophysiology of rigidity requires
more direct measures of muscle spindle function in man.

1   Supported by Swedish Medical Research Council Grant No. B 70–14 X–2881–01.

With the introduction of the microneurographic technique of HAGBARTH and VALLBO [1968] such a method became available, and in the present investigation a comparison is made between the inflow of nerve impulses from the muscle spindles of healthy subjects and rigid patients. The method permits both single- and multi-unit recordings but we have decided to use multi-unit recordings for the initial analysis for the following reasons: (a) multi-unit recordings are technically easier to perform, especially in patients with motor disorders causing involuntary muscle contractions, and (b) as multi-unit recordings from muscle nerves are dominated by I_A afferent units (see below), each multi-unit record will be a measure of the average behaviour of the spindles supplied by the impaled fascicle. In contrast, large samples of single units are necessary before statistically reliable comparisons can be made between different groups of persons.

This report will first describe the findings leading to the conclusion that a multi-unit recording from a muscle nerve can be used as an index of the average I_A afferent activity, and then discuss some of our results from parkinsonian patients with bearing on the pathophysiology of rigidity.

Material and Methods

Normal subjects. The characteristics of the multi-unit record in response to passive stretch, voluntary contraction, etc., are based on recordings in well over 50 muscle nerve fascicles in 8 healthy adult subjects. Nerve block experiments were performed successfully on 2 occasions in one subject.

Parkinsonian patients. Multi-unit recordings were made from 32 muscle nerve fascicles in 9 patients (47–58 years old), with Parkinson's disease since 1–8 years. In all cases rigidity was the dominant symptom, but in most of them some tremor and/or akinesia were also present.

Nerve recordings. All recordings on the parkinsonian patients and on most of the normal subjects were made in the median nerve at the elbow level where fascicles supplying the finger flexor or the pronator muscles in the forearm were impaled. Some of the recordings on the normal subjects were made in the peroneal nerve at the fibular head in fascicles supplying the anterior tibial or the peroneal muscles. No important differences were noted between the results obtained in the two nerves.

The general experimental procedure and the procedure employed for identifying an impaled nerve fascicle were the same as described previously [HAGBARTH and VALLBO, 1968; VALLBO, this volume]. When analyzing the results, the nerve signals were quantified with the aid of an electronic integrating circuit (time constant 0.06 sec) and in the figures, the 'mean voltage neurogram' obtained in this way is shown together with the original nerve record. During the experiment, the subject was usually sitting with a slightly abducted arm lying on a table where the hand and fingers were attached to an adaptable lever. Transducers connected to the lever measured angular movements either at the wrist or at the metacarpophalangeal joints, as well as the force developed in active pronation or finger flexion movements against resistance. Passive movements of fingers and hand were manually controlled. During recording from the peroneal nerve, the patient lay on an experimental table and corresponding movements and measurements were performed in the ankle joint.

The EMG activity from the muscles involved was recorded with either surface- or concentric needle electrodes.

Partial nerve blocks were made by injecting 15–30 ml 0.5% Lidocaine around the median nerve proximal to the recording site.

Results

A. Normal Subjects

1. General Characteristics of Multi-Unit Responses in Muscle Nerves

Usually, in a resting, relaxed subject, when recording from a fascicle innervating a muscle held at short or intermediate lengths, there was EMG silence and no steady-state nerve activity was seen. Such nerve activity, clearly discernible from the noise, occurred only if the muscle was stretched far beyond its resting length and, even then, the sustained nerve activity was less pronounced than during rather weak voluntary isometric contractions. In contrast to the low static sensitivity to stretch, the dynamic sensitivity was high and even at short muscle lengths weak taps against the muscle belly or small rapid passive joint movements caused transient bursts of impulses in the multi-unit neurogram. Similarly an electrically elicited muscle twitch gave rise to a burst of activity during the stretch phase. Isometric voluntary contractions always elicited typical changes in the multi-unit neurogram: a marked sustained increase of activity persisting during the whole contraction period without sign of adaptation or fatigue.

As discussed by HAGBARTH and VALLBO [1968], these findings can be explained if it is assumed that the multi-unit neurogram is dominated by impulses from primary spindle endings. The basis for this suggestion is illustrated in figure 1, where the similarities between the responses of a single I_A afferent unit and the multi-unit responses to passive stretch, isometric voluntary contraction and an electrically elicited twitch are clearly shown.

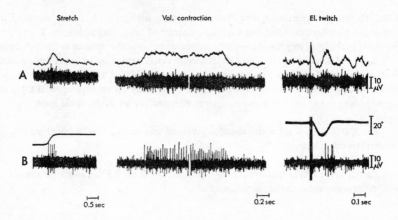

Fig. 1. Comparison between a multi-unit recording from a pure muscle nerve fascicle and a single-unit recording from a I_A afferent fibre in such a fascicle. *A* Multi-unit recording from a fascicle innervating the finger flexor muscles. Original nerve record shown in the *lower tracing. Upper tracing* is a mean voltage record obtained by integrating the original neurogram (time constant 0.06 sec). *B* Single I_A afferent unit *(lower tracing)* from finger flexor muscle (different subject from A). The *upper tracing* shows movements in the metacarpo-phalangeal joints and is applicable to both *A* and *B*. Note similarities between single- and multi-unit responses during passive stretch, voluntary isometric contraction and electrically elicited muscle twitch. The rhythmic response after the electrical twitch is due to bouncing of the plate supporting the arm.

2. Effects of Fusimotor Nerve Blocks on Multi-Unit Muscle Nerve Activity

a) Voluntary Contractions

When recording from a single I_A afferent unit, the sustained increase of activity seen during isometric voluntary contractions is taken as good evidence that the contraction is associated with increased fusimotor activity to the spindles [HAGBARTH and VALLBO, 1968; VALLBO, 1970]. If this view is correct and if the multi-unit neurogram is dominated by I_A afferent impulses, the increased nerve activity seen during isometric contractions should be substantially reduced by a fusimotor nerve block.

Such selective nerve blocks were performed on two occasions and, as shown in figure 2, the results were in good agreement with the postulate. Whereas in the control situation each isometric finger flexion caused a pronounced sustained increase in the multi-unit neurogram, after the nerve block contractions of similar strength gave rise only to slight increases of neural activity (fig. 2A). It was also shown that, before the nerve block, rapidly alter-

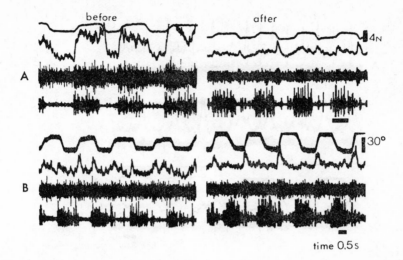

Fig. 2. Multi-unit responses recorded in a median nerve fascicle innervating the finger flexor muscles obtained before (left) and after (right) injecting 30 ml 0.5% Lidocaine around the nerve proximal to the recording site. Healthy subject. In *A* isometric, and in *B* isotonic (load 400 g) finger contractions were performed. *Upper tracings.* Mechanogram. In the isometric case shown as force and in the isotonic as movement (extension in the metacarpophalangeal joints shown as upward deflection). *Second tracings.* Integrated neurogram. *Third tracings.* Original nerve record. *Lower tracings.* Finger flexor EMG.

nating voluntary finger movements gave rise to one burst of impulses at the start of the flexor contraction and one during the stretch phase. After having blocked the fusimotor fibres, however, the contraction response was almost totally abolished but the stretch response remained at least as prominent as before (fig. 2B).

During the anaesthesia the maximal muscle strength of the finger flexors was reduced but in one experiment could be partly restored by high frequency vibration (≈ 100 c/sec) against the forearm muscles. It was also clear that the nerve block caused the subject to overestimate the muscle power evolved both during maximal and submaximal contractions. In conformity with this he reported that during the anaesthesia he had to develop a greater voluntary effort than before to be able to lift a test weight.

b) Passive Stimuli

During the nerve block, the effect of passive movements and local muscle taps were also tested and the results compared with those obtained in the

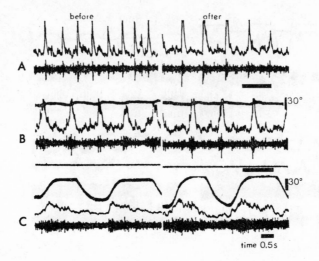

Fig.3. The effect of 0.5% Lidocaine on multi-unit nerve responses from a relaxed muscle. Same experiment as in figure 2. *A* Nerve responses to local taps against the muscle belly. *Upper tracings* show integrated and *lower* original nerve records. *B* Nerve responses to rapid small passive stretches. *Upper tracings* show movements in the metacarpo-phalangeal joints (extension upwards). *Second and third tracings* show integrated and original neurograms, respectively. *Lower tracings* show finger flexor EMG. *C* Nerve responses to slow, large passive movements. Same tracings as in *B* except that finger flexor EMG is not included.

control situation. As shown in figure 3, which is taken from the same experiment as figure 2, the nerve responses to taps against the muscle belly and small, rapid as well as large, slow, passive stretches were essentially unaltered by the anaesthesia.

B. Parkinsonian Patients

1. Muscle Nerve Activity at Rest

Although instructed to relax the arm completely, some degree of EMG activity was often present in the finger flexor or pronator muscles of the parkinsonian patients. In most cases a 'spontaneous' multi-unit nerve activity was simultaneously recorded in nerve fascicles innervating these muscles. The spontaneous nerve activity was sometimes seen when no EMG activity could be detected, but it could be temporarily abolished during the rising phase of an electrically induced muscle twitch. If the degree of rigidity varied, the spontaneous nerve activity usually varied in parallel with the EMG activity, e.g. it increased with increasing degree of rigidity. In the experiment illustrated in figure 4, the patient was quite rigid during the period shown in the left part of the figure and during this phase both EMG and nerve records showed more activity than a little later when the rigidity was less pronounced (right part of the figure).

2. Muscle Nerve Activity During Passive Movements and During Voluntary Contractions

Figure 5A shows that in parkinsonian patients small rapid *passive stretches* gave rise to fairly prominent bursts of impulses similar to those seen in controls (cf. fig. 3). However, during maintained stretch the responses were quite different. Whereas in the normal subject maintained stretch caused little or no sustained activity, in the rigid patient the increase of nerve activity had a tonic character and lasted throughout the whole stretch period (fig. 5B).

As when recording from a healthy subject, a *voluntary isometric contraction* performed by a rigid patient generally caused a sustained increase of multi-unit activity in the nerve supplying the contracting muscle. However, the increase was often weak or moderate in comparison with that seen in normals (fig. 5C). Due to the high background activity, the difference may have been more apparent than real but on the other hand the nerve response to voluntary contraction was sometimes weaker than to maintained stretch, which was not seen in normal subjects.

Discussion

A. Normal Subjects

The results of the partial nerve block experiments give strong support to the notion that multi-unit neurograms are dominated by I_A afferent activity. The result agrees with the fact that most single units encountered in muscle nerve fascicles can be identified as primary spindle afferents [HAGBARTH and VALLBO, 1968; VALLBO, 1970] but as each fascicle contains several fibre types with different functional characteristics (e.g. skeletomotor neurons, tendon organ afferent neurons, group II afferent neurons) the I_A dominance may nevertheless seem surprising. Although several factors no doubt contribute to give this effect [HAGBARTH and VALLBO, 1968] the main reason is probably

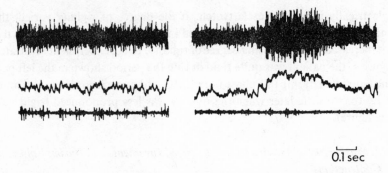

0.1 sec

Fig. 4. The effect of different degrees of rigidity on the spontaneous multi-unit activity recorded in a median nerve fascicle innervating the finger flexor muscles in a patient with Parkinson's disease. Left and right part of the figure were taken with an interval of less than 1 min. *Upper tracings.* Multi-unit nerve activity. *Middle tracings.* Integrated neurogram. *Lower tracings.* Finger flexor EMG. Note that nerve and EMG activity change in parallel. The prominent neural burst seen in the right part of the figure is a passive stretch response and was included to show that recording conditions had not deteriorated.

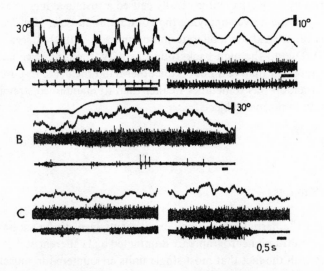

0,5 s

Fig. 5. Recording from median nerve fascicle innervating the finger flexor muscles in patient with Parkinson's disease. *A* Nerve responses to rapid small (left) and slow large (right) passive movements. *B* Nerve response during a large long-lasting passive stretch. *C* Nerve response during isometric voluntary finger flexions. *Upper tracings* (in *A* and *B*), movements in the metacarpo-phalangeal joints (extension upwards). *Second travings* in *A* and *B* and *upper* in *C*, integrated neurograms. *Third tracings* in *A* and *B* and *second* in *C*, original neurograms. *Lower tracings,* finger flexor EMG.

that the recording method itself involves a bias for activity in large diameter afferent axons with low threshold.

It has long been postulated that skeletomotor and fusimotor neurons are co-activated during voluntary muscle contractions and that the fusimotor activity prevents muscle spindle unloading during the contraction [for review see GRANIT, 1970; chapt. VII]. Previous microneurographic recordings in man have supported this hypothesis [HAGBARTH and VALLBO, 1968, 1969; VALLBO, 1970, and this volume] and it is further strengthened by the outcome of the partial nerve block experiments. The reduction of maximal muscle power noted after the nerve block may have been due to imperfect selectivity of the block, i.e. some skeletomotor neurons may have been anaesthetized together with the fusimotor nerve fibres. However, other factors could also contribute. The preliminary observation that high frequency vibration of the muscle partly restored maximal power could indicate that some 'servo loop assistance' is necessary for the maintenance of normal muscle strength.

In contrast to the marked effects during voluntary contractions, the partial nerve block did not appreciably change the afferent multi-unit responses to passive stretch. This suggests that in a completely relaxed muscle there is little or no fusimotor bias on the spindles—in other words, the so-called α–γ linkage seems to be quite firm also during voluntary relaxation.

B. Rigid Subjects

As all necessary controls have not yet been performed, the interpretation of our results from the rigid patients must still be regarded as preliminary. However, assuming that the properties of the spindles themselves are not greatly changed, it seems fairly safe to conclude that the results do not agree with the hypothesis that the involuntary contraction in rigidity is of the 'pure α type'. If this were the case, the contraction should unload the spindles and no afferent activity should be seen in the neurogram. Therefore, the finding that there is a 'spontaneous' static afferent activity from rigid muscles which increases with increasing degree of rigidity, suggests that both skeletomotor and fusimotor neurons are activated during the involuntary contraction.

The next step in the analysis would be to decide whether the *balance* between skeletomotor and fusimotor neuron activity is the same in rigidity as during normal voluntary contractions. During a normal voluntary isometric contraction, the fusimotor drive seems to increase largely in parallel with the strength of the extrafusal contraction [VALLBO, this volume]. Our data are at present insufficient for making a similar quantitative evaluation in patients. It should be pointed out, however, that even if it is assumed that the 'α–γ balance' is altered in patients with Parkinson's disease, the disturbance need not be the same during the *involuntary* contraction causing rigidity as when the patient makes a *voluntary* muscle contraction. For example, although one explanation for both the increased afferent activity during apparent EMG silence and the altered response to passive stretch could be a relative fusimotor dominance, it cannot be ruled out that the patient has better voluntary control over his skeletomotor than over his fusimotor neurons (as could be suggested from records like those shown in figure 5C). In view of such possibilities it seems important that a forthcoming analysis of muscle spindle behaviour in parkinsonian rigidity includes a comparison between normal and rigid muscles in similar states of contraction. In addition, it is desirable that the multi-unit recordings should be complemented by representative single-unit samples.

Author's address: Dr. B.G. WALLIN, Department of Clinical Neurophysiology, Academic Hospital, *S-750 14 Uppsala 14* (Sweden)

New Developments in Electromyography and Clinical Neurophysiology,
edited by J. E. Desmedt, vol. 3, pp. 273–276 (Karger, Basel 1973)

Central *versus* Proprioceptive Influences in Brisk Voluntary Movements[1]

H. C. HOPF, K. LOWITZSCH and H. J. SCHLEGEL

Department of Neurology, University Hospital, Göttingen

Inhibitory as well as excitatory reactions can be induced at the spinal level by activating the afferent fibres of the proprioceptive systems. These reactions can easily be demonstrated if the motoneurons are in a steady state of activation. This is the case, for example, when a muscle is kept in a certain position. Inhibition results in a reduction of the muscle activity and in a decrease of the monosynaptic and polysynaptic reflex discharges. Excitation causes the opposite reactions. As soon as the steady activation of the spinal motoneurons is influenced by supraspinal impulses the above mentioned mechanism is disturbed.

Until now it has been uncertain whether supraspinal structures or proprioceptive afferents are more effective in regulating the motoneuron output. This study examines what happens when supraspinal activation interferes with proprioceptive influences.

Methods

The experiments were performed on 19 healthy subjects. The EMG activity of the biceps and triceps brachii muscles was recorded simultaneously by an oscillograph. The mechanogram of the movement was taken from the forearm. The subjects were blindfolded throughout the experimental procedure and were instructed to adapt the muscle force relying only on verbal information about what was going on.

Supraspinal activation was brought about by a brisk voluntary movement. Proprioceptive stimulation was produced by several procedures, such as weight-loading the muscle, muscle vibration, direct electrical stimulation of the active muscle or of its antagonist, and also by eliciting a stretch reflex.

Weights between 0.25 and 5 kg were used. In particular, attention was payed to the

1   The investigations were supported by the Deutsche Forschungsgemeinschaft (SFB 33).

situation when the subject was told to adjust the muscle force to 0.5 kg but the weight to be lifted was 2–5 kg. For vibration the vibrator described by HAGBARTH and EKLUND [1966] was fixed over the distal tendon of the biceps. Single electrical stimuli were applied to the muscle and caused a movement of the forearm of about 2 cm. The stretch reflex was induced by dropping a weight of 0.25 kg from a distance of 30 cm against the forearm being elevated by the biceps brachii.

Results

1. A simple analysis of a brisk voluntary movement shows that the muscle is already active 50–100 msec before the arm contacts the weight. During this period no peripheral regulating influence occurs. Also the subsequent activation pattern remains unaltered for at least 150 msec even if the conditions require an increase or a decrease of activity.

2. Vibratory stimuli do not influence the pattern of voluntary activation, regardless of whether vibration starts within a variable time before the movement begins or during the initial phase of the movement (fig. 1).

3. Direct electrical stimulation of the active muscle is followed by a silent period lasting about 100 msec, if the muscle is kept under slight steady activation [GRANIT, 1950; HUFSCHMIDT, 1966]. In our experiments it was impossible to demonstrate any effect of muscle stimulation on the activation pattern of a brisk voluntary movement. Whether the agonist or the antagonist was stimulated made no difference (fig. 2).

4. A mechanically induced monosynaptic reflex discharge in the active muscle appeared only up to 40–50 msec before the initial activation and during the beginning of the after-activity. Yet the normal activation pattern was not disturbed (fig. 3).

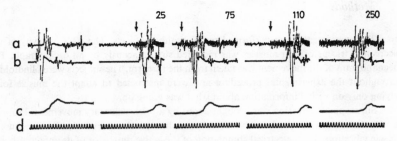

Fig. 1. Brisk voluntary movement and vibration. The normal innervation pattern is shown on the left. In the other records vibration is started (arrow) at variable intervals (number = msec) before the movement begins. *a* biceps brachii; *b* triceps brachii; *c* mechanogram; *d* time base 50 c/sec.

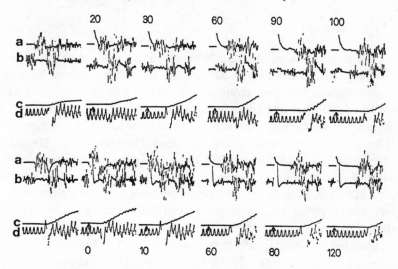

Fig.2. Brisk voluntary movement and direct electrical stimulation of the active muscle (above) and of its antagonist (below). The stimulus is applied (arrow) at variable intervals before the initial muscle activity. Technical details are as in figure 1.

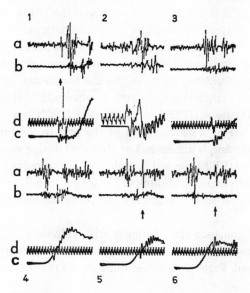

Fig.3. Brisk voluntary movement and the monosynaptic stretch reflex. A reflex discharge (arrow) can be observed only when the interval between the reflex activity and the start of voluntary activity is more than 40 msec or when the reflex activity is in time with the start of the afteractivity. *a–d* as in figure 1. The short biphasic deflection of trace d indicates the movement when the forearm is pushed down by the dropping weight.

Discussion

The innervation pattern of a brisk movement starts with synchronous muscular activity lasting about 50–70 msec. It is often preceded by a short initial inhibition. The first activation is followed by a second inhibition, which lasts up to 100 msec and appears simultaneously with the activation of the antagonist. Subsequently after-activity can be recorded which corresponds to the force necessary for maintaining the intended position [HOPF, HANDWERKER, HAUSMANNS and POLZIEN, 1967]. If this activation pattern were mediated by the γ loop one would expect additional proprioceptive impulses to have a strong influence. If, however, the pattern were due to direct activation of the spinal motoneurons from supraspinal levels the voluntary activity might be expected to predominate over the peripheral input.

Our results demonstrate that during a brisk voluntary movement: (1) there is a considerable delay of 150 msec or more in the efficacy of the regulating mechanisms; (2) the activation pattern is not altered by vibratory stimuli; (3) the inhibitory effect of direct electrical muscle stimulation is suppressed, and (4) the monosynaptic reflex discharge is suppressed during the first inhibition, the first activation, the beginning of second inhibition, and the beginning of the after-activity of the muscular activation pattern.

Some other findings should be mentioned in this context. The contraction of a muscle is adjusted exactly to the expected force requirement before the movement is initiated, i.e., a considerable time before the muscle can learn by experience what activity is truly adequate [HOPF, HANDWERKER, HAUSMANNS and POLZIEN, 1967]. HAMMOND [1956] and HAGBARTH [1967] reported that the stretch reflex, induced by a sudden push of the arm in the same direction as the original movement, is diminished or abolished during voluntary muscle activity. SOMMER [1939] observed that the unloading reflex disappeared during voluntary activation. According to HAGBARTH and EKLUND [1966] the effect of vibration on slow movements can be counteracted by voluntary effort.

From these findings we conclude that during a brisk voluntary movement the impulses from supraspinal structures dominate the level of excitation of the spinal motoneuron pool. Perhaps this does not concern the period of after-activity.

Author's address: Prof. H. C. HOPF, Neurologische Universitätsklinik, v. Siebold-Strasse 5, *D-34 Göttingen* (FRG)

Proprioceptive Reflexes and the H-Reflex

New Developments in Electromyography and Chimical Neurophysiology,
edited by J. E. Desmedt, vol. 3, pp. 277–293 (Karger, Basel 1973)

Methodology of the Hoffmann Reflex in Man

M. Hugon

Department of Neurophysiology, University of Provence, Saint-Jérôme, Marseilles

Electrical methods of eliciting the monosynaptic reflex from the calf
muscles, nowadays known as the Hoffmann- or H-reflex have been described
by Hoffmann [1922], Magladery *et al.* [1952], Paillard [1955] and others.
The present study attempts to better define the parameters of the H-reflex and
the methods used for its study. The presentation of quantitative data will also
be discussed.

The H-reflex is a brief contraction of the calf muscle and we are concerned
here only with its electrical aspects. The general arrangement for testing the
H-reflex is presented in figure 1.

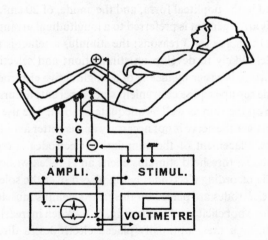

Fig. 1. Experimental set up for studying the H-reflex. Recording sites in the soleus
muscle (S) and in the gastrocnemius (G). The supports immobilizing the knee are not illustrated.

Position of the Subject

A semi-reclining position is preferred by most subjects who can then remain comfortable for a prolonged period of examination. The head and arms are supported, to eliminate variations in reflex excitability which may be caused by changes in head position affecting the labyrinthine influence on the spinal cord, as well as by contractions of neck and shoulder muscles which can amount to unwanted Jendrassik manoeuvres. The foot maintained on a foot-plate is flexed dorsally to a position which places the soleus muscle under adequate stretch so that the Achilles tendon reflex is easily elicited. The foot-plate is then locked at that angle. The knee is flexed to about 120° and maintained at that angle. In this way the soleus inserted on the foreleg is stretched optimally for the activation of monosynaptic reflexes. The gastrocnemius inserted on the femur bone is relaxed by the knee flexion, thereby reducing the depressive influence which the afferents from that muscle may have on the soleus reflex loop. Under these conditions, the contraction of the soleus is isometric while that of the gastrocnemius is isotonic.

Stimulation

The stimulating electrodes are placed on the dorso-ventral axis; the cathode, of 2 cm² is placed in the popliteal fossa, and the anode, of 20 cm², is placed on the patella. This arrangement is preferred to a longitudinal arrangement of electrodes along the nerve for 3 reasons: the stimulus artefact is reduced; anodal block is less likely to develop on stimulation; and selective stimulation of the nerve trunk is easier by means of a single active electrode. The SIMON [1962] electrode set-up is quite convenient since it permits accurate placement of the cathode on the skin as well as adequate fixation. The use of a needle electrode inserted near the nerve is not necessary and is better avoided with non-patient subjects. Placement of the stimulating electrodes is considered to be adequate when a threshold stimulus elicits an H-reflex without any direct M-response, the recording electrodes being placed over the soleus muscle. When recording electrodes are placed on the gastrocnemius muscles, the same threshold stimuli do not elicit any H-reflex (fig. 2, 3). When increasing the stimulus delivered through these optimally placed electrodes, a direct M-response will be evoked, first in the soleus and then in the gastrocnemius. Another point is that, when recording from the soleus muscle, the M- and H-responses will have a similar configuration for stimulation at the optimal

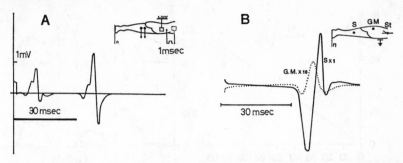

Fig. 2. Responses to popliteal stimulation in the triceps surae. *A* Bipolar recording with skin electrodes optimally placed on the soleus muscle. Notice the similarity of the wave form of the M- and H-responses. *B* Monopolar recording of the H-reflex in relaxed muscles triceps surae, with the active recording electrode placed either on the soleus (continuous line) or on the gastrocnemius (dotted line); in the latter case the amplification was increased by a factor of 10. It is likely that the small (H' response recorded monopolarly over the gastrocnemius is in fact a genuine soleus response.

position in the popliteal fossa. The latter in fact permits a roughly selective activation of the nerve fibers of the soleus. When the stimulating electrodes are placed more medially or more laterally in the popliteal fossa, an M-response appears first in the gastrocnemius and only for stronger stimuli in the soleus (fig. 3).

The stimulus duration should be 1 msec, which makes it more selective for the afferent axons with long utilization times. Shorter durations would favour the activation of α motor axons [see VEALE *et al.*, this volume]. To insure a stable intensity of the stimulus during the course of the experiment, one should reduce the skin resistance, use non-polarizable electrodes and, if possible, use a constant current stimulator with high internal resistance.

The frequency of stimulation is low enough for the third H-reflex of a series to be more than four-fifths the height of the first reflex. Stimuli are given every 5 sec, or sometimes at longer intervals: a shorter interval should not be used. Long intervals are necessary to avoid the 'late depression' described by MAGLADERY (called by PAILLARD: zone V). When looking for experimental changes in H-reflex amplitude, it is necessary to adjust the stimulus intensity to obtain an H-reflex of 50% of the maximum H-reflex amplitude in the subject investigated, thereby permitting both increases and decreases of the response in the course of the experiment [COQUERY, 1962].

The number of tests necessary to determine a mean value and its standard deviation depends on the variability encountered in any given subject. About 20 measurements are usually sufficient, and certainly at least 10 should be

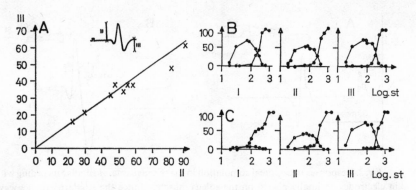

Fig. 3. Evidence for the rough proportionality between amplitudes of the different components of the H-reflex. *A* Amplitude of the negative component of the H-reflex (abscissa) plotted as a function of the amplitude of the larger positive component of the same response (ordinate) in one subject. Notice the proportional relation. *B, C* Recruitment curves of the M- and H-responses plotted in different ways. The recording is bipolar from soleus in *B*, and monopolar from the same skin position in *C*. The H-reflexes are all produced by the soleus, but the M-responses are contributed only by soleus in *B* (selective bipolar recording) whereas a contribution from gastrocnemius must also be present in *C*. As a result the M recruitment curve is spurious in *C*. The 3 adjacent diagrams were plotted using the amplitudes of the first, second and third components of the triphasic wave form, respectively.

used. For a quick test, 7 measurements can be taken, and the 2 extreme values omitted when calculating the mean. Our results were obtained from a population of 20 young female adults, and each mean value was calculated on the basis of at least 10 measurements.

The recording electrodes are placed 3 cm apart on the skin over the soleus muscle which produces the H-reflex. They are placed along the mid-dorsal line of the leg, about 4 cm above the point where the 2 heads of the gastrocnemius join the Achilles tendon. The active electrode is proximal. The amplitude of the reflex is measured on the largest deflexion. We found a fairly constant proportionality between the sizes of the negative and the positive componets of the response, and we generally measure the larger of the two (fig. 3 C).

Electrical Response and Mechanical Response

Contraction in the calf muscles can be recorded under isometric conditions by a strain gauge fixed under the support of the metatarsal region. Figure 5 shows that the force recorded is proportional to the corresponding

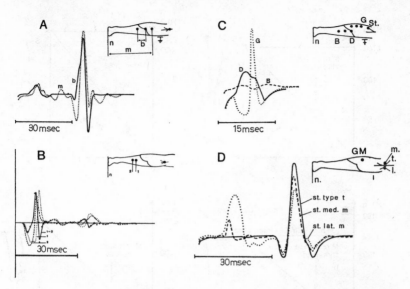

Fig.4. Problem of selectivity in recording M- and H-responses from triceps surae. *A* Monopolar (....) and bipolar (——) recording (see inset diagram of the leg) of the same H-reflex elicited by popliteal stimulation. The bipolar recording in optimal position eliminates some of the irrelevant activities seen on the monopolar record. *B* Three superimposed traces of the same M-response recorded monopolarly from electrode No. 1 (....), from electrode No. 2 (————) and bipolarly from electrodes No. 1 and 2; the latter trace is exactly the difference between the first two (calculated substraction indicated by dots). *C* Estimation of diffusion of gastrocnemius M-response to the soleus recording site. The nerve to medial gastrocnemius is stimulated to elicit a large M-response recorded from electrode G. The same response is much attenuated at electrodes D and even more at B (monopolar recording). *D* M- and H-responses elicited from various electrode positions in the popliteal fossa. —— = optimal placement on the nerve fibres to soleus with no M-response present in spite of a large H-reflex. ———— = stimulating electrode moved slightly medially. .... = same moved slightly laterally. The H-reflex is a little smaller in both latter instances, but a direct M-response appears which is not produced by soleus (see text). It must be stressed that these trace were recorded monopolarly from a site over the gastrocnemius, thus emphasizing the importance of optimal placement of the stimulating electrode in relation to the soleus nerve. Other records document the necessity for selective recording from soleus.

H- or M-response, whether recorded by monopolar electrodes or bipolar (Fig. 4). The relationship no longer obtains if the stimulus evokes both M- and H-responses, whose mechanical effects summate while the electrical signs remain distinct. These results apply only to the conditions considered above and they would not be obtained with different positions of the limb segments

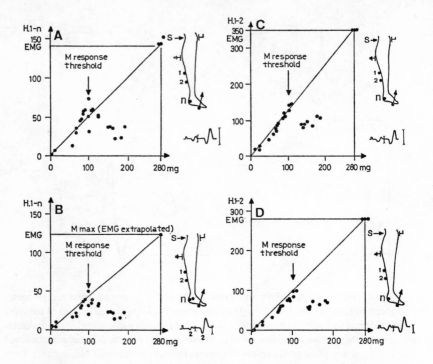

Fig. 5. Relation between EMG and mechanical responses in triceps surae. Popliteal stimulation elicits either pure H-reflex or both M- and H-responses depending on the intensity. Abscissa: mechanical force in mg. Ordinate: amplitude of the H-reflex EMG recorded either monopolarly (*A, B*) or bipolarly (*C, D*) over soleus. H-responses are measured peak-to-peak in *A* and *C*, while only for the main component in *B* and *D*. The scatter of the data is less with bipolar recording. The vertical arrows indicate the M-response threshold, on the right side of which the myogram increases while the H-reflex decreases.

and a different degree of stretch on the soleus. The linear relationship between electrical and mechanical responses means that the EMG of each reflex does provide a fairly good estimate of the strength of the reflex. The maximal M response represents excitation of all the motor fibres of the nerve to soleus and, therefore, the activity of 100% of the soleus motoneurones. The maximal M-response therefore represents a standard by which the H-reflexes are to be compared in different subjects. It can be assumed, for instance, that an H-reflex equivalent to 25% of the maximum M-response represents activity in 25% of the soleus motoneurones relevant to soleus. Measuring H- and M-responses by comparison with the maximal M-response in this way means that responses which have a different absolute value can, nevertheless, be compared on the basis of their relative values. However, this procedure can only be employed if the maximal M-response represents activity in the soleus muscle alone. This can be ascertained by checking whether the M- and H-responses do indeed present a similar configuration. If they do not, one should probably change the positions of the recording electrodes to minimize pick-up of the M-response produced in the nearby gastrocnemius muscle.

The Soleus Electrical Response and the Muscle Architecture

Dissection of soleus muscles obtained at autopsy indicates that the muscle fibres are short, about 40 mm in the adult. The end-plates stained by the acetylthiocholine method of Koelle-Couteaux appear located in the middle part of the muscle fibres and only one end-plate is seen on each muscle fibre [cf. DESMEDT, 1958] (fig. 6). The action potential elicited in a muscle fibre by the motor axon starts at the end-plate and propagates on either side towards the tendon ends. The corresponding electrical field produced in the extracellular space is triphasic [cf. LORENTE DE NÓ, 1947]. With synchronous activation of a bundle or sheet of muscle fibres arranged side by side, these electrical fields summate consistently and can be recorded at a distance. With monopolar recording with respect to a distant reference electrode, the wave form of the H-reflex is indeed triphasic [fig. 6; HUGON, 1962] which suggests that most soleus muscle fibres are probably activated simultaneously. If the recording electrode is close to the motor end-plates, on the surface of the muscle or in its belly (fig. 6) the initial positive deflexion is absent and the electrical response starts with a negative component (fig. 7). If the recording electrode is near the insertion of the muscle into its tendon, the terminal positive phase is reduced or absent (fig. 7C, 8A) [DESMEDT, 1958].

The electrical field is strong in the vicinity of the fibres and weaker further away. Figure 8A shows the importance of this gradation. One consequence is that threshold responses can be detected more readily with intramuscular needle electrodes than with surface electrodes on the skin. A fine intramuscular electrode, 100 μm in diameter and insulated except at the tip, does not distort the distribution of potentials and will record the potential field produced by the muscle fibres adjacent to it. A larger needle of 250 μm will reveal both the general electric field and the local activity of individual motor units (fig. 8B). The unit activities are distributed over a period of about 12 msec, which gives some indication of the desynchronization of soleus motor units in the H-reflex, and also in the T-reflex.

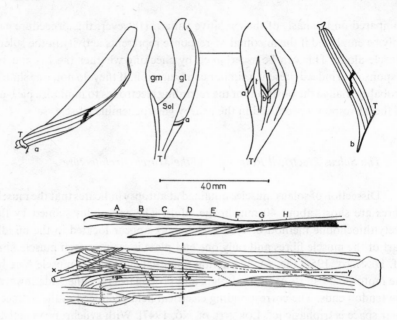

Fig.6. Anatomical structure of the triceps surae. Above and centre: posterior and anterior aspects of the triceps. T = achilles tendon. a and b = lines of sampling of the muscle shown on a larger scale at the left (a) and right (b). The 40-mm scale applies to these enlarged pictures. The area of the motor end-plates can be seen halfway along the muscle fibres (method of Koelle-Couteaux). Below: general posterior view of the limb muscles. gm = skin projection of the point of entry of the nerve into the medial gastrocnemius muscle. gl = the projection for the gastrocnemius nerve; Sa = the projection for the anterior soleus nerve; Sp = the projection for the posterior soleus nerve; Jp = the projection for the nerve to tibialis posterior. Longtitudinal section between X and Y: general orientation of fibres in the gastrocnemius and soleus muscles. Note the relative positions of the proximal insertions of the gastrocnemius and soleus muscles.

The same technique shows that the synchronization of units in the direct M-response is roughly the same, which suggests that the desynchronization is due to the intrinsic pattern of motor innervation in this muscle. In other words the synchronization in the efferent H-reflex volley is as good as in the direct α motor axon volley elicited by a brief stimulus [cf. GASSEL and OTT, this volume].

The field potential change begins and ends at the same time throughout the soleus muscle (fig. 7). The true latency of the H-reflex (or the M- or T-response) should be measured to the time of onset of the change in field potential, that is the moment when depolarization begins in the region of the motor end-plates (see also fig. 8A). Measurements based on later components of

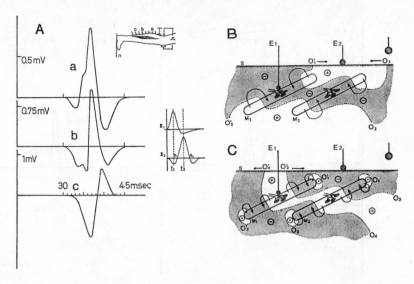

Fig.7. Potential field in soleus during an H-reflex. *A* Same H-reflex recorded monop-olarly from the skin over the soleus at the points a, b and c (see inset). Positive is downwards. In the triphasic wave form, the first phase corresponds to the synchronous development of depolarization in the stimulated fibres, the second phase represents this depolarization underneath the focal electrode, and the last phase corresponds to the end of depolarization. The last positive component is missing in the c derivation which is over the extreme end of the muscle. B Diagram of the principle of the triphasic field change at the moment t1 in the small middle diagram. The isoelectric boundaries O1 and O2 separate the positive from the negative zones in the volume conductor. For an active monopolar recording in E1 the EMG is negative and maximal, with no initial positive phase. At the same moment, the EMG is positive in E2. *C* The same diagram for the instant t2. The wave of depolarization has reached the tendon ends of the muscle fibres. In the centre, repolarization has begun. For an active monopolar recording in E1 the EMG has become positive, and of low ampli-tude. In E2 the EMG has become negative and now reaches its maximum, which is small by comparison with the maximum in E1. Eventually E2 becomes weakly positive, when the isoelectric surface O3–O'3 has reached the ends of the muscle fibres.

the electric response are complicated by muscle conduction time and field distortion which are, however, constant for a given muscle.

Records are usually made with bipolar electrodes. The resultant electrical response represents the difference between the 2 monopolar recordings cor-responding to each of the electrodes with reference to a common indifferent electrode (fig. 4B). The bipolar recording eliminates distant activities and emphasizes the activity from the nearby sources. When the electrodes are closer together the selectivity of the method is increased but activities of more

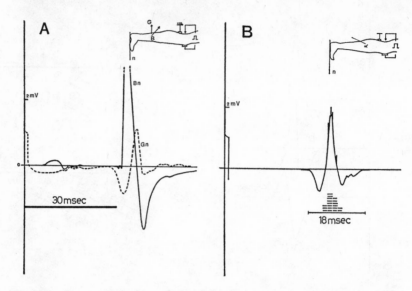

Fig.8. H-reflex recorded with surface and depth electrodes in soleus muscle. *A* The same response recorded by a depth electrode (Bn) and by a focal surface electrode (Gn). Compare with E1 and E2 of figure 6. Electrode B is situated in the region of the motor end-plates. *B* H-reflex recorded with a fine intramuscular electrode, inserted at a little distance from the motor end-plate region. The EMG potential appears as a triphasic wave with superimposed activity of muscle fibres immediately adjacent to the electrode tip. These motor unit potentials show that the H-reflex takes place within a period of about 12 msec. The synchronization of the H-reflex is equivalent to that of the M-response recorded in the same way, and it must reflect the pattern of neuromuscular activation in soleus.

distant motor units can then be ignored. For practical purposes a useful compromise is achieved by using recording electrodes of 1 cm² placed 3 cm apart on the skin, in the optimal position considered above which permits rather selective pick-up of the soleus activity with little contribution from the gastrocnemius. It may be pointed out in passing that bipolar electrodes placed over the gastrocnemius cannot be so selective as they will pick up activity from the soleus located underneath.

Recruitment Curves for M- and H-Responses

The recruitment curves relate the changes in amplitude of the M- and H-responses as a function of the intensity of the electrical stimulus. Under the experimental conditions described above, stimulation in the popliteal

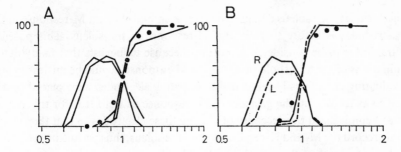

Fig. 9. Proposed design for the representation of recruitment curves. Abscissa: stimulus intensity plotted on a logarithmic scale; the intensity evoking an M-response of amplitude 50% of maximum M is taken as unity. Ordinate: voltage of the largest component of the M- or H-responses plotted on a linear scale, with 100 representing the amplitude of the maximum M-response. *A* Two trials on the same subject on one side. The stability of recording conditions is indicated by the similarity of M recruitment curves. The similarity of the H recruitment curves is taken to indicate comparable excitability along the H-reflex loop in these 2 trials. *B* Comparison of right (R) and left (L) sides in the same normal subject. The asymmetry of H-reflex recruitment curves is considered significant in view of the similarity between the M recruitment curves on the 2 sides.

fossa will produce an H-reflex in the soleus and an M-response in both the soleus and the gastrocnemius. We have found that a reflex does not appear in the gastrocnemius unless some voluntary contraction background is present in that muscle, either tonic or phasic. However, other authors have not always found this [KOTS and KRINSKI, 1967].

a) Recruitment curves for direct M-responses. With a cathode alongside the nerve to soleus an M-response is obtained, and the recruitment curve is S-shaped (fig. 9). This shape can be explained from the distribution of α fibres in the nerve trunk. The largest fibres are excited first, but they are few in number and the M-response is of small amplitude. As the stimulus increases, fibres of higher threshold are excited, and these are more numerous, so that the M-response increases rapidly. Lastly, the smallest α fibres, of higher threshold, are excited, but these are again rather less numerous, and the curve rises more slowly again. Figure 10 provides a diagrammatic illustration of this: the figures are based on observations on the cat by ECCLES and SHERRINGTON [1932]. The cumulative curve of recruitment for the α fibres of the soleus (cS) follows an S-shape which results by integration of the histogram of nerve diameters. The gastrocnemius fibres are larger in the histogram and the cumulative recruitment curve (cG) is shifted along the abscissa. If the same situation is applied in man a stimulating cathode lying fairly proximally over the common trunk to the soleus and gastrocnemius should elicit an M-response in the gastrocnemius before there is any M-response in the soleus. This possible situation is to be considered important because it suggests that many determinations of the 'threshold of the direct M-response' would be misleading for evaluating the soleus H-reflex if the recording electrodes were placed on the leg so as to pick up the gastrocnemius response as well. It is for this reason that I emphasized the necessity of placing these electrodes so that they would only record the M- and H-responses from the soleus, thus permitting meaningful comparisons of their respective thresholds.

The experimental conditions vary from one subject to another and from one experiment to another using the same subject, just because of small differences in the placement of electrodes. Using several experiments as a basis for each set of figures eliminates these variables and shows up only the significant results, as in the following procedure (fig. 9).

1. The maximal M-response is taken as 100% and other responses are expressed relative to this maximum.

2. The stimulus which evokes an M-response of 50% is taken as unity and other values for the stimulus are expressed relative to this.

3. The amplitude of the M-responses is plotted on an ordinate with a linear scale and the corresponding stimulus amplitudes appear on the abscissa on a logarithmic scale.

4. The dimensional ratio between the 1–10 scale on the logarithmic abscissa and the 0–100 scale on the linear ordinate is fixed arbitrarily at 2.66. If these conventions are always used it is possible to compare the M-curves obtained in the course of different experiments and such curves are, in fact, closely similar (fig. 11).

The use of a logarithmic scale is suggested because of the following: the threshold stimulus current eliciting an M-response and the stimulus current necessary to elicit a maximum M-response present a roughly constant ratio of 2. The constancy of this intensity ratio means that the distance along the abscissa between the threshold and the maximum intensities for M would be fairly constant when plotted on a logarithmic scale, even though their absolute values (and thus also their difference) would vary among the subjects and among the experiments. This rather constant ratio is not surprising as it must result from the roughly standard Gaussian scatter of α motor axon diameters in the nerve to soleus. On the other hand, the scaling of the abscissa which is based on taking as unity the stimulus intensity eliciting a 50-percent M-response can be justified by the steep slope of the M recruitment curve at that point. It does, in fact, insure that the recruitment curves from different

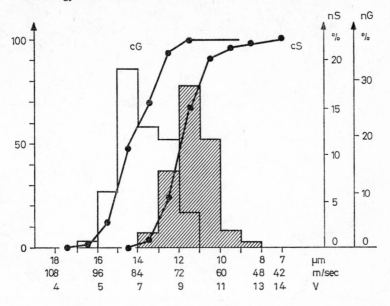

Fig. 10. Tentative explanation of the M recruitment curve by the distribution of diameters of efferent α axons. The diagram is based on the data of Eccles and Sherrington [1930] for axon diameter in the cat. Abscissa: fibre diameter with corresponding conduction velocity and voltage of a threshold electrical stimulus to the nerve trunk. The scales are supposed to be roughly proportional. Ordinate on the right, percentage α efferent axons in the histograms for the soleus (nS) and the gastrocnemius (nG) nerves. Ordinate on the left, percentage of α efferent axons recruited in either nerves when the voltage of the electrical stimulus is increased; 100% means that all axons in the nerve have been activated. This theoretical recruitment curve assumes that the threshold for each axon is proportional to its diameter. The recruitment curve would also describe the muscle response in first approximation if the motor units thus recruited are of comparable size. Notice that, in the cat, the α efferents to gastrocnemius have a slightly larger diameter than those to soleus. It is not known whether this also applies to man.

subjects would be shifted only minimally along the abscissa with respect to one another.

b) Recruitment curves for the H-reflex. The H-curves are defined by the threshold for eliciting the reflex, by the maximum H amplitude obtainable, and by the characteristics of its extinction when stimuli of greater intensity produce occlusion between antidromic and reflex orthodromic impulses in the same α motoneurones. The H-reflex begins to disappear as the M-response increases and it is completely absent for a stimulus which produces a maximum M-response. Persistence of a response is attributed to antidromic excitation of the initial segment of the motoneurone axon (F-response). The H-reflex

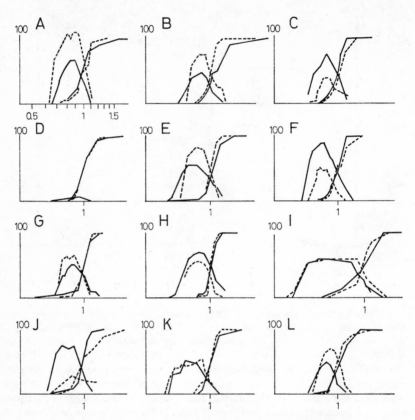

Fig. 11. H and M recruitment curves in 12 normal adult subjects. Bipolar recording with surface electrode from soleus. Same conventions for diagrams as in figure 9. —— = right side. - - - - = left side. In view of the constancy of the M-curves, the asymmetry of the H-curves is significant: it is not constitutional but represents variability in monosynaptic excitability. The curves D, I and J, repeated several times, apparently represent unusual anatomical arrangements, probably atypical distribution of fibre size.

appears to contain the reflex activities of motor units supplied by α fibres of high threshold since it is necessary to achieve a maximal M-response in order to occlude the H-reflex completely.

The H maximum is defined by its mean value which I found as 52% of M maximum for 52 curves obtained from 12 normal adults examined on both sides. Its distribution was such that 42 of 52 values for H maximum lay between 35 and 75% of M maximum (fig. 12 B).

The summit of the H-curve often has the form of a plateau: the H-reflex stops increasing before the M-response has begun to increase. It appears that

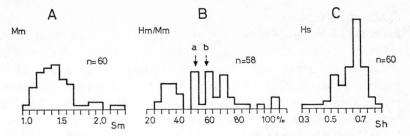

Fig. 12. Quantitative data on the H- and M-responses, based on 60 recruitment curves recorded on both limbs in 12 adult normal subjects. The ordinates represent the number of trials. *A* Stimulus intensity required to elicit a maximum M-response, the intensity for eliciting a 50-percent maximum M-response being taken as unity. *B* Ratio between maximum amplitude of H-reflex and maximum amplitude of the corresponding M-response, on a percentage scale. *C* Threshold stimulus intensity required to elicit an H-reflex, the intensity eliciting a 50-percent maximum M-response being taken as unity.

antidromic activity or recurrent activity (refractoriness; Renshaw inhibition) is not responsible for this failure to further increase.

The threshold values for H were obtained, in 51 cases from 60, by stimuli between 0.5 and 0.75 with a modal value of 0.65 as expressed with respect to the stimulus intensity eliciting a direct M-response of 50% of its maximum amplitude (see above) (fig. 12C). In the same subject, the maximum of the H-reflex can vary even though the H-reflex threshold would be the same. Thus, there is little correlation between the threshold excitability of the reflex and the maximal excitability of the same reflex.

Occasional peculiarities may be seen: in one subject there was no H-reflex on the right or left and it appeared only with voluntary contraction of the soleus. In another subject the recruitment curve for H was very far to the right of the curve for M; further studies of this case suggested that the curve for M soleus was in fact contaminated by a gastrocnemius M-response, elicited at low threshold and developing into a very large maximum M-response (fig. 4D). This gastrocnemius interference of course upsets the scaling system which implies that both M- and H-responses be produced exclusively by the soleus muscle, as discussed above.

When genuinely recorded, the H-reflex appears for a stimulus intensity of 0.6 along our conventional scale. Its maximum amplitude reaches about 50% of the maximum M amplitude. The H-reflex amplitude starts diminishing when the threshold for the soleus M-response is reached.

The M recruitment curve should be reproducible if it is assumed (see above) that it indeed reflects the scatter of electrical excitabilities in a standard, roughly Gaussian, population of soleus motor axons. If this be accepted, one could say that when identical M recruitment curves are recorded in separate experiments on a given subject, the technical conditions for stimulation and recording have been fairly similar. Furthermore, if similar H recruitment curves are also observed the excitability of the H-reflex loop must have remained the same. On the other hand, if the M recruitment curves are similar but the H recruitment curves differ significantly, this would indicate a change in the H-reflex loop excitability. The latter could be assessed by changes in H threshold stimulus and in H-reflex maximum amplitude. All this is readily performed with the proposed method of presenting the results.

Suggestions for a Practical Protocol of H-Reflex Testing

When the subject is seated and in a state of complete relaxation and with the electrodes and limb positions arranged, the following procedures should be carried out in turn:

1. Discover the position for the stimulating cathode which can produce an H-reflex without any sign of an M-response, from either soleus or gastrocnemius. Use threshold stimulation, with a frequency of 0.5 stimuli/sec.

2. Lower the frequency of repetition to 1 in 5 sec or more to study the H-reflex.

3. Determine the stimulus which will evoke a maximum M-response at which the H will be completely occluded. Measure M (3 estimates).

4. Diminish the stimulus and measure M (3 estimates) and H (5 estimates) at each level.

5. Determine the stimulus just necessary to obtain the maximal M-response again (3 estimates). Measure the corresponding H (5 estimates).

6. Diminish the stimulus in about 5 steps of one-tenth of its value which will generally bring the stimulus intensity below the threshold of H. At each step of stimulus intensity measure M 3 times and H 10 times.

In this protocol it is necessary to make a total of about 100 measurements, each taking about 5 sec, so that the whole period of observation will last about 12 min.

Conclusions

Methodology and parameters of the H-reflex in man have been presented critically and the significance of the responses currently recorded has been discussed on the basis of new evidence. One of the points deserving emphasis is that the placement of the recording electrodes should be such as to insure selective pick-up of the direct M-response of the soleus muscle, with little or no contamination of the record by the direct M-response from gastrocnemius. In most studies the M-response threshold stimulus and its maximum amplitude are used as references to evaluate the H-reflex [PAILLARD, 1955]. When the muscles are completely relaxed, the H-reflex elicited by electrical stimulation in the popliteal fossa only involves the soleus muscle. It is only when the gastrocnemius has a background of voluntary contraction that a reflex response appears in that muscle.

The H-reflex recorded from a relaxed subject will thus, in fact, be a soleus reflex even for different placements of recording electrodes on the posterior aspect of the foreleg. The situation is not so straightforward for the assessment of M-responses which appear both in the gastrocnemius and in the soleus. One must, therefore, place the recording electrodes over the soleus muscle in such a way as to minimize or eliminate the pick-up of the gastrocnemius M-response. Under such conditions, which are easily achieved after a few trials, the wave forms of the M- and H-response will be similar, as they now both reflect electrical responses of the soleus activated by a fairly synchronous volley in its α motor axons (fig. 2A). It then becomes possible to quantitate much more accurately the H-reflex features and recruitment curve in relation to the genuine soleus M-response since a major spurious source of inconsistency has been eliminated. With a new method for displaying the results in a special diagram the values of stimulus (specified on the basis of the soleus M-response) can be scaled so that the M and H recruitment curves would nearly coincide in different experiments on the same subject and in different subjects (fig. 9, 11). The suggested protocol permits evaluation of the necessary data for these curves in about 12 min, even though the stimuli are applied at intervals of no less than 5 sec (minimum required to avoid depression of the H-reflex) and averages of several responses rather than isolated values are plotted.

The presentation of such detailed procedures may be useful to help improve current methods of recording M- and H-responses in man and to standardize the methods for displaying the results. Special methodologies can, of course, be derived easily from this basic protocol in order to explore specific problems.

Author's address: Prof. MAURICE HUGON, Laboratoire de Neurophysiologie, Faculté des Sciences Saint-Jérôme, Traverse de la Barasse, *F- 13 Marseille 13ᵉ* (France)

New Developments in Electromyography and Clinical Neurophysiology,
edited by J.E. Desmedt, vol. 3, pp. 294–307 (Karger, Basel 1973)

Maturation of Human Reflexes[1]

Studies of Electrically Evoked Reflexes in Newborns, Infants and Children

R. F. MAYER and R. S. MOSSER

Department of Neurology, University of Maryland School of Medicine, Baltimore, Md.

Maturation of reflexes from fetal to adult life raises important problems. Studies have been carried out both in man and animals concerning the anatomical development of the nervous system [CAJAL, 1909; ARIËNS KAPPERS, HUBER and CROSBY, 1936]. Extensive studies of human fetal and newborn behavioral reflexes have been reported [LANGWORTHY, 1933; WINDLE, 1940; CARMICHAEL, 1954; HOOKER, 1952, 1954]. Many of these studies have shown a cephalocaudal and proximal-distal development of the nervous system [KINGSBURY, 1932]. Recent studies in animals indicate that the cervical areas of spinal cord develop sooner than the lumbar areas [EKHOLM, 1967; MELLSTRÖM and SKOGLUND, 1969; SCHEIBEL and SCHEIBEL, 1971]. In man and animals the diameters of axons in peripheral nerves are small at birth and gradually enlarge reaching adult values by the age of 5 years in man [REXED, 1944; NYSTROM and SKOGLUND, 1966; GUTRECHT and DYCK, 1970; EKHOLM, 1967] and the internodal lengths also increase [VIZOSO and YOUNG, 1948; GUTRECHT and DYCK, 1970]. Most of the early studies concerned with maturation of the nervous system have been anatomical or behavioral, but during the past 20 years there have been numerous neurophysiological studies as well. EEG and cerebral-evoked potential studies of cerebral maturation have been summarized by PURPURA and SCHADÉ [1964] and MINKOWSKI [1967]. SKOGLUND [1966, 1969] summarized maturation of spinal reflexes in animals. This chapter surveys the maturation of spinal reflexes in man. At present, there is limited information on cutaneous polysynaptic and long loop reflexes, and most studies have been concerned with the monosynaptic H-reflex

1   Supported in part by a research grant (No. MH1 7006) from the National Institutes of Health, United States Public Health Service.

[HOFFMANN, 1918, 1922; MAGLADERY and McDOUGAL, 1950] that will be considered from birth to adulthood.

Conduction Velocity in Peripheral Nerves

In 1960, THOMAS and LAMBERT reported that the velocities in motor fibers in the ulnar nerve of newborns were about one-half those of normal young adults. In premature infants the mean velocity was reported to be 21 m/sec and in full-term newborns 28. By the age of 3 years, most of the values were in the low adult range and at 5 years, the velocities were the same as in adults. In premature infants SCHULTE, MICHAELIS, LINKE and NOLTE [1968] found motor nerve conduction velocities as low as 12 m/sec (ulnar) and 6 m/sec (tibial) at 25 weeks of gestation and motor nerve conduction velocity appears to correlate with gestational age rather than with birth weight. The rate of maturation may vary in different sensory and motor nerves, but the fastest velocities in newborn nerves are nearly similar (25–30 m/sec). Additional studies are needed concerning the maturation of sensory fibers [see DESMEDT, NOEL, DEBECKER and NAMÉCHE, this series, volume 2]. The velocities in human newborn nerves are greater than those reported in newborn kittens.

The H-Reflex

Low intensity electric stimulation of the posterior tibial nerve evokes the classical H-reflex response in the calf muscle [HOFFMANN, 1918, 1922; see HUGON, this volume]. In the normal relaxed adult subject, reflex responses with H-reflex properties may also be recorded in other muscles of the limbs, but not in the small muscles of the hand or foot. However, in adults with an upper motoneuron lesion, H-reflexes can be recorded also in small hand and foot muscles [TEASDALL, PARK, LANGUTH and MAGLADERY, 1952; GARCIA-MULLIN and MAYER, 1969].

THOMAS and LAMBERT [1960] reported that H-reflexes could be evoked in small hand muscles in normal prematures, newborns and infants of less than 6 months of age, but no H-reflexes were recorded in 60 subjects aged 2–14 years. They related this to the immaturity of the nervous system at birth. Persistence of the H-reflex in hand muscles beyond the age of 1 year has been interpreted as evidence of CNS dysfunction [HODES, GRIBETZ and HODES, 1962; FRENCH, CLARK, BUTLER and TEASDALL, 1961; STIMSON, KHEDER,

HICKS and ORLANDO, 1969]. Additional studies of the H-reflex in normal newborns have provided further information [HODES and GRIBETZ, 1962; HODES and DEMENT, 1964; BLOM, HAGBARTH and SKOGLUND, 1964; MAYER and MOSSER, 1967, 1969a].

Threshold

In awake and quiet human newborns, H-reflexes are easily evoked in calf, foot, forearm and hand muscles by an electric stimulus which is subthreshold for motor axons and thus elicits no direct M response. In most newborns a large nearly maximal H can be seen before any M response occurs. As the intensity of the stimulus is increased, the M response enlarges and the H-reflex decreases. With maximal stimuli for motor fibers the H-reflex disappears as in the adult [see HUGON, this volume]. This requires a large stimulus since the electric threshold is greater in newborn peripheral nerve than in adults.

Latency

The latency of the H-reflex recorded in calf muscles of newborns is approximately half that recorded in adults (table I) [MAYER and MOSSER, 1969a]. The overall latency decreases slightly during the first year as the conduction velocity in peripheral nerves increases. Thereafter, the latency increases with age and growth in length of the nerves and does not reach adult values until after puberty and cessation of growth of the extremity.

Table I. H-reflexes in the triceps surae muscle (mean values and range)

Age	Latency, msec	Estimated Velocity, m/sec[1]	Amplitude, mV	H/M ratio
1–3 days	15.7 (13–17)	28.5 (24.8–37.0)	5.8 (2.2–8.2)	54.7 (41–67)
1–5 months	14.3 (14–15)	37.3 (32.1–47.9)	5.4 (2.3–9.4)	–
6–12 months	14.9 (13.5–16.5)	45.8 (40.7–51.6)	6.4 (2.8–12.2)	–
1–6 years	16.8 (14.0–19.5)	59.5 (46.9–83.5)	11.6 (4.3–20.0)	65.2 (54–73)
20–50 years	28.8 (26–32)	82.6 afferent 81.5 efferent	13.1 (3–19)	55.1 (27–75)

1 Overall conduction velocity.

The latencies of the H-reflexes elicited in hypothenar muscles by ulnar nerve stimulation at the wrist, are greater (17–18 msec) at birth than at 6–12 months (15–16 msec). Similar observations have been made of the H-reflex recorded in plantar foot muscles when the posterior tibial nerve is stimulated at the ankle: 23–24 msec at birth and 18–19 msec at 6–12 months.

THOMAS and LAMBERT [1960] measured the velocity of afferent fibers in the ulnar nerve subserving the H-reflex in newborns and found a mean value 30.1 m/sec. This velocity was roughly 10% faster than that in efferent fibers (27.5 m/sec) but the difference was not statistically significant. A similar relationship between the velocities in afferent and efferent fibers in adult man has been observed [DAWSON, 1956; MAYER and MAWDSLEY, 1965]. The velocities in the fibers which subserve the H-reflex increase with age at a rate similar to other peripheral nerve fibers. The adult velocities in the sciatic nerve to calf muscles are greater than those in the ulnar nerve or posterior tibial nerve to foot muscles; the adult values in the sciatic nerve are reached at a somewhat later age [MAYER and MOSSER, 1969a].

Amplitude

The H-reflex is characteristically a triphasic response of short duration suggestive of a synchronous discharge. In newborns, the H-reflex in calf muscles is significantly smaller than in infants aged 1–6 years or in adults (table I) [MAYER and MOSSER, 1969a]. In the awake quiet newborn, there is a slight fluctuation in the amplitude of the responses evoked every 10–30 sec as occurs in adults.

The amplitude of the awake newborn H-reflex recorded in many muscles is variable, as in adults. Factors which increase the amplitude are mild active contraction of the muscle, passive stretch of the muscle, reinforcement maneuvers such as biting and grasping and labyrinthine vestibular stimulation. Factors which decrease the response are active contraction of antagonists or strong contraction of the active muscle, passive shortening of the muscle, strong electric stimuli, rapid rate of stimulation (especially more than 1/sec), tapping of the tendon jerk or over the recording muscle and strong flexion and extension of the neck or other muscles. Consideration of these factors is important in any study of the H-reflex.

Post-Tetanic Potentiation

Following tetanic stimulation, e. g. at 400–500/sec, for 10–20 sec, there is augmentation of the H-reflex in adult muscles [HAGBARTH, 1962; CORRIE and HARDIN, 1964; MAYER and MAWDSLEY, 1965]. BLOM, HAGBARTH and SKOG-

LUND [1964] studied such post-tetanic potentiation (PTP) in human infants. They reported PTP in older infants, aged 11 days to 2.5 months but stated it was difficult to produce this in two 2-day-old newborns. MAYER and MOSSER [1969a] were unable to demonstrate PTP of the H-reflex in calf muscles in 5 newborns less than 3 days old but did observe it in infants aged 8 days to 6 months. However, in some newborns PTP could be recorded in small hand and foot muscles. PTP can only occur if a functional subliminal fringe is present. The lack of PTP of the monosynaptic reflex in newborn kittens has been interpreted either by the smallness of the subliminal fringe, or by a deficiency in transmitter mobilization during repetitive stimulation in the (immature) presynaptic terminals [ECCLES and WILLIS, 1965; SKOGLUND, 1960c, 1966, 1969; WILSON, 1962].

Excitability Cycles

The excitability of α motoneurons in man can be tested by studying the recovery cycle of the H-reflex after a conditioning electric stimulus (fig. 1) [LANGUTH, TEASDALL and MAGLADERY, 1952; PAILLARD, 1955; ZANDER OLSEN and DIAMANTOPOULOS, 1967; GARCIA-MULLIN and MAYER, 1969, 1972; GASSEL, 1970]. Similar studies have been reported in newborns, infants and children [MAYER and MOSSER, 1969a]. In this report excitability cycles of H-reflexes were recorded in calf, hand and foot muscles. The curves recorded at various ages are shown in figure 2. The excitability cycles in calf muscles of newborns showed no evidence of 'early facilitation' of the test response with stimuli of equal intensity separated by 2.5–10 msec. However 'early facilitation' was observed in calf muscles of infants aged 8 days to 17 months. It was also present in hand and foot muscles of newborn infants. Following

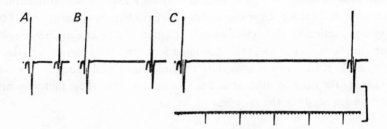

Fig. 1. Recordings of H-reflexes from calf muscles in an awake newborn using paired stimuli. The test response follows the conditioning response by 100 msec *(A)*, 200 msec *(B)* and 500 msec *(C)*. Note the rapid recovery of the test responses. Calibration: 10 and 100 msec, 2 mV.

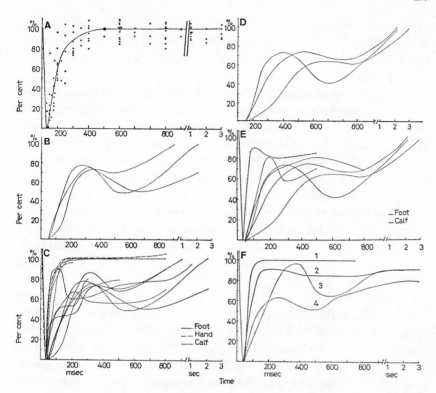

Fig. 2. Excitability cycles of H-reflexes recorded in calf muscles, or hand and foot muscles where indicated, of humans from birth to adulthood. The amplitude of the test response measured as a percentage of the conditioning response is plotted against the time between stimuli. *(A)* Birth, *(B)* 6 months, *(C)* 6–12 months, *(D)* 12–18 months, *(E)* 1–3 years, *(F)* Adults, 1–2, Spastic paraparesis, 3–4, Normals. Note the change in the cycles recorded from calf muscles at and after 6 months and the similarity of the cycles recorded in adults with spastic paraparesis and those in newborns. Each cycle or point represents the mean of the responses recorded from different subjects [MAYER and MOSSER, 1969a].

the 'early depression' at 10–20 msec which was frequently very short (30 msec) and in some incomplete, the test response became greater than 50% of the conditioning response at 100 msec and this has been termed 'rapid recovery'. The response returned to approximately 100% of control by 200–300 msec, 'complete recovery', and persisted for 2–3 sec. No 'late depression' occurred between 500 and 800 msec as in adult curves. This cycle in newborns is monophasic and differs from the adult biphasic curve. The cycles recorded from hand and foot muscles in newborns were similar to those in the calf except for the 'early facilitation'. The rate of recovery of the test H-reflex was similar in hand, foot and calf muscles and there was no indication that recovery was faster in upper than in lower extremities.

When the conditioning shock was subthreshold for H-reflexes and the test stimulus produced a maximal H without a direct M response, the excitability cycles recorded in calf muscles of newborns differed from those shown in figure 2. The curves were more variable. Some showed slight facilitation between 100 and 300 msec and other showed mild depression at 400–800 msec, that recovered by 3 sec. This conditioning stimulus had little but variable effect on the amplitude of the test response and the effect could not be predicted nor correlated with any specific factors in newborns. Studies in adults have suggested that the predominant influence of a subthreshold conditioning shock on the H-reflex is depression of the response but an intercurrent facilitation at 50–300 msec occurs in some [TABORIKOVA and SAX, 1969; TABORIKOVA, this volume].

There was little change in the excitability cycle from birth through the first 5 months of life. Although there were variations from one infant to the next, it was not until the age of 6–12 months that definite changes occurred. First, the recovery of the test response recorded in calf muscles was less than 50% at 100 msec, 'slow recovery' (fig. 2, table II). The test response was usually less than 80% of the control by 300 msec, 'incomplete recovery'. In most subjects, 'late depression' at 500–700 msec was not observed and the test response returned to 100% by 2–3 sec. There was some variation in the cycles recorded in normal infants aged 6–12 months in that the cycles were either of the newborn or the adult pattern.

After the first year of life, the excitability cycles continued to show 'early facilitation' and slow and incomplete recovery. However, the curves became biphasic and the 'late depression' observed at 400–800 msec recovered by 2–3 sec. This curve pattern persists throughout normal development in children and in adults (fig. 2, table II).

In hand and foot muscles the H-reflexes were small and the cycles recorded after 6 months were more variable than those in newborns. In the hand, the curves were generally monophasic and the recovery was rapid, slow, complete or incomplete in different subjects (table II). In the foot the curves were usually rapid and complete in recovery and there was less change with maturation than in hand or calf muscles. H-reflexes were not present in hand and foot muscles after the first year of life except in one normal subject aged 13 months (fig. 2). In adult patients with chronic upper motoneuron lesions,

Table II. Motoneuron excitability in newborns, infants and children

Age	Site recorded	Early facili-tation	Excitability cycle	Post-tetanic potentiation
1–3 days	hand	+	RR, C, monophasic	±
	foot	+	RR, C, monophasic	±
	calf	0	RR, C, monophasic	0
4 days to 5 months	hand	+	RR, C, monophasic	+
	foot	+	RR, C, monophasic	+
	calf	+	RR, C, monophasic	+
6–12 months	hand	+	RR, SR, I, C, monophasic	+
	foot	+	RR, C, monophasic	+
	calf	+	SR, I, monophasic	+
1–50 years	hand	0	0	0
	foot	0	0	0
	calf	+	SR, I, biphasic	+

+ = present; 0 = absent; RR = rapid recovery, > 50% at 100 msec; SR = slow recovery, < 50% at 100 msec; C = recovery near 100%; I = recovery incomplete, < 80%.

H-reflexes can be recorded in small hand muscles and their excitability cycles are similar to those recorded in newborns [GARCIA-MULLIN and MAYER, 1972].

Additional studies have been carried out on 10 normal infants aged 5–12 months [MAYER and MOSSER, 1969 b] in hand, calf and in some foot muscles (table III). In the calf, the cycles change and recovery is usually slow and incomplete. However, in foot muscles there is little change in the curves until the response is no longer recordable. The H-reflex in hand muscles is more variable and recovery may be incomplete or complete. The cycles differ from one infant to the next suggesting that the rate of maturation also differs in different individuals. Also the maturation of the reflexes in different segments in the same infant varied so that in one infant aged 10 months, the cycle in calf muscles was of the 'adult' type, in the foot it was of the 'newborn' type and in the hand it was of the 'infantile' type. In another infant aged 11 months, the cycle in the hand was of the 'newborn' type while in the calf it was 'infantile'. In another aged 12 months, the response in the hand was of the 'adult' type (absent), in the calf it was 'infantile' and in the foot it was of the 'newborn' type.

Table III. Maturation of H-reflexes in infants

Age, months	Calf	Foot	Hand
5	RR, C, monophasic	NT	small, variable
6	SR, C, monophasic	NT	*absent*
8	RR, I, monophasic	RR, C, monophasic	SR, I, small
8	SR	present	NT
9	SR, C, monophasic	RR	SR, small
10	SR, I, monophasic	RR	small, variable
10	SR, I, biphasic	RR, C, monophasic	SR, I
11	SR, I, monophasic	NT	*absent*
11	SR, I, monophasic	NT	RR, C, monophasic
12	SR, I, monophasic	RR, I, monophasic	*absent*

NT = not tested; RR = rapid recovery, > 50% at 100 msec; SR = slow recovery, < 50% at 100 msec; C = recovery near 100%; I = recovery incomplete and less than 80%.

These observations provide additional data that maturation proceeds in a rostral-caudal direction and includes H-reflexes in the hand before those in the foot. However, changes occur in the development of cervical and lumbo-sacral segments in man at the same time and either might precede the other in some subjects. Thus maturation also differs within the same spinal segment as in sacral 1 where H-reflexes subserving calf muscles develop more rapidly than those of foot muscles.

M Response and H/M Ratio

The amplitudes of the direct M response in hand, foot or calf muscles in newborns vary but are generally one-third to one-half those recorded in adults [THOMAS and LAMBERT, 1960; BORENSTEIN, 1968; MAYER and MOSSER, 1969a]. The amplitudes of the H- and M responses increase with age and reach adult values by the age of 10–12 years. With paired maximal electric stimuli for motor fibers, the recovery of the test M response in the calf muscles of newborns (the relative refractory period at neuromuscular junction) occurs within 30 msec and this is similar to that in children and adults [DESMEDT, 1958; LAMBERT, 1966].

The maximum M response in muscles represents the activity of all motor units in the muscle while the maximum H-reflex recorded with the same electrodes represents the total activity of the motor units activated by I_A afferents. The ratio of H to M is an index of motoneuron excitability and of what portion of the motoneuron pool can be activated by this reflex [ANGEL and HOFMANN, 1963; LANDAU and CLARE, 1964; TABORIKOVA and SAX, 1968; GARCIA-MULLIN and MAYER, 1969]. The ratios determined in calf muscles of awake newborns are listed in table I. These are similar to those recorded in children and adults using the same techniques. The ratio determined in newborns is not suggestive of increased motoneuron excitability as reported in adults with chronic hemiplegia [ANGEL and HOFMANN, 1963; LANDAU and CLARE, 1964; GARCIA-MULLIN and MAYER, 1972].

Antidromic Motor Responses – F Responses

In adults, F responses are distinguished from H-reflexes by the following characteristics: (1) the threshold for electrical stimulation is the same as that of motor fibers and higher than that of afferent fibers which subserve the H-reflex; (2) conduction velocity in antidromic fibers is the same as that in orthodromic fibers to the muscle; (3) there is neither PTP nor augmentation during mild active contraction of the recording muscle, and (4) the excitability cycles of F responses are different from those of H-reflexes, since the F responses are altered only while the paired stimuli are separated by 3–20 msec [THORNE, 1965; MAYER and FELDMAN, 1967]. F responses may be blocked by spinal anesthesia, proximal whole nerve blocks and disease processes which alter proximal nerve, plexus or motor roots but are not blocked by deafferentation.

In newborns, MAYER and MOSSER [1969a] reported that F responses could be recorded especially in hand and foot muscles. The responses were small and variable in amplitude (100–400 µV), they did not occur with every stimulus and had the same characteristics as those recorded in children and adults. The latency of the F response in newborns was the same as the monosynaptic H-reflex to the same muscles (16–19 msec in hypothenar muscles and 22–25 msec in plantar foot muscles). No evidence of altered motoneuron excitability, increased or decreased, was observed in newborns tested antidromically. THORNE [1965], who studied F responses in adults both with normal and abnormal upper motoneuron function, concluded that the main change in the latter was a superimposed true reflex component on the F response and this was mediated by both low and high-threshold muscle afferents. Additional studies of this response using single motor units in infants are needed.

Sleep

In normal human adults, reflex activity is depressed during sleep [LEE and KLEITMAN, 1923; TUTTLE, 1924]. HODES and DEMENT [1964] have reported that H-reflexes show a variable decline in amplitude (30% or less) as the normal adult passes from wakefulness into slow wave (EEG phases 2–4) sleep. During the 'ascendant' stage 1 of sleep, the H-reflexes are reduced by 90% or more of their values in the slow wave stages or they may be totally absent. The electrically evoked reflexes were smaller and less variable in sleep with rapid eye movements (REM sleep) than in this stage with minimal eye movement. They observed these changes in the responses recorded in calf and plantar foot muscles, but since they did not distinguish between H and F responses in the foot, it is likely that both were depressed during sleep. We observed in several normal adults that in REM sleep both the H-reflex in calf muscles and the F response in foot were absent.

In newborns and infants aged 2–4 months, HODES and GRIBETZ [1962] have shown that H-reflexes recorded in forearm and hand muscles were reduced as the infants fell asleep. The reflexes were restored by spontaneous arousal or by waking. The fact that newborns pass from wakefulness to sleep and even REM sleep very quickly may explain why some investigators [LANGWORTHY, 1933] could not elicit ankle jerks in all newborns, while we have been able to record H-reflexes in calf muscles and evoke ankle jerks in all wakeful newborns tested. The knee jerk is markedly decreased or obliterated in newborn infants (EMG quadriceps muscle) during REM sleep [PRECHTL, VLACH, LENARD and GRANT, 1967]. Although the amplitude of a single test H-reflex is changed very little during drowsiness or light sleep in newborns and infants, the excitability cycle is changed. Recovery is much slower than during wakefulness, since the test response is usually 5–20% of the conditioning one by 100 msec; at 200–300 msec recovery is only 30–50% and in some there is a 'secondary depression' from 400–800 msec. In others, recovery is gradual but is usually still incomplete by 3 sec. During deeper stages of sleep the amplitudes of the responses also become depressed and the recovery cycle very slow. In some, the 'early depression' persists for over 100 msec and recovery is slow and incomplete. The H-reflex is very depressed or obliterated in all newborns during REM sleep at a time when the ankle jerks are also depressed or absent. H-reflexes recorded in hand and foot muscles are altered in the same way. During deep sleep it was also observed that the amplitude of F responses was decreased and the variability of the responses greater. F responses could not be evoked during REM sleep when H-reflexes were absent.

Since the changes in the excitability of H-reflexes during sleep are similar in newborns, infants and adults, it is likely that the pathways necessary for this depression of the monosynaptic reflex are functioning at birth. Studies in man have suggested that tracts in the anterior funiculus and anterior half of the lateral funiculus carry the impulses [SHIMIZU, YAMADA, YAMAMOTO, FUJIKI and KANEKO, 1966]. In animals, it has been concluded that during REM sleep supraspinal inhibitory volleys from medial and descending vestibular nuclei induce presynaptic depolarization of group I afferents and hence phasic depression of monosynaptic reflexes [POMPEIANO, 1967]. Alpha motoneurons are tonically inhibited postsynaptically throughout REM sleep possibly by reticulospinal inhibitory volleys and γ-motoneurons appear to be more strongly inhibited during this phase than α-motoneurons [POMPEIANO, 1966]. The slow recovery of the excitability cycles, especially the prolonged 'early depression' during sleep in newborns, may in part be related to depression of γ-motoneurons as well as α-motoneurons.

Exteroceptive Reflexes

Exteroceptive reflexes have been studied polygraphically in premature [WEINMANN, MEITINGER and VLACH, 1969] and term infants [PRECHTL, VLACH, LENARD and GRANT, 1967]. When tactile stimuli are applied to circumscribed skin areas, local reflex activity is elicited in single muscles or small groups of muscles adjacent to or beneath the site of stimulation. These polysynaptic exteroceptive reflexes become weak or are abolished during slow sleep and may be unchanged or only slightly reduced during REM sleep. Changes in cardiac and respiratory rhythm and EEG arousal effects were demonstrated by PRECHTL, VLACH, LENARD and GRANT [1967] in response to tactile stimuli. but not in response to a tendon stretch stimulus.

Polysynaptic reflexes have been extensively used in nursery examinations to account for most of the behavior of newborn infants. SCHULTE, LINKE, MICHAELIS and NOLTE [1968] confirmed the increase in flexor activity that accompanies the Moro reflex and demonstrated that the amount and duration of extensor activity remains stable during early maturation. We are not familiar with studies to demonstrate the time of disappearance of polysynaptic reflexes electrophysiologically. We have observed that infants with low thresholds for monosynaptic reflexes fail to habituate to polysynaptic reflexes and that in this respect habituation as a criterion of polysynaptic maturation parallels the development of a more adult type of H-reflex recovery cycle. However, additional studies of polysynaptic reflexes in newborns are necessary.

Discussion

From the studies of electrically evoked responses in newborns, certain functional characteristics of the spinal cord at this stage of development can be postulated. The lack of both 'early facilitation' in the excitability cycle and PTP of the H-reflex in calf muscles of newborns 1–3 days old suggests a small functional subliminal fringe. This may be related to immaturity of the afferent fibers, that have slow conduction velocities, or to their connections in the spinal cord. Since these phenomena change within a few days after birth, during which time there is little change in the conduction velocity of peripheral fibers, they most likely result from immaturity of the afferent terminals.

The rapid and complete recovery of the excitability cycles of H-reflexes at birth as well as the presence of H-reflexes in hand and foot muscles suggest increased excitability of α-motoneurons. This increase in motoneuron excitability is not reflected however in larger H/M ratios or F responses as seen in adults with chronic spastic hemiplegia [GARCIA-MULLIN and MAYER, 1972]. These findings taken together suggest that the increased excitability is presynaptic and that the motoneuron pool is not enlarged but is small. This increase in excitability may be comparable in part to the observations that in the monkey excitatory synapses located on the dendrites of the motoneurons develop first and the inhibitory connections develop later [BODIAN, 1966]. Studies in the cat have correlated the development of motor activity with maturation of motoneuron dendrite bundles [SCHEIBEL and SCHEIBEL, 1971] and synaptic boutons [CONRADI and SKOGLUND, 1969].

The rapid H-reflex recovery cycle in newborns may in part result from a lack of inhibition of the spindle discharge on motoneurons as well as active γ-motoneurons. The muscle spindles and fusimotor system are functioning in human newborns since the tendon jerks are present and very active. This differs from the newborn cat in whom γ driving has been reported to be absent [SKOGLUND, 1960a] and the recovery cycle of the monosynaptic reflex slow [SKOGLUND, 1960a, b].

The excitability of motoneurons changes especially between the ages of 6 and 12 months. Although this follows a rostrocaudal and proximodistal course, maturation proceeds at different rates even within the same spinal segments. Since the spinal reflex activity in newborns changes to that of the adult at about the same time that myelination becomes complete in spinal tracts [LANGWORTHY, 1933; DEKABAN, 1959], it has been postulated that the increased motoneuron excitability at birth results from an imbalance of facilitatory and inhibitory effects at the spinal motoneuron level [MAYER and

MOSSER, 1969a]. This may result from poorly myelinated corticospinal and rubrospinal tracts and well myelinated vestibulospinal, reticulospinal, spinocerebellar and posterior column tracts.

However the change in the excitability H cycle from the newborn to the adult occurs during the period when nerve conduction velocity increases, although the velocity is still less than that in adults. In the kitten it has been reported that reflex activity becomes similar to that in the adult when the velocity of afferent fibers is approximately one-third of normal [SKOGLUND, 1960c, 1969] while in human newborns the velocity is about half that in adults and reflex activity is still immature. It is hoped that with additional techniques and studies more significant data can be obtained concerning how the human nervous system matures and functions.

Author's address: Prof. R. F. MAYER, Department of Neurology, University of Maryland Medical School, *Baltimore, MD 21201* (USA)

New Developments in Electromyography and Clinical Neurophysiology,
edited by J. E. Desmedt, vol. 3, pp. 308–316 (Karger, Basel 1973)

An Electrophysiological Study of the Organization of Innervation of the Tendon Jerk in Humans[1]

M. M. GASSEL and K. H. OTT

San Francisco, Calif.

Introduction

A high degree of localization of the stretch reflex has been reported. Limitation of reflex contraction to one head of the quadriceps or gastrocnemius on stretching of this part of the muscle only [LIDDELL and SHERRINGTON, 1924] and even reflex restriction to small muscle strips of 2–3 mm in diameter on discrete stretching of these strips of muscle has been reported [COHEN, 1953]. However there are conditions in which the reflex contraction is more widespread. Increased afferent transmitter mobilization by post-tetanic potentiation, increase in the level of excitability of the motoneuron pool by turning the face of the animal toward the tested limb or by the crossed extension reflex, and fast stretch rates result in a more widespread reflex response.

Clearly, there must be a discrete anatomo-physiological substrate for the localization of response. The muscle spindles from a localized section of muscle must be in a particular effective relationship to the motor units within this region. Alternatively, the potential for more diffuse monosynaptic activation of motoneurons must also be present.

There is an electrophysiological correlate to the anatomy of nerve supply and end plate distribution in a muscle [GASSEL, 1963, 1964]. The latency of response to proximal stimulation of the muscle nerve is measured by recording evoked potentials from the muscle at increasing distances from the fixed point of stimulation. There is a progressive increase in latency of the evoked potentials in more distal recordings from muscles with a longitudinal orientation of

1   This work was supported in part by NIH Grant NB-07562, and Fellowships FR-05355-07 and NB-5101–13.

end plates, such as the rectus femoris, in which longer branches of the nerve supply distal parts of the muscle. On the other hand, the latency is independent of electrode distance in the biceps brachii in which the innervation zone is concentrated in a horizontal band in the middle of the muscle where the main nerve trunk terminates.

The distribution of muscle spindles is not as well structured as that of motor end plates, with somewhat greater dispersion through the muscle mass [GREGOR, 1904; BARKER and CHIN, 1960; CHIN, COPE and PANG, 1962]. However, the greatest density of neuromuscular spindles occurs in the neighborhood of the main intramuscular nerve trunks; and as already indicated, the distribution of the trunks follows a different pattern in the biceps brachii and rectus femoris.

This report investigates the organization of innervation of the monosynaptic reflex excited by sudden stretch (tendon tap) in muscles of different fascicular arrangement in humans. The relationship between stretch receptors and motor units in the biceps brachii and rectus femoris is studied by relating the latency of the direct motor (M) response to that of the monosynaptic reflex, electrically and mechanically evoked in proximal, mid and distal parts of the muscle.

Methods and Material

The techniques of direct stimulation of the femoral nerve, of the trunks of the brachial plexus, and of reflex evocation have been described in detail [GASSEL, 1963, 1964; GASSEL and DIAMANTOPOULOS, 1966]. The femoral nerve was stimulated percutaneously at the inguinal ligament immediately lateral to the femoral artery by means of a bipolar surface electrode (DISA 13 K 62). The trunks of the brachial plexus were stimulated at Erb's point. The stimulus was a square wave pulse, 0.15 msec in duration led from a DISA Multistim through a double-shielded output transformer. The output impedance was 15 KΩ and the maximal-current output 15 mA.

The tendon jerk reflex was elicited by striking the appropriate tendon with a conventional reflex hammer equipped with a microswitch. The microswitch was embedded in the head of the hammer and activated the sweep generator of the oscilloscope 0.3 msec after contact with the skin.

The electrical activity of the biceps brachii and rectus femoris was recorded by 3 concentric needle electrodes with an external diameter of 0.45 mm and a leading-off surface of 0.07 mm^2. The electrodes were placed in the proximal, middle and distal parts of the tested muscle. Simultaneous recordings from the 3 electrodes were amplified and displayed on the 3 independent CRT of the electromyograph (DISA).

Five to 10 photographs were made of maximal direct M responses followed by maximal monosynaptic reflexes recorded from the same intramuscular electrodes. Submaximal stimulation of the femoral nerve elicited a monosynaptic reflex in 7 subjects and the latency of the electrically evoked reflex was also recorded at each electrode position in the rectus femoris.

The latency measurements were made to the start of the action potential. The distance from the stimulating electrode to each recording electrode was measured with a tape measure in the case of the rectus femoris, and with obstetrical forceps in the instance of the biceps brachii [GASSEL, 1964]. The data was analyzed by the IBM Model 40 computer at the University of California Medical Center, San Francisco. 35 normal subjects ranging in age from 18 to 68 years were studied. There were 25 investigations of the rectus femoris and 23 of the biceps brachii.

Results

The latency of the knee jerk (KJ) reflex and that of the direct M response recorded in the rectus femoris are plotted against the distance of the recording electrodes from the stimulating electrode (fig. 1 and 2). The latency of both the KJ reflex and the direct M response increases with the distance of the recording electrode along the rectus femoris. The slope of the latencies of the KJ reflex is somewhat greater than that of the direct M response (table I). The increasing latency of the M response with distance is related to the delay for conduction in the longer branches which supply the distal parts of the muscle. This latency is independent of subject size. On the other hand, latency of the KJ reflex is not only a function of electrode distance in the rectus femoris but also is related to the size of the subject (fig. 3).

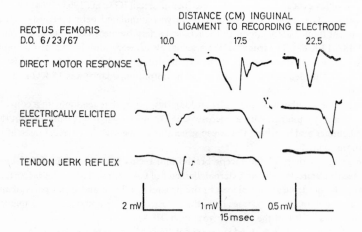

Fig. 1. Illustration of evoked potentials in the rectus femoris.

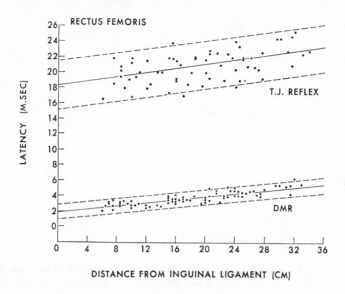

Fig. 2. Latency of the knee jerk (T.J.) reflexes and of the direct motor responses (DMR), collected from normal subjects, are plotted against the distance of the recording electrodes from the stimulating electrode at the inguinal ligament. The best straight line described by the coordinates is computed by the method of least squares. The distribution of points above and below this line was considered to be Gaussian, and the standard deviation was computed. Twice the standard deviations above and below the best straight line are indicated by broken lines. The data from which the graph is drawn are tabulated in table II.

The variable of subject size was further evaluated by separately computing the slope of latencies for each individual subject by the method of least
squares, and then calculating the mean of the slopes and the standard deviation
(table II). It is clear that the slope of the latencies is reduced relative to that
based upon the pooled data and not significantly different from that of the M
response similarly computed.

The difference between the latency of the KJ reflex and the distal motor
latency (KJ minus M response) is plotted as a function of distance of the
recording electrodes from the stimulating electrode in figure 4. This graph
represents the time taken for the reflex to traverse the entire afferent limb and
the proximal efferent limb up to the point of stimulation of the femoral nerve
adjacent to the inguinal ligament. The best straight line has a slightly positive
slope, again related to subject size. There is essentially null slope when computing the slope of latency against electrode distance for each individual and
averaging this value (table II).

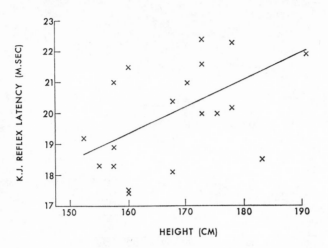

Fig. 3. The latencies of the knee jerk reflex to a recording electrode 15 cm from the point of stimulation of the femoral nerve is plotted against the height of the subject. The best straight line is drawn by the method of least squares.

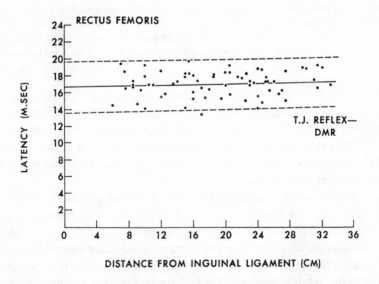

Fig. 4. Latency of the direct motor response (DMR) subtracted from that of the knee jerk (T. J.) reflex at each electrode position in the rectus femoris and plotted against distance from the stimulating electrode at the inguinal ligament. The best straight line is computed by the method of least squares. Twice the standard deviations are indicated by broken lines.

Table I. Data computed from pooled values from normal subjects

		Tendon jerk reflex	Direct motor response	Tendon jerk reflex—direct motor response	Electrically evoked mono-synaptic reflex	Electrically evoked monosynaptic reflex—direct motor response
Rectus femoris	m	0.140	0.100	0.021	0.060	0.015
	b	18.3	1.91	16.7	14.3	11.7
	SD	1.55	0.500	1.50	1.45	1.30
Biceps brachii	m	0.120	0.053	0.032		
	b	14.0	4.00	10.0		
	SD	0.920	0.491	0.823		

m = slope of best straight line calculated by method of least squares; b = describes the intercept at the ordinate similarly calculated; SD = standard deviation; Latency = m × distance + b ± 2SD.

Table II. Average slopes of latencies

		Tendon jerk reflex	Direct motor response	Tendon jerk reflex—direct motor response	Electrically evoked mono-synaptic reflex	Electrically evoked monosynaptic reflex—direct motor response
Rectus femoris	m	0.128	0.114	0.011	0.107	0.017
	SD	0.056	0.033	0.048	0.064	0.040
	N	21	25	23	7	7
Biceps brachii	m	0.047	0.020	0.033		
	SD	0.046	0.046	0.046		
	N	19	23	19		

The slope of the latencies for each individual was calculated by the method of least squares. The mean (m) of the slopes (in msec/cm) an[d] the standard deviation (SD) was then determined. N = number of subjects.

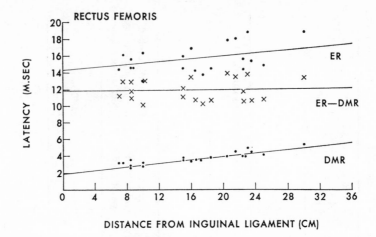

Fig. 5. The latencies of the electrically evoked monosynaptic reflex (ER) and of the direct motor response (DMR) on stimulating the femoral nerve at the inguinal ligament is plotted against the distance to the recording electrodes in the rectus femoris. The latency of the DMR subtracted from that of the ER at each recording electrode is plotted. The best straight line is computed by the method of least squares.

An electrically evoked reflex (ER) was elicited on submaximal stimulation of the femoral nerve in 7 subjects. The latency of the reflex is plotted against distance from the stimulating electrode at the inguinal ligament to the recording electrodes in the rectus femoris (fig. 5). The latency of the ER increased *pari passu* with that of the M response. When the latency of the M response was subtracted from that of the ER at each recording electrode and plotted against the electrode distance the best straight line through the coordinates had an essentially null slope (m = 0.017 with a SD of 0.040; table II). In each instance the KJ reflex was elicited and the latency of the ER subtracted from the KJ. This value represents primarily the latency for the stretch receptor volley to reach the inguinal ligament; a small portion of this latency is attributable to the fact that the mechanically evoked reflex occurs after temporal summation of the afferent volley in the anterior horn whereas the ER does not [LLOYD, 1943]. The latency of this distal afferent volley (KJ minus ER) was found to average 2.0 msec (SD = 1.3) greater than the distal efferent volley (direct M response).

The latencies of the M response from the biceps brachii show a slightly positive slope with distance when the best straight line is plotted by the method of least squares, based upon data pooled from all subjects (fig. 6). The slope is

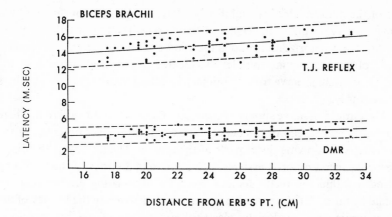

Fig.6. The latencies of the biceps jerk (T.J. reflex) and the direct motor response (DMR), evoked on stimulation at Erb's point, are plotted against distance to the recording electrodes in the biceps brachii. The criteria of graphing are similar to those described in figure 2 and based upon data graphed in table I.

essentially null when computed for each subject and then averaged (table II; m = 0.020, SD = 0.046), indicating that the latency is independent of distance of the electrode along the biceps.

The latencies of the biceps jerk reflex based upon pooled data did increase with distance, but when the slope of the latencies of each individual subject was averaged, the latency was again independent of distance of the electrode along the biceps (table II). Similarly, plotting the latency of biceps jerk reflex minus the M response at each electrode against the electrode distance for each subject resulted in a family of nearly horizontal lines (table II).

The latency of the biceps jerk was found to increase with subject size in a manner analogous to that illustrated in figure 3; and it is noteworthy that the slope of the latencies of the biceps jerk with distance based upon pooled data is greater than that of the M response, indicating that the tendon jerk latency is more sensitive to subject height than the M response.

Discussion

When the influence of subject height is discounted, the latency of the tendon jerk in the biceps brachii and that of the direct M response on stimulation at Erb's point is independent of electrode position in proximal or distal parts of the muscle. This invariability is related to the entry of the musculo-cutaneous nerve trunk in the center of the muscle. The motor end plates are concentrated in a horizontal band in the middle of the muscle and the greatest density of neuromuscular spindles are located in the neighborhood of the main intramuscular nerve trunk, both factors acting to excite all muscle fibers synchronously.

The pattern of nerve supply of the rectus femoris is quite different. There is a longitudinal spread of the nerve entry zone with longer branches of the femoral nerve supplying distal parts of the muscle. The latency of the direct M response increases progressively with increasing distance of the recording electrodes within the rectus femoris from the stimulating electrode at the inguinal ligament. The progressive increase with distance in the latency of the monosynaptic reflex evoked by electrical stimulation of the femoral nerve proximal to the muscle mass is, not surprisingly, attributable to the increasing latency of the M response. However the latency of the KJ reflex also increased *pari passu* with the M response and consequently the KJ minus the M response at each electrode was also independent of distance, when corrected for the factor of subject size.

There is a number of possible explanations for the apparent lack of influence of dispersion of the neuromuscular stretch receptors in proximal and distal parts of the rectus femoris muscle on the latency of the KJ reflex. A leveling mechanism acting in the periphery could serve to equalize in time the arrival of proximal and distal stretch receptor contributions to the afferent volley. After tendon tap the wave of mechanical tension passing up the rectus femoris could fire the distal spindles before the proximal spindles. The receptors throughout the muscle, in this case, would discharge serially rather than simultaneously. Indeed, FULTON and LIDDELL [1925] attributed the relative asynchrony of nerve impulses associated with the tendon jerk to the time taken for conduction of the wave of increased tension down the muscle. LLOYD [1943] found a discrepancy of 0.2–0.3 msec between the afferent volley evoked by stretch and that by electrical stimulation of the preterminal nerve bundles of the medial gastrocnemius of the cat. He ascribed part of this 'excess latency' to the time 'for transmission of the tension wave from the free end of the tendon to the site of the first-responding receptors'. We have recorded the wave of tension after tendon tap in proximal and distal parts of the muscle [see GASSEL and DIAMANTOPOULOS, 1966]. A dampened oscillation traversed a distance of 15 cm in 1–2 msec in 4 of 6 subjects. We have no direct evidence on which to relate serial spindle activation to the wave of tension induced by tendon tap.

It is of interest that the latency of the ER subtracted from that of the tendon jerk (TJ) reflex was found to be an average of 2 msec (SD = 1.3) longer than that of the distal motor latency. The value TJ minus ER represents primarily the latency for the stretch receptor volley to reach the stimulating electrode adjacent the inguinal ligament. A small part is attributable to the central temporal summation associated with the mechanically evoked but not the electrically evoked reflex. The longer latency of the TJ minus ER than the M response, in spite of the wide terminal branching of the motor axons, may be advanced to support the hypothesis of excess latency associated with the spread of a tension wave after the tendon tap.

On the other hand, there is also an anatomical and physiological basis for an effective common afferent input to all parts of the muscle during sudden muscle stretch. There are several times more motor units than spindles and the afferent fibers must distribute to several motor units. Indeed, spindle afferents from one muscle influence agonist and antagonist motoneurons widely, although autogenic excitation is most potent [ECCLES, ECCLES and LUNDBERG, 1957]. It has recently been concluded, from a study of evoked excitatory postsynaptic potentials (EPSP) by computer techniques, that almost all of the 300 motoneurones of the medial gastrocnemius receive afferent imput from each I_A fiber, although the size of EPSP varied considerably in the different motoneurons [MENDELL and HENNEMAN, 1968].

It is therefore possible, and quite consistent with our results, that virtually simultaneous activation of a localized group of neuromuscular spindles effectively activates a widespread reflex response.

Summary

The latency of the tendon jerk to proximal, mid and distal parts of the biceps brachii is equal, a finding consistent with the localization of the nerve supply to a transverse band in the central area of the muscle. In the rectus femoris with a longitudinal orientation of nerve supply and longer nerve branches innervating distal parts of the muscle, it was found that differences in the latency of the tendon jerk were attributable solely to differences in efferent innervation rather than to differences in afferent innervation. The muscle, therefore, acted as though there was an effective, common afferent input to all parts of the muscle during sudden muscle stretch.

Author's address: Dr. M. M. GASSEL, 425 Warren Drive, *San Francisco, CA 94131* (USA)

New Developments in Electromyography and Clinical Neurophysiology,
edited by J. E. Desmedt, vol. 3, pp. 318–322 (Karger, Basel 1973)

A Study of the F Response by Single Fibre Electromyography[1]

J. V. TRONTELJ

Institute of Clinical Neurophysiology, The University Hospitals of Ljubljana,
Ljubljana

There has been some discussion in the literature about the nature of the
F response, which appears in certain muscles upon electrical stimulation of
their nerves as a variable deflection following the M response at a latency
which approximately corresponds to that of the monosynaptic reflex (H-
reflex). The F response has been interpreted by several authors as a monosynaptic or polysynaptic reflex response [HOFFMANN, 1922; MAGLADERY and
MCDOUGAL, 1950; LIBERSON, 1962] in which the afferent fibres conduct more
slowly than the motor fibres, and was even used to estimate the conduction
velocity of muscle afferents [LIBERSON, ZALIS, GRATZER and GRABINSKI, 1964;
ANGEL and ALSTON, 1964].

Using voluntary contraction of the stimulated muscle and a conditioning
tetanic stimulus, HAGBARTH [1962] was able to convert the F response into a
typical H-reflex. Therefore he admitted the possibility that the F response
might be a rudimentary H-reflex lacking pre- or postsynaptic facilitation or
partly blocked by the antidromic volley.

DAWSON and MERTON [1956], however, could find no difference between
the afferent and efferent conduction velocities for the F response and suggested
that it represents recurrent discharges of antidromically activated motoneurones, as described in the cat by RENSHAW [1941]. This view has been experimentally supported by a single motor unit study in normal man [THORNE,
1965] and by a study of the F response in deafferented human limbs [MAYER
and FELDMAN, 1967], and it seems to be prevailing at the present time.

1   This work was supported by Boris Kidrič Fund of Slovenia (grant No. 306/20:71)
and by the US Department of Health, Education and Welfare, Social and Rehabilitation
Service (grant No. 19-P-58397-F-01).

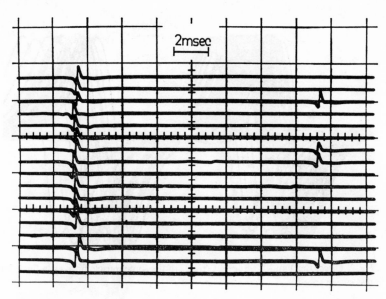

Fig. 1. Direct (left) and F responses (right) of a single muscle fibre in m. abductor pollicis brevis.

GASSEL and WIESENDANGER [1965] have shown, however, that the analogous responses in cat plantar muscles are composed of a recurrent, a monosynaptic and a polysynaptic reflex component. This led to an assumption that human F responses may also contain both a recurrent and a monosynaptic reflex component, and possibly also a polysynaptic component [GASSEL, 1969].

Using single fibre EMG as described by EKSTEDT [1964] and STÅLBERG [1966], one can successfully isolate and identify responses of single motoneurones even from a synchronous discharge of a large part of a motoneurone pool. This method was applied to study the F responses in normal man.

Figure 1 shows a typical recording from a single muscle fibre in m. abductor pollicis brevis while repetitively stimulating the median nerve at the wrist. Some of the direct or M responses were followed by an F response. Stimulus strength was just at the threshold for that motoneurone. Whenever the M response failed, the F response also failed. In other words, the F responses required the direct activation of the motor axon, which is in contrast to H-reflex responses, and can be taken as an evidence of their recurrent origin.

Recording from single muscle fibres allows considerable precision of latency measurements: up to 5 μsec (fig. 2). The mean variation of latencies

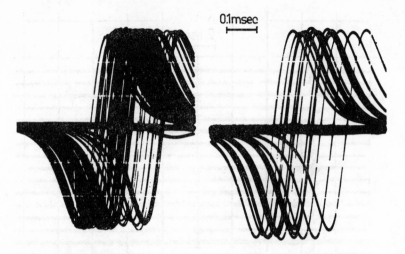

Fig. 2. Direct (left) and F responses (right) of the same muscle fibre as in fig. 1, but at expanded sweep speed.

of 100–300 successive M responses, expressed as standard deviation, was 59 $\pm$ 15 μsec in 15 motoneurones of various muscles; it was mainly due to the variation of neuromuscular transmission time and to the inconstancy of the time and point at which the stimuli triggered off the propagated nerve action potentials [STÅLBERG and TRONTELJ, 1970]. Latency variation of the F responses following the same M responses was only slightly larger: 73 $\pm$ 27 μsec. This is in contrast to the ratio between the M and H responses of 18 soleus motoneurones, which was 57 $\pm$ 23 : 155 $\pm$ 62 μsec (fig. 3).

There was a high correlation between the latencies of consecutive M and F responses (mean correlation coefficient for 15 motoneurones was + 0.76, p < 0.001). This can be interpreted as an evidence of the initiation of both responses in the same nerve fibre.

It can be safely concluded, therefore, that the described F responses really are recurrent discharges.

Such responses could be readily detected in many muscles of both the upper and lower limbs and were quite common even in the soleus (fig. 4). Under circumstances a considerable part of the H-wave as recorded over the soleus may be composed of recurrent responses.

On the other hand, typical reflex responses resembling the H-reflex in their latency and the size of its variation, as well as in their being suppressed by increase in stimulus strength were sometimes found in the muscles of the hand and forearm (fig. 5), especially on voluntary contraction.

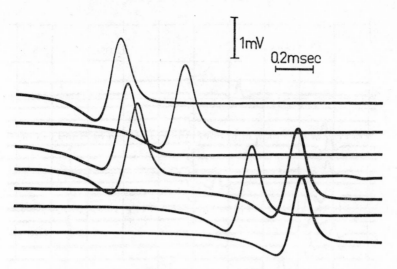

Fig. 3. Successive H-reflex responses of a muscle fibre in the soleus showing typical amount of latency variation. The beams were delayed 31 msec after the stimulus.

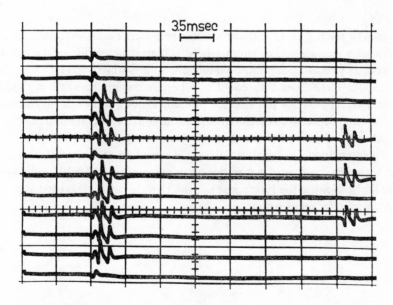

Fig. 4. Recurrent responses of a soleus motoneurone (represented in the recording by a pair of muscle fibres).

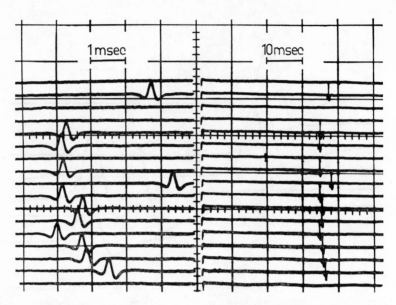

Fig.5. Reflex responses of a muscle fibre in m. abductor digiti quinti. The left beams are delayed 31 msec after the stimulus.

It is of interest to note that recurrent responses could be facilitated by Jendrassik's manoeuvre, by subthreshold voluntary activation of the motoneurone, as well as by subthreshold afferent volleys which also facilitate the H-reflex. In some cases they could also be blocked by the same methods. Facilitation may be due to increased probability for antidromic invasion of the soma, and depression to a too strong facilitation of soma invasion resulting in shortening of the axon-soma delay [ECCLES, 1955].

To summarize, electrical stimulation of mixed neves in normal man can elicit both reflex and recurrent responses in most muscles of both upper and lower limbs. Neither the F response nor the H-wave can be regarded as pure recurrent or pure reflex responses.

Author's address: Dr. J. V. TRONTELJ, Institute of Clinical Neurophysiology, The University Hospitals of Ljubljana, Holzapflova ulica, *61000 Ljubljana 5* (Yugoslavia)

New Developments in Electromyography and Clinical Neurophysiology,
edited by J. E. Desmedt, vol. 3, pp. 323–327 (Karger, Basel 1973)

The F Response after Transverse Myelotomy

O. E. MIGLIETTA

Department of Physical Medicine and Rehabilitation, New York Medical College,
New York, N. Y.

Stimulation of some mixed nerves in man will evoke the appearance of some late electrical responses designated as H and F, following the direct muscle or M response.

While there is general agreement on the reflex nature of the H response, there has been much discussion as to the origin and significance of the F response in man. This potential, which can normally be recorded from small foot and hand muscles, has been considered and treated as a reflex response by MAGLADERY and McDOUGAL [1950], HAGBARTH [1962], LIBERSON, ZALIS, GRATZER and GRABINSKI [1966], resulting apparently from stimulation of afferent fibers in the peripheral nerve, with polysynaptic connections and slower conductions in afferent than efferent fibers. DAWSON and MERTON [1956], however, reported that conduction was the same in afferent and efferent pathways of the F response and they suggested that it resulted from backfiring of motoneurons following their antidromic activation by centripetal volleys. Similar conclusions were reached by THORNE [1965] in normal subjects and additional evidence for the latter view has been provided by GASSEL and WIESENDANGER [1965] and McLEOD and WRAY [1966] in animal experiments. The present paper reports on the behavior of the F response following surgical deafferentation of the lower extremities in four paraplegic patients. This study supports the view that in man the F response results from antidromic activation of the motoneurons.

Material and Methods

Four adult males with traumatic spinal cord injury of 2–5 years duration were studied. The spinal cord injury ranged in location from the lower cervical to the mid-thoracic level.

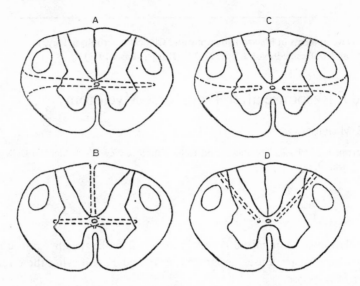

Fig. 1. A BISCHOF's longitudinal myelotomy. *B* POURPRÉ's 'myelotomie en croix'.
C ROTHBALLER's bilateral longitudinal myelotomy. *D* ROTHBALLER's dorsolateral myelotomy.

All patients were severely spastic. Phasic spasms, usually flexor, but sometimes extensor or even involving more elaborate pedaling movements, dominated the picture. Spasticity problems had been treated with varying degrees of success by physical and pharmacological means, but these measures ultimately proved inadequate and all four patients were operated upon by Dr. A. ROTHBALLER. Neurosurgical intervention for the relief of spasticity in paraplegics dates back to FOERSTER [1913], MUNRO [1945] and MACCARTHY and KIEFER [1949]. These were attempts to interrupt what was conceived of as a pathological reflex arc, attacking the afferent limb by posterior rhizotomy, or the efferent limb through anterior rhizotomy or cordectomy. Attempts to accomplish more selective sections or lesions within the spinal cord itself began with the introduction of longitudinal myelotomy [BISCHOF 1951]. In this a longitudinal incision is made through the equator of the spinal cord in the coronal plane, cutting about three-quarters of the way through the cord, in order to interrupt most connections between the dorsal and ventral halves of the spinal cord bilaterally (fig. 1 A). In 1959, a somewhat similar end was achieved with a different technical approach by POURPRÉ [1960] with his 'myelotomie en croix' (fig. 1 B). In 1961, ROTHBALLER introduced a bilateral longitudinal myelotomy, equatorial in location (fig. 1 C) and in 1966 modified this as a bilateral dorsolateral longitudinal myelotomy (fig. 1 D) [ROTHBALLER, 1969]. The aim of such operations is to prevent sensory or reflexogenic impulses from impinging upon the anterior horn cells. The dorsolateral longitudinal approach attempts to do this with transecting lesions in the zone immediately below the dorsal grey column. Hopefully it leaves uninterrupted any connections between the lateral corticospinal (pyramidal) tract and the motor unit, thus preserving residual voluntary motor power or at least its potential for return in patients with partial cord transections.

Electrophysiological studies were carried out in the lower extremities before operation and at intervals of 3, 8 and 12 weeks post-operatively. Clonus and H-reflexes were elicited from the gastrocnemius-soleus muscle group by using standard techniques [MAGLADERY, PORTER, PARK and TEASDALE, 1951; MIGLIETTA, 1964]. The M and F responses were recorded from the small flexor muscles of the foot. Surface electrodes were used and all electrical responses were followed and measured from a Tektronix Storage Scope. Latencies were measured to the onset of the first deflection of the action potential, following percutaneous supramaximal stimulation of the posterior tibial nerve at the popliteal fossa and at the ankle. The conduction velocity was calculated in the usual way by dividing the distance between the stimulating cathodes by the difference in latency of the responses at each site.

A bipolar stimulator was used and the stimuli were rectangular pulses 0.1–0.5 msec in duration, delivered through a ground isolated output. During testing the patient lay supine with the leg adequately supported.

Results

Before myelotomy. Sustained clonus was present in all patients at a frequency of 6–7/sec. The H-reflex was easily elicited with subthreshold stimuli for the motor fibers, increased gradually with increasing pulse intensity and disappeared with maximal stimulation of motor fibers. Its latency averaged 30 msec (fig. 2A).

A late response (fig. 2B) was picked up from the small foot muscles in all patients and had all the characteristics associated with the F response: it appeared late after the direct M response; it gradually increased and remained

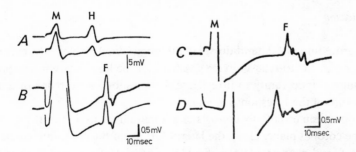

Fig. 2. A, B Normal adult man. *A* Gastrocnemius-soleus group with M and H responses recorded during submaximal stimulation of the posterior tibial nerve. *B* F response recorded from small flexor muscles of the foot on supramaximal stimulation of the posterior tibial nerve at the ankle. *C, D* Patient after myelotomy. M and F responses recorded from small flexor muscles of the foot on stimulation of the posterior tibial nerve at the ankle in *C* and at the popliteal fossa in *D*. Distance between the two points of stimulation 370 mm.

present with supramaximal stimulation of the posterior tibial nerve; its amplitude was rather low and variable in occurrence, shape and occasionally in latency; its latency shortened with proximal stimulation at the popliteal fossa. F response latency varied from 50 to 60 msec with stimulation at the ankle.

Conduction velocity of the posterior tibial nerve averaged 46.5 m/sec (44.4–48.5 m/sec). This was similar to the average conduction velocity of 45.2 m/sec (43.5–47.8 m/sec) of the volley proceeding in a proximal direction and giving rise to the F response.

After myelotomy. All 4 patients had excellent clinical results from the operation and made an uneventful recovery. There was immediate and total disappearance of the spasticity including flexor and extensor spasms, as well as improvement in postural deformities. All deep tendon reflexes disappeared completely and never returned. Clonus and H-reflex were abolished and remained absent. Conduction velocity of the posterior tibial nerve was unchanged and the F responses were as easily elicitable and had the same characteristics as before the operation.

The responses of one of the patients recorded after myelotomy are shown in figure 2C and D. The difference in latency between distal (fig. 2C) and proximal (fig. 2D) stimulation of the posterior tibial nerve is 8 msec for both the descending volley giving rise to the direct M response, and for the ascending volley giving rise to the F response. With a distance of 370 mm between the two points of stimulation, the conduction velocity along the fibers evoking the M and F responses is identical at 46 m/sec.

Discussion

The procedure of bilateral dorsolateral longitudinal myelotomy, carried out in four patients to relieve severe spasticity of the lower extremities, provided a unique opportunity to investigate the question of the reflex or antidromic nature of the F response in man.

The operation essentially consists of a longitudinal section of the spinal cord in the coronal plane, along the lateral surface of the cord, immediately below the dorsal root entry zone (fig. 1D). The aim of the operation is to reduce sensory impulses impinging upon the anterior horn cells. The success of the procedure was evident by the marked clinical improvement, by the abolition of the deep tendon reflexes, clonus and H-reflex in all 4 patients.

The F response, on the other hand, continued to be easily recorded from the small muscles of the foot. The persistence of the response in the deafferented limbs clearly supports the view that in man the F response results from antidromic activation of motoneurons in the spinal cord as originally suggested by DAWSON and MERTON [1956] and subsequently supported by THORNE [1965], GASSEL and WIESENDANGER [1965] and McLEOD and WRAY [1966]. The orthodromic and antidromic conducted volleys were found to travel at similar speeds. Similar conclusions were reached by DAWSON and MERTON [1956] and by THORNE [1965]. The latter author, recording from single motor units in normal subjects, showed that the F response occurs only when a motor nerve fiber is excited, giving an early direct response first and that the velocity of the centrifugal and centripetal impulses were identical.

In view of the present findings in man and previous experiments in man and animals, it appears reasonable to conclude that: (1) The F response represents the discharge of motoneurons in the cord following their antidromic activation; (2) the term 'reflex' for the F response should not be used, to avoid misunderstanding with a true reflex mechanism; and (3) F response latencies should not be used as an indirect test to determine the conduction velocity of sensory fibers.

Summary

The F response was recorded from the small muscles of the foot in 4 paraplegic patients. Deafferentation of both lower extremities by dorsolateral myelotomy was successfully carried out to relieve their severe spasticity. Deep tendon reflexes, clonus, flexor responses and H-reflex were abolished. The F response persisted and was as easily elicited as before the operation. These findings in man support the view that the F response is due to antidromic stimulation of the spinal motoneurons in the cord.

Author's address: Prof. O. E. MIGLIETTA, M.D., New York Medical College, Flower and Fifth Avenue Hospitals, Fifth Avenue and 106th Street, *New York, NY 10029* (USA)

New Developments in Electromyography and Clinical Neurophysiology,
edited by J.E. Desmedt, vol. 3, pp. 328–335 (Karger, Basel 1973)

Supraspinal Influences on H-Reflexes

HELENA TÁBOŘÍKOVÁ

Department of Physiology, School of Medicine, State University of New York,
Buffalo, N.Y.

When the popliteal nerve is weakly stimulated at the fossa poplitea
(fig. 1A), there is a monosynaptic reflex discharge into the soleus muscle
(fig. 1B) [HOFFMANN, 1922; MAGLADERY, 1955; PAILLARD, 1955; HUGON, this
volume]. The group I_A fibres from the muscle spindles of triceps are solely
responsible for this reflex, which was named the H-reflex by MAGLADERY
[1955]. When the applied stimulus evokes a maximal discharge of these fibres,
the reflex usually does not activate all the motor units of the soleus muscle.
By collision technique it has been shown [TÁBOŘÍKOVÁ and SAX, 1968] that the
activated percentage ranges from 24 to 100, with a usual value around 50%.
This size at the middle of the motoneurone pool places it in the most sensitive
range for signalling by the size of the H-reflex an increase or decrease in the
α-motoneuronal excitability. Furthermore, the H-reflex has an advantage over
the tendon reflex because this latter is complicated by its muscle spindle initiation that is under control of the fusimotor system.

Provided that the stimulating electrode is placed on the skin in a fixed
relationship to the underlying popliteal nerve, a given stimulus will result in a
constant monosynaptic bombardment of the α-motoneuronal pool by I_A
impulses. Variations in the size of the resulting H-reflex provide a reliable
criterion of the net level of the other excitatory or inhibitory influences playing
on those motoneurones. Only two disturbing factors have to be taken into
account: presynaptic inhibitory action on the I_A synapses [DEVANANDAN,
ECCLES and YOKOTA, 1965a,b; DECANDIA, PROVINI and TÁBOŘÍKOVÁ, 1967a,
b; DELWAIDE, 1971], and a change in the transmitter output with repetitive
stimulation, which with I_A synapses is usually a diminution due to transmitter
depletion [CURTIS and ECCLES, 1960].

Potentially the utilization of H-reflexes should provide a most valuable method for revealing the supraspinal influences that are playing upon moto-neurones and that are often greatly changed in conditions giving disorders of movement (spasticity, tremors, etc.). These supraspinal influences may form a background of supraspinal activity, or they may be induced by various procedures at spinal level that activate ascending pathways to the brain that in turn evoke descending discharges, as for example occurs in spino-bulbar-spinal reflexes.

Supraspinal Influences as Background

Originally it was reported that, in contrast to tendon reflexes, there was no enhancement of H-reflexes by the Jendrassik manoeuvre [SOMMER, 1940; HOFFMANN, 1951; BULLER and DORNHORST, 1957], and PAILLARD [1955] found enhancement to be present only sometimes and then to a small degree. It was, therefore, proposed that the Jendrassik manoeuvre caused selective activation of γ-motoneurones with a resulting fusimotor discharge and increase in the sensitivity of the annulospiral endings giving enhanced tendon reflexes. However, a careful reinvestigation [GASSEL and DIAMANTOPOULOS, 1964; LANDAU and CLARE, 1964] has shown that the H-reflex invariably was poten-tiated in the Jendrassik manoeuvre, both in normal subjects and in patients with parkinsonism and spasticity, though as a rule to a lesser degree than the ankle jerk. It even occurred when there was blockage of the fusimotor fibres by procaine. It appears that, though fusimotor bias contributes to the poten-tiation of the ankle jerk, supraspinal influences must also be acting directly on α-motoneurones and increasing their excitability, in addition to the supraspinal action on the fusimotor system. It should be pointed out that Jendrassik enhancement cannot occur for those infrequent H-reflexes that involve virtually 100% of the motoneurone pool [TÁBOŘÍKOVÁ and SAX, 1968]. Such a case apparently is illustrated by DELWAIDE [1971, fig. 20], there being a large enhancement when the H-reflex was reduced.

Mild voluntary contraction of the triceps muscle increases the H-reflex [HOFFMANN, 1922; PAILLARD, 1955; GOTTLIEB, AGARWAL and STARK, 1970; GOTTLIEB and AGARWAL, 1971; DELWAIDE, 1971], which is simply explained by the summation on the triceps motoneurones of the excitatory influences from the pyramidal tract with the I_A impulses of the H-reflex. Conversely, mild contraction of antagonists usually depresses the H-reflex [GOTTLIEB, AGARWAL and STARK, 1970; GOTTLIEB and AGARWAL, 1971], which presumably is attributable to pyramidal inhibition of the triceps motoneurones. It has been shown that the best potentiation of tendon jerks occurs when there is simultaneously a Jendrassik manoeuvre and a mild voluntary contraction [OTT and GASSEL, 1969].

The general tonic excitatory influences from higher centres can be appreciated when H-reflexes are reduced or abolished during drowsiness or sleep of the subject, as has been reported by COQUERY [1962] and PAILLARD [1959]. There have been parallel investigations on cats [GASSEL and POMPEIANO, 1965].

Supraspinal Influences as Long-Loop Reflexes

SHIMAMURA, MORI, MATSUSHIMA and FUJIMORI [1964] calculated that in man a long-loop reflex for the hind limb (tibial nerve to m. tibialis anterior via a spino-bulbo-spinal pathway) would have a latency of about 60–85 msec. This calculated value was in good agreement with the observed times, 63–85 msec, for reflex responses of tibialis anterior evoked by stimulation of the tibial nerve at a strength adequate to evoke a maximal reflex. In 9 subjects DELWAIDE [1971] has observed very delayed reflex responses of the triceps muscle with an intensity of stimulation that may be below the threshold for evoking a H-reflex. In figure 1 C (LLR) the latency was 145 msec, and it decreased to 125 msec (fig. 1 D) when the stimulus was increased to produce a large H-reflex, and even a small M-response. It seems highly probable that, as DELWAIDE suggests, this is an example of a long-loop reflex with a pathway up to the brain stem or even higher.

It is unusual to observe such overt examples of long-loop reflexes in response to tibial nerve stimulation at a strength subliminal for evoking H-reflexes. However subliminal activation of triceps motoneurones can often be demonstrated by testing with an H-reflex at appropriate intervals [PAILLARD, 1955; TÁBOŘÍKOVÁ and SAX, 1968; DELWAIDE, 1971]. In figure 1 E are curves from 4 experiments (cf. fig. 2 and 3 of TÁBOŘÍKOVÁ and SAX, 1968]. Superimposed on the prolonged depression given in one experiment (solid line) there was in the other 3 experiments a delayed wave (60–300 msec) of relative and even of absolute facilitation. In figure 1 F there is a similar series of superimposed curves [DELWAIDE, 1971: fig. 10A].

In the usual experimental arrangement, with the conditioning and testing stimuli at a strength for setting up a large H-reflex [cf. MAGLADERY, 1955; PAILLARD, 1955] the situation is complicated by the delayed input to the

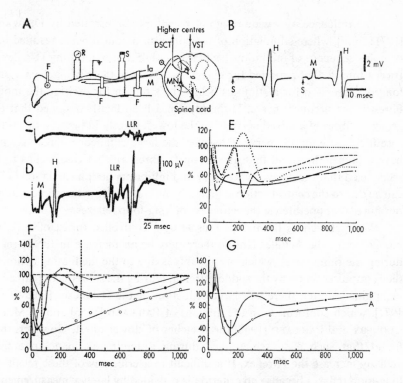

Fig. 1. A Diagram of experimental arrangements and pathways of H-reflex. S and R are stimulating and recording arrangements, and F indicates fixation devices. The H-reflex pathway is illustrated by a single I_A and M fibre and pathways to and from higher centres are shown (cf. fig. 2). In *B* are two specimen H-reflexes, the stimulus in one being above threshold for the motor fibres, hence the early M-response [TÁBOŘÍKOVÁ and SAX, 1968]. In *C*, stimulus was below threshold for H-reflex but there was a later long-loop reflex (LLR). In *D*, a stronger stimulus evoked an initial M-response and H-reflex and a much larger LLR [DELWAIDE, 1971]. In *E* are time courses of sizes of a testing H-reflex after conditioning by an H-stimulus below threshold for evoking an H-reflex as in *C* [TÁBOŘÍKOVÁ and SAX, 1969]. In *F* is a similar set of curves in another investigation [DELWAIDE, 1971]. *G* is similar to *F*, but conditioning is by stimulation of peroneal nerve at a strength 5% above the motor threshold [DELWAIDE, 1971].

spinal cord from the various receptors stimulated by the muscle contraction resulting from the H-reflex and even from the M-response. Under such conditions there will be delayed bombardments on the α-motoneurones of triceps through pathways at spinal level superimposed upon the bombardments from supraspinal levels.

The influence of a wide range of receptors is illustrated by DELWAIDE [1971: fig. 7], where stimulation of the common peroneal nerve resulted in a complex sequence of inhibition-facilitation-inhibition of a testing H-reflex in triceps muscle that persisted for over 1 sec (fig. 1 G), there being even a prolonged depression for a stimulus below the strength for exciting the motor fibres in the peroneal nerve. Hitherto, it has been tacitly assumed that the group I_A fibres of a mixed nerve have the lowest threshold because this is the case for a pure muscle nerve. However, the large cutaneous fibres may not have a higher threshold than the group I_A fibres. In that case, it has to be recognized that the delayed excitatory and inhibitory actions in figure 1 C–G can be due to the central action initiated by impulses in low-threshold cutaneous fibres that operate via the pathways of long-loop reflexes.

When stretching the triceps muscle by a controlled thrust of a Teflon plunger onto the Achilles tendon there may be an increase in the H-reflex during the ramp onset, which presumably is due to the facilitating action of the I_A impulses set up by the sudden stretch. During the subsequent plateau of maintained stretch there is a large reduction of the H-reflexes [DELWAIDE, 1971], which conforms with the findings of PAILLARD [1955] and of MARK, COQUERY and PAILLARD [1968] for stretches of slower onset. However, they found that such a stretch applied during a Jendrassik manoeuvre may actually increase the H-reflex. It is difficult to ascribe any of these results to long-loop reflexes because no criterion is provided by latency measurements. A preliminary attempt has been made to carry out very rapid dorsiflexions and plantarflexions of the ankle by a catapult rubber device and to test the motoneuronal excitability by H-reflexes at various times during the movement and the subsequent plateau position [TÁBOŘÍKOVÁ, DECANDIA and PROVINI, 1966; ECCLES, 1966]. Dorsiflexion sometimes set up an initial brief excitation as reported by DELWAIDE [1971], but there was always the sustained depression during the plateau. On the other hand, with plantar flexion there was an initial depression on which was superimposed a brief excitatory phase usually from 100 to 300 msec, much as in figure 1 E, F. Probably long-loop reflex responses contribute to the complex time course of the actions on the testing H-reflex. The receptors of the ankle joint would be expected to be powerfully excited by these rapid angulations, but cutaneous and muscle afferents would also be excited. This investigation was made to test for the central actions of a sudden natural movement. However much more sophisticated instrumentation is required if it is to be systematically carried out.

When a vibrator is applied to a muscle tendon, there usually develops in the normal subject a slowly rising contraction of that muscle [HAGBARTH and

EKLUND, 1966, 1968; DEGAIL, LANCE and NEILSON, 1966; LANCE, DEGAIL and NEILSON, 1966], which is attributed to reflex discharges initiated by the group I_A fibres of that muscle [cf. MATTHEWS, 1966]. This prolonged incrementing reflex is absent after spinal cord section, but this is attributed to interruption of the tonic excitatory action descending principally in the vestibulospinal fibres, but also in reticulospinal fibres [GILLIES, BURKE and LANCE, 1971 a, b]. It is, however, not clear that this tonic vibration reflex is exclusively due to a monosynaptic I_A action on α-motoneurones that have a background facilitation by descending influences from the brain stem. It is not possible so to explain the continuance of the reflex discharge for up to 1 sec after cessation of the vibration [LANCE, DEGAIL and NEILSON, 1966, fig. 1; MATTHEWS, 1966, fig. 1]. It must also be recognized that the vibration of the tendon in the human subject will excite cutaneous receptors as well as group I_A and these may contribute to the response, particularly if it has a long-loop component, as seems likely from its slowly incrementing character and its after-discharge. For our present purpose it is important that H-reflexes are depressed during the vibration [HAGBARTH and EKLUND, 1966; DEGAIL, LANCE and NEILSON, 1966] and this effect has been extensively investigated by DELWAIDE [1971]. It is probable that, in part, this depression is due to supraspinal influences that were postulated above [cf. DEGAIL, LANCE and NEILSON, 1966; LANCE, 1970] as being concerned in the tonic vibratory reflexes. However, much of the depression is probably effected at a purely spinal level by two mechanisms: the continued activation of the I_A fibre would result in transmitter depletion [CURTIS and ECCLES, 1960] with a resulting depression of the testing H-reflex, and probably the continued afferent input into the spinal cord would give presynaptic inhibition of the group I_A terminals on the motoneurones [DEVANANDAN, ECCLES and YOKOTA, 1965a, b; DECANDIA, PROVINI and TÁBOŘÍKOVÁ, 1967a, b; LANCE, 1970; DELWAIDE, 1971].

Possible Supraspinal Pathways for Long-Loop Reflexes Influencing H-Reflexes

In figure 2 are shown diagrammatically the simple monosynaptic H-reflex pathway and superimposed thereon spinal and supraspinal pathways. Some diagrammatic simplification is achieved by having the ascending and the descending components of long-loop reflexes shown separately in figure 2 A and B, respectively. Many of the pathways have been discovered in the cat and monkey [JANSEN and BRODAL, 1954; OSCARSSON, 1965; BRODAL, 1967;

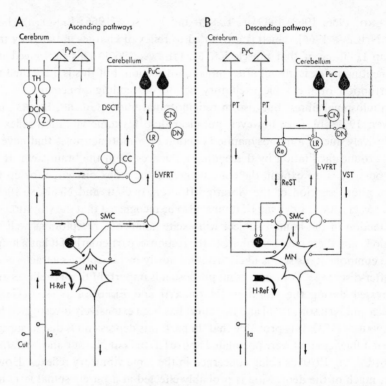

Fig. 2. Ascending (*A*) and descending (*B*) pathways possibly involved in supraspinal influences on H-reflexes (H-Ref). PyC, pyramidal cells; PuC, Purkyně cells; Cut, cutaneous fibre; I_A, I_A afferent fibre; DCN is dorsal column nuclei with Z component; TH, ventrobasal thalamus. All other symbols are explained in the text.

ECCLES, ITO and SZENTÁGOTHAI, 1967], but are presumed to exist in man. Because of their insignificance in man [NATHAN and SMITH, 1955] the red nucleus and rubrospinal tract are omitted. In outline both the I_A and cutaneous ascending pathways activate Clarke's column cells (CC) and so go via the direct spino-cerebellar tract (DSCT) to the cerebellum. Furthermore, the cutaneous input ascends to the cerebrum via 2 relays and the I_A by 3 relays. The cutaneous input also excites via interneurones of the spinal motor centre (SMC) the cells of origin of the bVFRT that ascends to the lateral reticular nucleus (LR) and so to the cerebellum, as well as to the cerebellar nuclei (CN). In figure 2*B* the pathways down are from the cerebrum via the pyramidal tract (PT) to the motoneurones (MN) and spinal interneurones (SMC) and also from the cerebrum to reticular nuclei (Re) and so down the reticulospinal tract (ReST) to the spinal motor centre and eventually to motoneurones as shown. From the cerebellum the Purkyně cells project to the cerebellar nuclei and to Deiter's nucleus (DN) and so via the ReST and the vestibulospinal tract (VST) to the spinal cord and thence to motoneurones either directly or via interneurones (SMC).

Several complicated circuits can occur even with this greatly simplified diagrammatic representation. These diagrams serve merely to illustrate the levels of complexity that can be involved as soon as supraspinal mechanisms are brought into the picture. And they must be brought in. Current investigations with adequate stimulation have shown that small muscle stretches or taps on the pads of the cat's foot evoke powerful responses from cells in the cerebellum and cerebellar nuclei, which should result in excitatory and inhibitory influences on motoneurones via the pathways shown in figure 2 B [ECCLES, SABAH, SCHMIDT and TÁBOŘÍKOVÁ, 1971; FABER, ISHIKAWA and ROWE, 1971; ECCLES, SABAH and TÁBOŘÍKOVÁ, 1972]. Furthermore, it has also been shown that, in accord with the pathways in figure 2, cutaneous stimulation activates pyramidal cells in the monkey's sensorimotor cortex in the region projecting via the pyramidal tract to motoneurones that cause movement related to the area of cutaneous stimulation [ASANUMA and ROSÉN, 1972; ROSÉN and ASANUMA, 1972]. Finally, PHILLIPS, POWELL and WIESENDANGER [1971] find that the group I_A afferents from the baboon forelimb project to area 3 a in close proximity to the motor cortex.

Author's address: Dr. HELENA TÁBOŘÍKOVÁ, Department of Physiology, School of Medicine, State University of New York, *Buffalo, NY 14214* (USA)

New Developments in Electromyography and Clinical Neurophysiology,
edited by J. E. Desmedt, vol. 3, pp. 336–341 (Karger Basel 1973)

Vestibular Influences on Proprioceptive Reflexes of the Lower Limb in Normal Man

P. J. DELWAIDE and P. DELBECQ

Section of Neurology, Department of Medicine, University of Liège, Liège

The vestibular system is uniquely involved in proprioceptive integrations and in the control of muscular tone and balance. Its role should be particularly significant in bipedal species and in man. It thus appears important to examine the interaction between descending vestibular influences and the spinal pathways involved in segmental proprioceptive reflexes. This problem has not yet received much attention in man [BENSON, 1959]. This paper considers the influence of vestibulo-spinal pathways on T- and H-reflexes of the soleus muscle and on the T-reflex of the short biceps.

Material and Method

28 normal adult volunteers of both sexes, aged 18–24 years, were submitted to one or more experimental sessions. They were sitting in a special chair with the lower limb maintained in a restraining device [DELWAIDE et al., 1969]. Standard methods were used to elicit the T- and H-reflexes of soleus with controlled stimuli [see HUGON, this volume; DELWAIDE, this volume]. The T-reflex of the short biceps was elicited by mechanical taps on the muscle tendon at its insertion on the peroneal head [HUGON, 1967]. Intervals of 8 sec separated each stimulus. As a rule the amplitude of 10 successive reflexes was averaged. Each session lasted about 2 h. Vestibular stimulation was achieved either by the careful injection of 30 ml water at 20 °C into the ear canal on one side (caloric stimulation), or by delivering an electric current of 3–5 mA for 300 msec through a skin silver plate (2.5 cm diameter) on the mastoid bone, the reference anode being placed on the contralateral mastoid or ipsilateral arm (electric stimulation). The stimulator was a Grass Model S88 with a constant current output. The intensity of the vestibular stimulation was adjusted so that it elicited a definite subjective impression of body displacement. The intensity should not be such as to produce a withdrawal response of the patient. No evident movement of the subject was noticed for the stimulations used in this work.

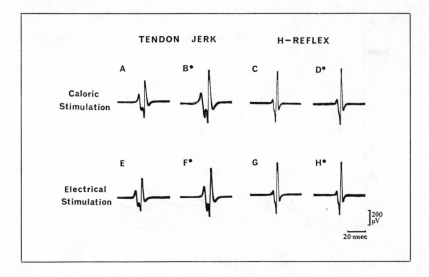

Fig. 1. Effects of caloric vestibular stimulations (A–D) and of electrical vestibular stimulations (E–H) on soleus tendon jerk (left side) and on H-reflex (right side). The responses evoked during vestibular stimulations are marked with a point. By both stimulations the tendon jerks are obviously increased, while H-reflexes are not.

Results

Vestibular Caloric Stimulation

The injection of cold water into the external ear canal lasts about 20–30 sec. The maximum T-reflex of soleus is definitely increased with respect to control about 10 sec after the beginning of the injection, it remains increased for several seconds, and in one case for as long as 2 min, after the end of the injection (fig. 1 A, B). The T-reflex can be increased by about 50–100% and the potentiation is more important when the caloric stimulation in stronger (colder water or larger amount injected). A similar potentiation is observed for caloric stimulation of either the ipsilateral or the contralateral ear canal. The time course of the effect is parallel to that of the sensation of dizziness reported by the subject. The H-reflex recorded under similar conditions is not significantly potentiated (fig. 1 C, D). The size of the control H-reflex was carefully adjusted to about half the maximum H-reflex amplitude in the same muscle so that the lack of potentiation effect of vestibular stimulation cannot be ascribed to saturation of H-reflex loop [cf. HUGON, this volume].

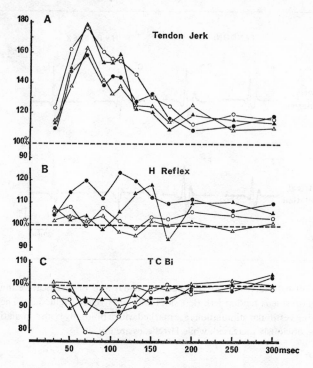

Fig. 2. Comparison of effects of electrical vestibular stimulation on soleus tendon jerk (A), H-reflex (B) and T-reflex of the short biceps (C) in 4 subjects. On the ordinate, results are expressed as a percentage of control values.

Vestibular Electric Stimulation

The electric current (see Method) is delivered for 300 msec during which time one soleus reflex is elicited at a chosen interval after the beginning of the stimulation. As a rule the maximum soleus T-reflex is potentiated during the vestibular stimulation and the peak of the effect is recorded for intervals of 70–90 msec (fig. 1 E, F). Pooled results obtained in 3 normal subjects are presented in figure 2A. The effect presents a characteristic time course with an early peak and a subsequent decay; the latter may probably be ascribed to adaptation of the peripheral vestibular system to the applied electric current.

The size of the maximum effect is similar to, or slightly less than, that obtained in the caloric tests for the intensities of electric current used. It is of interest that roughly similar potentiations of the soleus T-reflex are re-

corded, no matter on which side of the head the cathodal current is delivered
(fig. 3 A, B). The similarity between caloric and electric vestibular stimulations
further appears when the H-reflex is studied since no significant potentiation
thereof has been observed (fig. 1 G, H and 2 B). Note that the ordinate of
figure 2 B has been made larger than that of figure 2 A in order to present in
more detail the variations of H-reflex amplitude. These changes in H-reflex
are below 10% in 2 cases and vary up to about 20% in the other 2 experiments
which are far less important than the changes recorded for the T-reflex in the
same experiments (fig. 2 A).

Effects on the Biceps T-Reflexes

It is of interest to enquire whether the above results depend on a general
facilitation of the tendon jerks of the lower limb (such as in the Jendrassik
manoeuver) or rather involve the extensor reflexes only. The same experimental program has thus been used with the maximum T-reflex of the short biceps.
Vestibular stimulations, either caloric or electric, have not increased the
biceps reflex in any of the 6 experiments performed. Reductions by up to 10%
of the amplitude of the reflex were observed in a number of trials (fig. 2 C).
More results are needed before one could decide whether a mild genuine
inhibitory effect can result from vestibular stimulation.

Discussion

One of the significant results obtained is the similarity between the
potentiation effects on the soleus T-reflex of both caloric and electric vestibular
stimulations. The caloric stimulation would appear at first sight to achieve
a fairly specific activation of the vestibular sense organs. The DC electric
current applied to the mastoid might involve other structures in the head but
the data would suggest that activation of the vestibular system is certainly
one of the main effects produced. This observation is of interest since the
electric stimulus is more flexible to apply and better controlled, especially for
repeated stimulations.

Another finding is that similar facilitating effects are recorded, no matter
which side is stimulated (fig. 3). This was rather unexpected, since activation
of the vestibular system on either side tends to produce opposite shifts of
balance and displacements of the head. However, acute experiments on the

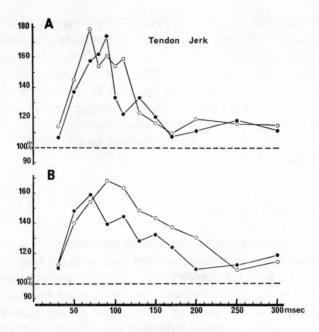

Fig. 3. Curves of T-reflex facilitation plotted respectively after ipsi- (●) and contro-lateral (○) vestibular stimulations. On the ordinate, results are expressed as a percentage of control values.

cat have indicated that in many instances the vestibulospinal tracts from both sides produce qualitatively similar effects on hind-limb segmental reflex path-ways [Hongo *et al.*, 1971; Aoyama *et al.*, 1971].

The vestibular potentiation of the soleus T-reflex is not accompanied by any consistent parallel changes of the amplitude of the soleus H-reflex (fig. 1, 2). This indicates that the excitability of the soleus α motoneurons is not in-creased to a significant extent, at least when it is tested by the I_A afferents from soleus activated in H-reflex testing. Along this line one should seek to account for the rather selective potentiation of the soleus T-reflex by a vestibulospinal effect which would be exerted either on the soleus fusimotor system or on premotoneuronal mechanisms modulating presynaptic inhibition of the I_A afferents. Other types of potentiation of the T-reflex without parallel changes of the H-reflex have already been interpreted by a γ loop effect whereby soleus spindles are made more excitable without there being necessarily a marked change in I_A afferents and α motoneuron excitability [Paillard, 1955; Rush-worth, 1960, 1964]. The alternative possibility of an effect organized uphill from motoneurons and involving presynaptic inhibition deserves careful consideration in view of recent evidence that changes in the I_A afferent may occur [DELWAIDE, 1970 and this volume; BURKE and LANCE, this volume]. In the decerebrate cat, BARNES and POMPEIANO [1970] found that vestibular stimulation elicits dorsal root potentials related to primary afferent depolarization (PAD) in the group I afferents of extensor muscles of the hindlimb, an effect which is parallel to the vestibular potentiation of monosynaptic extensor reflexes. Further studies are required to explore whether such a mechanism is also in operation in man. If so there should be no great difficulty in explaining why the T-reflex is potentiated while the H-reflex is not, even though they both involve a I_A afferent volley from soleus. It is wellk nown that the I_A volley is much more synchronized for the electrical stimulus evoking the H-reflex than for the Achilles tendon percussion evoking the T-reflex [GASSEL and DIAMANTOPOULOS, 1965]. Thus, with a strong presynaptic inhibitory background on I_A afferents, the more prolonged afferent discharge of the T-reflex would be at a disadvantage. By contrast if we suppose that the vestibulospinal influence reduces such presynaptic inhibitory mechanism on I_A afferents, the resulting potentiation would be more important for the T than for the H-reflex of soleus.

The present study also indicates that the T-reflex of the short biceps, a flexor of the knee, is not potentiated and actually sometimes slightly reduced (fig. 2C) by a vestibular stimulation which potentiates the soleus T-reflex. This suggests that the descending vestibular effects are not generalized, but instead differentially involve the flexor and extensor proprioceptive reflex pathways.

Acknowledgement

We wish to thank Prof. J. E. DESMEDT (Brussels) for his advice during the preparation of this paper.

Author's address: Dr. P. J. DELWAIDE, Hôpital de Bavière, bd. de la Constitution, B-4000 Liège (Belgium)

New Developments in Electromyography and Clinical Neurophysiology,
edited by J. E. Desmedt, vol. 3, pp. 342–359 (Karger, Basel 1973)

An Objective Technique for the Analysis of the Clinical Effectiveness and Physiology of Action of Drugs in Man

M. M. GASSEL

San Francisco, Calif.

Neuropharmacological techniques for predicting clinical effectiveness, based upon animal experiments, have been found wanting. The objective criteria for studying drug effect in animals frequently lack established clinical validity, e. g. criteria used for establishing activity as an analgesic or a 'muscle relaxant'. Animal models reproducing clinical disorders have been difficult to construct, and have provided only a modest basis for predicting clinical pharmacology. On the other hand, clinical judgment alone has proved to be inadequate in evaluating drug effect. Patients' subjective reports of change are frequently invalid, and objective techniques currently in use, such as EMG techniques and mechanical devices for recording resistance to muscle stretch, are inaccurate. Clinical analysis of drug action is also generally overly circumscribed, without evaluation of the frequently significant contribution of sedation, tranquilization or analgesia to the overall effectiveness.

This study describes a technique for comprehensive evaluation of drug effects in man, using valid objective criteria in a controlled setting.

The stretch reflex is evaluated both before and at intervals after the drug. The mechanically evoked monosynaptic stretch reflex (ankle jerk), the electrically evoked monosynaptic H-reflex and its recovery curve using paired stimuli are all studied in the same population of triceps surae motoneurones. Changes in excitability of the stretch reflex are also evaluated objectively in patients with spasticity by increase or decrease in the occurrence of clonus after a drug. The activity of the flexion reflex afferent system is evaluated by changes in excitability evoked in the same triceps surae motoneurones on stimulating cutaneous afferents. The flexion reflex is also evaluated objectively in spastic patients by ipsilateral or crossed effects evoked by a regular shock stimulus to the skin. An evaluation is made of drug effect on the response of spinal reflexes to facilitory influences.

The effects of sedation or tranquilization are evaluated by subjective reports of change and a timed test of intellectual performance. A comparison is made with doses of barbiturates to judge further the relationship of sedation to the observed changes. Analgesic effects are studied by a novel technique; simultaneous evaluation is made of clinical relief of pain and change in pain experience produced by a drug.

Reflex alterations are related to changes in psyche, to the experience of pain, to subjective reports of change, and to clinical effects evaluated at regular intervals during the investigation. The results of the investigations are analyzed and graphs provide a pattern of temporal changes showing the onset, height of effect and duration of action, an evaluation of the site of action, and specificity of drug effect.

Striking and reliable changes in the various parameters are produced in normal subjects and patients with barbiturates, diazepam (Valium), during inhalation of tobacco smoke, by intravenous nicotine, orphenadrine (Norflex), and with nefopam, an analogue of orphenadrine. It has been possible to analyze the therapeutic action of drugs in addition to those purported to be effective in disorders of the motor system, including sedatives and hypnotics, tranquilizers, stimulants and analgesics.

Methods

Subjects lie prone with the leg secured in a metal frame which fixes the position of thigh, knee, and ankle (fig. 1). A spring device returns the foot to a standard position after reflex elicitation. The frame has been designed to maintain immobilization comfortably over long periods, frequently 4–5 h [GASSEL, 1970a; GASSEL and OTT, 1970a]. Attached to the leg frame is a solenoid-driven tendon hammer that delivers a blow of constant magnitude at regular frequency to the Achilles tendon, adequate to evoke a maximal reflex. The evoked reflex potential is recorded with stainless steel electrodes, 2.5 cm in length, bent at right angles 1 cm from the end, with the distal 1 cm fixed in place subcutaneously over the triceps surae muscle. The reflex is amplified and recorded on film or measured from a storage oscilloscope. A monosynaptic H-reflex is also evoked in the same population of triceps surae motoneurones by electrical stimulation of the posterior tibial nerve in the popliteal fossa. The stimulating electrodes are fixed in place subcutaneously in optimal position to evoke a maximal H-reflex with a minimal direct M response. The regularity of stimulating and recording conditions over long periods (3–5 h) has been demonstrated by objective means, i.e. measurement of evoked physiological potentials (nerve or muscle action potentials and direct measurement of current delivery). The experimental protocol has been described in detail [GASSEL and DIAMANTOPOULOS, 1964; GASSEL and OTT, 1970a].

The recovery of the H-reflex was studied before and after the drug with the interval between the conditioning and the test reflex increased in 8 steps from 50 msec to 5 sec. The invariable shocks to the posterior tibial nerve were adjusted to evoke maximal H-reflexes. A statistically significant evaluation was made as at least 5, generally 10, but up to 40 responses were averaged in each determination; the number related to the variability of the reflex. The mean, SD, variance, and standard error of the mean were computed on an Olivetti model 101. The changes in reflex recovery, produced by a drug, at any interval between conditioning and test reflex was usually representative of the overall drug effect; however, the changes at the conditioning-test interval at the peak of intercurrent facilitation were frequently most sensitive. Reflex excitability and recovery at the peak of intercurrent facilitation, ascertained from the initial recovery curve, were selected for detailed study at frequent intervals (see fig. 2, 3). The mean of the conditioning and the test H-reflexes and of the ankle jerk reflexes were then graphed, with amplitude in mV on the ordinate, and time in minutes on the abscissa, before and after the drug. Initially a graph was also made expressing the amplitude of the test reflex as percent of the conditioning H-reflex. However, this ratio had little relationship to demonstrable clinical changes, e. g. when sedatives, such as diazepam, depressed reflex excitability (the conditioning H-reflex) to a greater degree than reflex recovery (the test reflex) at some time in the course of action. This resulted in an apparent enhancement of reflex recovery with, however, a clinical decrease in spasticity.

The activity of the flexion reflex before and at intervals after the drug was evaluated by recording changes in excitability of the monosynaptic reflex evoked in the same population of triceps surae motoneurones by stimulating cutaneous afferents [GASSEL and OTT, 1970a]. A regulated system is described for delivering an unvarying, single-shock stimulus to the skin of the dorsal and plantar surfaces of the ipsilateral distal foot via subcutaneous electrodes fixed in place to condition the reflex.

Evidence has been advanced from animal studies which suggests that an important criterion of a 'muscle relaxant' is its ability to block facilitation of spinal reflexes evoked by supraspinal stimulation [GINZEL, 1966a, b]. Alterations in response of spinal reflexes to facilitory influences are evaluated. Facilitation of H- and T-reflexes by the Jendrassik maneuver is studied, employing clenching of the hands or sniffing. The degree and pattern of facilitation is judged before, and at intervals after, administration of a drug.

Subjective reports and standard tests engaging attention and of intellectual performance are obtained at intervals during the investigation in order to judge the relationship between sedation and reflex alterations. The test of intellectual performance involves reciting in order the following four entities: (a) Methodist Episcopalian; (b) Royal Irish Constabulary; (c) serial subtraction (100 minus 7), and (d) Babcock sentence. The minimum time taken and accuracy of the performance is recorded after a period of preparation in establishing stable control values. A comparison is made with the effect of doses of barbiturates.

Studies also were performed evaluating, simultaneously, clinical relief of pain and change in pain experience, judged according to a novel experimental protocol. The clinical and experimental relief of pain were closely related. This evidence supplements that of reflex alterations, psychic changes and the clinical evaluation studied at intervals after the drug.

Changes in stretch and flexion reflexes were also evaluated objectively in patients with spasticity by recording clonus evoked by the reflexes (fig. 6), and by observable alterations in the flexion reflex elicited by a regular shock stimulus to the skin.

Changes in clinical status, intellect, and subjective reports, periodically evaluated, are recorded on the graph in relationship to reflex changes.

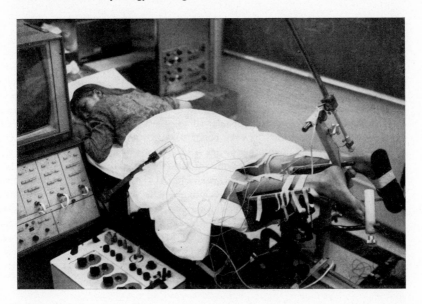

Fig. 1. Photograph of subject during investigation, showing frame for maintaining position, with automatic solenoid hammer.

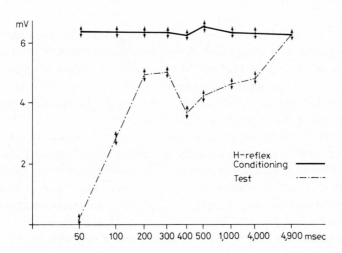

Fig. 2. The recovery curve of the H-reflex in a normal subject, employing maximal reflexes. *Ordinate:* amplitude of the reflex in mV. Vertical bar above or below the mean is equal to the standard error of the mean. *Abscissa:* time interval in msec between the conditioning and test stimuli.

Artifacts of the Procedure

A methodological feature of great importance in the protracted study of drug effects is the stability of stimulating and recording conditions. An apparatus has been designed to fix the position of the thigh, leg and ankle, and for delivering a regular blow to the Achilles tendon adequate to evoke a maximal reflex (see above). Comfortable immobilization has been maintained for the period of study frequently lasting 4–5 h. The design of electrodes and the technique for delivering a regular stimulus and for maintaining fixed recording conditions has been reported [GASSEL and DIAMANTOPOULOS, 1966; GASSEL and OTT, 1970a]. Constant EMG monitoring of relaxation is made from the same recording electrodes over the triceps surae. Sources of error in studies of reflex excitability in man have been reported [OTT and GASSEL, 1969; GASSEL and DIAMANTOPOULOS, 1965; GASSEL, 1969; see also HUGON, this volume]. Features compromising the study of reflex recovery, particularly in patients with spasticity or Parkinson's disease are soon to be published [GASSEL, DIAMANTOPOULOS and HUGHES, 1972]. Spurious and unrepresentative changes in reflex recovery in these patients are not uncommon, and some abnormal features of the recovery curves reported in these patients need to be reviewed. It is essential that there is monitoring for relaxation and absence of EMG activity, for the occurrence of clonus, and for flexor withdrawal associated with reflex elicitation.

Monosynaptic Reflex Excitability and Recovery

Rationale

The monosynaptic H-reflex evoked in the triceps surae has been identified as the equivalent of the ankle jerk that depends on muscle stretch receptors whose sensitivity is regulated by the fusimotor system [PAILLARD, 1955]. The H-reflex, that bypasses the muscle spindle, provides a measure of central excitability and the relationship of the H-reflex to the ankle jerk provides a means of evaluating fusimotor activity.

The H- and T-reflexes are usually abolished together in patients with absent reflexes and enhanced conjointly in patients with hyperactive reflexes; those normal subject with hypoactive tendon reflexes have low amplitude and high threshold H-reflexes. There are, however, physiological states and pathological conditions in which there is a predominant or even exclusive loss of the mechanically evoked T-reflex. The dissociation has been reported in acute spinal shock in humans [DIAMANTOPOULOS and ZANDER OLSEN, 1967; WEAVER, LANDAU and HIGGINS, 1963], in the course of therapy with chlorproethazine [MATTHEWS, 1965], and vincristine [SANDLER, TOBIN and HENDERSON, 1969], during spindle bursts of synchronized sleep, and in paradoxical sleep between the rapid eye movement phases in cats [GASSEL and

POMPEIANO, 1965], and in the stage of recovery after procaine nerve block in humans [GASSEL and DIAMANTOPOULOS, 1964]. These changes have been attributed to a specific depression of fusimotor function. However, there are basic differences in the nature of the afferent and efferent nerve volleys associated with these reflexes. Therefore, inferences regarding fusimotor function, based upon the relationship of changes in the two reflexes are not always justified [GASSEL and DIAMANTOPOULOS, 1965].

There are important clinical correlations with the reflex changes. In states characterized by predominant depression of the T-reflex, as in paraplegia or hemiplegia during the shock phase, signs of spasticity such as clonus or resistance to passive stretch are reduced or absent. However, the converse is not true and there are rare instances, both clinical [KREMER, 1958] and experimental [CANNON, MAGOUN and WINDLE, 1944] in which tendon jerks are hyperactive while resistance to passive movement is decreased. It is, of course, understood that single-shock electrical stimulation or tendon tap is not a 'natural' stimulus. The monosynaptic reflexes so evoked are best considered as the special response of the α-motoneurones to contrived and abnormally abrupt test volleys. Clinical applications and limitations of reflex studies in man have been critically reviewed [GASSEL, 1969].

A basic assumption involved in the study of monosynaptic reflexes as a test of motoneurone excitability is that presynaptic afferent polarization is a constant. However, hyperpolarization of presynaptic afferent terminals occurs during post-tetanic facilitation of the monosynaptic reflex, and presynaptic depolarization has been reported during the reflex depression of the rapid eye movement phase of desynchronized sleep [GASSEL, MARCHIAFAVA and POMPEIANO, 1965] or with the monosynaptic reflex depression following succinylcholine administration [COOK, NEILSON and BROOKHART, 1965]. There are also considerable differences in the pattern of presynaptic inhibition in, for example, decerebrate and spinal states [CARPENTER, ENGBERG, FUNKENSTEIN and LUNDBERG, 1964]. In decerebrate cats an increase of the monosynaptic reflex is not inevitably related with motoneurone excitability judged by the occurrence of spontaneous discharge of the same group of motoneurones [GRANIT and JOB, 1952]. Clearly motoneurone excitability is not reflected definitively by any single method of study. An alternative technique of evaluating central excitability by changes in discharge of motoneurones has recently been developed for use in clinical neurophysiology [GASSEL and OTT, 1970b].

The present analysis of drug action was accompanied by serial evaluation of clinical changes (contralateral side in paraplegic subjects) and objective evidence of change (increase or decrease of clonic beats) in spastic patients with hyperactive stretch reflexes. The evidence was supplemented by study of reflex recovery and of the activity of the flexion reflex. The reflex changes produced by various drugs were found to be reliable and generally consistent with clinical effects.

The recovery curve of the H-reflex with paired stimuli is useful for investigating pharmacological agents. A stimulus to the posterior tibial nerve which evokes a maximal H-reflex is followed, at intervals from 1 msec to 10 sec, by an identical stimulus. The size of the second reflex relative to the first is taken as a measure of the change in excitability of the motoneurone pool. The mean amplitudes of both reflexes are plotted as a function of the interval between the reflexes [cf. HUGON, this volume].

The recovery curve shows a phase of facilitation from about 50–400 msec that peaks at 200 msec [HOFFMANN, 1924; MAGLADERY, TEASDALL, PARK and PORTER, 1951; PAILLARD, 1955; DIAMANTOPOULOS and ZANDER OLSEN, 1967]. It appears to be superimposed on a long-lasting depression, beginning a few msec after the conditioning volley and fading gradually at 2–5 sec. This phase of intercurrent facilitation appears to be the most sensitive parameter in disease states. It is enhanced and occurs with shorter latency in spasticity [MAGLADERY, 1955]; it is heightened in Parkinson's disease, and decreased somewhat after successful thalamotomy [ZANDER OLSEN and DIAMANTOPOULOS, 1967; YAP, 1967]; and it is increased in chorea and dystonia musculorum deformans [COOPER, 1966; TAKAMORI, 1967]. The 'intercurrent facilitation' is abolished during spinal shock in man but then increases progressively with the return of muscle tone, finally becoming enhanced with the development of spasticity [DIAMANTOPOULOS and ZANDER OLSEN, 1967]. The origin of the intercurrent facilitation is complex and to some extent problematical [see critical review, GASSEL, 1970b]. However, it is an index of disease states and it will be shown that it is also a sensitive criterion of drug action.

Depression of Reflex Excitability and Recovery

1. Nicotine

Inhalation of tobacco smoke (10 studies) and intravenous administration of 0.8–1 mg nicotine (4 studies) caused a marked inhibition of H-reflex excitability and recovery (fig. 3). The subjects were nonsmokers or had stopped smoking at least 24 h before the study. There was predominant de-

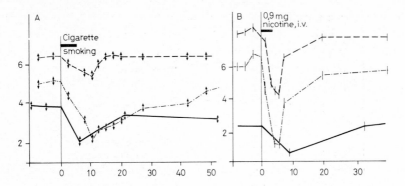

Fig. 3. Effect of smoking a tobacco cigarette (A) and i.v. injection of nicotine (B) in normal subjects. The graph shows changes in ankle jerk amplitude and in H-reflex recovery at height of intercurrent facilitation (mV, *ordinate*) as a function of time (min, *abscissa*). Tendon jerk, *continuous line.* Conditioning H-reflex, *interrupted line.* Test H-reflex (300 msec interval (A), 200 msec interval (B), *dots and dashes.* Vertical bar above and below the mean is equal to 1 standard deviation. Control values before the drug are indicated on the left.

pression of motoneurone recovery (i.e., the test reflex) in most studies and this depression was frequently the first evidence of effect.

Subjective changes reported concomitant with the reflex changes included a feeling of tranquility, extreme fatigue and nausea, a 'relaxation of muscles', and even flaccidity. Deterioration on intellectual testing occurred transiently with the trough of reflex depression in 3 of 14 studies. There was evidence of altered pain experience at this time [GASSEL, to be published]. The dramatic changes in the experimental system produced by nicotine are to be contrasted with the results of experiments employing techniques for evaluating changes in passive resistance to muscle stretch [cf. WEBSTER, 1964].

Inhalation of tobacco smoke is an efficient way of administering nicotine. It has been calculated that the equivalent of about 1 mg of nicotine delivered intravenously can be derived in man by smoking a cigarette [SILVETTE, HOFF, LARSON and HAAG, 1962; CLARK and RAND, 1964]. Nicotine is responsible for the pharmacological activity attributable to alkaloids in tobacco smoke [CLARK, RAND and VANOV, 1965].

Experimental depression of the knee jerk by nicotine was reported by SCHWEITZER and WRIGHT [1938]. A depression by tobacco smoke is seen in cats [CLARK and RAND, 1964] and man. The inhibitory action of nicotine on monosynaptic reflexes appears to be related primarily to its excitatory action on the inhibitory Renshaw interneurone having a cholinergic synapse [CURTIS,

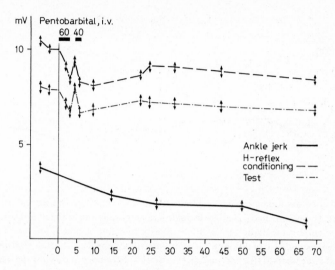

Fig. 4. Graph of the effects of 100 mg i. v. pentobarbital in a normal subject. *Ordinate:* mean amplitude of the conditioning and test H-reflexes at the height of intercurrent facilitation (300 msec) and of the ankle jerk. Vertical bars above and below the mean indicate the standard error of the mean. *Abscissa:* time in minutes after the start of the injection. Control values are indicated on the left.

ECCLES and ECCLES, 1957]. The effects of nicotine were prevented experimentally by beta erythroidine which blocks nicotinic excitation of the Renshaw cell. However, the action of nicotine is complex and additional mechanisms may be directly or indirectly involved in the reflex depression. Nicotine excites a variety of sensory receptors, including the muscle spindle, baroreceptors and chemoreceptors, and the midbrain reticular formation [GINZEL, 1967].

2. Pentobarbital Sodium

The effects of pentobarbital sodium (Nembutal), 50–100 mg, by slow i. v. injection, were studied in 5 normal subjects and 6 patients with spasticity. Figure 4 shows the effects in one normal subject; some reflex depression occurred after the injection of 60 mg of pentobarbital in a 2-min period, and the injection was stopped temporarily. There was slight restitution of reflexes in the next minute. At 3 min after the start of the injection the subject was drowsy and the time taken for intellectual testing increased from 25–150 sec. An additional 40 mg of pentobarbital produced an additional trough of depression after 2 min, followed by partial recovery. At 6 min after the injection there was slurring of speech, diplopia, and the subject was unable to complete the test of intellectual performance. Slight slowing in the latter persisted for 30 min and mild sedation remained for over 2 h.

Pentobarbital generally produced drowsiness, a considerable deterioration in intellectual performance, slurring of speech and occasionally diplopia. This was evident shortly after the injection and was most pronounced at 5–15 min. Some small defect in intellectual performance was frequently still demonstrable at 30–45 min after the injection and mild sedation continued for a few hours.

The monosynaptic reflexes were depressed *pari passu* with the intellectual changes, although the reflex depression outlasted the evidence of intellectual deterioration and subjective reports of significant sedation. Some reflex depression frequently continued for 2.5–3 h, up to the end of the experiment. The most sensitive index of drug effect was depression of H-reflex recovery and of the T-reflex. Depression of the ankle jerk was marked especially later in the study. It is noteworthy in this regard that fusimotor depression has been demonstrated experimentally during barbiturate induced EEG synchronization [HONGO, KUBOTA and SHIMAZU, 1963].

There was evident decrease in spasticity and clonus associated with depression of reflex excitability and recovery in 3 of 6 patients investigated.

The effects of intravenous, intramuscular, and oral administration of diazepam and pentobarbital have also been tested [GASSEL, DIAMANTOPOULOS and HUGHES, 1972]. Diazepam produces a similar pattern of depression of spinal reflex excitability and recovery with concomitant intellectual obfuscation. The depressant effect is global rather than specific to spinal motoneurones, and the activity of such drugs is most appropriately classified as sedative rather than 'muscle relaxant'.

Facilitation of Spinal Motoneurone Excitability and Recovery and Its Clinical Correlate

Depression of reflex excitability and recovery by a drug frequently had a clinical correlate in decrease in the clasp-knife response and in clonus in patients with spasticity. In order to investigate further validity of the objective criteria it was necessary to test the converse proposition: that a drug facilitating spinal motoneurone excitability or recovery would aggravate the clinical state of spasticity. The occasion for testing this was presented by a drug extensively studied as a potential muscle relaxant in clinical trials. The

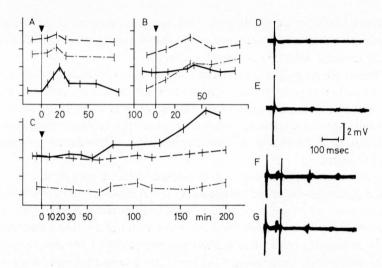

Fig. 5. Effect of R-738 administered intravenously (10 mg) in *A*, intramuscularly (10 mg) in *B* and orally (in a dose of 90 mg) in *C*. *A–C* Normal subject. *Ordinate*, reflex amplitude in mV. Vertical bar about the mean values is equal to 1 standard error of the mean. *Abscissa*, time in min after administration of the drug. Continuous line refers to the ankle jerk. For testing H-reflexes, two stimuli were delivered at an interval of 200 msec and the amplitude of the first H-reflex is indicated by an interrupted line while that of the second testing response appears as a dot-dash line. *D* to *G* illustration of reflex facilitation and increase in clonus after R-738. *D* and *E* ankle jerk. *F* and *G* Two H-reflexes elicited at an interval of 100 msec. *D* and *F* are before, and *E* and *G* are 100 min after the administration of 60 mg R-738 orally.

drug, Nefopam (R-738), is an analogue of orphenadrine and consists of a cyclization of the diphenhydramine molecule.

R-738 was found to facilitate monosynaptic reflex excitability and recovery in all of 21 normal subjects studied. The onset, peak effect and duration of drug action were related to the mode of administration. Intravenous injection of 10 mg of R-738 enhanced reflexes from about 4–45 min, with peak effect at 15 min (fig. 5A). There was sometimes a late phase of depression of reflex excitability from 45–80 min after the injection, especially evident in spastic patients. After i. m. injection of 10 mg of R-738, reflex facilitation was frequently first apparent at 5–7 min and the peak of action occurred at 30–40 min (fig. 5B). There was slow return to control levels at 1–2 h with generally little late depression. The first objective evidence of drug effect after 60 or 90 mg of R-738 per os was at 45–60 min (fig. 5C). The peak of effect was at 1.5–2 h, followed by a slow return to control levels at 3–4 h.

The T-reflex was usually enhanced to a greater extent than the electrically evoked reflex and this difference was especially evident with oral therapy, suggesting fusimotor driving.

The sensitivity of the technique was demonstrated by investigation of a patient with spasticity and an idiosyncratic reaction to a number of drugs. There was a prolonged period of facilitation of spinal reflexes and a long duration and intense psychic effect after i.v. administration. The reflex enhancement was still present 2 h after the injection. The patient was later found to have a history of increased sensitivity and prolonged effects from diphenhydramine (Benadryl), meprobamate and meperidine (Demerol). R-738 is chemically related to diphenhydramine.

R-738 usually induced a feeling of relaxation without drowsiness or deterioration in intellectual performance. There was thus a striking discrepancy between the psychic tranquillity and the excitatory effects on spinal motoneurones. There was also a decrease in pain experience evaluated by our method and a close correspondence between obtundation of clinical and experimental pain, evaluated at intervals in the patients.

The onset, peak of action, and duration and intensity of drug effect were related to the mode of administration. There was a close relationship between objective evidence of drug action judged by changes in motoneurone excitability and recovery, subjective changes in psyche, and alterations in pain experience.

The effectiveness of R-738 was studied in patients with spasticity of spinal or cerebral origin. There was generally aggravation of the clinical status associated with objective evidence of facilitation of reflex excitability and recovery. There was increase in activity of the flexion reflex and an increase in spasticity and clonus (fig. 5). There was sometimes later improvement in the clinical state during depression of reflex excitability and recovery at about one hour after i.v. injection of R-738. The clinical aggravation of spasticity after oral or i.m. administration of R-738 was generally prolonged.

R-738 is an effective remedy in the painful low-back syndrome, largely associated with intervertebral disc disease. Amelioration of back pain and increase in mobility occurred in 19 of 21 patients. Clinical improvement and increase in mobility occurred during an interval when reflex excitability and recovery were augmented, stretch and flexion reflexes hyperactive, alterations in psyche minimal, and the experience of experimental pain decreased. It was concluded that R-738 is an analgesic with interesting features [GASSEL, DIAMANTOPOULOS and HUGHES, 1972].

Control Studies with Intravenous Saline Placebo

The effect of intravenous saline placebo was investigated in 7 patients with spasticity and 7 normal subjects. There was little or no change over prolonged periods of observation in 4 patients and in 4 normal subjects. An increase or decrease in reflex excitability occurred during the injection in two normal subjects and two patients; the change was prolonged, persisting for 15 min with slow return to control in one normal subject, and continuing for 1.5 h in a patient with a paraplegic syndrome. A decrease in motoneurone excitability occurred at 40–55 min after the injection in one patient. The change was correlated with decrease in spasticity, clonus, and in the activity of the flexion reflex. A lesser increase or decrease in excitability occurred in two other studies. These changes represent spontaneous alterations in excitability and should be recognized in any protracted evaluation of drug action. The regularity of the pattern of change in reflex excitability after drugs, such as nicotine, diazepam, pentobarbital, and orphenadrine derivatives, and the differences in onset and duration of effects after oral, i.m. or i.v. administration, support the utility of the technique for evaluation of drug action.

Study of Effect of Drugs on Responsiveness of spinal reflexes
To Supraspinal Facilitation

There was no change in the pattern or degree of facilitation of reflexes evoked by the Jendrassik maneuver (clenching of fists or sniffing) during the height of reflex facilitation by R-738 or reflex inhibition by diazepam or pentobarbital. It is, therefore, impossible to support, in man, the proposed thesis based upon animal experiments [GINZEL, 1966a, b] concerning abolition of supraspinal facilitation of spinal reflexes as an important criterion of the action of R-738 and of 'muscle relaxants' in general.

Flexion Reflex Activity

Rationale for Technique of Study

Hyperactivity of the flexion reflex afferent system is an established component of spasticity, and indeed the distressing spasms associated with this hyperactivity are an important clinical problem. It can be argued, with reason, that this is the system most specifically involved with muscle spasm associated with the painful low-back and, indeed, with painful muscle spasm of almost any etiology.

In cats a relatively synchronized reflex in the anterior tibial muscle is evoked by stimulation of the proximal stump of a transected posterior tibial nerve, and the reflex changes have been demonstrated during phases of sleep in the chronic animal [GASSEL, MARCHIAFAVA and POMPEIANO, 1965]. In man, however, a polysynaptic reflex sufficiently regular to be useful in evaluating drug effect is unusual. An alternative and more analytic method of study has been developed [GASSEL and OTT, 1970a] that evaluates changes produced by stimulating skin afferents in the excitability of triceps surae motoneurones. The conditioning shock to the dorsal or plantar surface of the distal foot is followed at intervals from 10 msec to 10 sec by a test ankle jerk. The size of the conditioned T-reflex relative to that of the unconditioned T-reflex is a measure of the change in excitability of the motoneurone pool. The mean T-reflex amplitude is plotted as a function of the interval between the conditioning ipsilateral skin stimulus and T-reflex. Investigations of a large series of normal subjects have shown excitability changes at 40–90 msec with 'local sign', judged to be mediated by group II afferent skin fibers. There is facilitation of the ankle jerk on stimulation of the dorsal skin surface of the distal foot and inhibition on stimulation of the planter surface (fig. 6). A late enhancement of excitability occurs at 100–250 msec in both cases. The change in excitability at this latency has been calculated to be transmitted by the smallest skin afferents. Long loop, spino-bulbo spinal reflexes, excited by the

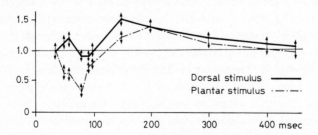

Fig.6. Illustration of the recovery curve of the ankle jerk conditioned by a stimulus delivered with subcutaneous electrodes to the dorsal and to the plantar surfaces of the ipsilateral distal foot. Each point represents the average value of 12 reflexes normalized to control and evaluated at each interval represented. The vertical bar is equal to one standard error of the mean. The time interval between the conditioning shock to the skin and the test reflex is plotted on the abscissa.

cutaneous shock, might also make a contribution to the late change in excitability. There is evidence of late effects but of reverse sign on stimulation of the skin of the distal foot in spinal man and striking abnormalities have been found in other patients with spasticity.

It was not practicable to record the entire graph of reflex changes after a drug, because of the limited time available to study adequately serial drug effects. Therefore changes in excitability were selected at two intervals, that at the peak of 'local sign' effect attributed to effects of stimulating group II skin afferents, and that at the height of the late facilitation, ascribed primarily to effects transmitted by the smallest skin afferents. In each subject the critical interval between conditioning stimulus and test reflex was determined before the drug and control values were established. Alterations in the extent or pattern of excitability change produced by the drug were evaluated at intervals. The drug action on activity of the flexion reflex was also studied objectively in spastic patients with exaggerated flexion reflexes. Changes in the ipsilateral or crossed effects evoked by an invariable cutaneous shock to the popliteal fossa were investigated at intervals during the study. There was a close relationship between the drug action on the flexion reflex judged by excitability changes in normal subjects, and on observable change in the flexion reflex in patients.

Results

Excitability changes evoked by stimulating cutaneous flexion reflex afferents were studied after i. v. pentobarbital sodium, 80–100 mg. Associated with the fall in control T-reflex excitability there was a decrease in the changes induced by the regular cutaneous stimulus without, however, a change in the sign (facilitation or inhibition) of changes evoked by dorsal or plantar stimulation respectively (fig. 7). There was evident decrease in the flexion reflex after pentobarbital injected intravenously in 2 of 6 patients with spasticity during depression of reflex excitability and recovery, and associated with sedation and drowsiness. Preliminary studies with diazepam indicate a similar depression of excitability change evoked by stimulating cutaneous reflex afferents [GASSEL, DIAMANTOPOULOS and HUGHES, 1972].

After R-738, the most striking change induced by cutaneous afferents occurred during the height of facilitation of monosynaptic reflexes. The reflex conditioned by cutaneous stimuli was facilitated to a lesser extent than the control reflex that resulted in an inverted effect from facilitation to de-

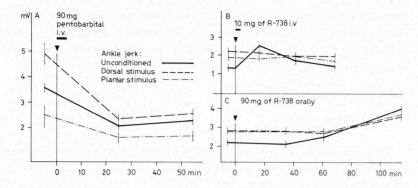

Fig.7. Action of drugs on excitability changes of T-reflex evoked by stimulating cutaneous afferents. *A* Effect of 90 mg pentobarbital sodium i.v. on early changes in excitability evoked by conditioning the ankle jerk with a shock at 60 msec to dorsal and plantar surfaces of ipsilateral foot (see *fig.7*). *Ordinate:* reflex amplitude in mV. Vertical bar above and below the mean is equal to 1 standard error of mean. *Abscissa:* time interval in min. after administering drug. Control values before pentobarbital are indicated on the left. *B, C* Effect of R-738 administered i.v. *(B)*, and orally *(C)*, on late excitability changes evoked by conditioning the ankle jerk with a shock to dorsal or plantar surface of ipsilateral foot. The peak of late facilitation studied in this subject occurred at 260 msec. Graphic symbols are the same as in *A*.

pression (fig. 7). The 'local sign' of excitability changes was maintained to some extent, i. e. stimulation to the dorsal skin of the foot continued to evoke relatively higher amplitude T-reflexes than plantar stimuli.

The change in T-response to cutaneous stimulation is a sensitive index of drug action. There is good correspondence with evidence of physiological effect based upon motoneurone excitability and recovery data regarding the onset, peak effect, and duration of action of R-738 (fig. 5 and 7).

The inversion of facilitation to depression of the ankle jerk, evoked by stimulating the skin of the ipsilateral foot during R-738 is noteworthy. Depression of the ipsilateral ankle jerk as a late effect of stimulating cutaneous afferents is precisely the picture found in paraplegic man and in some other spastic patients. Patients with spasticity studied with R-738 generally showed exaggeration of the flexion reflex to an unvarying cutaneous shock, during this interval of inversion. Thus R-738 facilitates the activity of the flexion reflex afferent system.

The change in excitability evoked by stimulating cutaneous flexion reflex afferents sometimes outlasted alteration in monosynaptic reflex excitability and recovery. This persistence is noteworthy in view of the occasional discrepancy observed between activity of the flexion and stretch reflexes in patients with spasticity, after drugs such as diazepam, pentobarbital and some orphenadrine derivatives. Hyperactivity of the flexion reflex outlasted that of the stretch reflexes in these instances.

Discussion

The importance of comprehensive evaluation of drug effects by an analytic system using objective criteria in a controlled setting is especially well emphasized by the study of R-738.

There were frequent and striking discrepancies between objective drug effect on reflex excitability described in this report and drug effect evaluated by EMG, the patients' subjective reports, or evaluation on clinical grounds alone.

There was reference to the disparity between subjective reports of relaxation, including those by patients with spasticity, at a time when spinal reflex excitability was grossly enhanced and the clinical state greatly aggravated. Indeed, subjects frequently stated explicitly that their 'muscles felt more relaxed' during the peak of facilitation. Clearly such reports need verification. Patients sometimes denied subjective change when the intellectual performance test showed evident deterioration. Subjects were sometimes anxious and tense at the start of the study and the tension had a counterpart in persistent motor unit activity in the EMG. R-738 produced a feeling of relaxation and a capacity to lie comfortably immobile for long periods. EMG activity stopped during the interval when spinal reflex excitability was grossly enhanced. Muscle spasm associated with painful musculoskeletal disorders could decrease with sedation, analgesic therapy, or as the result of a specific depressant action on spinal reflex activity (largely a function of the flexion reflex). EMG and integrated EMG provide no means of analyzing the action and are frequently misleading. The failure of clinical observation alone to provide a valid judgment of drug effectiveness needs no endorsement. The unequivocal evidence of facilitation of spinal reflexes by R-738 associated with aggravation of spasticity is the first such presented after extensive clinical trial. Indeed many studies concluded that the drug had 'muscle relaxant' propensities. The demonstration of its analgesic action is similarly the first such reported. Additional discrepancies in the action of some proprietary drugs are soon to be published [GASSEL, DIAMANTOPOULOS and HUGHES, 1972].

Summary and Conclusions

A new procedure is described, employing the methodology of clinical neurophysiology, for providing evidence of the therapeutic effectiveness of drugs in man. There is a comprehensive evaluation of drug effects using valid objective criteria in a controlled setting. Changes in stretch and flexion reflexes produced by a drug are studied objectively. Reflex alterations are related to changes in psyche, to a test of pain experience, to subjective reports of change and a test of intellectual performance, and to clinical effects evaluated at regular intervals during the investigation.

Stretch reflexes involving the same population of triceps surae motoneurones, are the mechanically evoked monosynaptic T-reflex (ankle jerk), the electrically evoked monosynaptic H-reflex, and the recovery curve of the H-reflexes (technique of paired stimuli). The activity of the flexion reflex afferent system is studied by changes in excitability evoked in the same group of triceps surae motoneurones on stimulating cutaneous afferents. The results are verified in patients with spasticity by recording the occurrence of clonus associated with the monosynaptic reflexes, and by judging the activity of the flexion reflex evoked by a regular shock stimulus delivered to the skin.

Objective evidence is presented of the action of drugs in man which either depress spinal stretch or flexion reflexes, such as nicotine and barbiturates, or drugs which facilitate spinal stretch or flexion reflexes, such as nefopam (R-738). Therapeutic effectiveness is analyzed, and discrepancies in conclusions are presented regarding drug action based upon preliminary animal experiments, subjective reports of change, evidence from EMG studies, and clinical observation alone.

A reliable pattern of changes has been demonstrated after oral, i.m. and i.v. administration. Idiosyncrasy in drug absorption or metabolism can be deduced by deviations from established patterns of effect. It has proved possible to predict the clinical utility of a wide range of drugs acting on the nervous system including sedatives and hypnotics, tranquillizers, stimulants and analgesics, from preliminary analysis of effects in normal subjects.

Author's address: Dr. M.M. Gassel, 425 Warren Drive, *San Francisco, CA 94131* (USA)

New Developments in Electromyography and Clinical Neurophysiology,
edited by J. E. Desmedt, vol. 3, pp. 360–366 (Karger, Basel (1973)

Diagnostic Use of Monosynaptic Reflexes in L₅ and S₁ Root Compression

J. DESCHUYTERE and N. ROSSELLE

EMG Laboratory and Department of Physical Medicine, University of Louvain, Louvain

It is well-known that conduction velocity (CV) is slowed in the chronically compressed segment of a peripheral nerve, for instance in the so-called peripheral entrapment neuropathies. When a spinal root is compressed, the CV may also be reduced in the afferent and/or efferent fibres which produces an increase in latency of the corresponding monosynaptic reflex [LIBERSON, 1963]. Reflex testing appears, indeed, as a useful diagnostic tool for supplementing routine EMG examination in compression syndromes involving spinal roots L_2–S_1 [DESCHUYTERE and ROSSELLE, 1970]. This paper further documents the use of monosynaptic reflexes in a large group of patients with recognized L_5 or S_1 root compression, the latter being established by clinical and EMG evidence. A matched group of normal subjects served as control. Special concern is given to the methods for eliciting reflexes from the extensor digitorum longus (EDL) muscle by stimulation of the peroneal nerve [DESCHUYTERE and ROSSELLE, 1971].

Material and Methods

50 normal subjects aged 24–64 years without any signs or symptoms of neuromuscular disorders served as controls for 50 patients aged 28–64 years. The patients had clinical and EMG evidence of compression of spinal root L_5 or S_1. The reflex responses were recorded with a concentric needle electrode inserted either in the EDL or in the triceps surae (soleus and medial gastrocnemius). The equipment for recording is a Medelec MS-3R. Rectangular electric pulses of 2-msec duration were delivered by an isolation unit at a chosen strength through percutaneous electrodes. The tibial nerve was stimulated in the middle part of the

popliteal fossa, and the common peroneal nerve in the distal lateral part of same. Latencies were measured from the stimulus artefact to the beginning of the reflex discharge. In 5 normal adults the nerve action potentials were recorded and averaged from the sciatic nerve high in the thigh in order to evaluate CV in afferent fibres. Motor CV were also estimated over the same segment.

Results

Normal subjects

The classical H-reflex of the triceps surae involves spinal root S_1 and can be used for estimating increases in latency in relation to S_1 root compression. The methodology of the H-reflex is dealt with by HUGON [this volume] and we only wish to emphasize that electrode placements should be such as to permit accurate determination of the latency of the earliest component of the H-reflex in the present application. Slight changes in stimulating electrode position in the popliteal fossa can modify the wave form of the muscle responses without altering the H-reflex latency. In our 50 normal subjects the H-reflex latency was between 25 and 30 msec, depending on the size of the person.

To test the L_5 root, we investigated the reflex response of the EDL muscle to stimulation of the common peroneal nerve. The position of the stimulating electrodes must be carefully adjusted and we found that I_A afferents from EDL appeared to be mostly localized in the ventromedial part of the common peroneal nerve in the distal part of the popliteal fossa. In almost every subject, a reflex response was elicited for stimulus intensities which were subliminal, liminal or slightly supraliminal, but always inframaximal for a direct M-response in EDL. In many instances a similar response with H-reflex features was also recorded from the tibialis anterior and the peroneus longus muscles.

When the stimulating electrodes were moved more proximally along the nerve trunk, the late response in EDL appeared with a shorter latency, thus in agreement with its central origin. Another point is that the H-reflex in EDL exhibits more variability in amplitude than the direct M-response in the same muscle, for a given stimulus intensity. We have also observed that when the stimulus is increased progressively, the M-response increases while the H-response may decrease, much in the same manner as the well-known soleus recruitment curves [HUGON, this volume].

When the position of the stimulating electrodes is slightly changed the configuration and size of the M- and H-responses may change, probably in relation to the number of afferent and efferent fibres activated. In a patient with marked wasting and weakness of the antero-lateral muscles of the legs subsequent to acute poliomyelitis, the H-reflex was elicited with unusual facility from EDL; this can presumably be explained by a more selective activation of the I_A afferents in the nerve trunk with reduced motor axons content. It is important to state, however, that when variations of wave form did occur as a result of changes in electrode position, the minimal latency of the H-reflex in EDL remained fairly constant. The latencies recorded were between 27 and 33 msec, depending upon the subject's size and age. Little difference (less than 1 msec) was found between the H-reflexes in EDL on the left and right side in any individual. It is worth trying to adjust the experimental conditions to obtain a better definition of the earliest components of the H-reflex. This can be tested from the shape of the maximum M-response which is generally similar to that of the reflex in the same muscle.

A most effective facilitation procedure for the H-reflex in EDL is by using paired stimuli at intervals of 170–250 msec. In this way the second stimulus occurs approximately at the time when the muscle relaxes after the contraction elicited by the first stimulus, such relaxation phases probably being associated with a burst of I_A afferent activity. Similar phase III facilitation was observed in the cat by BIANCONI et al. [1964]. We generally used repetitive activation at 4–5/sec which produces a state of cumulative facilitation; this method is easily applied for clinical testing of patients. We checked that the latency of the H-response was not modified by such facilitation processes which merely served to make the response better defined and larger.

Recording of the EDL muscle response to stimulation in the popliteal fossa and higher in the thigh served to estimate the motor CV. The sensory CV was determined by recording from the proximal sciatic nerve and averaging the nerve response elicited by a near-threshold stimulus to the common peroneal nerve at the popliteal fossa. Such stimuli elicited only a H-reflex in EDL and thus presumably only activated I_A afferents. We found that this sensory CV was faster than the motor CV for the same stretch of the nerve. Table I presents estimations of CV in afferent fibres based either on direct recording of the nerve potential or from the latency of the EDL H-reflex elicited by stimulation at 2 different levels along the peroneal nerve. The CV in afferent fibres is similar for the 2 methods and it is faster than the motor CV measured by recording EDL direct responses.

Table I

Stimulation site	Detection site	Distances, cm	Values of latencies, msec		
			M-wave	H-wave	NAP
Distally (deep peroneal nerve at capitulum fibulae)	EDL	14	4.5	35	–
Popliteal fossa (common peroneal nerve)	EDL	20	5.5	34	–
Thigh region (sciatic nerve)	EDL	42	10,5	29,5	–
Distally	thigh region	28	–	–	5.5
Popliteal fossa	thigh region	22	–	–	4.5

Patients with Root Compression

In 47 of the 50 patients studied it proved possible to elicit the H-reflex both in EDL and in triceps surae on both sides. In view of the influence of subject's size on the latency of reflex responses, it was preferred to compare reflex latencies on the normal and affected sides in each patient. The differences in reflex latencies are plotted in figure 1 with separate symbols for the EDL and triceps responses which correspond to the L₅ and S₁ root, respectively. Differences from 1 to 10 msec between the reflexes on the 2 sides were observed, and this is considered to be evidence of CV slowing in the affected root. It is interesting to note that increases in latency can be recorded in patients with no evidence of motor nerve fibre involvement, as judged from routine EMG and peripheral motor CV.

Conclusions

In this series of patients, the study of latencies of H-reflexes in the 2 hind limbs proved of significant value to detect slowing of CV in a compressed spinal root. The classical H-reflex of triceps surae has been used to test the S₁ root. For the L₅ root we used the late response with H-reflex features which is elicited in the EDL by stimulation of the common peroneal nerve. Observations on the latter reflex have been scarce although it is quite

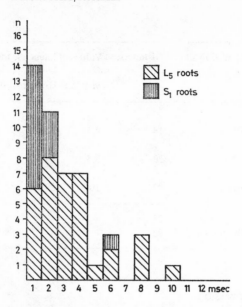

Fig. 1. 47 patients. Ordinate: number of patients (n). Abscissa: differences of reflexive latencies compared with contralateral side (msec).

interesting for diagnostic evaluations of the L_5 spinal root. We think the late response in EDL can be considered to be a monosynaptic reflex similar to the classical H-reflex for several reasons. The response could appear for stimulus intensities which were infra-maximal for the motor axons, except in several subjects in whom such H-responses could not be elicited without there being also a direct M-response. Furthermore, when the H-response was present in EDL, an increase in stimulus intensity recruited a larger M-response and decreased the H-response. Other arguments are the variability of the response amplitude and the fact that its latency decreases when the stimulus is delivered at a more proximal level.

It would be difficult to interpret these late responses in EDL as F-waves. The responses can be elicited for stimulus intensities which are smaller than those necessary to evoke a maximal direct M-response. Moreover, when the wave form of the late response is analyzed, it appears to include at times more motor units and other ones than those in the preceding direct M-response. This implies that the motor units discharging in the late response had not necessarily also been activated by a direct preceding volley in the motor axons.

TEASDALL *et al.* [1952] showed that true H-reflexes did occur in antero-lateral muscles of the foreleg, in patients with brain stem or spinal cord lesions. This implies that low-threshold I$_A$ afferents must be present in the peroneal nerve. The fact that H-reflexes are less readily elicited from these physiological flexor muscles deserves careful discussion. For one thing, it becomes increasingly apparent that I$_A$ afferents and reflex pathways are submitted to an inhibitory control from supraspinal structures, and removal of such controls may result in an increase of monosynaptic reflexes in patients with upper motoneurone lesions [see BURKE and LANCE; DEL-WAIDE; HAGBARTH, this volume]. Immaturity of such descending control may explain the presence of responses similar to H-reflexes in infants [see MAYER, this volume]. One might suppose that dorsiflexors of the foot are normally receiving a stronger descending control than calf muscles which would explain the different readiness in producing a H-reflex. These problems are still far from being solved and more than one descending pathway should be considered in this relation [see KUYPERS, this volume].

Some consideration should also be given to the respective diameters of afferent and efferent nerve fibres in the nerves to EDL and to triceps surae. The motor axons to both muscle groups present a similar CV in the thigh [GASSEL and TROJABORG, 1964]. The CV in the I$_A$ afferent was found, on the average, about 17 % higher than that of the motor axons over the same segment [DIAMANTOPOULOS and GASSEL, 1965]. We have also found that CV in I$_A$ afferents was faster than in the corresponding motor axons for the EDL. In animal experiments, REXED and THERMAN [1948] and LLOYD and CHANG [1948] noted that the largest afferents had a smaller diameter in flexor than in extensor muscles. Even if this were true in man for the afferents in peroneal and posterior tibial nerves, respectively, it would still remain that these peroneal I$_A$ afferents should conduct faster than the motor axons to EDL. On the other hand, it would appear likely that EDL has a larger proportion of fast (or phasic) motor units than the gastrocnemius muscle and, especially, the soleus which is mainly responsible for the H-reflex response [see HUGON, this volume]. Since motor axons supplying phasic motor units have a larger diameter [see BURKE, this volume], it could well be that in man the motor fibres to EDL would have a larger diameter than the motor fibres to triceps surae and soleus. These considerations of fibre size in the 2 nerves would perhaps explain why the direct M-response appears more readily than the H-reflex in EDL as compared to the calf muscles.

With the technique so far developed as it is today, in our estimation, 2 msec may be considered as evidence of root lesions provoked by chronic compression and 1 msec may be indicative of such lesions. Measurements of the latencies of the reflex responses are a valuable addition to EMG in the diagnosis and the follow-up of compression syndromes of the fifth lumbar and the first sacral roots.

Authors' address: Dr. J. DESCHUYTERE and Prof. N. ROSSELLE, EMG Laboratories, Academic Hospital, University of Louvain, Capucijnenvoer 35, *B-3000 Leuven* (Belgium)

New Developments in Electromyography and Clinical Neurophysiology,
edited by J. E. Desmedt, vol. 3, pp. 367–370 (Karger, Basel 1973)

The Influence of Diazepam and Chlorpromazine on the Achilles Tendon and H-Reflexes

C. H. M. BRUNIA

The Dr Hans Berger Clinic, Breda, and Department of Psychology, Tilburg University,
Tilburg

In previous experiments on a perceptual motor task, we found an increase
of Achilles tendon (T) reflex amplitudes with no simultaneous changes of
H-reflex amplitudes [BRUNIA, 1971], suggesting an increase of fusimotor
activity under these conditions [cf. PAILLARD, 1955]. BATHIEN and HUGELIN
[1969] however, using different tasks found an increase of both T- and H-
reflex amplitudes. Testing different levels of attention, BATHIEN [1971] concluded that with relatively simple tasks only fusimotor neurones would be
influenced, while during more complex tasks activation was stronger and the
α-motoneurones would be facilitated as well.

For the task we used a 'binary choice generator', an apparatus that produces tones of high and low pitches in a random succession. The subject has
to press a button with his right or left hand, depending on whether the pitch is
high or low. It is perhaps worthwhile to mention that sinus arrhythmia shows,
during this task, a significant decrease. This seems to be of some importance,
because this parameter is often used as an index of arousal or of mental load
[BRUNIA, 1970; BRUNIA and DIESVELDT, 1971].

Furthermore, it is clear that the task does not imply the stimuli, used to
evoke the reflexes, as has been done by REQUIN [1969].

One could speculate that even a task not requiring too much alertness
from normal subjects should also cause an increase of H-reflex amplitudes,
if the base level of excitability were lowered artificially. This was investigated
by giving subjects depressant drugs. At the same time we checked for any preferential influence of these drugs on the α or on the fusimotor system.

Experiment 1 was done with 24 normal subjects. Rest and task periods
of 4 min were alternated during 40 min: 12 subjects started with rest, the
others with a task period. 16 minutes after the beginning of the experiment,

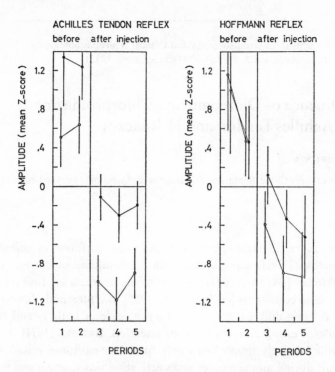

Fig. 1. Influence of 10 mg diazepam on T- and H-reflexes of 24 subjects. Length of experiment 40 min. Intravenous injection after 16 min. Mean Z-score of amplitudes during task (●) and at rest (○). The 0.95 level of confidence is indicated by vertical lines.

subjects received 10 mg diazepam i.v. The electrical stimulus was 1.2 times the threshold of the H-reflex, at a frequency of 1/3 sec. Technical details are the same as in previous experiments [BRUNIA, 1970, 1971]. The results (fig. 1) can be summarized as follows: (a) before injection, T-reflex amplitudes are significantly larger during task than at rest (p < 0.001). After diazepam the rest-task difference is significantly larger than before (0.001 < p < 0.01); (b) before injection, H-reflex amplitudes show no significant rest-task difference. After injection amplitudes are significantly larger during task (p < 0.001), and (c) T- and H-reflex amplitudes show an equally strong and significant decrease after injection, both at rest and during task.

Experiment 2 involved 10 subjects. All started with a rest period. They received 25 mg chlorpromazine i.v. 16 min after the beginning of the experiment. Further details are the same as in experiment 1. The results (fig. 2) can be summarized as follows: (a) T-reflex amplitudes are significantly larger

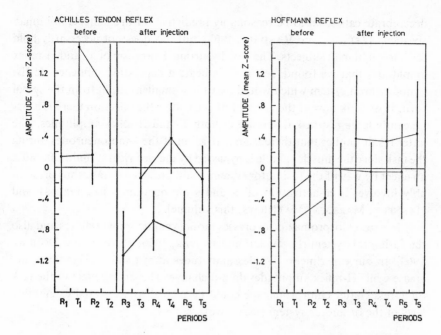

Fig.2. Influence of 25 mg chlorpromazine on the T- and H-reflexes of 10 subjects. Length of experiment 40 min. Intravenous injection after 16 min. Mean Z-scores of amplitudes during task (●) and at rest (○). The 0.95 level of confidence is indicated by vertical lines.

during task than at rest, both before and after injection (before: $p < 0.05$, after: $0.0005 < p < 0.005$); (b) H-reflex amplitudes show no significant rest task differences, and (c) after injection, T-reflex amplitudes are significantly smaller, only at rest ($0.001 < p < 0.01$). No significant changes of H-reflex amplitudes are seen.

Discussion

Diazepam has a stronger depressant effect on polysynaptic than on monosynaptic reflexes. It is more effective in decerebrate than in spinal animals [NGAI, TSENG and WANG, 1966]. Midcollicular decerebration rigidity is affected by a smaller dose than anemic decerebration [SCHALLEK, in ZBINDEN and RANDALL, 1967]. This points to a larger influence on the fusimotor neurones than on the α-motoneurones [cf. GRANIT, 1970]. Experiments with decerebrate cats seem to be reasonably predictive for drug effects in human spasticity [ZBINDEN and RANDALL, 1967], but this does not necessarily hold for normal human subjects. The equally strong depression of T- and H-reflex amplitudes that we found, seems to indicate a depressing influence on the α-motoneuron system which would result in a smaller output from the spinal cord. If we look now at the effect of the task after the injection there is a significantly large rest-task difference of both T- and H-reflex amplitudes. The latter result implies that there is an activation of the α-motoneurones during the task since the muscle spindle is bypassed by the electrical stimulus. Because an effect by way of the fusimotor system is still present, the task thus presumably involves a co-activation of α- and γ-motoneurones [see GRIMBY and HANNERZ; MARSDEN; MATTHEWS; this volume].

In cats, chlorpromazine depresses monosynaptic spinal reflexes and also the fusimotor system [HENATSCH and INGVAR, 1956; HUDSON and DOMINO, 1963]. In our experiments T-reflex amplitudes at rest show a significant decrease while H-reflex amplitudes do not change. The strong effect of the task on T-reflex after chlorpromazine could be related to a lowered excitability level of the fusimotor system (see above).

Summary

At a normal resting level of CNS excitability, the perceptual motor task used increased the size of the T-reflex without changing the H-reflex. When the central excitability was reduced by an i.v. injection of diazepam, the task now increased the size of both T- and H-reflexes. It is argued that the psychological task would induce an activation of both α- and γ-motoneurones. Furthermore, diazepam reduced the amplitudes of T- and H-reflexes roughly equally, thus suggesting a depression of α-motoneurones. Chlorpromazine lowered the T-reflex only which suggests a depression restricted to the fusimotor system.

Author's address: Dr. C. H. M. BRUNIA, Department of Psychology, Tilburg University, Tilburg (The Netherlands)

New Developments in Electromyography and Clinical Neurophysiology,
edited by J. E. Desmedt, vol. 3, pp. 371–374 (Karger, Basel 1973)

Diagnostic Application of a Battery of Central Motoneuron Electromyographic Tests

S. L. Visser and W. G. Buist

Laboratory of EEG and Clinical Neurophysiology, Free University, Valerius Clinic,
Amsterdam

Besides the study of simple monosynaptic reflexes, several other methods have been described which examine the function of the central motoneuron. Mostly these are based on multisynaptic reflexes or reactions. In disorders of the peripheral motoneuron, EMG changes are usually clearly recognizable; those observed in central motoneuron disorders are much less so. Also, it is difficult to differentiate between psychogenic disorder (hysteria, aggravation, non-cooperation) and organic disorder of the central motoneuron. It is, therefore, necessary to apply a battery of tests in order to obtain a positive indication of pathology. This study investigates the values of the different tests, and the combinations of tests which give the greatest reliability.

Methods

A series of tests were selected in order to evaluate their combined significance for practical diagnosis [Buist, 1970; Buist, Visser and Folkerts, 1972].

A. Knee tendon reflex (KTR). EMG recording of this reflex permits quantification [Bergman, Hirschberg and Nathanson, 1955] and facilitates comparison with other results. The influence of Jendrassik's manoeuvre can be traced as well as the phenomen of irradiation to other ipsi- and contralateral muscles [Hoefer, 1949; Hoefer and Putnam, 1940a, b; Broman, 1949]. The reflex was elicited by mechanical tap, and surface EMG was recorded from m. rectus femoris. The maximal size of the reflex was measured with and without the Jendrassik manoeuvre.

B. Duration of after-contraction [Hufschmidt and Schwind, 1960; Steinbrecher, 1959, 1965] which we studied in the rectus femoris with surface EMG electrodes. Its duration was expressed in msec.

C. Tonic vibration reflex (TVR) [Eklund and Hagbarth, 1965, 1966; deGail, Lance and Neilson, 1966]. We tested this also in rectus femoris and the reaction was scored in degrees of arc of displacement of the leg with respect to the thigh.

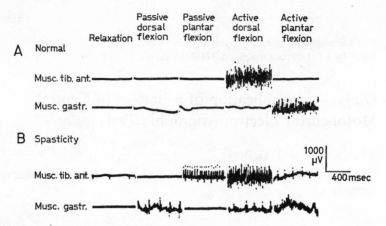

Fig.1. EMG during relaxation, passive stretching and shortening and voluntary contraction of an agonist and antagonist. Recorded with concentric needle electrodes in m. tibialis anterior and gastrocnemius-soleus. *A* Normal adult. *B* Spastic patient.

D. EMG during relaxation (in agonist and antagonist), active contraction and passive stretching and shortening (fig. 1). Important earlier studies on the diagnostic application of this examination anticipated much of our present knowledge [WERTHEIM SALOMONSON, 1920; VON WEIZSÄCKER, 1921; PRITCHARD, 1930; LINDSLEY, 1936; HOEFER and PUTNAM, 1940a, b; see also THIEBAUT and ISCH, 1952; LEFEVRE and SCHERRER, 1952; STEINBRECHER, 1965; RONDOT, 1968]. Moreover, it is possible to analyze abnormal movements, e.g. tremor, clonus, chorea etc. [BORNSTEIN and SAENGER, 1914; FOERSTER and ALTENBURGER, 1933; HOEFER and PUTNAM, 1940b].

The EMG pattern was evaluated as isoelectric, or with presence of motor unit potentials (MUAP) and the presence of irregularly grouped potentials, or tremor was noted. This was done at relaxation, in antagonist muscle during contraction of the agonist, at passive stretching and passive shortening.

E. Hoffmann or H-reflex [see HUGON, this volume].

F. Achilles tendon reflex (ATR) using mechanical stimulation and surface EMG recording. From these records we measured: the maximal size in mV of the ATR and its ratio (A/M) to the M response in Hoffmann test, before and during a Jendrassik manoeuvre. The maximal amplitude of the H-reflex with and without Jendrassik was also measured, as well as the H/M ratio. The presence of clonus and of irradiation to contralateral limb during H or ATR testing was noted.

G. Silent period [see ANGEL, this volume] which we tested in the gastrocnemius-soleus group using electrical stimulation of the posterior tibial nerve and surface EMG recording.

Results

With these tests, 39 normal subjects (mean age 43.4), 27 patients with a spastic syndrome (mean age 53.4) and 23 patients with a parkinsonian syn-

Table I. Abnormal and normal values of some relevant variables in spastic patients

	Spastic patients	Normal range
KTR with Jendrassik, mV	2.0 – 5.4	0.3 – 2.9
A/M ratio	0.08– 0.54	0.01– 0.17
H/M ratio	0.26– 0.74	0.06– 0.30
TVR, degrees	0 –16.8	9.5 –33.7

drome (mean age 64.6) were examined. It was statistically verified which of the 24 items contributed in a significant way to differentiation of these categories. Multivariate analysis (Atlas-computer, London University) was used. With the aid of canonic analyses the value of the EMG variables for the discrimination between the three groups of test subjects has been investigated. The most relevant variables with a coefficient of correlation > 0.25 are: EMG pattern at relaxation with respect to irregular grouped action potentials or tremor; KJR (id with Jendrassik); TVR; stretch reflex at passive stretching; ATR, especially A/M ratio; H-reflex, especially H/M ratio and the effect of Jendrassik manoeuvre.

Discriminant analysis (multiple regression analysis) showed that the variables which discriminated optimally between the normal and the spastic syndromes were the stretch reflex, KJR with Jendrassik and the TVR.

Tremor (tremor frequency) and stretch reflex also gave sufficient differentiation. For discrimination between spastic and parkinsonian syndromes the variables tremor and TVR gave satisfactory results.

The protocol for such testing can be summarized as follows: (a) examination with surface electrodes of KTR, ATR, and H-reflex; (b) examination with needle electrodes of an agonist and antagonist during relaxation and slow passive stretching and (c) if necessary, completed by examination of TVR.

Spastic patients have an increased KTR with Jendrassik, A/M ratio, H/M ratio and mostly a positive stretching reflex. The TVR is reduced (table I). Patients with parkinsonian syndrome have no electrical silence during relaxation, spontaneous action potentials, grouped action potentials and/or tremor being found at rest. There is mostly a positive stretching reflex.

This battery of tests can be applied in routine clinical EMG diagnosis. Moreover, there are several applications such as the evaluation of the influence of drugs upon central motoneuron disorder.

In 14 patients with a parkinsonian syndrome, EMG studies were made before and during each of the first 3 weeks of a course of L-DOPA given orally (mean dosage in weeks 1–3 was 1057, 2275 and 3289 mg daily). Pathological EMG activity during rest, passive stretching and shortening and reciprocal inhibition showed unmistakable improvement in the course of medication (Spearman test $p < 0.05$). The diagnostically less important EMG variables (H-reflex and silent period) remained unchanged. Since no over-all 'desynchronization' could be demonstrated in the EEG simultaneously recorded, this EMG improvement is probably a more specific effect than a general activation of the CNS [Visser, 1971; Visser and Postma, 1971].

The influence of a new experimental drug (MK-130) was studied in 10 patients with multiple sclerosis. Clinically, the drug had a slight positive influence on spasticity, but muscle weakness also improved. These trends were partly supported by EMG data showing a decrease in amplitude of the monosynaptic reflex responses (H, ATR, KTR) [Dubbelman and Visser, unpublished information].

Author's address: Dr. S. L. Visser, Laboratory of EEG and Clinical Neurophysiology, Free University, Valerius Clinic, *Amsterdam* (The Netherlands)

Servo-Control in Coordination of Movement

New Developments in Electromyography and Clinical Neurophysiology,
edited by J. E. Desmedt, vol. 3, pp. 375–382 (Karger, Basel 1973)

Servo Control, the Stretch Reflex and Movement in Man

C. D. MARSDEN

The Bethlem and Maudsley Hospitals, and King's College Hospital, Institute of
Psychiatry, De Crespigny Park, London

It is generally accepted that the muscle spindle machinery possesses the
properties of a *length servo* system, i.e. a self-regulating closed-loop mechanism using negative feedback from the spindles to maintain a constant muscle
length. Such a system automatically compensates for changes in load during
steady muscle contractions. Direct evidence for such motor behaviour in man
comes from investigation of the 'silent period' produced by unloading a
muscle during a steady effort or, conversely, stretching a muscle during a
steady muscle contraction.

Unloading the spindles in a contracting human muscle causes a period of
silence in its EMG, whether the unloading is achieved by superimposing a
muscle twitch evoked by an electric shock to its nerve [MERTON, 1951], or by
sudden release of the external support against which the muscle is contracting
[ANGEL, EPPLER and IANNONE, 1965; see ANGEL; STRUPPLER, this volume].
In both instances the period of silence in the EMG is interpreted as due to a
pause in the firing of the spindles of the muscle although, in the case of the
superimposed twitch, autogenic inhibition by stimulation of Golgi tendon
organs and other effects also have to be taken into account. The converse
experiment also gives the expected answer, namely, that sudden stretching of a
muscle during a steady voluntary effort calls forth an opposing muscle
contraction which, from its brief and constant latency, is clearly a stretch
reflex [HAMMOND, 1960].

Both types of experiment confirm that the spindle machinery in human
muscles resists sudden changes in load during steady voluntary efforts. During
such isometric contractions the muscle spindles must continue to discharge
to be silenced by unloading, and recently direct evidence of this has been
discovered by VALLBO [1970; fig. 6]. Recording from fine electrodes inserted

into the human median nerve, he isolated a single spindle afferent which was silent at rest but which began to discharge at the onset of isometric contraction of the flexors of the index finger and ceased at its relaxation. Since some degree of shortening of the muscle against its own internal elasticity must have occurred, this observation confirms the prediction that fusimotor neurones discharge during willed muscle contraction so as to offset the unloading of the spindle during such an *isometric* contraction.

The behaviour of muscle spindles when man exerts a steady muscle contraction, as inferred from the loading and unloading experiments and as observed directly by VALLBO [this volume] confirm that they act as a positional length servo system which resists externally imposed postural changes.

So far the spindle machinery has only been examined with regard to its behaviour during steady isometric contraction. When the muscle is allowed to shorten to produce free unrestrained movement, muscle spindles might continue to be biased by fusimotor contraction so as to offset their progressive unloading during the movement [MATTHEWS, 1964]. Such linkage of γ-moto-neurone discharge to α-motoneurone activation–α-γ *linkage* [GRANIT, 1968]– would preserve the capacity of the spindle to compensate for fluctuations in load during free movement. However, such behaviour has not been previously demonstrated in man. If spindles do, indeed, remain equally sensitive through-out the range of the free contraction of a muscle, then the spindle machinery possesses the characteristics for action not as a simple *length servo* system, but as a *follow-up servo*. Here varying spindle length by altering fusimotor bias could change muscle length; the spindles would not in fact record length, but the difference in length between muscle and spindle, so acting as misalignment detectors. The question is whether or not the γ loop can drive α-motoneurones to initiate and control free movements as proposed by MERTON [1953]. This postulate has been criticised on a number of grounds on the basis of animal experiments, but, as we shall see, many of the objections derive from the peculiar circumstances of study of anaesthetised or decerebrate animals, in which the spindle system is studied isolated from many of its control mech-anisms which operate in intact man. The true test of the 'follow-up servo' theory of muscle contraction has to be applied to willed free movement in normal man, or, if possible, conscious animals.

The *follow-up servo* theory leads to a number of predictions that can be tested experimentally in man. At the National Hospital for Nervous Diseases (London), Dr. P. A. MERTON, Mr. H. B. MORTON and I have studied the effects of altering the resistance to movement of the top joint of the thumb. The results of a number of experiments will be summarised here, for they are to be presented in full elsewhere. The method of study has been demonstrated to the Physiological Society in March 1971 [MARSDEN, MERTON and MORTON, 1971].

The top joint of the thumb was chosen, for it is a simple hinge joint operated by a muscle which lies in the forearm: flexor pollicis longus (FPL). The subject's hand and forearm lay supported half-way between full pronation and supination with the palm facing inward. The proximal phalanx of the thumb was gripped firmly by a metal screw clamp so that contraction of the thumb flexors only caused flexion of its top joint. The pad of the thumb pressed on an aluminium joy-stick incorporating strain gauges (Ether 3 A-1 A-35 OP) to record force of contraction. The joy-stick was bolted to the spindle of a printed electric motor (Printed Motors Ltd., type G9M4) which allowed free rotation of the top joint of the thumb from full extension to 90° of flexion. The other end of the printed motor spindle was attached to the spindle of a sensitive potentiometer (Penny & Giles, type 11/1) so as to record movement at the top joint of the thumb.

The output of the potentiometer was used to deflect a spot horizontally on a CRT (Tektronix 502 A) to indicate thumb position to the subject. In tracking tasks, he made flexion movements of 20° from a backstop, tracking a second CRT spot which took 1.2 sec to make the equivalent excursion.

The EMG of the active muscle (FPL) was led off by surface electrodes on the lower forearm, and rectified or, on occasions, rectified and integrated.

The position signal and treated EMG were averaged in a *Biomac* which was triggered after a delay by movement of the thumb from the backstop.

1. The Stretch Reflex and Silent Period during Movement

If the muscle spindles remain sensitive during movement, then it should be possible to demonstrate a stretch reflex to sudden loading, or a silent period on unloading during movement. This proved to be the case.

The subject steadily flexed the top joint of the thumb through 20°, matching his performance against the tracking spot on the CRT. The current applied to the printed motor provided a constant force against which the movement was carried out. In randomly presented trials, the current was abruptly increased or dropped when the thumb had flexed roughly 9°. The subject was unaware of when this was to happen, for movements against a constant force were interspersed randomly with movements with loading or unloading. The change in thumb position and the full-wave rectified EMG recorded from the surface of FPL were averaged in separate channels of a

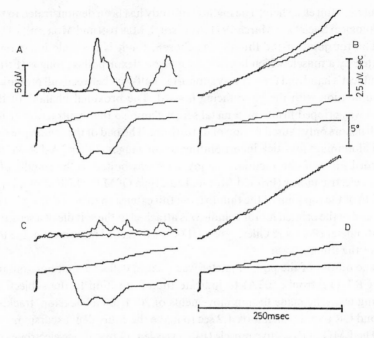

Fig. 1. A The stretch reflex in flexor pollicis longus, to a 5.5° extension movement of the terminal phalanx, starting 50 msec after the start of the Biomac sweep. The average of 16 responses in the rectified EMG is superimposed on the average of 16 controls. The averages of the corresponding positional records are shown below. *B* The average of 16 movements against constant force, superimposed on the average for sixteen in which extra resistance was encountered 50 msec after the start of the sweep. In this experiment the rectified EMG was integrated before averaging and the start of the response is indicated by the change of slope. *C* and *D* The same as *A* and *B*, respectively, but with the thumb anaesthetised by a Xylocaine ring block. (No similar effect is seen if the opposite thumb is anaesthetised.) Subject C. D. M. In all records the Biomac sweep is triggered when the thumb has flexed 8°. The sweep duration is 250 msec. During this time the subject's positional display was extinguished in all trials, so that vision played no part in any of the responses [MARSDEN, MERTON and MORTON, 1971].

Biomac. The *Biomac* sweep was triggered when the thumb had flexed 8°, and unloading occurred 50 msec after the start of the *Biomac* sweep.

Loading reversed the direction of thumb movement and the resulting muscle stretch elicited a stretch reflex (fig. 1 A). The latency, measured from the instant of applying the extra force, was some 50 msec. The nature of this stretch reflex requires comment. It is not equivalent to the constant short latency diphasic wave produced by a tendon tap. The latency of a monosynaptic response in the human long thumb flexor is estimated at about 22 msec. We have never obtained both a short latency tendon jerk and a long latency stretch reflex in the same record, as was demonstrated by HAMMOND [1960] during sudden loading of human biceps during steady contractions.

Unloading caused acceleration of thumb movement and a period of silence in the EMG followed some 50 msec after the load was released. This silent period lasted some 50 msec.

These results indicate that spindles are operative during movement, despite the unloading caused by muscle shortening. The presence of a stretch reflex is evidence that the spindles are not far behind the shortening muscle, while the silent period also indicates that the spindles are keeping up with the muscle as it contracts. Fusimotor discharge must have increased to offset the unloading produced by unrestrained contraction of the muscle.

2. Servo Action during Movement

The experimental situation was the same as before, the subject making steady flexions of the top joint of the thumb against a constant force, monitoring his performance against a tracking spot on the CRT. In random trials, his movement was halted (not reversed) when the thumb had flexed 9° by applying current to the printed motor so as to mimic a stiff spring. This ensured that no muscle stretch occurred when the load was added; in fact, the muscle must have continued to shorten against the elasticity of its tendon etc. when the movement of the printed motor was halted; this continued shortening should silence rather than excite its spindles.

Figure 1 B shows the results of averaging 16 movements against a constant force, superimposed on 16 in which the extra resistance was added. Inspection of the position trace confirms that the extra resistance merely halted the movement and caused no muscle stretch. The integrated EMG records show that the subject responded to the added resistance (shown by the change of slope compared with the averaged control record) some 50 msec afterwards. In the best subjects this early response was seen as soon as 40–50 msec after the resisting force started to increase. Its size varied from subject to subject, and from run to run in the individual subject. (It may be that it can be trained during repeated movements, but we offer no further evidence on this point here.) This compensatory response is, presumably, too early to be voluntary. In separate experiments the subject was instructed to contract FPL as hard and as quickly as he could on perceiving the added resistance (either via his digital senses or visually); the EMG evidence of the pull never occurred earlier hatn 150–170 msec after the extra resistance was applied.

The early compensating response to extra resistance is not a stretch reflex, for the muscle continues to shorten at a reduced rate. It must be a response of the type predicted by the *follow-up servo* theory of muscle contraction. If this compensating increase is due to spindle drive to α-motoneurones, then it should occur after the same delay as is needed for the stretch reflex to become operative. The latencies of the reflex responses to loading and unloading are the same as that of the response to halting movement, about 50 msec, so the first evidence of muscular compensation for added resistance to movement occurs at the time the stretch reflex becomes operative. We conclude that this *servo response* is due to operation of the stretch reflex as a *follow-up servo* system.

3. *The Effect of Anaesthesia of the Thumb on the Servo Response and Stretch Reflex during Free Movement*

The conventional statement of the servo hypothesis leads to the expectation that anaesthesia of the thumb (which ought not directly to affect the muscle-spindle-based machinery) should not affect the *servo response* to halting a movement. The experiments described previously were, therefore, repeated before and after anaesthesia of the thumb (achieved either by Xylocaine ring block at the base of the thumb, or by occlusion of the circulation for an hour or more with a tight rubber band at the base of the thumb, or with a sphygmomanometer cuff at the wrist).

The result was unexpected (fig. 1 D). With the thumb anaesthetised the early compensatory response in the EMG to halting the thumb's movement was abolished. As there is no reason whatever to think that any of the methods used to anaesthetise the thumb can have directly affected the active muscle itself, which lies in the forearm well clear of the field of operations, we must conclude that either our interpretation of this *servo response* as mediated by muscle spindles was incorrect, or conventional physiology was wrong. We then examined the effect of thumb anaesthesia on the stretch reflex and on the silent period to unloading. The results were dramatic. Anaesthesia of the thumb considerably reduced, and on occasions, abolished the stretch reflex (fig. 1 C) and silent period. Against conventional concepts of spindle function, we find that loss of afferent information from skin and joint in man profoundly affects the operation of the spindle machinery, which, to a degree, becomes inoperative.

This observation has a number of consequences, not least of which is that studies of spindle function in animals, where limbs are rendered anaesthetic by denervation and tendons severed from joints, can give no information on how the system behaves when all its control mechanisms are operative in intact animals or man. The *follow-up servo* theory has been criticised on the grounds that a maximal spindle stimulus in the experimental animal evokes only a modest reflex contraction, certainly much less than the muscle's maximal force [MATTHEWS, 1966]. Loop gain appears, from these experiments, to be insufficient for an effective *follow-up servo* system. Again, it has been demonstrated in the baboon that the distribution of spindle feedback is not confined to functionally related motoneurones [CLOUGH, KERNELL and PHILLIPS, 1968] which suggests that the stretch reflex does not possess the precision of action necessary for function as a *follow-up servo*. However, in such animal experiments the muscle's tendon has to be separated from its attachments and joint and skin nerves are severed, or the animal is anaesthetised. Our experiments clearly indicate that in man the loop gain of the stretch reflex of the spindle depends on inputs from the digit, so experiments on isolated muscles in animals cannot indicate loop gain, or even distribution of spindle input, in the intact animal.

The question arises as to what structures in the thumb are responsible for affecting the reflex responses of spindles in the long thumb flexor. Is it skin receptors, joint receptors, or both? We can offer no certain evidence of this as yet. If the apparatus is rearranged so that the thumb pulls the joy-stick down via a plate cemented to the nail-bed, the servo response can be demonstrated when movement of thumb flexion is halted, as before. If the terminal phalanx of the thumb is then anaesthetised (by a rubber band applied just distal to the interphalangeal joint), leaving sensation of movement at the joint unimpaired, then this servo response is attenuated. However, if the whole thumb is then anaesthetised, the servo response is now abolished. The conclusion is that both information from nail-bed, or from the pad of the thumb instead, and from joint receptors is required for servo action.

Our experiments show, for the first time, a response of the type predicted by the *follow-up servo* hypothesis. However, they do not tell us that the muscle's contraction was driven via the stretch reflex. Provided fusimotor discharge is accurately linked to that of direct cortico-motoneurone activity, the spindle machinery would aid the muscle to overcome resistance to movement, without, necessarily, being responsible for the initial contraction. We cannot judge, from these results, how much of the contraction of the muscle is driven by this servo system, and how much is due to activity of direct cortico-motoneurone pathways. It must be presumed that corticofugal impulses reach both α- and γ-motoneurones during willed movement, for both are accessible to electric shocks applied to the motor cortex of the baboon [KOEZE, PHILLIPS and SHERIDAN, 1968]. The relative power of the two cortical projections to α- and γ-motoneurones, and their temporal relations will decide the extent to which a muscle contraction is initiated and maintained via the stretch reflex. Direct recording of spindle afferent discharge, by the technique used by VALLBO, may answer the question of the role of the spindle servo in willed movement.

Author's address: Dr. C. D. MARSDEN, The Bethlem and Maudsley Hospitals, and King's College Hospital, Institute of Psychiatry, De Crespigny Park, *London, SE 5* (England)

New Developments in Electromyography and Clinical Neurophysiology,
edited by J. E. Desmedt, vol. 3, pp. 383–403 (Karger, Basel 1973)

Effect of the Peripheral and Central 'Sensory' Component in the Calibration of Position

A. W. MONSTER, R. HERMAN and N. R. ALTLAND

Department of Rehabilitation Medicine, College of Medicine, Temple University
Philadelphia, Pa.

Effective coordinated motor behavior depends on the presence and availability of a precise system of spatial referents. The essential knowledge of initial and final position of each anatomical part of the body, at any one time, is obtained by the integration of references from each joint within the proprioceptive spatial schema, which is itself within an invariant perceptual space. This integration process is functionally possible because: (1) the anatomical features of the body structure are highly predictable; (2) a number of sensory modalities continuously contribute temporal-spatial information concerning the various displacements of joints, tactile feedback and changes in the state of muscles (peripheral component), and (3) other internal feedback loops, derived from motor collaterals at cortical and subcortical levels, most likely also contribute important information (central component). Part of this information reaches consciousness. The peripheral component can be separated into two integral parts, namely, response-related feedback at the time of the motor act, and 'knowledge of result' [BILODEAU, 1966] following its completion. SPERRY [1950] first introduced the concept of a central component, also called 'sense of innervation' and this concept has recently gained anatomical support [KUYPERS, 1960; TOWE and JABBUR, 1961; WIESENDANGER and TARNECKI, 1966; OSCARSSON, 1970]. It has been found that in general the prominence of internal loops appears to grow ever greater as the nervous system phylogenetically increases in size and complexity, and as the subdivisions of the nervous system, originally concerned with transmitting feedback information generated by muscular contraction, gradually acquire more and more neurons concerned with internal feedback functions [CHANG, 1950; HASSLER, 1966; EVARTS, BIZZI, BURKE, DELONG and THACH, 1971].

Functional integration of the peripheral component seems to be assured by the presence of a fast-conducting pathway from the periphery (dorsal column-medial lemniscal system) which projects to the highest levels of the neuromotor system, specifically the postcentral gyrus of the sensorimotor cortex. It also projects to structures presumably involved in the integration and execution of complex coordinated motor activity, such as the cerebellum. Neuroanatomical and electrophysiological studies have shown internal feedback loops that transmit information from motor to sensory, motor to motor and sensory to motor areas. This systemic organization enables us to locate any part of our body with our eyes closed, direct our gaze at a specific part, skillfully position body segments with respect to the environment (e. g. the hand for the purpose of picking up an object), or temporarily attach spatial references to relevant aspects of the environment with respect to parts of our body. Accomplishing this, with a minimum of conscious involvement toward the details of the integration process itself, is essential. 'Obtrusiveness into consciousness is not necessarily an advantage' [SPERRY, 1969]. When we wish to pick up some object we certainly do not want to be concerned with all the changes in joints or adjustments in muscle tension. This is automatically taken care of by various control mechanisms.

Systematic study of the mechanisms subserving 'position sense' have traditionally emphasized the significance of the peripheral sensory component (peripheral hypothesis), particularly the receptor organs of the joint capsule and pericapsular tissue together with their afferent projections [PROVINS, 1958; MOUNTCASTLE and POWELL, 1959; LLOYD, 1960; MERTON, 1964; LLOYD and CALDWELL, 1965]. Kinesthesis, the perception of change in position, has been studied much more extensively than the calibration of anatomical positions within the spatial reference scheme [For a review and evaluation of various kinesthetic judgement criteria, see HOWARD and TEMPLETON, 1966, chap. 4]. It is, however, the latter aspect of position sense that 'places the study of kinesthetic sensation at the level of the mechanisms which confer upon it its local sign' [PAILLARD and BROUCHON, 1968].

Validity of the peripheral hypothesis has recently been questioned as a result of studies on functional losses from partial and complete deafferentations [FOERSTER, 1927; LASSEK, 1953; GELFAN and CARTER, 1967; KONORSKI, 1967; TAUB and BERMAN, 1968], from surgical intervention of central connections [JACOBSEN, 1932; SPERRY, 1947; WEISS, 1950; BUCY, 1957; MYERS, SPERRY and MCCURDY, 1962], from rare cases of selective spinal lesion [LASHLEY, 1917], as well as from various kinds of irreversible [NATHAN and SEARS, 1960] and reversible (cooling) sensory deprivation [CHASE, CULLEN, SULLIVAN and OMMAYA, 1965]. Experiments on sensory disarrangement in humans, utilizing primarily transformations on input to distance receptors, in effect vision and auditory sense, have also been informative [HELD, 1961; FREEDMAN and WILSON, 1967]. Illusions created by obstructing movement in the absence of peripheral sensations have been studied as well [HELMHOLTZ, 1867; KORNMÜLLER, 1931; MERTON, 1964], but unfortunately not in a quantitative manner. Analysis of the difference in position calibration under conditions of active (i. e. voluntary) and passive positioning has received some attention due to the need to understand the significance of the central (voluntary) component [GOLDSCHEIDER, 1889; LLOYD and CALDWELL, 1965; PAILLARD and BROUCHON, 1968]. The source and nature of this difference have become particularly relevant issues in view of the importance assigned by some investigators to the relationship between informational flow rate, cue discrepancy and the mechanisms of motor learning and motor-sensory compensation [VON HOLST, 1954; PAIL-LARD, 1960; HARRIS, 1963; HELD and FREEDMAN, 1965]. All these experimental and clinical data have led to the general conclusion that: 'sensory defects lead to motor dysfunction and *vice versa,* while major sensory defects in a single modality lead to observable consequences for other modalities' [FREEDMAN, 1968].

Mechanisms of position calibration, therefore, most likely involve both central and peripheral components. The significance of these components varies with the task to be accomplished. To quantify precisely the functional dependency under well defined physiological circumstances, particularly in skilled voluntary positioning, a number of techniques were applied to isolate the information contributed by each sensory component. A comparison was made between active and passive positioning. The effect of a selective nerve block on position calibration with and without an external load was studied, and analysis performed of the effect of reflexive and voluntary inputs on 'position sense' under isometric conditions. The voluntary control was utilized to evaluate the quantitative outcome of the complex of information resulting from the various experimental conditions, as illustrated in figure 1. More specifically, we have addressed ourselves to the following questions: (1) Can we identify in this quantitative outcome the role played by individual sensory modalities, particularly joints? (2) How do muscle receptors affect the calibration of position under normal functional conditions, and what is the effect of external stimuli, such as

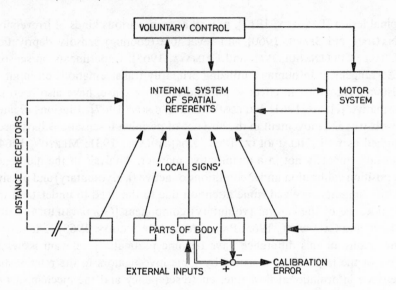

Fig. 1. Block diagram for calibration test of spatial referents system.

load? and (3) What role, if any, is played by the central component (corollary discharge, internal feedback) and what is its functional and anatomical implication?

Method

Two functionally identical, symmetrical parts of the human body, the ankle joints, were used to test the position calibration of one joint within the system of spatial referents. The method utilized is similar to the one used previously by SLINGER and HORSLEY [1906] and PAILLARD and BROUCHON [1968] for the upper extremity, and by LLOYD and CALDWELL [1965] for the lower extremity.

The subject is seated in a chair of adjustable height with the lower part of his trunk and thigh portion of his legs firmly secured to the chair. The feet are placed in two parallel rotary cradles. The height of the chair and the horizontal distances are adjusted so that the knee angle is at 45°. The center of rotation of each ankle joint is made to coincide with the axis of rotation of its cradle. The foot is initially in the relaxed position. The cradles have a moment of inertia that is small with respect to that of the foot. The shoes are padded with a soft material to minimize tactile cues during rotation and to give the impression that both feet are essentially bare and similar. Strain gage and position transducers are attached to the shaft of each cradle. In addition, the right cradle is attached to an electric servo motor. The motor can be controlled either as a position servo or as a force servo, the latter being the normal mode of operation during the experiment (except in the isometric case) since it is necessary to control the force independent of the joint angle.

Both feet are initially lined up in the relaxed position and in the isotonic case the reference foot is moved either passively or actively. The final reference position is then estimated actively with the other foot after a designated time delay, varying between 0 and 15 sec. After 15 sec most dynamic somatosensory receptors (and possibly their 'final target' neurons) have returned to a steady firing rate [MOUNTCASTLE, POGGIO and WERNER, 1963; AMASSIAN, 1970; GRANIT, 1970; VALLBO, 1970]. Neither visual nor auditory feedback is present except during the initial training period. The subject signals the beginning and end of each trial with a finger movement. Time intervals between positioning and estimation are varied at random. A passive movement is imposed by applying a specific loading pattern to the reference foot. This moves the ankle at a well defined speed to the point where the load is balanced by the passive resistance. The described technique leads to good reproducibility since it becomes obvious if any voluntary or reflex activity is present either during the imposed change in position or at the final reference position. The reference is always estimated actively with the other foot.

Controlled mechanical stimuli, such as a phasic stretch and vibration, can be applied to the tendon of the reference foot in order to induce a condition of misalignment or to interfere with the calibration of the existing spatial relationship between both feet by creating an illusion of movement. The mechanical stimulator is attached to one of the cradles and is used both for tapping and for vibrating the Achilles tendon at a reproducible location.

Selective blocks of the tibial nerve were done with 0.5–0.75% xylocaine [RUSHWORTH, 1960], and its effectiveness tested with mechanical taps to the tendon and sensorimotor examination below the level of the block. A few joint blocks were done as well (both capsule and ligaments) but the results are preliminary to date. The trajectories of each trial (position of reference foot, error between both feet and the load applied to the reference foot) are recorded on magnetic tape (HP 3960) and a number of parameters (particularly the estimation error) are analyzed with a computer (DEC, PDP-12). Figure 2 shows trajectories of typical responses seen during passive positioning (A1–E1), active positioning (A2–D2), positioning with short time delays (A2–B2) and positioning after selective nerve block (C2).

Results

Testing the position calibration was accomplished by means of introducing a sensation of change in the reference position and then estimating this new position.

Passive Position Calibration

It is clearly illustrated in figure 3 that passive positioning of the reference (indicated by open circles) leads to a much wider spread in the estimation error than active positioning (indicated by filled-in circles). This

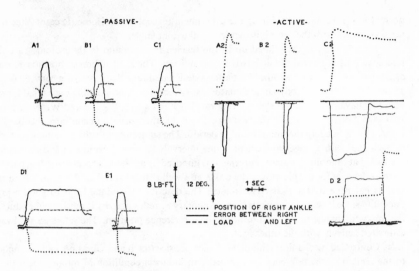

Fig.2. Typical trajectories of position calibration tests under different experimental conditions.

implies that during active positioning additional information is made available to the spatial referents system, leading to increased reproducibility. A relation between the error and the reference position was present in most subjects independent of both the foot used as the reference and the direction of the movement. Similar findings were reported by LLOYD and CALDWELL [1965], but primarily during passive positioning. It is felt that this tendency to underestimate large excursions from the relaxed position is an experimental artifact due to an inability to satisfy the condition of strict anatomical constraint. It occurs neither when both feet are positioned simultaneously, nor when positioned by slow oscillating movement. To correct for this systematic shift a position-dependent zero was established. This 'zero' refers to that value of the average error occurring as a result of active positioning with zero time delay between positioning and estimation. The indication 'under' or 'over' relates to that computed average in all illustrations.

Figure 3 is misleading since systematic errors due to the following important parameters are not indicated:

1. The time delay (T) between positioning and estimation was not constant for all trials but was taken from a uniform distribution (T = 0, 1, 5, 7 and 12 sec.) T was measured from the biginning of the position change of the reference and corrected for different speeds of positioning.

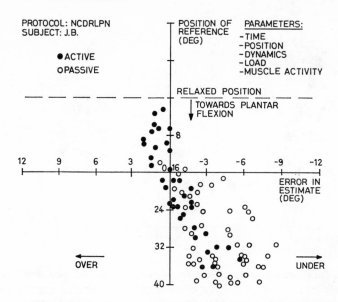

Fig. 3. Distribution of errors for passive and active positioning of the reference and different final positions of the reference.

2. The average speed of *passive* positioning was generally slower than that of *active* positioning, such as to elicit no reflex activity during large (30-40°) imposed deflections from the relaxed position.

The nature of the temporal factor (effect of the value of T) is illustrated more clearly in figure 4. The error is consequently found to be monotonically dependent on T, shifting from an initial overestimate towards an underestimate. Similar results were obtained for the upper extremity by PAILLARD and BROUCHON [1968], and attributed to inputs from dynamic joint afferents [also see BLOCH, 1890].

The dynamic event is further characterized by the large increase in the overestimate for small values of T, when the speed of passive positioning is changed from slow to fast (a 250-percent increase in speed in this case). For illustration, the decay time of a ventro-basal thalamic target cell driven by wrist extension in a macaque monkey is indicated by a dotted line[MOUNT-CASTLE, POGGIO and WERNER, 1963]. The neuronal decay time shows a striking similarity to the temporal behavior of the estimation error. The relation between the size of the overestimate and the speed of passive positioning at time T = 2 sec (location of the small arrow in fig. 4), is further

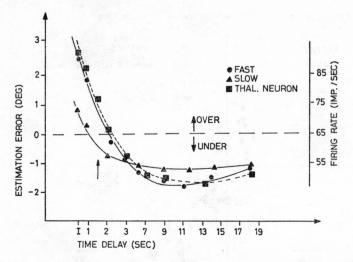

Fig. 4. Effect of decay time of the dynamic component on the estimation error when the response is elicited by passive positioning of the reference at two different speeds (fast and slow). The decay time of a target neuron of a wrist joint afferent, located in the ventrobasal complex of the thalamus of a macaque monkey (thal. neuron) is shown for comparison [derived from MOUNTCASTLE, POGGIO and WERNER, 1963].

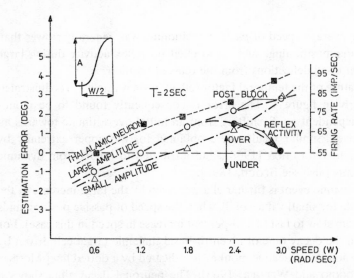

Fig. 5. Effect of passive positioning speed on the size of the error when estimated after a delay of 2 sec. The dynamic sensitivity of a target neuron of a wrist joint afferent, located in the ventrobasal complex of the thalamus of a macaque monkey (labeled thalamic neuron), is shown for comparison [derived from MOUNTCASTLE, POGGIO and WERNER, 1963].

illustrated by figure 5 and found to be quite linear for the range of speeds indicated. The sinusoidal loading pattern used to passively position the reference is shown in the left upper portion of figure 5. Since the acceleration and deceleration, at the beginning and end of movement respectively, are smaller and better defined, the use of a sinusoidal pattern was felt to be preferable to the use of a ramp. The existence of a significant amount of dynamic sensitivity explains the lack of reproducibility in the size of the error that was observed during earlier experiments using manual positioning of the reference. It was noted that if the amplitude of the sinusoid (A in fig. 5) was increased sufficiently, the tendency of the overestimate to increase with increasing the positioning speed would invert itself and the error would start to decrease again as shown by the two filled-in circles marked 'reflex activity' at 2.4 and 3.0 rad/sec. Further study of the relation between the loading and positioning trajectory revealed that an apparently significant amount of reflex activity was elicited at the higher speeds, an observation which was easily verified by recording the surface EMG from the m. triceps surae. A selective block of the tibial nerve eliminated the decrease in the dynamic sensitivity and extended the linear relationship between the error and the speed of stretch (the two circles marked 'post-block' in fig. 5). This was somewhat surprising, since the fiber size distribution of fusimotor fibers and dynamic joint afferents partly overlap, and therefore the block should affect both [BOYD and DAVEY, 1968]. The block had no effect at the slow speed of stretch. The simultaneous occurrence of myotatic reflex activity and a decrease in the overestimate is of interest when comparing these findings with the case of active positioning. An attempt was made to block the dynamic receptors of the joint capsule by means of xylocaine injected into the joint space and the supporting ligaments of the ankle. This was done in an effort to further demonstrate the articular nature of the source of the overestimate. However, complications such as pain, psychosomatic reaction and in one subject, development of a tremor in the frequency range of physiological tremor have thus far prohibited reliable verification.

Active Position Calibration

In the analysis of active positioning a distinction was made between two ways of estimating the new position: (1) starting from the same initial position of both feet (filled-in characters in fig. 6, 'normal'), and (2) starting

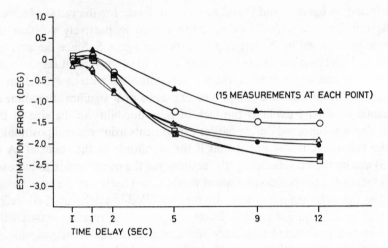

Fig. 6. Effect of decay time of the dynamic spatial component on the estimation error when the response is elicited by active positioning of the reference. Two conditions are shown for 3 subjects (JB, AP and AB): (a) estimating of the absolute position of the reference (normal), and (b) estimating of the incremental change; the change in the reference is made starting with some initial error between reference and estimating foot (incr.). ○ = JB (Incr.), △ = AP (Incr.), □ = AB (Incr.), ● = JB (Normal), ▲ = AP (Normal), ■ = AB (Normal).

from a variable initial error between both feet using an approximately fixed incremental change (open characters in fig. 6, 'incr.').

The findings of the first analysis show the same basic decay time as seen during passive positioning,except for a complete absence of the overestimate immediately following the dynamic event. Fast and slow positioning lead to a similar curve. Sometimes a slight increase in error was noted approximately 1-2 sec following positioning of the reference. This phenomenon has also been observed by BOWDITCH and SOUTHARD [1880], and might be reflexive in nature as was indicated subjectively by some subjects. Positioning of the reference followed by its immediate estimation often caused the reference foot to be slightly retracted during estimation (see fig. 2, A2 and B2), creating an overestimate. The error increase after 1-2 sec could also be due to a shorter decay time of the process(es) affecting the active calibration error immediately following the dynamic event as compared to those operating under the passive condition.

The purpose of the second experiment, the incremental mode, was to test the validity of the hypothesis that the increase in spatial resolution for active as compared to passive positioning is related mainly to the dynamic event, i. e. the position change. This would also indicate the role played by sensory inputs primarily related to the absolute value of the reference position within the normal range of motion of the joint, such as passive resistance and/or cutaneous sensations. The variation of the error in the 'incremental' mode tended to be somewhat larger as compared with the 'normal' mode, and increased with increasing time delay. This phenomenon is possibly due to central decay of the magnitude of the reference increment. It is clear, however, that the curves of the normal and incremental mode shown in figure 6, are basically the same. The size and polarity of the error at the large delays ($T \geq 12$) are similar to the values obtained as a result of passive positioning. The variation in the error under static conditions was found to be generally slightly larger than immediately following the dynamic event, an observation also made by PAILLARD and BROUCHON [1968] and by LASHLEY [1917]. The latter author also pointed out that the quicker the movement, the more accurately it is made. This can be attributed to the greater variability of joint receptor afferents during static conditions as compared to dynamic conditions [WERNER and MOUNT-CASTLE, 1963]. Differences in the average steady-state error for different subjects is partly due to the way the 'zero' is calculated but mainly inherent to the quality of the position sense for this joint. Threshold for passive movement is approximately 1.4° as compared to 0.3 and 0.5° for elbow and shoulder [GOLDSCHEIDER, 1889]. Comparison of dynamic sensitivity among these joints also shows the ankle to have relatively low dynamic sensitivity. Statistical significance might be improved by screening of subjects, use of other joints or further restriction in the number of uncontrolled variables, such as normal muscle tone of the subjects, attentiveness and lenght of the sessions.

Effects of Load on Position Calibration

Precise calibration of coordinated movement and of position can be obtained under conditions of minimal load and when overcoming some opposing force or balancing a load in the direction of movement. What central and/or peripheral mechanisms regulate this smooth adjustment to demand and how and where these processes are represented is controversial to date [JACKSON, 1876; SHERRINGTON, 1900; BATES, 1947]. There seems to be evidence, however, that at the level of the pyramidal tract neurons, we are dealing with a 'final' integrated output primarily representing a force demand [EVARTS, 1968]. How this temporal-spatial pattern is synthesized with the help of the spatial referents system is currently a question of great significance. It was expected that evaluation of the effect of load on position calibration might tell us something about the basic principles of motor regulation, particularly the various hypotheses concerning α-γ relations [MATTHEWS, 1964; GRANIT, 1968]. It might also provide indications concerning some of the essential differences in motor control strategy under isotonic and isometric conditions if combined with selective nerve blocks ,as proposed by PHILLIPS [1971].

It makes considerable difference, of course, if the load is applied to the reference foot or to the estimating foot. An error resulting from the latter method is essentially due to incomplete load compensation, when expressing the magnitude of the perceived sensation, but is subject to the same errors that were noted in the 'no-load' condition. The no-load condition is shown in figure 7 ,with dots. A plus (+) indicates that the load is applied in the direction of movement (e. g. towards plantar flexion if the reference is moved to plantar flexion) and to minus (–) when opposing the movement (e. g. towards dorsiflexion if reference is moved to plantar flexion). Note that $T = 12$, therefore, all measurements in figure 7 were made in the steady state. Since the magnitude of the reference itself is not affected by the load, the gain in the reflex loop of the estimating foot plays a primary role in explaining the resulting error. The sensitivity of this loop (S_E^A) is approximately the same in both directions and of the order of a few tenths of a degree per lb. ft. of torque. That the relationship between error and load is linear is unlikely, but needs to be tested and compared with muscle proprioceptor activity, using the technique of HAGBARTH and VALLBO [1967]. Also note the similarity between the passive resistance curve and the systematic shift in the error as a function of position; this position-dependent gradient is indicated with S_E^P. When the direction of movement of the reference is varied in both directions (e. g. dorsiflexion as well as plantar flexion), a discrete change is noted in the magnitude of the error, around the relaxed position. This also occurs in the no-load condition and for smaller values of T, but is generally less pronounced.

The results seen in the case where the load is applied to the reference are rather surprising (fig. 8). No correlation is noted between the polarity of the load during both conditions shown (load to dorsiflexion [DF] and to plantar flexion [PF]) and the spatial shift of the error. Good correspondence between the two curves is seen immediately following the dynamic

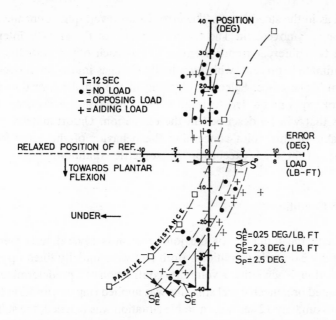

Fig. 7. Effect of a load applied to the estimating foot on the steady-state estimation error with active positioning of the reference.

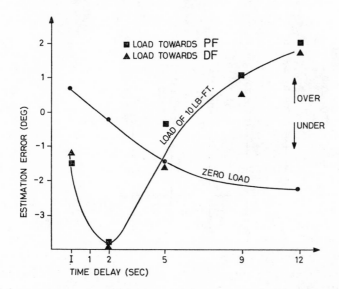

Fig. 8. Dynamic behavior of the estimation error when a load is applied to the reference position with active positioning of the reference.

event, as well as in the steady state. The error is, however, quite dependent on the direction of movement. The temporal nature of the error is interesting in that two different trends seem to follow each other. Initially, a large underestimate is present, which gradually changes into a small overestimate when T increases. Selective blocking of the tibial nerve during further control experiments led to large overestimates and a number of complications that will be described in the discussion. Unfortunately, no data is available at this point concerning the behavior of the error for different values of the load, except for the isometric condition.

Isometric Condition

To eliminate the influence of the joint receptors several tests were carried out under isometric conditions. Both feet were initially lined up in the relaxed position. Then either a voluntary contraction of a predetermined force was initiated or a mechanical stimulus was applied (tap or vibration to tendon). Approximately 12 sec after a stable situation was reached, the subject estimated the position of the foot as it was perceived. Going from left to right in figure 9: a maintained voluntary force towards dorsiflexion is

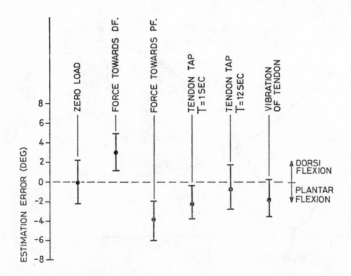

Fig. 9. Steady-state errors under isometric conditions of the reference but in the presence of a voluntary or reflex induced bias between reference and estimating foot.

sensed as a position change in that direction; similarly, a force towards plantar flexion leads to an estimation error in that direction. Recall that this direction sensitivity was not observed under isotonic condition! A phasic stretch induced a reflex response also leading to a dynamic error in the direction of plantar flexion ($T = 1$); the error gradually decayed back to zero ($T = 12$). The tonic vibration reflex (TVR) [EKLUND and HAGBARTH, 1966], normally induced a contraction of the extensors and led to an estimation error towards plantar flexion.

Both the phasic and particularly the tonic reflexes led to relatively large calibration errors in comparison with the amount of force that was elicited. Distinct sensations of position change were perceived during vibration but not under the other experimental conditions.

Discussion

The quality of the perceived calibration standard was characterized by various systematic errors that were related to a number of well defined parameters, such as time, joint angle, load and muscle stretch. It was noted that there exists a certain amount of variation under presumably identical conditions. This variation is either due to the 'sensory stage' of the perceptual process (for instance, the anatomical configuration of the ankle joint) and consequently independent of the motives of the observer, or results from the 'decision stage', determined by the observer's bias. The two stages cannot be differentiated with the present technique, since this would call for a correlative analysis of the subjective response and the activity evoked in particular neuronal elements at a succession of levels in the nervous system. The striking similarity in temporal behavior of the firing frequency of a thalamic neuron originating from a wrist joint afferent in a monkey and the decay in the position calibration error, is an example of such an approach. It also indicates that subsequent transformations at higher levels occur along linear coordinates. An essential prerequisite for such a correlative approach is that there must be a well defined and measurable correspondence between the end result of the perceptual process and the neuronal activities studied.

It must be emphasized that the mechanisms involved in the conscious perception of position may not be identical to those utilizing this information for the control of motor outflow. As early as 1917, LASHLEY stated: 'In many cases the perception and control of movement might be one and the same phenomenon, but certain experimental data and facts of everyday experience indicate a certain degree of independence of the accuracy of movement from any stimulation resulting from the movement itself.' What is perceived is a trade-off between possibly conflicting information from peripheral and most likely central components. We perceive an actual veridical process if the nervous system has the ability to evaluate the available information and can, if necessary, rearrange central/peripheral relationships instead of persisting in an illusion. This brings up the two essential questions: (1) Can we isolate what central and peripheral components affect the position calibration under the experimental conditions described? and (2) By what mechanisms and at what level do these effects impress their input on the measured calibration error?

For a systematic analysis of the error behavior, it is necessary to differentiate between the following two error categories: (1) those due to an actual incorrect perception, and (2) those beyond the resolution of the sensory inputs converging on the perceptual mechanism. This resolution is not constant but depends on the state of this mechanism. Stimuli that go unperceived can have a definite effect on the measured calibration error but are limited in magnitude by the resolution of the perceptual mechanism, which has its final resolution in the sensory stages of the modalities converging on it. Based on the data presented by GOLDSCHEIDER [1898] for the ankle joint, virtually all tendencies shown in figures 4-9 are beyond the perceptual resolution for *passive* movement. This explains the relatively large variation around average errors that showed great consistency when the test was repeated. It also indicates that other factors must play an essential role in magnifying this resolution and in assuring reliability at the time of voluntary movement. This is particularly critical for the most proximal joints, and perhaps that is why these are the most accurate ones. In view of these considerations it is logical to first consider the implications of that experiment where the sensation complex to both feet is most similar, i.e. active positioning without an external load.

There is no doubt that active movement leads to accurate localization of a body part, but only during a time period extending a few seconds beyond the occurrence of the dynamic event. During this time period a nearly perfectly symmetrical similarity is present between the neuronal activity evoked by both events, i.e. positioning and estimating. The assumption that the increased resolution is based simply on template matching of comparable sensations, obviously does not explain what happens when the sensation complex is different. We know from both LASHLEY's

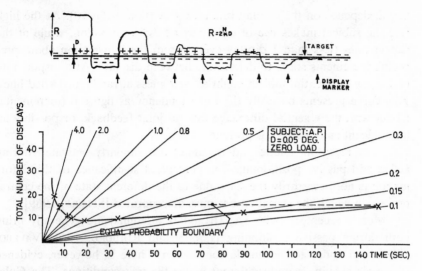

Fig. 10. Evaluation of the number of incremental changes in position necessary to reach a specific target position (upper half). Only the polarity of the error is fed back to the subject (visually) at specific increment of time and at a rate varying from 4/sec to 1 every 10 sec. Since the initial error (R) is only a few degrees, the resulting curve (lower half) is taken as a measure of central decay of the spatial information from the preceding increments.

experiments and from everyday clinical observations, that distinct sensations of movement can be created by means of voluntary contraction in the absence of any passive sensation. This sensation is apparently independent of the fact if the movement actually takes place [LASHLEY, 1917; MERTON, 1964]. Good spatial resolution for position increments can be maintained for short periods of time following positioning, and calibration of the initial position through the use of another modality such as vision, as long as no unpredictable resistance is present. The temporal decay of the position calibration was also noted by other investigators [GOLD-SCHEIDER, 1898; PAILLARD and BROUCHON, 1968], but no data are available for the case of active positioning in the absence of joint sensation. A possible way to simulate this condition is by using movement changes below the level of joint resolution as described in the following experiment: a subject is asked to resolve a positional difference of a few hundredths of a degree by successive incremental movements. This is accomplished on the basis of information about the polarity of his position error, displayed to him visually at a regular rate (upper portion of fig. 10). Optimal performance, i.e. a minimum number of displays to reach the target, depends on the display rate (lower portion of fig. 10). At the high rate the subject makes use of only part of the information, while at the slower rates (less than 1 display every 2 sec) the information about preceding increments starts to deteriorate between displays. At a display rate of 1 every 10 sec, the subject might as well guess at random (dashed line). This curve presents basically the same tendency as figure 6 (incremental mode) with the essential difference that no joint feedback, or possibly an insignificant amount of it, is present.

Joint receptors (static and dynamic) are similarly activated during active and passive positioning. The perceptual significance of this information is not necessarily the same due to the different state of the central process in which the sensation becomes incorporated [SPERRY, 1969].

Muscle receptors (with IA, IB and II afferents) are generally firing both during passive positioning [JANSEN and RUDJORD, 1965; VALLBO, 1970], and under active muscle contraction. There is, however, evidence that their activity is quite different under the two conditions. The Golgi tendon organs (IB afferents) are quite insensitive to passive stretch but have a low threshold for active contraction [HOUK and HENNEMAN, 1967]. These receptors signal muscle tension, which is determined primarily by the state of contraction of the muscle and not by the position of the joint. Two types of intrafusal fibers have been recognized: the nuclear bag and nuclear chain fibers, as well as two kinds of receptors, the primary and secondary, centrally innervated by fusimotor or γ fibers [for review, see MATTHEWS, 1964]. Spindles are under fusimotor control during active contraction, anticipating or keeping up with the muscle shortening, while the stretch-induced reflex response is the primary response during passive positioning. No correlation appears to exist between muscle stretch and fusimotor drive but a relation between joint afferents and static and dynamic muscle reflexes has been proposed [FREEMAN and WYKE, 1966]. No perceptual information is derived from the spindles or tendon organs during passive stretch [MERTON, 1964; GELFAN and CARTER, 1967], but that does not necessarily deny their role in the position calibration of voluntary movement, a fact that was overlooked by LASHLEY. This was pointed out by PAILLARD and BROUCHON [1968] who hypothesized that the information derived from the pattern of innervation is possibly susceptible to intervention via more peripheral routes. They attributed this intervention to the γ-dynamic system. The possible significance of the γ-spindle system for motor learning has been brought up by a number of investigators [see for instance: BUCHWALD, STANDISH, ELDRED and HALAS, 1964] since its discovery by LEKSELL [1945]. Selective blocking of the motor nerve as a means of verifying this 'γ-dynamic' hypothesis is of limited use since the block prevents all spindle firing, static as well as dynamic, during contraction of the extrafusal fibers due to lack of fusimotor support. It consequently decreases the gain in the myotatic servo loop effectively to zero. Since in our experiments movements were made from the relaxed position towards plantar or dorsiflexion, an opposing force (the passive resistance) drives the reference foot back in the absence of adequate load compensation. The result is a gradual decay of the reference position in the order of a few degrees, such as shown in figure 2, C2. If we now keep in mind that the order of magnitude of the tendencies in figures 4-9 is of the order of a few degrees and so is the perceptual threshold, it is clear that even though this method does lead to overestimation for large values of T, it certainly does not lead to the kind of curves shown in figure 4. We intend to repeat the experiment either with another joint or by compensating for passive resistance and gravity forces. The effect of the block was, however, quite pronounced in the test shown in figure 10. Large overestimates and inability to reach the target position resulted as soon as the tendon tap response disappeared. This observation supports the idea that spindles play an important role in position calibration but the evidence is, of course, inconclusive in view of the limited selectiveness of the block (cutaneous as well as other small motor fibers are attacked), the uncertainty with respect to joint inputs and to a lesser extent by the nature of the positioning task itself. It is interesting that extraocular muscles, which are not subjected to varying external conditions, such as load, have stretch receptors. The afferents of these receptors project to the cerebellum, like those of skeletal muscles, but do not create a stretch reflex [KELLER and ROBINSON, 1971].

Muscle contraction may play a minor role in the perception of the joint angle, as suggested by the fact that the discharge rate of slowly adapting receptors of the joint capsule is affected by contraction of muscles which tense it [SKOGLUND, 1956]. This is most likely the source of the dipping effect observed in passive positioning at high speeds (fig. 5), and the calibration error resulting from tendon taps (fig. 9). In both cases the fast stretch produced a muscle contraction and possibly recruited dynamic joint fibers through capsule deformation. This is further supported by the fact that it was only present if reflex activity was elicited and then outlasted the contraction for up to 12 sec. The dip at high speed of passive stretch could also be partly due to two other factors: (1) the fact that the speed of stretch is decreased as a result of the reflex activity, and

(2) the saturation of the dynamic joint receptors at these stimulus intensities [STEVENS, 1957]. That the presence of active muscle contraction can affect passive position sense was also reported by BROWN, LEE and RING [1954], in studies of the effect of capsule blocks on position sense. Strictly speaking, passive positioning, in the actively contracted state is impossible and, therefore, their conclusions must be questioned. LOEB [1890] claimed that movements were relatively underestimated when they occurred under conditions of slight contraction and overestimated when under conditions of greater contraction. This agrees with the findings in figure 8, for the steady-state condition. It is a challenging problem how these and other observations can be related to established neurophysiological facts.

In the case of active positioning in the presence of a load, we can explain the independence of the transient error from the direction of the load (fig. 8) if we accept the hypothesis of corollary discharge intervention through spindle afferents. When the load is in the plantar flexion direction, it is supported by the flexor musculature. Movement towards plantar flexion will automatically lead to a transient period of overextension of the intrafusal fibers with respect to the extrafusal fibers. When the load is towards dorsiflexion, however, the load is supported by the extensors and movement to plantar flexion will result in a short period of overcontraction of the intrafusal fibers with respect to the extrafusal fibers also signalling an underestimate. Movement in the direction of dorsiflexion similarly leads to a large transient overestimate, independent of the direction of the load. This also accounts for the error observed when maintaining a steady voluntary force under the isometric condition, since the directions of force and movement are the same. If we base the steady-state load sensitivity for the isometric condition on the values from figure 7 ($S_E^A = 0.25°$/lb. ft.) a load of 12 lb. ft. (as used in fig. 9) corresponds to an error of approximately 3°, which is quite close to the actual error.

Notice that the time constant of the transient effect in fig. 8 is only about 2 sec and of the same order of magnitude as the time period over which an improvement in calibration error is observed during active positioning in figure 4. It also corresponds to the optimal display rate in figure 10. This is, however, conjectural and a number of controlled experiments need to be carried out to obtain further and more conclusive evidence.

The calibration error observed during vibration of the tendon is of a different nature, since it leads to a distinct sensation of position change under the isometric condition and in the absence of a voluntarily initiated response. It was established that for a given frequency the sensation complex was quite dependent on the stimulus location on the tendon as well as the joint angle. Moderate amplitudes induced a normal tonic reflex in the direction of plantar flexion, stronger stimuli led to a flexor-withdrawal type response [EKLUND and HAGBARTH, 1966; MATTHEWS, 1966]. An effect on postural stability through the vestibular system was also reported. The fact that the contraction develops rather slowly (up to 30 sec depending on various parameters), suggests that polysynaptic and most likely central routes are involved. It was also noted that the strength of the sensation of movement was mostly independent of the magnitude of the extensor contraction, indicating other receptors might be involved, i. e. the dynamic joint receptors. When more mechanical energy reaches the ankle joint, the sensation of movement is stronger. Under isotonic condition, the reflexively induced position change can be overcome voluntarily, but to what extent there is an interference with position calibration is unknown so far. MATTHEWS [personal communication] reported errors of the order of 10° in tests on the position calibration of the forearm. This is not impossible in view of the large magnitude of the effect of the dynamic joint receptors; peak-to-bottom in figure 4 is about 5°.

Summarizing the discussion, it is felt that many questions remain about the mechanisms involved in position calibration. The significance of the joint afferents under passive as well as active conditions appears to be considerable and techniques need to be perfected to isolate this peripheral component. Combining the psychophysical measurements with direct monitoring of the various muscle afferents is another prerequisite to obtaining conclusive evidence in humans under functional circumstances. The hypothesis that position calibration during voluntary movement is dominated by intervention of the corollary discharge by feedback from muscle spindles [PAILLARD and BROUCHON, 1968] is supported by our findings, but the evidence is still quite incomplete.

Acknowledgement

The authors acknowledge the cooperation of Dr. C. S. TENG in doing the various nerve blocks and for taking an active interest in the experiments. This work is supported by grant 23P-55115 of the United States Social and Rehabilitation Services.

Author's address: Dr. A. W. MONSTER, Krusen Research Center, Moss Rehabilitation Hospital, 12th Street and Tabor Road, *Philadelphia, PA 19141* (USA)

New Developments in Electromyography and Clinical Neurophysiology,
edited by J. E. Desmedt, vol. 3, pp. 404–417 (Karger, Basel 1973)

Servo Control of the Intercostal Muscles

T. A. SEARS

Department of Neurophysiology, Institute of Neurology, London

In this review I have brought together the results of neurophysiological
research on breathing in animals and in man which have also shed some light
on the more general problem of the nervous regulation of movement. The
review has two aims. Firstly to give an account of animal experiments which
validate the hypothesis that breathing movements are controlled through a
servo mechanism which utilizes the length stabilizing properties of the stretch
reflex. Secondly to describe the testing of this hypothesis by experiments on
conscious man and to describe the re-assessment of the significance of servo
control in the light of the results obtained. The account is confined to work on
the intercostal muscles. For work on the diaphragm the reader should consult
the excellent accounts by VON EULER [1966a, b; 1970].

Animal Experiments

There is insufficient space to review in detail the relevant work on the
muscle spindles and stretch reflex of limb muscles but recent comprehensive
accounts are available [MATTHEWS, 1964, 1970, 1971; GRANIT, 1955, 1970].
In summary, the salient points that bear on the problem of respiratory muscle
control are: that the stretch reflex represents a sustained motor output from
the CNS in response to a steady afferent barrage induced by muscle stretch.
The essential afferent discharge is thought to be conveyed in the fast-conducting muscle spindle, primary afferent fibres which monosynaptically excite
α motoneurones innervating the same and synergic muscles. Following the
demonstration by JANSEN and MATTHEWS [1962] that the static and dynamic
responses of muscle spindles to stretch are under independent control from

the CNS, MATTHEWS [1962] described two types of fusimotor fibre, static and dynamic. Because dynamic fusimotor fibres neither appreciably modified the response of a muscle spindle to steady stretch [MATTHEWS, 1962; APPELBERG, BESSOU and LAPORTE, 1966] nor offset the unloading of spindles during muscle shortening [LENNESTRAND and THODEN, 1968], their excitation cannot represent the 'demand' for movement as conceived in the theory of the 'length follow-up servo' mechanism of MERTON [1951, 1953]. Thus LENNESTRAND and THODEN [1968] have concluded that the demand for movement must be fed to the servo mechanism via the static fusimotor fibres which do have the required properties. The innervation of spindle secondary endings by static fusimotor fibres [APPELBERG, BESSOU and LAPORTE, 1966] suggests a role also for these endings in the stretch reflex. Studying the effect of vibration on the stretch reflex, MATTHEWS [1969] has concluded that spindle secondaries indeed provide a supporting excitatory drive to motoneurones and he has speculated that they may do so by inhibition of the disynaptic pathway mediating autogenetic inhibition [MATTHEWS, 1970]. However, GRILLNER [1970] has recently produced evidence opposing this interpretation.

Strictly speaking, the *length follow-up servo* hypothesis requires that the command for movement is fed entirely via the fusimotor neurones (the 'indirect' route). It seems often to be ignored that ELDRED, GRANIT and MERTON [1953], GRANIT, HOLMGREN and MERTON [1955] clearly emphasized that different movements are likely to be compounded from signals fed 'directly' and 'indirectly' to the α motoneurones in varying combination. Reviewing studies on the Magnus- de Kleyn reflexes, the cutaneous reflexes [ELDRED and HAGBARTH, 1954] and electrical stimulation in the brain stem and cerebellum [GRANIT and KAADA, 1952], GRANIT [1955] emphasized the importance of 'α-γ linkage' as the probable means by which the brain regulated movement. However, it was left unanswered as to whether the γ loop was actually utilized in natural movements initiated by the CNS such as in willed movement. One aspect of the loop was subsequently investigated by HAMMOND [1954] who demonstrated that the human biceps muscle under voluntary activation shows a response to rapid stretch which has the properties of a stretch reflex. This response consists of an increase in tension with a latency measured in the EMG of 50–60 msec, substantially shorter than the earliest voluntary response to a cutaneous stimulus to the limb but much longer than the latency to the biceps jerk. HAMMOND [1956] also made the important discovery that the stretch reflex, which is a *spinally* mediated reflex automatically providing compensation against loads through its length compensating properties, nevertheless can be caused to occur, or not to occur, according to the instructions given to the subject (prior instruction). Later, ANGEL, EPPLER and IANNONE [1966] demonstrated that a sudden release of a muscle in steady postural contraction evoked a brief latency silent period in the EMG. They attributed this to the abrupt withdrawal of monosynaptic excitation from the α motoneurones, the inference being that the spindles were not unloaded by the control postural contraction [see ANGEL; STRUPPLER et al., this volume].

My own interest in the possibility of servo control of respiratory muscles began with observations on the effects of sectioning thoracic dorsal roots on the Breuer reflexes. When the inspiratory-inhibitory, expiratory-excitatory reflex of Breuer was elicited by tracheal closure at the height of inspiration, the normal *recruitment* in frequency and number of expiratory motor units during the prolonged expiratory pause was reduced or abolished by sectioning the dorsal roots in the same segment [SEARS, 1958] signifying that the force-controlling properties of the Breuer reflex depend in part on the *segmental* afferent flow. Subsequently NATHAN and SEARS [1960] described a temporary paralysis of the diaphragm and intercostal muscles which followed sectioning of cervical and thoracic dorsal roots respectively for the relief of intractable pain. They suggested that the paralysis may have been due to opening of a 'γ loop' in respiratory muscles although, at that time, none of the supporting data was available for respiratory muscles. Such evidence was subsequently obtained, in two parallel investigations. Firstly, intracellular recording from respiratory motoneurones was used to determine the nature both of the *central* 'demand' signals for respiratory movement and the segmental afferent inflows which act upon the motoneurones. Secondly, anatomical and physiological studies were carried out to define the innervation and mechanical properties of the intercostal muscles and their spindles. Thus ANDERSEN and SEARS [1964] established that intercostal muscles comprise both fast (mean rise time 25 msec) and slow (rise time 45 msec) motor units innervated by the low threshold (large diameter) group of motor fibres in the intercostal nerves; it had also been demonstrated by deafferentation studies that the motor fibres show a bimodal distribution [SEARS, 1964a]. As in the classic experiment of KUFFLER, HUNT and QUILLIAM [1951] on limb muscles, excitation of the small diameter motor fibres conducting in the range 15–30 m/sec did not produce recordable external tension, so it was concluded that these must innervate the intrafusal muscle fibres of the spindles found abundantly distributed in intercostal muscles [ANDERSEN and SEARS, 1964]. By recording directly from fine intercostal nerve filaments, it was demonstrated that during spontaneous breathing, intercostal fusimotor neurones fire with a respiratory rhythm, not only reciprocally between inspiratory and expiratory fusimotor neurones, but also in tight α-γ linkage. As with α motoneurones, the recruitment in frequency and in number of fusimotor neurones occurred in inspiration for external intercostal muscles and in expiration for internal intercostal muscles, with complete or relative inhibition of such activity in the respective opposite phases of the respiratory cycle [SEARS, 1963a, b, 1964b]. The preliminary communication of this work to the Australian Physiological Society [SEARS, 1962] provided the first description of supraspinal activation in α-γ linkage for a natural movement 'demanded' by the CNS. At the same time CRITCHLOW and VON EULER [1962, 1963] showed that the spindles of the external intercostal muscles fired during the *active* phase of contraction of these muscles, instead of being silenced by the unloading which occurred when they applied procaine to the ventral roots to block the fusimotor fibres [c.f. MATTHEWS and RUSHWORTH, 1957]. They interpreted this result as being due to a combination of rhythmic supraspinal and segmental reflex drives on the fusimotor neurones but emphasized the need for direct recording. This they subsequently did also by recording directly from intercostal nerve filaments [EKLUND, VON EULER and RUTKOWSKI, 1963, 1964]. Both groups of workers showed that in procedures which affected breathing such as hyper- and hypo-ventilation, in hypoxia and during elicitation of the Breuer reflexes, α and fusimotor neurones always responded together in tight α-γ linkage. A significant proprioceptive control of 'tonic' fusimotor neurones, its responsiveness to tonic neck reflexes [c.f. MASSION, MEULDERS and COLLE, 1960] and some modification of the α-γ linkage in response to stimulation of the anterior lobe of the cerebellum were all important features of γ control emphasized by the work of EKLUND, VON EULER and RUTKOWSKI [1964], CORDA, EKLUND and VON EULER [1965] and CORDA, VON EULER and LENNESTRAND [1966]. In particular, VON EULER and PERETTI [1966] established that four different modes of fusimotor control can be identified. Tonic-dynamic, tonic-static, rhythmic dynamic, and rhythmic static. Through this multiplicity of control is permitted a great flexibility in the co-ordination of respiratory and postural control of the intercostal muscles [VON EULER, 1966a, b].

In discussing the significance of a central respiratory drive in α-γ linkage I drew on the length follow-up servo hypothesis and on my studies employing intracellular recording from intercostal motoneurones [SEARS, 1963a]. These had shown that the largest diameter afferent fibres present in the muscular branches of the intercostal nerves [SEARS, 1964a] monosynaptically excite α intercostal (and abdominal) motoneurones [ECCLES, SEARS and SHEALY, 1962; SEARS, 1963b, 1964c]. Contrary to expectation, no reciprocal 'direct' inhibition between inspiratory and expiratory muscle spindles and their respective antagonistic motoneurones was demonstrable. In this respect intercostal muscles resemble the hip adductors in not giving 'direct' inhibition to their antagonists [ECCLES and LUNDBERG, 1958]. Although the mean amplitude of the EPSP was small (approx. 1.0 mV) they could be as large as 5.0 mV in individual motoneurones. Expiratory motoneurones also received weak monosynaptic excitation from adjacent segments. In contradistinction to most lumbar motoneurones [see CURTIS and ECCLES, 1960], the monosynaptic EPSP of expiratory motoneurones showed a considerable amplitude potentiation with increasing frequency so that above 60 c/sec the temporal summation of successive EPSPs caused a sustained depolarization of α motoneurones. For a motoneurone close to threshold, such as those to which the demand to breathe is fed in α-γ linkage this additional synaptic excitation evoked repetitive firing of up to 40 impulses/sec which was sustained throughout the period of stimulation [SEARS, 1963b, 1964b]. These responses to repetitive stimulation of the I_A pathway, provided a crude model for the probable form of the postsynaptic potential in α motoneurones evoked by the spindle input under rhythmic fusimotor drive in α-γ linkage. What then is the form of the direct input to α motoneurones? During spontaneous breathing, intracellular recording from thoracic motoneurones revealed slow, rhythmic changes in potential in phase with breathing. These were shown to be due to alternating, excitatory and inhibitory synaptic drives distributed reciprocally to inspiratory and expiratory motoneurones. The rhythmic synaptic drives showed an interdependence in which increased excitation of inspiratory motoneurones was associated with a greater inhibition of expiratory motoneurones and *vice versa* [ECCLES, SEARS and SHEALEY, 1962; SEARS, 1963b, 1964e]. They were named 'central respiratory drive potentials' (CRDPs) to connate their central origin and shown to depend on synaptic drives originating at supraspinal levels. These data from the animal experiments allowed the following interpretation. 'Failure of the extrafusal muscle fibres to shorten at the same rate as the intrafusal fibres, due to resistance to air flow to and from the lungs, would lead to an increased discharge from the muscle spindles with the consequence that the monosynaptic pathway would exercise an increased excitatory drive on the intercostal motoneurones. There would be summation with the "central respiratory drive potential" and so an increase in the discharge frequency of active motoneurones and a recruitment of others into activity' [SEARS, 1963a].

Collectively, the experiments described above provided as much, if not more direct evidence than is available for the control of limb movement, that respiratory movements of the ribs are controlled as conceived in the follow-up length servo hypothesis modified, however, by the finding that the demand for breathning is fed to the motoneurones in tight α-γ linkage.

The effects of dorsal root section on the response to mechanical loading of the respiratory system [SEARS, 1958, 1963a, 1964b; VON EULER and FRITTS, 1963; CORDA, EKLUND and VON EULER, 1965] and spontaneous breathing [NATHAN and SEARS, 1960] can thus be explained in terms of interruption of the afferent limb of the γ loop.

Human Experiments

The question which now arises is: 'can servo control of the intercostal muscles be demonstrated in conscious man?' To investigate this we initially developed techniques for studying the effects of resistive loading on the electrical activity of intercostal muscles during spontaneous breathing [NEWSOM DAVIS, SEARS, STAGG and TAYLOR, 1965] but for the reasons which will become clear later these gave equivocal answers to the question [NEWSOM DAVIS, SEARS, STAGG and TAYLOR, 1966]. Evidence was, therefore, sought for servo control during willed respiratory movement [NEWSOM DAVIS and SEARS, 1967, 1970]. The experiments were analagous to those used on the human biceps muscle by HAMMOND [1954, 1956, 1960; see also HAMMOND, MERTON and SUTTON, 1956; ANGEL, EPPLER and IANNONE, 1965]. We studied electromyographically the effects on contracting intercostal muscles of sudden alterations in mechanical load as subjects (with open glottis) either held lung volume constant (equivalent to holding a limb in steady postural contraction supporting its mass against gravity) or changed it at a constant rate (equivalent to moving the limb at constant velocity).

Brief (10–100 msec), timed pulses of pressure (0–80 cm H_2O) were delivered to the mouth through an electrically operated, two-way valve and used suddenly to oppose (load) or to assist (unload) contractions of the intercostal muscles; positive pressures assisted inspiratory (external) and opposed expiratory (internal) intercostal muscles and negative pressures had the opposite effect. Resistive loading and unloading were also used. EMGs were recorded through pairs of fine wires [TAYLOR, 1960] and quantified by 'gated' integration for epochs of usually 50 or 500 msec duration [NEWSOM DAVIS, SEARS, STAGG and TAYLOR, 1965]. The height of each epoch gave the integral for that period while its slope (estimated from the 500 msec epoch) gave a measurement of the level of activity against time. Online averaging of the data, usually on 10 experimental runs, was carried out with an averaging computer.

Air flow was recorded with a pneumotachograph and integrated to give a direct display as percent vital capacity (% VC) on a digital voltmeter. Air flow could be held constant either by watching the scale of a flow meter or, in the case of expiratory muscles, by singing a note at attempted constant pitch and intensity [see SEARS and NEWSOM DAVIS, 1968].

The results of a typical loading experiment are illustrated in figure 1 which also shows, by comparison with photographically superimposed traces of the CRO (upper set), the derivation of the computer averaged records (lower set). Ten experimental runs were averaged. In figure 1 A the subject expired at a flow rate of 0.13 l/sec recording being initiated when he reached 20% VC. This control record (T.8 internal intercostal muscle) shows the relatively constant level of expiratory activity which contributed to the development of the total pressure required to overcome both the elastic recoil pressure at that volume and the low resistive flow pressure [see BOUHUYS, PROCTOR and MEAD, 1966; NEWSOM DAVIS and SEARS, 1970, for detailed discussion]. Following sudden loading of the expiratory muscles by an opposing pressure pulse of $+60$ cm H_2O (fig. 3 B) there is a brief latency inhibition (inhibitory response IR) in the EMG as seen by the plateau in the E(500) trace and the reduced height of the 6th epoch in the E(50) trace, succeeded by a dramatic increase in activity in the 7th epoch, the excitatory response (ER). The IR and ER are more prominently seen at the higher flow rate (fig. 1 C). Identical results were obtained in loading experiments on inspiratory intercostal muscles.

The lung volume at which a given region of the intercostal muscles becomes active, i.e. the volume threshold, differs in different parts of the chest wall [SEARS and NEWSOM DAVIS, 1968]. This may now be understood in terms of the differing lung volumes at which the excitatory synaptic drives have brought the membrane potentials of the motoneurones in question to critical firing potential as discussed earlier for the CRDP. Low threshold sites are those which become active as soon as one attempts to change one's lung volume away from the 'relaxation volume' (the volume at which for a given posture the elastic recoil forces of the rib cage and the lung are in equilibrium). Any additional excitation of the motoneurones, such as from spindle primary afferent fibres, should recruit motoneurones close to firing threshold into activity. This expectation is realized in figure 2 A which illustrates a loading experiment on inspiratory muscles. The extra mechanical load of an opposing pressure, recruited activity (7th epoch) at a lung volume which during the control movement was just subthreshold for activity at the recording site. This indicates a close functional relationship between spindles responding to the load and to the spatial distribution of α-γ linked demand signals, a control

Load : exp T8

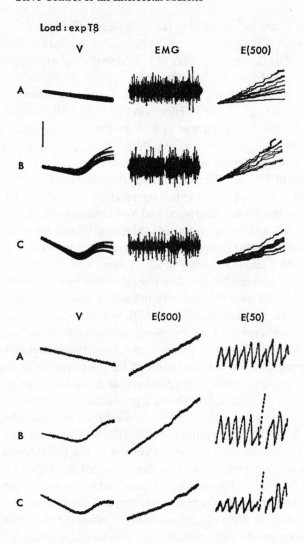

Fig. 1. Derivation of averaged responses during loading of expiratory intercostal muscles (81 CS). Subject expired at flow rate 0.13 l/sec in *A* and *B*, and 0.28 l/sec in *C*. Recording for 500 msec initiated at lung volume 20% VC. *A* Control. *B* and *C* Opposing pressure of $+60$ cm H_2O applied 200 msec after onset of trace, for 100 msec. Upper three records superimposed, lower three, averaged traces for 10 individual experimental runs. Lung volume (V = % vital capacity) electromyogram (EMG) and integrated EMG reset at 50 (E 50) or 500 (E 500) msec. Calibration 2% VC; trace duration 500 msec. From NEWSOM DAVIS and SEARS [1970], reproduced by permission of the Journal of Physiology.

feature which is important in maintaining the shell-like configuration of the rib cage [see DA SILVA, 1971]. Of interest in this connection are VALLBO's findings [1970] from spindle afferent recordings of a close spatial relationship between activities in the skeletomotor and fusimotor systems. When the same opposing pressure was applied at higher lung volumes, at which activity was then present in the control records (first 5 epochs in E[50] of fig. 2 B and C), the excitatory response in the 7th epoch was preceded by a striking inhibitory response in the 6th epoch. This pattern of response to loading was highly repeatable. In a large number of experiments the minimal latency of the inhibitory response was about 20 msec. That of the excitatory response was 50–60 msec, which is short compared to the voluntary reaction time of about 140 msec in these muscles [DRAPER, LADEFOGED and WHITTERIDGE, 1960].

As shown in figure 3, sudden unloading of contracting intercostal muscles by an assisting pressure resulted in a reduction or total abolition of the EMG, i. e. a silent period with a minimal latency of 22–25 msec. The greater the assisting pressure the more prominent the silent period as seen by the decreased slope of the 6th epoch in figure 3 C. Resistive unloading was equally, if not more effective, in evoking a silent period (fig. 3 D). We attributed the silent period to the sudden withdrawal of monosynaptic excitation from the intercostal motoneurones, the evidence in support of this from the animal experiments being as strong as for the explanation of the silent period in limb muscles. The silent period was especially prominent at those muscle lengths (lung volumes) at which the highest intrathoracic pressures can be developed, such as at low lung volumes (e. g., 15% VC) for inspiratory muscles and high lung volumes (e. g., 80% VC) for expiratory muscles [COOK, MEAD and ORZALESI, 1964]. Nevertheless, the most important feature of the unloading experiments was that a silent period could be demonstrated throughout the vital capacity. Thus the relevant muscle spindles could not have been unloaded by the contractions underlying the control movement and must have been in receipt of a static fusimotor drive at least sufficient to offset the shortening. Indeed, our demonstration of a silent period at the extremes of shortening for both inspiratory and expiratory muscle spindles shows that this postulated static fusimotor drive to the respective muscle spindles must have increased towards either extreme of the vital capacity. This conclusion is further supported by the behaviour of the excitatory response to loading at different lung volumes.

The ER of the intercostal muscles occurring 50–60 msec after the load, closely resembles in its latency and behaviour the EMG correlate of the biceps muscle stretch reflex [see HAMMOND, MERTON and SUTTON, 1956]. Thus it was

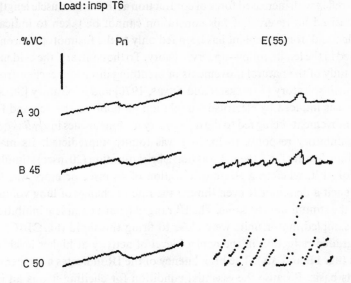

Load : insp T6

Fig.2. Averaged responses (10 runs) to loading of inspiratory intercostal (61 CS) at increasing lung volumes. Inspiratory flow, 0.3 l/sec. Load -31 cm H_2O, duration 60 msec in all three experiments. Loading initiated at 30% VC in *A*, just below volume threshold for activity at this site (under control experimental conditions), just above threshold in *B*, and well above at 50% VC in *C*. Calibration, 5% VC; trace duration 500 msec. From NEWSOM DAVIS and SEARS [1970], reproduced by permission of the Journal of Physiology.

enhanced by the *prior* instruction 'resist' and reduced or abolished by the prior instruction to 'let go' when the subjects *perceived* the stimulus. We therefore interpreted the ER as the response of a servo mechanism serving to stabilize the demanded movement against changes in mechanical load [c.f. HAMMOND, MERTON and SUTTON, 1956]. The ER was shown to increase either with increasing lung volume for inspiratory muscles (fig. 2) or with decreasing lung volume for expiratory muscles [fig. 6 and 7 in NEWSOM DAVIS and SEARS, 1970]. Thus the conclusion again could be drawn that a fusimotor drive must have increased to offset the spindle unloading with shortening, although this experiment could not distinguish between static and dynamic fusimotor activation. The increased sensitivity of the load compensation reflex with increased muscle shortening presumably reflects the need for the gain of the reflex to increase in order to control the augmenting elastic recoil forces as lung volume changes towards either extreme of the vital capacity. [MARSDEN, MERTON and MORTON, 1972, have recently described an increase in gain of the stretch reflex with increased force of contraction at the same muscle length.] It is emphasized however, that this conclusion cannot be taken to indicate that the 'demand' for movement has been fed only to the fusimotor neurones as conceived in the length follow-up servo theory. To the contrary, the evidence from the study of the natural movements of breathing discussed earlier, from the study of inspiratory [ANDERSEN and SEARS, 1970] and expiratory [SEARS, 1966] apneusis induced by electrical stimulation all point to the demand for breathing movements being fed to the respiratory motoneurones in α-γ *linkage*.

The inhibitory response to loading was totally unpredicted. Its main distinguishing feature from the ER was that a brief duration (20 msec) stimulus readily evoked it, whereas a stimulus duration of 40 msec or more was required to elicit a definite ER even though the rate of change of lung volume caused by the stimuli was the same. The IR ranged from a complete inhibition when the sampled motor units were close to firing threshold (fig. 2B, C) to simply a graded reduction in the control level of activity at higher levels of activation (fig. 1B and 3A). The brief latency of the IR indicates a segmental reflex as its basis. Because the essential condition for eliciting it was an increase in tension in already contracting muscles, evidence was adduced that the IR is due to autogenic inhibition from the tendon organs known to be present in the intercostal muscles [BARKER, 1962; CRITCHLOW and VON EULER, 1963] especially proximally and in the levator costae muscle [GODWIN-AUSTEN, 1969]. From the work of JANSEN and RUDJORD [1964] and HOUK and HENNEMAN [1967], it would be expected that tendon organs would be firing during actively produced movement and hence would exert an inhibitory effect on intercostal motoneurones. But we never saw any early facilitatory effect of unloading as would be expected on the basis of disinhibition. Hence, we concluded that the autogenetic inhibitory pathway to intercostal motoneurones [SEARS, 1964c] is inhibited in the control movement [cf. HUFSCHMIDT, 1966]. This would require that throughout the course of a learnt (practiced) movement against a predictable load, such as the control movement in our experiment, an appropriately phased centrifugal control would need to be exerted on the disynaptic pathway mediating autogenetic inhibition. Presynaptic inhibition specifically of group I_B fibres by the cortico-spinal tract [ANDERSEN, ECCLES and SEARS, 1964] could serve this function and it is noteworthy that negative dorsal root potentials are evoked in thoracic roots by stimulation of the trunk area of the sensori-motor cortex [AMINOFF and SEARS, 1971].

What is the functional significance of the IR of intercostal muscles, i.e. of autogenetic inhibition? According to the servo theory, the length stabilizing properties of the stretch reflex are utilized in a length or follow-up length

Unload: insp T6

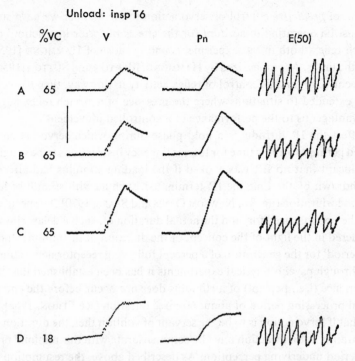

Fig. 3. Averaged responses (10 runs) of an inspiratory intercostal (6ICS) to unloading. *A–C* Unloading under static conditions at 65% VC; load +20, +40, +60 cm H_2O, respectively, duration 100 msec. *D* Resistive unloading; subject maintaining mouth pressure of −40 cm H_2O against large airway resistance, and with low inspiratory flow rate. At 200 msec from onset of recording, connection to low-resistance pathway for 100 msec period. Calibration, 5% VC; trace duration 500 msec. From NEWSOM DAVIS and SEARS [1970], reproduced by permission of the Journal of Physiology.

servo mechanism to provide automatic compensation for changes in load. However, our results suggest that in *conscious* man, automatic compensation is not immediately allowed in the contingency of an unexpected load by the occurrence of autogenetic inhibition. Under these circumstance we conceive that the difference between the actual tension and that predicted, constituting the unmatched fraction of the tendon organ input (tension error), exerts autogenetic inhibition on the motoneurones. Thus servo action would occur only within prescribed tension limits and outside them, would actually be dampened or curtailed altogether.

Because the unloading experiments demonstrate that the excitatory effect of the spindle primary afferents at least is exerted at motoneuronal level,

some form of *predictive* control of servo action in the manner we have suggested must be operating to account for the effects of 'prior instruction' on
the stretch reflex, both in our experiments and in those of HAMMOND [1956,
1960] on the limbs. As emphasized by HAMMOND, MERTON and SUTTON [1956],
'if the executive organ has control of reflex activity in this way, then its scope
is clearly extended to situations where the presence of a stretch reflex might
be disadvantageous to the performance of a controlled movement'.

In effect the IR introduces a hold phase during which servo action is
dampened perhaps to allow time for the contingency instructions to be sought.
It is significant that no ER was evoked if the loading stimulus had already
been withdrawn by the time the IR terminated, such as with stimuli of less
than 40-msec duration (fig. 8 b, NEWSOM DAVIS and SEARS, 1970]. To speculate
further, the possible need for, and the actual duration of, such a delay, should
be considered in the light of the concept of the duration of the minimal 'processing period' for the generation of a percept following receptor nerve stimulation. Through psycho-physical experiments it has been established that the
first experience (i.e., percept) of a stimulus does not occur before the end of
a minimal processing period of about 60-msec duration [see EFRON, 1966]. It
follows that if servo action is to be the servant of volition then the duration of
any constraints upon its action must be commensurate with this minimal processing period underlying perception. As described above, the resumption of
servo control (in accordance with prior instruction) occurs at about 60 msec,
the earliest time at which an unexpected load could be expected to be perceived
through proprioception. However, from studies of minimal reaction times it
is known that willed movement in response to perception of visual, auditory
or cutaneous stimuli does not occur until considerably later, in fact with a
latency of about 140 msec in the case of the intercostal muscles. Nevertheless,
the reflex action which is allowed, or not allowed, after a delay of about 60 msec
is appropriately matched in sign and intensity to conform with the willed
movement that follows perception of the unexpected load. This harmonious
blending of reflex and voluntary illustrates how the conscious nervous system
uses past experience, i.e. learning, to optimize present behaviour.

These considerations linking the servo control of movement with load
perception invite a final speculation. The spinal interneurones mediating
autogenetic inhibition would, under centrifugal control as proposed above, be
the first central neurones to detect the tension error signal. Thus they could
play a role in the perception of unexpected loads, but not directly, for the
evidence at present is against the involvement of Golgi tendon organs in proprioception [see MERTON, 1964; SEARS, 1971]. However, they could operate

indirectly, perhaps by signalling to higher levels and exerting a control over afferent signals which project into consciousness from the periphery, or that occur in relation to the 'sense of effort' through which the required motor action is estimated [MERTON, 1964].

In conclusion, it is suggested that in normal *conscious* man the muscle spindle dependent stretch reflex mechanism is not allowed automatically to compensate for unexpected loads, as conceived in the servo theory, due to the operation of autogenetic inhibition. As a corollary, in the execution of learnt movements against predictable loads, the spindle input constitutes a 'learnt' input to the α motoneurones dependent on a programmed pattern of command signals fed in α-γ linkage to the respective motoneurones.

Author's address: Dr. T.A. SEARS, Department of Neurophysiology, Institute of Neurology, Queen Square, *London, WC 1 3 BG* (England)

New Developments in Electromyography and Clinical Neurophysiology,
edited by J. E. Desmedt, vol. 3, pp. 418–427 (Karger, Basel 1973)

Coordination of Posture and Movement[1]

G. L. GOTTLIEB and G. C. AGARWAL

Biomedical Engineering Department, Rush-Presbyterian-St. Luke's Medical Center,
and College of Engineering, University of Illinois at Chicago Circle, Chicago

The myotatic reflex traverses the largest and fastest of the peripheral nerve fibers, both sensory and motor, and but a single synapse within the spinal cord. It is easily elicited by a brisk tap to a tendon and it possesses an electrical counterpart, the Hoffmann reflex (H-reflex) which is also easily evoked in man in m. soleus [HOFFMANN, 1922; PAILLARD, 1955]. These facts have made it one of the most exhaustively studied of the spinal mechanisms. However, it is only within the past decade that we have begun to understand how the reflex is actually able to play the important role in motor coordination that so long has been suggested for it.

Questions arise from the fact that, although it is trivial to excite a reflex muscle twitch by tendon tap or mild electrical shock, such stimuli are really 'tricks' we play on the nervous system to cajole and coerce it into revealing its secrets. Both stimuli are so brief (a tendon tap lasts 10–20 msec and a shock less than 1.5 msec) that they are vanished long before any muscle response appears (the latency of the EMG of the reflex is 30–40 msec). It seems unlikely that the forces of evolution would have taken the trouble to provide us with so powerful a mechanism for causing a muscle to contract in response to stretch, if it were to be activated in only this manner. [For discussion of other features of active stretch reflexes see HAGBARTH; BURKE and LANCE; HOMMA, this volume.]

In the paragraphs that follow, we shall present evidence, both from our own and from other laboratories, to support a hypothesis which explains the apparent contradiction between what we expect the important functions of the

1   This work was partially supported by an NSF grant GK-17581 and by a Presbyterian-St. Luke's Hospital grant-in-aid.

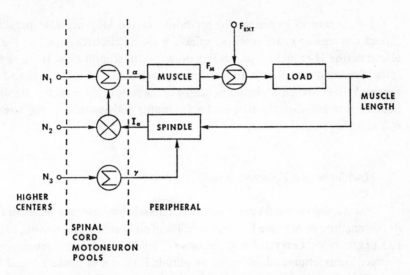

Fig. 1. A simplified block diagram of the peripheral motor system. Three classes of controlling signals from higher centers are shown: N_1 signals directly excite the α-moto-neuron pools, N_2 signals regulate the sensitivity of the monosynaptic reflex arc and N_3 signals determine fusimotor activity which influences spindle performance.

myotatic reflex to be, and what we have been able to demonstrate for it. The hypothesis is that the sensitivity of the myotatic reflex, that is the threshold and amplitude of response to a given stimulus (in engineering terms, the loop gain), is one of the controlled variables of the CNS. The reflex is continuously modulated, turned on and turned off, facilitated and inhibited, in accord with the ongoing demands of both posture and intentional movements.

Figure 1 will be useful in providing a basic orientation to the subsequent discussion. It shows a simplified block diagram of the myotatic reflex arc. Voluntary activation of the muscle can be accomplished either by direct excitation of the α-motoneuron pools (N_1), or excitation by way of the fusi-motor fibers (N_3), the so-called 'γ loop'. Both pathways are known to be capable of eliciting muscle contraction and under appropriate circumstances are known to be responsible for initiating contraction [ELDRED, GRANIT and MERTON, 1953; GRANIT, 1970]. In addition, we show a third centrally orig-inating input (N_2) which is meant to denote a site for central control over the sensitivity of the α-motoneuron pools to afferent signals. No denial of the monosynaptic nature of this pathway is intended, nor any implication as to the mechanism of this control, for such information is not yet available to us.

Two means of experimental excitation of the loop are also possible. Stretch (increasing x) activates the spindle while an electrical shock to the I_A afferents (the H-reflex) by-passes this organ. Both stimuli must traverse the spinal cord and therefore both may provide a measure of reflex excitability. However, the H-reflex is a better indicator of the state of the spinal excitability because it is less directly affected by fusimotor activity than is the stretch reflex.

Modulation of a Voluntary Agonist

A sustained voluntary contraction of a muscle increases the sensitivity of its motoneuron pool to the I_A afferent volley of an H-reflex [HOFFMANN, 1922; PAILLARD, 1955; GOTTLIEB and AGARWAL, 1971 b]. The static relationship between contraction and the reflex amplitude is shown in figure 2A and we use it to define a baseline against which to observe transient influences in the relationship which occur during phasic movements.

We can observe three phenomena during a phasic voluntary effort. Muscle tension (measured as foot torque for rotations about the ankle joint) is the ultimate mechanical output. The EMG is a useful measure of the aggregate α-motoneuron activity producing that tension, and the amplitude of the H-wave (the EMG component of the H-reflex) provides a relative indication of the sensitivity of the I_A afferent-α efferent pathway. We use the term 'effort' rather than 'movement' to describe what is a more or less isometric contraction. Strapping the foot to a rigid plate is probably insufficient to prevent movement from being sensed by the spindles but there is no gross rotation of the ankle joint. This is extremely important if one wishes to compare amplitudes of EMG responses because simple joint rotation will produce large changes in the tension and EMG relations. The origin of these changes is still unknown [INMAN, RALSTON, SAUNDERS, FEINSTEIN and WRIGHT, 1952; LIBERSON, DONDEY and ASA, 1962].

If a person exerting a steady dorsal effort suddenly makes a rapid plantar flexion, the two earliest signs of this act are a rise in the amplitude of the H-reflex in the soleus muscle [KOTS, 1969; COQUERY and COULMANCE, 1971] and a decrease in the anterior tibial EMG [HUFSCHMIDT and HUFSCHMIDT, 1954]. The former effect is observed 50–100 msec before, and the latter effect about 50 msec before the appearance of the soleus EMG. This increase in reflex excitability is also observable in the tendon jerk [GURFINKEL, KOTS and SHIK, 1965; GURFINKEL and PAL'TSEV, 1965; COQUERY and COULMANCE, 1971].

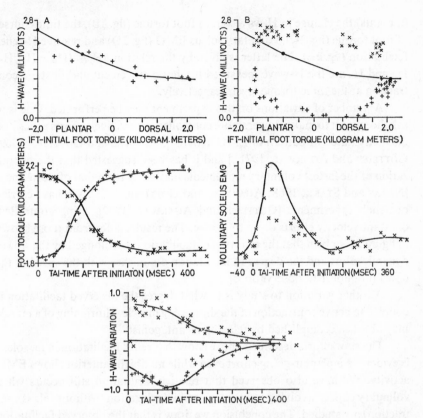

Fig. 2. Variation in H-wave in brisk effort. *A* The variation in H-wave with static initial foot torque (IFT). *B* The H-wave measured during brisk step efforts. In this and subsequent parts, x points are measurements made during plantar flexion efforts and + points are measurements made during dorsiflexion efforts. *C* The foot torque, measured at the instants when the H-wave was recorded, plotted *versus* time after initiation (TAI) of the movements. *D* The voluntary soleus EMG, rectified, filtered and averaged over 20 msec just before the stimulus. *E* H-wave variation during tracking plotted against the time after initiation of the movement. Positive values indicate facilitation and negative values indicate inhibition.

In a brisk effort, the degree of facilitation is extremely strong and reaches a maximum before the appearance of the soleus EMG [GOTTLIEB, AGARWAL and STARK, 1970]. Subsequently, a burst of EMG activity appears in the soleus and contraction follows. During the dynamic phase of the contraction, the relative facilitation of the H-reflex diminishes and is reduced to the static level by the completion of the effort. This is illustrated in figure 2 which shows (x points) the change in H-wave versus foot torque (fig. 2B), the time courses of foot torque (fig. 2C), voluntary soleus EMG (fig. 2D) and relative H-reflex facilitation (fig. 2E). The latter is given by the relation $\Delta H = (H_p - H_s)/H_s$; H_p and H_s are the H-wave measured in phasic movement and in static contraction at the same torque level, respectively.

A number of variations on this experiment may be performed. One is to ask the subject to make slower, more controlled efforts. Such efforts generate a distinctly different pattern of EMG activity [AGARWAL and GOTTLIEB, 1969; GOTTLIEB and AGARWAL, 1971a] and it has been suggested that the organization of the fastest voluntary movements may differ from that of slower ones [NAVAS and STARK, 1968; AGARWAL and GOTTLIEB, 1971]. We have carried out such experiments [GOTTLIEB and AGARWAL, 1972], using controlled, constant velocity efforts of up to 1.5 sec. The results differ from those shown in figure 2E only in that the duration of facilitation is prolonged by the slower dynamic effort and the degree of facilitation diminishes with the speed of the effort [unpublished observations].

Another variation to study is to what degree the observed facilitation is coupled to active contraction of the shortening muscle. Shortening of a muscle may also be accomplished by relaxing its antagonist.

The slow efforts described above show that reflex facilitation in m. soleus is evident in a plantar-going effort even while m. tibialis anterior shows EMG activity. We have also observed that relative facilitation still occurs when voluntary efforts involving only anterior tibial relaxation (without soleus contraction) are studied. The conclusion we draw is that the observed facilitation is not simply a phenomenon associated with activation of a motoneuron pool, but is a manifestation of a parameter (the reflex loop gain) which is independently regulated by the CNS.

Modulation of a Voluntary Antagonist

In the antagonist of a voluntary movement, the H-reflex is inhibited during, but not prior to the activation of the voluntary agonist [KOTS, 1969; GOTTLIEB, AGARWAL and STARK, 1970; COQUERY and COULMANCE, 1971]. Figure 2 shows these effects (+ points). The amplitude of the H-wave *versus* foot torque is shown in figure 2B, the time course of the effort in figure 2C and the relative inhibition in figure 2E. The EMG of the voluntary agonist m. tibialis anterior is not shown but does not differ significantly from the pattern of figure 2D.

As before, neither slowing the effort nor using only voluntary antagonist relaxation alters the basic pattern of this effect.

It is significant that reduction of the H-reflex in m. soleus is evident even while that muscle still shows EMG activity. This implies that the reduction in the reflex amplitude is not accomplished by inhibition of the soleus moto-neuron pool but by presynaptic or by dendritic mechanisms [RALL, BURKE, SMITH, NELSON and FRANK, 1967]. Contraction of m. tibialis anterior does enhance the degree and duration of H-reflex diminution but the reduction is apparent, even in the absence of such activation. Our conclusion here is that collaterals from the α motor fibers of the shortening muscle, the I_A afferents of the shortening muscle activated over the γ loop and the I_B afferents of the lengthening muscle may all act to inhibit the H-reflex in the lengthening muscle, but at least part of the observed effect must originate centrally. The function of this inhibition is presumably to prevent reflex opposition to a voluntary movement.

Modulation at Other Sites

It is not necessary to look only at the active joint of a voluntary movement to observe modulation of reflex sensitivity. One of the longest known mani-festations of modulation is the Jendrassik maneuver, a common clinical technique for demonstrating intact reflexes in seemingly areflexic patients. The maneuver usually consists of a vigorous contraction of the arm or hands just as the reflex is elicited.

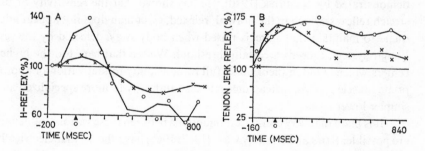

Fig.3. Relative modulation of reflex sensitivity during a Jendrassik maneuver (0 points) and an inverse Jendrassik maneuver (x points): *A* H-reflex. *B* Tendon jerk reflex. Each point is the average of 6 responses. Start of maneuver (at time = 0) detected from force measurements.

We have examined the influence of a variety of isometric contractions on both H-reflex and tendon jerk reflex. Figure 3 A shows the effects of two types of efforts on the H-reflex. These efforts are (1) a quick, sustained pull by the two opposed arms, with the hands grasping a U-shaped bar and (2) the inverse of that maneuver, i.e. a sudden relaxation of the previously contracted, opposed arms. Figure 3 B shows the effects of these two types of efforts on the tendon jerk reflex.

The common effect of these efforts (and of plantar flexion or dorsiflexion of the contralateral foot) is a brief, 400 msec period of considerable facilitation of the reflexes at the spinal cord. Following this interval, the spinal pathway shows a period of inhibition but the tendon jerk shows continued facilitation. This prolonged facilitation must be mediated by fusimotor activation of the spindles.

A voluntary movement, by producing forceful interactions with the external world, is likely to disturb the body's posture. The precise nature of these disturbances are not necessarily easy to predict, depending as they do upon complex physical relations between the limbs, the body and the object of our effort. If we recognize that a generalized facilitation of stretch reflexes is an effective way of providing a more rigid skeletal framework, the non-specific effect of these diverse efforts becomes quite reasonable. Making the reflexes more sensitive to stretch prepares the body for a possible, even likely, but not completely predictable disturbance without extraneous or possibly inappropriate recruitment of muscular effort. If and when these disturbances are realized, low level peripheral postural mechanisms are tuned and alerted to respond.

A related manifestation of these physiological mechanisms has been demonstrated by NASHNER [1970]. He has shown that the sensitivity of the stretch reflexes, which in the normal, relaxed, erect man does not appear adequate to maintain that state, is raised when body sway, detected by the vestibular system, exceeds a certain threshold. We see that here too the higher centers which could in theory take full responsibility for activation of appropriate muscle groups, instead delegate that authority to more specialized and simpler lower levels.

It is plain that very complex interactions between these various influences are possible. KOTS and MART'YANOV [1967, 1968] have shown that electrically elicited vestibular stimulation in man will facilitate the soleus H-reflex. They subsequently showed [KOTS and MART'YANOV, 1968] that a simultaneous voluntary activation of m. soleus, or of its antagonist m. tibialis anterior appears to 'disconnect' this vestibular influence.

We may tentatively conclude from this is that there appear to be a sequence of hierarchical neural mechanisms, each capable of adjusting and tuning those below it. All these higher centers must ultimately exert their influence at the periphery. Considered in this way, it may be more appropriate to consider the basic peripheral element of the motor system as, not the motor unit or the 'final common pathway', the α-motoneuron, but the entire system of spinal reflex loops.

An Hypothesis on Voluntary Motor Organization

Investigators of the myotatic reflex in animal preparations have generally sought to stabilize spinal parameters by preventing transient, spontaneous influences descending from above. Such studies inevitably tend to emphasize the static properties of the reflex, as a mechanism for maintaining a centrally determined muscle length in the face of external disturbances. What we have described above are the strategies employed in recruiting this mechanism into aiding and abetting precisely what it seemed it was intended to prevent, i.e. change of muscle length.

MERTON's speculative servo hypothesis [1953] provided a framework for much of our subsequent thinking over the last 20 years as to the functional role of this mechanism under dynamic, time varying conditions. GRANIT [1955] expanded upon this idea to postulate an α-γ linkage for cooperative recruitment of the α-motor and fusimotor pathways.

VALLBO [1970] has shown that during voluntary isometric contraction, the spindle primaries are active. This must certainly imply concomitant fusimotor activity and, since we have seen that during such efforts the sensitivity of the α motor pools to these afferent signals is heightened, the functional significance of this reflex loop in the regulation of normal voluntary movements is almost certainly considerable [see also C. D. MARSDEN, this volume].

We must be very cautious, however, in interpreting these findings in conventional servo-analytic terms. We generally think of three benefits accruing from increasing the gain of a closed-loop system. These are a reduction in the steady state error, a faster transient response and reduced sensitivity to changes in system parameters. The costs for these improvements are likely to be an increase in energy expenditure and a greater tendency towards instability.

The problem of instability is a very real one. The pathological phenomenon of clonus, observed in people with spasticity, is probably caused by the uncontrolled elevation of the closed-loop gain resulting in a 'relaxation oscillator' type of behavior. It is interesting to observe that a person can voluntarily stop these oscillations by contracting the offending muscle [AGARWAL and GOTTLIEB, 1972] which reduces the loop gain at the muscle [GOTTLIEB and AGARWAL, 1971a].

It is very difficult, however, to evaluate the relative importance of the other benefits and costs. It is likely that reducing energy expenditure is a far more important design parameter in the evolution of the motor system than it is in most engineering design tasks. But such a criterion as steady state accuracy is probably of little importance since there are other feedback loops operating through our awareness, e.g. visual and tactile feedback, which would play an overriding role.

The speed of the transient response is also not necessarily an important factor. During voluntary movements, any such benefits afforded by increased gain could be achieved equally well, in theory, by suitable alterations (from above) in the activation patterns of the α and γ-motoneuron pools themselves.

We might consider rather, that the myotatic reflex be looked upon as a short-term local representative from the distant higher centers. In repose, the reflex is not adequately sensitive to respond to any but the most abrupt disturbances. When a voluntary movement is made, however, a series of predictive commands descends the cord. The α-motoneuron pool of the agonist muscle is activated in a manner anticipated to produce the necessary forces. More or less simultaneously, the fusimotor fibers activate the agonist spindles according to the expected amount of muscle shortening. The primary afferents remain active throughout the movement and their sensory input, acting through the elevated gain setting of the reflex path, contributes to proper behavior from the α fibers.

If these predictions of force and shortening are in error, however, simple compensations may be initiated at the segmental level far more rapidly than they can be generated at the clever but remote higher centers. These compensations involve nothing more than the recruitment or release of parts of the agonist α-motoneuron pool. This first line of defense against flawed commands must suffice until the error is detected and more suitable commands received from above.

At joints removed from the actual movement, reflex sensitivities are also briefly raised so that such disturbances to our postural equilibrium as may be caused by the voluntary movement will be detected and opposed at the lower levels of the neural hierarchy.

Looked at in this manner the myotatic reflex is recruited, by need or anticipated need, to detect and provide regulation for disturbances in planned motor activity. It acts to provide a simpler, low level of control in brief intervals before higher level control can be asserted.

Author's address: Dr. G. L. GOTTLIEB, Biomedical Engineering Department, Presbyterian-St. Luke's Hospital, 1753 West Congress Parkway, *Chicago, IL 60612* (USA)

Reflex Effects of Muscle Vibration

New Developments in Electromyography and Clinical Neurophysiology, edited by J.E. Desmedt, vol. 3, pp. 428–443 (Karger, Basel 1973)

The Effect of Muscle Vibration in Normal Man and in Patients with Motor Disorders

K.-E. HAGBARTH

Department of Clinical Neurophysiology, University Hospital, Uppsala

I. Afferent Inflow
Induced by Muscle Vibration in Normal Man

A. Vibration-Induced Spindle Discharge Imitating the Effect of Fusimotor Activation

As shown in recordings from muscle afferents both in cat [GRANIT and HENATSCH, 1956; BIANCONI and VAN DER MEULEN, 1963; BROWN, ENGBERG and MATTHEWS, 1967] and man [HAGBARTH and VALLBO, 1968] the primary endings of the muscle spindles are easily excited by means of vibration applied to the tendon of a muscle. Thus, one can use the vibration technique to evoke a steady high-rate discharge in the stretch reflex afferents, that imitates a powerful static response in these fibres. As judged by recordings from muscle nerves in healthy subjects, passive sustained elongation of a relaxed muscle is not sufficient to produce any potent static spindle discharge. The vibration-induced discharge rather imitates the static fusimotor activation of the primary endings that normally occurs under isometric voluntary contraction [HAGBARTH and VALLBO, 1968].

A tendon tap in man normally evokes a massive synchronous afferent volley from the spindle primaries, resulting in a phasic stretch reflex with accompanying transient reciprocal inhibition of antagonistic motoneurons. With the vibratory stimulation one can avoid such dynamic compulsory motor responses and instead investigate how a sustained inflow from the stretch receptors, similar to that occurring during voluntary contractions, affects muscle tone.

B. Technique for Vibratory Stimulation
of Spindle Primaries

According to BROWN, ENGBERG and MATTHEWS [1967], virtually all
spindle primaries in the cat soleus muscle can be selectively driven by high
frequency vibration of very low amplitude (about 50 μm), longitudinally applied to the soleus tendon. In the intact man it is not possible to obtain such an
efficient and selective stimulation. Regardless of whether one uses vibrators
with large contact areas over the muscle bellies [e.g. DE GAIL, LANCE and
NEILSON, 1966] or smaller vibrators pressed against the muscle tendon
[EKLUND and HAGBARTH, 1966] vibration amplitudes of 0.5–1 mm (peak-to-
peak displacement, perpendicular to the contact surface) are required to get
a sufficient spread of the vibration waves within the large human extremity
muscles. It has been shown in recordings from whole muscle nerve fascicles
that with a small vibrator securely fastened to the muscle tendon the vibration-
induced barrage of impulses from the intramuscular stretch receptors increases
in strength with increasing amplitude (up to 2 mm) and increasing frequency
(up to 200/sec) of the mechanical oscillations. Such recordings also show that
the efficiency of the vibratory stimulus increases with increasing passive elongation of the muscle [HAGBARTH and VALLBO, 1968] and that the barrage of
impulses abruptly ceases as soon as the vibrator is moved away from the
muscle tendon to, for example, neighbouring soft tissues or muscles or tendons on the opposite side of the limb. This does not imply that the stimulus
cannot spread to antagonistic muscle groups but it indicates that, as long as
the oscillation amplitude does not exceed 2 mm, tendon vibration is not a
potent stimulus for spindles in antagonistic muscles, in particular not when
these muscles are slack and short.

In cats, the upper frequency limit of the spindle primaries can be as high
as 800/sec [VON EULER and PERETTI, 1966] and vibration frequencies around
400/sec are often used to obtain optimal stimulation of spindle primaries with
minimal involvement of secondary endings and Golgi tendon organs [cf.
BROWN, ENGBERG and MATTHEWS, 1967]. Little is known about the upper
frequency limit of human primary spindle endings but it may well be lower
than in the cat. During weak or moderate isometric contraction, frequencies
up to 75–100/sec have been observed [VALLBO, 1970a, b]. Only a few recordings
have been made from single IA afferents during tendon vibration in relaxed
subjects. The units observed followed vibration frequencies up to 75/sec; at
higher frequencies they tended to respond to each second or each third beat
[unpublished observations]. Still, it seems that also in man the stimulation of

the primary becomes more selective with increasing vibration frequency. Thus, as recently shown [EKLUND, 1971 b] vibratory stimulation around 40/sec applied to the muscle belly tends to bring in autogenetic inhibitory effects, presumably depending upon a proportionally stronger involvement of secondary spindle endings and/or Golgi tendon organs.

It follows that a vibratory stimulus within the range of 100–200/sec and 1–2 mm applied to the tendon would be an appropriate stimulus if one wants to activate the primary endings in a human muscle with minimal involvement of other intramuscular stretch receptors and minimal spread of the stimulus to antagonists. Such vibration has also been found to be particularly potent in eliciting reflex autogenetic excitatory effects, i.e. a motor response of the type to be expected from the primary endings. Vibrators oscillating at about 150/sec with an amplitude of about 1.5 mm, fairly independent of the external load, are easily made of compact cylindrical DC motors equipped with an appropriately large excenter on the axis. Such small vibrators, attached transversally with rubber bands over human muscle tendons have the advantage of being easy to handle in clinical practice and are, therefore, well suited as standard tools for neurologists and physiotherapists interested in the clinical application of the vibration technique. A description of the physical properties of such vibrators has recently been given [EKLUND, 1971 a].

C. Afferent 'Busy Line' and Presynaptic Inhibition

Great attention has been paid to the fact that tendon jerks and phasic stretch reflexes are suppressed during muscle vibration [DE GAIL, LANCE and NEILSON, 1966; HAGBARTH and EKLUND, 1966a; LANCE, DE GAIL and NEILSON, 1966; RUSHWORTH and YOUNG, 1966; DELWAIDE, 1971 and this volume; LANCE et al., this volume]. A main reason for this suppression may well be that the primary endings are so engaged by the vibratory stimulus that they are not able to respond efficiently to a transient superimposed muscle stretch. In fact, multi-unit recordings from human muscle nerves, the afferent response to a tendon tap, which normally appears as a distinct shower of impulses, stands out quite vaguely against the barrage of impulses produced by vibration [unpublished observations].

It seems probable that an afferent 'busy line' phenomenon also contributes to the suppression of the H-reflexes during vibration, but there is also evidence that the vibration-induced afferent inflow may cause spinal presynaptic inhibition of the monosynaptic pathway [LANCE, NEILSON and TASSINARI, 1967; GILLIES, LANCE, NEILSON and TASSINARI, 1969; YAMANAKA, 1964; DELWAIDE, 1971].

D. The Effect of Vibration Depending upon Muscle Length and State of Contraction

The effect of the vibratory stimulus upon the intramuscular receptors does not depend only on the characteristics of the vibrator itself and the way it is applied. As mentioned above, the initial length of the muscle stimulated is an important factor since the spindle response to the stimulus increases with increasing passive elongation of the muscle [GRANIT and HENATSCH, 1956; HAGBARTH and VALLBO, 1968]. The relative insensitivity of the endings in a passively shortened muscle can, however, be compensated for by a weak voluntary contraction, presumably indicating that fusimotor drive then helps to sensitize the spindles. It is still questionable whether under normal conditions there may also occur changes in fusimotor tonus in a completely relaxed muscle, i.e. changes that can alter spindle sensitivity without accompanying EMG signs of muscle activity. Recent microneurographic findings in healthy alert subjects indicate that normally the α–γ linkage is quite firm [HAGBARTH and VALLBO, 1968; HAGBARTH and VALLBO, 1969; VALLBO, 1970a, b] and that during complete relaxation (EMG silence) the primary endings have little or no fusimotor bias [WALLIN, HONGELL and HAGBARTH, this volume].

It should also be pointed out that as the muscle starts to develop reflex tension in response to the stimulus, the muscular 'stiffening' probably helps to enhance the spread of the vibration waves to more distant spindle endings in the muscle belly. On the other hand, on the assumption that the vibration-induced contraction is of a pure α type, the extrafusal shortening will tend to unload the spindles, and this might reduce their sensitivity to the externally applied oscillations [MATTHEWS, 1967; cf. BROWN, ENGBERG and MATTHEWS, 1967]. Finally, the active tension must be presumed to stimulate Golgi tendon organs [HOUK and HENNEMAN, 1967; cf. STUART, MOSHER and GERLACH, 1971] and with their autogenetic inhibitory effect they will tend to curtail the autogenetic excitatory effect from the spindles.

These examples of secondarily induced changes in the sensory inflow during continued vibration should be kept in mind as we now proceed to describe the motor and perceptual effects of muscle vibration in healthy subjects and in patients with various types of central motor disorders.

II. Motor and Perceptual Effects of Muscle Vibration in Normal Man

A. The Tonic Vibration Reflex (TVR)

The vibration-induced afferent inflow from the spindles imitates the effect of fusimotor activation; in a similar way, the TVR imitates the tonic stretch reflex normally operating during voluntary isometric contractions. Thus, in all healthy subjects, adequately applied muscle vibration within the range of 100–200/sec causes a reflex change of muscle tonus that is restricted to the muscles vibrated and their antagonists and that accords with the pattern of the stretch reflex: autogenetic excitation and reciprocal inhibition. These excitability changes can be demonstrated in various ways.

1. Subject Remaining Passive During the Test

There are now many reports confirming that a vibratory stimulus of the standard type applied to an initially relaxed skeletal muscle tends to produce an involuntary tonic contraction in this muscle [e. g. RUSHWORTH and YOUNG, 1966; MARSDEN, MEADOWS and HODGSON, 1969] a contraction that increases in strength with increasing muscle length [EKLUND and HAGBARTH, 1966; JOHNSTON, BISHOP and COFFEY, 1970] and that stays as long as vibration continues and the subject remains passive. Under 'isotonic' conditions when the muscle is free to shorten against gravity, a slow joint movement occurs as an increasing number of motor units are gradually recruited. It often takes 15–30 sec before a steady contraction is reached. If then some extra load is applied that tends to stretch the muscle, the contraction immediately increases like a true 'load compensating response' [HAGBARTH and EKLUND, 1966a]. The time course of the TVR is different under isometric conditions when muscle shortening is prevented by joint fixation. The plateau contraction is reached more rapidly and often in two stages, an initial rapid rise of tension is followed by a second slower phase [EKLUND, to be published].

The slow development of the TVR during isotonic conditions deserves some comment. It is hard to exclude that it may be due, in part, to a gradually increasing efficiency and spread of the vibration waves in the muscle as the active tension slowly increases during shortening. This cannot be the whole explanation, however, for a similar slowly rising contraction also appears in response to selective electrical repetitive stimulation of the group I muscle afferents [DE GAIL, LANCE and NEILSON, 1966; LANG and VALLBO, 1967; MARSDEN, MEADOWS and HODGSON, 1969]. The slow rise of the response has been considered as evidence that the tonic stretch reflex belongs to the polysynaptic reflexes [GRANIT, 1970] and the following sentence of GRANIT is, in this connection, worth quoting: 'It seems likely that in man the stream of impulses from vibrated primaries is hitting an "unprepared" central organization for mobilizing "tone", thus providing the experimenter with a unique opportunity of seeing a process *in statu nascendi* which otherwise he encounters only as a fully developed state in the decerebrate animal.'

It is not only the development of the TVR that is slow under isotonic conditions; the motor effect of the stimulus declines also slowly as vibration suddenly. When the contraction itself has ceased, a fascilitatory effect still remains, as shown by the fact that a second period of vibration, initiated after a pause of 10–15 sec, gives a more rapid rise of tension than did the initial vibration reflex is in a similar way enhanced in muscles which have just relaxed after a period of strong isometric contraction [EKLUND and HAGBARTH, 1966].

The H-reflexes provide no reliable index of the motoneuron excitability changes occurring during vibration [e. g. BARNES and POMPEIANO, 1970; cf. also COOK, NEILSON and BROOKHART, 1965]. As previously mentioned, the H-reflexes are suppressed in the muscle stimulated, and also in the antagonists [HAGBARTH and EKLUND, 1966a; RUSHWORTH and YOUNG, 1966]. The latter antagonistic effect, cannot be due to an afferent 'busy line' phenomenon but 'presynaptic inhibition' might be involved [LANCE, NEILSON and TASSINARI, 1967; DELWAIDE, 1971] besides a true reciprocal inhibition of the antagonistic motoneurons. The vibration-induced reciprocal inhibition is responsible for the fact that when two identical vibrators are simultaneously applied to two antagonists, the effects tend to cancel each other [LANCE, DE GAIL and NEILSON, 1966].

Two identical, simultaneously running vibrators are useful also when testing the symmetry of the vibration reflexes. Normally, the vibration reflexes are of similar strength in the right and left limbs but asymmetries can easily be induced by caloric vestibular stimulation or merely by passively turning the head of the subject to one side. In other words, the tonus shifts induced by tonic neck reflexes and tonic vestibular reflexes are brought to light with the vibration technique [EKLUND and HAGBARTH, 1966]. A more generalized enhancement of the TVR occurs when the subject is cold, and the Jendrassik's manoeuvre also has a certain potentiating effect [EKLUND and HAGBARTH, 1966; MARSDEN, MEADOWS and HODGSON, 1969]. Warming the subject [EKLUND and HAGBARTH, 1966], and drugs such as barbiturates and diazepam, give a suppression of the vibration reflexes [LANCE, DE GAIL and NEILSON, 1966]. It should be noted that many of these manoeuvres which affect the strength of the tonic reflexes leave the responses to tendon taps unaffected, indicating that the TVR can provide a better index of muscle tone than phasic myotatic reflexes.

2. Vibration Affecting Voluntary Effort Required to Activate the Muscles

The central excitability changes induced by muscle vibration are experienced by the subject as a rather peculiar phenomenon since sustained contractions and tonus shifts appear without any sense of effort. When he tries to move his limb during the test he notices that voluntary joint movements in one direction are aided by the vibration, whereas movements in the opposite direction meet with a certain 'resistance'. Vibration of the antigravity elbow flexors makes the forearm feel lighter than normal and voluntary contractions in the elbow flexors tend to become stronger than the subject intended, whereas contractions in the elbow extensors tend to become correspondingly weaker. This holds true for both sustained voluntary contractions and fast contractions occurring during rapid alternating movements [HAGBARTH and EKLUND, 1966a; EKLUND and HAGBARTH, 1966]. When the subject becomes aware of this 'falseness' in his motor control, however, he is fully able to compensate for it. Thus, maximal voluntary power is not affected by vibration and if the subject is allowed to watch the position of his arm, the force signal or the integrated EMG signals, he can easily adjust the strength so as to oppose the effect of the vibration. When the test is performed isometrically, however, he cannot compensate fast enough to avoid the initial rise of tension as vibration starts, nor can he avoid a short period of overcompensation when vibration suddenly stops [EKLUND and HAGBARTH, 1966].

The effects described mainly concern voluntary contractions in the large extremity muscles such as those which serve the purpose of keeping a limb lifted against gravity, exerting a pressure against an external resistance, or moving the limb up and down. It is noteworthy that other types of voluntary motor acts, like refined manipulatory finger movements under tactile and visual control, seem to be comparatively little affected by vibration of the prime movers [HAGBARTH and EKLUND, 1969]. Thus, most healthy subjects notice no difficulty in performing handwriting whilst a vibrator is operating on the finger (and hand) extensors, but they do notice that a greater central command is needed to flex a limb against gravity while the extensor muscles are vibrated. As recently shown [GOODWIN, McCLOSKEY and MITCHELL, 1971] the greater central command needed to contract an arm muscle while vibrating on an antagonist is reflected in enhanced heart rate and ventilation.

B. Vibration-Induced Falling (VIF)

Quite different and more spectacular motor responses appear when vibration is applied to certain leg and trunk muscles in blindfolded subjects standing with the feet together, as during the Romberg test [EKLUND, 1969]. Vibration symmetrically applied over the erector spinae muscles, the knee flexors, the calf muscles and their antagonists is particularly efficient in producing involuntary displacements of the center of gravity, so that the subject tends to fall forwards or backwards depending upon the site of the stimulus. These falling reactions induced by vibration of muscles actively engaged in the maintenance of equilibrium cannot be explained, like the TVR responses, in terms of tonus shifts restricted to the muscles vibrated and their antagonists. It seems more likely that vibration-induced muscle afferent signals arriving at supraspinal equilibrium centers are responsible for the unintentional displacement of the center of gravity. The 'false' setting of the equilibrium centers can, like the TVR responses, be compensated for by voluntary effort [cf. EKLUND and LÖFSTEDT, 1970], and when the subject keeps his eyes open during the test the falling reactions are greatly reduced in strength.

C. Vibration and Position Sense

When the TVR is allowed to develop under isometric conditions, it often happens that a blindfolded subject gets the illusion that a slow joint movement occurs, while in fact the joint is kept quite still. Such illusions may be quite strong, even though in some subjects they seem to be totally missing. GOODWIN, MCCLOSKEY, MATTHEWS and MITCHELL [1971], describe how a vibration-induced isometric contraction in the elbow flexors leads to a false perception of the arm extending itself.

It has been shown in both the lower [EKLUND, 1972] and the upper extremities [GOODWIN, MCCLOSKEY, MATTHEWS and MITCHELL, 1971] that the subjects fail to perceive the full extent of an involuntary vibrationinduced position or movement. These findings are compatible with the idea that the spindle discharge set up by vibration is wrongly perceived as being due to a stretch or load of the muscle stimulated, thus giving a tendency to underestimate the vibration-induced muscle shortening.

III. Muscle Vibration in Patients with Central Motor Disorders

Different types of abnormal motor responses to muscle vibration of the standard type are encountered in patients with central motor disorders. Various motor syndromes (spasticity, weakness, rigidity, tremor, hyperkinesia) can be accentuated by muscle vibration and, thus, the vibration technique can be used for diagnostic purposes to reveal mild or latent motor dysfunctions [HAGBARTH and EKLUND, 1966b]. In some patients with spastic pareses, however, appropriately applied vibration can also be used to improve motor power and active range of movement [HAGBARTH and EKLUND, 1966b, 1969] and other beneficial effects of vibration has been described in children with cerebral palsy and disturbed 'body image' functions [EKLUND and STEEN, 1969]. In the following section, particular attention will be paid to these possible diagnostic and physiotherapeutic applications of the vibration technique. Some attempts will be made to interpret the outcome of the vibration tests in neurophysiological terms, even though, at the present stage of our knowledge, such interpretations are apt to be rather hypothetical. Generally speaking, however, the normal ability to compensate for the TVR phenomenon and the ability to carry out accurate manipulatory finger movements while a vibrator is operating on the prime finger movers indicates an efficient central command over the tonic proprioceptive reflexes and also commands which allow coordinated movements even though a heavy afferent 'noise' interferes with the normal sensory feedback from the muscles. Apparently, these compensatory functions are often missing in patients with central motor disorders.

A. Spastic Pareses

1. Strength and Time Course of the TVR
On the assumption that the exaggeration of the phasic stretch reflexes in spastic muscles depends upon a fusimotor sensitization of the primary spindle endings, one might expect such muscles to give abnormally strong responses to vibration. This is often but not always the case. Especially in spinal forms of spasticity it is not unusual to find that the TVR is weak or absent even though the tendon jerks are exaggerated [DE GAIL, LANCE and NEILSON, 1966; LANCE, DE GAIL and NEILSON, 1966]. This may be taken as an indication that the TVR utilizes central paths or motoneurons other than the monosynaptic reflex, but alternative interpretations are possible. One should recall that the vibratory

stimulus is not selective for the primary endings and thus an absent TVR in a spastic muscle may well be due to an abnormal release of vibration-induced autogenetic inhibitory reflexes or skin reflexes that oppose or conceal the TVR. In paraplegic patients with exaggerated withdrawal reflexes, the vibratory stimulus may cause a generalized flexor reflex that totally overrides or conceals any concurrent TVR response. Furthermore, 'inverse TVR responses' are occasionally seen [HAGBARTH and EKLUND, 1968, 1969]. In patients with grossly overactive stretch reflexes in the calf muscles, for instance, these muscles may contract in response to vibration of the antagonist ankle flexors, presumably due to spread of the vibratory stimulus.

In most cases of mild or moderate spasticity of the cerebral type, however, the spastic muscles respond more briskly to vibration than do healthy relaxed muscles. In particular, the vibration reflexes in the spastic muscles have a more abrupt onset [HAGBARTH and EKLUND, 1968; JOHNSTON, BISHOP and COFFEY, 1970; LANCE, BURKE and ANDREWS, this volume]. A tendon jerk followed by a silent period may appear as vibration suddenly starts and the EMG activity then rises rapidly. As in healthy subjects, the plateau level of the TVR, is quite dependent upon the initial passive length of the muscle: the vibration-induced tension increases with increasing muscle length [HAGBARTH and EKLUND, 1968; BURKE and LANCE, this volume]. When vibration suddenly stops, the tonic muscle activity usually ceases within 1 or 2 sec. On the whole, the vibration reflexes in human spastic muscles often have a time course similar to that described for the soleus muscle of the decerebrate cat [MATTHEWS, 1966]. It is still uncertain to what extent the abrupt onset depends on a higher degree of 'readiness' in the central paths mediating the tonic stretch reflex [GRANIT, 1970]. Influences from other receptors than the primary endings may also play a role in determining the time course of the TVR [GILLIES, BURKE and LANCE, 1971 a].

2. Abnormal radiation of the TVR

In many spastic patients the vibration reflexes spread to functionally allied muscles acting at neighbouring joints, a phenomenon which is not seen in healthy subjects. Thus, vibration applied on the volar side of the wrist may cause not only flexion and pronation movement of the hand in hemiplegic patients but also elbow flexion and adduction of the whole arm [HAGBARTH and EKLUND, 1968]. In some patients, the whole limb can be reflexly extended or flexed depending on whether vibration is applied to one of the flexor or the extensor muscles. Such reactions are hard to explain in terms of stimulus spread [LANCE and DE GAIL, 1965]; they are presumably due to an abnormal central spread of the excitatory (and reciprocal inhibitory) inflow from the spindles.

3. Vibration Affecting T- and H-Reflexes

As in normal man, muscle vibration suppresses the tendon jerks, probably depending in part on a peripheral 'busy line' phenomenon and in part upon 'presynaptic inhibition' of the monosynaptic path. According to LANCE, NEILSON and TASSINARI [1967], DELWAIDE [1971] and LANCE et al. [this volume], the autogenetic vibration-induced suppression of the T- and H-reflexes is less pronounced in spastic than in normal muscles, suggesting that 'presynaptic inhibition' is less efficient in the patients. This in turn, has led to the notion that a deficiency in these inhibitory mechanisms may contribute to the hyperreflexia in spasticity [DELWAIDE, this volume].

4. Vibration-Induced Clonus

In agreement with the rule that vibration tends to accentuate various motor dysfunctions, it can, in spastic patients, cause sustained rhythmical alternating contractions in antagonistic muscle groups, that very much resemble ordinary clonus [HAGBARTH and EKLUND, 1968; KANDA, HOMMA and WATANABE, this volume].

In patients exhibiting spontaneous clonus or clonus elicitable by fast passive joint movements, vibration has a clear potentiating effect that often remains for a while after the vibrator has stopped. Vibration can, however, induce clonus in spastic muscles when other more common provoking methods fail to do so. The clinical implications of this fact are limited, for vibration-induced clonus occasionally appears also in healthy subjects, especially in the soleus muscle [HAGBARTH and EKLUND, 1966a].

5. Vibration-Induced Compulsory Changes in Posture, Changes in Maximal Voluntary Power and Active Range of Movement

Many spastic patients have an impaired voluntary control over the tonic vibration reflexes [HAGBARTH and EKLUND, 1968, 1969]. There is no constant correlation between this impairment and the plateau strength of the TVR as tested during rest, and the impairment can also be quite pronounced in patients with no apparent initial signs of muscle weakness. When the phenomenon is pronounced, the vibration causes a compulsory change in posture that the patient cannot overcome, and it even happens that his voluntary attempt to counteract the vibration-induced movement results in an enhancement of the reflex and, thus, a movement opposite to that intended. A more sensitive test, however, is to apply vibration while the patient performs slow, alternating flexion and extension movements. Vibration of the flexors will then cause a contraction in these muscles that increases with increasing muscle length as the patient tries to extend the arm, and EMG recordings verify that at the same time he loses voluntary power in the antagonistic extensors so that full range of extension is prohibited. Extensor vibration may have exactly the reverse effect and prohibit a full-range flexion of the joint.

In patients with initial muscular weakness, the vibration is often about equally efficient in prohibiting active movements opposing the TVR pattern as it is in potentiating movements in the direction of the TVR. Again, there is no consistent relation between the strength of this potentiating effect and the strength of the TVR in itself, as tested during rest. Thus, in some patients with pronounced weakness and weak or absent vibration responses the combined effect of vibration and voluntary effort can result in muscle contractions which are quite powerful [HAGBARTH and EKLUND, 1966b]. At the same time as vibration potentiates voluntary power in the muscle vibrated, it helps to inhibit the antagonistic hypertonus, thus facilitating active movements in the TVR direction even further. The potentiating effect of vibration upon voluntary power often remains for 5–10 sec after the end of stimulation.

In typical cases of spastic hemiplegia the hypertonus prevails in the flexors of the arm and the extensors of the leg whereas the weakness prevails in the antagonists to these antigravity muscles. In such patients with the characteristic flexion posture in the arm, vibration of the elbow extensors can help to restore a more normal balance between the flexor and extensor tone and between flexor and extensor power. In other words, appropriately applied vibration can help patients of this type to 'break away' from the dystonic posture in which they tend to be locked and the active range of movement can be considerably increased.

Since the acute beneficial effects of muscle vibration can be striking, there is reason to believe that appropriately applied vibrators automatically turned on (for instance by EMG signals) to assist certain movement patterns can help certain patients in their daily activities [HEDBERG, OLDBERG and TOVE, 1967]. The practical applicability of such devices has, however, not yet been tested.

6. Long-Term Therapeutic Effects

After training sessions with vibration-supported active movements many patients report that for a few hours they feel more relaxed and less handicapped by their spasticity. Vibrators are now being tested in routine therapeutic work at various rehabilitation clinics and many physiotherapists seem convinced that repeated sessions of vibration treatment may give long-term beneficial effects [HAGBARTH and EKLUND, 1969]. KNUTSSON, LINDBLOM and ODÉN [1970] measured maximal voluntary power and range of movement before and after a 3-week period of vibration treatment in 23 hemiparesthetic patients. Their statistical analysis revealed no significant long-term improvement of maximal voluntary power, but a slight increase in range of movement. Still, the subjective effect as experienced by most of the patients and as reported by the physiotherapist was a marked improvement. Encouraging observations as regards long-term improvements of motor control after vibration treatment has been reported from screening tests performed on children with cerebral palsies [EKLUND and STEEN, 1969]. More information about the results of prolonged vibration treatment is needed, but even at this stage it seems that the muscle vibrators can be useful tools in the rehabilitation of selected patients.

7. Concluding Remarks on TVR in Spasticity

Much work remains to be done with muscle vibration tests in *spastic patients*. From a pathophysiological point of view it is of interest to note how the reciprocally organized TVR can either potentiate or relieve an abnormal spastic posture depending upon where the vibrator is applied. There is reason to believe that at least in some spastic patients there is an abnormal fusimotor-maintained discharge in the I_A afferents that contributes to the hypertonus and that also tends to lower the contraction power in the antagonists. It has been shown, for instance, that depression of the stretch reflex by local cooling can greatly increase the contraction power of the antagonists [KNUTSSON, 1970]. Local cooling of a spastic muscle combined with vibration over the antagonist should accordingly be a most efficient way of changing the balance between flexor and extensor power.

As yet, little is known about the clinical value of classifying spastic patients according to the way they respond to muscle vibration. Systematic quantitative studies of the various effects of muscle vibration described have to be carried out in clinically well-defined groups of patients with spasticity due to circumscribed lesions at various levels of the neuraxes. Animal experiments can give valuable hints as to which central pathways are involved in the control of the TVR. Thus, GILLIES, BURKE and LANCE [1971 b] have recently presented evidence that descending pathways of great importance for the maintenance of the TVR in the cat are the vestibulo-spinal and pontine-reticulospinal tracts.

By studying not only the TVR but also the electrically elicited tonic stretch reflex [LANG and VALLBO, 1967] in spastic patients, one might acquire information as to what extent the abnormal vibration responses in spasticity depend on an abnormal sensitivity of the muscle spindles. MARSDEN, MEADOWS and HODGSON [1969] recently used this technique in healthy subjects to show that the normal ability to suppress the TVR is not dependent upon changes in fusimotor outflow to the spindles. It is conceivable, however, that the potentiating effect of vibration upon voluntary power in spastic pareses depends not only on the synergetic effect of the reflex and voluntary activation of the motor units, but also on fusimotor induced sensitization of the spindle endings during the voluntary effort.

The vibration-induced falling reactions have not yet been thoroughly studied in spastic patients, but preliminary screening tests have revealed no obvious correlation between the strength of the calf muscle TVR (as tested on a sitting or lying subject) and the backward-falling reaction elicited by calf muscle vibration during standing.

B. Parkinsonian Rigidity and Tremor

LANCE, DE GAIL and NEILSON [1966] and HAGBARTH and EKLUND [1968] found TVR of normal strength and time course in the rigid muscles of patients with Parkinson's disease. Since the TVR is often superimposed upon a pre-existing hyperactivity in the rigid muscles, however, the plateau tension developed during vibration tends to be higher than in healthy subjects. HAGBARTH and EKLUND noted, for instance, that vibration on the palmar side of the wrist tends to cause a slow pronation movement which the patient may have difficulty in preventing voluntarily. Variable results were obtained in EMG recordings from rigid antagonists to the muscles vibrated. A vibration-induced reciprocal inhibition may be seen, but in other instances vibration causes a more widespread increase of rigidity involving also the antagonists.

As in spasticity, more striking effects of muscle vibration can be seen if the tests are preformed during the execution of willed motor acts [HAGBARTH and EKLUND, 1968]. Thus, the motor handicap exhibited by a parkinsonian patient trying to perform rapid alternating joint movements is often accentuated during vibration of one of the prime movers. Vibration of the wrist flexors and extensors during handwriting occasionally gives a striking effect, in that it potentiates or reveals a tendency for micrographia. The writing test

Table I. Abnormal motor responses to muscle vibration

1. *Tonus shifts* (excitation and rec. inhibition) *radiate* to muscles acting on neighbouring joints.	Pyramidal disorders with spasticity
2. *Fast development.* Clonus.	
3. *Maximal voluntary power affected.* Unable to compensate for the tonus shifts.	
4. Potentiation of *tremor* and *choreoathetosis.*	Extrapyramidal and cerebellar disorders
5. Coordination in skilled motor acts *(handwriting)* seriously affected.	

may also clearly reveal and document another effect of the vibration, namely its tendency to potentiate or uncover the parkinsonian tremor.

Vibration on the calf muscles during standing usually evokes a compulsory backward-falling reaction, which the patient cannot prevent even with the eyes open, and which is characterized by a lack of compensating arm and body movements to avoid falling. Vibration of the thigh flexors or the erector spinae muscles during walking may elicit compulsory running [unpublished observations].

C. Cerebellar Syndromes and Choreoathetosis

According to DE GAIL, LANCE and NEILSON [1966] and HAGBARTH and EKLUND [1968], cerebellar lesions are often accompanied by diminished or absent TVR, as tested during rest. In some patients with cerebellar syndromes, however, the vibration has a marked accentuating effect upon the 'intention' tremor, and muscle coordination in handwriting or during fast alternating movements (adiadochokinesis) may be seriously upset by the vibratory stimulus. In a few patients with cerebellar syndromes it has also been observed that even though the TVR in itself is weak, it can affect maximal voluntary power in a similar way as in spastic pareses.

Vibration on the calf muscles during standing with eyes open may, as in parkinsonian patients, cause compulsory backward falling reactions, but unlike the parkinsonian patients, those with cerebellar disturbances often exhibit a large variety of ineffective compensatory arm and body movements during the test.

Like parkinsonian tremor and cerebellar intention tremor, choreoa-thetotic movements in patients with, for instance, Huntington's chorea tend to be accentuated by vibration, primarily in the limb vibrated. Even in early stages of Huntington's disease, vibration can provoke marked disturbances in motor coordination during, for instance, the handwriting test.

Many of the vibration effects described in patients with *cerebellar* and *extrapyramidal disorders* are hard to interpret in strict neurophysiological terms. They are of clinical diagnostic interest, however, since they can serve to reveal mild or latent motor disorders. It is also of clinical interest to note that the effects of muscle vibration can be quite different in patients with very similar clinical syndromes. Thus, in some cases of cerebellar disorders, vibration reveals a latent 'intention' tremor, whereas in other cases with an overt tremor, vibration has little or no potentiating effect. Does this indicate that there are fundamentally different types of cerebellar 'intention' tremors and, if so, would it be of any clinical value to differentiate between them?

Table I gives a schematic summary of the main types of abnormal responses to vibration that any clinician equipped with a vibrator can find in patients with different types of central motor disorders.

Author's address: Prof. K.-E. HAGBARTH, Department of Clinical Neurophysiology, University Hospital, *S-71450 Uppsala* (Sweden)

New Developments in Electromyography and Clinical Neurophysiology, edited by J. E. Desmedt, vol. 3, pp. 444–462 (Karger, Basel 1973)

The Reflex Effects of Muscle Vibration

Studies of Tendon Jerk Irradiation, Phasic Reflex Inhibition and the Tonic Vibration Reflex

J. W. LANCE, D. BURKE and C. J. ANDREWS

Division of Neurology, The Prince Henry Hospital and The School of Medicine, University of New South Wales, Sydney

Clinical observations on the alteration of spinal reflexes by disease of the nervous system have preceded by almost a century the ability of the physiologist to explain them. The absence of a tonic stretch reflex in normal relaxed man and in the anaesthetized animal has hampered the physiologist in his attempts to investigate the basis of muscle tone. He has had to make a choice between studying tonic reflexes in an artificial preparation, the decerebrate animal, or exciting a synchronous volley through the monosynaptic reflex arc by electrical stimulation of muscle nerve or dorsal root in the hope that this would reflect the sustained activity of tonic mechanisms. That this is not always the case can be seen by the discrepancy between the briskness of the tendon jerks and postural performance in normal man as well as in patients with disorders of the cerebellum, basal ganglia or upper motoneurone.

A new tool for the investigation of tonic mechanisms in normal man and the intact cat has come to hand with the discovery that the primary spindle endings are extremely sensitive to vibration [ECHLIN and FESSARD, 1938; GRANIT and HENATSCH, 1956; BIANCONI and VAN DER MEULEN, 1963; CROWE and MATTHEWS, 1964] and that continued vibration of the muscle belly or tendon will induce a tonic contraction of the muscle vibrated. This tonic vibration reflex (TVR) thus acts as a model for the tonic stretch reflex (TSR), although deprived of group II afferent input because the secondary spindle ending is relatively insensitive to vibration [BROWN, ENGBERG and MATTHEWS, 1967].

The discovery of the TVR, the recognition of separate dynamic and static fusimotor fibres and the acquisition of further knowledge about the reflex effects of group II and group I$_B$ afferents now permit more precise analysis of spinal mechanisms than was previously possible. The stretch reflex arc can be analyzed in terms of a synchronous discharge such as the monosynaptic reflex (MSR), tendon jerk or H-reflex, and a continuing discharge, such as the TVR or TSR, both of which may be divided into dynamic and static components.

Reflex Irradiation

A normal subject with active reflexes, or a spastic patient, will respond to percussion of the Achilles tendon by contraction of the hamstrings as well as the calf muscles, and the hamstrings may also contract as the knee jerk is elicited. LOMBARD [1889] noted that 'not infrequently the flexors as well as the extensors contracted in response to the blow on the ligamentum patellae'. Percussion of one leg may cause contraction of muscles in the opposite limb as in the 'crossed adductor response', and a sharp tap on the radius or shoulder tip may be followed by a reflex contraction of all major muscle groups in the upper limb on the same side, and occasionally on the opposite side as well.

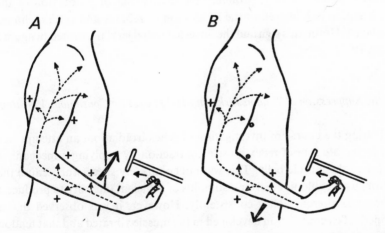

Fig. 1. A Muscle contractions produced by tapping the radius in a subject with brisk reflexes. The propagation of a vibration wave through the limb initiates reflex contractions (+) in biceps, triceps, brachioradialis, finger flexors and extensors. The limb moves in the direction of the stronger muscles so that the elbow and fingers flex, the reaction of a normal 'supinator' or radial jerk. *B* The mechanism of the 'inverted supinator jerk'. When reflex arcs employing the 5th and 6th cervical segments are interrupted by disease, the biceps and brachioradialis response to the vibration wave set up by radial tap is absent (o). Other muscles respond normally, causing the elbow to extend while fingers flex [from LANCE, 1970].

This phenomenon, known as reflex irradiation, was ascribed by SHERRINGTON [1898] to jarring of afferent fibres at, or just peripheral to, the dorsal root ganglion. In 1944, WARTENBERG suggested that the contraction of muscles remote from the site of impact was caused by the transmission of a mechanical insult through the body. However, this postulate did little to alter the prevailing view that reflex irradiation resulted from intraspinal dissemination of afferent impulses through a complex synaptic network.

Subsequent studies from this laboratory have shown that transmission through skeletal structures of a vibration wave set up by percussion of bone or tendon could adequately explain reflex irradiation [LANCE, 1965; LANCE and DE GAIL, 1965]. Interference with the reflex arc of the percussed muscle by ischaemia and by procaine infiltration did not abolish the irradiated reflexes, and the latencies of these irradiated reflexes were found to be delayed by the time taken for the vibration wave to spread from the point of percussion to each muscle.

Thus the propagation of the vibration wave is a means of imparting a sinusoidal stretch to muscle spindles at a distance from the point at which the stimulus is applied, and is probably the mechanism of production of the normal tendon jerk as well as the irradiation of reflexes and certain clinical signs such as Hoffman's sign and the 'inverted radial jerk' illustrated in figure 1 [LANCE, 1970].

The Suppression of Tendon Jerks and H-Reflexes by Continuous Vibration

During the course of investigation of reflex irradiation, an attempt was made to initiate a brief reflex muscle contraction by applying a mechanical vibrator to the muscle belly. This was not completely successful because the vibrators used could not deliver a stimulus of sufficient intensity to produce a synchronous afferent volley consistently. However, it was observed that a sustained reflex contraction developed in the muscle vibrated and that tendon jerks or H-reflexes were suppressed or abolished while the vibrator was applied to the appropriate limb, as shown in figure 2 [LANCE, 1965; DE GAIL, LANCE and NEILSON, 1966]. Reflex suppression was subsequently confirmed by RUSHWORTH and YOUNG [1966], MARSDEN, MEADOWS and HODGSON [1969] and DELWAIDE [1969, 1971]. After cessation of vibration, the amplitude of the tendon jerk or H-reflexes takes about 5 sec to return to normal. By recording the afferent volley of the H-reflex from the human sciatic nerve with needle electrodes, it was found that the reduction in amplitude of the afferent volley

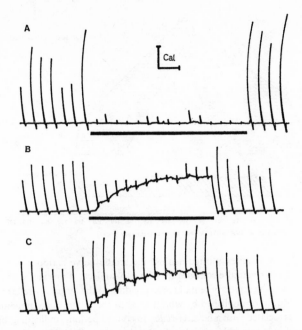

Fig. 2. Effects of continuous muscle vibration: suppression of tendon jerks, and the tonic vibration reflex in a normal subject. *A* Vibration of the quadriceps muscle while knee jerks are elicited every 5 sec. Knee jerks are depressed during the period of vibration (black bar) even without the development of a tonic contraction, probably because of the spread of the vibration wave to flexor muscles. *B* Suppression of knee jerks accompanying a tonic contraction induced by vibration. *C* Voluntary contraction of quadriceps in same subject as *B*, without suppression of knee jerks. Calibration: vertical, 0.4 kg for *A*, 0.6 kg for *B* and *C*. Horizontal, 10 sec [from DE GAIL, LANCE and NEILSON, 1966].

caused by occlusion of I_A afferent fibres from vibration-induced activity was insufficient to account for the degree of suppression of the H-reflex as shown in figure 3 [LANCE, NEILSON and TASSINARI, 1968; GILLIES, LANCE and TASSINARI, 1970]. A central inhibitory process was therefore postulated to account for the phasic reflex suppression.

Subsequent experiments in the cat [GILLIES, LANCE, NEILSON and TASSINARI, 1969] were designed to investigate the proposed central mechanism. The application of a vibrator to the intact limb of the cat inhibited the MSR of triceps surae much as it does in man. Cutaneous receptors were shown to play little role in this inhibitory phenomenon since vibration of the flayed limb still inhibited the MSR while vibration of flaps of skin with intact nerve supply had no effect on the MSR. Although the MSR was profoundly sup-

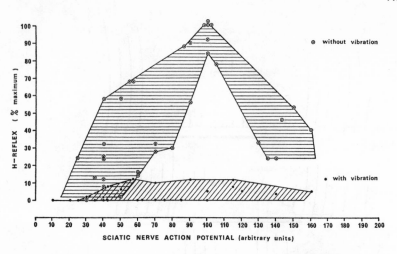

Fig. 3. Amplitude of the H-reflex plotted against the sciatic nerve action potential recorded simultaneously by perineural electrodes in 3 human subjects. For each subject, 100 arbitrary units represents the sciatic potential corresponding to the largest H-reflex recorded. The H-reflex is reduced or abolished by vibration while the sciatic nerve potential remains of an amplitude associated with a large H-reflex in control periods, indicating that suppression of H-reflexes is not caused by peripheral occlusion ('busy line') [from GILLIES, LANCE and TASSINARI, 1970].

pressed during vibration, the F response did not alter, suggesting that the inhibitory process was taking place at a presynaptic level. This was substantiated by the finding that the administration of picrotoxin abolished the inhibitory effect of vibration on the MSR. Furthermore muscle vibration evoked a dorsal root potential, the time course of which paralleled that of the MSR suppression, and, during vibration, primary afferent depolarization of I_A afferent terminals could be demonstrated, using an extracellular electrode positioned in the spinal cord in the vicinity of the motor nucleus of the gastrocnemius-soleus muscle (fig. 4 C–D). Stimulation at this site evoked a double ventral root volley, the first peak of which was a direct motoneurone discharge and the second of which was relayed through the MSR arc. Muscle vibration suppressed the second of these volleys, but did not alter the first (fig. 4 A–B). Thus it may be concluded that muscle vibration inhibits the MSR due to presynaptic inhibition of I_A afferent terminals.

Vibration is a powerful stimulus to the spindle primary ending [BROWN, ENGBERG and MATTHEWS, 1967], but it may also activate other receptors such as Pacinian corpuscles and cutaneous receptors, although these latter appear

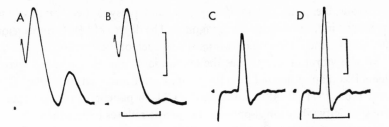

Fig.4. A, B Ventral root recording of direct motoneurone volley and monosynaptic reflex in the cat, showing selective suppression of the latter by vibration of the limb, in *B*. *C, D* Antidromic volley evoked by intraspinal stimulation of group I fibres, recorded from nerve to soleus in the cat. Augmentation of the response during vibration indicates primary afferent depolarisation [from GILLIES, LANCE, NEILSON and TASSINARI, 1969].

to have little effect on the MSR [MCINTYRE and PROSKE, 1968]. Activity in group I_A afferent fibres may evoke presynaptic inhibition of I_A afferent terminals [COOK, NEILSON and BROOKHART, 1965], and it seems probable, therefore, that muscle vibration inhibits the MSR due to presynaptic inhibition of I_A afferent terminals evoked by vibration-induced activity in I_A afferent fibres. Similar conclusions have been reached by DELWAIDE [1971] based on experiments in man [see DELWAIDE, this volume].

BARNES and POMPEIANO [1970a, b] confirmed that muscle vibration induces presynaptic inhibition of I_A afferent terminals due to activity in I_A afferent fibres. They found that vibration of a flexor muscle inhibited presynaptically the MSR of the antagonistic extensor muscle and that vibration of lateral gastrocnemius-soleus evoked presynaptic inhibition of the medial gastrocnemius reflex pathway. However, this was insufficient to suppress the MSR of medial gastrocnemius in the latter instance, presumably because of overwhelming vibration-induced I_A postsynaptic facilitation.

It seems probable then that in the cat (and presumably also in man) the application of a vibrator to the muscle belly or tendon suppresses the MSR of that muscle because vibration spreads to and thus also stimulates the primary spindle endings of antagonistic muscles. It would therefore be expected that if the vibrator were applied directly to the antagonistic muscle, or even to more distant muscles of the limb, phasic reflexes would be inhibited at least as readily. Such is indeed the case [RUSHWORTH and YOUNG, 1966; BURKE and ASHBY, 1972]. Thus in normal man when the MSR is elicited during vibration applied to the limb over the homonymous muscle it is testing a complex interaction of group I_A synaptic effects – presynaptic inhibition and postsynaptic facilitation from the agonist, and presynaptic inhibition and postsynaptic inhibition from the antagonist. The degree of MSR suppression, if any, will depend on the relative intensity of each of these actions.

Muscle vibration suppresses the tendon jerks and H-reflexes of spastic patients less than in normal man [DE GAIL, LANCE and NEILSON, 1966; DEL-WAIDE, 1969, 1971; BURKE and ASHBY, 1972]. In patients with moderate to severe spasticity vibration suppresses the phasic reflexes by less than 30%. In hemiparetic patients the affected side may be contrasted with the 'normal' side, on which phasic reflexes are always reduced to less than 50% and may be completely abolished by muscle vibration. The small inhibitory effect observed in spastic limbs is always maximal when the vibration is applied to the homonymous muscle. Vibration applied to the antagonistic flexor muscle has little effect on extensor MSR and vibration applied to more distant muscles across the knee joint fails to suppress the test reflex.

An inhibitory mechanism underlying the normal vibration-induced phasic reflex suppression does not appear to be operating in spasticity. These results are best explained by postulating that presynaptic inhibitory mechanisms, deprived of supraspinal support, are suppressed in spasticity. It is suggested that the absence of these inhibitory mechanisms plays a significant role in the hyperreflexia of spasticity.

The Tonic Vibration Reflex in Normal Man

Vibration-induced activity in group I_A fibres produces a tonic contraction of the muscle vibrated which was described independently by LANCE [1965] and EKLUND and HAGBARTH [1965]. LANCE reported: 'A vibrator was applied over the quadriceps muscle in twelve subjects while the knee jerk was elicited at intervals of 5 or 10 seconds. With the exception of one patient with absent knee jerks, all subjects usually experienced a tonic contraction of the quadriceps. This increased in strength until a plateau was reached and ceased immediately the vibrator was switched off. The force exerted by the knee jerk was reduced to one-third in two subjects without loss of muscle power by infiltration of the quadriceps with procaine. After this procedure tonic contraction was also reduced to one-third. While tonic contraction was in progress, it could be reduced by the subject consciously relaxing the muscle, but resumed its former level when the subject's attention was distracted. Force of contraction increased as the frequency of vibration was increased, reaching its maximum at about 80 c/sec.' HAGBARTH and EKLUND [1966a] applied a

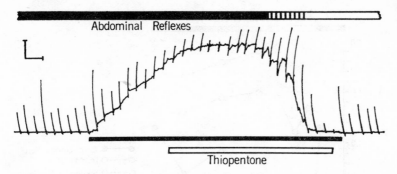

Fig.5. Abolition of quadriceps TVR in normal subject by intravenous thiopentone 400 mg with preservation of knee jerk. Recording from force transducer strapped to ankle shows knee jerks elicited at intervals of 5 sec, which become superimposed on TVR as this develops. Black bar indicates duration of vibration at 50/sec. Calibration: vertical 0.4 kg, horizontal 10 sec [from DE GAIL, LANCE and NEILSON, 1966].

vibrator at 100 c/sec to the muscle tendon and noted that the reflex contraction increased as the vibrated muscle was stretched. They found that a dilute solution of lidocaine applied to the muscle nerve abolished the activity in the vibrated muscle while voluntary power was preserved. EKLUND and HAG-BARTH [1966] gave the euphonious name of tonic vibration reflex to the phenomenon and showed that it was influenced by the degree of muscle stretch, head position, caloric stimulation of the labyrinths and by changes in body temperature.

DE GAIL, LANCE and NEILSON [1966] observed that the TVR could not be recorded from a muscle when the tendon jerk of that muscle was absent as a result of compression of the appropriate posterior roots, and was absent below the level of spinal cord transection in paraplegic patients. The vibration-induced tonic contraction could be simulated by tetanic stimulation of the medial popliteal nerve at an intensity which was subthreshold for efferent fibres. The TVR was found to be more susceptible to the action of barbiturates and other polysynaptic blocking agents than the tendon jerk (fig.5, 6). The tonic contraction of one muscle could be abolished by the simultaneous vibration of its antagonist and was potentiated by a preceding tetanus to the muscle nerve [LANCE, DE GAIL and NEILSON, 1966].

The effect of vibration in initiating a reflex tonic contraction was confirmed by RUSHWORTH and YOUNG [1966], MARSDEN, MEADOWS and HODGSON [1969] and DELWAIDE [1971]. MARSDEN and his colleagues showed that the tonic contraction induced by electrical stimulation of I_A afferent fibres

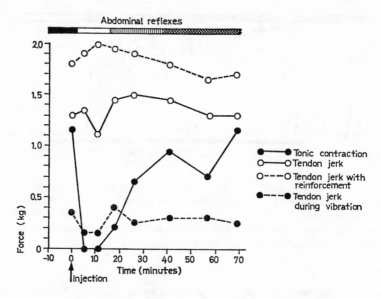

Fig. 6. Effect of a polysynaptic blocking agent, in this case Ciba 28,882 Ba, in abolishing the TVR of human quadriceps without any significant depression of the knee jerk. The TVR and abdominal reflexes recover together [from DE GAIL, LANCE and NEILSON, 1966].

could be inhibited at will in the same manner as the TVR, suggesting that the inhibition did not depend upon the withdrawal of γ efferent effects on the spindle.

The TVR of normal man develops some seconds after the start of muscle vibration and then increases slowly over a period of 20–60 sec. The latency of onset is variable, probably because of the ability of the normal subject to inhibit the response. In some tense subjects a phasic spike may be seen in the EMG and tension records 50–100 msec after the first beats of the vibrator. Phasic reflex contraction then subsides and the tonic contraction slowly augments. The TVR is well maintained as long as vibration persists and ceases 0.5–2.0 sec after the vibrator is removed [LANCE, DE GAIL and NEILSON, 1966; BURKE, ANDREWS and LANCE, 1972]. The force of the TVR is augmented by reinforcement, using the Jendrassik manoeuvre or pinna twist, and declines from the enhanced level over a period of 10 sec. Although it is possible to drive spindle primary endings maximally in the experimental animal by a single vibrator [BROWN, ENGBERG and MATTHEWS, 1967], this does not apply in man, since the application of 2 vibrators, one to the muscle belly and one to the tendon, will evoke a stronger reflex contraction than either one alone. The

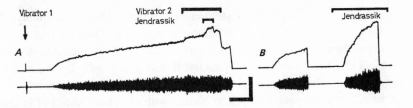

Fig. 7. The augmentation of the quadriceps TVR of normal man by the use of 2 vibrators and reinforcement. *A* The TVR is elicited first by vibration of the ligamentum patellae, then increments with the application of a second vibrator to the muscle belly, and again when the Jendrassik manoeuvre is performed. *B* TVR in the same subject with and without reinforcement. Lower trace, EMG [from BURKE, ANDREWS and LANCE, 1972].

performance of the Jendrassik manoeuvre during such vibration may increase reflex tension still further (fig. 7).

When the TVR is studied during an isometric contraction, it increases with increasing muscle stretch. Similarly, the reflex EMG is usually greatest when the muscle is at its most stretched point during slow sinusoidal passive movements of the limb. If the muscle is stretched suddenly, the TVR may vanish, probably because of cortical inhibition. When the TVR is evoked during reinforcement, increasing muscle length has little effect on the amplitude of the reflex contraction, suggesting that reinforcement is capable of compensating for the shorter muscle length in the position of least stretch [BURKE, ANDREWS and LANCE, 1972].

The TVR in Disorders of Muscle Tone

Cerebellar Disorders

LANCE, DE GAIL and NEILSON [1966] noticed that the TVR was diminished in 7 patients with disease of the cerebellum, more so on the side of greater deficit. HAGBARTH and EKLUND [1968] also found the TVR to be weaker than normal in 5 patients with a cerebellar syndrome but in only one of two patients with unilateral disease was it less on the affected side. It is probable that the TVR is normally potentiated in man by the substantial group I_A afferent projection to the cerebellum since stimulation of the deep cerebellar nuclei in the cat is known to influence both γ- and α-motoneurones. Because the method of eliciting the TVR in the experimental animal employs a frequency and amplitude of vibration which stimulates primary endings maximally, the TVR becomes independent of fusimotor innervation. Thus the fact that removal of the cerebellum in the decerebrate or anaesthetized cat has no effect on the TVR [MATTHEWS, 1966; GILLIES, BURKE and LANCE, 1971a] does not negate the observation in intact conscious man. Rather, the differing results in the different preparations are consistent with the observations of GILMAN [1969] that chronic cerebellar ablation produces hypotonia due to a decrease in fusimotor drive.

Parkinson's Disease

The TVR in Parkinson's disease does not differ significantly from that of normal subjects [LANCE, DE GAIL and NEILSON, 1966; HAGBARTH and EKLUND, 1968; BURKE, ANDREWS and LANCE, 1972].

Spasticity

HAGBARTH and EKLUND [1966b, 1968] and DE GAIL, LANCE and NEILSON [1966] found that muscle vibration evokes a TVR in spastic patients, although it is usually of lower amplitude than in normal subjects. The TVR is absent in patients with complete spinal transection. In hemiparetic patients, the TVR as recorded by a force transducer is usually of lower amplitude than it is on the clinically normal side, but this is difficult to assess because the tension recording is the resultant between the forces exerted by antagonistic muscles and the reciprocal relationship between antagonistic muscle groups is diminished in spastic states.

HAGBARTH and EKLUND [1968] reported that the TVR of spastic subjects is often of more abrupt onset and cessation than that obtained in normal subjects. This has been confirmed using isometric recording [BURKE, ANDREWS and LANCE, 1972]. There is much variability, however, even in a single subject, depending on the frequency of vibration and the mode of application of the vibration to the limb (fig. 8). There does not appear to be any consistent difference in the TVR in cases of spasticity of different aetiologies; the differences in the form of the TVR found in any one subject are greater than the differences between patients, and it is therefore concluded that the shape of the TVR is of little help in distinguishing the type or site of lesion. However, the TVR is usually absent in patients with complete spinal transection,

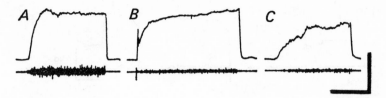

Fig. 8. The TVR of spastic man. The 3 tension recordings illustrated are all from the quadriceps of the same patient. Lower trace, EMG. *A* Low frequency vibration (50 c/sec) of the ligamentum patellae. *B* High frequency vibration (200 c/sec) of ligamentum patellae. *C* Low frequency vibration (50/sec) of quadriceps muscle. Calibration: vertical, 4 kg (measured by force transducer at ankle); 1 μV for *A*, 2 μV for *B* and *C*; horizontal, 10 sec [from BURKE, ANDREWS and LANCE, 1972].

presumably due to severance of the anterior columns [c. f. GILLIES, BURKE and LANCE, 1971a].

In some spastic patients reinforcement by the Jendrassik manoeuvre potentiates the TVR. In all patients studied, particularly those with incomplete cord lesions, the clasp-knife phenomenon could be readily elicited in the quadriceps stretch reflex, indicating involvement of the dorsal reticulospinal system in the dorsal quadrant of the spinal cord [BURKE, KNOWLES, ANDREWS and ASHBY, 1972]. It is therefore likely that the descending pathways mediating the Jendrassik manoeuvre traverse the ventral quadrant of the spinal cord.

Sometimes vibration may induce flexor spasms, possibly due to cutaneous stimulation. Occasionally spread of vibration to the flexor muscle may produce an overpowering TVR of that muscle rather than of the extensor to which the vibration is applied. As noted by HAGBARTH and EKLUND, the ability of the spastic patient to suppress the TVR voluntarily is impaired, but the force produced by voluntary contraction of paretic muscles is potentiated by vibration of that muscle, so that it may be useful in the rehabilitation of hemiplegic patients.

A TVR cannot be elicited from the quadriceps muscles of spastic limbs unless the muscle is subjected to a certain degree of initial stretch, as in the decerebrate or the anaesthetized cat [MATTHEWS, 1966; GILLIES, BURKE and LANCE, 1971a]. A small TVR may be recorded at 30–45° knee flexion, and this increases in size with increasing stretch of the quadriceps, although it tends to reach a plateau at about 90° (fig. 9). Inhibition of the TVR with increasing muscle stretch, as found with the quadriceps stretch reflex, does not occur whether the TVR is recorded isometrically or during a stretching

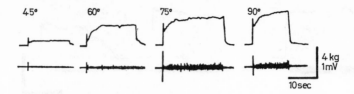

Fig. 9. The effect of muscle stretch on the TVR of spastic man. Increasing muscle stretch increases the size of the TVR, the effect tending to plateau between 75 and 90°. Lower trace, EMG [from BURKE, ANDREWS and LANCE, 1972].

movement. This property is important if the TVR is to be used therapeutically since the force generated by the TVR would automatically increase in response to an undesirable flexion movement.

The Tonic Vibration Reflex in the Cat

MATTHEWS [1966] found that the TVR was present in the decerebrate cat with an initial phasic spike which was followed by a plateau of tension maintained for the period of vibration. The TVR was abolished by section of the spinal cord at the first cervical segment, thus demonstrating that the reflex depended upon brain-stem structures. The TVR was not affected by ablation of the cerebellum. The plateau tension of the TVR was linearly related to the frequency of vibration. These observations were confirmed by GILLIES, BURKE and LANCE [1971a] who found that the TVR was not affected by section of the dorsal quadrant of the spinal cord but was reduced to 20–40% of the control level by a discrete lesion in the ventral column of the cervical spinal cord (fig. 10) and to 20% or less by ablation of the lateral vestibular nucleus on the same side. After these lesions, the tonic plateau of the reflex was more severely affected than the initial phasic spike, which was reduced only to 60–80% of its former level. When the vestibulospinal tract was stimulated caudal to the ablated lateral vestibular nucleus, the TVR was restored to its former level (fig. 11). Exploration of the brain stem disclosed that the TVR of triceps surae could be inhibited consistently from the medullary reticular formation and potentiated from the caudal part of the pons and the lateral vestibular nucleus (fig. 12) [GILLIES, BURKE and LANCE, 1971b].

The areas of facilitation and inhibition found were smaller than those described earlier by MAGOUN and RHINES [1947], which was attributed by

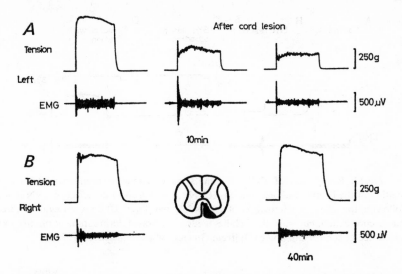

Fig. 10. Effect of lesion in left ventral quadrant of spinal cord. *A* EMG and tension record of TVR before, 10 and 40 min after lesion of ipsilateral ventral quadrant at the fourth cervical segment. *B* Control recordings from contralateral side and the extent of the lesion confirmed by histology. The lesion reduced the plateau of the TVR proportionately more than the initial phasic spike [from GILLIES, BURKE and LANCE, 1971a].

GILLIES and colleagues to the fact that the frequency and amplitude of vibration employed would have discharged all primary spindle endings, thus making the TVR independent of γ efferent drive. More recently ANDREWS, KNOWLES and HANCOCK [1972], working in this laboratory, have shown that the activity of brain-stem areas is highly sensitive to barbiturate anaesthesia. In animals where the total dosage of barbiturate was low (25–30 mg/kg), the areas of the brain stem affecting the TVR were remarkably similar to those found by MAGOUN and RHINES [1947] to alter tendon jerks and stretch reflexes, and by GERNANDT and THULIN [1955] to alter the MSR, with a central inhibitory zone (fig. 13, section 1) surrounded by a lateral facilitatory area (fig. 13, sections 2, 3). These areas produced a bilateral effect non-specifically on flexor and extensor muscles but a small area was found at the caudal part of the inhibitory area, which gave rise to reciprocal enhancement and suppression of flexors and extensors. Between the inhibitory and facilitatory areas lay 'mixed areas', first described by GERNANDT and THULIN [1955], from which facilitation was elicited by direct stimulation, which altered to an inhibitory response when the motor cortex contralateral to the TVR being tested was

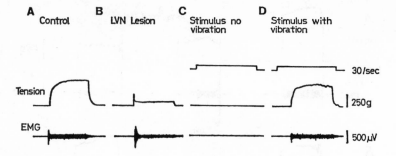

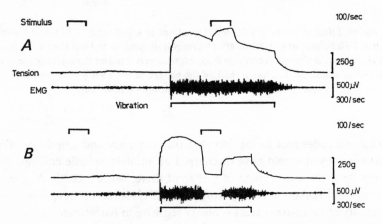

Fig.11. Restoration of TVR by stimulation of lateral vestibulospinal tract after ablation of the lateral vestibular nucleus (LVN). *A* and *B* show the effect of LVN lesion. Following the ablation, stimulation of vestibulospinal fibres (C) does not evoke a direct muscle contraction, but when the Achilles tendon is vibrated during stimulation, the TVR is recorded at control amplitude (D) [from GILLIES, BURKE and LANCE, 1971 a].

Fig. 12. Effect of brain-stem stimulation on the TVR of the decerebrate cat. The reflex was facilitated from the pontine reticular formation (A) and inhibited from the medial medullary reticular formation (B). The control stimulus did not evoke muscle contraction. Duration of electrical stimulation (2 sec), and of vibration indicated by markers [from GILLIES, BURKE and LANCE, 1971 b].

stimulated at the same time. It is probable, therefore, that each part of the reticular formation contains both facilitatory and inhibitory neurones and that the effect obtained by a direct stimulus simply samples the net effect of the area being tested. Inhibitory neurones can be potentiated from the motor cortex, but the response from facilitatory neurones is independent of cortical stimulation.

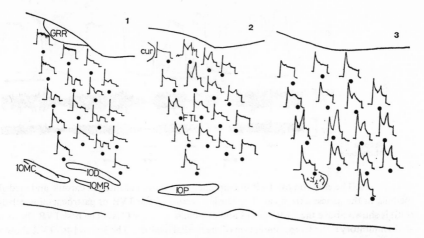

Fig. 13. Sagittal planes of brain stem 1, 2 and 3 mm from midline demonstrating the effect of stimulation, at the point shown, on the TVR of gastrocnemius-soleus in the anaesthetized cat. The tension trace of the TVR is superimposed on each point of stimulation. Inhibition of the TVR is observed 1 mm from the midline and facilitation 2 mm and 3 mm from the midline. Points are separated by 1 mm.

cur,   cuneate nucleus, rostral division;
GRR,   gracile nucleus, rostral division;
IOD,   dorsal accessory nucleus of the inferior olive;
IOMC, medial accessory inferior olive, caudal division;
IOMR, medial accessory inferior olive, rostral division;
IOP,   principal nucleus of the inferior olive;
FTL,   lateral tegmental field [from ANDREWS, KNOWLES and HANCOCK, 1972].

ASHBY, ANDREWS, KNOWLES and LANCE [1972] found that the TVR of gastrocnemius-soleus was inhibited by stimulation of the contralateral pericruciate area and internal capsule. This effect persisted after section of the pyramidal tract in the medulla although the cortical area from which inhibition could be obtained was reduced in size and a higher stimulus intensity was required for a comparable degree of inhibition. The extrapyramidal tract responsible for inhibition of the TVR traverses the internal capsule and the medial part of the midbrain dorsal to the cerebral peduncle. Recently AN-DREWS, KNOWLES and LANCE [1972] have shown in lightly anaesthetized cats that the inhibitory effect on the TVR from the medial medulla is potentiated by stimulation of the motor cortex or internal capsule contralateral to the TVR being examined (fig. 14).

The cortical area capable of inhibiting the TVR directly and the cortical area projecting to the medial medullary reticular formation overlap consider-

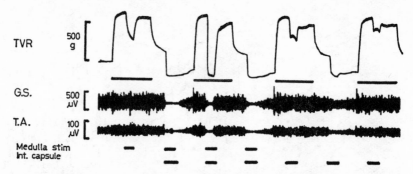

Fig. 14. The effect on the TVR of combined stimulation of internal capsule and medial medulla in the anaesthetized cat. The tension record of the TVR of gastrocnemius-soleus (GS) is shown above the EMG of GS and of tibialis anterior (TA). The first TVR shows a slight inhibitory effect from stimulation of the medial medulla. The 3rd and 4th TVR show a similar slight inhibitory effect from the contralateral internal capsule. Combination of both stimuli (second trace) shows a total inhibition of the TVR and EMG of GS and TA [from ANDREWS, KNOWLES and HANCOCK, 1972].

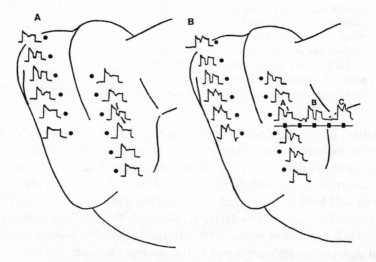

Fig. 15. The cortical areas producing inhibitory effects on stimulation. *A* The results of direct stimulation of the motor cortex with the TVR of the contralateral GS superimposed on the points of stimulation. *B* The results of combined stimulation of motor cortex and medial medulla similarly superimposed. At one point, 3 responses have been mounted to demonstrate the direct effect of cortical stimulation (A), the combined effect of stimulating cortex and medulla (B), and the effect of medullary stimulation alone (C) [from ANDREWS, KNOWLES and LANCE, 1972].

ably (fig. 15). The inhibition of the TVR resulting from stimulation of the medial medullary reticular formation is not influenced by simultaneous stimulation of the caudate nucleus or the red nucleus contralateral to the TVR under test, or the motor cortex ipsilateral to the TVR. It is apparent, therefore, that although direct stimulation of the medial medulla gives rise to bilateral inhibitory effects, it is given specific direction from the motor cortex contralateral to the TVR being examined. It should be noted that the inhibitory effects on the TVR produced from the motor cortex or internal capsule affect both flexors and extensors equally and occur at a stimulus intensity which is subliminal for movement. If the stimulus current is doubled, flexor movements are produced from these sites.

Stimulation of the red nucleus contralateral to the limb being tested facilitated the TVR in both flexor and extensor muscles. The caudate nucleus could not be shown to exert any effect on the TVR, either directly or via the brain-stem reticular formation [ANDREWS, KNOWLES and LANCE, 1972].

MCCULLOCH, GRAF and MAGOUN [1946] found that the application of strychnine to a strip anterior to the motor cortex in the monkey evoked neuronal discharge in the ipsilateral medial medulla but they could not find clear evidence of discharge in the contralateral medulla. The present studies demonstrate that the effect on the medulla of stimulating the motor cortex is bilateral and that the effect of stimulating the medullary centres directly on the TVR of flexor and extensor muscles is bilateral. In spite of this, the TVR of flexors and extensors in a given limb can be inhibited only from the contralateral cortex. It is highly likely that interruption of this cortico-bulbo-reticular pathway could account for the release of tonic reflexes which is associated with hemiplegia in most patients.

Summary

The recent appreciation of the sensitivity of the spindle primary ending to a vibratory stimulus has made possible the more accurate analysis of postural mechanisms in the intact normal subject. Clinical phenomena such as reflex irradiation, 'inversion' of tendon reflexes and even the normal tendon jerk may be explained by the transmission of a vibration wave set up by percussion. The application of a vibrator to muscle belly or tendon of normal subjects, evokes a reflex tonic contraction of that muscle (TVR) and simultaneously suppresses tendon jerks or H-reflexes. Studies in both man and cat have shown that this phasic reflex suppression is caused by 'presynaptic' inhibition.

The TVR of normal man usually augments slowly over a period of 20–60 sec, can be inhibited voluntarily, is inhibited by vibration of the antagonistic muscle, and can be increased in amplitude by reinforcement or a preceding tetanic stimulation of the muscle nerve. The TVR increases as muscle length is increased and is also influenced by tonic neck and labyrinthine reflexes.

The TVR in Parkinson's disease does not differ from that of normal subjects, but in spasticity it is usually of more abrupt onset and cessation, and cannot be inhibited voluntarily as readily as in the normal subject. There does not appear to be any consistent difference in the TVR in cases of spasticity of different aetiologies. The TVR is diminished or absent in patients with damage to the spinal cord.

The TVR in the cat can be inhibited by stimulation of the contralateral motor cortex, internal capsule and medial medullary reticular formation. It is facilitated by the lateral vestibular nucleus and vestibulospinal tract, the lateral medullary reticular formation and the contralateral red nucleus. Inhibition or facilitation obtained from the medulla is bilateral and effects both flexors and extensors, with the exception of a small area in the caudal part of the medial medulla from which reciprocal effects may be obtained. Stimulation of the motor cortex potentiates the inhibition obtained from the medial medulla of either side but the effect is apparent only on the contralateral TVR. The facilitatory effects from the lateral medulla are unaltered by cortical stimulation. Mild facilitation obtained from 'mixed areas' in the medulla can be converted to profound inhibition of the TVR by concomitant stimulation of the contralateral motor cortex.

It appears that the corticoreticular pathway relays solely upon inhibitory neurones in the reticular formation and thus interruption of this pathway rather than the pyramidal tract which it accompanies could be responsible for the increased muscle tone of hemiplegic patients.

Acknowledgements

This research programme has been supported by grants from the National Health and Medical Research Council of Australia, the Adolph Basser Trust, Mr. and Mrs. EDWIN STREET, Ciba Co. Ltd., and Merck, Sharp & Dohme Ltd. Mr. K. NORCROSS and Mr. N. SKUSE have provided able technical assistance.

Figures are reproduced with the permission of Butterworths, London, and the editors of Brain, the Journal of Neurology, Neurosurgery and Psychiatry, Journal of Physiology and Journal of Neurophysiology.

Author's address: Prof. JAMES W. LANCE, Department of Neurology, The Prince Henry Hospital, Anzac Parade, Little Bay, *Sydney, N.S.W. 2036* (Australia)

New Developments in Electromyography and Clinical Neurophysiology,
edited by J. E. Desmedt, vol. 3, pp. 463–468 (Karger, Basel 1973)

A Survey of Japanese Research on Muscle Vibration

S. HOMMA

Department of Physiology, School of Medicine, Chiba University, Chiba

Many studies on muscle vibration have been reported in the last decade. Since the phenomenon of 'tonic vibration reflex' (TVR) was first reported by HAGBARTH and EKLUND [1966] many Japanese workers have studied the TVR and in 1968 a research group was organized in Japan with (since 1970) the support of the Ministry of Health and Welfare. A symposium on TVR was held at the 23rd Congress of the Japan EMG Society, November 23–24, 1970, Nagasaki (President, S. NAGAI). This chapter presents a selection of the research work chosen by the reviewer. The symposium was published in full [HORI, 1971].

I. Basic Research

A. Some Principles of the TVR (Work of S. HOMMA, Department of Physiology, School of Medicine, Chiba University, Chiba, Japan)

Motoneuronal spike potentials recorded at the ventral root during sinusoidal stretch of the soleus and the gastrocnemius muscle were recorded in acute experiments on cats and computer processed. Analysis of nonsequential interval histograms showed that they fired at intervals which were distributed around the integer multiples of the vibration frequency [HOMMA, KANDA and WATANABE, 1971a]. Intracellular experiments showed that α-motoneurones were excited only when the depolarization threshold was attained by the synchronously arriving EPSP whose frequency coincided with that of the vibration [HOMMA, ISHIKAWA and STUART, 1970]. It was also found by data analysis of the human EMG that the same principle of 'integer multiple' existed during the human TVR.

The relationship between vibratory excursion (ΔL) and vibratory frequency (f) was investigated in the cat [Homma, Ishikawa and Watanabe, 1967; Homma, Kobayashi and Watanabe, 1970]. This ΔL-f relation was also investigated in man. In both cases the TVR was found to be elicited most effectively by vibration at about 100/sec [Homma, Kanda and Watanabe, 1971 b].

B. A Device for the Recording of TVR Tension (Work of S. Tsukahara, Department of Physiology, School of Medicine, Fukushima University, Fukushima, Japan

Slow flexion of m. biceps brachii during vibration was investigated with a special device [Suzuki, 1961]. At moderate intensities of vibration the muscle contracted with a very slow increment. They found that the slope of the increment was proportional to the vibratory intensity and also found that vibration at about 100/sec was most effective. The increment could not be abolished by local anaesthesia of the skin. They supposed the contraction to be reflex in nature because flexion disappeared during nerve block.

II. Clinical Research

A. TVR Studies from a Neurological Point of View (Work of S. Watanabe, Department of Internal Medicine, School of Medicine, Chiba University, Chiba, Japan)

It has been shown in normal subjects that the H-reflex is inhibited by vibration [Yamanaka, 1964]. Comparison of two successive H-reflexes offers a convenient method of studying the effect of vibration in normal subjects and in patients [Matsuda, 1969].

Figure 1 shows the results of an experiment where the posterior tibial nerve was stimulated with pairs of stimuli repeated every 10 sec. It was noticed that the ratio of amplitudes of H-reflexes to the first and second shock was markedly increased during vibration although direct plotting of the two waves shows apparently almost similar time courses of suppression. This increase of ratio was not seen with vibration of the antagonist muscle although the direct plotting showed a similar suppressive time

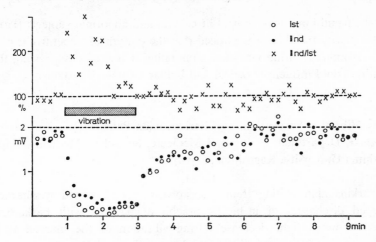

Fig. 1. The effect of vibration on H-reflexes in a normal young subject. Pairs of stimuli 200 msec apart were given to the posterior tibial nerve every 10 sec. Vibration (horizontal hatched bar) was applied to the ipsilateral Achilles tendon (187/sec, 80 G) and H-reflexes recorded from m. triceps surae. Open circles and dots indicate amplitudes of H-reflex to the first and second shocks respectively; crosses are the second H-reflex expressed as a percentage of the first.

course during vibration. Through studies on cases of Marie's cerebellar ataxia, traumatic cerebellar ataxia, olivo-ponto-cerebellar atrophy, Wallenberg's syndrome, vertebral artery insufficiency and spinal cord injury, it was observed that H-reflex ratio did not change in the patients with cerebellar disorder, and that is decreased in spinal injury whereas it increased markedly in the normal subject as shown in the figure.

It was concluded that the effect of vibration on the 'recovering' H-reflex seemed to be exerted through a rather complex mechanism involving some rivalry between excitatory and inhibitory effects, probably under supraspinal control.

B. Estimation of the Level of Central Excitatory State in Vibratory Stimulation of the Muscles (Work of M. OKAMOTO, Y. SAITO, S. KAWAI and H. HORI, Department of Surgery, Nara Medical College, Nara, Japan)

Fluctuation in the size of the H-reflex during repetitive stimulation with varied interval has been investigated as a criterion of supraspinal excitatory and inhibitory control in normal and abnormal subjects [HORI, 1959; OKADA, 1962]. It was noticed that the pattern of 'repetitive evoked EMG' is altered during vibratory stimulation of muscles, suggesting that vibration could influence cortical and bulbar excitatory systems.

C. Vibration Susceptibility of Parkinsonian Patients (Work of M. TA-KEDA, Department of Internal Medicine, School of Medicine, Kago-shima University, Kagoshima, Japan)

Parkinsonian 'resting' tremor is known to be suppressed by vibration. Grouped discharges of m. biceps brachii of patients with Parkinson's disease showed slight decrease during vibration of the muscle. Hand tremor of the ipsilateral side also decreased. Participation of muscle afferents and subcortical nuclei in the parkinsonian tremor is suggested.

D. Clinical Application of TVR (Work of Y. YAMANAKA, Department of Orthopedic Surgery, School of Medicine, Chiba University, Chiba, Japan)

Percutaneous vibration at the belly of a muscle has the following effects:

1. Vibration at Low Frequency (Below 50/sec)
(a) Depression of the monosynaptic reflex was demonstrated by the study of H-reflex amplitude in m. triceps surae [YAMANAKA, 1964]. Skin afferents were known to participate partly in the above inhibition because the depressant effect of vibration upon H-reflex size was slightly weakened by anesthesia of the skin area where vibration was applied.

(b) Peripheral blood flow was studied in the index finger with a photo-electric plethysmograph during vibration to the ipsilateral forearm. Increase in height of the pulse wave could be observed with concurrent increase in the temperature of the skin and muscle of the forearm.

2. Vibration at High Frequency (Above 50/sec)
(a) Gradual increment of TVR was apparent in this frequency range.
(b) Post-vibratory facilitation: increase of H-reflex size was observed immediately after vibration.

Table I. Effects of vibration therapy as an indication of diagnosis

Classification of effect	Effective frequency range, per sec	Indication	Mode of Vibratory Application
Inhibition by vibration	20–50	spastic and rigid motor disorder	(1) posture of preferably low muscle tone is recommended (2) vibration at muscle belly for 5 min or more
Effect upon the autonomic nervous system	20–50	disuse atrophy	(1) concurrent application with thermal and electric stimulation gave better prognosis (2) vibration at muscle belly
TVR	80–120	neurogenic motor disorder	(1) concurrent voluntary contraction should be done to increase muscle tone (2) vibration at tendon region
Post-vibratory facilitation	100–250	flaccid paralysis and disuse atrophy	(1) posture by which patients can increase muscle tone (2) vibration longer than 5 min

3. Clinical Application

Vibration therapy was tried on patients with cerebral paralysis of hemiplegic and athetoid types. Pegboard tests were used to assess the therapeutic effect. Vibration was more effective upon the athetoid than the hemiplegic type and differences were seen in the best frequency range (table I).

E. Effect of Vibration upon Jaw Muscle Disorder of Children with Cerebral Paralysis (Work of S. ISHIDA, Sode-ga-ura Therapeutic Institute for Retarded Children, Chiba, Japan)

More than 30 children with cerebral paralysis of the tetraplegic type, all with disorder of jaw opening and closing movements, received vibration therapy at 25/sec for 60 days (5 min/day). An improvement was found in EMG pattern, speech control, deviation of the lip position and ptyalism. There was, however, no appreciable change in the masticatory pressure.

F. Vibration as a Neurological Test (Work of K. IOKU, Department of Neurosurgery, School of Medicine, Osaka University, Osaka, Japan)

H-reflexes are difficult to elicit in the muscles of the upper extremities. However, the H-reflex was recordable from m. opponens pollicis in some of the patients with vestibular disorder (16 cases of whiplash injury and 1 acoustic neurinoma) during vibration of the muscle.

Romberg's sign was also studied in the same patients before, during and after vibratory stimulation to the neck or to the Achilles tendon with the aid of a modified X-Y recorder. It was concluded that, in patients with tonic neck reflex, Romberg's sign was facilitated by vibration and that, in the patients with cerebellar and bulbar disorder, the relevant symptoms were temporarily improved by vibration.

G. Use of Muscle Vibration in Rehabilitation Medicine (Work of M. IKEDA and T. TAMAOKI, Division of Rehabilitation, Institute for Balneotherapy, Tohoku University, Naruko, Japan)

Scissor leg gait due to pain of the spastic adductor muscle of the hip was found to be improved when vibration was applied to the abductor muscle of hip. The effect was explained by the vibratory inhibition upon the antagonist muscle groups.

Author's address: Dr. S. HOMMA, Department of Physiology, School of Medicine, Chiba University, *Chiba* (Japan)

New Developments in Electromyography and Clinical Neurophysiology,
edited by J. E. Desmedt, vol. 3, pp. 469–474 (Karger, Basel 1973)

Vibration Reflex in Spastic Patients

K. KANDA, S. HOMMA and S. WATANABE

Department of Physiology, School of Medicine, Chiba University, Chiba

It is well-known that, both in man and in the decerebrate cat, a slowly rising autogenetic reflex is elicited by muscle vibration [GRANIT, 1970; HAG-BARTH and EKLUND, 1966a,b; HOMMA, KANDA and WATANABE, 1971b]. In anesthetized cats and monkeys we found that the firing intervals of the moto-neuron coincided in a statistically significant way with some multiples of the period of the vibration used [HOMMA, KANDA and WATANABE, 1971a]. In the human tonic vibration reflex (TVR) the spike intervals of m. quadriceps femoris have a distribution which conforms to the same principle although this is not the case during voluntary contraction [HOMMA, KANDA and WATANABE, 1971c].

This transformation in the spinal cord of I_A input into a motoneuron output at a frequency which is lower and in integral ratio to the input can be discussed under the term 'decoding ratio'. This refers to the special pattern of spinal integration during forced vibration of muscles. It was also demonstrated by intracellular experiments that such motoneurons fired only at the peaks of the depolarizing postsynaptic potentials (EPSP) the frequency of which was always identical to that of the applied vibration [HOMMA, ISHIKAWA and STUART, 1970].

This paper deals with spastic patients and it will be shown that the 'preferred firing interval' of the spastic motoneurons were also harmonics of the vibratory cycle. The frequency of clonus which was present during vibration was also found to be some multiple of the vibratory frequency but was thought to be elicited through a slightly different mechanism from the one operating in TVR. It is proposed that such clonus elicited by vibration be called 'clonic vibration reflex' (CVR).

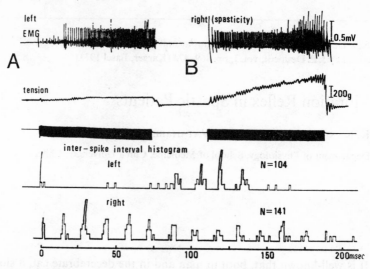

Fig. 1. Comparison of TVR in the normal and spastic m. quadriceps femoris in hemi-plegia. *A* Simultaneous records of EMG spikes, TVR force (tension) and monitored vibration (during 15 sec), respectively, from the normal left side. *B* The same records from the spastic right side. Bottom traces are non-sequential interval histograms made from the normal and spastic EMG spikes shown above.

I. TVR in Spastic M. Quadriceps Femoris

A patient recovering from hemiplegia was used as the subject. Poly-urethane-insulated wires were inserted into the spastic right m. quadriceps femoris. Unitary EMG spikes were recorded bipolarly through balanced pre-amplifiers. EMG, TVR force and monitored vibration were recorded on a multi-channel tape-recorder, TEAC 351-F (TEAC Co., Tokyo). The EMG spikes were later processed by a small computer, ATAC 501-20 (Nihon-kohden, K.K., Tokyo) and non-sequential inter-spike interval histograms were displayed on a strip chart recorder by using slow-speed analog read-out of the computer. A pneumatic type of vibrator [HOMMA and ZENITANI, 1971] was generally used.

Figure 1 shows simultaneous records from the normal and spastic sides when the patellar tendon receives vibration at 67/sec. During vibration, unitary EMG spikes are elicited with parallel increment in the TVR tension. Increments in both EMG and TVR force are observed. Recruitment of other motor units is observed on both sides. The non-sequential interval histograms show that the peaks of intervals are found around harmonics of the vibratory frequency. Although the intensity of vibration is adjusted tŏ be similar on both

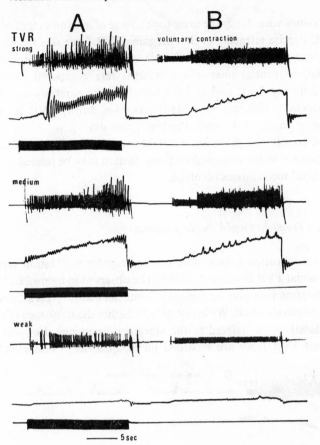

Fig. 2. TVR and voluntary contraction of the spastic m. quadriceps femoris in hemi-plegia. *A* Records during TVR when weak, medium and strong vibrations, respectively (67/sec) are applied to the patellar tendon. *B* Records during voluntary 'retracing' of TVR time course.

sides, the size of TVR force is quite different. Another important difference is that, in the spastic muscle, the harmonics are found in wider ranges than those for the normal muscle. The spastic neuron thus appears capable of firing at shorter intervals than the normal one. In other words the 'decoding ratio' is smaller on the spastic side.

When the intensity of vibration is increased at different levels the spike intervals of EMG are shortened and more units are recruited. In figure 2, gradual recruitment of EMG spikes and parallel increment of TVR force can be observed in all three stages. Figure 2B shows records when the patient is requested to produce voluntarily the same time course of tension as that of the preceding TVR. Precise retracing of the foregoing TVR force was facilitated by the use of a storage oscilloscope (Tektronix type 564) to be monitored by the subject. Non-sequential interval histograms were compared between spikes recorded during TVR and spikes elicited by the carefully matched voluntary contraction. Histograms made from spikes during TVR showed regular peaking as in figure 1, while the histogram during voluntary contraction showed no separate peaks but a roughly normal distribution. Such a remarkable difference in the motoneuron firing pattern may be related to the difference in central mechanisms involved.

II. Vibration Frequency and Vibratory Clonus

More intense vibration causes more marked regular undulation superimposed on the usual TVR increment. This can be observed in figure 2B. This clonus-like undulation may even be present in some normal subjects although its amplitude is relatively small. We investigated whether the frequency of this clonus-like undulation was related to the vibratory frequency. In figure 3, EMG spikes and TVR force are recorded on m. quadriceps femoris of a

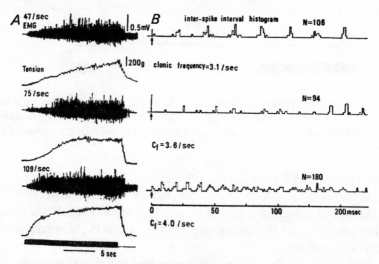

Fig. 3. TVR with superimposed vibratory clonus in m. quadriceps femoris of a normal subject. Frequencies of vibration used as indicated in A; duration of vibration is indicated by heavy bar. Non-sequential interval histograms correspond to the records beside them. Each clonic frequency (C_f) measured from the tension oscillation is shown in B.

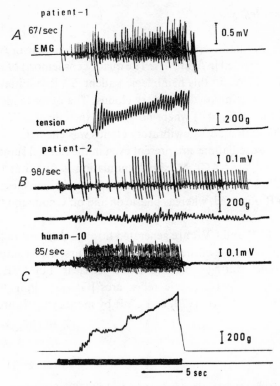

Fig. 4. Two types of TVR observed in spastic patients, and a normal TVR. EMG spikes in the upper and tension in the lower trace of each set. *A* TVR of a spastic patient with superimposed vibratory clonus. *B* TVR of another spastic patient with vibratory clonus only. This patient had a spinal cord tumor. *C* Record from the normal subject. Vibratory frequencies are indicated. Thick horizontal line at the bottom indicates duration of vibration.

normal subject when the patellar tendon is vibrated at 47, 75 and 109/sec. Relatively clear oscillation is superimposed in this subject. Interval histograms are computed for each EMG spike recording and the preferred firing interval is estimated for each vibratory frequency. The mean frequency (C_f) of the superimposed clonus-like oscillation is measured and found to be 3.1, 3.6 and 4.0/sec for vibration frequencies (V_f) of 47, 75 and 109/sec, respectively. The approximate ratios (1/n) of frequencies in the input and output sides, ($C_f = [1/n] \, V_f$) are 1/15, 1/20 and 1/27, respectively. C_f is changed by V_f. A reason for such deviation from the 'integer principle' mentioned above may be applicable to the very low frequency of the vibratory clonus which inevitably needs a very large number for the decoding ratio.

III. Clonic Vibration Reflex

TVR of the spastic m. quadriceps femoris are known to appear in two different modes. One is illustrated in figure 4A. The quadriceps femoris of the spastic side is vibrated at 67/sec in this hemiplegic patient. TVR is initiated tonically with very large superimposed vibratory clonus. The other mode is found in a spinal cord tumor (fig. 4B) where no gradual increment of TVR force could be found although moderate vibratory clonus was elicited.

Both TVR and clonic oscillations are present even in the normal subject but the latter are very small (fig. 4C). Maximal undulation is found at the beginning of TVR both in spastic and normal subjects. A steady increment is characteristic for all TVR observed, whereas phasic on and off is common for the CVR.

It is suspected that TVR and CVR are generated through different reflex arcs within the spinal cord because of their different time courses. The nature of these two types of reflex during vibration is not yet clear, but TVR is considered to be effected by polysynaptic reflex arcs [GRANIT, PHILLIPS, SKOGLUND and STEG, 1957; GRANIT, 1970] and CVR by monosynaptic arcs.

Summary

The patellar tendon was vibrated at various frequencies to elicit TVR of m. quadriceps femoris in normal and spastic subjects. EMG spikes and TVR force were recorded. EMG spikes were later processed by the computer and non-sequential interval histograms were made.

1. Intervals of EMG spikes are preferentially distributed about harmonics of the vibratory frequency. The ratio between vibration frequency, V_f and the α-motoneuron firing rate (MN_f) is referred to as the 'decoding ratio'.

Hence $MN_f = \frac{1}{n}V_f$, where n is an integer larger than 1.

2. Distribution of intervals of EMG spikes during voluntary contraction does not show any discrete peak or peaks but shows normal distribution.

3. TVR is superimposed upon by CVR which becomes exaggerated in the spastic patient. Frequency of CVR is influenced by the frequency of vibration.

4. In some spastic patients CVR can be observed in spite of the absence of TVR.

5. TVR and CVR are probably generated through different reflex arcs within the spinal cord.

Author's address: Prof. S. HOMMA, Department of Physiology, Chiba University, Medical School, *Chiba* (Japan)

Pathophysiology of Spasticity

New Developments in Electromyography and Clinical Neurophysiology,
edited by J. E. Desmedt, vol. 3, pp. 475–495 (Karger, Basel 1973)

Studies of the Reflex Effects of Primary and Secondary Spindle Endings in Spasticity

D. BURKE and J. W. LANCE

Division of Neurology, The Prince Henry Hospital and The School of Medicine, University of New South Wales, Sydney

Since the observations of SHERRINGTON [1898, 1909], the analogy between spasticity and the decerebrate state has been drawn by many authors, most notably by WALSHE [1919, 1929]. WALSHE considered decerebrate rigidity a suitable experimental prototype for spasticity, especially that associated with capsular hemiplegia, but also that occurring in 'paraplegia-in-extension'. Spasticity and decerebrate rigidity he thought of as being 'physiologically identical', although spasticity 'is but a fractional expression of decerebrate rigidity'. In the light of recent advances in neurophysiology, the analogy between the decerebrate and spastic states has probably outlived its usefulness, although it is still emphasized by many authors such as RUSHWORTH [1960, 1964a, 1966], THOMAS [1961], JANSEN [1962] and BRODAL [1962]. BRODAL stated: 'It is unfortunate that the experimentally produced condition in animals which appears to correspond to spasticity in man has been termed decerebrate *rigidity*. However this discrepancy in nomenclature should not prevent us from recognizing the essential identity of the facts behind the two terms.' One of the purposes of this paper is to point out discrepancies other than that of nomenclature.

While muscle vibration may be used as a relatively selective tool for the study of the reflex effects of group IA afferent fibres from spindle primary endings, muscle stretch activates both the primary and the secondary spindle endings, and these receptors differ in their sensitivity to the velocity and the length of the stretching movement [MATTHEWS, 1964]. In investigating the parts played by these spindle receptors in the phenomena of spasticity, the response of the stretch reflex of spastic man to increasing velocities of stretch and to changes in muscle length has been analyzed. The reflex effects of muscle vibration in spastic man have been studied in parallel so that the tonic stretch reflex could be compared with the tonic vibration reflex (TVR), and so that the effects of vibration on phasic reflexes could be determined. These studies have been reported in another chapter in this volume.

Experiments have been carried out in patients with spinal and with cerebral spasticity. Some patients with complete spinal transection were tested so that segmental reflex mechanisms could be more readily eluci- dated. Only patients with stable spasticity were studied but, in spite of this, it was often necessary to discard the first few stretching movements be- cause of initial reflex variability. Thereafter, a relatively stable level could be maintained for prolonged periods. In the various forms of spasticity few qualitative differences in stretch reflex behaviour were found, although marked quantitative differences were noted. This was not unexpected, since the studies involved features common to all forms of spasticity, and were not designed to highlight differences in other forms of reflex behaviour.

Stretch Reflexes of Lower Limb Muscles

The stretch reflexes of the hamstrings and quadriceps femoris muscles have been analyzed using both linear and sinusoidal stretching movements, the leg being moved manually at the knee. The excursion of the knee was measured using a goniometer, the output of which was differentiated to produce a voltage proportional to angular velocity. So that reflex activity could be recorded uncontaminated by factors due to limb mechanics and the active contractile and passive visco-elastic properties of muscles, the surface EMG has been recorded. In the initial experiments the EMG potentials were integrated to provide a measure of the reflex activity. Where appropriate these recordings have been supplemented by clinical obser- vations of muscle tone and by recordings of the resistance to passive stretch using a force transducer.

A. The Response to the Velocity of Stretch

In both the hamstrings and quadriceps, increasing the velocity of the stretching movement resulted in a greater stretch reflex response [BURKE, GILLIES and LANCE, 1970, 1971]. On cessation of movement reflex EMG

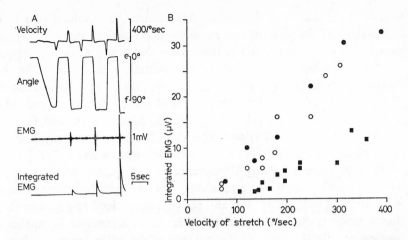

Fig. 1. A Sensitivity of the quadriceps stretch reflex to the velocity of stretch; the faster the velocity of stretch, the larger the reflex EMG. EMG subsides completely on cessation of movement: there is no response to static stretch. Stretching is indicated by a downward deflection of the angle record. *B* The EMG: velocity relationship. In the hamstrings muscles of each of the 3 patients illustrated, above a threshold velocity, the relationship between the velocity of the stretching movement and the reflex EMG is approximately linear. [From BURKE, GILLIES and LANCE, 1971.]

subsided, there being little or no response to maintained static stretch (fig. 1 A). These findings differ from those found in the decerebrate cat, but they are not surprising since spasticity is often defined in terms of a velocity-dependent increase in muscle tone. The stretch reflex of spastic man is thus a dynamic stretch reflex, as has been noted by earlier authors [RONDOT, DALLOZ and TARDIEU, 1958; RUSHWORTH, 1960; SHIMAZU, HONGO, KUBOTA and NARABAYASHI, 1962; HERMAN, 1970]. There is no electrical activity in spastic muscle at rest [HOEFER and PUTNAM, 1940; HOEFER, 1941, 1952; WEDDELL, FEINSTEIN and PATTLE, 1944], and slow stretching may fail to evoke a reflex response [RONDOT, DALLOZ and TARDIEU, 1958; FOLEY, 1961; HERMAN, 1970]. For many patients, a threshold velocity can be defined below which no EMG is produced.

In both the quadriceps and hamstrings, the relationship between the velocity of stretch and the integrated EMG response was found to be approximately linear over a wide range of velocities (fig. 1 B), a finding which had not previously been reported. The dynamic nature of the stretch reflex of any one patient can therefore be quantified in terms of this EMG: velocity relationship. Any change in muscle tone will be reflected by a change in this relationship, seen in the threshold velocity or in the slope of the relationship. Provided that standard surface electrode placements are used, recordings made from day to day are quite comparable [BUSKIRK and KOMI, 1970], so that alterations in spasticity, be they due to medication or to the passage of time, may be readily appreciated. These methods of testing have proved useful in the objective assessment of response to medication [cf. JONES, BURKE, MAROSSZEKY and GILLIES, 1970; ASHBY, BURKE, RAO and JONES, 1972] and in the analysis of the mechanism of drug action [BURKE, ANDREWS and KNOWLES, 1971].

The prominent velocity dependence of the spastic stretch reflex is consistent with the dynamic sensitivity of the spindle primary ending [MATTHEWS, 1963, 1964]. The absence of a maintained response to static stretch suggests that, in spasticity, either the motoneurone is capable of reacting only to the dynamic components of the reflex input or, alternatively, that the input itself is predominately dynamic, static stretch producing little maintained afferent activity. One would expect the latter to be correct if one accepts the postulate of RUSHWORTH [1960, 1964a, b] and of JANSEN [1962] that the dynamic nature of the spastic stretch reflex results from dynamic fusimotor drive. However, whether dynamic fusimotor neurones are *hyperactive* or not in human spasticity is at this stage conjectural. Elucidation must await the application of the microneurographic techniques of HAGBARTH, VALLBO and colleagues.

B. The Response to Changes in Muscle Length

Because the reflex response occurs during the stretching movement and subsides on cessation of movement, the effects of increasing muscle length have been studied using two forms of stretching: linear stretching movements in which the movement is divided into 3 sequential steps of equal amplitude, each performed at the same velocity of stretch; and small sinusoidal oscillations of the limb performed at different centres of oscillation.

The effect of increasing muscle stretch on the stretch reflex was found to differ in the quadriceps and hamstrings. With linear stretching movements the reflex response of quadriceps was greatest in the first step of the movement, but decreased and disappeared in subsequent steps (fig. 2A). With sinusoidal oscillations of small amplitude little reflex EMG could be evoked when the centre of oscillation was in the stretched position

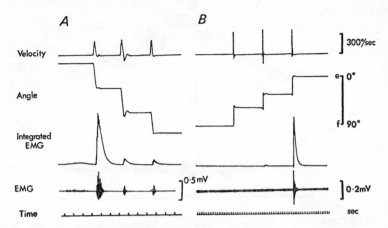

Fig. 2. A Diminution in the reflex response of quadriceps with increasing muscle length. The stretching movement (downwards on angle record) has been divided into 3 sequential steps of equal amplitude, performed at the same velocity of stretch. Resulting EMG activity is maximal in the first step, and further stretching evokes only small responses. [From BURKE, GILLIES and LANCE, 1970.] *B* Increase in the reflex response of hamstrings with increasing muscle length. A significant reflex response is produced only by the last step of the stretching movement which approaches the fully extended position of the knee joint (the most stretched position for the hamstrings). Before the stretch reflex had fatigued to a stable level, EMG was evoked by earlier steps of the stretching movement, but the response increased with subsequent steps. [From BURKE, GILLIES and LANCE, 1971.]

of the quadriceps, but the EMG increased progressively as the limb was moved back to less stretched positions, becoming maximal at the fully extended position of the knee joint, the least stretched position of the quadriceps (fig. 3). It thus appears that the effect of increasing muscle stretch on the quadriceps stretch reflex is inhibitory.

The response of the hamstrings to changes in muscle length was the reverse. With linear stretching movements EMG became maximal in the last step of the movement, which approached the position of greatest stretch (fig. 2 B). Similarly the response to small sinusoidal oscillations of the limb was maximal when the limb was oscillating at the position of greatest stretch of the hamstrings, but diminished and disappeared when the centre of oscillation was moved to a less stretched position.

Thus the effect of changes in muscle length on the stretch reflexes is inhibitory in the quadriceps and facilitatory in the hamstrings. To ensure that afferents from joint receptors do not play a significant role in the inhibitory effect of muscle stretch on the quadriceps stretch reflex, the

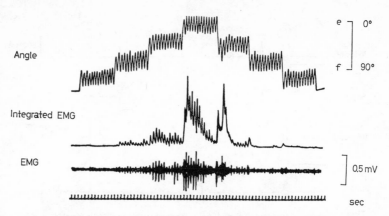

Angle

Integrated EMG

EMG

e ⌐ 0°

f ⌐ 90°

0.5 mV

sec

Fig. 3. Diminution in the reflex response of quadriceps with increasing muscle length. Small sinusoidal oscillations of the limb are performed while the centre of oscillation is systematically varied. The reflex response becomes maximal when the limb is oscillating about the fully extended position of the knee joint (the least stretched position of the quadriceps), but diminishes and disappears as the centre of oscillation is moved to a more flexed (more stretched) position. [From BURKE, GILLIES and LANCE, 1970.]

knee joint receptors were blocked by ischaemia, maintained until joint proprioception was abolished. Reflex EMG was then of smaller amplitude but the inhibitory effect of muscle stretch could be demonstrated as readily, and a clasp-knife phenomenon could still be elicited clinically.

In order to confirm these results, experiments were carried out to determine at which phase of a sinusoidal stretching cycle the reflex responses of quadriceps and hamstrings become maximal [BURKE, ANDREWS and GILLIES, 1971]. The EMG potentials produced by sinusoidal stretching movements were amplified and full-wave rectified, and the responses to a number of stretching cycles were averaged on a fixed programme averaging computer which was triggered when the goniometer output exceeded a preset trigger voltage. Joint position was averaged simultaneously so that precise phase relationships could be calculated. Accurate sinusoidal stretching movements were generated manually by matching the excursion of the limb with a sine wave of known frequency produced by a function generator.

The reflex responses were largely confined to the stretching phases of the sinusoidal cycle. Peak EMG of hamstrings occurred late in the stretching phase, slightly in advance of the position of greatest stretch (fig. 4B). The 'phase lead' increased as the rate of sinusoidal stretching

Fig.4. Phase relationships of the reflex response to muscle length during sinusoidal stretching at 0.5/sec. *A* Traces are from the quadriceps, in which the reflex response becomes maximal early in the stretching phase, at a position advanced 100° on the position of greatest stretch. *B* Hamstrings. The reflex response occurs late in the stretching phase, leading the position of greatest stretch by approximately 30°. 40 stretching cycles have been averaged in each record. Upward deflection of the (lower) angle recording indicates stretch of the appropriate muscle (calibration, 100°). [From BURKE, ANDREWS and GILLIES, 1971.]

increased, so that the peak of the reflex activity approached the middle of the stretching phase, the position of greatest velocity of stretch (fig. 5). Reflex activity of quadriceps occurred in the first half of the stretching phase in advance of the position of greatest velocity (fig. 4A). This phase lead was not altered significantly by altering the rate of stretching.

Considered with the results reported above, the phase relationships of the hamstrings stretch reflex may be interpreted in terms of a velocity-dependent reflex in which the effect of muscle stretch is facilitatory. Thus hamstrings EMG tends to peak near the position of greatest stretch at low rates of stretching, but approaches the position of greatest velocity as the rate of stretching increases. Such an interpretation cannot be invoked to explain the phase relationships of the quadriceps stretch reflex, since even extreme velocity sensitivity will not produce maximum EMG in the first half of the stretching phase unless the effect of muscle stretch is inhibitory. Thus the most satisfactory explanation of these experiments is again that the effect of muscle stretch is inhibitory in the quadriceps but facilitatory in the hamstrings.

The spindle secondary ending responds selectively to changes in muscle length [MATTHEWS, 1964]. In spinal preparations, the reflex effects of its group II afferent fibres are believed to be those of the so-called 'flexor reflex afferents' (FRA), in that they produce autogenic inhibition of extensor reflexes and autogenic facilitation of flexor reflexes [ECCLES and LUNDBERG, 1959a; LAPORTE and BESSOU, 1959]. The responses of the stretch reflexes of the spastic quadriceps and hamstrings muscles to changes in muscle length are thus compatible with the properties of the secondary

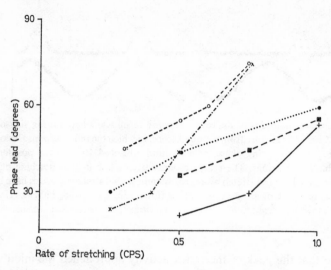

Fig. 5. The effect of increasing rate of stretching on the phase relationships of the hamstrings stretch reflex. For each of the 5 patients illustrated an increased rate of stretching results in an increased lead of the reflex response on the position of greatest stretch. [From BURKE, ANDREWS and GILLIES, 1971.]

ending and the synaptic action of its group II afferent pathway in spinal animals. Could any other stretch receptors be responsible for these effects?

Recent studies of Golgi tendon organ behaviour suggest that these receptors are best thought of as contraction receptors since their response to passive changes in muscle length is comparatively small [HOUK and HENNEMAN, 1967], especially in the absence of muscle contraction [STUART, MOSHER and GERLACH, 1971]. Prolonged inhibition of the quadriceps stretch reflex is therefore difficult to explain on the basis of tendon organs since their activity, by inhibiting muscle contraction, would abolish their adequate stimulus. Moreover, flexor tendon organs have an autogenic inhibitory effect on flexor motoneurones as extensor tendon organs have on extensor motoneurones [BIANCONI, GRANIT and REIS, 1964 a, b; GREEN and KELLERTH, 1967], so it is difficult to explain the differences between the stretch reflexes of quadriceps and hamstrings by the activation of tendon organs.

Could quadriceps primary spindle endings be responsible for the inhibition of the quadriceps stretch reflex through a presynaptic inhibitory mechanism? Certainly in normal man muscle vibration may induce such

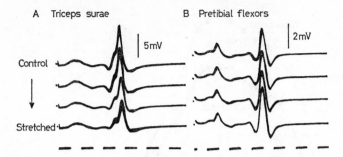

Fig.6. A The effect of static muscle stretch on the H-reflex of triceps surae. The triceps surae has been stretched from the control position (upper trace) in steps of 10°. The most stretched position is the lowest trace. There is progressive suppression of the H-reflex, but the M response is unchanged. Five sweeps have been superimposed at each muscle length. Time marker: 100/sec. *B* The effect of static muscle stretch on the H-reflex of the pretibial flexors. Increasing stretch of the pretibial flexors increases the H-reflex. Five traces superimposed at each position. Time marker: 100/sec. [From BURKE, ANDREWS and ASHBY, 1971.]

an intense IA afferent discharge that the monosynaptic reflexes of extensor muscles are suppressed by the resulting presynaptic inhibition [GILLIES, LANCE, NEILSON and TASSINARI, 1969; BARNES and POMPEIANO, 1970a, b; DELWAIDE, 1971]. However, it is probable that presynaptic inhibitory mechanisms are suppressed in spasticity [DELWAIDE, 1969, 1971; BURKE and ASHBY, 1972]. Furthermore, in both normal and spastic man the tonic vibration reflex increases as muscle stretch increases [EKLUND and HAGBARTH, 1966; HAGBARTH and EKLUND, 1968; BURKE, ANDREWS and LANCE, 1972], due to an increase in group IA firing rate [HAGBARTH and VALLBO, 1967, 1968]. It therefore appears unlikely that tonic reflex pathways can be suppressed by this mechanism.

Thus of all stretch receptors, the spindle secondary ending is the receptor most likely to be responsible for the differing effects of muscle stretch on the quadriceps and hamstrings stretch reflexes. To test this hypothesis further an attempt was made to isolate the responses to changes in muscle length uncontaminated by dynamic factors. Accordingly a study has been made of the effect of static muscle stretch on test H-reflexes of patients with spinal spasticity [BURKE, ANDREWS and ASHBY, 1971].

In spastic man, conditioning stretch of the triceps surae, induced by passive dorsiflexion of the ankle, was found to inhibit the H-reflex of triceps surae (fig. 6A and 7). Muscle stretch was the adequate stimulus for this inhibition since it was readily reproduced by pressure on the Achilles

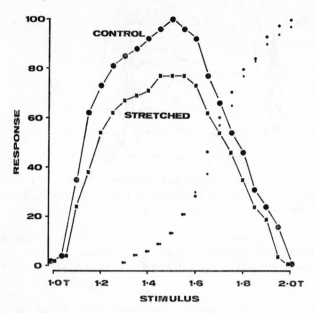

Fig. 7. The effect of muscle stretch on the recruitment curve of the H-reflex of triceps surae. With a muscle stretch of 10° the amplitude of the H-reflex (larger symbols) diminishes but that of the M response (smaller symbols) does not change. The suppression of the H-reflex is present at all stimulus intensities. Five traces have been averaged at each stimulus intensity. The amplitudes of the H and M responses are expressed as a percentage of their maximal values. The stimulus intensity is recorded in multiples of that stimulus threshold (T) for the H-reflex. [From BURKE, ANDREWS and ASHBY, 1971.]

tendon, the position of the ankle remaining constant. Stimulation of cutaneous receptors by squeezing the Achilles tendon rather than stretching it did not result in inhibition. It may be concluded that the endorgans responsible were muscle stretch receptors. The autogenic inhibitory effect of muscle stretch was abolished by infiltration of the triceps surae with dilute solutions of procaine which blocked the ankle jerk without altering the direct motor response. If dilute procaine preferentially blocks small nerve fibres, the abolition of the inhibitory effect may be attributed to reduction in the sensitivity of spindle receptors by fusimotor blockade, or to inactivation of small afferent fibres.

During selective block of large afferent fibres induced by ischaemia, muscle stretch still inhibited the H-reflex of triceps surae. The degree of inhibition was, if anything, more obvious during ischaemia than before it, presumably due to abolition of a facilitatory group IA effect. It seems likely then that the inhibitory effect of muscle stretch was mediated by small afferents, probably of muscle spindle origin. Such criteria are fulfilled by the group II afferents from the secondary spindle ending. If these were the responsible afferents, the effect of muscle stretch on flexor reflexes should be facilitatory and indeed, passive plantar-flexion of the ankle, stretching the pretibial flexors, was found to increase the H-reflex of the pretibial flexors in those patients in whom a consistently reproducible H-reflex could be evoked (fig. 6B).

Thus it seems probable that, while the dynamic sensitivity of the primary ending determines the velocity-dependence of the spastic stretch reflex, the responses to changes in muscle length are primarily the result of the reflex effects of the secondary ending. Presumably in lower limb muscles the static response of the primary ending is dominated by that of the secondary ending.

C. The Clasp-Knife Phenomenon

In view of the above, it would be appropriate at this stage to re-examine the postulated mechanisms of the clasp-knife phenomenon. First the phenomenon should be defined (fig. 8).

The clasp-knife phenomenon is said to occur in the stretch reflexes of antigravity muscles, being best developed in the quadriceps. The resistance

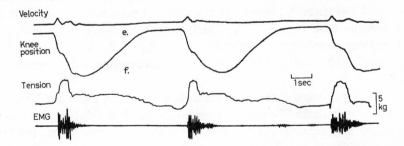

Fig. 8. The clasp-knife phenomenon of the quadriceps stretch reflex. The stretch reflex was elicited by flexing the knee, from the fully extended position (e.) to 90° flexion (f.). The stretch reflex is felt by the examiner as a 'catch' in the movement and is seen in the tension and EMG records. The resistance to stretch slows the stretching movement, as is seen in the velocity and goniometer (knee position) records. As the resistance is overcome the movement proceeds, and the reflex response diminishes. This is felt by the examiner as a 'give' in resistance. [From BURKE, GILLIES and LANCE, 1970.]

to passive stretch is said to increase to a critical tension at which the stretch reflex is inhibited so that further stretching may be accomplished without opposition [PATTON, 1965]. As a result the sensation of resistance experienced by the examiner has been likened to that experienced when closing the blade of a pen-knife. Generally the clasp-knife phenomenon is considered the equivalent of the lengthening reaction described by SHERRINGTON in the decerebrate cat [WALSHE, 1919, 1929; JANSEN, 1962; RUSHWORTH, 1964b; PATTON, 1965], although his original descriptions of the lengthening reaction distinguished two distinct phenomena, one occurring in the decerebrate cat and one in the chronic spinal dog [SHERRINGTON, 1909]. The reaction in the spinal dog was described by SHERRINGTON as follows: if the examiner bends the knee 'against the knee extensor's contraction he feels the opposition offered by the extensor give way almost abruptly at a certain pressure; the knee can then be flexed without opposition'. The essential feature appears to be abolition of resistance *during the stretching movement,* so this phenomenon corresponds well with the clasp-knife phenomenon of spastic man.

In the decerebrate cat, SHERRINGTON [1909] described a different phenomenon: 'starting with the knee in a posture of extension due to its extensor's rigidity, the observer forces it into flexion. On cessation of the flexion force the knee is found to remain in the new posture'. The essential features of this phenomenon are that *on cessation of movement* the resistance to stretch diminishes, but is *not* abolished so that some tone remains to maintain the new posture, and that further flexion meets with renewed resistance. Obviously, in the different preparations, two distinct phenomena have been described and, since their mechanisms probably differ, they should not be equated. In this paper the term 'lengthening reaction' will be restricted to that phenomenon occurring in the decerebrate cat.

The dependence of the clasp-knife phenomenon on a threshold tension is often emphasized: 'if excessive tensions are reached, inhibition of motoneurones may supervene and resistance collapses' [RUSHWORTH, 1964a]. The inhibition is usually attributed to the autogenic inhibitory effects of Golgi tendon organs activated by excessive muscle tension [JANSEN, 1962; PATTON, 1965]. LANDAU [1969] has postulated that, in addition to tendon organ effects, muscle contraction unloads spindle receptors, decreasing the excitatory afferent input. However, it is doubtful whether significant spindle shortening could occur during a stretching movement. Most authors would agree that the clasp-knife phenomenon is a tendon organ phenomenon induced by the excessive muscle tensions generated by the stretch reflex. Tendon organs have a very low threshold to muscle contraction [HOUK and HENNEMAN, 1967; STUART, GOSLOW, MOSHER and REINKING, 1970; STUART, MOSHER and GERLACH, 1971], so that they will be recruited as their motor units are recruited into the developing contraction, rather than *en masse* at some arbitrary total muscle tension.

Before one attempts to establish which receptor is responsible for the clasp-knife phenomenon, one must first establish that it is in fact an autogenic inhibitory phenomenon. Muscle tone depends upon complex mechanisms of which reflex activity is but one part. The mechanics of joint and limb, and the active contractile and passive visco-elastic properties of muscle also contribute to the resistance which the examiner appreciates as muscle tone. These non-reflex properties of muscle produce a resistance which increases with muscle length and increases with the velocity of stretch [RACK, 1966], and indeed it is possible that they may alter in the pathological states of spasticity [HERMAN, 1970]. It has been suggested by HERMAN [1970] that the dynamic sensitivity of the reflex and non-reflex components of muscle tone determine the occurrence of the sensation described as the clasp-knife phenomenon; that the clasp-knife phenomenon is not an autogenic inhibitory phenomenon, but a pseudo-inhibition produced by a fall in muscle tone when the resistance to stretch slows the stretching movement.

In the quadriceps of spastic man this does not appear to be so. Clinical observations reveal a marked decrease in muscle tone once the limb has been flexed past the position of the clasp-knife sensation [BURKE, GILLIES and LANCE, 1970]. This decrease in muscle tone persists as long as the flexed posture is maintained and indeed the limb may be oscillated freely in this region without any appreciable change in tone or reflex EMG. However, if the knee joint is extended so that the limb enters the region where the stretch reflex is active, similar movements evoke marked resistance and prominent reflex EMG (cf. fig. 3). As a result, if the knee jerk is elicited with the patient lying on a couch with knees flexed to 45°, as is usually done in normal clinical examination, the response may be clonic, because during relaxation the lengthening of the muscle is sufficient to elicit a brisk reflex response. If however the tendon is tapped when the knee joint is flexed to 90°, provided the initial excursion of the limb does not extend into the region where the stretch reflex is active, a pendular knee jerk may be evoked (fig. 9), because the lengthening phase after contraction now fails to evoke a reflex response. The clasp-knife phenomenon

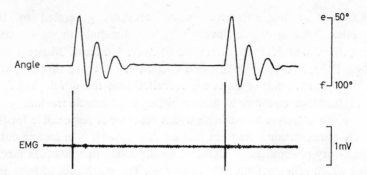

Fig. 9. Pendular knee jerk recorded in a spastic patient: With the knee at 90° flexion, a tendon tap evokes a reflex EMG response which extends the knee to approximately 50°. The limb oscillates like a damped pendulum because in this position of the knee joint the stretch reflex is suppressed so that the stretching phase of the oscillation fails to evoke a reflex response. [From BURKE, GILLIES and LANCE, 1970.]

of the quadriceps stretch reflex thus appears to be accompanied by a genuine suppression of the tonic stretch reflex, which persists as long as muscle stretch is maintained. Furthermore, although the hamstrings stretch reflex appears to be as sensitive to dynamic factors as the quadriceps stretch reflex, a clasp-knife sensation is not elicited on stretching the hamstrings. Muscle tone and reflex response become maximal at the fully stretched position of the hamstrings. The absence of a clasp-knife phenomenon in the hamstrings stretch reflex is readily explained since the autogenic effect of spindle secondary endings on flexor motoneurones is facilitatory.

Stretch Reflexes of Upper Limb Muscles

The reflex pattern of upper limb muscles was studied mainly in cases of hemiplegic spasticity [ASHBY and BURKE, 1971] using the same methods as for lower limb reflexes. In both triceps and biceps brachii a prominent dynamic stretch reflex was found. As in lower limb muscles, a threshold velocity could often be defined for the stretch reflex, as also reported by SIMONS and BINGEL [1971], and the reflex response appeared to be linearly related to the velocity of stretch. In the biceps of some patients a maintained response to static stretch could be elicited but this was not as prominent as the dynamic stretch reflex.

In both biceps and triceps, whether studied using stepwise linear stretching movements or small amplitude sinusoidal movements, increasing muscle stretch resulted in an increasing reflex response. The effect of muscle stretch was facilitatory in both muscles. This was confirmed by determining at which phase of a sinusoidal stretching cycle reflex EMG became maximal. In both muscles peak EMG activity occurred in the second half of the stretching phase of the cycle, and moved closer to the position of greatest velocity as the rate of stretching increased.

It appears then that in both muscles the net reflex effect of stretch receptor responses to changes in muscle length is facilitatory. It is therefore likely that if group II afferents have an inhibitory action on motoneurones of one of the upper limb muscles, this inhibitory effect is dominated by facilitatory effects of the group IA-mediated primary ending response to changes in muscle length.

It has been reported that a clasp-knife phenomenon may be elicited in upper limb stretch reflexes. In this study we were unable to demonstrate an autogenic inhibitory effect of muscle stretch and, on careful clinical testing, reflex tension did not disappear even in the most stretched positions provided that an adequate velocity of movement was maintained. Clearly then this clasp-knife sensation must be differentiated from the clasp-knife phenomenon of the quadriceps stretch reflex, in which muscle tone is greatly diminished or abolished past a certain degree of muscle stretch.

The apparent clasp-knife phenomenon reported in upper limb stretch reflexes may be adequately explained by the dynamic sensitivity of the reflex, as postulated by HERMAN [1970]. A fast stretching movement evokes a large reflex response which slows the stretching movement thus diminishing the dynamic stimulus and the resulting resistance. Moreover as the end of the stretching movement is approached the examiner tends to slow the stretching movement and thus does not maintain the same dynamic stimulus. The resistance opposing muscle stretch therefore falls, imparting a clasp-knife sensation. If, however, the amount of EMG per unit velocity at differing muscle lengths is calculated, there is greater reflex response at greater muscle lengths. It is suggested, therefore, that the term 'clasp-knife phenomenon' be reserved for those instances in which autogenic inhibition can be demonstrated, as in the quadriceps stretch reflex, and that the term 'pseudo clasp-knife reaction' be used to describe the sensation found in other muscles such as those of upper limb.

Supraspinal Control of the Reflex Effects of Muscle Stretch

The reflex effects of group II afferents, other FRA, and IB afferents, may be inhibited by pathways arising in the brain stem [ECCLES and LUNDBERG, 1959 b], and it has been suggested that such inhibition is one of the factors underlying decerebrate rigidity [GILLNER and UDO, 1970]. However, these reticulospinal pathways are not the only ones influencing transmission from the FRA. Stimulation of the pyramidal system increases the reflex effects of the FRA [LUNDBERG and VOORHOEVE, 1962], probably by an action on the interneurones of FRA pathways [LUNDBERG, NORRSELL and VOORHOEVE, 1962]. A lesion of pyramidal pathways would thus remove a supraspinal action facilitating transmission from the FRA. In spasticity, since the reflex effects of the secondary spindle ending appear to be demonstrable as the clasp-knife phenomenon, the removal of the pyramidal influence on FRA transmission must be compensated for by lesions in pathways inhibiting FRA transmission. The major system controlling transmission from the FRA is the 'dorsal reticulospinal system', which arises medially from ponto-medullary reticular formation and traverses the dorsolateral funiculus of the spinal cord, exerting an inhibitory action on the first order interneurone of the FRA pathway [HOLMQVIST and LUNDBERG, 1959, 1961; ENGBERG, LUNDBERG and RYALL, 1968 a, b]. In the decerebrate cat this system appears to 'switch off' the reflex pathways from the FRA at an interneuronal level.

Although SHERRINGTON [1909] described the lengthening reaction in the stretch reflex of extensor muscles of the decerebrate cat, he subsequently reported that 'so long as the stretch increases the reflex continues to increase' [LIDDELL and SHERRINGTON, 1924]. This has been confirmed by GRANIT [1958] who measured the reflex response under static conditions, and by MATTHEWS [1959] who measured the reflex response during slow stretching. This behaviour clearly differentiates the decerebrate stretch reflex from the spastic stretch reflex, and it appears probable that these differences arise from suppression of FRA pathways in the decerebrate state and release of these pathways in the spastic state. To investigate this possibility attempts have been made to reproduce the length-dependent reflex responses typical of spastic man in the decerebrate cat using acute or chronic spinal cord lesions and acute brain-stem lesions [BURKE, KNOWLES, ANDREWS and ASHBY, 1972]. Initially the reflex behaviour of 'intact' decerebrate cats was studied, all experiments being done using experimental methods similar to those used in spastic man.

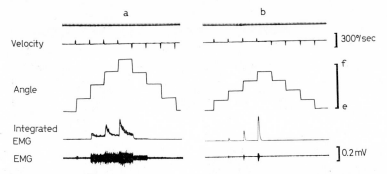

Fig. 10. Stretch reflex of the decerebrate cat. Two types of decerebrate preparation are illustrated: the intercollicular decerebrate (*a*) and the high decerebrate (*b*). In *b*, the stretch reflex consists of only a dynamic response to muscle stretch. In *a*, there is a prominent dynamic response, but this subsides to a plateau, static level on maintained stretch. In both preparations, the reflex response increases with increasing muscle stretch. Stretch is an upward deflection of the angle record. Time marker 1 sec/division. Amplitude of stretch (f) 100° for both preparations (e = 0°). [From BURKE, KNOWLES, ANDREWS and ASHBY, 1972.]

In agreement with earlier authors, it was found that reflex activity was more readily elicited from the extensor muscles of the decerebrate cat than from the flexors. When the level of midbrain section was intercollicular (as used by SHERRINGTON), dynamic and static components of the stretch reflex could be easily distinguished, although the prominence of either component could be altered by varying the level of midbrain section. Both components increased as muscle stretch increased, seen using linear stretching movements in figure 10, and using small sinusoidal oscillations in figure 11 A. With intercollicular decerebration the reflex activity of the dynamic component of the stretch reflex subsided on cessation of movement to a lower static level during maintained stretch (fig. 10 A). No autogenic inhibitory effect of muscle stretch could be demonstrated, in agreement with the earlier authors. This was not surprising since the pathways likely to produce such autogenic inhibition are suppressed in the decerebrate state. In fact, MATTHEWS [1969] has suggested that the autogenic effects of spindle secondary endings are facilitatory to extensors of the decerebrate cat, the reverse of the spinal pattern. The acceptance of this postulate would only strengthen the conclusion that in the decerebrate state there is no autogenic inhibitory activity capable of explaining the lengthening reaction. What then is the lengthening reaction described by SHERRINGTON in the stretch reflex of intercollicular decerebrate cats? It

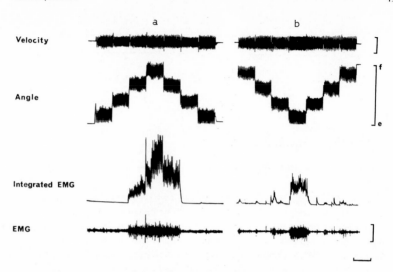

Fig. 11. Release of flexor reflex afferents in the decerebrate cat by a lesion of the dorsal quadrant of the spinal cord. The length-dependent facilitation of the quadriceps stretch reflex characteristic of the decerebrate cat (*a*) is altered to length-dependent inhibition characteristic of the spinal cat (*b*) by sectioning the dorsal quadrant of the spinal cord. *a* Control record. The effect of muscle stretch is facilitatory. Note that stretching is an upward movement of the angle record. *b* After ipsilateral dorsal quadrant lesion in the same cat. The effect of muscle stretch is inhibitory, as in chronic spinal man. Calibrations: velocity 300°/sec; angle e = 0°; f = 100°; EMG 0.5 mV; time 10 sec. [From BURKE, KNOWLES, ANDREWS and ASHBY, 1972.]

is suggested that the lengthening reaction is not an autogenic inhibitory phenomenon, but is due to the passive decline of the dynamic response to muscle stretch to a plateau static level on cessation of movement. This conclusion strengthens the view that the lengthening reaction and the clasp-knife phenomenon are quite dissimilar, and that the decerebrate cat is a poor model for spastic man.

In those chronic spinal cats which developed increased tone in the extensor muscles of the hind limb, the reflex pattern found was similar to that described earlier in spastic man. Increasing muscle stretch inhibited the stretch reflex of quadriceps, but facilitated the stretch reflex of hamstrings. Using acute and chronic partial spinal cord lesions it was found that the only consistent lesion required to reverse the length-dependent facilitation of the quadriceps stretch reflex of the decerebrate cat to the inhibition of the chronic spinal cat was a lesion in the dorsal half of the lateral column (fig. 11). Such lesions also released flexor re-

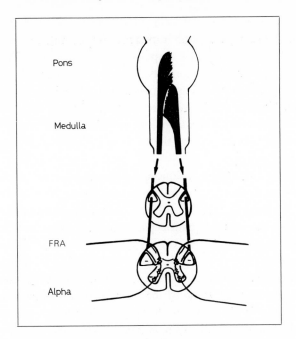

Fig. 12. Control of flexor reflex afferent pathways from the brain stem by the dorsal reticulospinal system. The two components of this system are illustrated: that arising from pontine reticular formation and controlling the inhibitory actions of the FRA, and that arising from medullary reticular formation and controlling the facilitatory actions of the FRA. Both components are bilateral. [From BURKE, KNOWLES, ANDREWS and ASHBY, 1972.]

sponses to noxious cutaneous stimuli, and a stretch reflex appeared in the hamstrings. Midline transverse lesions in the caudal brain stem also released the inhibitory effect of muscle stretch (and the inhibitory effects of other FRA) on the quadriceps stretch reflex, much as spinal lesions had done. However, the release of the flexor-facilitatory effects did not occur after lesions in the caudal pons, and only followed similar lesions in the medulla. Thus with higher lesions it was possible to release inhibitory actions on the quadriceps without also releasing the flexor-facilitatory effects.

The pathways defined are shown in figure 12. They appear identical to those described by LUNDBERG and colleagues who studied the synaptic actions evoked by electrical stimulation of muscular, cutaneous and articular nerves. It thus appears that an understanding of the control of flexor reflex afferent actions may be used to explain differences in different forms of spasticity [BURKE, KNOWLES, ANDREWS and ASHBY, 1972].

In spasticity due to spinal lesions the clasp-knife phenomenon may be demonstrable if the quadriceps retains sufficient muscle tone, and in addition there will be a heightened flexor response to noxious stimuli. With the complete release of flexor reflex afferents following spinal transection there may be such suppression of extensor tone that paraplegia-in-flexion develops, unless skin, bladder and bowels are adequately cared for. With adequate patient care there is less FRA stimulation, with less extensor inhibition and less flexor facilitation, so that the basic spinal pattern of paraplegia-in-extension may emerge.

In spasticity due to brain-stem lesions a pattern similar to that of spinal lesions may occur if there is medullary involvement. With lesions largely restricted to the pons or higher centres, a clasp-knife phenomenon may be just as prominent in the quadriceps as in the spinal patients (extensor-inhibitory action), but the flexor responses to noxious stimuli (flexor-facilitatory actions) will be less prominent. With cortical or internal capsular lesions, which are unilateral, partial suppression of brain-stem centres may be adequate to allow some of the phenomena of spinal spasticity, such as the clasp-knife phenomenon, to appear while others are still at least partially inhibited. Alternatively it is possible that the degree of inhibitory control of the FRA exerted by the dorsal reticulospinal system is normally small in intact man, as has been found in the cat by HOLM-QVIST and LUNDBERG [1961], so that there may be no need to postulate a lesion of this system in spasticity due to cortical lesions. The clasp-knife phenomenon could result from a 'normal' degree of group II reflex activity superimposed on a hyperactive stretch reflex.

It is suggested, therefore, that the level of activity in the dorsal reticulo-spinal system plays a major role in determining the clinical manifestations of the spastic state, and in differentiating the features of cerebral and spinal spasticity.

Summary

The stretch reflexes of spastic patients have been analyzed in an attempt to determine the roles played by the primary and secondary spindle endings.

In lower limb muscles the reflex response increases as the velocity of stretch increases, but there is little maintained response to static stretch. These properties are consistent with the dynamic sensitivity of the spindle primary ending. In the extensor muscles of the lower limb muscle stretch inhibits the reflex response, but in flexor muscles the reflex response increases as muscle stretch increases. Consistent results were found using linear or sinusoidal stretching movements, or when the effect of static stretch on the H-reflex was studied. This pattern of response to changes in muscle length is attributed to activation of spindle secondary endings, and in the light of these experiments the mechanism of the clasp-knife phenomenon is reassessed.

In the upper limbs, the responses to muscle stretch of both the biceps and the triceps brachii increase with increasing stretch and a true clasp-knife phenomenon as found in the lower limb cannot be elicited. The sensation of an apparent decrease in resistance to stretch may be explained in these muscles by the extreme sensitivity to the velocity of stretch. If the reflex effect of the secondary ending is inhibitory to one of these upper limb muscles then it appears probable that such effects are dominated by the facilitatory action of the primary ending.

The stretch reflex of intercollicular decerebrate cats consists of both dynamic and static components, both of which increase as muscle stretch increases. It is therefore suggested that the lengthening reaction is not an autogenic inhibitory phenomenon, as is the clasp-knife phenomenon, but depends on subsidence of the dynamic response to muscle stretch on the cessation of movement. The facilitatory effect of muscle stretch on extensor reflexes could be converted to inhibition, as found in spastic man, by selective lesions in the spinal cord or brain stem. The 'dorsal reticulospinal system', previously described by LUNDBERG and colleagues, could thus be defined, and it is suggested that this system plays a prominent part in determining the clinical manifestations of different forms of spasticity.

Acknowledgements

This research has been supported by grants from the National Health and Medical Research Council of Australia, the Adolph Basser Trust, Mr. and Mrs. EDWIN STREET, Ciba Co. Ltd., and Merck, Sharp and Dohme Ltd. Mr. K. NORCROSS and Mr. N. SKUSE have provided able technical assistance.

Figures are reproduced with the permission of the editors of Brain, the Journal of Neurology, Neurosurgery and Psychiatry and Archives of Neurology.

Author's address: Dr. D. BURKE, Division of Neurology, The Prince Henry Hospital Little Bay, *Sydney, N.S.W. 2036* (Australia)

New Developments in Electromyography and Clinical Neurophysiology,
edited by J. E. Desmedt, vol. 3, pp. 496–507 (Karger, Basel 1973)

The Role of the Fusimotor System in Spasticity and Parkinsonian Rigidity

P. Dietrichson

Institute of Neurophysiology, University of Oslo, and Department of Neurology,
Ullevål Hospital, Oslo

The increased stretch reflex in spasticity presents itself as an increased resistance to a quick stretch of the muscle and as an increased tendon reflex. In parkinsonian rigidity the disordered stretch reflex appears as an increased resistance to a slow and maintained muscle stretch and it is independent of the speed of the stretch. The main purpose of the present article is to determine whether the two types of augmented stretch reflex in spasticity and in parkinsonian rigidity, respectively, might be related to increased fusimotor activity. Furthermore the duality of γ motor innervation of the muscle spindles is now well documented [cf. Matthews, this volume] and it is indeed tempting to enquire whether either the dynamic or the static fusimotor systems could be differentially involved in the pathophysiological mechanisms underlying the two types of augmented stretch reflex in these conditions. In order to explore both the dynamic and the static fusimotor systems in spastic and parkinsonian patients it is necessary to study both phasic stimulations such as tendon taps and single nerve stimuli, and also prolonged tonic stimulation such as maintained stretch. We did not investigate muscle vibration [cf. Hagbarth, this volume; Lance et al., this volume].

The early clinical studies of the possible significance of the fusimotor system in spasticity and rigidity led to contradictory results. Direct recording from the spindle loop, as in physiological animal experiments, was not possible in humans until 1968 when Hagbarth and Vallbo introduced microneurography of human muscle axons in intact nerve. The indirect methods on patients involve: recording the silent period; comparing the mechanically and electrically elicited stretch reflex; comparing the maximal electrically elicited stretch reflex to the maximal direct muscle re-

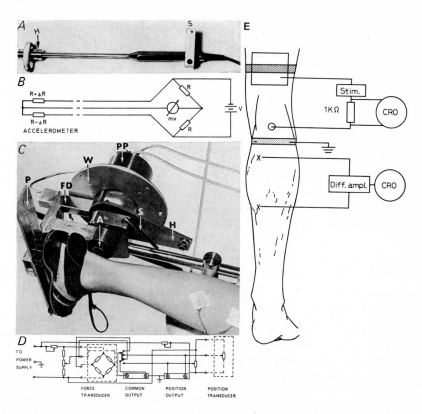

Fig. 1. A Mechanical stimulation. The housing (H) for the oil immersed silicone beam of the accelerometer is in the head of the hammer. The support (S) in the rear end is for calibration. Length of the reflex hammer = 24.5 cm; weight 302 g. *B* Diagram of the accelerometer and the Wheatstone's bridge arrangement. *C, D* Myograph (C) allowing both isometric and isotonic recording. The handle (H) can bring the force detector (FD) into contact with the pedal (P), and can be fixed by the wrench (S). The housing for the position potentiometer (PP) is mounted concentrically around the axis of the myograph (A). W is the pulley used for loading. When the myograph is used for the opposite leg, the pedal is positioned upside down. *D* The measuring circuit of the myograph. E Electrical stimulation and recording; posterior view of the right leg with the arrangement of the stimulating and recording electrodes for study of the ankle reflex.

sponse; studying the effect of reinforcement manoeuvres; performing fusi-motor blockade.

In the present study these indirect approaches were used with improved methodologies [DIETRICHSON, 1971 a, b, c; DIETRICHSON and SØRBYE, 1971].

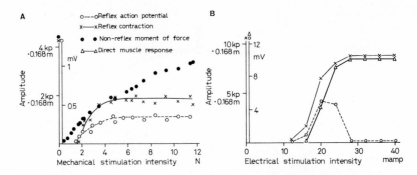

Fig. 2. Relation between stimulus intensity and size of the ankle reflex. Mechanical (A) and electrical (B) stimulation of the ankle reflex. Relation between stimulus intensity and reflex response both as reflex action potential and isometric reflex contraction.

Methods

Visual observation of the phasic stretch reflex response, described 100 years ago as the movement of the leg following a tap to the infrapatellar tendon [ERB, 1875; WESTPHAL, 1875] is still in use in the clinic. For experimental purposes a method is required by which both the size of mechanical and electrical stimuli and the amplitude of the mechanical and electrical responses can be quantitatively recorded. Particular emphasis was placed upon recording of the size of the mechanical stimulus, as the mechanical threshold of the stretch reflex permits evaluation of the phasic sensitivity of the muscle spindle [MATTHEWS, 1959].

For this purpose, a reflex hammer similar to that introduced by SØRBYE [1966] was used; a transducer in the hammer head signalled the retardation experienced by the hammer head when striking the tendon (fig. 1 A, B). It was found that the output signal of the hammer had a linear relationship to the non-reflex tension increase imposed upon the tendon-muscle system by the tap (fig. 2A). The hammer signal could therefore be used as an indication of muscle stretch.

The H-reflex was elicited by electrical stimulation of I_A afferents in the posterior tibial nerve. The strength of the electrical stimulus was estimated in mA by the voltage drop across a serial resistor in the stimulating circuit (fig. 1E). In normal subjects, the stimulating current amounted to about 40 mA. The size of the H- and M-responses evoked by the step-by-step increases in stimulus intensity is demonstrated in figure 2B.

The reflex potential was recorded by surface electrodes over the triceps surae muscle and the best recording electrode placement was determined.

The reflex contraction was recorded by a myograph allowing isotonic as well as isometric recordings (fig. 1 C, D). With some modifications this myograph was also used in testing the sensitivity of the muscle spindles to tonic stretch. The method described is easy to use, is not painful for the patient and permits accurate recording of ankle reflex with phasic mechanical stimulation and single electrical stimuli.

We examined groups of 15 spastics, 15 parkinsonian rigid patients and 15 normal subjects.

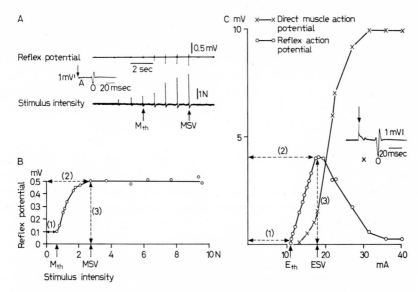

Fig. 3. A, B Mechanical stimulation. *A* Tendon taps of increasing intensities as indicated by increasing retardation force measured in Newtons (N) of hammer head (lower channel). Corresponding amplitude of reflex action potential shown in upper channel. Threshold and maximal stimulus intensity marked by arrows. Insert between the two channels: reflex action potential. Reflex latency measured from moment of tendon tap (arrow) to onset of reflex potential. Parkinsonian patient. *B* Relation between stimulus intensity (retardation force) and amplitude of the T-reflex. Taps are give to the most sensitive area of the tendon (see text). *C* Electrical stimulation. Relation between electrical stimulation intensity and amplitude of the H-reflex: (1) the stimulus threshold; (2) the electrical stimulus value required to elicit the maximal reflex response (3). The direct muscle response in this parkinsonian patient has a slightly higher stimulus threshold than the reflex response. By progressive increase of the stimulus intensity the direct response reaches a constant plateau. Insert (right) M- and H-response. Arrow points to the moment of stimulation.

The tendon jerk. With the patient prone, taps of varying intensity were applied to the most sensitive part of the Achilles tendon by means of the accelerometer hammer (fig. 3A). There was an increase in amplitude of the reflex potential on increasing intensities of the tendon tap (fig. 3B). The slope of the curve varied from subject to subject. On reaching a certain value, the reflex amplitude remained approximately constant in spite of a further increase of the tendon tap intensity. Thus, the following three parameters for the ankle reflex were obtained (numbers in brackets on fig. 3B): (1) the mechanical threshold value; (2) the mechanical stimulus required to elicit a maximal reflex potential, and (3) the maximal amplitude of the mechanically elicited reflex muscle action potential.

The electrically induced H-reflex. Electrical stimuli of increasing intensities were applied to the mixed tibial nerve in the popliteal fossa. The following three parameters for the electrically elicited ankle reflex were obtained (numbers in brackets in fig. 3C): (1) the H-reflex threshold; (2) the saturation intensity, and (3) the maximum amplitude of the H-reflex.

Results

Phasic Ankle Reflex in Spasticity and Parkinsonian Rigidity

Figure 4 presents for the 3 groups of subjects individual values of the threshold stimulus in relation to the maximum reflex response, for mechanical stimulation in A and for electrical stimulation in B. There is a clear separation of the data recorded in the normals and the spastics, only for the mechanically evoked T-reflex (fig. 4A).

Such a difference in mechanical sensitivity in spastics could be explained either by increased fusimotor innervation or by increased excitability of the motoneurone pool. A differentiation between these two possible mechanisms may be obtained by comparing the mechanically and electrically evoked stretch reflexes. Increased fusimotor activity would not, in principle, augment the H-reflex since the elicited afferent volley would not be influenced by the fusimotor activity.

Comparing the maximal amplitudes of the T- and H-reflexes, T/H ratio showed in normals a mean ratio of 0.29 (fig. 5). This low ratio indicates a more efficient and synchronous activation of I_A afferents by the electrical stimulus than by mechanical stimulation. In spastic patients the mean ratio was signi-

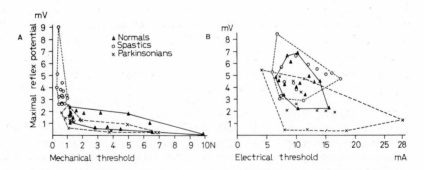

Fig. 4. Relation between mechanical threshold and maximal reflex amplitude in the individual 15 normal, 15 spastic and 15 parkinsonian subjects. The values in each group encircled. The same for electrical stimulation. Note separation between normals and spastics on mechanical stimulation only.

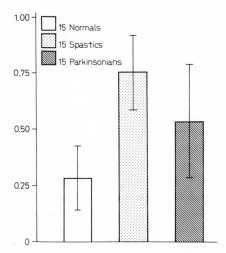

Fig. 5. T/H, mean ratio between the maximal amplitude of the mechanically and the electrically elicited ankle reflex.

ficantly higher (0.76) (p <0.01) which would suggest a lowered threshold of the muscle spindle receptors. If the spasticity was due to augmented α moto-neurone excitability both the T- and H-reflexes would be increased, leaving the T/H ratio unchanged from the normal value, which evidently is not the case. Also in parkinsonian patients the mean ratio (0.53) was higher than the normal value (p <0.01) (fig. 5).

A second method to test the possible role of fusimotor activity consists of a selective blockade of the fusimotor fibres by local anaesthesia of the muscle nerve. Such a selective blockade which would reduce the mechanically evoked stretch reflex, leaving the electrically evoked reflex intact, was attempted in 8 spastic patients. In all these patients a selective fusimotor blockade was obtained a few minutes after a successful injection, and it lasted for 5–35 min (mean = 10 min) after which there was also blocking of the α motor and I_A fibres. Within the same period, the spasticity decreased, as judged from the lowered resistance to quick passive movements of the ankle joint. An example is presented in figure 6 B, where an initially high ratio of 0.90 was reduced to a normal value with no effect on the direct and reflex muscle potential (lower record). This finding confirms RUSHWORTH's interpretation of early procaine experiments [RUSHWORTH, 1960] and supports the notion that an excess fusimotor innervation plays a role in the increased phasic stretch reflexes in spasticity.

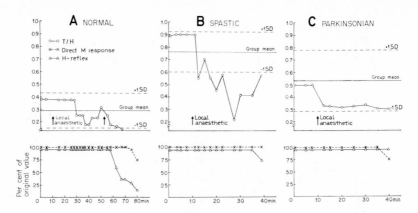

Fig.6. Fusimotor blockade. Effects on the maximal amplitude of the T- and H-reflexes by local anaesthesia of the tibial nerve (upper diagram) in a normal subject (A), a spastic (B) and a parkinsonian patient (C). Arrows indicate injection of 2.5 ml 0.25% Citanest ®. The selective fusimotor anaesthesia was controlled by unaltered amplitude of the direct M-response and of the H-reflex (lower diagram). Mean T/H ± 1 SD in 15 normals, 15 spastics and 15 parkinsonians indicated by horizontal unbroken and broken lines, respectively. Note short-lived normalization of the spastic ankle reflex in *B*.

Regarding the parkinsonian group, selective nerve block reduced the two highest T/H ratios (fig. 6C). Finally, the results suggest that fusimotor innervation also contributes to the normal ankle reflex, at least in some subjects (fig. 6A).

A higher ratio of the maximal reflex response to the maximal direct response (H/M) in spastics as compared to normals suggests that, in addition to excess fusimotor innervation, spasticity also involves increased α motoneurone excitability due to excitatory sources other than I_A afferents.

The wide variability in findings with the use of phasic stimulation in parkinsonian rigidity prevented definite conclusions regarding the role of fusimotor innervation in that condition. Such phasic stimulation is possibly not the most suitable mode of stimulation for testing possible changes in the fusimotor system in parkinsonian rigidity.

Tonic Ankle Reflex in Parkinsonian Rigidity and in Spasticity

Rigid muscles in parkinsonian patients exhibit increased resistance to slow and sustained stretching. The increased resistance is dependent on intact

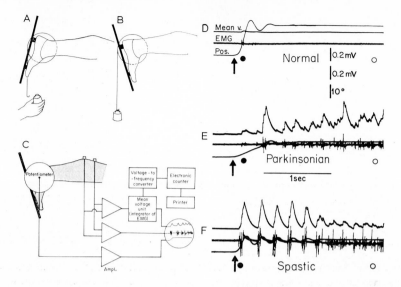

Fig. 7. A–C Tonic reflex recording. Position of the foot before (A) and after (B) passive dorsal flexion in the ankle joint. *C* Diagram of the experimental set-up for myographic (potentiometer) and EMG recording of the tonic ankle reflex response to a constant and sustained mechanical stimulus. *D–F* Tonic stretch reflexes. *D* Normal subject; *E* Parkinsonian, and *F* spastic patients. Onset of muscle stretch is indicated by arrow and by the upward deflection of the myograph position signal (Pos.). The reflex action potentials (EMG) and their integrated voltage (mean v.) recorded on the two upper channels demonstrate phasic (filled circles) as well as tonic components of the reflex response (open circles). The tonic reflex response was measured 3 sec after the onset of the stretch when the oscillations were moderate. In the parkinsonian and the spastic records the stable increase in muscle length was smaller and the tonic electric reflex response was larger than in the normal record.

dorsal roots and is commonly referred to as increased tonic stretch reflexes. One possible explanation of the increased tonic ankle reflexes would be an excess of fusimotor innervation [JANSEN, 1962]. This hypothesis predicts: (1) an increased reflex response to a given sustained mechanical stimulus, and (2) a reduction of the increased reflex response by a selective local anaesthesia of the fusimotor nerve fibres.

The aim of the present investigation is to clarify to what extent increased fusimotor innervation can explain the increased tonic stretch reflexes in parkinsonian rigid muscles.

The size of the tonic ankle reflex response to a constant submaximal stimulus was recorded electrically and mechanically (fig. 7). The electrical response was measured by integration, voltage-to-frequency conversion and

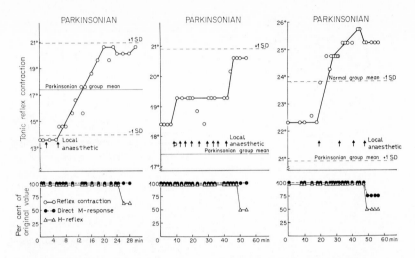

Fig. 8. Selective nerve blockade. Effects of local anaesthesia of the tibial nerve on increased tonic ankle reflex contraction in the triceps surae muscle in three parkinsonian patients (upper diagrams). Arrows indicate injection of 1.5–3 ml 0.25% Citanest. The period of selective nerve blockade was determined by unaltered amplitudes of direct and reflex muscle potential to electrical nerve stimulation (lower diagram). Mean and/or ± SD of tonic ankle reflex contraction to a constant mechanical stimulus in 20 parkinsonian patients and 10 normal subjects is indicated by horizontal unbroken or broken lines. Note decrease of the increased reflex response in the three patients.

counting. Mechanically the response was measured by the extension of the triceps surae, as indicated by the potentiometer of a myograph, giving the degree of dorsiflexion of the ankle joint at different loads. Ten normal controls, 20 parkinsonian patients, and 10 spastics were examined.

In normal subjects a maintained muscle stretch did not elicit a tonic reflex response (fig. 7 D). In all parkinsonians the same muscle stretch elicited definite tonic reflex responses (E). In spastics, the mean value of the responses was intermediate to the mean values obtained from the two former groups (F). A peripheral blockade of the fusimotor fibres was attempted in the tibial nerve in three parkinsonian patients (fig. 8). Control measurements to ensure the absence of any blocking effect on the alpha motor and I_A fibres were made by recording the direct M-response and the H-reflex. A weak solution (0.25%) of the local anaesthetic (α-propylamino-2-methyl-propionanilidi chloridum, Citanest® Astra) was injected around the tibial nerve in the popliteal fossa. The mechanical response to the sustained stretch of the triceps surae muscle (10 kg weight on the myograph), the M- and H-responses to a maximal shock to the mixed tibial nerve, as well as the rigidity were tested frequently. The injections were given in portions of 1.5–3 ml.

In all three parkinsonian patients (fig. 8) this procedure caused a gradual decrease of the reflex tonic contraction to lengthening of the muscle. This suggests that the increased tonic reflex response depends on the integrity of small-sized nerve fibres: either static fusimotor fibres or group II afferents. Since increased discharges from the secondary endings are caused by increased static fusimotor innervation, the latter seems to be important for the maintenance of rigidity. Recent observations of WALLIN, HONGELL and HAGBARTH [this volume], based on microneurography of human muscle afferents, confirm that the muscles in parkinsonians have an increased static sensitivity to maintain passive muscle stretch, suggesting an increased fusimotor innervation in these patients.

However, other possible causes for rigidity in parkinsonian patients are not excluded by the present experiments. The following mechanisms may operate in addition to the increased static fusimotor activity: (1) increased supraspinal input to tonic α motoneurones [STEG, 1964], and (2) increased excitability of interneurones, either mediating the tonic stretch reflex or inhibiting Renshaw interneurones [WILSON, TALBOT and KATO, 1964] or presynaptic inhibition on I_A afferent terminals [LUND, LUNDBERG and VYKLICKY, 1965], and thereby enhancement of the stretch reflex by removal of inhibition.

The Silent Period in Spastic, Rigid and Normal Subjects during Isotonic and Isometric Muscle Contractions

The silent period during a superimposed twitch contraction has been explained in various ways [MERTON, 1951; see ANGEL and STRUPPLER et al., this volume]. The present series of experiments started with a study of the silent period but no conclusive results were obtained because of variability in the silent period. The use of this test in clinical patients was difficult because in the same individual there was a wide variation in its duration; this was partly due to interference with tremor in parkinsonians. Therefore, in the subsequent experiments parkinsonian patients without tremor were used and the methods were standardized (fig. 9). More reliable data were obtained by averaging a large number of observations in each individual by means of a digital computer (fig. 10). It was found that in spastics the mean duration of the silent period was longer than in normals, whereas in parkinsonians it did not differ significantly from normals. Some parallels can be drawn between this finding

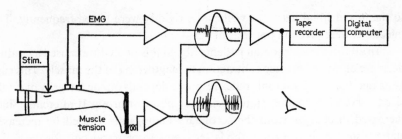

Fig. 9. Experimental arrangement for isometric recording of the silent period from the triceps surae muscle.

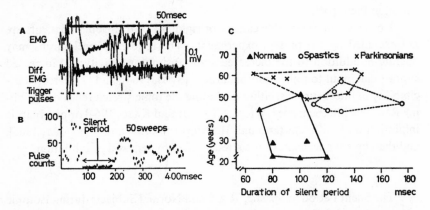

Fig. 10. A Single EMG trace from tape recorder, differentiated EMG and trigger pulses from the computer. The latter discriminated upward pulses higher than 37 μV of the differentiated EMG. *B* Magnified first part of computer histogram of 50 EMG traces, demonstrating a mean silent period of 115 msec. Normal subject. *C* The mean duration of the silent period (abscissa) in each individual whose age is plotted as ordinate. The values in each group encircled. No correlation between duration of the silent period and age.

in spastics and observations in animal experiments by JANSEN and RACK [1966] suggesting that the prolonged silent period in spastics may be interpreted as due to increased fusimotor innervation.

The study of the silent period during various twitch tensions was used to assess the role of autogenic inhibition in spasticity. The duration of the silent period, using isotonic and isometric contraction, was found to be independent of large variations in the twitch tension. This suggests that autogenic inhibition from the tension-sensitive tendon organs is not the main mechanism in the prolonged silent period in spastics.

Thus, the results obtained from the study of the silent period are also compatible with the notion of increased dynamic fusimotor innervation in spastics.

Author's address: D. P. DIETRICHSON, Department of Neurology, Ullevål Hospital, *Oslo 1* (Norway)

New Developments in Electromyography and Clinical Neurophysiology,
edited by J.E. Desmedt, vol. 3, pp. 508–522 (Karger, Basel 1973)

Human Monosynaptic Reflexes and Presynaptic Inhibition

An Interpretation of Spastic Hyperreflexia

P. J. DELWAIDE

Hôpital de Bavière, Liège

Over the years, the features of spastic hyperreflexia have suggested a
number of pathophysiological hypotheses derived from animal neurophysiology. As regards the γ loop, for example [MERTON, 1953; GRANIT, 1955], the
exaggerated monosynaptic reflexes can be accounted for by two different
mechanisms: (a) increased activity of the fusimotor system [RUSHWORTH,
1964], or (b) increased α motoneuron excitability [LANDAU and CLARE, 1964].

In spastic patients, BULLER [1957] found that the tendon response (T response) is more markedly increased than the H-reflex. Shortly afterward,
RUSHWORTH [1960] observed that procaine infiltration of nerve trunks, which
specifically paralyzes the fusimotor fibers [MATTHEWS and RUSHWORTH,
1958], reduces the hyperreflexia. These two findings seemed to suggest that
hyperreflexia results from γ hyperactivity.

Recent contrary evidence makes reassessment of these views necessary.
The schematic representations of γ loop organization have been complicated
by the results of animal studies showing that there is a very frequent involvement of inhibitory mechanisms situated upstream from the effector cell
[FRANK and FUORTES, 1957; ECCLES, 1964; LUNDBERG, 1967]. In addition,
experiments in man using vibratory stimulation, that causes a powerful and
rather specific activation of the primary endings of the muscle spindles
[BIANCONI and VAN DER MEULEN, 1963; MATTHEWS, 1966], have shown that
vibration exerts an inhibitory effect on monosynaptic reflexes [LANCE, 1965;
HAGBARTH and EKLUND, 1966; RUSHWORTH and YOUNG, 1966]. These results
are difficult to reconcile with the classical hypothesis of γ loop organization
which holds that the role of the I_A afferents is always facilitatory.

In our studies we sought first to determine how the inhibitory effect of
vibration in spastics differs from that seen in controls. The ensuing analysis

of vibratory inhibition yielded results which have helped to elucidate the pathophysiology of pyramidal hyperreflexia [DELWAIDE, 1971].

Material and Methods

Both the T- and H-reflexes were explored in 78 adult control subjects of various ages and in 58 spastic subjects. All subjects were tested in the sitting position. The foot was held rigid in a special apparatus [DELWAIDE, HUGON and THIRIET, 1970] and the knee was maintained immobile. Contractions were performed under isometric conditions. The angles formed by the leg and thigh and by the foot and leg were verified to be 130° and 110°, respectively, and the same values were used in all experiments.

T-reflexes were elicited with electromechanical hammer taps whose parameters were both adjustable and reproducible. The H-reflex was produced by 1-msec cathodal stimuli applied to the popliteal fossa with the electrode described by SIMON [1962]. The reflexogenic stimuli were delivered at a low frequency not exceeding 0.2/sec. EMG responses of the soleus muscle were recorded by two electrodes applied to the midline, a few cm from one another below the bulge of the gastrocnemius muscles.

Peak-to-peak measurements of the potentials were made. For every stimulation intensity, the amplitudes of 5–7 responses were averaged. To study the effect of a conditioning stimulus, test H-reflexes were selected corresponding to approximately 7 different stimulation intensities; in other words, the whole recruitment curve, and not merely one value from it, served as the test.

We used a vibrator derived from the Hagbarth model, consisting of a DC motor with an eccentric mass on its axis. The vibrator was fixed to the leg with elastic bandages at the level of the Achilles tendon. Except in special cases, a frequency of 100/sec was used.

Effects of Vibratory Stimulation on Monosynaptic Reflexes

The inhibition of the H- and T-reflexes by vitratory stimulation was described by YAMANAKA [1964], LANCE [1965], HAGBARTH and EKLUND [1966], LANCE, DE GAIL and NEILSON [1966] and RUSHWORTH and YOUNG [1966]. However, these authors paid little attention to the quantitative aspects of vibratory inhibition, which have therefore been investigated in the present study (fig. 1, 2 and 3).

Figure 1 illustrates the effect of vibration on the H-reflex. The control reflex (A) is inhibited when vibration acts on the Achilles tendon (B). After vibration has ceased (C), the amplitude of the reflex may be greater than that of the control (postvibratory potentiation). Figure 1, D indicates the time-course of the effect of vibration on the H-reflex: the amplitude of the control response shows relatively little fluctuation during the first 4 min and remains consistently in the neighborhood of 4.1 mV. As soon as vibration is applied, reflex amplitude is markedly reduced; upon cessation of the stimulus there is

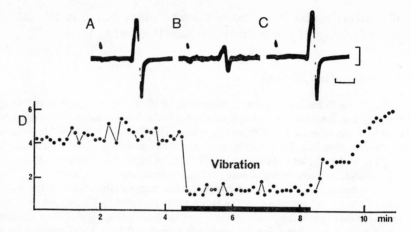

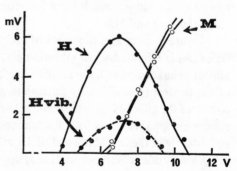

Fig. 1. Effects of vibratory stimulation. *A* Control H-reflex. *B* Vibration applied to the Achilles tendon. *C* H-reflex after the cessation of vibration. Calibrations, 1 mV and 20 msec. *D* Time course of effect of vibratory stimulation on the H-reflex. (Abscissa: duration of vibration is expressed as heavy black line.)

a gradual increase in amplitude. Throughout the vibration period the H-reflex values also fluctuate very little, thus providing consistent results.

Classical recruitment curves were established before and during vibration (fig. 2). When the direct motor (M) response values are identical to the control values, one may assume that the experimental conditions remained stable. In such cases it is reasonable to compare the amplitudes of the control and conditioned reflexes for every stimulation intensity. We propose expressing H-reflex inhibition by the percentage ratio between the maximal amplitudes of the 2 recruitment curves; for example, the vibratory effect of figure 2 would yield a numerical index of 30%

$$\frac{\text{H max. during vibration}}{\text{H max.}} \times 100$$

Fig. 2. Recruitment curves of the H-reflex. Solid line: control curve (H). Broken line: on the same coordinates, the recruitment curve obtained during vibration (H vib.). The M values are identical. Observe the reduction in amplitude of the H-reflex.

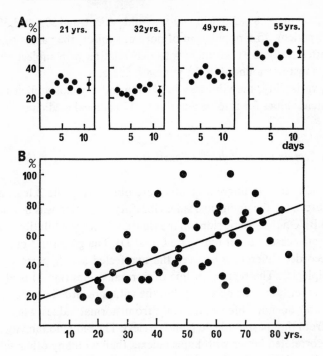

Fig.3. A Study of vibratory inhibition in 4 normal adults of various ages. *Ordinate:* intensity of inhibition, expressed as a percentage of control values (see fig.2). *Abscissa:* consecutive test days. Symbol with vertical bar = mean and 1 sD. *B* Intensity of vibratory inhibition in 52 normal subjects. *Ordinate:* intensity of inhibition, expressed as in *A* above. *Abscissa:* subject age (in years). The values in this normal population are scattered but show a correlation with subject age. The oblique line represents the regression line of normal values.

The influence of various experimental conditions was also studied, including the site of application of the stimulus, the effect of the Jendrassik maneuver, the occurrence of soleus contraction under isometric versus isotonic conditions, the vibration frequency and the initial degree of stretch of triceps surae.

The inhibition index was found to be similar in both legs of a given subject. Likewise, when vibratory inhibition of the H-reflex was measured on several consecutive days, comparable values were obtained as shown for 4 different subjects in figure 3A. From one subject to the next, however, the index varied. This can be seen from the mean values obtained in 4 subjects of various ages (fig. 3A): $29.6 \pm 5.3\%$ (1 SD) in a 21-year-old patient, $24.8 \pm 3.1\%$

in a 32-year-old patient; 35.8 ± 2.5% in a 49-year-old patient, and 50.8 ± 3.3% in a 55-year-old patient. In a population of normal subjects the values were found to be scattered (fig. 3B) but there was a correlation with subject age, inhibition being most pronounced in the youngest subjects.

Despite its variability, the vibratory inhibition index yields sufficiently high and consistent values for it to be a useful tool in clinical medicine.

Vibratory Effect on the H-Reflex in Spasticity

In hemiplegic patients, different results were obtained in the 2 legs, as illustrated in figure 4. On the unaffected side, vibratory inhibition was normal (compare A and B). On the spastic side (on the right), vibratory inhibition was reduced and in some cases absent (compare C and D). The graph in figure 5 illustrates the recorded differences between the unaffected and hyperreflexic sides in 15 hemiplegics. The regression line of normal values (fig. 3) is also plotted on this figure. It may be observed that the values recorded on the un-affected side (stars) do not differ statistically from normal values whereas those recorded from the spastic side do (solid circles). The reduction in vibra-tory inhibition correlated better with hyperreflexia than with any other sign of the pyramidal syndrome. In an attempt to determine whether this reduc-tion in inhibition was specific for pyramidal hyper-reflexia, we tested subjects free from neurologic disease but with very brisk reflexes (neurotics, hyper-thyroid subjects, etc.). Their inhibition indices (fig. 6) proved to be similar to those of controls. In neurologic patients, notably in parkinsonism, values were also in the normal range.

By testing patients with very mild forms of hemiplegia, we have also shown that the measurement of vibratory inhibition can serve as a sensitive and quantitatively accurate test of hyperreflexia in this condition. The dif-ference between the unaffected and spastic sides in hemiplegia seemed to us

Fig. 4. Comparison of vibratory inhibition on the unaffected (*A* and *B*) and spastic sides (*C* and *D*) of a hemiplegic patient. The value *(A)* of the unaffected side is inhibited during vibration *(B)* whereas the control value of the spastic side *(C)* remains unchanged during stimulation *(D)*. Calibrations, 200 μV and 20 msec.

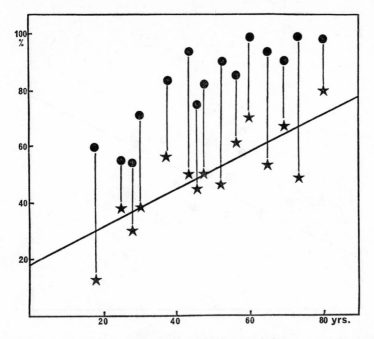

Fig.5. Comparison of vibratory inhibition in the unaffected and spastic sides in 15 hemiplegic patients. *Ordinate:* intensity of inhibition expressed as a percentage of control value. *Abscissa:* subject age (in years). The oblique line represents the regression line of normal values. Values from the unaffected side are shown as stars and from the spastic side as solid circles.

sufficiently clear-cut to constitute a useful ancillary examination in the clinical exploration of the pyramidal syndrome. In this chapter, however, we should merely like to stress the pathophysiologic significance of this difference. It seems logical to postulate that pyramidal hyperreflexia is related to a deficiency in the inhibitory mechanism normally activated by vibration. Let us, therefore, turn to an analysis of this mechanism.

Analysis of Vibratory Inhibition

When our study first began in 1968, no satisfactory explanation had yet been devised for the vibratory inhibition of monosynaptic reflexes in animals [GILLIES, LANCE, NEILSON and TASSINARI, 1969; BARNES and POMPEIANO, 1970] or in man.

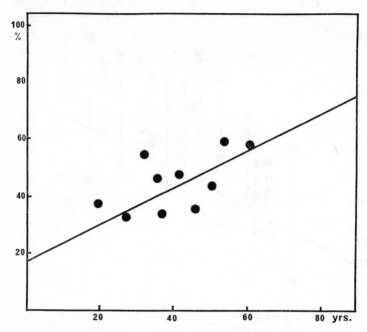

Fig.6. Measurement of vibratory inhibition in 10 subjects with no neurologic disease but with very brisk reflexes. *Ordinate:* intensity of inhibition expressed as a percentage of control value. *Abscissa:* subject age (in years). The oblique line represents the regression line of normal values.

The mechanism of action of this type of stimulation is difficult to analyze in clinical neurophysiology since, as figure 7A makes clear, vibration applied to the Achilles tendon stimulates not only the primary endings of the muscle spindles of the posterior compartment of the leg (1) but also, by diffusion, the receptors of the antagonist muscles of the anterior compartment (2). In addition, the cutaneous receptors of the leg (3) are vigorously activated during vibration. The role of numerous afferent impulses thus needs to be taken into account. In order to determine which of these inputs is responsible for the inhibitory effect, techniques were developed to permit more selective qualitative and quantitative activation of the peripheral receptors.

In the first stage of our study (fig. 7B), performed in collaboration with Prof. M. Hugon in Marseilles, we imposed a passive sinusoidal movement on the ankle. The mechanical apparatus used enabled us to control several parameters such as amplitude and frequency of movement. When the motor caused the ankle to pivot while the knee was immobilized, the muscles of the anterior

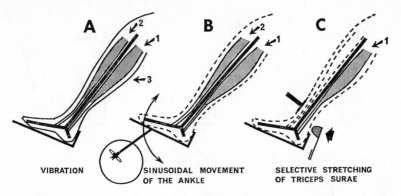

VIBRATION   SINUSOIDAL MOVEMENT   SELECTIVE STRETCHING
                   OF THE ANKLE   OF TRICEPS SURAE

Fig. 7. Different stages in the analysis of vibratory inhibition. Schematic representation of the various techniques used. *A* Vibration applied to the Achilles tendon stimulates the muscle spindles of soleus (1) and also those of tibialis anterior (2). Vibration also vigorously activates the cutaneous receptors (3). *B* The passive sinusoidal movement of the ankle alternately stimulates the receptors of soleus (1) and tibialis anterior (2). The cutaneous receptors are not stimulated by this maneuver. *C* A slow thrust against the Achilles tendon stimulates exclusively the spindles of soleus (1).

and posterior compartments of the leg were alternately stretched. Thus, the muscle spindles of the 2 antagonist muscle groups were activated for limited periods of the sinusoidal cycle. In most cases a frequency of 60/min and an angular movement of 6° were used. Under these conditions, cutaneous stimulation was greatly reduced or abolished.

Throughout the duration of ankle movement, the amplitude of the monosynaptic reflexes of soleus decreased, just as during vibratory stimulation. However, the inhibition was cyclically reinforced, being markedly greater when the muscles of the posterior compartment were stretched. An angular movement of equal amplitude could be imposed on the ankle from two different starting positions: either the foot in extension with tibialis anterior stretched and triceps surae relaxed, or the reverse with the foot in dorsal flexion. The muscle spindles of a stretched muscle are known to be more sensitive to further stretch. When the inhibition caused by the same angular movement of the ankle was measured from these two starting positions, it was observed that the intensity of inhibition was directly related to the initial degree of stretch of triceps surae.

Figure 8A–C gives the results obtained by raising the frequency of stretch without modifying the ankle movement. A rise in frequency clearly increased inhibition. The influence of the amplitude of the angular movement

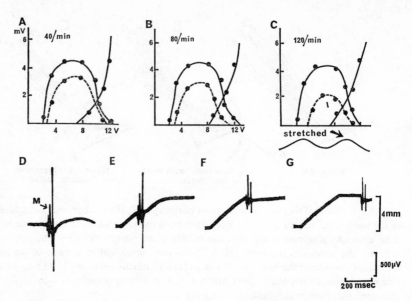

Fig. 8. Upper part *(A–C):* effects on recruitment curves of sinusoidal movement of the ankle at different frequencies. The control recruitment curve is shown as a solid line. The recruitment curve obtained during sinusoidal stretching is shown as a broken line. The values of the M curves are identical in all three cases. Note the increase in inhibition as frequency rises. Lower part *(D–G):* effect of gradual stretching of triceps surae. *D* The control response is preceded by an M response of small amplitude. *E–G* The H-reflex and the parameters of thrust are recorded on the same oscilloscope channel. *E* During the dynamic phase of stretch, the H-reflex is inhibited. The M response is identical to the control value. *F, G* Inhibition is reinforced during static stretch.

(shown as the abscissa) on the intensity of inhibition is illustrated in figure 9. The inhibition index rose gradually with the increase in amplitude. It is interesting to note that an inhibition of about 10% occurred even for a very slight angulation causing minimal stretch.

When the experimental conditions created by passive sinusoidal ankle movement produced significant inhibition, a vibratory stimulus applied at the same time to the Achilles tendon failed to increase the inhibition. In contrast, under the same conditions inhibition was clearly reinforced by cutaneous stimulation. This suggests that there are occlusive phenomena between the inhibitory mechanism of vibration and that of passive sinusoidal stretching.

The above technique thus indicates that cutaneous stimulation is not the determining factor in vibratory inhibition. Investigations of the soleus myotatic arc during sinusoidal movement suggest that the afferent impulses

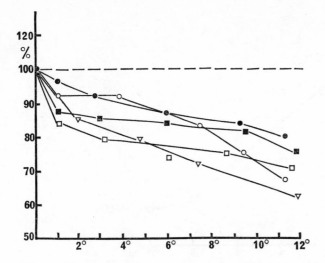

Fig.9. Sinusoidal movement of the ankle: reinforcement of inhibition of the H-reflex by angular movements of increasing amplitude. *Ordinate:* the intensity of inhibition is expressed as a percentage of the control value. *Abscissa:* amplitude of angular movement (in degrees).

responsible for inhibition arise in the posterior compartment of the leg, that they are influenced by very limited stretch (fig. 9), and that they are sensitive to the speed of stretching (fig. 8). Experimental conditions that facilitate the discharge of I_A fibers from triceps surae reinforce the inhibition of soleus monosynaptic reflexes. It is nevertheless possible that afferences from the anterior compartment play some role, particulary in view of the fact that inhibition persists during the phase of tibialis anterior stretch.

In a second series of experiments (fig. 7C), also performed in collaboration with Prof. HUGON, we developed a technique for selectively activating the muscle receptors of the posterior compartment. The leg and ankle were held rigid and the spindles activated by stretching the triceps surae. This was effected by means of a pneumatic piston which gradually applied pressure to the Achilles tendon, with both the speed and depth of thrust being adjustable. The ankle joint was immobilized in a special holder. Under these conditions, the tibialis anterior remained unaffected by the stretch. Strain gauges glued to an intermediate piece between the pneumatic piston and the Achilles tendon monitored the thrust parameters.

The excitability of the soleus myotatic arc was tested under 2 different conditions: during the dynamic phase of stretch and during the static phase in

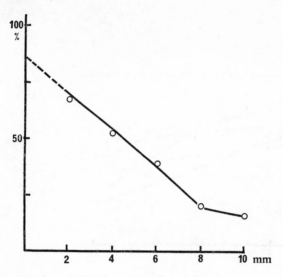

Fig. 10. Effects of depth of thrust against the Achilles tendon on the inhibition of the H-reflex. *Ordinate:* intensity of inhibition is expressed as a percentage of the control value. *Abscissa:* depth of thrust (in mm).

which the piston maintained a stretch of constant length. The lower part of figure 8 illustrates the changes in H-reflex amplitude during this maneuver. The experiment was conducted with a thrust of 4 mm over a period of 300 msec. Shown in figure 8 D is the control H-reflex. The same oscilloscope channel was used to represent both imposed movement and the reflex responses. H-reflex amplitude was reduced during the dynamic phase (E). The values of the M responses remained stable between D and E, indicating that the stimulation conditions did not vary. During the static phase (fig. 8 F, G), there was a marked reinforcement of inhibition lasting about 15 sec. During this phase, we studied the changes in H-reflex amplitude determined by increasingly deeper thrusts. Figure 10 shows control value (ordinate) plotted against the amplitude of the H-reflex expressed as a percentage of the depth of thrust against the Achilles tendon (abscissa). Inhibition was regularly reinforced from 2 to 8 mm of thrust. If we extrapolate the curve, we may assume that the inhibitory effect will already be significant for a very slight degree of stretch.

The controlled stretching of triceps surae specifically demonstrates that the inhibitory afferences originate in the posterior compartment of the leg; the fact that they can be triggered by very slight stretch argues for the involve-

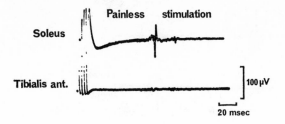

Fig. 11. Demonstration of a polysynaptic reflex in soleus by painless stimulation of the tibial nerve at the ankle with a volley of shocks (4).

ment of the primary spindle endings. This technique thus confirms the apparently paradoxical fact that the inhibition is triggered and increased by the discharge of I_A fibers from triceps surae.

However, on the basis of findings yielded by the two techniques described above, it is difficult to rule out the intervention of the I_B fibers or the group II afferents of muscular origin. Both types of fibers are classically considered to inhibit the soleus motoneuron pool by a postsynaptic inhibitory mechanism. They hence depress the excitability of the common final pathway. In order to complete our analysis of vibratory inhibition, we therefore needed to study the excitability of the soleus motoneuron pool itself using a pathway different from that involved in monosynaptic reflexes (i.e., via polysynaptic reflex). To this end, we sought a polysynaptic reflex capable of stimulating the soleus motor nucleus consistently. By applying a painless electrical stimulus to the tibial nerve at the ankle, we were able to evoke an EMG response exclusively in soleus (fig. 11). The amplitude and latency of this polysynaptic reflex varied with the experimental conditions but a sufficiently reproducible response could be obtained when these conditions were controlled. Since the amplitude of the reflex increased during voluntary contraction of soleus, the response did indeed seem to reflect an increase in motoneuron excitability. In order to compare the results obtained by monosynaptic as opposed to polysynaptic reflex testing, it is important to determine whether both reflexes stimulate the same motoneurons. However, this is difficult to determine experimentally in man since even very weak stimulation of the H-reflex activates a large number of motoneurons. The experimental investigations of HENNEMAN, SOMJEN and CARPENTER [1965] and the clinical studies of ASHWORTH, GRIMBY and KUGELBERG [1967] indicate, despite the reservations of GRIMBY and HANNERZ [1968; this volume], that motoneurons are recruited in the same order according to their size.

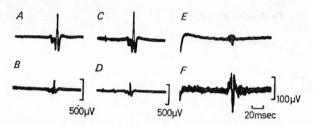

Fig. 12. Differential effect of vibratory stimulation on monosynaptic (A–D) and poly-synaptic (E–F) reflexes. The monosynaptic reflexes are profoundly inhibited (cf. A, C and B, D) whereas the polysynaptic reflex is enhanced (cf. E, F).

The polysynaptic reflex was elicited under two types of experimental con-ditions under which monosynaptic reflexes are clearly inhibited, namely dur-ing the application of vibration to the Achilles tendon and during the static phase of controlled stretching of triceps surae. Figure 12 illustrates the results obtained. Unlike the monosynaptic reflexes, that were inhibited, this poly-synaptic reflex showed an increase in amplitude. This suggests that the soleus motoneurons are not themselves depressed by vibration. Vibratory inhibition must selectively affect the myotatic arc upstream from the motoneurons.

Presynaptic Inhibition as a Controlling Factor in Myotatic Arc Excitability

The results of the analysis of vibratory inhibition are difficult to interpret within the framework of the classic schemas of γ loop organization, since they indicate: (1) that the inhibition of soleus monosynaptic reflexes is triggered by impulses originating in the muscles of the posterior compartment of the leg; (2) that inhibition is enhanced as I_A fiber discharge is increased by ex-perimental maneuvers, and (3) that inhibition occurs even when the excitability of the final common pathway is not depressed.

A postsynaptic inhibitory mechanism does not seem to be involved. If it were, motoneuron excitability would be reduced and the polysynaptic soleus reflex would also be inhibited during vibration. Such is not the case.

We suggest that the above findings can be explained by a presynaptic in-hibitory mechanism acting on the terminal arborizations of the triceps I_A fibers (fig. 13 A). This interpretation would explain why inhibition is restricted to the myotatic arc. Since inhibition occurs upon I_A fiber discharge, we hypo-

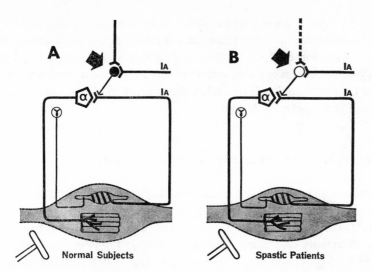

Fig. 13. Functional organization of the myotatic arc in normal subjects (A) and spastic patients (B). *A* A mechanism of presynaptic inhibition is added to the classical schema of the γ loop. The interneuron responsible for this inhibition (arrow) is activated by I_A afferents from homonymous muscles and by supraspinal pathways (vertical line). *B* In spasticity, the interneuron (arrow), is less active due to the reduction of supraspinal facilitatory influence (broken line).

thesize that the interneuron responsible for the presynaptic inhibition is activated by I_A fibers arising from homonymous muscles. In addition, since various supraspinal factors (e. g., the performance of mental arithmetic) are able to modify the intensity of vibratory inhibition, we suggest that this interneuron is likewise subject to the influence of supraspinal centers. The TVR is most likely mediated polysynaptically and follows a different pathway from that illustrated in figure 13 A.

A mechanism of presynaptic inhibition acting on the I_A afferents from soleus has, until recently [DELWAIDE, 1969; DELWAIDE and BONNET, 1969], received little attention from clinical neurophysiologists. Presynaptic inhibition of the I_A fibers from an extensor muscle following the discharge of the homonymous I_A fibers was described in the cat by ECCLES, ECCLES and MAGNI [1961] and ECCLES, MAGNI and WILLIS [1962], but these authors considered it of little quantitative significance. More recently, however, the studies of GILLIES, LANCE, NEILSON and TASSINARI [1969] and BARNES and POMPEIANO [1970] have shown that vibratory stimulation in the cat causes appreciable presynaptic inhibition of the I_A fibers from the extensor muscles.

According to the presynaptic inhibition hypothesis, the ultimate effect of the I_A afferences on the motoneurons is adjusted at a premotoneuronal level as a function of the preceding I_A discharge and supraspinal factors. Monosynaptic reflexes are thus unable to test true motoneuron excitability because the reflex amplitude depends both on this excitability and on the efficiency of presynaptic inhibition.

Pyramidal Hyper-reflexia and Presynaptic Inhibition

We have already mentioned that vibratory inhibition of the H-reflex is reduced in subjects with pyramidal hyperreflexia (fig. 4 and 5). It seems a reasonable conjecture that this departure from normal behavior is caused by a reduced activity of the interneuron responsible for vibratory inhibition under physiologic conditions. We thus propose (fig. 13 B) that the interneuron becomes less active because it is deprived of a supraspinal facilitatory influence. According to this hypothesis, the myotatic arc of the hyperreflexic patient is no longer subject to a tonic inhibitory control; in spasticity, therefore, all proprioceptive afferent impulses are able to gain direct access to the motoneurons and exert a motor influence.

Hyperreflexia is thus not necessarily related to increased γ activity. According to the presynaptic inhibition hypothesis, it is nevertheless still true that procedures which reduce γ activity (procaine infiltration, pharmacologic agents) will inhibit reflex responses.

In conclusion, we think that the fundamental mechanism of pyramidal hyper-reflexia consists of a reduction of the presynaptic inhibition which normally acts on the I_A afferents. This mechanism does not, however, exclude a possible contribution by other pathophysiologic factors to the pyramidal syndrome. Certain findings suggest that the reduction of presynaptic inhibition may represent a special case of a more general disturbance in spastic patients affecting the reactivity of the spinal interneurons [DELWAIDE, 1971].

Author's address: Dr. P. DELWAIDE, Section of Neurology, Institut de Médecine, Hospital de Bavière, B-4000 Liège (Belgium)

New Developments in Electromyography and Clinical Neurophysiology,
edited by J. E. Desmedt, vol. 3, pp. 523–537 (Karger, Basel 1973)

Renshaw Cell Activity in Normal and Spastic Man

J. L. VEALE, S. REES and R. F. MARK

Van Cleef Foundation Laboratory, Alfred Hospital and Department of Physiology,
Monash University, Melbourne

Renshaw cells are small interneurones of the ventromedial horn of the mammalian spinal cord that are monosynaptically excited by recurrent collaterals from nearby motoneurones [RENSHAW, 1941]. The synapse between motoneurone axons and these interneurones first attracted attention for pharmacological reasons because it seemed likely, and in fact proved to be correct, that the transmitter was acetylcholine [ECCLES, FATT and KOKETSU, 1954; CURTIS and RYALL, 1966]. The axons of Renshaw cells form inhibitory synapses on nearby motoneurones. Within a motor nucleus the frequency of firing of motoneurones, which is very much lower than that of other spinal cord neurones, may possibly be limited by recurrent inhibitory effects. It has also been postulated that these effects tend to confine activity within a motoneurone pool to its more active members in a manner similar to the lateral inhibition in sensory paths [GRANIT, HAASE and RUTLEDGE, 1960; BROOKS and WILSON, 1958; ECCLES and HOFF, 1932].

The integrative function of Renshaw cells has received much less attention until recently, when it was found that in addition to direct inhibitory effects on motoneurones, they can inhibit short-axon IA inhibitory interneurones on the pathway from the central processes of afferent fibres from muscle spindle afferents to the motoneurones of antagonist muscles [HULTBORN, JANKOWSKA and LINDSTROM, 1971 a, b and c]. This explains the occurrence of recurrent facilitatory effects that had previously been seen and shown to be accompanied by an increase in membrane resistance of motoneurones such as would be expected from suspension of continuous inhibitory bombardment [RENSHAW, 1946; WILSON, 1959; WILSON and BURGESS, 1962]. Thus, in addition to their direct inhibitory action,

Renshaw cells may cause an indirect disinhibition of motoneurones by their primary action on inhibitory interneurones. Furthermore, Renshaw cells may inhibit each other so that the effects of discharge of a motoneurone upon the excitability of its neighbours is not easily predictable [RYALL, 1970; RYALL and PIERCEY, 1971]. Although the territory of the recurrent motoneurone collateral is small, one branching axon of the Renshaw cell can extend 5.5 mm along the cat spinal cord to exert intersegmental influences [RYALL, PIERCEY and POLOSA, 1971].

The IA inhibitory interneurones provide the only known spinal mechanism for reciprocal innervation of muscles which allows contraction of a synergic muscle group to proceed with concomitant relaxation of antagonists. Renshaw cells, as inhibitors of the reciprocal inhibitory effects of muscle spindle afferents, could regulate the balance between simultaneous contraction of antagonist muscles about a joint such as is necessary for the supporting pillar function of a limb and the smooth relaxation of antagonists necessary for the flexion and extension of walking [HULTBORN, JANKOWSKA and LINDSTROM, 1971 c]. The finding that Renshaw cells appear to be inhibited during spinal stepping is in keeping with a role in switching from reciprocal to concurrent activation of antagonist muscles [SEVERIN, ORLOVSKII and SHIK, 1968]. Smooth movement it seems requires inhibition of recurrent disinhibition.

The control of Renshaw cell excitability at the segmental level is quite complex, involving afferent fibres from a variety of antagonistic and synergistic muscles [WILSON, TALBOT and DIEKE, 1960; RYALL and PIERCEY, 1971]. Renshaw cell discharge may also be modified by cerebellar or reticular stimulation [HAASE and VAN DER MEULEN, 1961] and, as has been shown in an investigation unfortunately carried out in acute experiments on cats without anaesthesia, by stimulation of the sensory-motor cortex [MACLEAN and LEFFMAN, 1967]. The effects are normally inhibitory but the detailed pattern of inhibition has not yet been shown to be reminiscent of that required for normal limb function. Decerebrate and spinal animals show very little difference in their Renshaw cell activity [HOLMQVIST and LUNDBERG, 1959]. Supraspinal control is therefore exerted from regions accessible to stimulation of the sensory-motor cortex as readily as from brain-stem structures.

There is almost no knowledge of Renshaw cell function in normal man. The silent period in the human myogram following electrical stimulation of the nerve to a voluntarily contracted muscle [HOFFMANN, 1919; HOFFMANN and KELLER, 1928] could be due to the spread of recurrent

inhibition in the spinal cord, following antidromic invasion of motoneurones [DENNY-BROWN, 1928] but it could also depend upon the
cessation of afferent activity from muscle spindle afferents during the brief
shortening of the extrafusal fibres [MERTON, 1951; PAILLARD, 1955]. So far
there is no conclusive proof one way or the other and probably both
mechanisms contribute.

The two postulated effects of Renshaw cell activity in the spinal cord
are considered to be disordered in clinical spasticity due to supraspinal
lesions. The possibility that loss of control of effects mediated by Renshaw
cells is a major part of the pathophysiology of spasticity is therefore high,
and worth investigating.

Experiments on man that may be directly compared with those of
Renshaw cell effects in the cat spinal cord require the ability to stimulate
motor axons in isolation so as to submit spinal motoneurones to antidromic and recurrent effects, without altering the reactivity of the cord in
other ways via the dorsal roots. In addition, a method is needed of testing
the excitability of motoneurones during recurrent conditioning with the
temporal resolution necessary to distinguish direct antidromic effect from
changes in reflex afferents consequent upon peripheral muscle contraction.
Both these requirements are satisfied by taking advantage of the differential
electrical thresholds of the M and H responses elicited from the triceps
surae muscle by stimulation of the posterior tibial nerve. If one uses 1 msec
duration square wave pulses of appropriate strength to the posterior tibial
nerve in the popliteal fossa, then an H-reflex response is elicited in isolation; whereas if pulses of 100 μsec duration are used, a direct M response is obtained at threshold. Figure 1B shows responses that can be
recorded, with two fixed pulse durations. The amplitudes of the responses,
M and H, have been plotted in figure 2A. For long-duration pulses
(1 msec) the sensory (IA) fibres appear more sensitive, as we obtain only
an H-reflex. As the strength is increased, more IA fibres are excited and the
H-reflex grows by recruitment of motor units. Threshold for motor fibres
is shown by the appearance of an M response, with a latency of 5 msec.
Further increase in stimulus strength gives recruitment of both the M and
H responses. For short pulses of 100 μsec duration the motor fibres are
more sensitive, as the M response appears first. With increasing strength
the M response grows by recruitment, and when the threshold for IA fibres
is reached, an H-reflex appears.

Figure 2B is a plot of a strength-duration curve for direct motor and
reflex responses. One does not always obtain such clear separation as

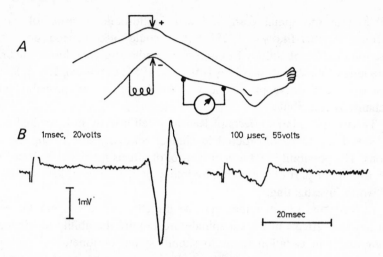

Fig. 1. A Diagram of stimulating and recording arrangement used in motor conditioning of the H-reflex. *B* Typical examples of surface myograms recorded from gastrocnemius-soleus following just suprathreshold stimulation of the posterior tibial nerve. Each trace is an average of 20 sweeps using the CAT. Stimuli applied 5 msec after start of each trace, with duration and strength as marked. The left trace shows an H-reflex only, and the right a direct M only.

shown in figure 2 B; only fractions of a volt may separate the thresholds of M and H responses. In some subjects one always obtains either an M or H response at threshold, irrespective of pulse duration, even though a differential sensitivity of M and H responses does still exist. It is possible that in these cases either the motor or the sensory fibres were significantly more accessible to the stimulating current. Some degree of searching in the popliteal fossa is generally required until a stimulating point that gives a clear separation of motor and sensory responses can be found.

The simplest explanation of the above results is that short pulses (100 μsec or less) preferentially stimulate motor axons and long pulses (1 msec or more) preferentially stimulate sensory axons. However, there are difficulties in using the H-reflex as an indicator of threshold of sensory fibres because of the unknown amount of central summation needed for motoneurone discharge. We therefore investigated the ulnar nerve, as it can be readily stimulated at two points along its general path as a mixed nerve (at the wrist and elbow), it has a purely sensory part in the fourth and fifth digits, and obvious regions of motor innervation. These experiments confirmed the differential sensitivity found for the posterior tibial

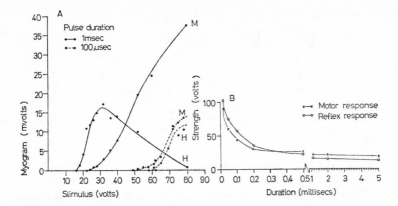

Fig. 2. A Curves of the M and H responses recorded from gastrocnemius-soleus following stimulation of the posterior tibial nerve, for two different pulse durations. *B* The strength-duration curves for motor and sensory (IA) fibres, as indicated by the appearance of an M (motor) or an H (reflex) response, respectively.

nerve. Figure 3 shows that for a volley involving roughly the same number of fibres in the ulnar nerve (neurogram, top traces), there is no recorded muscle activity with the longest pulse (1 msec), but considerable activity with the shortest pulse (20 μsec). Further work has shown that the entire neurogram for short pulses is accounted for by impulses in motor axons. A preliminary report of these studies has been published [VEALE, MARK and REES, 1971] and a full account is in preparation.

Utilizing this differential sensitivity in subjects with unequivocal separation of the M and H responses, it is possible to investigate the effects of an antidromic stimulation of motor axons (the conditioning stimulus) upon orthodromic H-reflex (the test reflex).

The subjects lay (for comfort) supine with the thigh and ankle of the leg under investigation supported so that the relaxed limb was raised off the surface of the couch (fig. 1 A). The skin of the limb was cleaned with chloroform, washed and dried carefully. Light abrasion was applied if necessary by either scrubbing with sand soap or gentle stroking of the skin with sterile needles. Recording electrode resistances were reduced to less than 10 K. Differential recording electrodes were applied over the gastro-cnemius muscle. The anode stimulating electrode was a copper sheet about 3 cm square applied over the kneecap. The cathode stimulating electrode was a rounded metal dome about 1 cm in diameter covered by saline-soaked cloth and mounted on a perspex rod whose position could be

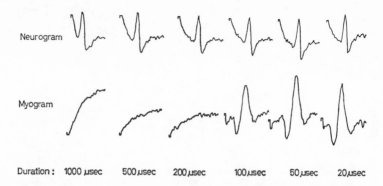

Duration: 1000 μsec   500 μsec   200 μsec   100 μsec   50 μsec   20 μsec

Fig. 3. Surface recordings taken from the ulnar nerve at the elbow (neurogram) and hypothenar muscles (myogram). Each pair of records shows 50 averages with the CAT of the results following stimulation of the ulnar nerve at the wrist, with pulse durations as shown. For each pair the strength was adjusted so as to produce neurograms of equal amplitude. The increase in effectiveness of motor stimulation with shortening of pulse duration is clearly seen.

freely adjusted. Subjects were not allowed to sleep during recording sequences.

The recorded electrical responses were amplified differentially with a Tektronix 122 Preamplifier, with a band width of 0.2–10 kc/sec, and after further conventional amplification, averaged on a Mnemotron CAT, Model 400B. Stimuli were timed using two Devices Digitimers and two Devices Mark IV Stimulators in series. One of these had been modified to produce square wave pulses down to 5 μsec duration, with a rise time of about 1 μsec. This provided the conditioning pulse, and generally a pulse duration of 50 μsec was used. The maximal voltage used was 100 V. The test stimulus was always 1 msec duration, and these stimuli were applied through the same electrode to a suitable spot over the posterior tibial nerve in the popliteal fossa. The short pulse (the conditioning stimulus) had to give an M response only with no H-reflex, and furthermore none obtainable when a weak to medium voluntary contraction of the triceps surae was made. The long pulse (the test stimulus) had to give an H-reflex with no direct M response. This necessitated working with relatively small responses in the muscles. In general, we tried to arrange strengths so that the amplitude of the M response was about a third of the H-reflex. This was not always possible: if the motor stimulus was too strong an H-reflex developed; if the sensory strength was lowered its variability became much larger.

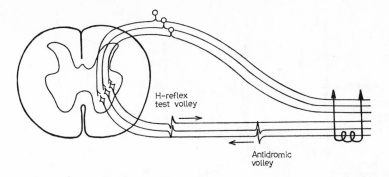

Fig.4. Diagram showing how collision in motor axons can still occur with negative conditioning, that is, with the motor volley released after the test volley which is inducing the H-reflex. If no suppression of the H-reflex occurs with negative times, there has been no collision in the motor axons and there is no commonality between motor axons carrying H-reflex action potentials and those carrying motor conditioning action potentials. In the example as drawn, one-third of the axons are common, and the H-reflex would be reduced to one-half.

We confined our investigations to the effects of the conditioning stimulus on a test reflex elicited up to 30 msec later. For longer times one could not be sure that any effects obtained were not secondary to the direct motor contraction (such as unloading spindles, stimulating Golgi tendon organs, etc.). It was also possible, if required, for any of the conditioning intervals to be negative, that is for the test stimulus to be applied before the conditioning stimulus.

All results presented were obtained by averaging 20–50 responses. With repetitions of an H-reflex there is an immediate problem in how fast a repetition rate is permissible. The excitability of the motoneurones following an H-reflex has not entirely returned to normal after 1 sec [TABORIKOVA and SAX, 1969]. With the number of repetitions needed for these investigations, slower repetition rates would make the experiments excessively tedious, and would be rather self-defeating, as recording conditions can change during a long experiment. The following procedure was adopted: a repetition rate of 1/sec was used for the test reflex, but stimulation was begun many seconds before averaging with the CAT was started. A sequence of 5 test stimuli (at one second intervals) was applied during each sweep of the CAT with averaging taking place for only a 100 msec period (over one-fifth of the sweep) with each stimulus. Four of these test stimuli were preceded by conditioning stimuli at four selected

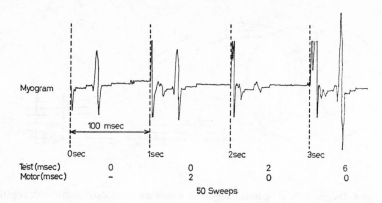

Fig.5. Averages of 50 sweeps of the CAT, with the sweeps split into 4 segments, each of 100-msec duration, and following the preceding segment at 1-sec intervals. In segment 1 the test stimulus alone has been given, and an H-reflex without an M response has occurred. In segment 2 a test stimulus has been given at the start, and after 2 msec a conditioning stimulus, which results in an M response. This is negative conditioning and there has been negligible effect on the H-reflex. In the third and fourth segments the conditioning stimulus occurs at the start of the segment and the test stimulus after 2 and 6 msec respectively. The inhibition at 2 msec and enhancement at 6 msec of the H-reflex can be clearly seen.

intervals. The test stimulus alone gave the standard against which the conditioned responses were expressed as a percentage. Frequently two of the conditioning intervals were kept constant throughout the whole recording session (which might last 1–2 h). The stability of the responses with time then gives an indication of the stability of the experiment as a whole.

Conditioning stimuli sent impulses antidromically up the motor axons, test stimuli sent impulses to the spinal cord via spindle afferents (fig. 4) and motoneurone excitability was judged by the amplitude of reflex contraction. Conduction time in the reflex pathway is about 15 msec in each direction, the total latency being about 30 msec. Therefore if the test stimulus is given first the conditioning stimulus could interact with the descending motor impulses by collision in motor axons even if it were not applied until 30 msec after the test stimulus. As the conditioning stimulus is applied relatively earlier the crossing point of orthodromic and antidromic volleys will move up the nerve until at some interval, the exact duration of which is not known, the antidromic volley will reach the motoneurones before the test volley in the spindle afferents reaches the spinal cord. An example of the results is shown in figure 5 in which each response is an average of 50 trials. The complete results are presented graphically in

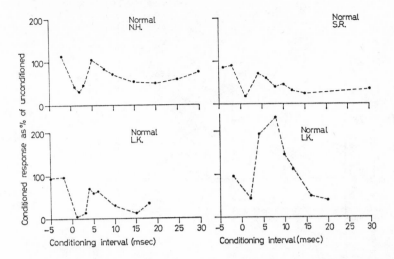

Fig.6. Four examples of the effects of motor conditioning on the H-reflex in normal subjects. The amplitudes of the conditioned reflexes have been expressed as a percentage of the unconditioned reflex. Note that two of the curves were obtained from the same subject (L.K.) but on different days.

figure 6 which shows the percentage changes in H-reflex amplitude (in terms of the unconditioned reflex response) with various intervals between conditioning and test stimuli. The size of the antidromic stimulus was kept constant.

The use of negative conditioning times enabled us to demonstrate that the conditioning and test stimuli were exciting two separate fractions of the triceps surae motoneurone pool. When the test stimulus preceded the conditioning stimulus by 2–5 msec, as shown in the graphs, the size of the H-reflex was virtually unchanged. If a reduction of the response had occurred it would have indicated that the antidromic motor action potentials of the conditioning stimulus had collided with the orthodromic motor action potentials elicited by the test stimulus and had consequently reduced the size of the recorded H-reflex (occlusion). Figure 4 clarifies this point. The conclusion that there was no collision in motor axons was consistent with the visual observation that the test volley was causing a contraction in one part of the triceps surae muscle and the motor (conditioning) volley a contraction in another part of this muscle.

Motor conditioning at 0 msec cannot be investigated, since the 2 stimuli were applied through one electrode. Inhibition of the H-reflex was

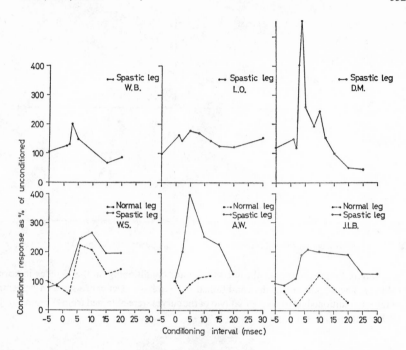

Fig. 7. Six examples of the effects of motor conditioning on the H-reflex in spastic subjects. The amplitudes of the conditioned reflexes have been expressed as a percentage of the unconditioned reflex. In three cases with unilateral symptoms, curves were obtained from the normal limb, and have been shown by a dashed curve for comparison.

evident at plus one msec and reached a point of maximal inhibition at 2–3 msec. Occasionally it fell to 10% of the unconditioned response but was more usually 30–50%. This inhibition was followed immediately by facilitation of the response, usually back to near the control level (80–120%) but occasionally up to 200%. The peak of facilitation occurred at about 5–8 msec. The amount of depression and facilitation could vary quite widely in the same subject on different days (fig. 6, subject L. K.), although the pattern of the response as described remained the same. This variation may depend upon the relative and absolute size of the test and conditioning volleys.

The period of facilitation of the response was followed by further inhibition which persisted in varying degrees for up to 30 msec – the limit of our testing period.

Similar experiments were performed on 10 patients with upper motoneurone signs in their lower limbs. These signs were secondary to cerebral haemorrhage, thrombosis or multiple sclerosis. Results from 6 of the 10 patients (aged 21–64 years) are presented in figure 7. In 3 of the patients the pathological signs were unilateral and their normal legs were studied as a control. In the spastic subjects there was a complete lack of the initial inhibition of the response. Facilitation was evident at 1–2 msec and reached a peak at 3–5 msec, which is earlier than in the normal subject. It was also enhanced, usually to 200 % but occasionally to 400–700 % and tended to be more prolonged. The inhibition which follows was not as pronounced.

The difficulties experienced in obtaining the curves in spastic patients and the relatively small number of subjects tested do not yet allow us to be sure that the curves presented are entirely characteristic of upper moto-neurone lesions.

Discussion

The antidromic volley in the motor axons could be exerting its effect on the H-reflex by 3 mechanisms.

1. The conditioning action potentials antidromically invade cell bodies.

2. The conditioning action potentials make the muscle contract (producing the M response) and thus set up an excitation of peripheral receptors. The central effects of these could explain the results.

3. The conditioning action potentials invade the collateral branches of the motor axons and stimulate Renshaw cells, whose activity is responsible for the subsequent effects.

The first possibility (1) can be rejected. The use of negative conditioning times has shown that the test and conditioning stimuli are exciting 2 separate fractions of the triceps surae motoneurone pool. This eliminates the possibility of the changes in the H-reflex being due to alterations in motoneuronal excitability secondary to their antidromic invasion.

The second explanation (2) can be excluded on a basis of timing. The conditioning stimulus has to propagate to the peripheral muscle and initiate a contraction (about 6 msec), and the contraction initiate a sensory discharge which then propagates back past the stimulating electrode in the popliteal fossa. Thus for conditioning intervals of less than about 12 msec, any action potentials generated in this way will arrive at the cord after the test stimulus, and thus could hardly influence the test response. Any sudden effects at 12 msec or so would be treated with caution, but there are none.

These arguments and experimental results exclude mechanisms (1) and (2) leaving only (3): that the antidromic volley has excited Renshaw cells via axon collaterals and that these in turn have induced complex excitability changes in adjacent motoneurones of the triceps surae motoneurone pool.

In normal subjects the effect of motoneurone discharge on the excitability of neighbouring cells follows the recovery curve of excitability of two H-reflexes elicited one after another at varying intervals as originally determined by PAILLARD [1955]. In his case the same motoneurones were tested; in our case, the amplitude of test H-reflexes measures fluctuations in excitability of motoneurones that have not recently discharged. If the recovery curves are so similar in shape it means that the synaptic action of recurrent motoneurone collaterals sets the pattern for the excitability cycle of motoneurones, as appears to be the case in other mammalian spinal motoneurones [ECCLES, 1955]. Early recurrent inhibition of extensor motoneurones is followed by an increase of excitability coming on about 3 msec later, which may or may not exceed control levels, and this is followed by a reinstatement of quite profound inhibition. After about 12 msec the motoneurones will begin to feel the altered afferent discharge consequent upon muscle contraction and its effects on muscle sense organs. The excitability curves become progressively more difficult to interpret and the later inhibition could have many causes. It seems almost certain that the brief elevation in excitability is due to recurrent disinhibition of motoneurones by inhibition, mediated by Renshaw cells, of the tonically active inhibitory interneurones. All the evidence in cats indicates that the inhibitory interneurones include those on the disynaptic inhibitory pathway from muscle spindles of antagonist muscles.

Similar excitability tests in patients with spastic limbs from upper motoneurone lesions show a great increase in recurrent excitatory effects and an absence of any recurrent inhibition. Therefore there has been a dissociation of the inhibitory and excitatory effects of motoneurone discharge, mediated presumably by Renshaw cell inhibition. Enhanced Renshaw cell activity could be responsible for the extra excitatory effects but this should also be associated with an increase in recurrent inhibition as well. Any influence on overall Renshaw cell activity thus should change

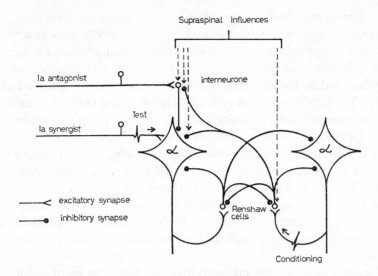

Fig.8. Schematic diagram of currently presumed inhibitory connections of the Renshaw cells. These go to other Renshaw cells, to α-motoneurones, and to inhibitory interneurones that are stimulated by I_A fibres from antagonists. Hypothetical but possible sites of influence on these mechanisms by supraspinal influences have been shown by dashed lines.

both actions equally. In spastic man the results show that the recurrent effects on interneurones affecting motoneurone excitability become stronger than the direct recurrent effects of motoneurone inhibition. Figure 8 is a diagram of the known connections of Renshaw cells in the cat spinal cord and it is apparent from consideration of the possible effects of facilitation and inhibition of the cells themselves that this could not account for our results. If recurrent effects occur at all the initial event should always be a monosynaptic inhibition as it appears in normal subjects. The only conclusion we can draw is that the inhibitory actions of Renshaw cell terminals on the motoneurones must be controlled separately from the inhibitory transmission onto the inhibitory interneurone. Two possibilities are an altered reactivity of the motoneurone to the inhibitory transmitter or a separate presynaptic inhibition of inhibitory terminals onto motoneurones.

The results of recurrent conditioning of spinal motoneurones in man show that presumed Renshaw cell inhibition and disinhibition appear to govern the early part of the excitability cycle of motoneurones. In spastic man recurrent inhibition is largely lost but recurrent facilitation is strong. Both these changes would tend to make for unchecked spread of activity in motoneurone pools leading to loss of fine control of graded muscular activity. Tendon jerks could also be enhanced by these changes. A normal volley initiated in the IA fibres will have some degree of temporal dispersion. The action potentials arriving first will excite (following suitable spatial summation) some motoneurones. Their action potentials will invade their collaterals to Renshaw cells and initiate the facilitation phase in the adjacent motoneurones of the pool. The remaining sensory action potentials in the latter part of the volley in the IA fibres will now encounter facilitated motoneurones, and hence the response will increase. This postulated mechanism is an example of positive feedback, since any effective α-motoneurone stimulation leads to increased sensitivity to further stimulation, and it might be involved in a number of pathological conditions.

Previous work on the mechanism of spasticity has emphasized the possible importance of overactivity of γ-motoneurones and hyperexcitability of muscle spindle afferents. The main evidence for this is that dorsal root section, or other interruption of afferent pathways from the limb into the cord, reduces spasticity [RUSHWORTH, 1960]. This does not argue against our proposition that recurrent effects mediated by Renshaw cells are important because deafferentation will also decrease the excitability of all cord mechanisms.

Furthermore if human Renshaw cells also inhibit IA inhibitory interneurones [HULTBORN, JANKOWSKA and LINDSTROM, 1971a] their augmented activity would block the reciprocal inhibition of stretch reflexes and shortening of any muscle would be opposed by the stretch reflexes in antagonists.

Summary

1. The electrical thresholds of sensory and motor nerve fibres in human nerve trunks differ according to the duration of the stimulating current pulse.

2. In the posterior tibial nerve, differential thresholds allow an antidromic volley in motor nerve fibres and an orthodromic volley in sensory fibres to be set up from the same stimulating electrode by changing the stimulus voltage and duration.

3. With near threshold stimulation the monosynaptic reflex response of triceps surae motoneurones is not occluded by an antidromic volley. Therefore, the volleys excite separate fractions of the motoneurone pool.

4. If the antidromic volley is timed to reach the spinal cord before the test afferent volley, the amplitude of the latter is modified in characteristic manner according to the exact timing of the two stimuli. Early inhibition of the reflex response is followed by a brief enhancement of excitability and a reinstatement of inhibition.

5. These excitability changes occurring in motoneurones that have not themselves discharged, and which have been induced by antidromic stimulation of neighbouring motor axons are probably due to the synaptic actions of Renshaw interneurones excited by recurrent axon collaterals.

6. In spasticity, early recurrent inhibition is lost and reflex enhancement is exaggerated.

7. The results can mostly be interpreted according to our knowledge of recurrent effects in the cat spinal cord, and the changes in spasticity could be an important pathophysiological mechanism.

Acknowledgements

We thank Miss LOUISE KEEGAN for her help in the laboratory. Dr. B.S. GILLIGAN kindly allowed us access to patients under his care. The work was done in the Van Cleef Laboratories and funded by the Van Cleef Foundation and a grant from the National Health and Medical Research Council of Australia. We thank Prof. A.K. MCINTYRE and Prof. R. PORTER for help with the manuscript.

Authors' address: Dr. J.L. VEALE, Van Cleef Foundation Laboratory, Alfred Hospital and Department of Physiology, Monash University, *Melbourne* (Australia). Dr. R.F. MARK, Department of Physiology, Monash University, *Melbourne* (Australia)

New Developments in Electromyography and Clinical Neurophysiology, edited by J. E. Desmedt, vol. 3, pp. 538–549 (Karger, Basel 1973)

Amplitude and Variability of Monosynaptic Reflexes Prior to Various Voluntary Movements in Normal and Spastic Man

E. Pierrot-Deseilligny and P. Lacert

Department of Neurological Rehabilitation, Hôpital de la Salpêtrière, Paris

Many studies have been conducted on the monosynaptic reflex (MSR) changes which occur either during [Paillard, 1955; Gottlieb, Agarwal and Stark, 1970] or before [Requin, 1969; Kots, 1969a, b; Coquery and Coulmance, 1971; Pierrot-Deseilligny, Lacert and Cathala, 1971] voluntary movement in man.

We have studied the changes in soleus MSR during the reaction time (RT) prior to a voluntary movement, thus during a period when the physiological mechanisms involved in the movement are being organized, special attention being paid to the following points:

1. A comparative study of the changes in the H- and T-reflexes which should provide some insight into the role of the γ loop during preparation for movement.

2. A comparison of voluntary contraction of different muscles, which elucidates the reflex changes associated with different types of movement. Subjects were thus asked to perform isotonic contractions, not only of the muscle tested, gastrocnemius-soleus (GS), but also of its antagonist, tibialis anterior (TA), of the contralateral GS and TA, and of the extensor carpi (EC) muscles.

3. A study of the variability of the H-reflex. The changes in this reflex as one approaches the voluntary movement clarify the physiological interpretation of the concomitant changes in amplitude.

4. A comparison of the results obtained under similar experimental conditions in spastic patients with lesions situated at different levels of the nervous system.

The experimental protocol followed has been described in detail previously [Pierrot-Deseilligny, Lacert and Cathala, 1971]. The reflex (C) evoked during the RT just prior to the requested movement was compared with the mean of control reflexes $(\overline{N})$ measured immediately before the test reflex; for each contraction we thus measured

$$\Delta H = \frac{1}{M}(C-\overline{N}) \times 100$$

ΔH represents the change in the reflex expressed as percent of the maximum M response. For each movement, we measured the time interval between the reflex evoked and the start of the following voluntary contraction, as detected by the appearance of EMG activity. The successive values of ΔH during a single test were grouped according to the corresponding values of this interval; the time interval was divided into 10-msec units and the mean and variance of ΔH were calculated for each such unit. Each type of trial (e. g., movement of the TA) from each population studied (e. g., normals, spastic patients with similarly located lesions) was treated separately. For each 10-msec unit, the following parameters were calculated: (a) the mean of ΔH; (b) the mean individual variance, defined as the mean scatter of ΔH around the mean for each individual, and (c) the inter-individual variance, defined as the scatter of the individual means around the group mean.

I. Normal Subjects

The results presented in this section were obtained from 10 normal adult subjects in the course of 47 recording sessions. During each session the subject performed the same type of movement (GS, TA or EC contraction) 375 times.

A. Phenomena Observed Prior to Voluntary Contraction in the Upper Extremities (EC)

The mean amplitude of the H-reflex (fig. 1 Bc) revealed facilitation beginning 100 msec before the movement and increasing as the start of contraction approached; however, reflex increase remained quite moderate. Individual variance (fig. 1 Cc) remained remarkably stable throughout the period preceding the start of contraction in the upper extremities and it also had the lowest values. Inter-individual variance (fig. 1 Dc) was very small.

Such results concerning voluntary movement at a great distance from the reflex arc being tested appear to provide a rough estimate of the nonspecific modifications which precede any type of motor activity.

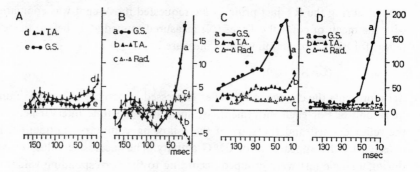

Fig. 1. A, B Changes in H-reflex amplitude in normal subjects prior to voluntary contraction of various muscles. Abscissa: time interval between the start of voluntary contraction and the reflex, which preceded the contraction. The onset of the voluntary contraction coincides with the ordinate. Ordinate: ΔH or variation in H-reflex amplitude expressed as a percentage of the maximum direct M response. *A* Contralateral movement. *B* Ipsilateral movement. *a* and *e:* contraction of gastrocnemius-soleus (GS); *b* and *d:* of tibialis anterior (TA); *c:* of extensor carpi radialis (rad.). The vertical bars represent 1 SD on either side of each symbol. *C, D* Changes in variance on the side ipsilateral to movement in normal subjects prior to voluntary contraction of different muscles. *C* The mean individual variance of ΔH is shown as the ordinate. *D* The inter-individual variance is shown as the ordinate. Each symbol represents the results obtained in a 10-msec bin.

B. Phenomena Observed on the Ipsilateral Side Prior to Voluntary Contraction of GS

The changes in mean amplitude of ΔH, mean individual variance and inter-individual variance make it possible to distinguish 3 periods in the time interval before contraction of the GS.

1. Period from 170–80 msec before Start of Movement

This period was marked by a depression of the H-reflex which reached a peak 80 msec before the movement (fig. 1 Ba). Such depression had not been observed by KOTS [1969a, b]. COQUERY and COULMANCE [1971], however, reported an inhibition phase which was less marked and earlier before the contraction than in our study. While inter-individual variance remained very small in this period (fig. 1 Da) individual variance increased slightly and reached distinctly higher values than in any other test situation (fig. 1 Ca).

This H-reflex depression may be compared to that found by REQUIN [1969], who studied the time interval between a warning signal and an execu-

tion signal and observed depression just prior to the latter. REQUIN and PAIL-
LARD [1969] suggested that this might result from a presynaptic inhibition
acting on the afferent I$_A$ pathway.

When tested at times equidistant from voluntary contraction, depression
of the H-reflex was observed only when the voluntary movement involved
the GS (fig. 1 B). The physiological significance of this phenomenon, if it was
indeed related to a presynaptic inhibition localized to the afferent pathway
leading to the motor nucleus about to be activated, might be twofold: firstly,
isolation of the α-motoneurones from postural afferent impulses originating
in the muscle spindles; secondly, protection of the motor nucleus from
excitability variations reaching it via the γ loop. We may add in this connection that the appearance of H-reflex inhibition coincides with a slight
facilitation of the tendon reflex (fig. 3 B), reflecting γ activity. This activity is
probably non-specific since it is also observed on the side contralateral to
movement.

2. Period from 80–30 msec before Start of Movement

During this period a progressive increase in H-reflex amplitude was observed. The decrease which had begun in the preceding period gradually
dissipated and gave way to a growing facilitation (fig. 1 Ba). Parallel to these
changes, there was a progressive and considerable rise in both individual
(fig. 1 Ca) and inter-individual (fig. 1 Da) variances. These phenomena might
be related to a modulation of α-motoneurone excitability by the pyramidal
system:

Among the physiological mechanisms that may underlie facilitation of
the H-reflex, one must consider a progressive increase in the excitability of the
soleus α-motoneurones which ultimately grows enough to reach discharge
threshold when voluntary movement begins. EVARTS [1966] has shown that
in the monkey the changes in pyramidal cell discharge occur 100 msec before
the appearance of EMG activity. Taking into account the time necessary for
conduction along the pyramidal and then the peripheral fibres, pyramidal influences would begin to reach the GS motoneurones approximately 90–80
msec before the start of movement, i.e. when we observe the onset of H-reflex
facilitation. It would not be surprising if these pyramidal influences were
facilitatory for the soleus motoneurones since in primates pyramidal stimulation produces complex effects, including a facilitation in this motor nucleus
[PRESTON and WHITLOCK, 1963].

This hypothesis also accounts for the concomitant changes in variance.
In our experiments, the mechanical characteristics of initial force and accele-

ration show considerable variation both during the same test and between subjects, as revealed by the marked variations in the initial EMG record. These variations, which result from a particular temporospatial recruitment of the α-motoneurones during each contraction, are ultimately linked to a given level of central activity; EVARTS [1968] has indicated a relationship between the level of pyramidal discharge and the instantaneous force of the corresponding muscular contration. The changes in pyramidal influences from one movement to the next – variations which cause the observed changes in the force and acceleration of the contraction – would begin to provide an additional source of individual and inter-individual variance 80 msec before contraction, as actually observed.

3. 30-msec Period Immediately Preceding Movement

During this period we observed, in agreement with KOTS [1969a, b] and COQUERY and COULMANCE [1971], a marked rise in H-reflex amplitude (fig. 1 Ba). In addition, individual variance (fig. 1 Ca) first became stabilized and then decreased, while inter-individual variance continued to rise (fig. 1 Da).

This facilitation of the H-reflex may result, we think, from a reduction of presynaptic inhibition in the afferent pathway in addition to a continuation of the α-motoneurone excitability increase from the preceding period. The following points support this hypothesis.

Immediately before GS voluntary contraction the H-reflex often increases considerably, reaching values far greater than maximum H in the relaxed muscle (fig. 2). This is in agreement with GOTTLIEB, AGARWAL and STARK [1970] who observed that the increase in H-reflex was maximal at the onset of the voluntary contraction, the same mechanisms being responsible for the facilitation observed immediately before and after the start of contraction. GOTTLIEB, AGARWAL and STARK [1970] have noted that H-reflex amplitude and the GS EMG have a different time course during the first 100 msec of contraction, from which they deduce that the H-reflex facilitation coinciding with the movement cannot result solely from modulation of α-moto-neurone excitability. They suggest that the facilitation may also be due to a lifting of presynaptic inhibition from the afferent reflex pathway.

The stabilization of individual variance followed by its decline also argues for a new factor intervening during the last 30 msec preceding voluntary movement. The drop in individual variance with a concomitant rise in the mean might result from motor nucleus saturation [COQUERY, MARK and PAILLARD, 1962; RUDOMIN and DUTTON, 1969a], but we do not think this likely. If the

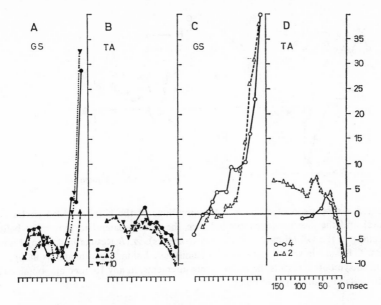

Fig. 2. Changes in H-reflex amplitude on the side ipsilateral to movement before voluntary contraction of the GS (A and C) and of the TA (B and D) in 5 normal subjects illustrating the extreme values for the population. Same co-ordinates as for figure 1 *A, B.* Each type of symbol (filled and open circles and triangles) represents a given numbered subject. Each symbol represents the mean of measurements obtained in the same test in a 10-msec bin.

subjects are considered individually, during the last 10 msec studied variance dropped in 2 subjects, as in the others, but their mean remained low. The studies of RUDOMIN and DUTTON [1969a, b] offer an alternative interpretation of the decrease in variance. These authors have shown that inhibition of the interneurones involved in presynaptic inhibition of the soleus I_A afferents causes a decline in MSR variability. Thus, the decrease we observe in individual variance might be related to central inhibition of the interneurones controlling presynaptic inhibition on the afferent reflex pathway.

The almost identical values of the H and tendon reflex obtained during the 40-msec period immediately preceding contraction (fig. 3 B) do not favour the hypothesis of MSR facilitation via the γ loop [PAILLARD, 1955]. These results agree with the finding of HAGBARTH and VALLBO [1969] that during voluntary isotonic contraction of soleus the increase in I_A discharge does not precede, but rather occurs during or slightly after the EMG activity.

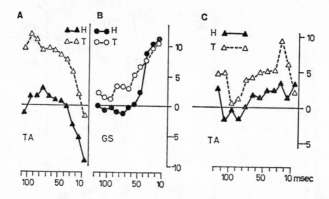

Fig.3. Comparison of changes in the amplitude of the H-reflex (filled circles and triangles) and the T-reflex (open symbols) on the side ipsilateral to movement before contraction of the GS (circles) and the TA (triangles). *A, B* Results in 4 normal adult subjects. *C* Results in spastic patients with hemisphere lesion. Co-ordinates as in figure 1 *A, B*. Each symbol represents the mean of measurements, in any 10-msec bin, obtained in subjects tested by either type of reflex.

C. Phenomena Observed on the Ipsilateral Side Prior to Voluntary Contraction of TA

Unlike KOTS [1969a] we observed progressive depression of the H-reflex during the 50 msec preceding movement of this muscle (fig. 1Bb), whereas individual variance increased progressively (fig. 1Cb). In this case, inter-individual variance remained very low (fig. 1Db).

This H-reflex decrease was less pronounced than the symmetrical facilitation which preceded contraction of the GS. Once again, our curves are in agreement with those of GOTTLIEB, AGARWAL and STARK [1970], who observed that the H-reflex was already somewhat depressed from the start of TA contraction and that the depression continued to increase during the first 100 msec of voluntary movement. This inhibition appearing before the start of contraction and continuing after its initiation may result from a reinforcement of presynaptic inhibition on the I_A afferent pathway [GOTTLIEB, AGARWAL and STARK, 1970].

The increase in individual variance observed during this period is in line with the latter hypothesis since RUDOMIN and DUTTON [1969a, b] showed that facilitation of the interneurones controlling presynaptic inhibition of the soleus I_A fibres increases MSR variability in the cat.

A comparative study of the H- and T-reflexes (fig. 3A) shows that the T-reflex curve runs parallel to the H-reflex curve but is displaced a fairly constant distance above it. The larger size of the T-reflex may reflect reinforcement of dynamic γ activity before voluntary contraction of the TA.

D. Individual Characteristics Observed

Figure 2 illustrates the variations observed between subjects. Each individual about to engage in voluntary motor activity presents a characteristic pattern of H-reflex changes and very similar results were obtained for each subject who repeated the same test on several occasions.

Results obtained in a given subject (ipsilateral to the movement) can also be compared depending on whether the contraction occurred in the GS or TA. This comparison is shown in figure 2, in which each symbol represents a different subject. The subjects with no inhibition phase before GS contraction (fig. 2C) showed facilitation prior to the inhibition which immediately preceded TA contraction (fig. 2D). Conversely, subjects with a marked inhibition phase prior to the facilitation preceding GS contraction (fig. 2A) had no facilitation before voluntary TA contraction (fig. 2B).

The H-reflex changes observed ipsilateral to movement thus appear to be the result of two factors: an effect associated with the type of voluntary movement under preparation during the 50-msec period immediately preceding contraction, and a change in excitability level of the reflex arc characteristic of each individual about to perform a voluntary movement of any sort, observed during the period preceding the last 50 msec.

E. Phenomena Observed Contralateral to Voluntary Movement

The H-reflex was facilitated to a slight extent on the contralateral side, regardless of whether the movement involved the GS or the TA. This facilitation does not appear to result only from a diffuse non-specific action. Contralaterally, during the 50 msec preceding GS contraction, the facilitation was less marked than that seen before contraction of the upper limb. However, the facilitation observed during the 50 msec preceding contraction of the TA was more pronounced (fig. 1A). The individual variance curves likewise differed and in fact diverged during the last 30 msec before the movement. We may conclude that each type of movement, when performed on one side, induces characteristic contralateral modifications.

II. Spastic Patients

Changes in H- and tendon reflexes were studied in a large number of spastic patients who were asked to perform a voluntary contraction on the affected side. However, in only 12 cases could the results be used because of the necessity of active participation by the patient in the experimental procedure and the large number of movements required for a valid test. All 12 subjects presented a unilateral nervous system lesion located as follows: in 6 patients, in a cerebral hemisphere; in 4 patients, in the brain stem; in 2 patients with a Brown-Séquard syndrome, in the spinal cord.

MSR changes preceding contraction of the GS could not be studied in these patients because of the proprioceptive silent period. In normal subjects, the soleus MSR gives rise to only a relative silent period in the EMG activity of the gastrocnemii reflected simply by a reduction in their activity, which makes it possible to detect the exact onset of the motoneurone excitation which will bring about GS contraction. This is not so in spastics, in whom the reflex response causes an absolute silent period in the EMG activity of both the gastrocnemii and the soleus muscles. Consequently in the spastic group only the MSR changes preceding contraction of the TA were studied. In all cases the contraction was performed on the spastic side. The patients tested had been previously chosen for their ability to perform a TA contraction of fair force with easily detectable EMG activity.

The results obtained in these patients differed from those observed in normal subjects; moreover, they varied greatly depending on whether the causal lesion was located in the cerebral hemisphere, the brain stem or the spinal cord.

A. Phenomena Observed on the Ipsilateral Side before Voluntary Contraction of TA in Spastic Patients with Cortico-Subcortical Lesions

In contrast to normals, these patients did not display progressive H-reflex inhibition during the last 50 msec before contraction (fig. 4Bb). This absence of inhibition was noted in all 6 spastic patients with lesions in the cerebral hemisphere. These patients thus formed a homogeneous population since throughout the period preceding contraction of the TA inter-individual variance remained less than 10% of the mean individual variance.

As discussed in normals, the hypothesis suggesting a presynaptic origin for the increasing H-reflex inhibition before TA contraction would account

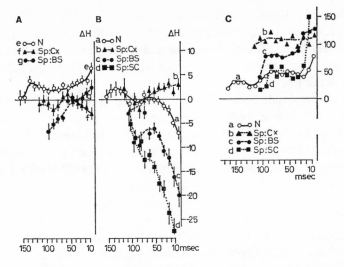

Fig. 4. Comparison of changes in H-reflex amplitude before a voluntary contraction of the tibialis anterior (TA) in normal and spastic subjects. Abscissa: interval between H-reflex and the start of voluntary contraction which coincides with the ordinate. Ordinate: ΔH or variation in H-reflex amplitude in *A* and *B*, and variance in *C*. The H-reflex is tested on the side ipsilateral to the voluntary movement in *B* and *C*, and on the side contralateral to the movement in *A*. *a* and *e:* normal adult subjects; *b* and *f:* spastic patients with hemisphere lesion; *c* and *g:* spastic patients with a brain-stem lesion; *d:* spastic patients with a spinal cord lesion.

for the rise in individual variance. In spastics with cortical lesions individual variance remains very stable (fig. 4 Cb). Along this line of reasoning the absence of H-reflex inhibition in these patients might well be due to a lack of reinforcement of presynaptic inhibition on the afferent pathway of the GS MSR which normally occurs prior to TA contraction. If this were true, it would indicate that the corticosubcortical structures destroyed in these patients are necessary for organization of the presynaptic inhibition pattern which leads to voluntary movement.

In order for dorsiflexion of the foot to take place during TA contraction, the resulting stretch of the GS must not trigger a myotatic reflex which would oppose the movement. One can thus understand the need for, and the functional significance of, the inhibition of the GS MSR which we observed ipsilateral to movement in normal subjects during the 50 msec preceding TA contraction. The absence of H-reflex inhibition in spastic patients with hemisphere lesions during this period accounts for the following clinical phenomenon, which was found in all patients: whereas the force developed by the TA contraction was relatively good, the amplitude and/or rapidity of dorsiflexion of the foot was very limited. These limitations undoubtedly result from the absence of inhibition of the GS myotatic reflex which normally occurs before contraction of the TA.

Lastly, the amplitude curves of the H- and T-reflexes ran parallel as in normals only until 40 msec before contraction. There was then a very clear inhibition of the T-reflex during the last 30 msec while the H-reflex remained stable (fig. 3 C). This probably reflects a decrease in dynamic γ activity which may thus partly compensate for the lack of inhibition of the GS MSR before TA contraction.

B. Phenomena Observed on the Ipsilateral Side before Voluntary Contraction of TA in Spastic Patients with Brain-Stem Lesions

In these 4 patients H-reflex inhibition increased as the start of contraction approached (fig. 4 Bc). This inhibition was more marked than in normals. In spastics with brain-stem lesions the entire curve of H-reflex amplitude variations before TA contraction was displaced below the corresponding curve for normal subjects (fig. 4 Ba).

Similar results were obtained in these 4 patients. All presented an increasing and marked inhibition of the H-reflex as contraction approached. In addition, regardless of the time interval between the reflex and the voluntary movement, the H-reflex showed consistent inhibition in all 4 subjects.

The progressive inhibition of the H-reflex during the period immediately preceding TA contraction was accompanied by a concomitant rise in individual variance (fig. 4 Cc). The mechanism of the decrease in H-reflex as one approaches TA contraction is probably similar in both cases. In these spastics with brain-stem lesions, TA contraction would be preceded by an increase in presynaptic inhibition on the afferent pathway of the GS MSR. This might indicate that, in contrast to hemispheric lesions, brain-stem lesions producing spastic hemiplegia do not cause profound disorganization of the modulations of presynaptic inhibition which precede voluntary movement. A similar conclusion may be drawn concerning hemisection of the spinal cord since, in these cases as well, a progressive inhibition of the H-reflex (fig. 4 Bd) and a rise in individual variance (fig. 4 Cd) occur as one approaches TA contraction.

It can be seen from figure 4B that in spastics with brain-stem lesions the progressive inhibition observed during the period immediately before contraction is superimposed on a background of early inhibition occurring before the movement and thereafter remaining more or less constant. Regardless of its mechanism, this 'additional' inhibition (relative to that seen in normals) might explain why, despite the exaggeration of the GS MSR, dorsiflexion of the foot in these patients is only slightly decreased in amplitude and speed.

C. Phenomena Observed Contralateral to Movement

In this test, again, the results differed greatly depending on the site of the lesion.

1. In patients with hemisphere lesions, the H-reflex was increasingly inhibited over the last 50 msec (fig. 4Af). Although less marked, this inhibition was not unlike that seen in normals ipsilateral to movement. This finding may be interpreted as the equivalent of the clinical phenomenon known as imitative synkinesis: the movement performed by the spastic side gives rise to the same movement on the healthy side.

2. In patients with brain-stem lesions, the H-reflex showed increasing facilitation as contraction approached. In the same manner as for the ipsilateral side, the entire curve of H-reflex amplitude variations before TA contraction was displaced below the corresponding curve for normals (fig. 4Ag). The 'additional' inhibition noted in these patients ipsilateral to movement would thus appear to be diffuse, affecting the GS monosynaptic reflex arc on both sides simultaneously.

Author's address: Dr. E. PIERROT-DESEILLIGNY, Laboratoire d'étude des handicaps moteurs, Service de Rééducation Neurologique, Hôpital de la Salpêtrière, 47, Bd de l'Hôpital, *F-75 Paris 13e* (France)

New Developments in Electromyography and Clinical Neurophysiology,
edited by J. E. Desmedt, vol. 3, pp. 550–555 (Karger, Basel 1973)

Supraspinal Control of the Changes Induced in H-Reflex by Cutaneous Stimulation, as Studied in Normal and Spastic Man

E. Pierrot-Deseilligny, B. Bussel and C. Morin

Laboratoire d'étude des handicaps moteurs, Service de Rééducation Neurologique, Hôpital de la Salpêtrière, Paris

The studies of Hagbarth [1960], Bathien and Hugon [1964] and Gassel and Ott [1970] have demonstrated the effect of cutaneous stimulation on the excitability of soleus motoneurons in normal man. Suprasegmental control of the interneurons activated by cutaneous afferents has, however, received very little attention in man. It seemed of interest to approach this problem in the following way: (1) by comparing the effect of cutaneous stimulation on the H-reflex in normal subjects and in patients with spastic hemiplegia with lesions at different levels of the CNS, and (2) by studying the effects of voluntary contraction on the H-reflex variations induced by cutaneous stimulation.

I. Study of the Effects of Cutaneous Stimulation on H-Reflex Amplitude in Normal Subjects and in Spastic Patients with Lesions at Different Levels of the CNS

This study was performed in 24 normal subjects and 19 patients with spastic hemiplegia. The site of the lesion was not the same in all patients: (1) 2 patients had the lesion in the paracentral lobule as verified during neurosurgery; (2) 12 patients also presenting a lesion of the cerebral hemisphere, had a superficial lesion involving the middle and lower part of the precentral gyrus or a deep lesion in the internal capsule, and (3) in 5 patients the lesion was situated in the upper portion of the brain stem (pons or mesencephalon).

The conditioning cutaneous stimulation, a volley of 10 shocks at 300 c/sec, was applied between two ring electrodes surrounding the base and tip of the small toe. Each H-reflex conditioned (C) by an ipsilateral cutaneous stimulus was compared with the control reflex (N) which immediately preceded it, and

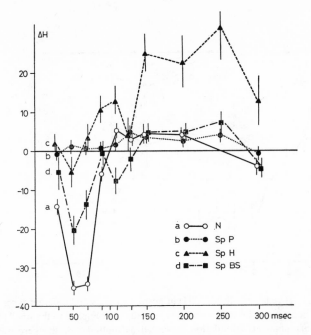

Fig. 1. Comparison of changes in H-reflex amplitude (ΔH) after cutaneous stimulation of the small toe in normal subjects and certain types of spastic patients. Abscissa: time interval between the start of cutaneous stimulation and the stimulus triggering the H-reflex. Ordinate: ΔH, representing H-reflex variation in response to cutaneous stimulation, expressed as a percentage of maximum H-reflex. Each symbol represents the mean of results obtained in each group of subjects for a given time interval. *(a)* normal subjects; *(b)* spastic patients with paracentral lobule lesions; *(c)* spastic patients with hemispheric lesions not reaching the paracentral lobule, and *(d)* spastic patients with brain-stem lesions. The vertical bars represent 1 SD on either side of each symbol.

$$\Delta H = \frac{C - N}{\text{maximum H}} \times 100$$

was calculated, where ΔH represents the variation of the reflex induced by cutaneous stimulation expressed as a percentage of the maximal H-reflex. The current intensity corresponding to the subjective perception threshold was estimated. Two types of tests were performed. (1) The cutaneous stimulus intensity was kept constant at twice the intensity required for the perception threshold (this is experienced by normal subjects as slightly painful), and the interval between the stimuli eliciting the H-reflex was gradually increased from 30 to 300 msec. (2) The interval between the two stimuli was kept constant at 70 msec and the cutaneous shock intensity was varied randomly.

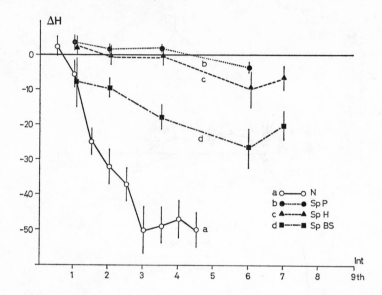

Fig. 2. Changes in H-reflex amplitude (ΔH) in normal subjects and different types of spastic patients after cutaneous stimuli of increasing intensity were applied to the small toe. Abscissa: intensity of cutaneous stimulation expressed as multiples of the perception threshold. Ordinate: same as for figure 1. *(a, b, c, d)* same as for figure 1. Each symbol represents the mean of results from each group of subjects for a given stimulus intensity. The interval between the onset of cutaneous stimulation and the stimulus triggering the H-reflex was maintained constant at 70 msec.

It may be seen from figure 1A that in normal subjects a slightly painful cutaneous stimulus gave rise, first, to an inhibition of the H-reflex which became maximal when the interval between the onset of skin shocks and the stimulus triggering the H-reflex reached 70 msec. The degree of inhibition increased with the intensity of the skin shocks (fig. 2A). These results are comparable to those obtained by other authors in humans [BATHIEN and HUGON, 1964; GASSEL and OTT, 1970] as well as in animals [HAGBARTH, 1952].

In spastic patients this inhibition was far less pronounced (fig. 1 B, C, D), even at high cutaneous stimulation intensities (fig. 2 B, C, D). These findings take on an even greater significance when one considers that in these patients, due to the frequent presence of associated sensory disturbances, any possible error in calculating the threshold tends towards overestimation. It will be noted that the degree of H-reflex inhibition seen after skin stimulation depended on the site of the lesion. There was no inhibition in subjects with a paracentral lobule lesion (fig. 1 B and 2 B), very slight inhibition in the remaining patients with a hemispheric lesion (fig. 1 C and 2C), and moderate inhibition in patients with a brain-stem lesion (fig. 1 D and 2D). In addition, it should be emphasized that, for each type of spastic patient, the degree of inhibition varied directly with the degree of motor recovery in the distal lower limb.

The absence of H-reflex inhibition after cutaneous stimulation in subjects with paracentral lobule lesions destroying the cortical area for motor representation of the lower extremity underscores the importance of the control exerted by the sensorimotor cortex in man on the spinal interneurons activated by cutaneous stimulation. This control may be exerted directly by the pyramidal pathway, since LUNDBERG, NORRSELL and VOORHOEVE [1962] have shown in the cat that stimulation of the pyramidal fibers has a facilitatory effect on the interneurons activated by the flexion reflex afferents, in particular the cutaneous afferents. The persistence of some H-reflex inhibition after toe stimulation in the other spastic patients might thus be due to a partial lesion of the pyramidal fibers corresponding to the lower limb in the cerebral hemisphere or brain stem. However, it is possible that, in man as well as in animals, cortical control over these spinal interneurons might be exerted indirectly via fibers not running exclusively in the pyramidal tract and modulating reticular function. This would explain why a fair degree of H-reflex inhibition persists after cutaneous stimulation in spastic patients with brain-stem lesions.

When the interval between toe stimulation and the stimulus triggering the H-reflex exceeded 70 msec, both normal subjects (fig. 1A) and spastic patients (fig. 1 B, C, D) exhibited H-reflex facilitation which was still present at 200 msec, and which disappeared at 300 msec. This late facilitation also exists in animals, where it has been shown to be due to the activity of an interneuron chain different from the chain responsible for early inhibition [LUNDBERG, 1969]. If, as is most likely, the same is true in man, our results might suggest that these two types of interneuron chain are not controlled in the same manner by suprasegmental structures. In our studies, the only abnormality seen at such an interval in spastic patients was a particularly marked and long-lasting facilitation (fig. 1 C) observed in those patients with hemispheric lesions which did not reach the paracentral lobule.

II. Effects of Voluntary Contraction on H-Reflex Variations Induced by Cutaneous Stimulation

This study was performed in 12 normal subjects and in 4 spastic patients who had recovered sufficient voluntary motor activity to be able to maintain

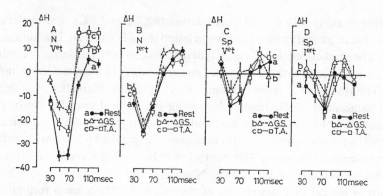

Fig. 3. Changes in H-reflex amplitude (ΔH) in normals (*A* and *B*) and in spastic patients (*C* and *D*) after cutaneous stimulation of the great toe (*B* and *D*) or the small toe (*A* and *C*) as measured at rest *(a)* and during sustained contraction of gastrocnemius-soleus *(b)* or tibialis anterior *(c)*. Same coordinates as for figure 1.

a long-lasting muscular contraction. The control and conditioned H-reflexes were evoked during sustained isometric contraction involving either the muscle being tested via the reflex, the gastrocnemius-soleus (GS), or its antagonist, the tibialis anterior (TA). The stimulus intensity used for triggering the H-reflex was adjusted such that the amplitude of the control responses was equal to half the maximal H-reflex obtained during the voluntary contraction. The skin stimulation was kept at twice subjective threshold and was applied to the small toe and then to the great toe in each subject.

When the stimulation was applied to the small toe in normal subjects (fig. 3A), its effect on the H-reflex was modified by voluntary contraction. The modification was identical regardless of whether contraction occurred in the GS (fig. 3Ab) or in the TA (fig. 3Ac). In both cases, there was a reduction in the early inhibition and an increase in the late facilitation observed at rest. However, when the conditioning stimulation was applied to the great toe, voluntary contraction of the GS or TA had no influence on the effect of stimulation on the H-reflex (fig. 3B).

Any interpretation of these phenomena must take into account an observation which to us seems fundamental, namely that the effect of voluntary contraction on the H-reflex changes induced by skin stimulation depends, not on the type of contraction performed (GS or TA), but on the cutaneous zone stimulated. This fact might be explained by the action of the pyramidal fibers, which are very likely brought into play during voluntary contraction. In cases of pyramidal-tract lesions, as in the spastic patients studied, voluntary contractions have only a slight and variable effect on the H-reflex changes induced by cutaneous stimulation, and in no case is their effect related to the toe stimulated (fig. 3 C and D).

It may appear paradoxical to assign two apparently opposed functions to the cortex via the pyramidal tract: activation of the spinal interneurons at rest, and inhibition of the same interneurons during movement. Nevertheless, it is conceivable that the cortex might act to facilitate certain well-adapted segmental reactions at rest while subordinating these reactions to movement during voluntary activity.

Author's address: Dr. E. Pierrot-Deseilligny, Laboratoire d'étude des handicaps moteurs, Service de Rééducation Neurologique, Hôpital de la Salpêtrière, 47, bd de l'Hôpital, *F-75 Paris 13e* (France)

New Developments in Electromyography and Clinical Neurophysiology,
edited by J. E. Desmedt, vol. 3, pp. 556–578 (Karger, Basel 1973)

A Systematic Analysis of Myotatic Reflex Activity in Human Spastic Muscle

R. HERMAN, W. FREEDMAN, A. W. MONSTER and Y. TAMAI

Department of Rehabilitation Medicine, Temple University School of Medicine, and Krusen Research Center, Moss Rehabilitation Hospital, Philadelphia, Pa.

In man, the dynamic behavior of the myotatic reflex system in spastic muscle has been attributed to increased activity of the primary endings of the muscle spindle and hence to release of the dynamic fusimotoneurones [JANSEN, 1962; BURKE, GILLIES and LANCE, 1970; HERMAN, 1970]. It has been implied that normal spindle behavior is essential for damping within the stretch reflex loop [PARTRIDGE and GLASER, 1960; BROWN and MATTHEWS, 1966]. Recently, there has been considerable interest in the role of the inherent rheologic (viscous, elastic, plastic) and contractile properties of muscle in the reflex control of position and/or force [MATTHEWS, 1959; PARTRIDGE, 1965]. This has led to studies of the input-output characteristics of the hyperactive myotatic system in different states of neuromuscular disease and dysfunction [PARTRIDGE and GLASER, 1960; JANSEN and RACK, 1966; POPPELE and TERZUOLO, 1968; HERMAN, 1970; ROSENTHAL, MCKEAN, ROBERTS and TERZUOLO, 1970]. Analysis of the effects of the relationship between muscle length and tension of the stretch reflex is complicated by the interaction of a number of essential components of the reflex system, e.g.: (1) patterns of afferent discharge; (2) time delays in the neural circuit; (3) character of motoneurone discharges (tonic or phasic), and (4) the inherent levels of muscle stiffness [PARTRIDGE, 1967; ANASTASIJEVIĆ, ANOJČIĆ, TODOROVIĆ and VUČO, 1969]. This complex interrelationship becomes evident when the dynamic reflex response of spastic muscle is modified by alterations in the rheologic and contractile properties of muscle (e.g. in contracture of spastic muscle).

The dynamic and static properties of the stretch reflex in subjects with spasticity has previously been tested utilizing ramp (linear) movements at various rates of stretch. However, since muscle stretch at a constant velocity is not similar to functioning physiological modes, ramp inputs are of only limited value in a systematic investigation of the dynamic properties of muscle.

Input functions, potentially capable of leading to a quantitative measure of the dynamic properties of the stretch reflex, have been applied to the extensor muscles of the cat [LIPPOLD, REDFEARN and VUČO, 1958; LENNERSTRAND and THODEN, 1968; POPPELE and TERZUOLO, 1968] and to various muscles in man [GOTTLIEB, AGARWAL and STARK, 1970; NASHNER, 1970]. In several of these investigations sinusoidal stretch at various frequencies has been used as the input to an assumed linearly functioning reflex system. However, attractive as this approach appears, it must be used with caution since criteria for linearity may not be satisfied (see Results, section IA). Nevertheless, sinusoidal stimulation under appropriate physiological conditions of amplitude and frequency is convenient for describing qualitatively the dynamic properties of the system. Observations of the frequency-dependent characteristics have provided both a sound basis for predicting the system's response to other dynamic signals (e.g. impulse input), and an evaluative means for studying the behavior of the reflex system to inputs which are more similar to physiological signals. This approach is also convenient for studying the dynamic properties of muscle (open-loop) to stretch with and without electrical stimulation applied directly to muscle or nerve.

This investigation, therefore, describes the response characteristics (output torque/input foot rotation) of the myostatic system and attempts to define some of its properties by analyzing the neural component (output EMG/input foot rotation) and non-neural component (output torque/input induced electrical stimulation and/or input foot rotation). In subjects with spasticity, gain (output/input) and phase angle (peak torque or peak EMG/input displacement) of the ankle myotatic reflex were analyzed across a wide range of frequencies (0.05–12/sec) and displacement amplitudes (1–10°). This data was then compared with data derived from impulse-type inputs.

Methodology

The ankle myotatic reflex system was investigated in normal subjects (n = 8), and subjects with spastic paraplegia (n = 5), and with hemiplegia (n = 12), chosen according to a classification system previously described [HERMAN, 1970]. Seven of the subjects were selected for nerve block experiments.

The ankle of each subject was aligned with the rotational axis of a servo-controlled positioning apparatus [HERMAN, SCHAUMBERG and REINER, 1967]. Data were recorded on an FM tape recorder (Honeywell 7600), monitored on a polygraph (Beckman Dynograph) and analyzed using a PDP-12 computer (Digital Equipment Corp.). In order to investigate the open-loop characteristics of m. triceps surae and its antagonist muscle, the reflex loop was completely blocked by a perfusion of xylocaine (2%) around the sciatic nerve.

The parameters of interest were foot position, net ankle torque, and EMG activity from m. soleus, medial gastrocnemius and tibialis anterior. EMG activity was recorded by means of wire electrodes, Teflon coated to 0.5 in of the tip inserted into the muscle bellies approximately 1 in apart.

The subject's foot was positioned at the desired angle of plantar flexion (PF) (mean positions 25, 15 and 5 °PF) to measure responses to tendon taps and sinusoidal displacements (frequency range 0.05–12/sec). Foot oscillation amplitudes ranged from 1 to 10° during sinusoidal movements. Open-loop muscle characteristics were determined passively or with sustained contraction, e. g. 25 impulses per second (imp/sec), or with a sinusoidally modulated impulse train, e. g. 25 ± 10 imp/sec. Each of these open-loop tests was performed with and without foot movement.

The data analysis provided gain (output/input) vs. frequency, and phase angle (output with respect to input) vs. frequency curves. EMG records were analyzed by dividing the input sinusoid of each cycle into 20 bins and measuring the amplitude of integrated EMG. The response time, rate of rise of integrated EMG and the magnitude and position of occurrence of peak integrated EMG were determined by this method.

Results

IA. Frequency-Response Characteristics of the Myotatic Reflex to Sinusoidal Inputs

The output torque/input displacement relationships of the total ankle myotatic reflex loop were determined both by varying the amplitude of sinusoidal movement (1–10° peak-to-peak) at any one frequency, and by increasing sinusoidal frequencies (0.05–12/sec) at any one amplitude of movement. In this presentation, frequency-response characteristics are often described by Bode plots which relate output torque to input rotation in the form of 'stiffness' or gain-frequency curves (expressed in dB = 20 log output/input) and of phase angle-frequency curves (relationship of the position of peak torque with respect to the position of peak displacement).

The response characteristics of the ankle myotatic reflex system as well as its components were analyzed primarily in a group of hemiplegic patients. This model was chosen for investigation because it presents *stable* responses to repetitive stretch.

Figure 1 illustrates a typical torque-frequency response curve of a grade III subject [HERMAN, 1970] at a 10° amplitude of input oscillation. Sinusoidal movements about a mean displacement of 5 °PF were used in order to reduce

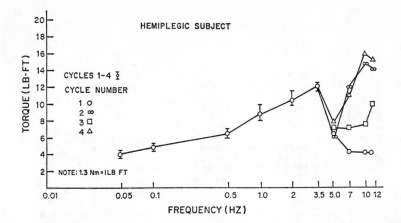

Fig. 1. Torque-frequency curve of the ankle myotatic reflex in a grade III hemiplegic subject. Note alterations in peak torque (above 5/sec) during the initial 4 cycles of a series of repetitive stretches (see text for explanation).

the influence of the torque-length properties of muscle which have been shown to exhibit little alteration between 10 °PF and 0° (neutral position of the ankle joint) [HERMAN and BRAGIN, 1967; table I]. A graded increase in torque values was observed from 0.05–3.5/sec, followed by a reduction at 5/sec. This rise in torque was associated with a transition from tonic to phasic EMG discharges in the extensor m. triceps surae, a decrease in EMG response times and an increase in the rate of rise of EMG [see below; also HERMAN, 1970]. Although the amplitude of integrated electrical discharges was considerably greater at slow rates of stretch, i.e. at 0.05–1/sec (fig. 2), rather small changes in torque occurred at these frequencies. This was apparently due to a tonic firing pattern which did not substantially support the development of active contraction, regardless of the rate of stretch [JOYCE, RACK and WESTBURY, 1969; HERMAN, 1970]. At 5/sec, despite an increase in the rate of rise of EMG activity and a decrease in the EMG response time (fig. 3), the magnitude of the torque response decreased, due in part to the position of peak EMG activity relative to the peak of the displacement sinusoid. Because the peak EMG discharge was in advance of peak displacement by only 50° (approximately 28 msec at 5/sec), torque was developed during the relaxation part of the cycle. These torque values fell below those obtained isometrically by an electrically in-duced twitch contraction [FREEDMAN, 1971; JOYCE, RACK and WESTBURY, 1969].

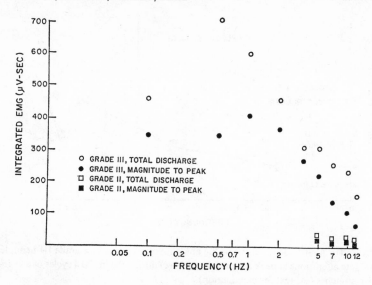

Fig. 2. Integrated EMG–frequency curve of the soleus muscles in both a grade II and grade III hemiplegic subject. The magnitude of integrated EMG at any one frequency is averaged over 10 cycles, and measured either through the entire cycle (total discharge) or to the peak of the EMG discharge. Note the 'roll off' at approximately 1/sec.

Figure 1 also illustrates marked variations between torque responses of the initial 4 sinusoidal cycles at frequencies of 7–12/sec. At 7/sec, despite considerable EMG discharge during the first cycle, the torque response was similar to values observed both during stretch of muscle at slow frequencies (e.g. 0.5/sec) and during tests in which the reflex loop was interrupted (by complete narcotization of the sciatic nerve). Thus, the EMG discharge during the first cycle was responsible for the augmented torque values developed during the second sinusoidal cycle. Subsequent torque responses represent EMG discharges of preceding cycles (e.g. the EMG discharge of the second cycle was uniformly suppressed, leading to a low torque response in the third cycle, while considerable EMG discharge during the third cycle influenced torque values of the fourth cycle). At frequencies of 9 or 10/sec, there was usually complete suppression of the second cycle EMG discharge, with limited EMG activity during each subsequent even cycle (i.e. cycles 4, 6, 8 and 10) interspersed between large phasic bursts during each odd cycle (i.e. cycles 3, 5, 7 and 9) (fig. 4). This pattern of electrical activity was associated with low torque values during the first cycle, followed by an increase during the second cycle, a somewhat lower response during the third cycle and an-

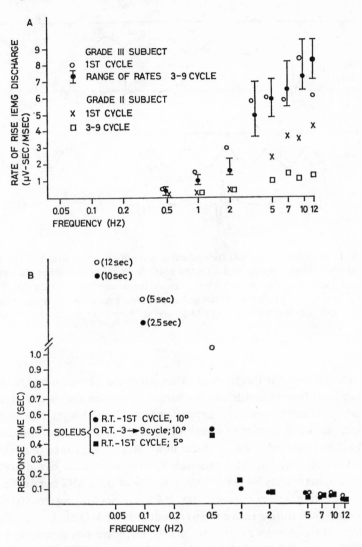

Fig. 3. A Rate of rise of integrated EMG as a function of frequency. In both a grade II and grade III subject, the rate of rise increases as the frequency increases. The first cycle response is usually higher than that of the average of rates across the third to the ninth cycle. *B* The response time (latency from the trough of the sinusoid to the initial EMG response) in a grade III subject as a function of the frequency of stretch. Note that the response time of the first cycle, as compared to the average of the third to the ninth cycles, is decreased and that there is essentially no difference between amplitudes of oscillation of 5 and 10°. At 1 and 2/sec, no data points for the average response times from third to the ninth cycle are indicated since EMG appears through the entire sinusoidal movement.

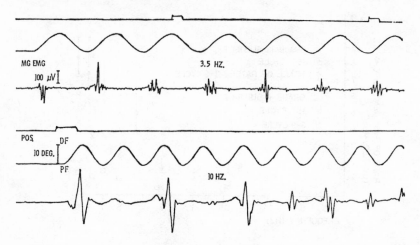

Fig. 4. Foot position and EMG (retouched) as a result of muscle stretch at two frequencies in a hemiplegic subject. At 3.5/sec the EMG discharge occurs during the movement toward dorsiflexion (DF) and an EMG burst occurs on each stretch cycle. At 10/sec the EMG activity occurs on alternate cycles and usually during movements towards plantar flexion (PF). Also note that the first cycle EMG burst at 10/sec is larger than the following bursts.

other increase during the fourth cycle. Alternate cycle inhibition was further pronounced at 12/sec in that EMG discharges were often completely absent during the even cycles, sometimes appearing as small discharges in m. soleus during later stages of a series of repetitive stretches. Torque responses were similar to those observed at 7 and 10/sec; however, the augmented third cycle response at 12/sec was due to the relatively short time duration of the stretch cycle (e.g. approximately 84 msec), the position of the EMG bursts (fig. 5) and the contraction time of muscle. Thus, at frequencies ranging from 0.05 to 5/sec, torque values during each cycle were induced by EMG discharges of the same cycle. At frequencies ranging from 7–12/sec, there was suppression of motor discharges, with the EMG discharge of one cycle influencing torque values of the following two cycles, the intensity of suppression being directly related to the frequency. Because of these findings, the frequency-response characteristics of the myotatic reflex have been illustrated at frequencies less than 5/sec. Despite varying conditions such as the amplitude of oscillation (e.g. 10, 5 and 2.5°) and the degree of spasticity (grade II, grade III), the ankle myotatic reflex in all hemiplegic subjects demonstrated uniform torque frequency curves.

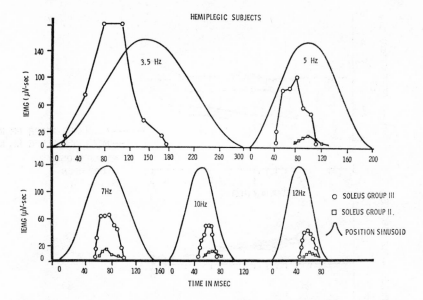

Fig.5. Amplitude of integrated EMG (IEMG) with respect to foot position during stretch at 5 frequencies. For ease of comparison, a sinusoid at each frequency, along with its time scale, is indicated. Note the decreasing amplitude of IEMG as frequency increases and the shift of the peak IEMG from leading the peak position (3.5/sec) to lagging (12/sec).

The frequency-response characteristics of the ankle myotatic reflex at various mean foot positions utilizing any one amplitude, and at various amplitudes, utilizing any one mean foot position are shown in figures 6 and 7. These figures illustrate the dynamic characteristics of the torque-length relationships and of the response of the system to increasing rates of stretch. Under both closed- and open-loop conditions, the relationship between output torque and input foot rotation is described by Bode plots for a group of 12 hemiplegic subjects whose hypertonia was classified as grade II or grade III.

Figure 6 demonstrates the gain and phase angle differences utilizing two mean displacements (e.g. 25 and 15 °PF). With the reflex loop closed, at any one frequency, the absolute value of the gain increased as a function of mean displacement. This was due primarily to the response of the passive properties (open-loop behavior) of muscle although the difference in gain between closed- and open-loop conditions at mean lengths of 25 and 15 °PF indicated the presence of enhanced reflex activity at 15 °PF. The phase angle plot in figure 6 shows torque leading displacement (ranging from 4–16°) at frequencies less

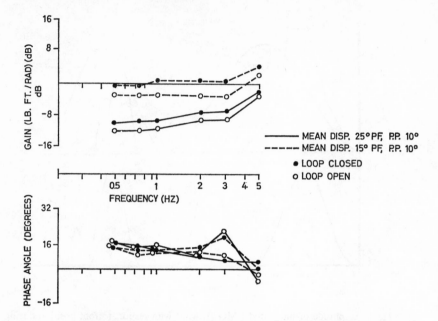

Fig.6. Frequency response curves in a hemiplegic subject at mean displacement of 25 and 15 °PF at a peak-to-peak (P.P.) amplitude of 10°. At each mean displacement, the responses of the total reflex system (●) and of muscle and connective tissue (○) are demonstrated. The gain curves show relative changes while the phase angle curves indicate absolute values.

than 2–3.5/sec. At 5/sec, there is a tendency towards phase lag. This finding was unexpected since peak EMG discharges were only in slight advance of peak sinusoidal displacement, suggesting that torque developed must have occurred during the relaxation part of the stretch cycle (fig. 5). Yet, the phase relationship between output torque and input displacement was limited to approximately ± 3°. Notice also the similarity in the slopes of the gain and phase angle curves in the presence and absence of reflex activity (fig. 6). These observations suggest that the rheologic properties of muscle influence greatly the behavior of the system. Further evidence for this concept will be presented below.

The dynamic properties of the intact myotatic reflex loop are presented in figure 7, for the entire sample of hemiplegic subjects. The ankle was modulated sinusoidally at frequencies ranging from 0.05–12/sec; however, the Bode plots indicate only a range to 5/sec for reasons already stated. Mean displacements of 25, 15 and 5 °PF and amplitudes of 10 and 5° peak-to-peak were utilized.

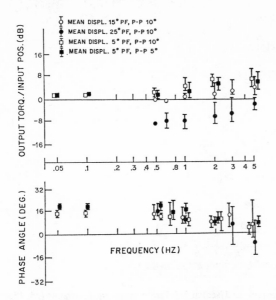

Fig. 7. Composite frequency response curves of the ankle myotatic reflex in hemiplegic subjects (n = 12) at 3 different mean displacements and at 2 amplitudes of oscillation. Gain curves increased slightly with increasing frequencies while phase angle advance decreased slightly from 2–5/sec.

The values for the gains were normalized at the lowest frequency for each mean displacement. The plots show remarkable uniformity at all mean lengths with respect to the shape of the curves. Although there is an increase in the absolute torque value from 25–5 °PF, the slopes of all the gain curves show a slight increase at the three mean displacements. The greater increase in gain between 25 and 15 °PF (as compared to the gain difference between 15 and 5 °PF) may be attributed to such factors as the limitation in time delay (response time) within the neural circuit, torque-displacement curves at 15 and 5 °PF, the time course of the active state, saturation levels of motoneurone output, lengthening of the series-elastic tissue and the interaction of increased rheologic properties with the hyper-reflexive properties of muscle [HERMAN, 1970]. On the other hand, *there was no dependency of the value of phase angle* on mean length or on the degree of reflex activity. A phase lead of approximately 6–16° was observed at all frequencies when mean displacements of 15 and 5 °PF were utilized. At 25 °PF, a slight phase lag (approximately 8°) was infrequently detected. This finding may be attributed to the influence of the lengthened flexor muscles and connective tissue upon net torque output.

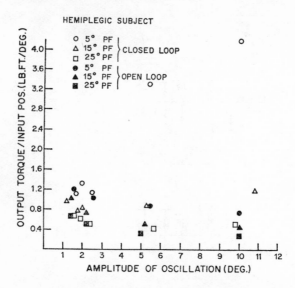

Fig. 8. Gain-amplitude curves (at 1/sec) at 3 different mean displacements (25, 15 and 5 °PF). With the reflex loop open (filled symbols), the gain decreases as amplitude increases. With the reflex loop closed (open symbols), gain decreases when no EMG is present. When EMG appears (between 6 and 10° at 25 °PF, 2 and 5° at 15 °PF, and 1–2° at 5 °PF), the gain increases as amplitude increases.

Figure 7 also demonstrates the dynamic behavior of the myotatic system when the amplitude of oscillation was varied about a selected mean length (e.g. the open and closed squares at 1/sec in fig. 7). When the amplitude of stretch was increased twofold (e.g. a change of peak-to-peak displacement from 5–10°), there was little change in the gain- or phase angle-frequency curves.

Further experiments were conducted to evaluate linearity in the myotatic system and its subsystems. Figure 8 illustrates that a test for linear transfer function is not satisfied for either the total system or for the two subsystems (i.e. the neural and non-neural subsystems). The test utilizes the premise that, at any given frequency, linearity is satisfied if the output response (e.g. torque) is linearly related to the input with no amplitude associated phase shift. The gain- and phase angle-frequency curves of muscle under both closed- and open-loop conditions suggest that the criteria for linearity have not been satisfied although very small amplitudes of movement were used. However, the curves do indicate that gain is a function of the innervation pattern of muscle. With the reflex loop *open*, gain was maximal with small amplitudes of stretch, decreasing with increasing amplitudes. This response pattern was also observed when EMG activity was absent with the reflex loop closed and when muscle (open-loop) was stimulated both by constant frequency pulses (e.g. 25 imp/sec) and by a sinusoidally varying impulse train (e.g. 25 $\pm$ 10 imp/ sec) (not illustrated). When reflex EMG activity was present, the gain increased with increasing amplitudes of stretch. These results may indicate that: (1) high gain values at small amplitudes (in muscle with the reflex loop open or in muscle with reflex loop closed, but with EMG absent) may be attributed to frictional elements within muscle, and (2) high gain values at large amplitudes of stretch (in muscle with active reflex loop behavior) are due to the recruiting, asynchronous character of motor discharges innervating muscle. Thus, when the reflex loop is intact, the character of motoneurone discharges (tonic or phasic) determines both the slope of the curve and the amplitude of the output response (fig. 8).

It must be emphasized that in these experiments in which responses were measured at a mean foot position of 5 °PF, torque developed during peak-to-peak oscillations of from 1–10° was independent of the torque-displacement characteristics of the system (table I).

IB. Frequency-Response Characteristics of Myotatic Reflex to Impulse Inputs

A well-known result from systems theory states that a linear system can be characterized by applying an input impulse function and measuring the system response; however, an impulse function is a mathematical device and cannot be physically realized. A realizable input function whose time duration is short with respect to the time duration of the output response may be considered to be an impulse-like signal; however, it must be stressed that biological systems within the range of physiological movements are not linear systems so that an impulse response approach is only approximate. ADOLPH [1959] suggested using a tendon tap as an impulse input to study reflex systems in order to determine the isometric system function; however, it is not clear whether he verified the assumption under which the tendon tap is considered to be an impulse function. In using the Achilles tendon tap as an impulse input to the myotatic reflex system, the data had to conform to the time constraints mentioned above.

In these experiments, the ankle torque wave form, which was elicited by an Achilles tendon tap, was decomposed into its component sinusoidal fre-

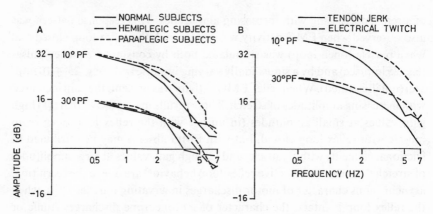

Fig. 9. A Mean amplitude-frequency curves of the tendon response for the entire population (n = 27) of subjects. Note that gain for each foot position is normalized at 0.5/sec to the values of the gain observed in hemiplegic subjects. *B* Amplitude-frequency curves of the tendon and electrically induced twitch responses. The increase in gain values at 10 °PF is due to enhanced reflex activity and to the torque-displacement characteristics of muscle. [The curves at each foot position are normalized to the value of the tendon response at 0.5/sec.]

quencies and plotted as an amplitude (in dB)-frequency plot. The average amplitude-frequency curves of the tendon jerk response in normal, hemiplegic and paraplegic man, are represented in figure 9A. Note that the amplitude curves decrease rapidly above 1/sec as the frequency increases. The minimum amplitude is usually reached at about 7/sec (the highest frequency indicated on the graphs). These amplitude-frequency curves are normalized to a convenient amplitude at 0.5/sec for ease in comparing the response at several different foot positions. Note that in figure 9A, the gain is considerably greater at 10 than at 30 °PF due to enhanced active reflex response (i.e. EMG output) and to the tension-length characteristics of muscle.

In order to verify that the system accepts the tendon tap as an impulse input, the twitch response of the chemically denervated m. triceps surae group to a supramaximal electrical stimulation (pulse duration 1.0 msec) was analyzed using the tendon jerk response analysis method. Figure 9B shows the amplitude-frequency curves for both the tendon jerk and supramaximal electrical twitch response at two different foot positions. The amplitude scale of the electrical twitch in the figure has been normalized to the tendon jerk amplitude response at 0.5/sec.

The analysis of the effect of direct electrical stimulation upon the m. triceps surae has provided further insight into the functional mechanisms of the ankle myotatic reflex system. Similar amplitude-frequency curves of the tendon jerk response (closed-loop) and of the direct electrical twitch response (open-loop) of muscle imply that the tendon jerk response is mainly a function of muscle tissue; i.e. the neural component of the myotatic reflex loop does not appreciably alter the shape of the contractile response of muscle at any given muscle length. The results obtained in studying the tendon response in the 3 groups of subjects supports this contention. This interpretation of the results is further substantiated by investigation of the frequency-response characteristics of muscle and connective tissue (see Results, section III).

II. The Frequency-Response Characteristics of the Neural Subsystem

The dynamic behavior of EMG response patterns (frequency sensitivity) during sinusoidal perturbation of the ankle appears strikingly similar to the EMG response patterns (velocity sensitivity) during ramp movements [HERMAN, 1970]. For example, at slow frequencies of stretch there was tonic firing throughout the stretch cycle. At higher frequencies (above 2/sec) phasic bursts were pronounced. With the increase of frequency, the duration of these bursts decreased and the magnitude of integrated EMG activity decreased, the curve 'rolling off' at approximately 0.5–1.0/sec in grade III subjects (fig. 2). The transition from tonic to phasic firing suggests a change from displacement-sensitive to velocity-sensitive behavior [LIPPOLD, REDFEARN and VUČO, 1958]. The contractile response of muscle varied according to the pattern of EMG discharges; that is, up to 1/sec, tonic firing patterns (e.g. motor discharge frequency at a rate of 6–12 imp/sec) did not alter appreciably either the peak torque or the degree of EMG activity (fig. 2). The lack of rate sensitivity of torque output with respect to tonic firing patterns was also observed by stimulating muscle (open-loop) during sinusoidal perturbation (fig. 10) or during ramp movements [JOYCE, RACK and WESTBURY, 1969; HERMAN, 1970]. On the other hand, phasic discharges were associated with increasing torque values as the frequency of stretch increased. Results of experiments (see below), utilizing a large amplitude, short duration, electrical pulse to muscle (open-loop) during ramp movement, also described increasing torque values as velocity increased (fig. 11, table I).

The response time, i.e. latency between the initiation of stretch (at the trough of the sinusoid) and the onset of EMG activity, as well as the rate of rise of EMG in both m. soleus and medial gastrocnemius were a function of the frequency of stretch. As frequency increased the response time decreased

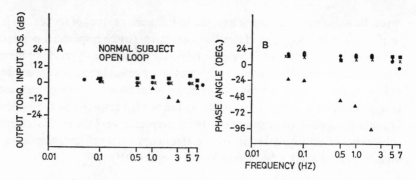

Fig. 10. Gain- and phase angle-frequency curves in a normal subject with complete local anesthetic block of the sciatic nerve. Note the similarity of the results for all experiments in which the foot was moving (●, ▦, ×). Only the curves for sinusoidal pulse modulation of isometrically contracting muscle differ from the near constant gain and phase angle values seen with the foot moving. Note also that gain is not altered as frequency of stretch increases when muscle is stimulated by a constant pulse frequency (●) or by a modulated pulse frequency (▦). See text for explanation. ●10 °PF − 10° foot moving-sust.
+ 0°
contraction 25 I.P.S. ▦ 10 °PF − 10° foot moving-sine exc. 25 ± 15 I.P.S. ▲5° PF isometric
+ 0°
muscle-sine exc. 25 ± 15 I.P.S. × 10 °PF − 10° passive movement (nerve blocked).
+ 0°

(fig. 3 B) and the rate of rise of integrated EMG increased (fig. 3A). There was a direct relationship between the development of torque and the response time and rate of rise of integrated EMG, similar to results obtained by HOMMA [1963] and HERMAN [1970]. HOMMA [1963] suggested that alterations in muscle spindle discharges as a function of the frequency of stimulation induced earlier postsynaptic potentials and ventral root spikes. With increasing stretch frequencies, summated EPSP responses progressively increased in slope and decreased in size and duration [HOMMA, ISHIKAWA and STUART, 1970]. According to these authors, when the rate of stretch was decreased, there was an increase in the response time. Such findings were attributed to a decrease in the rate of muscle receptor discharge, which would prolong the set-up time and increase central transmission times.

At any one frequency, EMG activity, i.e. the response time, rate of rise of integrated EMG and position of peak integrated EMG, during the initial cycle of a repetitive train of sinusoidal stretches, demonstrated enhanced dynamic sensitivity when compared to the responses of the subsequent cycles. Usually, during the first cycle, the response time was shorter and the rate of

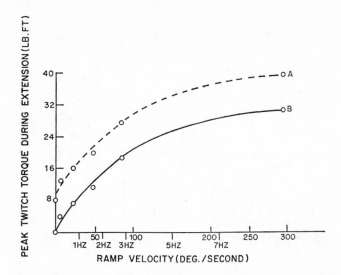

Fig. 11. Dynamic torque-velocity curve of chemically denervated m. triceps surae. Curve A includes the static torque at each point. Curve B displays only the dynamic torque characteristics. Note that the frequency markings on the abscissa indicate the maximum velocity achieved for a 10° oscillation at the sinusoidal frequency indicated.

rise of integrated EMG was greater than during the succeeding cycles (fig. 3). This behavior has been attributed to frictional forces developed in the intra-fusal muscle elements which are larger at the initiation of movement than during continuous movements, as well as to levels of hyperpolarization of spinal motoneurones [HOMMA, ISHIKAWA and STUART, 1970].

The dynamic sensitivity of the reflex loop can also be observed by relating the position of peak EMG output to peak displacement input, at increasing frequencies of stretch (fig. 5, 12). The position of peak EMG 'rolls off' at approximately 1–2/sec and lags displacement at 7–10/sec *(depending upon transmission time in the myotatic circuit)*. Note that the shapes of the frequency response curves of muscle (closed-loop) relating amplitude of integrated EMG and phase angle (peak EMG with respect to peak displacement) to frequency (fig. 3, 12), are similar in appearance to the shape of the frequency response curves of isometric muscle (open-loop) relating peak torque to peak displacement during electrical stimulation. Differences such as the 'corner' or 'roll off' frequency, are most likely due to variations in transmission lag time in the reflex loop (displacement input to EMG output) and in muscle (EMG input to torque output).

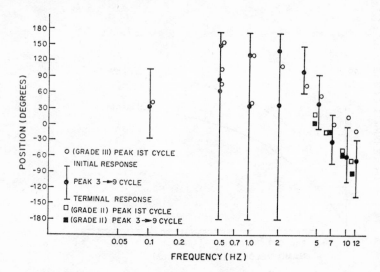

Fig. 12. Position of peak integrated EMG (with respect to the input sinusoid) as a function of frequency. Note the long-lasting duration of EMG firing which occurs at 0.5–2/sec and the generally greater lead angle of the first cycle response with respect to that of the remaining cycles of the input sinusoid. The grade II spastic subject did not exhibit appreciable EMG activity under 5/sec; however, position of peak EMG was similar to that observed in grade III subject.

The peak EMG output can be adjusted for the transmission time of the neural system. This calculation permits an estimation of the function of the muscle spindle – assuming nondynamic behavior of the motoneurone pool [ROBERTS, 1967; POPPELE and TERZUOLO, 1968]. Using a minimal transmission time as determined by tendon tap responses, the phase angle advance of peak EMG (2–12/sec) ranges from 120–160°. These findings support the contention that muscle spindles (i.e. the primary endings) are sensitive to both velocity and acceleration of stretch [MATTHEWS and STEIN, 1969; GRÜSSER and THIELE, 1968].

III. The Frequency-Response Characteristics of Muscle and Connective Tissue

The displacement input-torque output relationships of the ankle myotatic reflex system in man have been determined by utilizing two different input functions: (1) Achilles tendon tap (see section IB), and (2) variable frequency sinusoidal rotation of the foot at various stretch amplitudes (see section IA). As indicated in those sections, with the reflex loop closed, the results obtained from these two methods differed substantially with respect to the shape of the curves. The amplitude-frequency curves of the tendon response demonstrated decreasing amplitude components as frequency increased (fig. 9A); the sinusoidally induced gain-frequency curves showed a slight increase of gain over the frequency range (fig. 6, 7). These differences were reconciled by examining the frequency-response characteristics of muscle with the reflex loop open, the assumption being that the rheologic (visco-elastic-plastic) and the contractile properties of moving muscle may be responsible for the disparate reflex patterns. This would then explain the near constancy of the sinusoidally perturbed gain- and phase angle-frequency curves as opposed to the 'roll off' of the tendon jerk amplitude-frequency curve. This assumption was tested with two experimental procedures in normal subjects and in subjects with hemiplegic and paraplegic models of spasticity.

In experiment 1, similar to the investigations of the tendon jerk response, the output was measured at a fixed foot position, while in experiment 2, similar to the investigation of frequency response characteristics of the intact myotatic loop, the output was measured during sinusoidal movement of the ankle. Both experiments utilized similar mean displacements and mean tensions.

The study of muscle under open-loop conditions offers the advantage of observing the effects of various patterns of electrical stimuli on muscle being mindful of the fact that only frequency of stimulation (and not recruitment) is the controlling parameter. [Note that only relative amplitudes are demonstrated so that tendon jerk and modulation amplitudes could be normalized at 0.5/sec to some convenient value for the purpose of shape comparison in one subject and comparison of results among different subjects.]

Experiment 1

The patient's foot was rotated from 30–0° PF at different constant rates, i.e. 5, 23, 45, 85 and 300 degrees/sec. As previously demonstrated [HERMAN, 1970], in muscle devoid of reflex or voluntary innervation, there was no evidence of rate sensitivity; that is, peak tension at the end of stretch was not altered by increasing rates of stretch. During these movements the muscle was electrically stimulated by a single, supramaximal pulse (2.0 msec duration) applied at a predetermined angle. This was achieved by allowing the foot pad to mechanically close a microswitch at the selected angle during movement. As various rates of stretch were utilized, the resulting peak of the muscle contraction occurred at different angles of displacement (table I, B). In order to study the effect of velocity of muscle stretch on contraction, static or isometric twitch contractions were measured at the angle at which the peak occurred (table I, C). These findings were subtracted from the peak of the twitch toque which was obtained during ramp movement (table 1, A; fig. 11A). The resulting torque- velocity curve (fig. 11 B; table I, D) for the chemically denervated ankle musculature was used to adjust the electrical twitch amplitude-frequency curve for the effect of moving muscle. Since the amplitude-frequency response of the tendon jerk and twitch contractions were similar (fig. 9 B), the tendon jerk amplitude-frequency curve was corrected using similar values (fig. 13, A–B). The data derived from sinusoidal perturbation of the ankle joint with the reflex loop closed, was represented by curve C in figure 13. [Note that the curves are drawn from arbitrary gain values at 0.5/sec for ease of comparison.] The corrected tendon jerk amplitude-frequency curves and the sinusoidal gain-frequency curves were similar in shape, suggesting that the output characteristics of the stretch reflex are mediated by the properties of muscle (see below). Thus, if the above hypothesis were correct, the adjusted tendon jerk response curve would be similar in shape to the sinusoidal perturbation response curve with the reflex loop intact.

Experiment 2

In this series of experiments, sinusoidally varying inter-spike interval pulse trains employing various carrier frequencies, voltages and depths of frequency modulation, were utilized to stimulate the triceps surae both when the ankle was held fixed and when the ankle was moved sinusoidally from 10 to 0 °PF (fig. 10). A mean foot position of 5 °PF was selected since tension-length relationships were not altered appreciably by perturbations of 5° about this mean muscle length. When the ankle was maintained in a fixed position of 5 °PF, variations in the stimulus parameters did not cause a marked change in the gain- and phase angle-frequency curves [not illustrated]. Gain curves, normalized to a convenient value at the lowest frequency, show relative values, while phase angle curves demonstrate absolute values. Absolute gain, of course, would be modified by changes in the carrier frequency and voltage delivered to the muscle. The response characteristics of the isometrically contracting m. triceps surae, as represented by gain- and phase angle-frequency curves were similar to those exhibited by the m. triceps surae responding to a tendon tap or electrical twitch, i.e. the curves 'roll off' at approximately 0.5/sec. The frequency-response curves revealed a steady decrease in both gain and phase angle at frequencies past 0.5/sec. Output torque lagged input stimulation at all frequencies. In fact, phase angle determinations often showed

Table I. Torque-velocity relationships derived from induced twitch contractions during ramp movements. The dynamic response (D) is calculated by adjusting the total dynamic response (A) for the static response (C)

		Ramp speed, deg/sec				
		5	20	45	85	300
A	Dynamic twitch peak response, lb ft[1]	13	16	20	27	40
B	Angle at which dynamic twitch peak occurred, deg. PF	14.5	14.5	13	3	0
C	Static twitch peak response (lb ft) at similar angles (see B)	8	8.5	9	9	8.5
D	Adjusted twitch peak torque (lb ft) A–C	5	7.5	11	18	31.5

Note that from 14.5–0°, torque-length relationships were constant (line C).
[1] lb ft = 1.3 Newton meter.

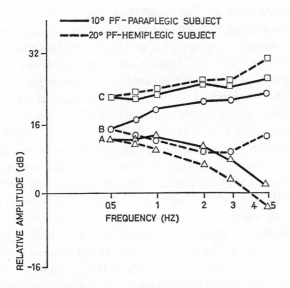

Fig. 13. Amplitude-frequency curves for a paraplegic subject (foot position 10 °PF) and for a hemiplegic subject (foot position 20 °PF). Curves: A, tendon tap responses (△); B, curve A corrected for foot movement (○) (see text); C, responses to sinusoidal foot movement input (□).

torque lagging displacement by more than 90° at 2/sec. In an isometrically contracting muscle, the marked lag in the relationship between peak torque and peak displacement is most likely due to time delays in impulse transmission and in contraction of muscle, while the changes in the slopes of the gain curves may be attributed to the number of effective pulses delivered to muscle [McKEAN, POPPELE, ROSENTHAL and TERZUOLO, 1970].

However, when sinusoidal displacement of the ankle coincided with the sinusoidal stimulation of muscle, the frequency response curves were modified markedly and appeared similar in shape to those observed in experiments with the reflex loop closed. Sinusoidally varying interspike interval trains (e.g. 25 ± 15 imp/sec) phase locked to the sinusoidal perturbation of the ankle (from 10–0 °PF) induced relatively flat gain- and phase-curves, with output torque leading input displacement (fig. 10). The degree of phase angle advance paralleled the phase angles observed with the reflex loop intact. Further, curves obtained with the reflex loop open or closed, resembled the frequency response curves of chemically denervated muscle which were determined by oscillating the ankle from 10–0 °PF (i.e. no nerve stimulation).

Thus, it appears that the contractile and rheologic properties of muscle couple to stabilize the total myotatic system, despite considerable neural transmission and contraction time delays. The coupling of these inherent muscle components of the myotatic reflex system seems to be responsible for the limited variation in the shapes of the gain curves, and for the constant absolute values of phase angle curves. This consistency is observed regardless of the magnitude of reflex activity, or of the delay in neural transmission (hence the position of peak EMG as related to peak displacement), or of the level of mean displacement and/or mean tension prior to oscillation of the ankle. In a previous paper [HERMAN, 1970], one of us indicated that a feedback control derived from an independently operated muscle spindle system allows for freedom and flexibility in control of imposed position, velocity and force and for prevention of oscillations within the reflex loop which might occur with excessive time delays following detection of error in position and/or force. It now appears likely that muscle dynamics are as, or more, important than the muscle spindle in protecting the system from instability and controlling the upright posture. With respect to the latter function, as the frequency and amplitude of the ankle joint movement range only to 0.5/sec and to ±2°, respectively, when the normal subject assumes a standing posture [NASHNER, 1970], the relatively high gain and phase advance (as compared to larger amplitudes of movement) of the ankle myotatic reflex is largely due to the inherent properties of muscle.

Conclusion

Utilizing sinusoidal and impulse inputs to the ankle myotatic reflex system, the frequency and amplitude responses of spastic muscles both with and without the reflex loop intact indicate that:

1. Relative gain and absolute phase angle values of the total reflex system are determined, for the most part, by the inherent rheologic and contractile properties of muscle.

In the total reflex system, the frequency response characteristics observed, i.e. a near flat gain and constant phase angle advance, are not the result of compensation for the responses of the neural and non-neural systems as was indicated by POPPELE and TERZUOLO [1968]. The 'roll off' of gain and phase with increasing frequency observed in an isometrically contracting muscle (open-loop), is considerably modified by imposing sinusoidal stretch (fig. 10). Thus, the decay in gain, which is a function of the number of impulses delivered to muscle, and the decay in phase angle, which is a function of the delay in contraction time, are altered by the rheologic and contractile properties of sinusoidally driven muscle.

Regardless of the subject group (normal, paraplegic, hemiplegic), of the degree of reflex activity (e.g. magnitude of integrated EMG), of the transmission time about the neural circuit (and, hence, the position of peak integrated EMG with respect to peak displacement), of the amplitude and frequency of oscillation and of mean displacement of the ankle joint, there is remarkable uniformity in the gain and phase plots. Further, evidence that torque leads displacement with and without the reflex loop intact, suggests that muscle properties can stabilize the myotatic reflex system [PARTRIDGE, 1967], a function often ascribed only to the primary endings of the muscle spindle [BROWN and MATTHEWS, 1966].

2. Increasing amplitudes of stretch at any one frequency do not elicit a linear response between torque output and displacement input. Muscle, without neural innervation demonstrates high gain values at small amplitudes of stretch and lower values at large amplitudes of stretch. Similar results have been noted in muscle and muscle spindles of experimental animals [HILL, 1968; MATTHEWS and STEIN, 1969]; this behavior is most likely due to frictional elements within muscle. Torque developed, either by tonic reflex firing or by induced frequency modulated pulse trains, is likewise not amplitude or rate dependent; e.g. gain-amplitude and gain-frequency curves indicate that muscle, influenced by constant frequency stimulation, resembles the behavior of muscle devoid of innervation (i.e. highest gain values at small amplitudes).

Lack of rate sensitivity during stretch has been recently emphasized [JOYCE, RACK and WESTBURY, 1969; HERMAN, 1970]; however, such results were not compared with responses to phasic patterns of motoneurone discharges. Torque developed, either by asynchronous, rapidly recruiting phasic reflex activity or by induced large amplitude, wide duration pulses, is rate and amplitude dependent [HERMAN, 1970; FREEDMAN, 1971, see table I, fig. 11]. Alterations in the force-velocity relationship during stretch may be ascribed to changes in the inherent visco-elastic properties during forcible contraction.

3. By adjusting the position of the peak EMG for transmission time of the monosynaptic reflex, and by assuming the spinal cord has a simple transfer function [ROBERTS, 1967; POPPELE and TERZUOLO, 1968], some inferences can be drawn with respect to muscle spindle function. Utilizing these criteria, the muscle spindle demonstrates a marked phase angle advance ranging from 120–160° across all frequencies; such findings support the contention that the muscle spindle responds to both velocity and acceleration of stretch [GRÜSSER and THIELE, 1968; MATTHEWS and STEIN, 1969].

4. Variations between the EMG responses during the first and subsequent cycles of a train of repetitive sinusoidal stretches are often observed; the response time of EMG, the rate of rise of EMG and the relationship between peak EMG to peak displacement all indicate increased reflex sensitivity during the first cycle (fig. 3, 12). Such results are attributed to static friction ('stiction') properties of intrafusal muscle fibers [LENNERSTRAND and THODEN, 1968] and to changes in levels of motoneurone hyperpolarization [HOMMA, ISHIKAWA and STUART, 1970].

5. Motoneurone discharges do not follow input frequencies from 7–12/sec. Complete suppression of motor discharges are observed during alternate stretch cycles at 12/sec. Such frequency-limiting behavior is ascribed to a number of central and peripheral mechanisms [HOMMA, ISHIKAWA and STUART, 1970].

Acknowledgement

This study was supported by grants No. 23P-55115 and 16P 56804 from the Social and Rehabilitation Services, Department of Health, Education and Welfare, Washington, D.C., USA.

The authors are grateful to Miss PATRICIA SAYLER for her editorial assistance, to Mrs. SARA MEEKS, Dr. NATHANIEL MAYER and Mr. FRED KUGLER for their technical help and valuable comments, and to Miss EILEEN McDONNELL for preparation of the manuscript.

Author's address: Dr. R. HERMAN, Department of Rehabilitation Medicine, Temple University School of Medicine, *Philadelphia, PA 19141* (USA)

New Developments in Electromyography and Clinical Neurophysiology,
edited by J. E. Desmedt, vol. 3, pp. 579–588 (Karger, Basel 1973)

Physiological Aspects of Hemiplegic and Paraplegic Spasticity[1]

R. HERMAN, W. FREEDMAN and S. M. MEEKS

Department of Rehabilitation Medicine, Temple University School of Medicine, and
Krusen Research Center, Moss Rehabilitation Hospital, Philadelphia, Pa.

The defining characteristic of clinical spasticity is excessive resistance of muscle to passive extension, the resistance increasing with increasing rates of extension [JANSEN, 1962; SHIMAZU, HONGO, KUBOTA, NARABAYASHI, 1962; HERMAN, 1970]. While this description may apply generally to one group of muscles (i. e. flexors or extensors) acting about a joint, it does not describe the behavior of all muscles controlling that joint. Neither is it suggestive of the integrative nature of the biological processes subserving the hyper-reflexive state.

The following comments attempt to clarify these issues by comparing the distinctive features of reflex activity in the spinal and hemiplegic models of spasticity. The discussion is based upon the response of spastic muscle to two different forms of natural stimuli: (1) controlled (rate, load) mechanical displacement of the joint utilizing physiological ranges of displacement and rates of stretch, and (2) small amplitude, high frequency vibration of tendon (or of muscle). The two stimulus categories ostensibly differ with respect to selective activation of the muscle receptors [BIANCONI and VAN DER MEULEN, 1963; MATTHEWS, 1966; STUART, MOSHER, GERLACH and REINKING, 1970]. The second type of stimulus predominantly elicits excitation of primary endings, while the first type elicits excitation of a number of receptors including the 'in parallel' primary and secondary endings of the muscle spindle, the 'in series' tension receptors of the Golgi tendon organs and the afferents from joint receptors [STUART, OTT, ISHIKAWA and ELDRED, 1965; GRANIT, 1970].

1   This study was supported by grants No. 23 P-55115 and 16 P-56804 from the Social
and Rehabilitation Services, Department of Health, Education and Welfare, Washington,
D. C., USA.

Autogenetic Responses

In subjects with complete spinal cord lesions, both rapid vibratory (e. g. 100/sec) and slow sinusoidal (e. g. 1/sec) stimuli to the extensor m. triceps surae, induce a slow progressive rise in EMG and net ankle torque [HERMAN and MECOMBER, 1971; see fig. 1]. A maximal response is often observed after 2 or 3 sec of vibratory stimulation and after several cycles of sinusoidal movement. On the other hand, in subjects with hemiplegia, there is a rapid rise of reflex activity in which a maximal response is often observed either within 0.1–0.5 sec after initiation of a vibratory stimulus or during the first cycle of a series of repetitive sinusoidal movements (fig. 1). It is reasonable to assume that vibratory stimulation in man alters myotatic reflex activity by enhancing primary muscle spindle discharges [HAGBARTH and EKLUND, 1966; HERMAN and MECOMBER, 1971]. Therefore, similarities in reflex behavior elicited by vibration and sinusoidal movement in either the hemiplegic or paraplegic group suggest that sinusoidal movements also enhance primary muscle spindle discharges.

Velocity sensitivity of the stretch reflex offers further evidence of the role of primary ending discharges in spasticity. Both ramp and sinusoidal stretch induce rate sensitive responses with respect to EMG response time, rate of rise of EMG and induced torque in the extensor muscles of the hemiplegic subject and in both extensor and flexor muscles of the paraplegic subject [HERMAN, 1970; BURKE, GILLIES and LANCE, 1971; HERMAN and MECOMBER, 1971].

The differences in dynamic reflex responses between the two models of spasticity (fig. 1) are attributed to the organizational pattern of the spinal cord. The slow rise of reflex response in subjects with spinal cord lesions can be ascribed to the progressive augmentation of the central excitatory state by cumulative events in the *interneuronal* pathways [GRANIT, 1970]. Since lesions of the spinal cord lead to removal of inhibitory influences upon segmental polysynaptic pathways, it is likely that primary ending discharges are transmitted through multisynaptic chains of the interneuronal pool [HERMAN, 1964; MAGLADERY and TEASDALL, 1958]. This concept is further supported by experimental evidence that spindle primaries can be modulated by polysynaptic circuits [GRANIT, 1970]. In contrast, the rapid build-up of reflex activity noted in spastic hemiplegia suggests that the primary ending discharges are transmitted largely through monosynaptic pathways. Thus, alterations in interneuronal behavior may explain the differences in autogenetic responses between the two models of clinical spasticity.

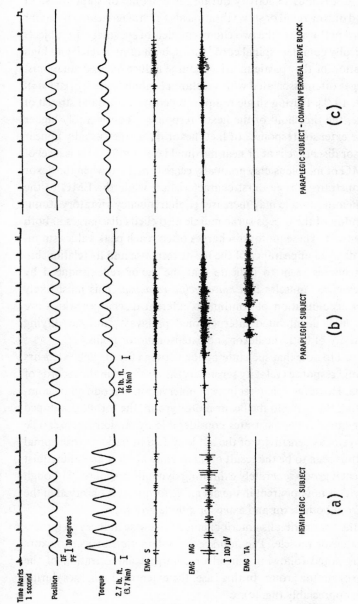

Fig. 1. Torque and EMG response of calf musculature to 1/sec sinusoidal displacement in a hemiplegic and a paraplegic subject. Note the following: (1) the EMG activity of the antagonist m. tibialis anterior is absent in the hemiplegic subject and present in the paraplegic subject. In the latter case, the threshold to firing is high as compared to the firing pattern of m. triceps surae; (2) torque and EMG response of the hemiplegic subject is maximal during first cycle while these responses in a subject with paraplegia progressively increase, reaching a maximum after 6 cycles, and (3) following a common peroneal nerve block the 'accumulative' pattern of EMG and torque remains unaltered while the magnitude of the response is decreased. The calibration levels for the paraplegic subject are identical before and after the common peroneal nerve block.

Reciprocal Innervation

The reciprocal behavior of agonist (e.g. triceps surae) and antagonist (e.g. tibialis anterior) muscle activity during the dynamic or static phase of ramp stretch and during rapid or slow vibration also characterizes the behavior of the spinal cord pathways in the two clinical models of spasticity. In subjects with physiologically complete spinal cord lesions, both ramp stretch and high frequency vibration of the spastic m. triceps surae induce marked antagonist (flexor) discharges often associated with substantial agonist activity [HERMAN and MECOMBER, 1971]. During single ramp or repetitive sinusoidal stretch of m. triceps surae, the threshold of the flexor response is considerably higher than that of the extensor response. In fact, flexor discharges usually appear when the extensor discharge is at or near maximal level (fig. 2). At this level of activity, the EMG of m. soleus may gradually cease (observed when the flexor discharge demonstrates phasic burst characteristics) while the EMG to the medial gastrocnemius muscle may increase. High frequency vibratory stimulation of the tendon of the triceps surae muscle also elicits discharges in both flexors and extensors. These motor discharges often reach peak values simultaneously. The delayed appearance of the flexor response and its relationship to extensor discharges seem to indicate that the flexor reflex induced by stretch of the extensor muscles is a *non-specific response*. This most likely results from an accumulation of continuous afferent discharges which are transmitted through disinhibited interneuronal pathways, thus modifying background activity of both the flexor and extensor motor pools.

It is of interest to note that such interaction between flexors and extensors limits the dynamic response (velocity sensitivity) as observed in the records of net ankle torque. However, when reciprocal innervation is modified by completely narcotizing the nerve to the flexor motor group, the torque developed by the extensor muscles demonstrates considerable dynamic response. The masking of the velocity sensitivity of the ankle system in patients with spinal cord lesions is thus seen to be the result of interacting agonist and antagonist ankle torque—each group separately exhibiting dynamic response. Although velocity sensitivity is not apparent in the net ankle torque, it is observed in the EMG records from both flexor and extensor muscle groups.

Stretch of the flexor (tibialis anterior) muscle likewise induces reciprocal discharges in extensor muscle. The non-specific interaction between flexors and extensors is noted following partial or complete narcotization of the nerve to the flexor motor group. In this case, the extensor discharges during stretch decrease appreciably (fig. 1c).

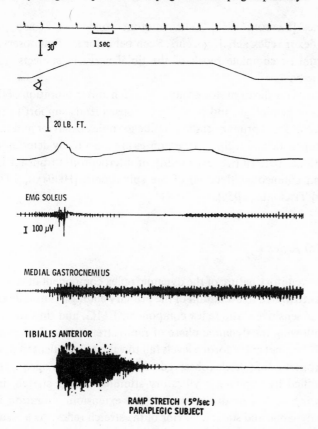

Fig. 2. Torque and EMG response to ramp stretch controlled at 15 °/sec in a paraplegic subject. The torque response 'collapses' during the ramp movement. This is associated with phasic discharges in m. tibialis anterior, suppression of EMG activity in the soleus and enhancement of EMG activity in the medial gastrocnemius. This is an illustration of a 'pseudo clasp-knife' response (see text).

In contrast, during the dynamic and static phases of ramp stretch and during slow sinusoidal stretch, there is *prolonged* reciprocal inhibition of antagonist flexors in hemiplegic subjects (fig. 1 a). This inhibition is augmented by high frequency vibratory stimulation of the triceps surae muscle. These findings are most likely due to excessive I_A activity from muscle receptors of m. triceps surae. The inhibitory nature of group I_A afferent discharges on the antagonist motoneurone pool has been well established in experimental literature. In general, motoneurones are inhibited when they receive impulses from group I_A fibers arising in muscles antagonistic to the muscle supplied by the motoneurones [PATTON, 1965]. When motoneurones lose I_A support, disinhibition of flexor reflex activity occurs. Such behavior is often observed following a partial or complete block of the tibial nerve in subjects with hemiplegic spasticity.

The behavior of the flexor motor groups to stretch and to vibration of the extensor muscles in hemiplegic and paraplegic subjects lends support to the view that I_A afferents from primary endings influence reflex behavior and that variations in autogenetic and reciprocal responses are most likely determined by the organizational pattern (i.e. excitability of interneurones transmitting muscle, joint and cutaneous afferents) of the spinal cord [HERMAN, 1970; MAGLADERY and TEASDALL, 1958].

Static Stretch Reflex

In man, the excitatory nature of the dynamic reflex component of spastic muscle, which is attributed to excitation of the primary endings, is usually not associated with a sensitive static reflex component. EMG, and thus torque, decay rapidly following the dynamic phase of ramp stretch at various rates. Often within 5–10 sec, net ankle torque levels fall to values equivalent to those of the passive properties of muscle (observed with the reflex loop open). This behavior is modified by applying a vibratory stimulus during stretch, immediately following stretch or during maintained extension. Vibration increases both the dynamic and static behavior of the stretch reflex. As a result, the magnitude of decay of EMG and torque is decreased and a sustained low amplitude tonic stretch reflex is observed [HERMAN and MECOMBER, 1971]. Such results strongly support the view that: (1) the decay in the stretch reflex is due to receptor adaptation (primary endings), and (2) the presence of a marked dynamic stretch reflex in the absence of a tonic stretch reflex is due to selective augmentation of the dynamic properties of the primary endings of the muscle spindle.

Pseudo Clasp-Knife Effect

The clasp-knife effect or lengthening reaction of spastic muscle is often considered to be one of the classical signs of clinical spasticity. This effect has been attributed to autogenetic inhibition of the stretch reflex in which resistance to movement decreases and a previously rigid extremity collapses readily

[PATTON, 1965]. However, in both hemiplegic and paraplegic spasticity, there is no experimental evidence of autogenetic inhibition. Reiterating a comment made by one of the authors in a previous publication [HERMAN, 1970]:

'When the stimulus performs work upon the spastic muscle in which the applied load and rate of stretch are unchanged (e. g. controlled load and rate of stretch), the development of tension as a function of length is below the threshold for a possible clasp-knife, or lengthening reaction if, in fact, such a phenomenon actually occurs. [However,]... when muscle performs work upon the stimulus, in which load and rate of stretch are uncontrolled, plastic changes in muscle do occur during stretch. Alterations in applied load and/or rate of stretch lead to diminution of motoneurone output, and hence, decay in tension during stretch. This change in tension during stretch may be classified as a *pseudo clasp-knife effect*. When muscle resistance overcomes the load and/or alters the rate of stretch, there is most likely a decrease in muscle spindle sensitivity (load, rate of stretch act as stimuli to muscle spindle) rather than an increased activity in tension receptors. The relationship between tension and electrical activity then seems to support the view that recruitment and frequency of motor unit firing corresponds to a particular intensity of afferent discharge which depends upon the rate of movement.'

These comments pertain specifically to hemiplegic spasticity; however, in subjects with spinal cord lesions, a sudden decrease in torque during stretch occurs only as a result of phasic discharges in the antagonist flexor muscles and/or alterations in the stimulus parameters, as described above (fig. 2). When flexor discharges during stretch of the extensors show continuous electrical activity rather than a single augmented EMG burst, induced torque rises without a clasp-knife effect.

Comparative Responses in Animal and Man

Alterations in both interneuronal and fusimotor activity appear to account for the observed differences in myotatic reflex activity in human subjects with either spinal or hemispheric lesions [MAGLADERY and TEASDALL, 1958; HERMAN, 1964; HERMAN, 1970; BURKE, GILLIES and LANCE, 1970, 1971] and in animals with either spinal or intercollicular lesions [TEASDALL, MAGLADERY and RAMEY, 1958; HOLMQVIST and LUNDBERG, 1959; ECCLES and LUNDBERG, 1959; KUNO and PEARL, 1960; JANSEN, 1965; GRILLNER, 1969; GRANIT, 1970] (table I).

In both decerebrate animals and hemiplegic subjects, flexor stretch reflex activity is absent. This may be ascribed to the lack of spontaneous dynamic and static fusimotor innervation of muscle spindles in flexor muscles. Extensor stretch reflexes, on the other hand, are pronounced. In the decerebrate cat, the response to both the dynamic and static phase of stretch is enhanced

Table I. Some comparative aspects of stretch reflex activity in cat and man

	Cat		Man	
	decerebrate	spinal	hemiplegic	spinal
Dynamic (Ext.)	+	+	+	+
Sensitivity (Fl.)	−	+	−	+[1]
Static (Ext.)	+	−	−	−
Sensitivity (Fl.)	−	−	−	−
TJ	*correlates* with dynamic but not static behavior	?	*correlates* with dynamic behavior	*no correlation—* due to accumulative effects and flexor discharge patterns
Vibration (Ext.)	peak EMG and tension within 0.5–1 sec	?	peak EMG and tension within 0.1–0.5 sec; (no antagonist firing)	peak EMG and tension within 3–4 sec; (antagonist firing)
Clasp knife (Ext.)	+ autogenetic inhibition	−	− (pseudo clasp-knife)	− (pseudo clasp-knife)

1 Present in approximately 40% of subjects.
+ = present.
− = absent.

[MATTHEWS, 1969]. This has been attributed to release of both dynamic and static fusimotoneurones. However, in the hemiplegic subject, an augmented response is observed only during the dynamic phase of stretch implying that only dynamic fusimotoneurones are released [SHIMAZU, HONGO, KUBOTA and NARABAYASHI, 1962; HERMAN and SCHAUMBERG, 1967]. In spinal lesions of both cat and man, dynamic sensitivity to stretch is detected in both extensors and flexors, appearing less consistently in flexor muscles [ALNAES, JANSEN and RUDJORD, 1965; BURKE, GILLIES and LANCE, 1970, 1971; see above]. As in hemiplegic subjects, there is no significant sensitivity to static stretch (i.e. no tonic stretch reflex appears).

Interneurones through which afferents are transmitted are known to be influenced by descending pathways from brain-stem centers. Alterations in these influences will modify interneuronal responses. In spinal preparations for example, there is release of tonic inhibition of flexor reflex interneurones mediating effects from group II and III muscle afferents, cutaneous and joint afferents (I_A afferents seemingly are independent of descending controls at the segmental level). Such changes following spinal lesions lead to enhanced excitability of polysynaptic reflexes. In man, polysynaptic activity is noted by observing both accumulative autogenetic responses and reciprocal excitation of flexors during slow stretch of extensors, and prolonged latency, low threshold, electrically induced H-reflexes following liminal stimulation of a peripheral nerve (e. g. tibial nerve) [MAGLADERY and TEASDALL, 1958]. Monosynaptic responses can also be revealed; however, they require higher intensity (e. g. rapid stretch or supraliminal stimulation) afferent support [HERMAN, 1964; FUJIMORI, KATO, MATSUSHIMA, MORI and SHIMAMURA, 1966].

In the decerebrate preparations, interneurones transmitting impulses from group I_B, II and III, afferents are tonically inhibited. However, strict adherence to this principle has recently been challenged by MATTHEWS [1969] and GRANIT [1970]. They suggest a facultative role for group II afferents (e. g. extensor group II afferents might facilitate extensor reflexes) due to 'switching' on and off of interneurones mediating autogenetic excitation. In subjects with hemiplegic lesions, afferent impulses (most likely I_A afferents) may be transmitted through both monosynaptic and polysynaptic pathways [HERMAN, 1964]. However, in contrast to observations of the spinal lesions, in hemiplegic subjects there is a relatively high threshold of interneuronal activity, often reaching a level corresponding to the threshold of the monosynaptic response.

Conclusion

The above findings in man support the contention that dynamic fusimotoneurones innervating the primary endings of the muscle spindle of both flexors and extensors in paraplegic subjects and of extensors in hemiplegic subjects are released. (It is difficult to evaluate the presence or absence of static fusimotor activity. The tonic stretch reflex, which most likely requires both support of static fusimotoneurones and suppression of autogenetic inhibitory reflex arcs [GRILLNER and UDO, 1970] is absent in both hemiplegic and paraplegic subjects. Furthermore, expected specific patterns of reciprocal innervation in spinal lesions, i. e. excitation of flexor motoneurones with concomitant inhibition of extensor motoneurones by group II afferents, is not explicit in our data.) While enhanced discharges from the primary endings of the muscle spindle are responsible for the dynamic behavior of muscle, the net output torque of myotatic systems governing any particular joint during passive movement is determined by a complex interaction between the degree of release of dynamic fusimotoneurones, the level of excitability of the motor pools and the manner in which I_A afferents are transmitted through the spinal cord, i.e. monosynaptically or polysynaptically.

These factors have been identified using controlled laboratory parameters such as applied load and velocity of passive muscle extension; however, in the clinical situation, interactions between the examiner's motor control system and that of his spastic patient obviate precise control of these parameters. Thus, clinical evaluation is further complicated by the art of the clinical examiner.

Acknowledgement

The authors are grateful to Dr. NATHANIEL MAYER and Dr. YASUHIKO TAMAI for their technical assistance and valuable comments, to Miss PATRICIA SAYLER for her editorial assistance, and to Miss EILEEN MCDONNELL for preparation of the manuscript.

Author's address: Dr. R. HERMAN, Department of Rehabilitation Medicine, Temple University, *Philadelphia, PA 19141* (USA)

Unloading Reflexes and the Silent Period

New Developments in Electromyography and Clinical Neurophysiology,
edited by J. E. Desmedt, vol. 3, pp. 589–602 (Karger, Basel 1973)

Studies of the Normal Human Silent Period

B. T. SHAHANI and R. R. YOUNG

Department of Neurology, Harvard Medical School and Massachusetts General Hospital, Boston, Mass.

The term 'silent period' refers to a transitory, relative or absolute decrease of EMG activity, evoked in the midst of an otherwise sustained contraction. In man, the tonic background EMG activity is usually produced by voluntary assumption of tension within a muscle. In animal studies, it has been produced by the evocation of tonic reflex contraction, for example, as the crossed extensor accompaniment of a prolonged flexor reflex. Though periods of electrical silence in EMG activity are often seen in association with intermittency of contraction, fatigue, tremor, clonus and so forth, the term silent period (SP) should be reserved for reflex pauses following a stimulus of some sort: an electrical shock to skin and peripheral nerve, a tendon tap or other means of eliciting phasic reflex activity or sudden decreases in the load against which the muscle is contracting.

The SP was discovered in man when HOFFMANN [1919] superimposed an electrically induced twitch upon voluntary EMG activity. Its reflex nature was quickly appreciated and over the years frequent attempts have been made to interpret the SP in terms of contemporary physiological mechanisms [DENNY-BROWN, 1928; GRANIT, 1955; MERTON, 1951, ANGEL, this volume] with the explanations becoming more complex as knowledge of muscle spindle, Golgi tendon organ and Renshaw cell behavior has increased. Simultaneously its behavior under various circumstances was used as a tool to dissect even more complex motor phenomena such as posture [DENNY-BROWN, 1928], normal control of movement [GRANIT, 1955; MERTON, 1951], and certain pathological entities such as spasticity, rigidity and tremor. ANGEL and STRUPPLER et al. [this volume] review the pathophysiology of these latter conditions and the mechanisms responsible for the SP produced by rapid unloading of a muscle during isometric contraction.

Most studies have dealt with the SP produced by electrical stimulation of peripheral nerves and the skin overlying them. Insufficient attention was paid to the possible role played by activity that such stimulation would induce in fibers other than muscle afferents and efferents. Because purely cutaneous stimulation has recently been shown to produce a SP and because the SP has always occupied such a central role in discussions of motor physiology, we plan to review these new findings as well as other observations of the normal behavior of the SP in an attempt to determine more precisely the role of various central and peripheral factors influencing motoneurone excitability in man.

Methods

Ten normal adult subjects were studied. EMG activity was recorded from adductor pollicis, abductor pollicis brevis, abductor digiti quinti, biceps brachii, tibialis anterior, soleus, masseter, and orbicularis oculi muscles, using concentric needle electrodes or clip electrodes [COPLAND and DAVIES, 1964] and was photographed on moving paper or with a Polaroid camera. Silent periods were evoked by electrical stimulation delivered through DISA 13 K 62 surface electrodes to the ulnar, median, superficial radial, medial popliteal, posterior tibial, sural or facial nerves, or the skin overlying the motor point of masseter or, through ring electrodes, to the digital nerves of the fingers. Stimuli were square waves 0.1–1.0 msec in duration and of nominal amplitude 10–50 V led from a DISA Multistim or from a Grass S 88 stimulator through a special SIU 5 isolation unit capable of 300 V output. SP were recorded during an evenly sustained isometric background contraction except in the case of orbicularis oculi where the tension produced is difficult to measure. Background (0.5–1.0 kg) and twitch tension were recorded primarily from adductor pollicis with the subject seated, the forearm supported and the hand fixed in a specially designed splint. The proximal phalanx of the thumb was connected to a steel bar on which two ceramic strain gauges (Ether 3A-1A 350P) were cemented. These made up two arms of a Wheatstone bridge circuit connected to a DC amplifier. Isotonic twitch contraction was recorded by interposing a spring between the thumb and strain gauge. All experiments were performed in a warm room (24–26 °C).

Results

As found by MERTON [1951], even a very small twitch is accompanied by a recognizable SP (fig. 1A and C), whereas a more characteristic, longer SP follows a supramaximal shock to the ulnar nerve (fig. 1B and D). With a maximal twitch, the duration of SP ranged in different subjects from 100 to 120 msec as measured from the stimulus artifact; it was fairly constant in any

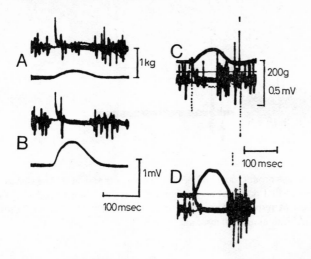

Fig. 1. The effect of alterations in stimulus strength upon the SP in adductor pollicis. *A, B* Isometric twitches. *A* Submaximal electric stimulus to the ulnar nerve. *B* Supramaximal stimulus. *C, D* Isotonic twitches (see methods). *C* Submaximal stimulus. *D* Supramaximal stimulus.

individual. Approximately 30 msec after the stimulus, the electrical silence is broken by a synchronous potential, the F response of MAGLADERY and MCDOUGAL [1950]. When the strength of stimulus was reduced (fig. 1 A and C), voluntary potentials appeared in the midst of the SP at 50–80 msec after the shock and occasionally between 30 and 50 msec. All these observations were the same under isometric and isotonic recording conditions.

However, there is a difference between these two conditions in the duration of the SP with submaximal twitches. With isometric recording the duration is the same as with the maximal twitches described above, whereas with isotonic twitch contractions the eventual resumption of voluntary activity takes place 15–25 msec earlier (fig. 1 C).

A SP is also produced by reflex contractions (fig. 2A). In figure 2B a single electrical stimulus, subthreshold for motor fibers, was delivered at the arrow to the medial popliteal nerve and the H-reflex response can be seen with a latency of 30 msec. When this is repeated during a background contraction in soleus, the electrical silence (fig. 2A) begins with the arrival of the H-reflex in the muscle rather than with the arrival of a direct muscle (M) response as noted previously with stimuli above threshold for motor fibres.

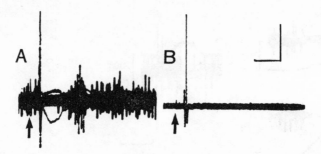

Fig. 2. The SP in soleus produced by an H-reflex discharge is seen in *A*. The electrical stimulus (arrow) to the medial popliteal nerve only activates I_A afferents producing the H-reflex without a direct M response. The same stimulus without the background EMG activity is seen in *B*. Calibrations are 100 msec and 500 μV.

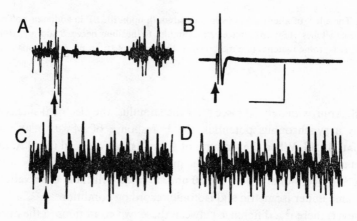

Fig. 3. Each of these represents EMG activity of 5 superimposed sweeps. *A* A SP seen in the background voluntary contraction in abductor pollicis brevis following stimulation of the median nerve at the wrist. *B* The response in orbicularis oculi following a maximal stimulus to the facial nerve. Voluntary contraction of the same muscle is seen in *D* while *C* represents superimposition of *B* and *D*. Note the very brief SP in *C* compared with *A*. Stimuli are delivered at the arrows. Calibrations are 50 msec for each, 2 mV for *B*, *C* and *D*.

SP studies in man have usually been restricted to muscles in the hand or, less often, the arm. Because muscles vary in anatomical configuration, size, speed of contraction and concentration of muscle spindles, clues to the nature of the SP may be obtained by studying it in a wide array of muscles. For example, when stimulating the facial nerve (fig. 3B) and recording the volun-

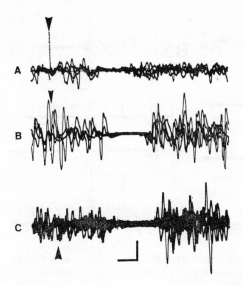

Fig.4. SP produced by cutaneous stimulation. *A* EMG of biceps brachii with stimulation of the superficial radial nerve at the wrist. *B* The same stimulus produces an SP in abductor pollicis brevis, and in *C* an SP in the latter muscle is produced by stimulation of the digital nerves of the middle finger. Stimuli are delivered at arrows. Calibrations are 20 msec and 500 μV.

tarily contracting orbicularis oculi (fig. 3D), only a very brief and relative electrical silence, lasting perhaps 30 msec, is seen (fig. 3C). This is quite different from the SP in abductor digiti minimi (fig. 3A) or masseter [see SHAHANI and YOUNG, this volume, and STRUPPLER, STRUPPLER and ADAMS, 1963] under the same circumstances.

Supramaximal stimuli delivered to a mixed nerve, as in MERTON's studies [1951] or those reported here, have potential effects other than those usually considered. In addition to producing impulses in motor fibers and in at least the larger diameter muscle afferents (I_A and I_B), such stimuli also activate larger cutaneous fibers running in the mixed nerve. Fortunately, one can stimulate such fibers in isolation in sensory nerves such as the superficial radial at the wrist or digital nerves in the fingers, and such stimuli also produce a SP in the background voluntary contraction (fig. 4, 5). This cutaneously evoked EMG silence is present from roughly 50–100 msec, after the stimulus in the upper extremity and from 80–140 msec, after the stimulus in the leg. There is also slight variation from muscle to muscle and nerve to nerve. This SP is produced in muscles proximal (fig. 4A, C) as well as distal (fig. 4B) to the

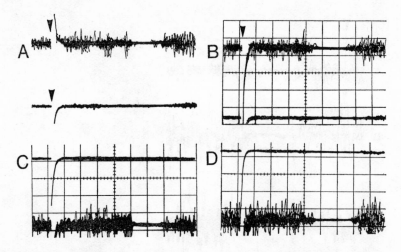

Fig. 5. A cutaneously evoked SP is seen with the same timing in each of a pair of antagonistic muscles. Simultaneous superimposed EMG recordings from soleus on the upper line of each pair and from tibialis anterior on the bottom line. In *A* and *B* voluntary activity is present in soleus, and in *C* and *D* the activity is in tibialis anterior. In *A* and *C* the stimulation is applied to the posterior tibial nerve at the medial malleolus, whereas in *B* and *D* the sural nerve is stimulated at the lateral malleolus. The stimulus artifacts, as at arrows in A, are clearly visible. Calibrations are 20 msec and 500 µV for one large division on the graticule.

site of stimulation. Furthermore, it occurs at approximately the same time in either of an antagonistic pair of muscles (fig. 5, compare A with C, and B with D) though the latency and duration varies in the same muscle depending on which nerve is stimulated (compare A and C with B and D). For example, the SP begins sooner in gastrocnemius or tibialis anterior following a stimulus at the ankle to sural nerve as compared with one to the tibial nerve (80 *versus* 100 msec). The SP in these muscles also has a longer duration with sural stimulation because the return of voluntary activity takes place at about the same time following the stimulus to either nerve.

Discussion

When a SP is produced by the method of MERTON [1951] used in this study, the stimulus to the mixed nerve sets up two volleys of impulses, one of which ascends to the spinal cord and the other descends to the muscle. The ascending volley in motor fibers either results (a) in a collision of ortho- and antidromic impulses somewhere in the motor axon between the cell body and the site of stimulation if that motoneurone was active within plus or minus the conduction time from cell body to stimulus site (of the order of 12 msec in the case of stimulation at wrist), or (b) antidromic firing of the motoneurone if it had not been active within plus or minus 12 msec of the stimulus. In addition, the antidromic motor impulses which either collide central to the origin of the recurrent axon collateral or invade the motoneurone would invade the collateral to activate Renshaw-like cells, if such exist in man. Allowing for conduction centrally in motor fibers, for conduction peripherally again to the muscle and assuming for the moment that the anterior horn cell will continue to fire at rates of roughly 25/sec, it can be shown that electrical silence will persist in the EMG from the arrival of the direct M response, some 3–4 msec after a stimulus at the wrist, for the next 45–65 msec even if one ignores possible Renshaw effects. The 'antidromic inhibition' described by RENSHAW [1941] in the cat could not have an effect on the human EMG peripherally for at least 30 msec after the stimulus; this time allows for conduction centrally, the onset of segmental inhibitory effects and the conduction peripherally of impulses which arose just before the recurrent collateral inhibition began. Therefore, the silence seen between the M and F responses (the latter occurring some 30 msec after the stimulus) is explicable purely on the basis of effects on conduction in motor axons and cannot be due to any changes in CNS activity. If antidromic firing is avoided by using a stimulus below threshold for motor fibers to produce a synchronous H-reflex, no EMG silence is seen during the first 30 msec after the stimulus (fig. 2).

The effect of the ascending volley produced in sensory fibers by the stimulation is more difficult to comprehend. The usual percutaneous stimulus would be expected to activate I_A and I_B fibers from primary endings on the spindle and Golgi tendon organs respectively, as well as large fibers from joint receptors, cutaneous and subcutaneous structures, and so forth. Again, allowing for conduction times, this volley of impulses could not produce an EMG effect in less than 30 msec after the stimulus, which is the latency of the H-reflex, taken to be the result of a volley in I_A fibers. This synchronous firing of much of the motoneurone pool would also be expected to produce, via the recurrent collateral pathway, considerable local Renshaw inhibition that would tend to produce EMG silence for some time after the arrival of the H-reflex in the muscle. Recent studies in the cat [RYALL, PIERCEY, POLOSA and GOLDFARB, 1972] have shown greater Renshaw cell activation following antidromic as opposed to orthodromic monosynaptic activation of the motoneurone pool. The effect of a volley in I_B, joint or cutaneous fibers is less clear. In the latter instance, at least, the SP produced (fig. 4 and 5) appears about 50 msec after the stimulus at the wrist and persists for another 50 msec or so. In summary, though the ascending volley in motor and sensory fibers may produce EMG silence by Renshaw or other mechanism from 30–75 msec (see below) after the stimulus, the first 30 msec of the SP is produced by peripheral antidromic effects and tends, therefore, to be less noticeable or absent when the stimulation of motor fibers is minimized or the conduction distance in them is shorter.

The next segment of the SP (from 30 to 50 or 60 msec after the stimulus) occurs too soon to be explicable on the basis of cutaneous effects (50–100 msec) or of effects produced by the volley descending from the site of stimulation to the muscle. In the mechanical trace (fig. 1) the twitch begins 12–15 msec after the stimulation or 9–12 msec after the arrival of the M response. A certain degree of extrafusal muscle contraction must occur before the spindles are unloaded or the Golgi tendon organs begin to discharge. Though I_A and I_B fibers are fast conducting, at least 30 msec must elapse in the case of human hand muscles between any peripheral action and the appearance in the EMG of increased or decreased activity related thereto. The effects of the descending volley on the SP cannot be expected before 50–60 msec elapse after the stimulus and the second segment of the SP must, therefore, be due to factors other than the modification of proprioceptive or cutaneous input produced by the stimulus. The persistence in humans of only the first 50 msec of the SP in de-afferented muscles [Shahani and Young, unpublished] supports this hypothesis and suggests that the initial volley in I_B, joint and other fibers plays relatively little role in at least the first half of the SP. The factors responsible for the silence between 30 and 50 or 60 msec must include antidromic collision and motoneurone firing which could produce relative silence, as noted above, up to 45 or 65 msec. One must also not forget possible effects of an antidromic volley in I_A fibers into the spindles themselves [Matthews, 1964]. In addition, a Renshaw effect, if present in humans, would be expected no earlier than 30 msec and would persist for at least the next 30 msec or so. Interestingly, if weaker stimuli are used, few motor axons are stimulated, less Renshaw effect would be expected, and under these conditions, a brief return of voluntary action potentials can be seen (fig. 1) at 50 to 60 or more msec. The second segment of the SP is probably due to a combination of some of the same peripheral antidromic effects described in discussions of the first segment as well as hypothetical Renshaw effects which may also spill over into the next segment.

The third segment of the SP also has a multifactorial basis. In upper extremity muscles, at least, the role of the cutaneous afferents merit serious consideration because, in isolation (fig. 4 and 5), they produce a SP between

50 and 100 msec after the stimulus. That interval comprises the latter half of the usual SP (the third segment in our terminology), and since we know that a single volley in radial nerve can produce it, we are forced to assume that a volley in all the large cutaneous fibers in the ulnar or median nerve at the wrist can also. HOFFMANN and associates described an SP in masseter following cutaneous stimulation of the oral mucosa [HOFFMANN and TÖNNIES, 1948] and in gastrocnemius and soleus following an electrical stimulus to the skin of the foot [HOFFMANN, SCHENCK and TÖNNIES, 1948]; SCHENCK and KOEHLER [1949] soon found that the latter stimulus could also produce an SP in both gastrocnemius and tibialis anterior. HAGBARTH [1960], using trains of noxious stimuli, found excitation in one group of leg muscles (usually flexors but at other times extensors, depending on the site of stimulation) and inhibition in the antagonist group. These silent periods, though clearly of cutaneous origin, were considered to be examples of reciprocal inhibition though at times no actual reflex was seen in the muscles antagonistic to those where the SP was seen. KUGELBERG, EKLUND and GRIMBY [1960], using the same techniques as HAGBARTH, reported much the same findings, including the observation that the SP began slightly before the associated reflex contraction. In addition to reciprocal inhibition, KUGELBERG *et al.* used the term 'ambivalent inhibition' to describe the simultaneous SP in antagonistic muscles.

In 1961, LIBERSON reported that cutaneous stimulation could produce an SP in many muscles – he cautioned against ignoring this factor – but subsequent authors in their discussions of the physiology of the SP [HERMAN, 1969; HUFSCHMIDT, 1966; MATTHEWS, 1964] were apparently not aware of the importance of cutaneous factors. HUGON [1967] and DELWAIDE, SCHWAB and YOUNG [1971] also described a cutaneously produced SP. Once again, physiological studies in man have demonstrated a new entity, the cutaneous SP, the basic neurophysiology of which remains to be worked out. The timing of the cutaneous SP (minimal latency in bulbar muscles with progressively longer latencies when recording from proximal arm, hand and leg) supports the contention that so-called 'long loop reflexes' [ECCLES, 1966; TABORIKOVA and SAX, 1969] may be involved, and other preliminary evidence suggests that the ipsilateral cerebellar hemisphere is necessary for its presence. Because it comes between 80 and 140 msec after stimulus in triceps surae, whereas the usual SP is up to 100 or 120 msec long, less contamination of the latter with cutaneous factors results, though an overlap is still present. Our impressions are that stimuli supramaximal for motor fibers are required to produce a clear-cut cutaneous SP and that lesser stimuli may tend to produce less cutaneous effects as compared to proprioceptive. Studies with quantitative evaluations of stimulus current are needed. Nevertheless, supramaximal stimuli, such as MERTON used, clearly produce cutaneous effects which cannot be ignored.

MERTON [1951, 1953] suggested that the latter half of the human SP was related to a response of a 'length servo' to an external disturbance causing shortening of the muscle which unloaded spindles, reduced their discharge ('spindle pause'), and thereby withdrew reflex excitation from motoneurones. The subsequently postulated closed loop feedback circuit provided a mechanism whereby the muscle could be set reflexly to any desired length, and it was postulated that voluntary muscle activity is produced by such a servo with the fusimotor fibers being the normal pathway for initiating voluntary contraction. This was an extremely fruitful hypothesis, giving rise as it did to the servo loop theory of motor control with many subsequent physiological experiments. However, since 1951 many studies, reviewed most recently by MATTHEWS [1964], BURKE [1971], and VALLBO and HAGBARTH [this volume] required the substitution of the concept of α-γ coactivation for MERTON's original theory of γ-firing leading α-firing around the servo loop during voluntary and other movements.

As noted above, the most constant period of electrical silence beginning about 50 msec after the stimulus lasts 30 or 40 msec and can be seen as long as there is any extrafusal contraction, whether isometric or isotonic (fig. 1). Since this period of silence is present even with minimal isotonic contractions, it may be related to modifications of the proprioceptive input dependent more upon changes in length of muscle than in tension. The 'pause' in I_A afferent discharge with unloading of the muscle spindles during the rising phase of the extrafusal twitch contraction which has been demonstrated in cats may, perhaps, be present in man as well. If so, this pause would withdraw excitatory input to the motoneurones during the time necessary to produce this part of the SP. The physiology of human muscle spindles is totally unknown, apart from recent data on afferent discharges in mixed nerves [HAGBARTH and VALLBO, this volume], and though much has been ascribed to their function and dysfunction in human health and disease, it is entirely speculative. Anatomical information about human muscle spindles is extant, however; no one has found typical spindles in facial muscles [SHAHANI and YOUNG, this volume]. Therefore, the absence of the third segment of the SP in orbicularis oculi at least supports our contention that it is related to changes produced in spindle discharge by the superimposed twitch.

The first segment of the SP is present in facial muscles but, as in masseter, it is slightly briefer than in more distal muscles (25 vs. 30 msec). The second segment of the SP is less obvious or absent in facial and masseter muscles.

Because of the shorter conduction distance, the antidromic effects discussed above would be shorter-lived and would not spill over into the second segment. There is also some question in man [RAMON y CAJAL, 1909] and animals [PORTER, 1965; SUMI, 1969] as to whether recurrent collaterals and Renshaw cells are present in masseter and other brain-stem motor nuclei. If in man they are not, as suggested by PENDERS and DELWAIDE [this volume], these findings in facial and masseter muscles would again support the hypothesis that Renshaw inhibition plays some role in the second segment of the SP. Muscle spindles have been demonstrated in masseter [COOPER, 1966] and the third segment of the SP is present there but not in facial muscles. There are, of course, other differences between facial and hand muscles, for example, the thin sheet-like nature of the former, the difficulty in measuring tension in the face and the shorter conduction distances and segregation of facial afferents in the trigeminal nerve from efferents in the facial nerve. The presence of a third segment of the SP in masseter and with weak isotonic contraction distally, and the failure to produce an SP in the face even with stimulation of both 5th and 7th cranial nerves [PENDERS, 1969] suggest, however, that the most important difference lies in the absence of spindles in the face. The absence of the third segment of the SP in the face, therefore, might support our hypothesis that the spindle pause plays a significant role in its origin.

To return to the peripheral limb musculature, an isometric twitch contraction of any amplitude or a maximal isotonic twitch produces an additional 15–25 msec of electrical silence compared with the SP produced by the submaximal isotonic twitch described above. This last portion of the SP seems, therefore, to be predicated upon the development of a certain tension within the muscle, suggesting that tension receptors such as Golgi tendon organs might be involved. MERTON [1951] felt it unlikely that tendon organs could account for any part of the SP seen with small isometric contractions because it was then thought that these receptors had relatively high thresholds for activation by tension. HOUK and HENNEMAN [1967] have shown in the cat, however, that tendon organs are quite sensitive to tension developed by active contraction of extrafusal fibers; for example, the contraction of a single fiber that produces 0.1 g tension may cause inhibitory afferent activity in I_B fibers at the rate of 25/sec. Human tendon organs may discharge, therefore, in proportion to tension developed by a muscle and, as far as timing goes, could provide inhibitory input to account for much of the last half of the SP including the latter 20 msec or so. HUFSCHMIDT [1966], for example, thinks that I_B input accounts for most of the SP. It was also shown that in cats, isotonic contractions which merely shortened a muscle without producing much ten-

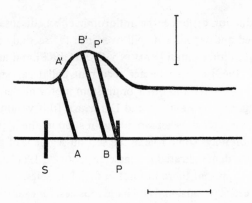

Fig.6. Schematic representation of factors presumably involved in SP (S–P where S is the stimulus). The first 50–60 msec (S–A) is related to various antidromic factors including Renshaw recurrent inhibition. At A' the muscle spindles begin to be unloaded and the Golgi tendon organs begin to be stimulated. At B' the spindle pause ceases, and tendon organ firing may carry on until P'. Voluntary activity resumes at P. Calibrations are 100 msec and 1 kg.

sion resulted in a feeble tendon organ response [HOUK and HENNEMAN, 1967] which might then fail (as in fig. 1 C) to produce the last 15–25 msec of the SP. HERMAN [1969] reported no difference in the length of the SP in triceps surae under isometric or isotonic conditions. He also found the SP duration to persist for 175 to more than 200 msec, and we cannot account for this discrepancy.

To account for the last half of the SP, at least the following factors must be considered: poorly understood cutaneous input, particularly with electrical stimuli supramaximal for motor fibers, muscle spindle unloading and Golgi tendon organ activation. The first, though of little interest to neurophysiologists, is clearly demonstrable in man, whereas the latter two have been extensively investigated in animal laboratories but remain *terra incognita* clinically. The time relationships between stimuli and EMG response may be difficult to visualize because of the delay around the loop from changes in firing of muscle receptors, spinal modification of resultant input, increase or decrease of motoneurone activity to eventual EMG changes. Figure 6 schematically portrays hypothetical roles played by the spindle pause (A'–B') and Golgi tendon organ activation (A'–P') in relation to the resulting EMG responses (A–B and B–P, respectively).

Clearly, multiple factors are involved in the SP, and it would be useful to simplify them if possible. By the use of phasic reflex contractions (tendon jerk or H-reflex) to produce the muscle twitch, antidromic factors may be eliminated along with some though not all of the cutaneous factors because the stimuli responsible for an H-reflex and, to a lesser extent, those producing a tendon jerk also produce phasic cutaneous input. Using these reflexes, however, one may be left with Renshaw and Golgi tendon organ inhibitory activity, a pause in excitatory afferents from spindle, as well as other mechanisms yet to be described. Submaximal isotonic contractions could minimize tendon organ activity but would introduce afferent input from joint receptors produced by the movement. The use of the unloading reflex to produce a SP [ANGEL, this volume] eliminates the putative burst of Renshaw and Golgi tendon organ activity, as well as other antidromic factors, but again adds quick joint movement to complicate the issue. The rebuttal that joint movement otherwise, during voluntary elbow flexion or passively during certain biceps contractions, produces no SP does not convince one that input from joint movement under different 'command' circumstances (i.e., the unloading reflex) may not affect the SP. Since MERTON's original hypothesis, the simplified servo theory of muscle contraction has had to be modified. Nevertheless, its basic truth remains irrefutable – the subject attempts, using complicated and not altogether conscious means, to do as instructed – to maintain a certain prescribed muscular tension. If this tension is exceeded by the superimposition of a twitch, a SP is required to allow the tension to fall to the preordained level. When the subject is instructed to produce a maximal contraction, there is no need for relaxation and only the early antidromic portion of the SP is seen. What the subject has been instructed to do determines the result one obtains, and differences in instructions account, at least in part, for the reported variability in the results of SP studies.

The mechanisms responsible for this reflex regulation, including the 'command program' given to the subject by the experimenter and delivered to the spinal cord from higher centers, and their ineffectiveness during maximal effort or quick phasic movements (which can break into the SP at any time), remain to be delineated. In the future, when this has been done, abnormalities in the SP associated with disease states may be used to infer something about the pathophysiology of the disease itself (rigidity, spasticity, tremor, etc.). Meanwhile, we can only make correlations between clinical states and findings observed, for example, in the SP or other reflex tests, in well controlled studies.

We remain in the natural history stage of Clinical Neurophysiology. Though it is difficult to resist the urge to jump to conclusions about mechanisms of human disease on the basis of comparison between a few human studies under poorly defined conditions and reports from animal experiments carried out under artificially simplified conditions, we must resist. We should concern ourselves with gathering data of a type which will remain useful as our understanding evolves.

Summary

Studies of the normal human SP produced in many different muscles by electrical stimuli have been carried out under carefully controlled conditions to analyze factors influencing motoneurone excitability in man. The first 50 msec of electrical silence appear to be produced by antidromic effects in motor axons, including possible Renshaw inhibition during the latter 20 msec or so. The last 50–70 msec of the SP is associated with changes in CNS activity produced by proprioceptive and cutaneous input from the twitch or its stimulus. Purely cutaneous input is shown to produce a SP from 50 to 100 or more msec after a single large stimulus, and the occurrence of a SP cannot, therefore, be taken to establish the existence of γ or other proprioceptive routes of motoneurone excitation. A pause in muscle spindle afferent input may account for much of the second half of the SP which is shown to be absent in facial muscles where spindles have not been seen. The last 20 msec of the SP, because it depends upon the degree of tension in the muscles, may relate to Golgi tendon organ activity. The importance of the hitherto ignored cutaneous mechanisms of the SP is stressed to caution against premature and egregious extrapolation from alterations in the SP in various pathological situations to explanations of even more complex phenomena such as spasticity, rigidity and tremor. There is obviously a variety of excitatory and inhibitory influences on spinal motoneurones, and it is important to include all these factors in any comprehensive theory of muscle control in man.

Acknowledgement

This work was supported by a grant from Mrs. W. B. LLOYD. We wish to thank Mr. FREDERICK E. BEW, Medical Research Council, for his help in building the myograph.

Authors' address: Dr. B. T. SHAHANI and Dr. R. R. YOUNG, Department of Neurology, Massachusetts General Hospital, Fruit street, *Boston, MA 02114* (USA)

New Developments in Electromyography and Clinical Neurophysiology,
edited by J. E. Desmedt, vol. 3, pp. 603–617 (Karger, Basel 1973)

The Unloading Reflex under Normal and Pathological Conditions in Man

A. STRUPPLER, D. BURG and F. ERBEL

Department of Neurology, Technical University of Munich, Munich

We have examined both the phasic stretch reflex and the unloading reflex under various experimental conditions. The two reflex responses are superimposed upon an isometric contraction to see what factors are responsible for each and how they are changed in various motor disorders. By comparing both types of responses we hoped to eliminate some of the confusion and contradiction in the literature on human reflexology.

The electrical activity produced by a muscle in sustained isometric contraction can be temporarily decreased, or even completely interrupted, by application of a quick, passive stretch which causes the muscle to twitch (silent period) [HOFFMANN, 1920]. A silent period is also elicited by unloading, thus suddenly allowing the limb to move so that the muscle reaches a new, shorter length. In the first case, the EMG silence follows a reflexly evoked burst of spinal motoneuron activity, after which the muscle returns to its original length. In the second case, the silent period is produced by simply allowing the muscle to shorten.

After the phasic stretch reflex several factors operate together to interrupt motoneuron firing:

1. The unloading of spindles during muscle shortening produced by the twitch, thus withdrawing *peripheral facilitatory* input.

2. Autogenetic inhibition by Golgi afferents, as muscle tension briefly increases against the external load or resistance.

3. Recurrent inhibition produced by the synchronized Renshaw cell activity that automatically follows the burst of motoneuron firing.

4. The refractory phase of motoneurons which is more effective in damping output directly after a population of neurons has just been discharged synchronously.

It is difficult to interpret changes in this type of silent period following a stretch-reflex, because the factors outlined above may vary independently, according to experimental conditions. For example, the degree of Renshaw inhibition and the refractory period of motoneurons are proportional to the number of motoneurons firing synchronously, while the effect of Golgi receptor activation depends upon the tension developed by the twitch. The cessation of spindle afferent firing is dependent partly upon the acceleration of the muscle shortening and on the level of fusimotor activity prior to the applied stretch. A further complication is introduced by an electrically induced twitch contraction, in which case the motoneurons are also invaded antidromically by the stimulus to the motor nerve. Recurrent inhibition is, of course, also increased in this instance as the antidromic spike reaches the recurrent collaterals [RENSHAW, 1941]. It is, thus, not surprising that the opinions concerning the cause of changes in the reflexly evoked silent period differ. Some authors stress the cessation of spindle afferent activity [MATTHEWS, 1933; MERTON, 1951; PAILLARD, 1955]; others favour autogenetic inhibition [HUFSCHMIDT, 1959], while ECCLES [1955] pointed out the importance of recurrent inhibition. Many of the problems inherent in the study of motoneuron activity by means of the post-twitch silent period are overcome by using the 'unloading' of an isometrically contracting muscle. Here, the only factor of importance (provided the release is quick enough) is the change in afferent drive to the motoneurons. SOMMER [1939] and HUFSCHMIDT [1959, 1966] considered that the EMG silence after unloading represented active inhibitory input from the tendon receptors. Others favour the interpretation that the pause is a consequence of the withdrawal of the Ia afferent drive which is believed to sum with the descending drive to maintain the contraction [STRUPPLER and SCHENCK, 1958; STRUPPLER, LANDAU and MEHLS, 1964, 1969; ANGEL, EPPLER and JANNONE, 1964; HOFMANN and ANGEL, 1967]. A study of unloading would thus give some information about the algebraic effects of the other inputs currently playing upon the motor cell, because this technique removes a major, segmental source of drive without stimulating any of the others.

Analysis of single motor units shows that the basic factor underlying the unloading reflex is a prolongation of the inter-spike interval in motoneuron discharges (fig. 1). After this phenomenon the motoneurons discharge synchronously (rebound). When the unloading is performed slowly enough the rebound may occur during the shortening, even while the acceleration continues. ALSTON, ANGEL, FINK and HOFMANN [1967] think

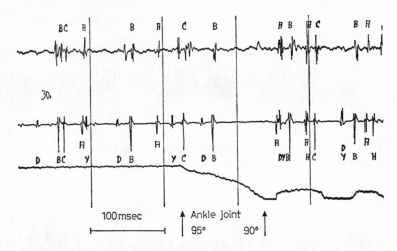

Fig. 1. Unloading reflex. Sudden removal of a load applied to the contracting tibialis anterior. Single motor units can be identified as indicated by letters. Note the prolongation of the inter-spike interval when the muscle is allowed to shorten. The upper and middle traces give the simultaneous recording of single motor units in the tibialis anterior muscle. The lower trace shows the angular displacement of the joint.

that the rebound after the pause represents a form of correction made through the γ efferents and via IA afferents to 'take up the slack' and prepare the muscle to maintain its new, shorter length. Spindles are not likely to be able to fire in such a way without a marked increase in their own fusimotor drive. An alternative proposal is that the rebound after the silent period in the unloading reflex could result from the hypersynchronous activation, from above, of a pool of motoneurons, all of which reach firing threshold simultaneously after their previous silence [ERBEL and STRUPPLER, 1970]. On this view, the suprasegmental drive primarily responsible for the resting isometric contraction simply takes over again, and resumption of IA drive would not be necessary to terminate the silent period.

Factors Regulating the Unloading Reflex

A. Parameters of Unloading

If the limb does move at a certain velocity in reaching a new position, the EMG silence may not be seen (fig. 2). For very small displacement, the

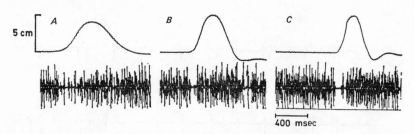

Fig. 2. Unloading reflex in the brachial muscle. By increasing the acceleration of muscle shortening, from A to C, the silent period is prolonged [from STRUPPLER, LANDAU and MEHLS, 1969].

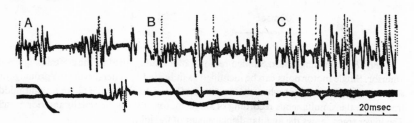

Fig. 3. Dependency of the silent period from the degree of muscle shortening. *Upper trace*, EMG of the brachial muscle; *middle trace*, EMG of the antagonistic triceps muscle; *lower trace* angular displacement. The shortening of the muscle is 15 mm in *A*, 2–3 mm in *B*, and 1 mm in *C* [from STRUPPLER, LANDAU and MEHLS, 1969].

acceleration must be rather rapid, and this requires a forceful contraction to begin with. Even with rather high force during the isometric contraction there is also a threshold for displacement, as very short movements will not produce a silent period, even if they are rapid. Such is not the case with the twitch contraction evoked reflexly or by nerve stimulation, as profound EMG silence follows such twitches, even when the limb is rigidly fixed so as to give an isometric contraction [DIETRICHSON, this volume]. Figure 3 illustrates these relationships in the unloading reflex. An important argument against Golgi afferent activity causing EMG silence after unloading is our finding that it makes no difference whether the test muscle shortens because a restraint is released or because an external force is applied in the direction of muscle force [STRUPPLER, LANDAU and MEHLS, 1964]. In contradiction to HUFSCHMIDT, we conclude that the tension-sensitive Golgi tendon organs cannot play an essential role in the unloading reflex. The acceleration and displacement parameters are roughly the same for both the flexors and extensors of the upper limbs, while, in the leg, the tibialis anterior muscle gives the clearest and most reproducible responses.

Soleus contractions are less sensitive to unloading. Neither arm muscles nor tibialis anterior show consistent H-reflexes, in contrast to soleus, which suggests somewhat different control mechanisms for neurons to these muscles [HOFMANN and ANGEL, 1967].

B. Mode of Innervation

When a muscle is unloaded during an isotonic contraction a much less pronounced silent period is observed. Isometric conditions are thus required to analyze changes in the duration or intensity of motoneuron inactivation. Another difficulty in interpretation can arise with the fatigued muscle, where the mode of innervation shifts from that of steady, asynchronous discharge to one of periodic synchronized burst firing which makes it difficult to evaluate the unloading silent period.

Effects of Training or 'Presetting'

When the subjects were instructed to react as soon as they sensed the removal of the weight, the duration of the silent period could be changed. For example a 'preset' to move the arm in the same direction as it moved when the weight was released, caused the silent period to shorten [HAGBARTH, 1967]. A possible explanation for this effect is that intention (preset) increases the direct, supraspinal activation of α-motoneurons, and in the unloading reflex this direct, descending input overrides the withdrawal of spindle afferent drive. Alternatively, it is possible that one effect of intention is to activate the intrafusal fibres themselves, so that they will not fall completely slack when the extrafusal fibres shorten. The pause in IA input, therefore, might be of shorter duration. If a counter-unloading move was intended, on the other hand, its execution at the time of the release of the weight was associated with a longer silent period.

Relationship to Activity in the Antagonistic Muscle

Sudden unloading of muscle contracting against a weight produced a movement which simultaneously shortened the agonist and stretched the antagonist. About 50 msec after the beginning of the silent period in the agonist [STRUPPLER and SCHENCK, 1958] a typical stretch reflex occurred in the antagonist, when measured in muscles of the upper extremities.

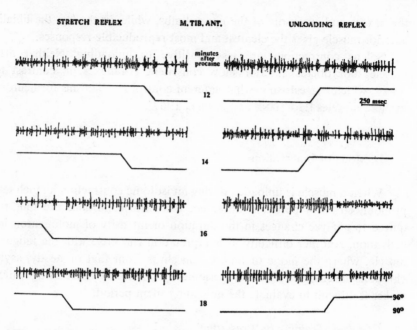

Fig. 4. Stretch reflex *(left)* and unloading reflex *(right)* in the tibialis anterior muscle during partial procaine block of the peroneal nerve. In each line the upper trace illustrates the EMG activity and the lower trace the angle position of the ankle joint. Note the disappearance of the reflexes 18 min after the application of procaine.

ERBEL [1969] and HERMAN and MAYER [1969] independently tested the silent period in soleus after blocking the peroneal nerve which supplies an antagonistic muscle. They observed that, when the block had suppressed the reflex activity in the tibialis anterior muscle, the unloading reflex remained unchanged in the soleus. This finding suggests that reciprocal inhibition from the tibialis anterior is not a causal factor in producing the silent period in the triceps surae muscle.

Dependency on the Gamma Loop

After complete deafferentation by means of posterior rhizotomy in man, a sustained voluntary contraction can still be performed but with ataxia, and no silent period follows a sudden removal of the external load [STRUPPLER and STRUPPLER, 1960]. Thus, with a purely suprasegmental mode of activation there is no longer a regulatory feedback from the γ loop and the muscle length does not affect the firing pattern of moto-

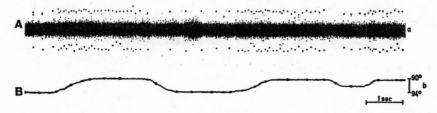

Fig.5. Soleus muscle spindle afferent during stretch and unloading. *A* Single unit recording from the tibial nerve. *B* Angle position of the ankle joint. Note steady firing frequency proportional to each of the three muscle lengths (static response) and the abrupt (dynamic) response to applied changes.

neurons. A similar effect can be seen with partial procaine block of a motor nerve [MATTHEWS and RUSHWORTH, 1957] where preservation of voluntary activity suggests that the large α fibres have been relatively spared by the local anaesthetic, and the attenuation of the silent period which then follows is presumed to reflect selective blockade of the small γ efferents [LANDAU, WEAVER and HORNBEIN, 1960]. When EMG records of single motor units are studied (fig. 4) it can be seen that, after partial procaine block, the inter-spike interval is only minimally affected by unloading. Thus, when the spindle sensors are not stretched by their own intrafusal muscle fibres to correspond with the length of the extrafusal fibres, the stronger suprasegmental drive now required to maintain motoneuron activity is relatively insensitive to segmental changes.

Recording from Single Spindle Afferents in Man

In 1953 ELDRED, GRANIT and MERTON demonstrated in animal experiments, the value of recording directly from single spindle afferents in the analysis of reflexes. VALLBO and HAGBARTH [1968] recently succeeded in direct recording of spindle afferents in man by means of a very precise microneurographic method and recorded the discharge patterns of single spindle afferents [VALLBO, this volume]. Using the same method, STRUPPLER and ERBEL [1972] investigated the effects of unloading in certain test muscles. Spindle afferents were identified by their discharge pattern during passive stretch and the other criteria given by VALLBO (fig. 5). When the isometrically contracted soleus muscle was unloaded while recording with a micro-electrode from single spindle afferents in the tibial nerve, a pause in the discharge was observed (fig. 6). This effect was most pronounced in units with high dynamic sensitivity. From these various lines of evidence

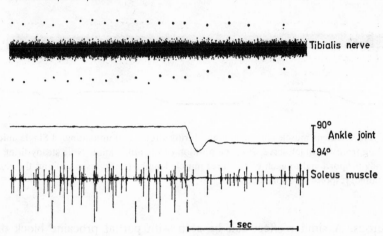

Fig.6. Unloading reflex in the soleus muscle. *Upper trace,* muscle spindle afferent in the tibial nerve; *middle trace,* position of the ankle joint; *lower trace,* EMG of the soleus muscle. There is a pause in the discharge of the spindle afferent during the shortening.

there can be little remaining doubt that the size of the unloading reflex is affected by a variety of converging inhibitory and excitatory proprioceptive and exteroceptive afferents on the α-motoneurons. The most important factor in the EMG pause during unloading must be the withdrawal of spindle afferent facilitation to the pool of motoneurons supplying the test muscle.

Investigation of Hypotonic Syndromes

1. Hypotonia of the peripheral type (tabes dorsalis, posterior rhizotomy). In this type of hypotonia the resistance to all forms of passive movements is decreased and there is also abnormal or excessive mobility at the joints. A voluntary muscle contraction can be maintained, but it is ataxic. Neither the stretch nor the unloading reflex can be elicited and it is thus clear that the spinal motoneuron pool is being driven exclusively by suprasegmental input. We, therefore, may assume that hypotonia reflects the loss of peripheral tonic drive. From this it is suggested that the normal downflow of excitation is sub-threshold for a portion of the motoneurons unless facilitatory feedback from the spindle afferents is also available to raise the excitability to firing levels [STRUPPLER and SCHENCK, 1958].

2. Interruption of vestibular afferents. Block of vestibular afferents causes a decrease in tonic innervation and reduced stretch reflexes, but does not affect unloading reflexes [RISTOW and STRUPPLER, 1964] so that hypotonia of this type does not appear to result primarily from a disturbance in fusimotor drive. One can assume a deficiency of direct α-motoneuron activation which would be consistent with animal experiments showing a monosynaptic, facilitatory path via the vestibulospinal tract to extensor α-motoneurons [GRILLNER, HONGO and LUND, 1970].

3. Cerebellar hypotonia. Our experiments include three patients, two having strictly unilateral neocerebellar lesions and a third having cerebellar syndromes predominantly on one side. In the first two cases one cerebellar hemisphere was removed surgically for a benign tumour. The third patient had multiple sclerosis with predominantly cerebellar syndrome on one side. All three showed typical clinical signs of hypotonia ipsilateral to the lesion as described so clearly by HOLMES [1939].

Methods

The testing apparatus was arranged for each subject so that recordings could be obtained from both arms at the same time. The patients were sitting comfortably in a chair with both upper arms extended horizontally in front on a board-support, with the forearms flexed at 90°. To both wrists a light bar was connected, to which a weight was attached by a cord running over a pulley. The patients were asked to keep this bar always perpendicular to the cord. Thus, in every case, we could ascertain that the same force was applied to both arms. The weight was held by an electromagnet and could be released by a switch. All experiments were performed with the subjects' eyes closed. We selected the brachial muscle for recording, as it is purely flexor, unlike the biceps muscle which has double function. Flexible wire electrodes were inserted into the agonistic and antagonistic muscles (brachial and triceps muscle of both sides) in order to record both sides simultaneously. The movements of the forearms were measured by means of a potentiometer attached by levers fixed to the elbows.

Results

In agreement with STRUPPLER and SCHENCK [1955] the results showed that the unloading reflex was clearly decreased on the side with clinical hypotonia. It can be seen in figure 7 that the firing rate of some units is not at all reduced during unloading, while others barely increase their interspike intervals. This finding should be contrasted to the normal unloading reflex side (fig. 7, lower traces). Thus, the unloading reflex is attenuated and incomplete on the side of neocerebellar lesions, and the question arises

UNLOADING REFLEX   BRACHIAL MUSCLE
cerebellar lesion (right side), normal (left side)

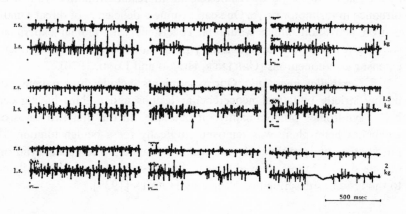

Fig. 7. Unloading reflex in the brachial muscle after ablation of the right cerebellar hemisphere. In each line the upper trace illustrates the unloading reflex on the affected side and the lower trace the unloading reflex on the normal side. Removal of the weight is marked by arrows.

whether the cerebellar lesion removes some direct descending influence on the spinal motor centres or whether the deficits are produced indirectly by interference with other facilitatory systems. In experimental animals the γ efferent drive can be greatly reduced by cerebellar ablation [VAN DER MEULEN and GILMAN, 1965; GILMAN and McDONALD, 1967b; GILMAN, 1969; GILMAN and EBEL, 1970], and it is reasonable to assume that the change in the silent period reflects the same abnormality responsible for the hypotonia; namely, a decrease in segmental IA facilitations produced by a corresponding loss of fusimotor drive from above. Some authors [TOWER, 1940; GOLDBERGER, 1969; GILMAN and MARCO, 1971] have also found after pyramidotomy a reduced resistance to passive muscle stretch, which was later shown to be a diminished response of muscle spindle primaries [GILMAN, MARCO and EBEL, 1971]. This type of hypotonia closely resembles the type seen with neocerebellar lesions and there is anatomical and electrophysiological evidence for extensive projections from the cerebellar dentate nucleus to the motor cortex. The pathways involved were recently reviewed by EVARTS and THACH [1969]. Several authors [WALKER, 1938; HASSLER, 1950, 1956; COMBS, 1959; UNO, YOSHIDA and HIROTA, 1970] pointed out that an important part of the cerebellar nuclear efferents influence the motor cortex via the specific relay nuclei of the ventro-lateral thalamus (VL). HASSLER [1956] emphasized the role of the posterior part of the VL mediating the cerebello-cortical impulses. Evidence that the VL neurons activate monosynaptically pyramidal tract neurons (PTN) was provided by studies of YOSHIDA, YAJIMA and UNO [1966], AKIMOTO and SAITO [1966]. Both the cerebellum [GILMAN and McDONALD 1967] and the pyramidal tract [GRANIT and KAADA, 1952; SHIMAZU, HONGO and KUBOTA, 1962; YOKOTA and VOORHOEVE, 1969] facilitate γ-motoneurons. Finally, KOEZE, PHILLIPS and SHERIDAN [1968] concluded that the action of PTN on muscle spindles and on motor units were capable of functional independency. We, therefore, favour the hypothesis that the neocerebellum may influence the muscle tone at least partly by PTN and the cortico-spinal tract.

Postsurgical hypotonia

In order to test this hypothesis we investigated the unloading reflex in 3 patients with stereotactic lesions in the posterior part of the thalamus and in the area of subthalamus where afferents from the brachium conjunctivum via the red nucleus pass through.

The first two patients had hemi-parkinsonism with pronounced tremor and no rigidity or akinesia. The third had bilateral parkinsonian tremor, the tremor being more pronounced on one side. After stereotactic surgery we observed a very slight hypotonia in the extremity relieved of its tremor. The criteria of hypotonia were those described by HOLMES [1939], and the findings closely resembled those in cerebellar lesions, except that there was little or no ataxia. It should be emphasized that the tremor was promptly and totally abolished in all three cases and that the hypotonia was not associated with any pathological reflexes or change in the phasic stretch reflex excitability.

In each patient the silent period produced by unloading of the voluntarily contracted muscle was clearly reduced contralateral to the stereotactic lesions. A typical example is illustrated in figure 8. On the operated side, the spike intervals vary considerably: some motor units continue firing during unloading, whereas others may only slightly decrease their discharge. In this same subject it was also found that the stretch reflex elicited during contraction of the muscle was followed by a shortened and incomplete silent period on the affected side. This phenomenon is shown in figure 9 in contrast with the normal side where a complete and long-lasting

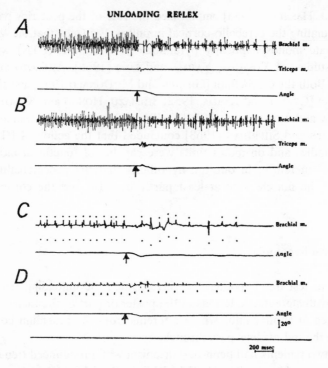

Fig. 8. Unloading reflex in patients with stereotactic lesions in and just below the posterior part of the ventro-lateral thalamus. On the affected side *(B, D)* there is no clearcut silent period following unloading of the muscle and as shown in single unit records there is only a small prolongation of the inter-spike interval. *A, C* normal side.

silent period followed the twitch contraction elicited by a tendon tap in the control brachial muscle. On the side contralateral to the stereotactic lesion the activity of motor units was only partly reduced.

Simultaneous recordings of the brachial muscle and the antagonistic triceps muscle on both sides revealed in most cases a much smaller antagonistic stretch reflex on the affected side compared to the normal. This finding is illustrated in figure 10 following unloading of the brachial muscle and in figure 9 following a stretch reflex. With HERMAN and MAYER [1969], we may suggest that this decreased triceps stretch response is a natural consequence of reduced disinhibition during an incomplete unloading response of the antagonistic muscle.

As stated earlier, the bulk of animal experimental work supports the hypothesis that cerebellar lesions produce hypotonia by affecting the level

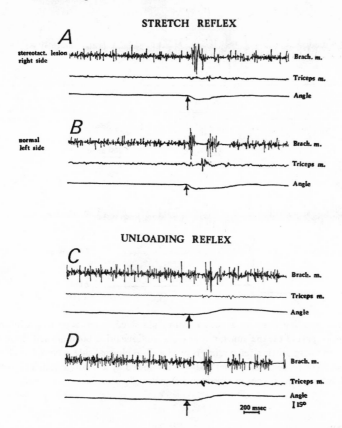

Fig. 9. Stretch reflex and unloading reflex in a patient with a stereotactic lesion in and below the posterior part of the ventrolateral thalamus of the left side. Note the EMG continues firing after both stretch and unloading of the muscle in *A* and *C*. Normal side in *B* and *D*.

of fusimotor activity and, thus, the sensitivity of the muscle sensors. Clinically one result is hypotonia, and we have now found a similar disturbance after a small stereotactic lesion that is believed to interrupt the outflow from neocerebellum via thalamus to corticospinal pathways. The reflex studies carried out in the two groups of patients show nearly identical changes, and we take the results in the stereotactically treated patients as further evidence in man that descending fusimotor activation receives an important facilitatory drive from the neocerebellum via its major ascending pathway to the cerebral cortex.

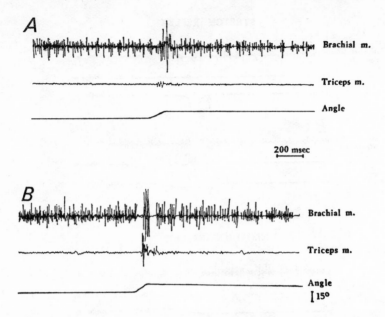

Fig. 10. A Unloading reflex in a patient with right-sided stereotactic lesion in and below the posterior part of the thalamus; motor units continue firing but a normal rebound occurs and a small antagonistic stretch reflex is produced. *B* Normal side, a complete silent period is induced. The rebound is normal and again followed by a silent period. The antagonistic stretch reflex in the triceps muscle is prominent.

Summary

To analyze certain aspects of central motor disorders in man we have studied the modifications of isometric contractions by phasic stretch or unloading of the muscle. Our technique involves simultaneous testing on both sides of the body. The unloading reflex could be elicited in the muscles of both arms and legs if the thresholds for displacement and acceleration were exceeded. Single motor unit recordings revealed that the unloading reflex was associated with a prolongation of the firing intervals of individual motoneurons usually followed by a spate of rapid firing called 'rebound'. If the muscle was unloaded slowly enough, the central drive could cause a rebound still during shortening of muscle. It was found that the unloading reflex could be modified by intention, by deafferentation, by differential procaine block of fusimotors, but not by block of the reflex arc in an antagonistic muscle. Microneurographic recording of single I_A afferents during unloading of the soleus muscle indicated that the beginning of the silent period coincided with the cessation of spindle afferent firing. In unilateral lesions of the cerebellar hemispheres some motoneurons keep firing, not only during unloading, but also following the phasic stretch reflex. In both cases there was an incomplete silent period. After thalamotomy (VL) and subthalamotomy for relief of parkinsonian tremor, the same abnormalities were noted. We, therefore, assume that a shift in tonic innervation from γ to α drive occurs when the cerebello-thalamo-fugal afferents to the cerebral cortex are interrupted, and the similarities between the two groups of patients suggest that the neocerebellar afferents can influence the γ drive via the corticospinal tract.

Blocking the vestibular nerve caused a slight hypotonia with a normal unloading reflex, but a reduction of stretch reflexes. In this case hypotonia may be caused by a diminished α drive. In deafferentation of tabes dorsalis the stretch and the unloading reflex were both reduced or even abolished. The reduced γ drive in this type of hypotonia can be compensated by an increased α drive.

Acknowledgement

We thank Professor W. W. HOFMANN for advice during the preparation of the manuscript.

This study was supported by the 'Deutsche Forschungsgemeinschaft'.

Author's address: Prof. ALBRECHT STRUPPLER, Department of Neurology, Technical University of Munich, Ismanigerstrasse 22, *D-8 München 80* (FRG)

New Developments in Electromyography and Clinical Neurophysiology,
edited by J. E. Desmedt, vol. 3, pp. 618–624 (Karger, Basel 1973)

Spasticity and Tremor

Effects on the Silent Period and the After-Volley

R. W. ANGEL

Department of Neurology, Stanford University School of Medicine, Stanford, Calif.

When a muscle is unloaded rapidly during an isometric contraction, the normal reaction is an electrical silent period, followed by a burst of large action potentials. This phenomenon, described by HANSEN and HOFFMANN in 1922, was soon applied to the study of human motor function [HANSEN and RECH, 1925; HOFFMANN, 1934; SCHWERIN, 1936; SOMMER, 1939], but the early investigations were done before the development of modern ideas concerning the muscle spindles and their role as governors of movement. In the light of current knowledge, the unloading technique has acquired further significance, and it continues to provide new information about motor activities.

The Silent Period

In clinical applications, a given muscle is unloaded several times under standard conditions. The resulting EMG patterns can be measured and compared with normal controls. The first part of the normal pattern is a sharp reduction in muscle action potentials. We assume that this 'silent period' is caused by sudden withdrawal of excitatory inflow from the muscle spindle afferents [STRUPPLER, LANDAU and MEHLS, 1964; ANGEL, EPPLER and IANNONE, 1965]. During voluntary contraction, the α-motoneurons are excited by nerve impulses from many sources, probably including afferent fibers from the muscle spindles. During some contractions, the spindle afferent facilitation may be necessary to continued firing of the motoneurons. In these cases, the spindle pause and reduction in I_A afferent activity that follows unloading will result in a silent period. According to such a working hypothesis, the occurrence of a silent period on the unloading test implies that the given contraction depends on spindle afferent excitation. Conversely, the absence of a silent period would imply that the motor discharge is relatively independent of spindle afferent activity.

The After-Volley: Possible Mechanisms

No definitive explanation is yet available for the motor activity following the silent period, but several facts are compatible with a spinal reflex mechanism [ALSTON, ANGEL, FINK and HOFMANN, 1967]. If acceleration of the limb is reduced experimentally by attaching a mass, the after-volley is reduced in size. This implies that the α-motor discharge is affected by proprioceptive feedback. Furthermore, if movement of the limb is arrested within 30 to 100 msec after unloading, the volley tends to occur about 20 msec after the block is applied. This suggests that the feedback is relayed to the muscle within a time which approximates the latency of a muscle stretch reflex.

Since the extrafusal fibers continue to shorten during the silent period, the after-volley cannot be an ordinary stretch reflex. If the α-motoneurons resume firing in response to a spindle afferent barrage, one must assume that the static γ-motoneurons are active during the silent period, since only the static fibers can activate the spindles during extrafusal shortening [LENNERSTRAND and THODEN, 1968].

Animal experiments suggest at least three mechanisms that might cause the γ-motoneurons to increase their activity during the silent period. The first possibility is a disinhibition resulting from contraction of the agonist muscle. It is debatable whether the muscles influence their own fusimotor neurons reflexly by changes of length or tension [GRANIT, PASCOE and STEG, 1957; HUNT and PAINTAL, 1958; ECCLES, ECCLES, IGGO and LUNDBERG, 1960]. However, some γ-motoneurons do appear to be inhibited by repetitive stimulation of ventral root fibers of α-motoneurons [BROWN, LAWRENCE and MATTHEWS, 1968; ELLAWAY, 1968; GRILLNER, 1969]. If the fusimotor neurons are inhibited by recurrent impulses from the α-motoneurons, it is plausible that α silence would produce disinhibition, permitting the gammas to resume activity during the silent period.

A second possible mechanism for resumed γ discharge is suggested by the observation that extension applied to the muscle causes inhibition of the fusimotor neurons [FROMM and HAASE, 1970]. If such stretch causes inhibition, it

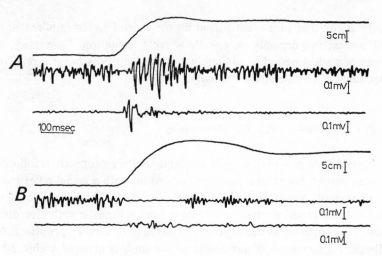

Fig.1. Unloading reflex. Top line: position of hand. Center: EMG from biceps. Lower line: EMG from triceps. As trace begins, patient is supporting 4.5 kg by contraction of biceps. At time shown by upward deflection of top line, weight is released, and elbow begins to flex. *A* Normal arm of patient with spastic hemiparesis. The silent period in biceps is followed by large after-volley. *B* Spastic arm of same patient. Unloading is followed by a silent period, but the after-volley is much smaller and delayed in onset.

is not unreasonable to propose that shortening may cause disinhibition. As the muscle shortens after unloading, the gammas might then resume firing and produce the after-volley through the stretch reflex loop.

A third possible mechanism which could activate the γ-motoneurons during the silent period is the fusimotor reflex elicited by stretching the antagonist [BARRIOS, HAASE and HEINRICH, 1967; HAASE and VOGEL, 1969]. In the cat, γ-motoneurons of the pretibial muscles are activated by tetanic stimulation of low threshold afferents of the ipsilateral gastrocnemius [SCHLEGEL and SONTAG, 1969, 1970]. Stretch of the antagonist muscle might thus activate the γ-motoneurons of the agonist.

There is no evidence that any of these mechanisms operate in man, but we assume, as a second working hypothesis, that the spindles do resume activity during the period of α silence and thereby produce the after-volley through the γ loop.

By whatever means the spindles might be activated, the resulting barrage of afferent impulses would be likely to excite a large α motor response. During the silent period, the motoneurons would receive little or no inhibition via the

Renshaw recurrent circuit or the Golgi tendon organ afferents. Hence, it is not surprising that the after-volley is often larger than the muscle action potentials before unloading.

Unloading Response in Patients with Spastic Hemiparesis

If the unloading response did have two distinct parts, each one mediated by a different neural mechanism, a disease process could affect either part separately. In patients with spastic hemiparesis due to lesions of the upper motoneuron, the affected biceps muscle shows an apparently normal silent period but a striking reduction of the after-volley as compared with the normal response on the opposite side [ANGEL, 1968]. Figure 1A shows the unloading reflex obtained on the normal side of a patient with spastic hemiparesis. The EMG from the normal biceps shows a definite silent period followed by a large after-volley. Figure 1B shows that a silent period is present on the affected side also, but the after-volley is strikingly reduced in size.

According to our first hypothesis, the presence of a silent period implies that spindle afferent excitation is necessary to sustain the base line contraction, in the spastic limb as well as in the normal. Hence, it appears that a lesion of the corticospinal tract does not abolish the role of spindle afferent impulses in sustaining a voluntary contraction. During the contraction, motoneurons on the paretic side are excited by inflow from the muscle spindles, and they cease firing when that inflow is withdrawn, just as on the normal side.

According to our second hypothesis, the reduced after-volley on the affected side points to impairment of a mechanism that normally reloads the spindles. On the further assumption that reloading is mediated by a fusimotor reflex, one might infer that this (hypothetical) reflex is impaired on the spastic side.

Figure 2 shows another finding in some patients with spastic hemiparesis. When flexion of the elbow is stopped experimentally, the normal biceps shows a large burst of action potentials, followed by a rhythmic alternation of silence and activity, at about 10/sec. In the affected limb, the arrest of movement is followed by a relatively small burst of activity and rhythmic oscillation at a much lower frequency, roughly 5/sec. Current knowledge provides no explanation for the difference of frequencies between the two sides. This phenomenon cannot be demonstrated in all patients, because the normal (or abnormal) muscle does not invariably show rhythmic oscillation when movement is arrested.

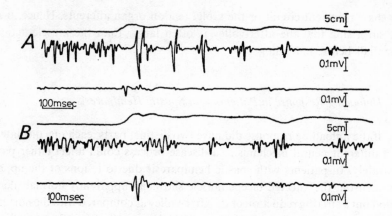

Fig. 2. Unloading reflex, same patient as in figure 1. Flexion of elbow was arrested artificially within 50 msec after unloading, as shown by shortened upswing of position trace. *A* Normal arm; the silent period is followed by rhythmic alternation of EMG volleys and silent periods, at about 10/sec. *B* Spastic arm; EMG volleys following the silent period are smaller and at slower frequency than those on normal side.

Unloading Response in Patients with Parkinson's Disease and Action Tremor

As in the case of pyramidal tract disease, patients with parkinsonian rigidity have a normal silent period on the unloading test [ANGEL, HOFMANN and EPPLER, 1966]. This is presumptive evidence that the rigid muscle has retained its sensitivity to small changes of length. If the silent period is the response of a servo mechanism, then parkinsonian rigidity does not necessarily interfere with function of that mechanism.

Since the resting tremor of parkinsonism tends to disappear during voluntary motor activity, it is not amenable to study by the unloading technique. On the other hand, many patients with parkinsonism have a form of tremor that appears and continues during voluntary innervation. LANCE, SCHWAB and PETERSON [1963] observed that this 'action tremor' was unaltered in frequency by an intrathecal infusion of procaine or a degree of ischemia which abolished the stretch reflex. They inferred that the tremor mechanism did not depend on afferents from the muscle spindles. We have reached the same conclusion by means of the unloading test. In three patients with action tremor, bursts of tremor activity were often seen immediately after unloading

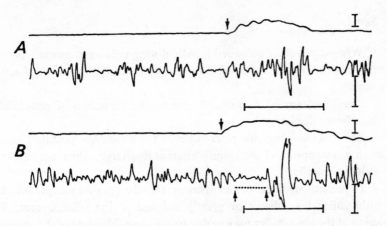

Fig.3. Unloading reflex in patient with action tremor. Upper line: velocity of hand. Lower line: EMG from biceps. *A* Before thalamotomy. As trace begins, patient is supporting 4.5 kg by contraction of biceps, which fires in discrete volleys. Weight is released at time (arrow) when a tremor volley is 'due'. Silent period does not occur. *B* After thalamotomy. Baseline EMG now shows continuous interference pattern, and unloading produces a silent period, seen above dashed line. Calibrations, 100 cm/sec, 0.2 mV, 125 msec.

(fig. 3A), at a time when normal subjects would show a silent period [ANGEL, AGUILAR and HOFMANN, 1969]. From this we inferred that the tremor volleys were relatively independent of proprioceptive feedback. After sterotaxic thalamotomy, each of these patients showed abolition of the contralateral action tremor. The separate bursts of motor activity formerly seen during voluntary contraction, were replaced by a normal interference pattern, and normal unloading reflexes were found (fig. 3B). The results suggest that surgery may have restored the normal dependence on proprioceptive feedback during voluntary innervation of muscle. They suggest also that action tremor is determined primarily by a central mechanism, rather than by oscillation of the segmental reflex. Current evidence suggests that the most probable pacemaker of tremor at rest is the dentato-thalamic system [cf. HASSLER, MUNDINGER and RIECHERT, 1970]. The disappearance of action tremor following thalamotomy indicates that its pacemaker may be the same. On the other hand, LANCE [1970] has postulated a central synchronization of motoneuron activity at about 10 impulses/sec, whether the motoneurons are thrown into activity by supraspinal or spinal reflex mechanisms.

Summary

1. When a muscle is unloaded rapidly during voluntary contraction, the normal response is a silent period in the EMG, followed by a burst of action potentials.

2. The silent period is probably due to the reduction of pre-existing spindle afferent discharge.

3. The mechanism of the after-volley is unknown. We postulate that it is due to a resumption of the spindle afferent discharge, which activates the α-motoneurons reflexly.

4. In patients with spastic hemiparesis, the silent period was present, but the after-volley was absent or greatly reduced in the affected arm. The reduction of the after-volley may be due to defective reloading of the spindles.

5. In patients with parkinsonism with action tremor, the unloading response was abnormal. Bursts of tremor activity were seen during the time when electrical silence would normally occur. This finding suggests that action tremor is relatively independent of spindle afferent activity. After the tremor was abolished by thalamotomy, the unloading response became normal.

Author's address: Dr. R. W. ANGEL, Veterans Administration Hospital, 3801 Miranda Avenue, *Palo Alto, CA 94304* (USA)

New Developments in Electromyography and Chimical Neurophysiology,
edited by J. E. Desmedt, vol. 3, pp. 625–628 (Karger, Basel 1973)

EMG Changes Induced by Sudden Stretch

C. Tardieu[1], J. C. Tabary and G. Tardieu

Hôpital Raymond Poincaré, Garches

EMG changes were correlated with displacement of the limb or finger
when a sudden stretch was applied during simple maintenance of a posture,
in 15 normal subjects [Tabary et al., 1965]. This approach differs from other
studies in which the limb was not free to move and the isometric force of the
muscle was recorded [Merton, 1951].

In one type of experiment (93 trials in 15 subjects) the displacement of
the index finger was recorded by filming a white spot at the finger-tip with
a fast camera (160 frames/sec). The subject was seated comfortably with the
hand resting on a support. The index finger was splinted so that only the metacarpophalangeal joint could move. Surface electrodes were used to record the

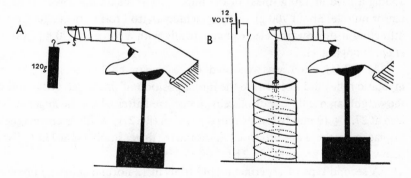

Fig. 1. Experimental set-up for sudden changes of the posture of the index finger.
A Weight addition. *B* Downward pull by electromagnetic device.

1   Maître de Recherche INSERM Contrat 712.156.

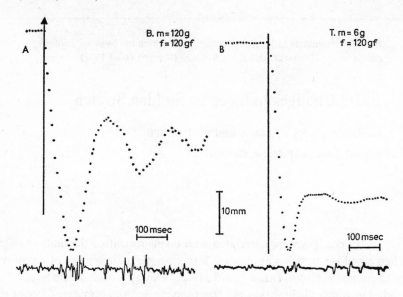

Fig. 2. Displacement of the index finger (dotted line) and EMG of the extensor of the index finger simultaneously recorded. A Sudden addition of a weight of 120 g. The index stabilizes about 25 mm below its initial position. *B* Same experiment but with electromagnetic pull. The index stabilizes about 40 mm below its initial position.

EMG of the index extensor muscle on the same film. The subject was instructed to keep the finger horizontal. The stretch was produced either by suddenly adding a load of 120 g (mass of the finger: 10 g; maximum force of the extensor muscle: about 700 g), or by attaching an iron rod to the finger tip and attracting it downwards electromagnetically with a force of 120 g [TARDIEU *et al.*, 1968] (fig. 1).

An EMG burst was recorded for each stretch but this well-known myotatic reflex did not restore the initial position of the finger. After adding the weight, an oscillatory displacement occurred after which the finger stabilized at 27.5 ± 19 mm below its initial position (fig. 2A). With electromagnetic displacement there was little oscillation and the finger stabilized at 37.8 ± 9 mm below its initial position (fig. 2B).

A second type of experiment (143 trials in 12 normal subjects) involved different kinds of stretches at the elbow while the EMG of the biceps and triceps muscles was recorded with surface electrodes. The lower forearm was held horizontal, keeping the upper arm alongside the body. The subject was asked to maintain this position. Four types of sudden change were tested:

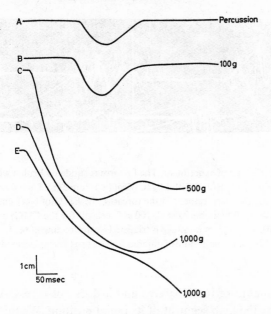

Fig.3. Displacements of the distal part of the forearm after percussion of distal part of the radius (A), loading of the wrist with 100 g (B), with 500 g (C) and 1,000 g (D and E) dropped from various heights.

percussion of the styloid process (which elicits a reflex discharge in the biceps after 18–20 msec); a weight of 100 g falling from a height of 100–10 cm onto the hand; a weight of 500 g falling from a height of 20–1 cm; a weight of 1 kg falling from a height of 10–1 cm (fig. 3). In each case, the EMG of the biceps showed a brief (reflex) burst, followed by a silent period. The EMG activity reappeared after latencies of 132 ± 29 msec, 140 ± 27 msec, 147 ± 26 msec and 160 ± 31 msec, respectively, for the 4 types of stretch considered (fig. 4).

When correlated with the displacement of the forearm these silent periods of the EMG cannot be accounted for by simple unloading according to the length servo hypothesis [MERTON, 1951, 1953]. The silence in the biceps EMG occurs while the muscle is still being lengthened. Furthermore, the biceps becomes active again at about the same delay after the stretch, whether the muscle is shortening (fig. 4A) or still lengthening (fig. 4B) at that time. Thus the sudden initial stretch would appear to elicit with almost constant timing the whole sequence of the initial burst, the silent period and the return of postural activity. This sequence of events is more complex than one would expect from a simple length servo control. The reflex response appears indeed

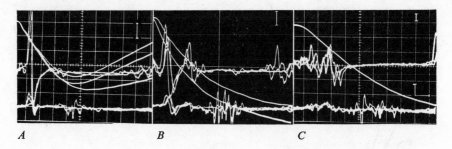

A   B   C

Fig.4. Subject C.T., second type of experiment. The forearm is suddenly loaded with 100 g dropped from 1 m (A), with 500 g (B) and with 1 kg (C) dropped from 2 cm. Each frame presents from above downwards the displacement of the forearm, the EMG of biceps and the EMG of triceps. Calibrations, 20 msec horizontal, 100 μV vertical (for the EMG) and 1 cm vertical (for displacement). The horizontal sweep is triggered by the loading of the forearm. Notice the similarity in EMG patterns in spite of differences between the displacements.

to be regulated by the amount of initial stretch and it does not necessarily achieve the resumption by the limb segment of its initial position. We think that a better understanding of the mechanisms involved is important for a reconsideration of the revalidation procedures in cerebral palsy.

Author's address: Mme C.TARDIEU, Hôpital Raymond Poincaré, *F-92 Garches* (France)

New Developments in Electromyography and Clinical Neurophysiology,
edited by J. E. Desmedt, vol. 3, pp. 629–634 (Karger, Basel 1973)

Analysis of the Shortening Reaction in Man

P. Rondot and S. Metral

Centre Hospitalier et Universitaire, and Centre Claude Bernard, Hôpital de la
Salpêtrière, Paris

An abnormal reaction of muscle occurring on passive shortening was reported by Westphal in 1877. He had already studied the inverse phenomenon,
that is, muscular contraction on extension of its tendon. He was thus surprised
to find that on occasions passive shortening of a muscle was accompanied by
a contraction which he named the 'paradoxer Muskelreaction'. At this time
he already clearly distinguished the defective muscular relaxation of certain
patients who proved to have great difficulty in not passively aiding imposed
movement. In the spinal animal, Sherrington [1909] described analogous
findings under the name of 'the shortening reaction'. Subsequently this reaction has been the subject of several communications [Wertheim Salomonson,
1914; Samojloff and Kisseleff, 1927; Rademaker, 1947; Broman, 1949;
Thiebaut and Isch, 1952; Schaltenbrand and Hufschmidt, 1957]. The
'Adaptationsspannung' of Förster [1921], and the 'local postural reflexes'
described by Foix and Thevenard [1929] in parkinsonian syndromes correspond very probably to phenomena of the same type. Rondot and Scherrer
[1966] studied the shortening reaction in patients with athetosis and dystonia,
in whom it is particularly marked, as it is in other lesions of the extrapyramidal
system. We believe the importance of the shortening reaction in the pathological physiology of these diseases is underestimated.

Methods

The shortening reaction has been studied by polygraphic records on the biceps and
triceps, with additional records from the infraspinatus, biceps, triceps and finger extensors.

Platinum surface electrodes were secured by bands over the muscles of the arm, or
stuck over the shoulder muscles. More recently we have used a multi-electrode placed into

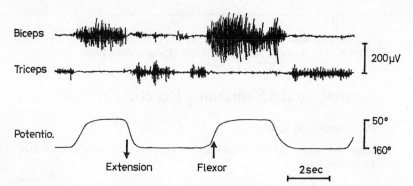

Fig. 1. EMG of biceps and triceps in a patient with athetosis during passive flexion and extension of the forearm. Activity appears in biceps during flexion and in the triceps during extension. This activity is increased at the end of movement, and persists in biceps and triceps while flexion and extension is maintained.

the extensor digitorum communis so that the firing rate of a motor unit can be studied during passive extension of the wrist and fingers.

The forearm is supported in a horizontal trough pivoted vertically on an axis through the elbow joint. A potentiometer placed on this pivot records the degree of movement.

The same system is used in the study of the unloading reaction. A traction of 5 kg is applied in the horizontal plane, and then suddenly released by cutting the wire by which the weight is supported.

We have studied 29 cases of athetosis, 25 cases of dystonia, 11 patients with Parkinson's disease and 1 case with the stiff-man syndrome.

Results

The Shortening Reaction

Both in patients with Parkinson's disease and with athetosis, the shortening reaction began during the course of the passive movement, and increased in amplitude as the degree of movement increased. It frequently persisted for several seconds after the movement had finished, while the muscle was passively held shortened (fig. 1). The recorded electrical activity may, particularly in patients with athetosis, increase at the end of the passive movement. This may be seen clinically as muscular spasm.

There may be no electrical activity in the stretched antagonist muscle, in patients with dystonia or athetosis. If there is motor unit activity, it characteristically appears at the beginning of the passive movement and then disap-

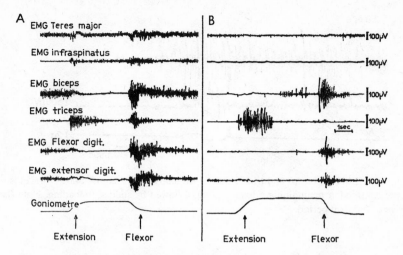

Fig. 2. Spread of the shortening reaction in a patient with the stiff-man syndrome. EMG of teres major, infraspinatus, biceps, triceps and of the flexors and extensors of the fingers of the right forelimb. Records before (A) and after (B) the injection of Diazepam.

pears during the course of stretch. Sometimes activity persists in the antagonist throughout the passive movement, but it frequently becomes less and disappears if the movement is repeated several times.

Characteristics of the Shortening Reaction

Shortening Reaction and Velocity of Movement

In contrast to what occurs during the stretch reflex in spasticity, the shortening reaction does not usually increase as the velocity of shortening increases. The inverse phenomenon is often observed, the integrated electrical activity decreasing as the angular velocity increases.

Involvement of other Muscles in the Shortening Reaction

The shortening reaction may spread to other muscles [RONDOT, 1968]. This spread, which is particularly marked in the muscles of patients with dystonia, often follows certain patterns. The activity in the teres major recorded during passive internal rotation of the shoulder spreads to the biceps.

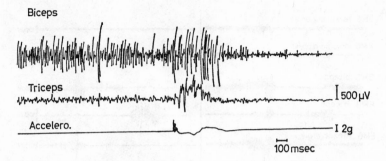

Fig. 3. EMG of biceps and triceps during an 'unloading reaction'. Start of trace: isometric contraction maintaining a load of 5 kg, followed by sudden unloading. Note the absence of a silent period.

The active recorded in infraspinatus during passive external rotation of the shoulder spreads to the triceps, and sometimes to the finger extensors. This spread of activity has seemed to be greatest in those patients in whom dystonia is most severe. A very marked example, illustrated in figure 2, was recorded from a patient with the stiff-man syndrome. The pattern of spread seemed to be related to the posture imposed by the dystonia. It should also be mentioned that if the dystonia was accompanied by a tremor, the electrical activity recorded during shortening reaction was also rhythmic.

The Shortening Reaction and Cutaneous Stimuli

If the shortening reaction is particularly marked, we have observed abnormal reactions to prolonged cutaneous stimuli. This is found frequently in patients with dystonia, less so in those with athetosis. These reactions are most brisk with the first stimulation, and then tend to attenuate with repeated stimuli. If the electrical activity accompanying shortening reaction is rhythmic, then cutaneous stimulation also evokes a rhythmic activity in the same muscles.

The Shortening Reaction and the Silent Period

We have recently studied the silent period by examining the unloading reaction in 3 athetotic and 2 dystonic patients, in whom the shortening reaction was particularly marked. In 4 of these 5 patients the reaction after the

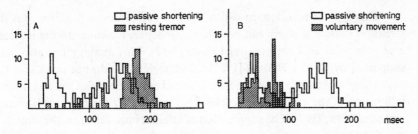

Fig.4. Resting tremor in a patient with a parkinsonian syndrome. Histogram of the discharge frequency of a single motor unit in extensor indicis. *A* During passive shortening and during resting tremor. *B* During passive shortening and during voluntary movement.

sudden release from a load was studied in the biceps. In 3 of them the traction was 5 kg but in one weaker patient only 2 kg was applied. In the fifth patient, the finger extensors were studied, loaded by a force of 1 kg. In 4 of 5 patients the silent period was absent (fig. 3). In the fifth, although it was present, it was less than 40 msec in duration. It should be noted that in this last case some spasticity was also present.

Discussion

The shortening reaction must be distinguished from poor voluntary control of muscular contraction. WESTPHAL [1877, 1880] was already aware of this difficulty. If the shortening reaction was voluntary, it would of course be surprising to meet it with such frequency in disease, and to find it on only one side in patients with hemiathetosis. In the case of Parkinson's disease with resting tremor, we have been able to distinguish clearly between the shortening reaction and voluntary movement [RONDOT and RENOU, 1972]. Single motor unit action potentials were recorded with a multi-electrode at rest, during isometric contraction, and during the shortening reaction (fig. 4). Voluntary contraction immediately modified the frequency of motor unit firing, so that it increased from 5 to 15–20/sec. During the shortening reaction, rhythmic activity continued, and the frequency was only modified by the appearance of a double discharge in each of the rhythmic bursts. In other words, the shortening in this particular case facilitates the tremor, an effect opposed to that of voluntary contraction. We think, therefore, that the shortening reaction is distinct from voluntary movement, and cannot be compared with it.

This type of reaction, in which the shortened muscle is active, but the stretched muscle is often, but not always, inactive, is similar to the inverted myotatic reflex of LAPORTE and LLOYD [1952]. This analogy has also been supported by DENNY-BROWN [1962]. Nevertheless, we do not think that the shortening reaction is related to an inverted myotatic reflex, as anaesthetic infiltration of the antagonist muscles does not modify the reaction [RONDOT and SCHERRER, 1966]. The suppression of afferent impulses from the stretched muscles should suppress the shortening reaction if the hypothesis that the latter is an inverted myotatic reflex were correct. We therefore consider the shortening reaction to be autogenic.

The afferent fibres for this reaction cannot be the I_A afferents, as there is no activity on stretching, and no silent period, which suggests that spindle activity is minimal or absent. Neither can the I_B fibres be concerned, as infiltration of the antagonist muscles does not abolish the reaction. It is difficult to be sure which pathways are involved in this reaction. Are group II fibres concerned? In support of this hypothesis may be mentioned certain points in common with the flexion reflex, in particular the response to proprioceptive and cutaneous stimulation. However, the flexors are not solely concerned with this reaction.

The frequency with which the shortening reaction is found in patients with athetosis and dystonia, and in some parkinsonian syndromes, suggests that it is involved in the pathophysiology of these conditions.

Summary

The shortening reaction is characterized by the appearance of electrical activity in a muscle as the insertions are approximated. It is observed most frequently in patients with athetosis, dystonia and in certain parkinsonian syndromes. It is an autogenic reaction which cannot be compared with an inverted myotatic reflex. The reaction spreads into other muscles following certain patterns. Frequently it is also accompanied by abnormally brisk responses to cutaneous stimulation.

Author's address: Prof. P. RONDOT, Hôpital de la Salpêtrière, 47, Bvd de l'Hôpital, F-75 Paris 13e (France)

New Developments in Electromyography and Clinical Neurophysiology,
edited by J.E. Desmedt, vol. 3, pp. 635–640 (Karger, Basel 1973)

Influence of Peripheral Stimulation on the Silent Period between bursts of Parkinsonian Tremor

G. RENOU, P. RONDOT and N. BATHIEN

Laboratoire de Physiologie, Centre Hospitalier-Universitaire St. Antoine, Paris
et Centre Cl. Bernard, Hôpital de la Salpêtrière, Paris

Previous studies have characterized motor unit discharges during parkinsonian tremor [RENOU, RONDOT and METRAL, 1970; RENOU, 1971]. One
particular pattern of discharge has been observed: when the tremor increases
in amplitude the same unit discharges twice in the same beat of tremor. The
frequency of discharge can reach 50/sec when there is a marked tremor. When
the tremor increases, it has also been shown that the silent period between
beats tends to last longer, thus reducing the duration of motor unit discharge
during the beat.

As units can discharge transiently at such a high frequency during the
beat a marked excitation of the motoneurone must occur at that moment.
The progressive lengthening of the silent period as the tremor increases in
amplitude suggests, on the other hand, an inhibitory process which would
increase in parallel with the tremor. We have attempted to investigate the
characteristics of the silent period separating tremor beats by studying the
effect of an electrical stimulus given at different phases of the tremor. Two
phenomena were studied: the variations in reflex response to stimulation
during tremor, and the changes in frequency and amplitude of tremor induced
by stimulation.

Methods

Sixteen patients with parkinsonian tremor affecting the lower limbs were studied. The
stimulus was a 1-msec electric pulse of variable intensity applied to the medial division of
the sciatic nerve in the popliteal fossa. Potentials were recorded from surface electrodes
over soleus, or from an EMG needle in this muscle. The monosynaptic H-reflex could thus
be studied, and the tremor recorded.

Stimulation was triggered at varying times in the tremor cycle. Variation of the intensity of stimulus to the sciatic nerve could produce either the H-reflex, or the direct motor (M) response. Direct stimulation of the soleus muscle was also used in each case. Repetitive stimulation to the saphenous nerve was tried but not explored in detail as changes in tremor were inconstant.

Results

A. Amplitude of the H-Reflex during the Tremor Cycle

With constant stimulus, the H-reflex varied considerably in amplitude during the cycle. The response was greatest if the stimulus was given at the beginning of a burst, but then decreased rapidly as the stimulus was progressively delayed. The response was very small between beats and increased rapidly at the beginning of the next (fig. 1 A). These results confirm those of EKBOM, JERNELIUS and KUGELBERG [1952] and of ISCH, ISCH-TREUSSARD and JESEL [1962].

B. The Effect of the Stimulus on the Frequency and Amplitude of Tremor

If a weak stimulus, capable of producing one-quarter to one-third of the maximal H-reflex, was given, no effect on the tremor was observed. As the intensity was increased the H-reflex (or the M response, according to the intensity) was followed by a prolonged silent period (fig. 1 B). A further burst of tremor appeared after about 250 msec. A greater number of motor unit potentials discharged in this burst than in those preceding the stimulus. The following cyclic discharges were at the same frequency as the tremor.

The silent period after this burst presented always the same duration whatever the time of the stimulus in relation to the tremor cycle. The silent period lasts about 250 msec after the stimulus if the response is an H-reflex, but the period is a little shorter by about 30 msec if the response elicited is an M response. Direct stimulation of the muscle also produces a silent period which is a little shorter than when the stimulus is given to the nerve. It appears that it is the muscular contraction itself which determines the silent period. This silent period lasts much longer than that obtained in the normal subject by giving a shock during tonic contraction, or by sudden unloading [see ANGEL; STRUPPLER *et al.*, this volume].

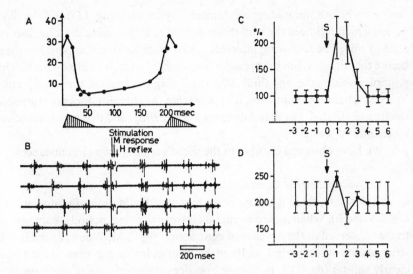

Fig.1. A Amplitude of the monosynaptic reflex in relation to the tremor cycle. Abscissa: on the lower line the beats of tremor are drawn diagrammatically in grey. Ordinate: the amplitude of the H-reflex response as a percentage of the direct motor (M) response. *B* Silent period after stimulation. Note the constancy of the duration of the silent period whatever the time of stimulation. Note also that the first beat after the silent period contains more motor unit potentials than the preceding beats. *C* Amplitude of the beats in relation to their rank before and after stimulation. Ordinate: Planimetric measures of integrated EMG activity expressed as a percentage of control values preceding the stimulus. These are mean values obtained from several patients. The vertical lines indicate the range of values and not the standard deviations. Abscissa: Rank of beats of tremor before and after stimulation. *D* Intervals between beats before and after stimulation. Ordinate: Mean values of intervals (msec) recorded from several patients. The vertical lines indicate the range of values, and not the standard deviations. Abscissa: Rank of beats of tremor before and after stimulation.

We have already noted that the first discharge which follows the silent period is made up of more motor unit potentials (fig. 1 B). Using planimetric measurements of the integrated EMG, we have observed that after the first burst the next two were also richer in motor unit potentials, although to a lesser extent. The facilitation thus appears to last more than 600 msec (fig. 1 C). A comparable facilitation is shown in figure 2A on the basis of a tonic contraction in a patient with Parkinson's disease.

The stimulus induced a silent period of constant duration whenever it was applied, thereby shifting the time of onset of the next burst of tremor. The frequency before stimulation was remarkably constant (fig. 1 D) and the silent period which followed the stimulus appeared to be an unusually long disturbance of this. The following intervals varied more or less, compared to those before the stimulus, but the previous frequency was attained after 4 beats. This pattern would make one think of a regulating mechanism transiently upset by a disturbance. In summary, it appears that the stimulus does not truly shift the tremor rhythm, but that it disturbs it transiently, and that some correcting mechanism operates.

We have observed changes in the silent period in two circumstances.

1. Tonic Voluntary Contraction

In this case, there is a state of tonic central facilitation of motoneurones (fig. 2A and B). When tonic contraction is weak the silent period is comparable to that observed in the absence of contraction (fig. 2A). When contraction is stronger the silent period, induced by a stimulus of the same intensity, is clearly shorter (fig. 2B). In neither case does the amplitude of the monosynaptic reflex change.

2. Vibration Applied to the Tendon

The effects of applying vibration at 100/sec to the Achilles tendon are shown in figure 2C and D. Control values before vibration are shown in figure 2C. The H-reflex response is then maximal. When vibration is applied (fig. 2D), its effects are seen in the well-known fall in amplitude of the phasic monosynaptic reflex. The silent period after the stimulus diminishes in proportion.

Discussion

In the patient with Parkinson's disease, the silence after stimulation can be compared to the silent period seen in the normal subject after an electrical stimulus to the motor nerve given during tonic voluntary contraction. Evidence of this lies in the similarity of the silent periods obtained in the same patient, in figure 2B (tonic contraction without marked tremor) and D (tremor during vibration).

This silent period appears abnormally long in patients with Parkinson's disease [HUFSCHMIDT, 1962; LIBERSON, 1962]. The features of the silent period seem to be related to the strength of the muscular contraction evoked by the stimulus delivered to the nerve, or to the muscle itself. One might consider

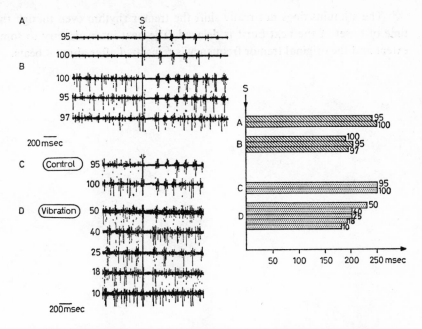

Fig.2. Left: The number before each line shows the amplitude of the H-reflex as a percentage of its maximum. *A* and *B* Tonic voluntary contraction; *C:* before vibration; *D:* recordings during vibration of increasing effects from above downwards. Right: Duration of silent period following stimulation corresponding to the 4 types of recording shown on the left. The number before each line indicates the amplitude of the H-reflex as a percentage of its maximum.

two possible mechanisms for this: either a pause in I_A facilitation through spindle unloading, or a Golgi autogenic inhibition. The first mechanism would appear unlikely as the sole explanation for the prolonged silent period in a parkinsonian patient. Activation of I_A afferents by vibration of the tendon produces a reduction, not a disappearance, of the silent period. We prefer to suggest an abnormally prolonged and intense central action of Golgi tendon afferents.

The end of the silent period in Parkinson patients is associated with a greater discharge of motor unit potentials in the next burst. This facilitation lasts for several hundred msec, as compared to about 100 msec in the normal subject. This facilitation is, of course, only expressed in a discontinuous fashion in the successive (increased) bursts of the tremor, the latter being separated by silent periods.

The stimulus does not really shift the tremor rhythm even though the time of onset of the next burst is delayed. The next intervals vary to some extent and the original tremor frequency is reinstated after about 4 beats.

Author's address: Dr. G. RENOU, Laboratoire de Physiologie, C.H.U. St-Antoine, 27, rue Chaligny, *F-75 Paris 12e* (France)

New Developments in Electromyography and Clinical Neurophysiology,
edited by J. E. Desmedt, vol. 3, pp. 641–648 (Karger, Basel 1973)

Blink Reflexes in Orbicularis Oculi[1]

B. T. SHAHANI and R. R. YOUNG

Department of Neurology, Harvard Medical School, and Massachusetts General
Hospital, Boston, Mass.

A blink is defined as (a) transient relaxation with disappearance of tonic
activity in muscles such as levator palpebrae, which are responsible for keeping the lids open, accompanied by (b) phasic activity in muscles such as orbicularis oculi, associated with the eye closure [COGAN, 1969]. *Reflex* blinks are
usually evoked by stimuli such as a mechanical tap or electrical current delivered to the skin in the periorbital region. Relatively little attention has been
paid to the reflex behavior of muscles other than orbicularis oculi in which the
EMG reflex blink presents two successive components.

One has a latency of 20–45 msec and is correlated with the clinically observed blink of the lids. It tends normally to habituate quickly (to decrease in
amplitude and duration, eventually perhaps to disappear) with repetitive
stimulation, as does the clinical blink. It appears bilaterally in response to a
unilateral or midline stimulus, represents prolonged and asynchronous firing
of motor units and can be evoked by a wide variety of stimuli including pin-
prick, a puff of air or corneal stimulation. Because of its appearance, adequate
stimuli and tendency to habituate it has always been considered a 'nociceptive'
or cutaneous reflex. Our demonstration [SHAHANI and YOUNG, 1972], that it
is the electrical counterpart of the movements which close the eye, supports
the notion that it serves to protect the globe. It is, clinically speaking, *the* blink
reflex and is clearly 'exteroceptive' or 'cutaneous' in nature. We have termed
it the second component of the orbicularis oculi reflex because it has a longer
latency than the one to be described next.

The other or first component has a latency of 10–15 msec, appears ipsilaterally following a unilateral stimulus, is brief and relatively synchronous
and habituates much more slowly than the second. The significance of this

1   This work was supported by a grant from Mrs. W. B. LLOYD.

brief burst of activity in terms of the protective function of the reflex as a whole is unexplained. More attention has been directed toward defining the adequate stimulus responsible for it [SHAHANI and YOUNG, 1972]. The afferent input necessary for both travels via the trigeminal nerve [KUGELBERG, 1952; TOKUNAGA, OKA, MURAO, YOKOI, OKUMURA, MYASHITA and YOSHITATSU, 1958]. Whereas the stimuli which evoke the second component are multiple and cutaneous in nature, the first component normally is produced only by a mechanical tap or electrical shock to the periorbital region.

Because of the latency and shape of the first component and the fact that it is most readily recorded from muscles beneath the skin which is tapped, KUGELBERG [1952] suggested it might well be proprioceptive or myotatic in nature, although he had no convincing evidence for that assumption. This hypothesis was subsequently assumed by others to have been proven [GAN-DIGLIO and FRA, 1967; RUSHWORTH, 1962], despite the fact that facial muscle spindles had not been found on careful examination in humans [KADANOFF, 1956; COOPER, 1966]; neither have typical spindles been demonstrable in the facial muscles of animals [BOWDEN and MAHRAN, 1956; BRUESCH, 1944].

Because of the similarity of the two reflex components in orbicularis oculi to those in tibialis anterior during cutaneous evoked flexor reflexes [SHAHANI and YOUNG, 1969, 1971], we proposed that the adequate stimuli for the first component in orbicularis oculi were also cutaneous or exteroceptive in nature. This hypothesis was supported by our findings [SHAHANI and YOUNG, 1968] that the first component can be produced by an electrical stimu-lation which does not produce vibration, stretching or deformation of any mechanoreceptors. This was also shown to be true when the elctrical pulses were delivered to discrete areas of facial skin beneath which neither facial muscles nor the putative facial muscle afferents could be found [SHAHANI and YOUNG, 1968]. These observations were subsequently confirmed by PENDERS and DELWAIDE [1969] in man and LINDQUIST and MÅRTENSSON in the cat [1970]. Other physiological data to support the contention that the first com-ponent is cutaneous rather than proprioceptive in nature have appeared [SHA-HANI, 1968, 1970; SHAHANI and YOUNG, 1972; PENDERS and DELWAIDE, this volume]. This contribution reviews some of the newer evidence which further substantiates the exteroceptive nature of the first component.

Methods

Details of the techniques can be obtained elsewhere [SHAHANI and YOUNG, 1972a, b]. In brief, reflexes were recorded with surface electrodes on the skin over orbicularis oculi in

Fig.1. Surface EMG activity recorded from orbicularis oculi in one normal subject; *A* is response to a tap on the glabella, *B* to an electrical stimulus to the glabella, *C* to electrical stimulus to the dorsum of the nose and *D* to an electrical stimulus to the ear. Calibrations are 30 msec and 200 µV.

man, with fine needles in that muscle in the macaque monkey and occasionally *vice versa*. Stimuli were painless small-aplitude square waves of less than 0.5-msec duration delivered through disc electrodes, or mechanical taps administered using a hand-held phonograph cartridge. When sound was used as a reflex stimulus, it was produced either by the delivery of a square pulse to earphones or by striking a metal plate held near the ear with the phonograph transducer. The oscilloscope sweep was triggered either by the stimulator or by the pulse from the transducer produced by the blow. The human data were recorded from more than 50 subjects, all normal with the exceptions noted. Three adult monkeys were studied under light phencyclidine anesthesia but without local anesthesia [SHAHANI and YOUNG, 1972b]. In the silent period studies, the nerve to the appropriate muscle was stimulated supramaximally during voluntary contraction of orbicularis oculi or masseter.

Results

Figure 1 A illustrates the two components in orbicularis oculi following a tap on the glabella. In figure 1 B the same two components are seen following an electrical stimulus to the same area. Note that the first component is even more synchronous and its latency a few msec shorter with electrical stimulation. The same two components are present in figure 1C when the stimulating electrodes are arranged sagittally over the dorsum of the nose from the tip to a point 2 cm proximally. The first component can even be recorded following stimulation to the skin over the auricle; although it is then smaller and less regular in appearance (fig. 1D). As noted before, similar reflexes with both components can be evoked by stimuli delivered to skin throughout the area supplied by the first division of the ipsilateral trigeminal nerve and even more widely under some circumstances. The ways in which these two components behave during habituation and dishabituation, sleep and so forth have been described elsewhere [RUSHWORTH, 1962; SHAHANI, 1968; SHAHANI and YOUNG, 1972]. The significant point here is that a first component follows electrical stimulation over many areas of the face.

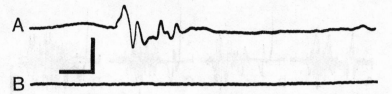

Fig. 2. EMG activity recorded from needles within the orbicularis oculi muscle of a monkey. *A* Response to a tap on a flap of skin elevated from the glabella. *B* No response to a similar tap on the structures deep to the flap. Calibrations are 5 msec and 200 μV.

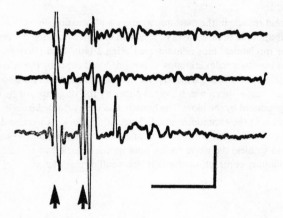

Fig. 3. Surface EMG activity recorded from orbicularis oculi of a patient with Tay-Sachs disease. The traces are sequential responses, reading from bottom up, to a loud noise. The two discrete reflex components may be seen above the arrows. Calibrations are 50 msec and 100 μV.

Blink reflexes following stimulation of an isolated flap of skin over the forehead were studied in monkeys. A tap delivered to a flap of skin with nerve and blood supply intact but otherwise isolated from the glabella produces a first component (fig. 2A) whereas the same tap delivered to the exposed subcutaneous tissue, muscle and periosteum fails to do so (fig. 2B). Electrical stimuli delivered to the skin also elicit a reflex, while stimuli administered directly to the exposed orbicularis oculi fail to do so.

In normal subjects a brief loud sound elicits a prolonged, asynchronous easily habituated blink reflex identical to the usual second component, except its latency is longer (48–63 msec). An early first component is not seen following auditory stimulation except when the subject is unusually prone to startle.

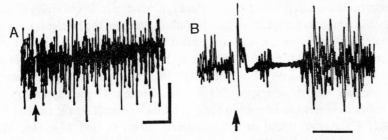

Fig.4. Surface EMG activity recorded from orbicularis oculi in a normal subject during voluntary closure of the lids. A supramaximal shock to the ipsilateral facial nerve was delivered at the arrow. Calibrations are 100 msec and 300 μV. Note the absence of a silent period. *B* EMG activity during voluntary contraction of masseter recorded from a coaxial needle electrode within the muscle. A supramaximal shock was delivered to the motor point of masseter at the arrow. Calibration is 50 msec. Notice the characteristic silent period.

Figure 3 illustrates blink reflexes recorded from a child with Tay-Sachs disease following auditory stimuli. Note the presence in successive traces of two separable components (arrows), the second beginning at 54–62 msec and the first at 30–32 msec. The usual electrical or mechanical periorbital stimuli in this patient evoked reflexes with two components at the normal latencies.

When one attempts to record a 'silent period' [MERTON, 1951; SHAHANI and YOUNG, this volume] from voluntarily contracting facial muscles by stimulating the facial nerve supramaximally (fig. 4A) it proves very difficult to do so. In contrast, a silent period is recorded from masseter following per-cutaneous stimulation at its motor point (fig. 4B). One does *not* find a period of EMG silence after stimulation of either facial or supraorbital nerve, whereas such silent periods are easily produced in skeletal muscle (including masseter) by stimulation of mixed or sensory nerves [SHAHANI and YOUNG, this volume, page 589].

Discussion

A first component can be evoked in orbicularis oculi by stimulation of the skin over a wide area of the face. These stimuli, whether mechanical or electri-cal, presumably give rise to impulses in a number of the afferent fibers running in the subjacent dermis and perhaps even deeper. Mechanical stimuli would be effective by deformation of mechanoreceptors, the production of generator potentials and eventual generation of conducted action potentials, whereas electrical stimulation would give rise to a synchronous afferent volley directly, albeit in a more heterogeneous group of fibres. Perhaps the briefer latency and greater synchrony of the first component following electrical stimulation is thereby explained. If muscle receptors such as spindles were present in facial muscles it should be possible to stimulate their afferent fibers percutaneously as one can with those responsible for the H-reflex. Though typical muscle spindles have not been found in facial muscles of man [GANDIGLIO and FRA, 1967; KADANOFF, 1956] or other animals [BOWDEN and MAHRAN, 1956; BRUESCH, 1944] the possibility of their existence is difficult to deny absolutely. They have been found in masseter, tongue and extraocular muscles [COOPER, 1953; COOPER and DANIEL, 1949]. As the stimulating electrodes are moved further from orbicularis oculi the likelihood of stimulating such muscle afferents becomes increasingly remote. When they are over various areas of the forehead or glabella, afferents from frontalis might be stimulated though if present they would be least likely at the midline or glabella. With them over the dorsum of the nose, afferents from procerus or compressor naris muscles might be stimulated though the nerves to these muscles run superiorly from the bridge of the nose in the first instance and inferolaterally in the second [GRAY, 1954], directions which would lead them away from the dorsum of the nose. With the electrodes over the ear, afferents from the auricularis muscles might be stimulated. Using these techniques, the probability of stimulating putative muscle afferents, activity in all of which should give rise to the same first component, becomes increasingly unlikely, but cannot be categorically denied. Other evidence, such as the fact that in man or monkey the same stimulus may evoke either a first component or a second component, or both depending on the level of alertness, habituation and so on [SHAHANI and YOUNG, 1972b; SHAHANI, 1968, 1970], also makes it difficult to imagine that the stimuli evoke activity in different receptors when a first component is produced as opposed to other times.

The most convincing evidence has been derived from stimulation of the monkey skin flap [SHAHANI and YOUNG, 1972b]. This activity does produce a normal first component whereas direct stimulation of subcutaneous tissue, muscle and periosteum does not, regardless of whether the stimuli are mechanical or electrical. In the monkey we found the first component easier to produce than the second. Perhaps this is because of the anesthesia used or because of species differences. Further studies need to be carried out, perhaps in man on flaps of skin during plastic surgical procedures. Meantime, the adequate stimulus for the first component of the monkey blink reflex is cutaneous in nature.

In addition to cutaneous ones, other sorts of 'exteroceptive' stimuli such as noise are effective though they tend to produce only a second component under normal circumstances. As PENDERS [1969] has described, we found the latency of what appears to be the second component longer with auditory stimuli, perhaps because of increased time spent on the afferent side of the reflex arc. When the two components are seen with auditory stimulation in the Tay-Sachs patient, they both arise at a longer latency than with skin stimuli. Whether or not these two components in pathological circumstances associated with hyperactive startle are analogous to the normal two components remains to be proven. If they are, the latency of the first is prolonged by about 20 msec, while that of the second is about 30 msec longer than normal. Nevertheless, under these abnormal circumstances, an earlier, more synchronous, less habituating reflex component can be produced in the orbicularis oculi muscle by an auditory stimulus which has no possibility of stimulating spindles or other proprioceptive stuctures.

Our inability to record a typical silent period from facial muscles confirms the observations of PENDERS [1969]. To be sure, the situation in the face with most of the afferents at least in one nerve (trigeminal), and all the efferents in another is different from other muscles and this is discussed further in another chapter [SHAHANI and YOUNG, this volume]. Nevertheless, PENDERS [1969] found that stimulation of neither trigeminal nor facial nerves produced a silent period in facial muscles and we would agree. He also found that a normal first component could be evoked so as to fall within 1 msec of a maximal antidromic volley in facial nerve [PENDERS and DELWAIDE, this volume]. There is no refractory period for this reflex component and PENDERS suggests this may reflect an absence of Renshaw recurrent inhibition in the facial nucleus. Whatever the significance of the absence of a silent period in terms of its usual evocation by proprioceptive (spindle, tendon organ) and cutaneous inputs, the physiology of facial muscles is clearly different from that of limb muscles or of masseter, a cranial muscle that does contain spindles. The situation vis-a-vis the nature of the first component has been clearly defined in the monkey: it is cutaneous just as is the second component. This is very likely true in man as well and the burden of proof now rests upon anyone who suggests that the first component is proprioceptive.

Summary

Reflex activity in orbicularis oculi in man and monkey following a mechanical or electrical stimulus to the periorbital regions consists of two discrete components. The second is clearly a cutaneous reflex and serves to close the eyelids. The first was considered to be a proprioceptive reflex though critical evaluation reveals its exteroceptive nature. It can be produced by electrical stimulation which cannot stimulate proprioceptive receptors directly. It can be produced by mechanical stimulation of skin alone, at least in the monkey. Furthermore, auditory evoked blink reflexes may also have two components in the presence of hyperactive startle reflexes such as in the patient with Tay-Sachs disease reported here. The physiological organization of facial reflexes is shown to be different from that of skeletal muscle reflexes as illustrated by studies of the 'silent period' following electrical stimulation of both types of muscle. Both components of the blink reflex should be considered cutaneous or exteroceptive in nature.

Authors' address: Dr. B. T. SHAHANI and Dr. R. R. YOUNG, Department of Neurology, Massachusetts General Hospital, Fruit Street, *Boston, MA 02114* (USA)

New Developments in Electromyography and Clinical Neurophysiology,
edited by J. E. Desmedt, vol. 3, pp. 649–657 (Karger, Basel 1973)

Physiologic Approach to the Human Blink Reflex

C. A. PENDERS and P. J. DELWAIDE

Department of Internal Medicine and Pathology, University of Liège, Liège

Electrical stimulation of the infraorbital [KUGELBERG, 1952] or supra-orbital [RUSHWORTH, 1962] nerve elicits a blink reflex which on EMG analysis proves to consist of two distinct components (fig. 1 a): an early unilateral component (R_1) and a late bilateral component (R_2). Since the initial studies of KUGELBERG [1952], R_1 has generally been considered a monosynaptic proprioceptive reflex while R_2 is believed to be a polysynaptic reflex of cutaneous origin.

Studies on the physiology of this interesting reflex have, however, been rare, and the nature of R_1 remains in dispute. RUSHWORTH [1962, 1968], STRUPPLER and DOBBLESTEIN [1963] and GANDIGLIO and FRA [1967] accept the hypothesis of KUGELBERG, whereas TOKUNAGA, OKA, MURAO, YOKOI, OKUMURA, HIRATA, MIGASHITA and YOSHITATSU [1958] and subsequently SHAHANI [1968, 1970], SHAHANI and YOUNG [1968] and PENDERS and DEL-WAIDE [1969 a, b] have proposed that this early component is also of cutaneous origin. Other issues are pending, for example, the organization of the moto-neuron pool of orbicularis oculi which appears different from the better known motor nuclei of the limbs.

Methods

48 healthy adult subjects were tested. They sat comfortably in an armchair with a headrest. They were asked to relax and keep their eyes half-closed.

Electrical stimulation was delivered via silver disc surface electrodes; the cathode was positioned near the point of emergence of the supraorbital nerve, with the anode approximately 1 cm above it. Painless stimuli lasting 0.7–1 msec were provided by a Grass S 8 stimulator feeding isolation transformers and constant current units.

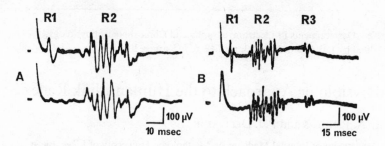

Fig. 1. Reflex responses to supraorbital electrical stimulation. *A* Usual response. *B* Third component occasionally seen in certain subjects. Above, ipsilateral recordings; below, contralateral recordings.

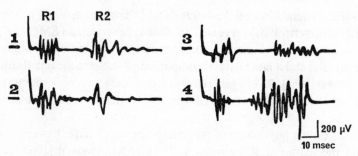

Fig. 2. Examples of polyphasic R_1 responses obtained in a single subject. The variations in morphology of R_1 are clearly visible.

The response also was recorded with silver surface electrodes, one of which was fixed to the lower eyelid and the other to the root of the nose. The signals were amplified by a Tektronix 565 oscilloscope and photographed with an Alvar Cathomatic Camera.

Results

Electrical stimulation of the supraorbital nerve readily elicited the ipsilateral reflex with two components described by KUGELBERG (fig. 1a). Occasionally, particularly when the amplitude of R_2 was large, a third component was observed around 75–90 msec (fig. 1b). This component, which has not been described in the literature, occurred in certain young subjects. Like R_2, it could be produced symmetrically in both orbicularis oculi muscles and elicited by stimulation anywhere on the face.

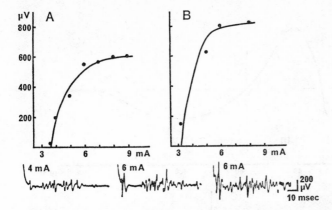

Fig. 3. R₁ recruitment curves. *A* standard conditions. *B* slight voluntary contraction of orbicularis oculi. The two left tracings were recorded under standard conditions; that on the right was recorded during slight voluntary contraction.

A. Features of the R₁ Response

The morphology of the early component was generally simple, being either diphasic or triphasic. Not infrequently, however, R₁ was frankly polyphasic, in which case it varied morphologically in the course of a single experiment (fig. 2).

The latency of R₁ was only relatively stable, since fluctuations sometimes exceeded 3 msec. In the same subject, the measurement of 100 responses yielded a variability of 8.2%, whereas under the same conditions variability is only 2% for the H-reflex.

The minimum stimulation intensity required to elicit the two reflex responses varied from one subject to another, although it ordinarily fell between 3 and 8 mA. The threshold was roughly comparable for R₁ and R₂.

The amplitude of the R₁ response increased with the intensity of stimulation (fig. 3 a). The maximum amplitude obtained was 10–30% of that of the maximal direct response elicited by stimulation of the facial nerve.

B. Analysis of the R₁ Response

Slight voluntary contraction of m. orbicularis oculi increased the amplitude of R₁, which rose to 40–60% of the maximal response obtained by stimulating the facial nerve (fig. 3 b).

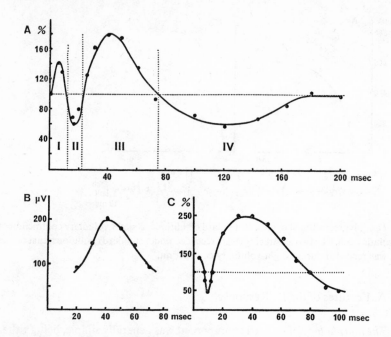

Fig. 4. R₁ recovery curves. *A* Both the conditioning and test shock were delivered to the supraorbital region at the same intensity (i.e., an intensity yielding 50% of the maximum R₁ response). *B* Conditioning and test shock delivered to the same region at the response threshold. *C* Conditioning shock delivered to cheek. Test shock as in *A*.

The Jendrassik maneuver had no effect on R₁ amplitude regardless of whether the reflex was elicited by electrical stimulation or percussion of the orbital region. The application of vibration (80–100 c/sec) to the forehead caused no marked reduction of R₁ amplitude, in contrast to its well-known ability to decrease the H-reflex. Vibration also failed to evoke in orbicularis oculi a tonic response comparable to the tonic vibration reflex described for soleus by EKLUND and HAGBARTH [1965].

The R₁ recovery curve obtained by delivering two stimuli at various intervals to the supraorbital region proved to be consistent in different subjects (fig. 4a) and it differs from the recovery cycle of the H-reflex [PENDERS and DELWAIDE, 1969a]; it can be subdivided into 4 phases:

1. Until the 10th msec there is an increase in the test response, becoming maximal around 5 msec. This explains why a short salvo at 200 or 300 c/sec produces an R₁ response of greater amplitude than a single shock.

2. Next, there is a brief phase of moderate depression.

3. This is followed by a potentiation phase with a maximum around 40 msec. The potentiation is independent of the size of R_2, as it shows no change when the conditioning shock is adjusted to the subjective threshold but subliminal for any response (fig. 4b). Similarly, potentiation persists even when the R_2 response is abolished by stimulation at a rate of 1 c/sec.

4. Lastly, there is a phase of variable depression that never exceeds 200 msec.

Comparable recovery curves are obtained by applying the conditioning stimulus to the territory of the maxillary nerve (V_2) or the mandibular nerve (V_3) or by contralateral stimulation (fig. 4c).

Thus, stimulation anywhere on the face can modulate the R_1 response. SHAHANI [1970] had already shown that this reflex can be elicited by stimulating any part of the forehead on the ipsilateral side. Using facilitatory maneuvers (slight voluntary contraction or paired stimuli 40 msec apart), we were sometimes even able to obtain R_1 after stimulation of the cheek or chin. Conversely, in 6 subjects the same facilitatory maneuvers produced an early response in the contralateral orbicularis oculi and even in orbicularis oris or in mentalis.

C. Analysis of the R_2 Response

The polyphasic R_2 response occurred bilaterally upon stimulation of the V_1, V_2 or V_3 territories. Its latency (25–40 msec) increased when weak stimulation was used or when a vibrator was applied to the face.

With a conditioning shock delivered to the face (V_1, V_2 or V_3), the recovery curve showed complete inhibition until 150–200 msec. When such a stimulus was applied to the upper limbs (e. g., to the median nerve), after a 50-msec latency period there was a marked depression in R_2 lasting approximately 100 msec. However, unlike GANDIGLIO and FRA [1967] we could not elicit R_2 by delivering a conditioning stimulus to the limbs even with a painful stimulus.

One important feature of R_2 which it shares with other polysynaptic reflexes, is that it is extinguished after repeated stimulation. This habituation phenomenon can be altered under certain psychological conditions: stress and calculation increase the amplitude of R_2 and the minimal frequency of stimulation to produce habituation. A similar effect is seen in neurosis. Old age, extrapyramidal syndrome (idiopathic or iatrogenic parkinsonism) also produce more difficult habituation [MESSINA, 1970; PENDERS and DELWAIDE, 1972].

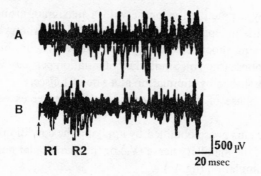

Fig.5. Absence of a true silent period in orbicularis oculi. Each tracing is obtained from the superposition of 3 CRT sweeps. *A* Sustained voluntary contraction. *B* Same as *A* with the addition of supraorbital electrical stimulation.

D. Organization of the Orbicularis Oculi Motoneuron Pool

Two important features of this motoneuron pool deserve mention: the absence of a true silent period and of Renshaw effects.

The absence of a true silent period (fig. 5) is tested during a sustained voluntary contraction, by eliciting a synchronous response from the muscle either by reflexogenic stimulation or by stimulation of the motor pathway. This phenomenon has been confirmed in numerous subjects; at the most, certain individuals show a slight reduction of voluntary activity immediately after the R_2 burst. It has been equally impossible to detect a silent period in other muscles of facial expression after stimulation of the facial nerve.

The absence of Renshaw recurrent inhibition can be tested easily since the afferent and efferent pathways in the blink reflex are anatomically distinct from one another and can thus be stimulated separately. Using the reflex to test motoneuron excitability after delivering a supramaximal volley to the facial nerve, we observed that the R_1 response is transmitted 100% 5–6 msec after the direct response (fig. 6). This means that antidromic stimulation of the motoneuron pool produces an extremely brief refractory period. This is incompatible with the duration (60–80 msec) of the Renshaw recurrent inhibition in spinal α-motoneurons [BROOKS and WILSON, 1959]. Conversely the excitability of the spinal γ-motoneuron (devoid of recurrent inhibition) becomes normal after 6 msec [HUNT and PAINTAL, 1958].

The presence of the 100-percent R_1 response 5 msec after the antidromic stimulus seems to indicate a lack of Renshaw cells in the blink reflex path-

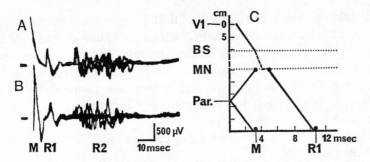

Fig.6. Absence of Renshaw activity in motoneuron pool of orbicularis oculi. *Left:* tracings obtained from the superposition of 3 CRT sweeps. *A* Supraorbital stimulation (control). *B* Same stimulation 1 msec after supramaximal stimulation of facial nerve at the parotid level. *C* Conduction diagram. Stimulation at parotid level (Par.) produces orthodromic volleys which yield motor response (M) and simultaneously, an antidromic volley which depolarizes the motoneuron (MN) and possible Renshaw cells. One msec later, supraorbital stimulation (V_1) was applied. Distances were estimated according to the data of KUGELBERG [1952]; the conduction speed determined is in agreement with that of KUGELBERG and of SHAHANI. The motoneuron is able to transmit the reflexogenic volley several msec after receiving antidromic stimulation.

way; cutaneous inhibition of such cells is a polysynaptic mechanism of 10–20 msec latency [WILSON, 1965] that cannot be considered in our experiments.

Discussion

The cutaneous origin of the early component of the blink reflex now appears to be demonstrated. The variability in morphology and particularly in latency displayed by R_1 is not characteristic of a monosynaptic proprioceptive reflex. According to BROWN and RUSHWORTH [this volume], the fluctuation of latency in some single motor units is no greater than that for the H-reflex. These authors, however, do not exclude the possibility that other units of the early component are evoked by afferents of cutaneous origin. The fact that R_1 and R_2 have comparable thresholds indicates that both responses depend on afferent fibers of comparable caliber and that, in any case, R_1 cannot be attributed to the excitation of large-caliber fibers such as the I_A fibers. Moreover, SHAHANI [1970] has stressed that it is possible to evoke R_1 by stimulation anywhere in the supraorbital region, and we have pointed out that stimulation even in another area of the face will sometimes elicit R_1. In addition, it has been demonstrated by SHAHANI [1970] that the early and late responses behave similarly during judicious procaine infiltration of the supraorbital region. Lastly, TOKUNAGA et al. [1958] have reported that the R_1 response is not abolished by destruction of the mesencephalic nucleus of the trigeminal nerve whereas it does disappear upon destruction of the principal sensory nucleus of the trigeminal nerve.

Besides this feature, the blink reflex presents other distinctive physiological characteristics.

The facts that the amplitude of R_1 can be modified by delivering a conditioning shock anywhere to the face and, conversely, that an R_1 can be elicited in other facial muscles by supraorbital stimulation, clearly demonstrate that close interconnections exist throughout the cutaneous region of the face. This area behaves physiologically as a homogeneous entity and fails to exhibit the traditional clear subdivisions described by anatomists for the three branches of the trigeminal nerve. However, the most distinctive organizational feature of the orbicularis oculi motoneuron pool is its lack of a γ system and of Renshaw effects.

Phase 3 of the R_1 recovery curve is characterized by marked potentiation. During this same interval the H-reflex, in contrast, shows a phase of complete inhibition related primarily to a pause in muscle spindle afferent activity [PAILLARD, 1955]. This pause, according to MERTON [1951], is the principal mechanism of the silent period, a phenomenon which does not exist in orbicularis oculi.

The inability of the Jendrassik maneuver or of vibration to affect the R_1 response provides further evidence that this motoneuron pool is devoid of a γ system. Moreover, histologists have not seen muscle spindles in the muscles of facial expression, and retrograde degeneration techniques have failed to demonstrate the presence of proprioceptive fibers at this level [HOSOKAWA, 1961].

Since the Renshaw cells are known to be partially responsible for the silent period [ECCLES, FATT and KOKETSU, 1954], it was interesting to investigate whether Renshaw activity could be demonstrated in orbicularis oculi where no such phenomenon exists. As described above, a supramaximal volley delivered to the motor pathway induced only a very brief refractoriness of the motoneuron pool, whereas postsynaptic inhibition of the Renshaw type is known to last 60–80 msec [BROOKS and WILSON, 1959]. The absence of Renshaw activity at this level thus seems to be clear. Let us recall that in the cat Renshaw effects have already been reported to be missing in the α-motoneurons of the intercostal muscles [EKLUND, VON EULER and RUTHOWSKI, 1964] and in the motoneuron pool of the hypoglossal nerve [PORTER, 1965].

Thus, the blink reflex consists of two reflexes of cutaneous origin which appear in response to a single stimulus. The R_1 component is distinctive for its short latency and for its resistance to habituation. An oligosynaptic reflex arc would account for these distinctive features.

The blink reflex, moreover, offers a good approach to the physiology of the muscles of facial expression in man. The present study demonstrates the unusual organization of these muscles and suggests that findings obtained in the limbs cannot be extrapolated without direct evidence to the muscles of facial expression.

Acknowledgements

Dr. PENDERS is Aspirant, and Dr. DELWAIDE is Chercheur qualifié of the Fonds National de la Recherche Scientifique (Belgium).

Author's address: Dr. C. PENDERS, Institut de Médecine, Hôpital de Bavière, *B-4000 Liège* (Belgium)

New Developments in Electromyography and Clinical Neurophysiology,
edited by J. E. Desmedt, vol. 3, pp. 658–659 (Karger, Basel 1973)

Some Comments on Blink Reflexes

J. Moldaver

Neuromuscular Laboratory, Hospital for Special Surgery, Cornell Medical Center,
New York, N.Y.

I wish to add some information which might be useful in clarifying the
mechanism of the 'blink' reflex. The reflex is now described as an exteroceptive
skin reflex of the orbicularis oculi [Young and Shahani, Penders and Delwaide, this volume]. I think that this muscle can also be the effector of a deep
or stretch reflex which can be elicited easily as follows. While retracting the
skin in the lateral corner of the eyelids towards the temporal region, a gentle
percussion on the pulling finger results in a contraction in the ipsilateral orbicularis oculi. This stretch reflex has a low threshold. It is depressed in lower
motoneurone lesions, as one would expect, but it is markedly increased in
upper motoneurone lesions.

When recording the EMG of the extra-ocular muscles in man [Breinin
and Moldaver, 1955] the levator palpebrae was also evaluated. It occurred
to me later on that it was difficult to understand why a very rapid muscle such
as the levator palpebrae should be the antagonist muscle of the orbicularis
oculi, which is a relatively slow muscle [Moldaver, 1964]. Actually, some
features of the orbicularis oculi appear unique among the skeletal muscles.
For instance, it has a medial and a peripheral component capable of having
2 different functions. Both the superior and inferior orbicularis oculi have 2
kinds of muscle fibers: the peripheral or orbital portion is reddish and thicker
and represents a slow muscle, while the central or palpebral component is
thinner and paler and is a fast muscle.

In the mechanism of blinking, the palpebral component, i.e. the thin,
rapid, white muscle, contracts first and the blinking is accompanied by a relaxation of the levator palpebrae. This follows the general law of reciprocal
innervation of antagonistic muscles. With forced closure of the eyelids, the
contraction of the orbital or red portion acts as a sphincter muscle. The 2

portions, fast and slow, can be activated either independently or simultaneously.

By using a small electrode on the skin and a current of threshold intensity, it is very simple to stimulate either of the 2 different motor points or the 2 independent small nerve branches and to observe the contractions of the 2 distinct portions of that muscle. I am wondering whether this little-known distinction explains some features of the blink reflexes, and namely the presence of several components.

It has often been stated that muscle spindles have not been found in human facial muscles and there is only a rare atypical spindle-like structure in some animals. I am questioning the fact as to whether enough specific attention has been paid to the palpebral or fast white muscle component for identification of spindle-like structures.

While the exteroceptive mechanism of the blink has been well documented in chapters of this volume, it may be wise to keep in mind the possibility of a proprioceptive component as well, which has so far not been critically excluded. The evidence that the orbicularis muscle includes 2 portions with different physiological features suggests that further studies are required to elucidate the functional organization of blink reflexes in man.

Author's address: Dr. J. MOLDAVER, 140 East 54th Street, *New York, NY 10022* (USA)

New Developments in Electromyography and Clinical Neurophysiology,
edited by J.E. Desmedt, vol. 3, pp. 660–665 (Karger, Basel 1973)

Reflex Latency Fluctuations in Human Single Motor Units

W. F. Brown and G. Rushworth

Department of Clinical Neurological Sciences, St. Joseph's Hospital, London, Ontario, and Unit of Clinical Neurophysiology, Department of Neurology, The Churchill Hospital, Oxford

It has long been known that a light tap to the supra-orbital region of the human face evoked a reflex closure of the eye on the same side, even in the blind [Overend, 1896]. Many facial reflexes were later described, that followed local stimulation at various sites in the facial region.

In 1945, Wartenberg described these reflexes as myotatic reflexes for he found that brief stretch of a facial fold was the most effective stimulus for evoking them. Kugelberg [1952] recorded the reflex responses from the orbicularis oculi muscle electromyographically, and found that the response to glabella stimulation consisted of two components; firstly, an early, synchronous reflex, that was relatively unaffected by repeated stimulation, both in its amplitude and its latency; secondly, at longer and variable latency, a very asynchronous response, that, in the normal subject, rapidly habituated on repeated stimulation.

Considering the features of the first component, and comparing its latency and pathways with the jaw jerk, Kugelberg suggested that it was a proprioceptive reflex, while the second component had features of a nociceptive reflex of cutaneous origin.

Rushworth [1962] found that procainization of the skin of the glabella did not affect the first component of the reflex. This component was also found to be selectively exaggerated on the side of a hemiplegia, as are the tendon reflexes, and in early and very late tetanus. For these further reasons it was suggested [Rushworth, 1966] that the first component reflex had its origin in facial muscle spindles.

It has been questioned whether these organs existed in muscles supplied by certain cranial nerves, especially the extra-ocular muscles, the intrinsic muscles of the tongue, the larynx and the face, but deductions from studies in non-primate species has often been misleading so far as man is concerned.

Typical muscle spindles have been found in the extra-ocular muscles of man and higher primates [Cooper and Daniel, 1949], the intrinsic muscles of the human and simian tongue [Cooper, 1953], human laryngeal muscles [Rudolph, 1961; Rossi and Cortesina, 1965] and in the mimetic musculature of the human face [Kadanoff, 1956]. All these muscles are non-load-bearing, working under isotonic conditions, and the function of these proprioceptive organs is, therefore, likely to be different from those within load-bearing muscles.

Unlike some animals, including the rhesus monkey, the close attachment of human facial muscle fibres and facial skin makes it very difficult to stimulate them separately. Many physiological studies of the facial reflexes in man have been carried out by evoking the reflexes electrically, using superficial stimulating electrodes of the DISA type (i.e., bipolar 5 mm salinesoaked discs, 25 mm apart). Whether such gross stimulation fires end-organs within skin or muscle, or both, is debatable.

Recently Bynke [1971] has stimulated the trigeminal nerve in patients undergoing neurosurgical operations on the posterior fossa, and has recorded the reflex response in the facial nerve. The central time of this reflex was only 1.5 msec and is, thus, almost certainly monosynaptic.

Trontelj [1968] found that the H-reflex in single motor units has a small variation of latency, ranging from 400–2,400 μsec in the alert human subject. A motoneurone bombarded monosynaptically just above its threshold is likely to respond with little latency variation, whereas a polysynaptic reflex will have much greater variations of latency.

We have studied, by means of single unit recordings, the reflex to supra-orbital nerve stimulation, and have compared the latency variation with that of the H-reflex, that is presumed to be monosynaptic, and with direct facial nerve stimulation.

Methods

Four normal subjects have been studied for the H-reflex and 10 normal subjects for the facial reflexes. Six patients have also been used as subjects. One patient had a meningioma in the middle fossa, which was explored surgically, and the opportunity was taken to stimulate the trigeminal sensory ganglion and sensory roots at the time of operation.

Four patients had Bell's palsy, and the normal side of the face was used for this study. One patient had a spastic quadriparesis, without involvement of the face.

In the case of the H-reflexes, stimulation of the medial popliteal nerve was performed through surface stimulating electrodes that were strapped in place as well as supported by hand. In the case of the facial reflexes, stimulation of the supra-orbital nerve was performed either monopolarly through a small needle which was inserted through the skin of the orbit to overlie the supra-orbital foramen, or bipolarly using small silver discs stuck to the skin over the orbit. The monopolar needle electrode was insulated with teflon down to its tip and the indifferent electrode was strapped to the skin of the neck.

Recordings have been taken from either soleus muscle or the lateral orbital part of the orbicularis oculi by means of bipolar co-axial needle electrodes, and low-threshold, single motor units have been identified by their all-or-nothing behaviour. Minimal electrical stimulation has been employed, using isolated square waves of 0.1 msec or less.

For the facial reflexes, the subject lay in a darkened room with the eyelids resting closed. A continuous record of orbicularis activity was taken simultaneously with the reflex responses, and records showing background motor unit activity were rejected.

The reflexes in single motor units have been photographed on expanded sweeps. Latency variations have been studied in two ways: firstly, as histograms, for which at least 100 reflexes have been collected; secondly, 50 trials of the reflex response have been photographed at greatly expanded sweeps, so that the latency variation was directly displayed and recorded.

Results

The H-reflex in single motor units was found to have a latency fluctuation of 400–2,400 μsec, with a mean of 950 μsec, in the series reported by TRONTELJ [1968]. We have three satisfactory single units whose latency fluctuations fell within the same range, one of these is expressed as a histogram (fig. 1 A). The total range of latency variation in this unit was 1,800 μsec.

It is noteworthy that the histogram demonstrates the fluctuations of latency about one point in time, and that the majority of responses occurred between 38.7 and 38.9 msec. The threshold motor unit illustrated here has a longer latency than the very earliest units which are evoked with larger stimuli to the medial popliteal nerve.

When single units were studied in the orbicularis oculi, it was found that liminal stimulation of the facial nerve produced direct responses with very little latency variation (fig. 2 A). The range of fluctuation of latency in 6 units was only 135–350 μsec in alert subjects.

Low intensity stimulation of the supra-orbital nerve produces an ipsilateral early reflex in the lateral orbital part of orbicularis oculi and, under favourable conditions, single motor unit responses can be isolated and studied. In 35 single units the latency range was 400–2,500 μsec (table I). A histogram of the latency variation in an alert subject is shown in figure 1 B. As with the

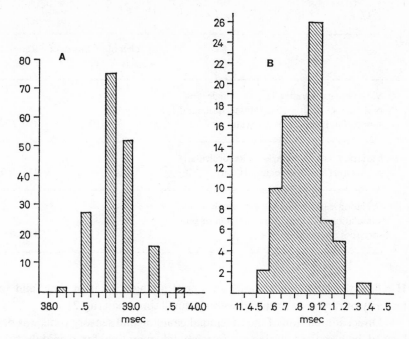

Fig. 1. A Latencies of a single soleus motor unit in the H-reflex. *B* Latencies of a single orbicularis oculi motor unit in the reflex evoked by electrical stimulation of the supra-orbital nerve. Alert subjects. The ordinate is number of trials. Note the normal distribution curve around one mean point.

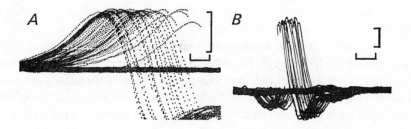

Fig. 2. A A single motor unit in orbicularis oculi. Direct stimulation of the facial nerve with very brief, just supraliminal electric pulses. Delay, 4.5 msec. 50 trials. Calibrations, 200 µV and 50 µsec. *B* Another single motor unit in orbicularis oculi. The reflex response was elicited by brief electric pulses, of near-threshold intensity, delivered to the supra-orbital nerve. Alert subject. 50 trials. Delay, 16.6 msec. Calibrations, 200 µV and 1 msec.

Table I.

	Number of subjects	Number of units	Range in μsec
Variation early reflex latency, single units orbicularis oculi, stimulation ipsilateral supraorbital nerve, alert subjects	15	35	400–2,500
Variation latency, single units orbicularis oculi, direct facial nerve stimulation	6	6	135–350
Variation early reflex latency, single units, stimulation of Vth sensory ganglion or sensory root at operation	1	3	150–400

H-reflex, the latency variations are about a single point in time, and the majority of responses occurred within a small time-span (fig. 2 B).

Direct stimulation of the trigeminal ganglion and sensory roots was performed in a patient during an intracranial operation for a middle fossa meningioma. The trigeminal nerve was well caudal to the pathological process. The patient was anaesthetized with nitrous oxide and halothane, and single units were isolated from the orbicularis oris. Three satisfactory single motor units were studied and their reflex latency to stimulation of the trigeminal ganglion and sensory roots had a fluctuation of only 150–400 μsec.

Discussion

The present results demonstrate that the reflex to supra-orbital nerve stimulation in alert subjects has a fluctuation of latency which is no greater than that for the H-reflex and, in an anaesthetized subject, the latency variation is extremely small. Furthermore, the histograms show a distribution of latencies which is unifocal. These findings suggest that the reflex is indeed monosynaptic, as in BYNKE's independent observations.

Muscle-spindles are the only sense organs known to connect monosynaptically with motoneurones and we would infer, therefore, that our results provide evidence that some muscle spindle afferents are involved in the early reflex component of the blink reflex to supra-orbital nerve stimulation;

of course, this does not exclude the possibility that other afferents also contribute to this reflex.

Recently, the plantar cushion reflex in the anaesthetized cat has been shown to be an oligosynaptic reflex [EGGER and WALL, 1971]. Their evidence is consistent with the reflex utilizing a trisynaptic, 4-neurone circuit, though it was possible to interpret the very earliest response as a disynaptic one. The reflex appears to originate in cutaneous pressure receptors.

We have found no latency fluctuations of single unit reflexes that would be consistent with this type of reflex in the human face in alert subjects, and the histograms in particular show only a single peak. Nevertheless, the possibility remains that the first component of the human blink reflex is itself a compound response, some units being monosynaptically activated by presumptive I_A fibres, while other units are evoked by afferents of cutaneous origin.

Authors' address: Dr. W.F. BROWN, Department of clinical Neurological Sciences, St. Joseph Hospital, *London, Ontario, 439–2231* (Canada), and Dr. GEOFFREY RUSHWORTH, Department of Neurology, Churchill Hospital, *Headington, Oxford OX3 7 LF* (England)

New Developments in Electromyography and Clinical Neurophysiology,
edited by J. E. Desmedt, vol. 3, pp. 666–672 (Karger, Basel 1973)

The Blink Reflex Induced by Photic Stimuli

Parameters, Thresholds, and Reflex Times[1]

H. C. Hopf, J. Bier, B. Breuer and W. Scheerer

Department of Neurology, University Hospital, Göttingen

The blink reflex in response to stimulation of the visual system comprises several components. The purpose of our study was to determine adequate stimulating parameters as well as the blink reflex thresholds and latencies under varying conditions.

Methods

The experimental arrangement is illustrated in figure 1. Iluminated circles of variable diameter, brightness, and exposure time were projected on a matt screen 25 cm in front of the subject's eye. The diameter of the stimulating area corresponded to visual angles of 2° 20', 7°, 14° and 23°, respectively. The exposure time was varied between 2 and 64 msec in 6 single steps (2, 4, 8, 16, 32, 64 msec).

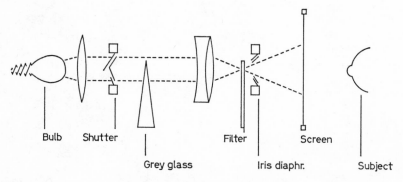

Bulb Shutter Grey glass Filter Screen Iris diaphr. Subject

Fig. 1. Schematic drawing of the experimental equipement.

1   The investigations were supported by the Deutsche Forschungsgemeinschaft (SFB 33).

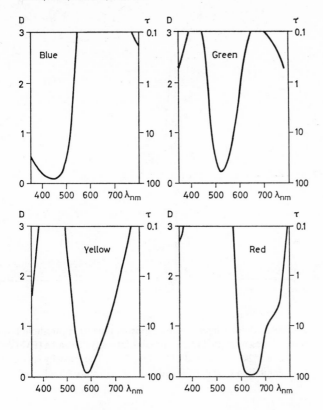

Fig. 2. Transmission maxima (t) and band width of the 4 filters.

White light and coloured light were tested. The white light was produced by a special bulb with a fairly flat energy output between 400 and 700 nm. When testing coloured stimuli a filter was interposed in the beam of the white light. The maximal transmission of the 4 filters used was at 655 nm (red), 580 nm (yellow), 522 nm (green) and 463 nm (blue) (fig. 2).

A glass wedge was used to adjust the stimulus intensity without changing the spectral composition of the light. The intensity is given in stimulation units. One stimulation unit corresponds to 3.97 asb. Before starting the experiment the subject was kept in a dark room for at least 15 min in order to attain a fair measure of dark-adaptation. The blink reaction of the eye lids was recorded electromyographically.

Results

The threshold of the blink reflex was determined by gradually increasing the stimulus intensity. The threshold intensity was taken as that level of illumi-

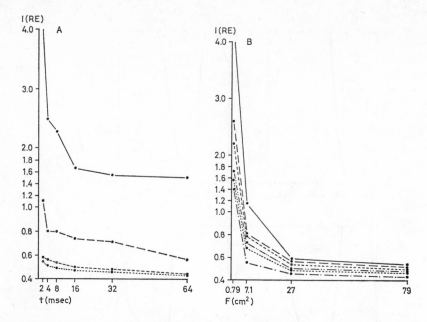

Fig. 3. A Relation between stimulus intensity and exposure time (t) at threshold level. *B* Diameter of the illuminated area as related to the stimulus intensity. The area (F) in cm² corresponds to a visual angle of 2°20', 7°, 14° and 23°, respectively. The intensity is given in stimulation units (RE). 1 stimulation unit corresponds to 3.97 apostilb. Number of subjects = 20.

nation at which 5 successive stimulus presentations gave at least p 3ositive responses. The threshold intensity also depends upon the exposure time and the stimulus area. In figure 3A stimulus intensity is plotted against exposure time. The relation between stimulus intensity and stimulus area is demonstrated in figure 3B. The lowest threshold of 1.75 asb (i.e., 0.44 stimulation units) was found with the largest stimulus area and the longest exposure time.

The threshold intensities to coloured stimuli differ markedly from each other. The highest values were observed with red light and the lowest values with yellow light. However, the yellow threshold intensity was even lower than that determined with white light (fig. 4).

The latency of the reflex as measured at threshold intensity and using the 20° stimulation area was 108.4 msec with the 2 msec exposure time and 122.0 msec with the 64-msec stimulus duration. The latencies showed considerable variation (75–245 msec). At higher stimulus intensities the latency decreased to a mean of 85.2 msec and 82.0 msec, respectively (range 75–95 msec) (fig. 5A).

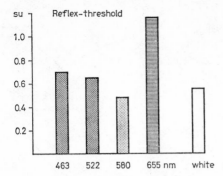

Fig.4. Threshold intensities in response to coloured stimuli. The wave length in nm corresponds to the maximal transmission of the filters. N = 12. su = stimulation units, defined in figure 3.

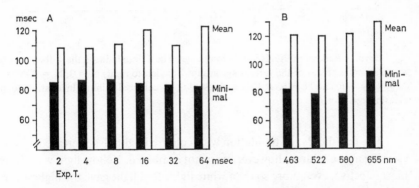

Fig.5. A Reflex latencies in response to white light using variable exposure times (EXP.T.). N = 18. *B* Reflex latencies following coloured stimuli. N = 5. Mean values are reproduced by the white columns and the minimal latencies by the black columns.

When coloured stimuli were tested, the reflex latencies were shortest with the yellow and the green lights (79 msec). Longest latency responses were observed with red light (95 msec) (fig. 5B).

The age factor plays an important role in regard to the reflex threshold and to the reflex latency. In 2- to 4-year-old children the threshold to white light was high (16.8 stimulation units). The lowest threshold values were found in 20- to 30-year-old adults (0.41 stimulation units). With further increase of age the thresholds again increased (10.1 stimulation units at an age of over 60 years) (fig. 6A).

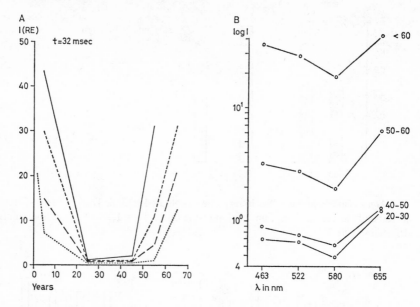

Fig. 6. A Reflex thresholds in relation to age. The 4 lines indicate the 4 different sizes of the stimulus area. For the various age groups from left to right: $N = 3$, $N = 5$, $N = 6$, $N = 7$, $N = 6$, and $N = 6$. *B* Reflex thresholds to coloured stimuli in relation to increasing age. $N = 6$.

The thresholds to stimulation with coloured light also increased with increasing age. Children, however, were not examined. Yellow light was always most effective, even more so than white light. Red light gave the highest thresholds (fig. 6B).

The course of the reflex latencies in relation to age was similar to that of the reflex thresholds. In young children (2–4 years) long latencies of 129.7 msec were observed ($N = 3$). In young adults (20–30 years) the reflex response appeared after 81.8 msec ($N = 6$). In subjects over 60 years the latency increased to 100.7 msec ($N = 6$). At threshold intensity the values were 181.0, 113.1 and 148.7 msec, respectively (fig. 7).

Conclusions

The blink reflex induced by photic stimuli is mediated by both the photopic and the scotopic systems. This can be derived from the exponential decrement of the stimulus intensity with increase of the exposure time. This curve

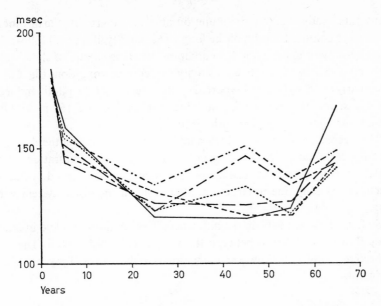

Fig. 7. Reflex latencies in relation to age. The values correspond to stimulus intensities of threshold level. The 6 different lines indicate the 6 different exposure times. For further details see figure 6A.

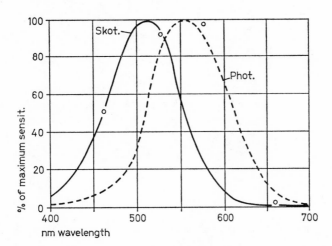

Fig. 8. The relative thresholds of the blink reaction as determined with coloured stimuli (open circles) and the curve of relative sensitivity of the photopic and the scotopic systems.

flattens out already at 32-msec stimulus duration. Furthermore, the threshold intensities to coloured stimuli of the long wave band (yellow, red) follow the relative sensitivity under photopic conditions whereas stimuli of short wave length (blue) follow the relative sensitivity for scotopic conditions (fig. 8).

As early as 1898, GARTEN reported on the latencies of the photic-induced blink reflex. In 9 subjects he found values between 70 and 161 msec. Our findings are in good agreement with his figures.

The decrease in latency and threshold of the reflex during the development from childhood to adolescence points to the factor of maturation of the reflex conducting pathways. The increase in latency and threshold in elderly subjects is probably due to structural alterations in the visual system with ageing.

So far, we cannot explain the differences in reflex threshold and in reflex latency that were observed between the various coloured stimuli. The low threshold to yellow light, however, corresponds to the results of psychological experiments.

Author's address: Prof. H. C. HOPF, Neurologische Universitätsklinik, v. Siebold-Strasse 5, *D-34 Göttingen* (FRG)

New Developments in Electromyography and Clinical Neurophysiology, edited by J. E. Desmedt, vol. 3, pp. 673–677 (Karger, Basel 1973)

Habituation of the Blink Reflex

Role of Selective Attention[1]

M. Gregorič

Institute of Clinical Neurophysiology, The University Hospitals of Ljubljana, Ljubljana

Habituation is a wide-spread phenomenon and has been defined as a gradual quantitative diminution of response to repeated uniform stimuli [Harris, 1943; Griffin and Pearson, 1968]. Habituation of the blink reflex can be observed electromyographically as diminution of the size particularly of its second component, R_2, which is generally interpreted as a polysynaptic defence reflex to noxious stimuli [Kugelberg, 1952; Rushworth, 1962; Shahani, 1970]. The speed and extent of habituation depends on the physical parameters of stimulation, particularly on the strength of stimuli and the rate of their presentation [Thompson and Spencer, 1966].

This study attempts to determine the basic characteristics in normal man of the course of habituation of the second component of the blink reflex, at a given set of physical parameters of stimulation, as well as the variability of habituation from one subject to another, and also in the same subject when controllable experimental conditions are kept constant.

Quantitative changes in the polysynaptic EMG component, R_2, of repetitively elicited blink reflex were studied in 15 normal subjects. Slightly painful electrical stimuli were applied to the skin over the right eye in series of 100 at a rate of 1/sec. Series of stimuli were repeated after rest periods of 5 min. The EMG responses were detected by a pair of surface electrodes placed on the lower margin of the right orbit. After suitable amplification they were rectified and integrated by a hybrid unit connected to an averaging computer. The results were displayed and printed out in the form of sequential histograms of magnitudes of individual EMG responses.

1   This work was supported by Boris Kidrič Fund of Slovenia (grant No. 306/20: 71) and by the US Department of Health, Education and Welfare, Social Rehabilitation Service (grant No. 19-P-58397-F-01).

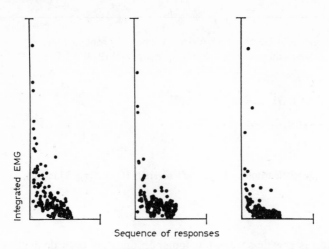

Fig. 1. Sequential histograms of magnitudes of the R₂ EMG component in the blink reflex. Three successive series separated by resting periods of 5 min. Notice increasingly rapid decrement though the recovery of the initial responses is almost complete.

The course of habituation of R_2 was usually in the form of an exponential curve. A large initial decrease was followed by fluctuation with a tendency towards slower further diminution of the size of responses. Interindividual differences in the course of habituation were large and instead of an exponential curve, irregular fluctuation, sometimes with certain periodicity, was frequently observed. When repeating the series of stimuli in the same subject, a phenomenon called potentiation of habituation was observed in most cases. In successive series the course of habituation was more and more rapid, even though the resting periods between the series resulted in a more or less complete recovery of the initial size of the response (fig. 1). However, in some subjects the variation of responses within the series and from one series to another was so large as to completely override the effects of habituation and potentiation of habituation. The variation observed in these cases was as large as the previously mentioned interindividual variation. We presumed that the variation of responses and the course of habituation is not only determined by the general reactivity of the nervous system in its fluctuation between sleep and vigilance, between relaxed wakefulness and alert attentiveness, but also depends to a considerable extent on the more brief and transient changes in mental state.

We were particularly interested to see whether and how habituation is influenced by the changes in selective attention when the level of non-specific attention or vigilance remains unchanged. In order to investigate this problem, several experimental situations were introduced which served to focus subjects' attention to the stimuli or to distract it from the stimuli. In one of these, the stimuli were preceded by a weak tone, and the subject was told to press a key as soon after the tone as he could. In another series he was asked to ignore the tone and to press the key after the stimulus. The sequence of two series was alternated. In most of the cases habituation was less pronounced and variation of the responses larger when subjects were pressing the key on tone, i.e. to an irrelevant stimulus (fig. 2 A). During another series of stimuli the subject was solving a simple arithmetic problem. The histogram of the responses always showed markedly slowed or even completely absent habituation. Consequently, the mean values of the responses were rather high in this series; in half the subjects they exceeded the control values by more than 100% (fig. 2 B). A third way used to distract subjects' attention was to present a short, not very exciting motion picture. In this series, too, habituation was less pronounced than in the control series or in the series with key-pressing on stimuli, although it was not as markedly reduced as in the series with calculation (fig. 2 C).

To summarize these findings, in all three experimental situations distraction of attention resulted in slower and less effective habituation, and in more pronounced variation of the responses. In contrast to this, habituation was more rapid when subject's attention was focused to the stimuli.

Changes in the level of vigilance do not seem to have played any significant role in the experimental situations of this study. Perhaps the level of vigilance was slightly increased during calculation.

During distraction from the stimuli, the blink reflex [LARSSON, 1960] and the abdominal skin reflex [HAGBARTH and KUGELBERG, 1958], have shown smaller responses and we have found in most of our own series that the first responses (not influenced by habituation) were also smaller. However, when stimulation in this series was continued at constant intervals there was even sometimes an increase instead of the usual decrease in size of the responses (fig. 2). Cerebral evoked potential studies in man have shown that the habituation decrement is related to the selective inattention associated with distraction [GARCÍA-AUSTT, BOGACZ and VANZULLI, 1964]. This is in contrast to our findings quoted above which suggest that, for the blink reflex, habituation would be learning not to respond to the stimulus.

The blink reflex is regarded as the most sensitive motor part of the orientation or startle reaction [LANDIS and HUNT, 1939; LARSSON, 1967; GOGAN, 1970]. Less effective habituation during distraction could be ex-

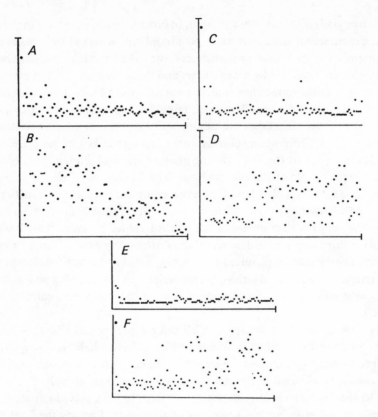

Fig. 2. Course of habituation of blink R_2 component in experimental situations involving distraction from the stimuli eliciting the blink reflex. Abscissa: rank of sequential blink responses elicited at 1/sec. Ordinate: size of each R_2 response (integrated EMG). The first response of each habituation sequence is indicated by a larger dot. *A* and *B* The subject presses a key either to the stimuli eliciting the blink reflex (*A*, mean R_2 amplitude 9.8 ± 4.7 arbitrary units) or to a tone (*B*, mean size 23.8 ± 12.9). *C* and *D* The subject pays attention to the stimuli (*C*, mean size of R_2 9.0 ± 5.9) or attempts to solve an arithemetic problem (*D*, mean size of R_2 21.3 ± 10.9). *E* and *F*. The subject pays attention to the stimuli (E, mean size of R_2 5.3 ± 2.8) or looks at a motion picture (*F*, mean size 15.3 ± 10.2).

plained by a reduction of inhibitory influences from higher levels. Habituation is slow or absent when the cerebral cortex is destroyed or by-passed [GLASER and GRIFFIN, 1962; GRIFFIN and PEARSON, 1968]. Recently we found in 4 patients with dementia due to arteriosclerosis that habituation of the R_2 blink component occurred more slowly than in normal subjects. Habituation may be regarded as a purposeful reaction in the sense that it suppresses a reflex response to a slight stimulus which has proved innocuous [KUGELBERG, 1952]. In contrast to distraction of attention, a decrease of vigilance potentiates habituation, and may result in the activation of other inhibitory mechanisms, possibly originating from the reticular formation itself.

If habituation of the blink reflex is compared to habituation of the flexion reflex [DIMITRIJEVIĆ, FAGANEL, GREGORIČ, NATHAN and TRONTELJ, 1972] it becomes obvious that the former is much more changeable and probably subjected to a more intricate control by higher levels. Habituation of the flexion reflex may be an essentially segmental process [DIMITRIJEVIĆ and NATHAN, 1970; THOMPSON and SPENCER, 1966] and the effect of distraction could in this case be explained as dishabituation due to a superimposed facilitation as the inattentive subject might be repetitively startled by stimulus presentations.

Author's address: Dr. M. GREGORIČ, Klinični center, Holzapflova ulica, *61000 Ljubljana 5* (Yugoslavia)

New Developments in Electromyography and Clinical Neurophysiology,
edited by J. E. Desmedt, vol. 3, pp. 678–681 (Karger, Basel 1973)

The Blink Reflex in Bell's Palsy

E. SCHENCK and F. MANZ

Departments of Psychiatry and Neurology, University of Freiburg, Freiburg

In patients with Bell's palsy the blink reflex, with its R_1 and R_2 components, may be abolished or reduced, and its latency is sometimes increased due to the pathological process in the facial nerve [OKA, TOKUNAGA, MURAO, YOKOI, OKUMURA, HIRATA, MIYASHITA and YOSHITATSU, 1958; RUSHWORTH. 1962; GANDIGLIO and FRA, 1967; BENDER, MAYNARD and HASTINGS, 1969], In 3 of the 4 cases reported by KIMURA, POWERS and VAN ALLEN [1969] the R_1 blink component returned between the 8th and 13th days after onset of the palsy, the latency being extended by a maximum of 6 msec; the 4 cases recovered from the paralysis with satisfactory results.

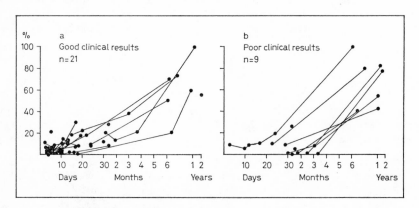

Fig. 1. Changes in amplitude of the R_2 component of the blink reflex after onset of Bell's palsy. The amplitudes are expressed as percentage of amplitude on the normal side in the same patient. Abscissa: time after onset of paralysis. *A* Patients with good clinical recovery. *B* Patients with poor clinical recovery. Data for the same patient are joined by a straight line.

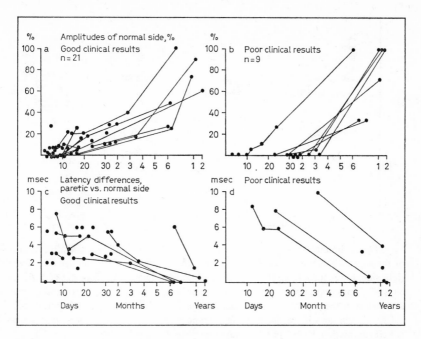

Fig. 2. Changes in the R_1 component of the blink reflex after onset of Bell's palsy. *A* and *B* Changes in amplitude of R_1 expressed in percentage of the R_1 amplitude on the normal side of the same patient. Data for the same patient are joined by a straight line. Abscissa: time after onset of the paralysis. *A* Patients with good clinical recovery. *B* Patients with poor clinical recovery. *C* and *D* Changes in the latency difference of R_1 on the paretic side with respect to the normal side after onset of the paralysis. Abscissa: time after onset. *C* Patients with good clinical recovery. *D* Patients with poor clinical recovery.

We have performed a similar investigation in Bell's palsy by determining the amplitudes of both blink reflex components and the latency of the R_1 early component. So far, our sample includes 30 patients with a definite clinical outcome. 21 patients show a favorable status, whereas 9 show incomplete recovery with a significant residual paresis and/or strong associated movements of the face. Nine patients were examined only once, 7 twice, 7 three times each, and the remaining 7 patients 4 times or more.

EMG recordings were made with skin electrodes from both m. orbicularis oculi simultaneously. Electrical stimuli of 0.2 msec pulse duration were applied consecutively and separately to the supraorbitalis nerve on both sides. Stimulation was irregular in order to avoid habituation of R_2. A mini-

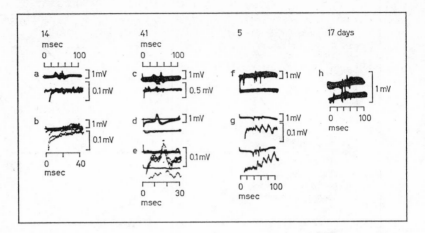

Fig. 3. Simultaneous EMG recordings from both m. orbicularis oculi after stimulation of the supraorbitalis nerve on one side. In each pair of recordings, upper trace from the right side, lower trace from the left side. *A–E* Patient D.S., studied 14 days (*A*, *B*) and 41 days (*C–E*) after onset of paralysis. *F–H* Patient K.B., studied 5 days (*F*, *G*) and 17 days (*H*) after onset of the paralysis. *A* Left stimulation, very small R_2 on the left (paralyzed) side. *B* Left stimulation with recording on faster time base to show absence of R_1 on the left side. *C* Left stimulation, both R_1 and R_2 have reappeared on the left side. R_1 on the left is shown on a fast time base in *E*. This can be compared with R_1 elicited on the right side by right stimulation in *D*. *F* Right stimulation eliciting normal R_1 and R_2 on the right (normal) side. *G* Two pairs of tracings under identical conditions with left stimulation. R_2 is present on the left side, but with small amplitude and increased latency. *H* Left stimulation, reappearance of the two components on the left side.

mum of 20 stimuli was given on each side. Amplitudes were estimated from the strongest reflexes only.

Figure 1 shows the amplitudes of the R_2 component as percentages of those on the unaffected side (100%), as a function of time after the onset of paralysis. In most cases of the good recovery group, R_2 remains detectable during the first 6–10 days, contrary to the less favorable cases. However, reflex amplitudes are reduced to about 5%. This amplitude recovers relatively fast, frequently before the 12th day. In only 1 of the 9 unsatisfactory cases did R_2 behave similarly. This patient had only a slight residual paresis but disturbing associated movements of the face.

Contrary to R_2, the R_1 components are often entirely absent during the first 14 days, even in the group with a good prognosis (fig. 2 *A*, *B*). In one case it could not be elicited even after 18 days. A favorable prognosis is indicated when the R_1 can be demonstrated during the first few days or when it reappears during this period. However, this is rarely observed. In only one case of the bad recovery group did a small R_1 reappear after only 12 days.

The increase in latency of the R_1 in comparison to the unaffected side is shown in figure 2 C, D. In the group with favorable prognosis the latency difference does not exceed 6 msec. Even in this group the latency difference disappeared only slowly, a result which is in agreement with the findings of KIMURA, POWERS and VAN ALLEN [1969]. Latency differences exceeding 8 msec were found in the group with a bad prognosis.

Figure 3 shows original recordings of 2 typical cases with good clinical results, to illustrate the dissociation between R_2 and R_1 in the first few days.

Summary

The prognosis of Bell's palsy is favorable if the blink reflex R_1 component is at all recognizable during the first 10 days or if it reappears during this time. This, however, is rarely observed. Most patients in whom the R_2 component can be elicited during this period also have a positive prognosis. The prognosis is not good when both R_1 and R_2 remain absent during the first 2 or 3 weeks. The greatest increase in latency of R_1, i.e. to 8 msec or more, was found in the unsatisfactory cases.

Author's address: Prof. Dr. E. SCHENCK, Psychiatrische und Nervenklinik der Universität, und Priv.-Doz. Dr. F. MANZ, Neurologische Universitätsklinik mit Abteilung für Neurophysiologie, *D-7800 Freiburg i. Br.* (FRG)

New Developments in Electromyography and Clinical Neurophysiology,
edited by J. E. Desmedt, vol. 3, pp. 682–691 (Karger, Basel 1973)

The Blink Reflex as a Test for Brain-Stem and Higher Central Nervous System Function

J. KIMURA

Department of Internal Medicine (Neurology), Faculty of Medicine, University of
Manitoba, and Department of EEG, Winnipeg General Hospital, Winnipeg, Manitoba

A unilateral electrical stimulation of the supraorbital nerve evokes the
early R₁ reflex (latency: 10.6 ± 2.5 msec) of m. orbicularis oculi on the
side ipsilateral to the shock and the *late R₂ reflex* on both sides [KUGEL-
BERG, 1952; TOKUNAGA, OKA, MURAO, OKUMURA, HIRATA, MIYASHITA and
YOSHITATSU, 1958; RUSHWORTH, 1962; GANDIGLIO and FRA, 1967;
BENDER, MAYNARD and HASTINGS, 1969; KIMURA, POWERS and VAN
ALLEN, 1969; SHAHANI, 1970]. In analogy to the pupillary light reflex,
the ipsilateral component of R_2 will be referred to as the *direct R_2* (latency:
31 ± 10 msec) and the contralateral component, as the *consensual R_2*
(latency: 32 ± 11 msec). Thus, stimulation on the right evokes a *right* R_1,
a *right* direct R_2 and a *left* consensual R_2. (The consensual R_2 is named
after the side of recording and not after the side of stimulation.) In addition
to the normal range of latencies (mean value ± 3 SD) as mentioned above
[KIMURA, POWERS and VAN ALLEN, 1969], the normal range of variation
between the two sides of one individual has also been established. For R_1,
a latency difference between the right and left sides exceeding 1.2 msec is
considered abnormal. For R_2, two types of comparison are possible.
Firstly, a latency difference between a direct and a consensual R_2 simultaneously evoked by unilateral stimulation should not exceed 5 msec.
Secondly, a latency difference between bilateral R_2 evoked by right-sided
stimulation and corresponding reflexes subsequently evoked by left-sided
stimulation should be less than 8 msec [KIMURA, RODNITZKY and VAN
ALLEN, 1970].

On the basis of a delayed R_1, it would be impossible to determine
which part of the reflex arc is involved. Analysis of R_2 is useful in distinguishing an afferent delay from an efferent delay. For example, a bi-

lateral (right direct and left consensual) delay of R_2 to right-sided stimulation implicates a lesion somewhere along the afferent (initial common) path on the right (right trigeminal nerve, right side of the pons, or right spinal tract and nucleus of the trigeminal nerve in the lateral medulla). On the other hand, a unilateral (right direct and right consensual) delay of R_2 on the right regardless of the side of stimulation implicates a lesion somewhere along the efferent (final common) path on the right (right facial nerve or right side of the pons at this level).

Studying patients with multiple sclerosis [KIMURA, 1970] we noted, early in 1968, that alteration of the blink reflex might be due to lesions in the brain stem and not necessarily due to lesions of the facial and/or trigeminal nerves [KIMURA, 1971a; LYON and VAN ALLEN, 1972a]. Since then a total of 170 patients with various brain-stem lesions (48 vascular, 13 neoplastic, 13 anoxic or traumatic, 13 with extrinsic posterior fossa tumors, and 83 with multiple sclerosis including 32 patients with no clinical signs of brain-stem involvement) and 26 patients with degenerative disorders (18 with amyotrophic lateral sclerosis, 8 with spinocerebellar degeneration including Friedreich's ataxia, and 10 with Parkinson's disease) have been studied at the University of Iowa and the University of Manitoba.

The blink reflex obtained from the 170 patients was analyzed according to the clinical localization and etiology of the brain-stem lesions. R_1 (fig. 1) was abnormal bilaterally or unilaterally on the side of the lesion in most patients with pontine (35 of 39 cases) and multilevel brain-stem lesions (36 of 41 cases) and in 7 of 15 patients with mesencephalic lesions. It was also abnormal in most patients with cerebellar or cerebello-pontine angle tumors (7 of 8 cases). On the other hand, R_1 was normal in most patients with medullary lesions (26 of 30 cases). Of the 32 patients with multiple sclerosis who showed no clinical evidence of brain-stem involvement, R_1 was unilaterally delayed in 11 and normal in the others. [A similar finding is reported by NAMEROW in this volume.]

R_2 was abnormal in a majority of patients with pontine (24 of 39 cases) and multilevel brain-stem lesions (28 of 41 cases). It was also altered in 7 of 8 patients with lateral medullary lesions involving the spinal tract and nucleus of the trigeminal nerve, all of whom showed an afferent delay of R_2 in the presence of a normal R_1. R_2 was normal in most patients with medial medullary (13 of 22 cases) and mesencephalic lesions (14 of 15 cases).

These findings thus indicate that a delay of R_1 is relatively specific to pontine involvement due to intrinsic lesions [KIMURA, 1970, 1971b;

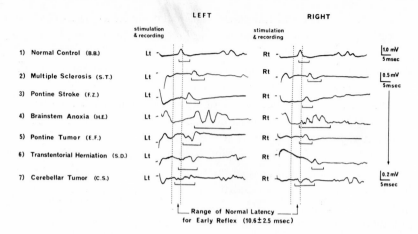

Fig.1. The early R_1 reflex (horizontal brackets) in a normal control (case 1) and in 6 patients (cases 2–7) with pontine involvement due to various intrinsic or extrinsic lesions. Note a bilateral delay in cases 2 and 3, and a unilateral delay on the left in cases 4 and 5, and on the right in case 6. Note also abnormally multiphasic wave forms of the early reflex bilaterally in cases 4 and 7, and unilaterally on the left in cases 5 and 6. The range of normal latency of the early reflex is enclosed by dotted lines [KIMURA, POWERS and VAN ALLEN, 1969].

NAMEROW and ETEMADI, 1970; LYON and VAN ALLEN, 1972b] or extrinsic lesions [KIMURA and LYON, 1972b]. While R_2 may also be altered by pontine lesions, it is more consistently affected by lateral medullary lesions [KIMURA and LYON, 1972a].

Of the 41 patients with multilevel brain-stem lesions, 8 were in unrousable coma and 5 had the syndrome of akinetic mutism when the reflex study was performed. In all the comatose patients and in 4 of 5 patients with akinetic mutism, R_2 was either totally absent or, at best, minimal in amplitude regardless of the side of stimulation even when a shock of very high intensity was used. A local block of the reflex pathway in the brain stem does not seem to account for this finding since R_1, that is more consistently affected by brain-stem lesions, was normal or nearly normal at least on one side in 6 of these patients [KIMURA, 1971b]. Besides, R_2 reappeared if the patients recovered from coma (fig. 2). A reversible functional suppression of R_1 was also seen (fig. 2), but less commonly. It is thus likely that the absence of R_2 in coma is due to diffuse suppression or inactivity of the multisynaptic reticular system regardless of the etiology of the coma [KIMURA, 1971b].

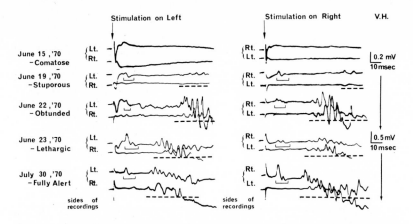

Fig. 2. The blink reflex in a patient recovering from coma (herpes simplex encephalitis). The stimulus was delivered to the supraorbital nerve first on the right and then on the left. The intensity of the shock was advanced to a level above which the evoked potential remained constant. The reflex responses were simultaneously recorded from ipsilateral (upper tracing in each pair) and contralateral (lower tracing in each pair) orbicularis oculi. Note a unilateral early R_1 reflex (horizontal brackets) recorded only in the upper tracing and bilateral late R_2 reflexes (broken lines) recorded in both the upper and lower tracings in each pair. Both R_1 and R_2 were absent on June 9 (not shown) and on June 15 when the patient was comatose. R_1 was normal but R_2 was markedly delayed and diminished on June 19 when she was responsive but stuporous. Note the progressive recovery in amplitude and latency of R_2 contemporaneous with her improvement to full alertness on July 30.

This view is consistent with a more recent study at the University of Iowa [LYON, KIMURA and MCCORMICK, 1972] with 48 other comatose patients (21 supratentorial mass lesions, 4 brain-stem lesions, 21 metabolic and 2 anoxic encephalopathy). R_2 was absent or minimal in amplitude in all 48 patients while R_1 was normal in coma in all the cases with supratentorial mass lesions with the exception of 2 patients who had advanced transtentorial herniation. Alteration of R_1 indicated primary or secondary structural changes in the pons, although reversible functional or pharmacologic block of pontine conduction was also documented.

Total absence of R_2 is not specific to comatose states as its absence may be noted, although less consistently, in various other conditions. During our investigation (fig. 3) R_2 was absent while R_1 remained normal or nearly normal, in 2 alert but immobile patients with features of the 'locked in' syndrome, in an alert and ambulatory patient with pseudobulbar palsy, and in 5 alert patients given therapeutic dosages of diazepam

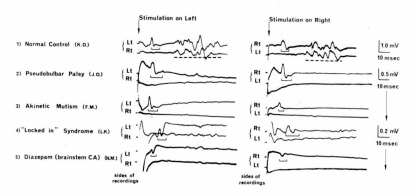

Fig.3. Absence of the late R_2 reflex in various neurological disorders. The stimulus was delivered to the supraorbital nerve first on the right and then on the left. When R_2 was absent with an ordinary shock of 10–20 mA and 0.1-msec duration, the intensity was slowly advanced up to 40 mA and 0.5-msec duration. The reflex responses were simultaneously recorded from ipsilateral (upper tracing in each pair) and contralateral (lower tracing in each pair) orbicularis oculi. Note a unilateral early R_1 reflex (horizontal brackets) recorded only in the upper tracing and a bilateral R_2 (broken lines) recorded in both the upper and lower tracings in case 1 (normal control). Note also virtual absence of R_2 regardless of the side of stimulation in cases 2–5. The early reflex was normal in cases 2, 3 and 5 and significantly delayed bilaterally in case 4.

(Valium) that presumably blocked the multisynaptic reflex arc already depressed by brain-stem lesions. In a case of the Wallenberg syndrome, also, R_2 was completely blocked when the side of the lesion was stimulated; it was normal when the uninvolved side was stimulated.

Effects of sleep on the blink reflex [SHAHANI, 1968] were studied in 8 normal subjects [KIMURA and HARADA, 1972]. In response to shocks of such intensity *(So)* that reflex responses remained constant in wakefulness, R_1 was obtained in 38, 16, 0, 5 and 41 % of the trials in stages I, II, III, IV and REM, respectively (fig. 4). The corresponding percentages for the late reflex were 81, 32, 17, 29 and 63 %. Statistical analysis of these and of the generally much higher percentages obtained by shocks of *S10* *(So* plus 10 mA) or *Sp* (paired *So* separated by 5 msec) intensity reveals that excitability of both R_1 and R_2 is higher in stage REM ($p < 0.01$) than in stages II–IV (with no significant difference between stages I and REM and among stages II, III and IV). This is in agreement with the observations of FERRARI and MESSINA [1972]. Furthermore, some of the R_2 obtained during REM sleep (fig. 4) were normal in latency (35 % of the responses obtained during this stage) and/or very prolonged in duration (16 %). These features

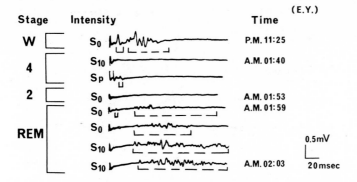

Fig.4. Right blink reflex evoked by shocks of *So* (above which responses remained constant in wakefulness) *S 10* (*So* plus 10 mA) or *Sp* (paired *So*) delivered to the right supra-orbital nerve. The top tracing was obtained with the subject fully awake immediately prior to the sleep session. The others were consecutive recordings showing marked suppression of both the early R_1 (small horizontal brackets) and late R_2 (brackets with broken lines) reflexes in stages IV and II and release from the suppression in stage REM. Note a relatively normal latency and a very prolonged duration of R_2 reflex in stage REM. Well developed R_1 (not shown) also appeared during stage REM with shocks of *Sp* intensity.

also appeared in stage I but not in stages II–IV ($p < 0.01$) when R_2, if present, was usually delayed or very delayed in latency and normal, short or slightly prolonged in duration.

EEG changes (high-voltage slow waves, spindles or alpha rhythm) in response to shocks, on the other hand, appeared only in 38% of the shocks delivered during stage REM. This is significantly less ($p < 0.01$) when compared to the corresponding percentages in stages I, II or III (74, 90 and 84%, respectively) and at about the same level with stage IV (45%).

It is noteworthy that if shocks of sufficiently high intensity are delivered, both R_1 and R_2 are frequently elicitable even in stages II–IV when their depression appears most profound, while in coma R_2 is totally absent regardless of stimulus intensity. It is interesting to note that neither the oligosynaptic nor the multisynaptic component of the blink reflex in man is influenced by sleep in the same way as the H-reflex in man or multisynaptic spinal reflexes in cats which become more depressed during desynchronized sleep than during synchronized sleep [GASSEL, MARCHIAFAVA and POMPEIANO, 1964; HODES and DEMENT, 1964; MORRISON and POMPEIANO, 1965].

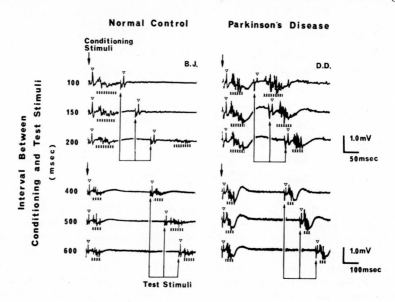

Fig. 5. Blink reflex evoked in the right orbicularis oculi by paired stimuli to the right supraorbital nerve. Arrows point to conditioning and test stimuli and broken lines underline the late R_2 reflexes of the conditioning and the test responses. Conditioning and test early R_1 reflexes (indicated by small triangles over the peaks) are seen between stimulus artefacts and corresponding R_2 in each tracing. Note inhibition of the test R_2 (in the presence of relatively constant test R_1) at the time intervals of 100 and 150 msec (period of complete depression) in a normal control and the absence of this inhibition in a parkinsonian patient.

As for the 26 patients with degenerative disorders, R_1 was normal in all but 2 cases of amyotrophic lateral sclerosis which exhibited borderline values. The average latency of R_1 was 11.1 msec for amyotrophic lateral sclerosis, 11.5 msec for spinocerebellar degeneration and 11.0 msec for Parkinson's disease. R_2 was slightly delayed in 2 cases of amyotrophic lateral sclerosis and in 2 cases of spinocerebellar degeneration, but was normal in the remaining cases. The average latency of R_2 in each category was 35, 37 and 29 msec, respectively.

In the 10 cases of Parkinson's disease (unilateral study in 5, bilateral in 5), and in 12 age-matched normal controls (unilateral study in 7, bilateral in 5), the recovery cycle of the blink reflex was studied by means of the paired shock technique with time intervals between the conditioning and test stimuli ranging from 5–800 msec (fig. 5, 6). For R_1, the amplitude of the test response was converted into a percentage of the amplitude of

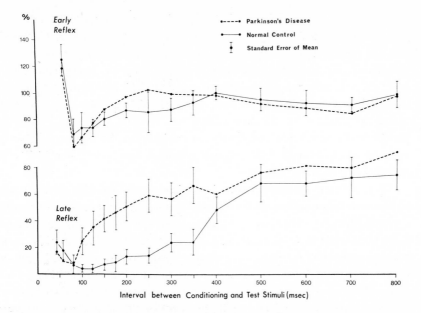

Fig.6. Recovery cycle of the early R_1 and the late R_2 reflexes in 10 parkinsonian patients and 12 age-matched normal controls (the test responses converted into a percentage of the corresponding conditioning responses). Note marked inhibition of R_2 of the test response at time intervals ranging from 80–400 msec in normals and a significant upward shift of the corresponding recovery curve in parkinsonian patients indicating disinhibition at time intervals of 100, 125, 150 and 175 msec ($p < 0.05$) and 200, 250, 300 and 350 msec ($p < 0.01$). Note also initial facilitation followed by relatively slight inhibition of R_1 of the test response beginning at the time interval of 80 msec in both normal subjects and parkinsonian patients. The test R_1 was difficult to measure at the time intervals of 20 and 40 msec because of its overlap with the R_2 of the conditioning response.

the corresponding conditioning response. For R_2 instead of the amplitude, the average amplitude X duration of response was used as a rough evaluation of response area to calculate the percentage.

When stimuli of adequate intensity (16–26 mA and 0.1 msec duration) were delivered in pairs, every 30 sec or longer, R_2 of the conditioning response was nearly constant and exhibited no evidence of habituation, either in Parkinson's disease or in normal controls. R_2 of the test response in normals, however, was virtually absent (less than 10% of the corresponding conditioning response) at time intervals of 80–175 msec between shocks (period of complete depression) and then recovered gradually to reach 50% at 400 msec and 75% at 800 msec intervals. In most cases inhibition was less complete at intervals less than 80 msec but the test R_2 was difficult to measure at intervals less than 40 msec because of its overlap with the conditioning R_2. In the parkinsonian patients, complete depression was noted only at the time intervals of 60 and 80 msec. Initial recovery of R_2 of the test response was rapid and reached 50% at 200 msec. This was followed by a more gradual return to reach 90% at 800 msec interval. Statistical analysis of the two curves reveals that this disinhibition in parkinsonian patients is significant at time intervals between conditioning and test shocks of 100, 125, 150 and 175 msec ($p < 0.05$), and 200, 250, 300 and 350 msec ($p < 0.01$).

As for R_1, those showing considerable random variation in amplitude to repeated single shocks were excluded from either group. In both normal controls (unilateral study in 4, bilateral in 3), and in parkinsonian patients (unilateral study in 3, bilateral in 5), initial unstable facilitation of R_1 up to the time interval of 60 msec was followed by relatively consistent inhibition at 80–125 msec intervals. Although the level of this inhibition varied markedly in degree from one individual to another, the test response was on the average reduced in amplitude to 60 or 70% of the conditioning response. At the time interval of 150 msec, the amplitude of the test R_1 then recovered to 80% after which it remained around the 80–100 percent level with some fluctuation until 800 msec. Although the percentage of recovery of R_1 is also slightly higher in Parkinson's disease than in normals at time intervals ranging from 125–350 msec, statistical analysis of these curves reveals no significant difference between the two groups ($p > 0.10$). It is clear, however, that the recovery cycle of R_1 differs not only from that of R_2 but also from that of the H-reflex [OLSEN and DIAMANTOPOULOS, 1967; TAKAMORI, 1967; YAP, 1967; LANDAU, 1969; PENDERS and DEL-WAIDE, 1969a, b].

The oligosynaptic early R_1 reflex in normals shows a relatively slight change following initial facilitation while the multisynaptic late R_2 reflex is profoundly affected over a much longer period although both reflexes are maximally inhibited at time intervals of 80–125 msec. This indicates an inhibition or block, at least in part, at the interneurons rather than at the motoneurons of the facial nucleus which is the final common path for both R_1 and R_2. Reduction of the R_2 depression noted in the parkinsonian patients in this study suggests that the underlying pathophysiology in parkinsonism involves interneurons rather than the motoneuron. This is consistent with the previous observations that in this disease habituation of the blink reflex [RUSHWORTH, 1967; PEARCE, HOSAN and GALLAGHER,

1968 and of the spinal reflex produced by cutaneous stimulation [DEL-WAIDE, SCHWAB and YOUNG, 1971] is reduced.

The blink reflex is a useful test not only for measuring brain-stem conductions but also for evaluating the influence of the higher nervous system upon brain-stem function.

Acknowledgements

This work was supported by the Medical Research Council of Canada (Grant MA-4000) and the Multiple Sclerosis Society of Canada. Part of this work was done when the author was associated with the Department of Neurology and Neurosensory Center (University of Iowa) supported by Grant NS 03354 of the National Institute of Neurological Disease and Stroke of the Department of Health, Education and Welfare of the United States of America.

Author's address: Dr. JUN KIMURA, EEG Department, Winnipeg General Hospital, 700 William Avenue, *Winnipeg 3, Manitoba* (Canada)

New Developments in Electromyography and Clinical Neurophysiology,
edited by J. E. Desmedt, vol. 3, pp. 692–696 (Karger, Basel 1973)

Observations of the Blink Reflex in Multiple Sclerosis [1]

N. S. NAMEROW

Department of Neurology, School of Medicine, University of California, Los Angeles,
Calif.

The blink reflex is of particular interest for the evaluation of the
pontine region, an area frequently affected by multiple sclerosis [NAMEROW,
1970].

Method

The supraorbital branch of the trigeminal nerve was stimulated percutaneously with
a pulse duration of 0.25 msec (Grass Model 8 stimulator). A recording needle electrode was
placed into the lower portion of m. orbicularis oculi at the center of the orbit while the
indifferent electrode was placed laterally over the malar eminence. During the recording
session, each subject was reclining in a bed in a quiet, isolated room with eyes gently closed.
Patients were aroused between successive reflex measurements, thereby avoiding sleep. A
Tektronic Memoscope was used to store and photograph each response.

31 patients with multiple sclerosis and 20 normal subjects, matched for age and sex
to the patient group, were studied. Patients were divided into 3 groups: those who had never
experienced pontine symptoms of facial pain, paresthesias, myokymia or weakness of
'peripheral' type, tongue paresthesias or jaw weakness; those who had experienced such
symptoms in the past or at the time of study; those with signs of an internuclear ophthalmoplegia. Only patients who had been closely followed by the UCLA Multiple Sclerosis
Clinic throughout the course of their illness were selected for study (mean duration of care,
5 years). Each patient had frequent serial neurological examinations and, on each occasion,
positive and negative findings were carefully documented. Therefore, information concerning pontine symptoms was reasonably accurate.

1   This research was supported by National Multiple Sclerosis Society Grant No. 516-C-3
and United States Public Health Grant No. NSO8711.

Table I. Reflex latencies from 20 normal subjects (40 tested reflexes)

	Early R_1 reflex response, msec	Difference $(L–R)^1$, msec	Late R_2 reflex response, msec	Difference $(L–R)^1$, msec
Mean	10.0	0.39	28.6	2.05
Standard deviation	1.18	0.22	4.3	1.61
Range	10 ± 2.4	≤ 0.8	29 ± 7.5	≤ 5

1 (L–R) = left minus right.

Results

The normal blink reflex with the ipsilateral early (R_1) and bilateral late (R_2) responses on stimulation of the supraorbital nerve is well-known. On the basis of our normative values (table I) an R_1 with a latency greater than 12.5 msec and an R_2 with a latency greater than 37.5 msec could be considered abnormally delayed. In addition, a latency difference between left and right R_1 of more than 0.85 msec and left and right R_2 of more than 5.5 msec also could be considered as abnormal. Amplitudes were considered as an unreliable index and were not used.

Multiple sclerosis can produce patchy or irregular demyelination so that one facial side can be apparently normal while the other shows obvious pathology. One can therefore test in each patient two separate facial halves or

Table II. Reflex latencies from 31 MS patients (62 tested reflexes)

	Pontine symptoms and/or signs				With signs of internuclear ophthalmoplegia	
	Without		With			
Reflex response	early	late	early	late	early	late
Mean	11.1	34.3	14.3	39.7	17.6	51.5
Standard deviation	2.4	9.3	4.8	13.3	3.9	12.5
Range	14.2 ± 5.8	43.5 ± 18.5	17 ± 8	50 ± 25	17.9 ± 7.1	52.5 ± 22.5

Table III. Clinical status and results of reflex latencies from 31 MS patients (62 tested reflexes)

Clinical status	Number of tested reflexes from number of patients () = Number of patients	Early reflex response		Late reflex response	
		normal	delayed	normal	delayed
Without pontine symptoms or signs	38 (14 without symptoms 10 with unilateral symptoms)	31	7	27	11
With pontine symptoms and/or signs	24 (10 with unilateral 7 with bilateral symptoms)	12	12	11	13
With signs of internuclear ophthalmoplegia	16 (8 with bilateral signs)	2	14	1	15

reflex arcs. 17 patients had pontine symptoms and signs as described and 7 of these were bilaterally affected, as were all patients with internuclear ophthalmoplegia.

Table II illustrates the latency of R_1 and R_2 related to the patient's clinical status. In patients without symptoms or signs the mean of both R_1 and R_2 was within the range of normal. In those with pontine symptoms there was an increase in the mean latencies into the abnormal range and this was even more prominent for those with signs of an internuclear ophthalmoplegia.

Table III shows the number of normal and abnormal reflex responses as related to the patient's clinical status. The patients with internuclear ophthalmoplegia were most consistently abnormal, in addition to giving the largest delays in latency. This would imply that either the more severely affected patients demonstrated this sign, or more likely, that the median longitudinal fasciculis is structurally important in mediating the reflex.

Fig. 1. A Section through pontine region demonstrating large demyelinated plaque at root of the 5th cranial nerve. This patient had no signs or symptoms referable to this particular lesion. *B* Blink reflex in patient in acute exacerbation with severe brain-stem and pontine symptoms and signs. Upper sweep, right eye. Lower sweep, left eye. Sweep speed: 10 msec/division.

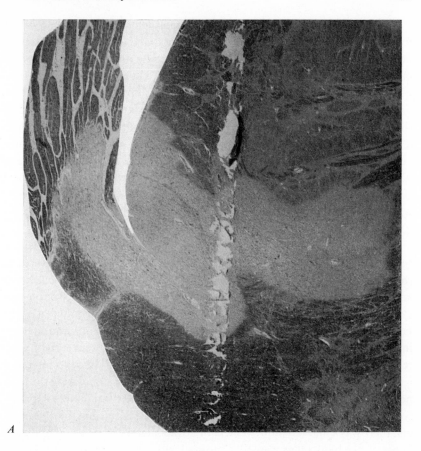

A

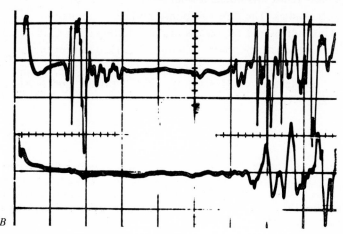

B

The patients with no symptoms or signs referable to pontine lesions frequently had an abnormal R_2 response; we found an abnormal latency in nearly 30% of the tested sides that were clinically normal. Silent lesions in multiple sclerosis have not been well recognized, but are not unusual to those who have seen cases at autopsy [NAMEROW and THOMPSON, 1969]. A similar finding is reported by KIMURA in this volume. An example of such a clinically silent lesion is shown in figure 1 A. This is a section of the pontine region from a young multiple sclerosis patient who committed suicide 19 days after a neurological examination had given virtually normal results. Yet, this large demyelinated plaque at the root of the 5th nerve was not associated with symptoms that one might attribute to such a lesion. This is the type of lesion which could have produced an abnormal reflex response in a patient without complaints referable to this area.

Like KIMURA [1970], we have noticed improvement in reflex latencies as patients evolve through an attack of multiple sclerosis. This is not always the case, however. One recent patient had a typical course of multiple sclerosis and was admitted following an acute exacerbation with primarily brain-stem symptoms. She became aphonic, quadriplegic and demonstrated a severe peripheral facial paresis. She had a bilateral 6th nerve palsy and a severe internuclear ophthalmoplegia. Figure 1 B shows the patient's right blink reflex at the time of admission. The R_1 response was delayed to 15 msec with a marked delay bilaterally of R_2. The direct response on stimulating the facial nerve was 3.5 msec. Three months later, the patient had considerably improved so that facial strength had returned to normal, speech was but mildly slurred and there was a marked improvement in extraocular movements. Despite this, however, the blink reflex was still abnormal with values essentially unchanged from the earlier studies. The direct response on facial stimulation was also unchanged. This patient and the results of reflex testing show that virtually normal clinical function can be observed in spite of marked slowing of neuronal conduction. These results can be compared with data on cortical evoked responses in multiple sclerosis [NAMEROW, 1968; DESMEDT and NOEL, this series, vol. 2, page 363].

Author's address: Dr. N.S. NAMEROW, Department of Neurology, UCLA School of Medicine, Los Angeles, CA 90024 (USA)

New Developments in Electromyography and Clinical Neurophysiology, edited by J. E. Desmedt, vol. 3, pp. 697–712 (Karger, Basel 1973)

The Clinical Significance of Exteroceptive Reflexes

R. R. YOUNG

Department of Neurology, Harvard Medical School and Massachusetts General Hospital, Boston, Mass.

Studies of the neurophysiology of movement have focused on the organization of proprioceptive reflexes elicited by afferent input from receptors within muscles, tendons, joints and, in most categories, the vestibular system. For example, the study by MELVILL-JONES and WATT [1971a] suggests that certain reflex behavior in the human leg may be related to impulses originating in the otolith apparatus. Special emphasis in both animal and human neurophysiology has been on investigations of monosynaptic reflexes, in part at least because of their relative simplicity. Monosynaptic reflexes likely comprise a small fraction of all proprioceptive reflex activity under natural circumstances, and from many normal, athletic and agile individuals no tendon jerks can be elicited; nevertheless, clinicians as well as physiologists tend to be fascinated, perhaps unduly, with deep tendon reflexes and the H-reflex. On the other hand, the tonic stretch reflex, which is proprioceptive but, as far as can be told, not monosynaptic, would appear to be one of the fundamental building blocks of posture.

Segmental afferent input, particularly the proprioceptive input, was at one time thought to be crucially important for motility, so much so that it was generally accepted that the forelimb of a monkey, following complete posterior rhizotomy, was profoundly ataxic proximally and virtually paralyzed, particularly as regards the performance of independent voluntary movements such as reaching or grasping [MOTT and SHERRINGTON, 1895; TWITCHELL, 1954]. Subsequent studies of stepping in the deafferented spinal cat [BROWN,

1911] and of the scratch reflex (see below) required modification of these ideas. Later, observations in man [LASHLEY, 1917] and monkey [BOSSOM and OM-MAYA, 1968; KNAPP et al., 1963] revealed quite useful and well-controlled movements to be present in deafferented limbs. This has tended to direct attention (1) toward 'control signals' or 'commands' which descend to the spinal cord, either with fully formed programs for movement or to trigger and modulate such programs stored locally ('centrally patterned movements') [DELONG, 1971], and (2) away from the role played by local spinal afferent input and the reflexes related thereto. While dorsal root activity is clearly not prepotent, it must nonetheless figure prominently in the control of movement. Our concern will be with this afferent activity, particularly as it relates to non-proprioceptive reflexes.

Types of Reflexes

Any rigid categorization of reflexes will finally prove artificial and unsatisfactory though it may be useful for discussion. In addition to proprioceptive reflexes and the exteroceptive ones discussed below, we may designate two other types: *interoceptive* and *teleceptive*. In the former, chemoreceptors, pressure receptors, and other visceral afferents provide information which is used in regulation of various smooth muscle, gland and other effector organs controlled by the 2 divisions of the autonomic system. Autonomic physiology, particularly in the areas of vasomotor, respiratory and sudomotor control and their disorders, constitutes a very active field of clinical investigation with a literature which the average clinical neurophysiologist tends to overlook. For example, it is well-known that tetanus toxin blocks postsynaptic inhibition on mammalian spinal motoneurones [BROOKS et al., 1957]; hence, the overactivity of these cells in the form of the spasms of clinical tetanus is presumed to relate to normal excitatory activity in the presence of compromised inhibition. However, only recently have studies of the autonomic physiology of such patients documented similar overactivity of the sympathetic nervous system in clinical tetanus [CORBETT et al., 1969]. This suggests that tetanus toxin also affects central autonomic neurones, and since many of the complications (including some of the mortality) encountered in severe tetanus appear to be due to this sympathetic overactivity, its treatment by a combination of autonomic blocking agents (propranolol and bethanidine [PRYS-ROBERTS et al., 1969]) is important. Other interoceptive reflexes which are well known to clinicians include carotid sinus and oculo-cardiac reflexes and the automatic or reflex contraction of the 'spastic' bladder. Interoceptive spinal input also has profound influence on somatic musculature as evidenced by the 'board-like' abdomen, where peritoneal irritation and pain produce reflex contrac-tion of abdominal musculature. Much hunting, feeding, evacuatory and sexual behavior provide examples of more complex interoceptive control of posture and motility.

Teleceptive reflexes utilize the eyes, ears, and olfactory-taste afferent systems. The pupillary response to light, blink to visual threat or a loud noise, and opto-kinetic nystagmus represent simpler reflexes of this type. Many of these reflexes, however, tend to be very complicated and 'high level'; some are reflex only in the broadest sense of a sensory input which produces an appro-priate response. Much behavioral research dealing with reaction times uses visual or auditory input while activity in the olfactory-taste afferents obviously has to do with emotional and visceral reflexes. Clinically speaking, conditions such as Creutzfeldt-Jakob and Tay-Sachs diseases, in which startle [SHIMA-MURA, this volume] to a light flash or loud noise is unusually prominent, ap-pear to demonstrate abnormal excitability of some of these teleceptive re-flexes which may also be more than usually evident in normal persons under conditions of extreme apprehension or fright. BICKFORD [1972], in his studies of 'microreflexes', has contributed greatly to our understanding of photo-myoclonus as seen on the EEG, another example of an exaggerated startle-like teleceptive reflex.

In addition to his studies of proprioception, SHERRINGTON and his col-leagues also dealt with *nociceptive* or flexor withdrawal reflexes [SHERRINGTON, 1906a, 1910] and the scratch reflex [SHERRINGTON, 1906b]. The latter is the classic example of a prepatterned integrated movement (with considerable dexterity even in a deafferented limb) released or activated by cutaneous stimuli. DENNY-BROWN wrote extensively of the effects of body surface con-tact as examples of non-proprioceptive control of the motor system. In his discussions of abnormal movements and postures, he stressed tactile explora-tory and avoiding reactions, their release by cerebral lesions, and the import-ance of light contact sensation in the organization of equilibrated reactions at supraspinal levels; he wrote [1966] 'the whole motor apparatus is subservient to exteroceptive effects in the cerebral cortex'. In fact, in his work with MOTT more than 75 years ago, SHERRINGTON found that 'afferent impulses, both from the skin and from the muscles, *especially the former*, are necessary for the carrying out of highest level movements'. They reported little impairment of function in the forelimb of a monkey from which the only afferent input was in cutaneous fibers from the hand, whereas in other animals the hand was useless if muscle afferents were intact but cutaneous input from the hand was no longer present [MOTT and SHERRINGTON, 1895].

Exteroceptive Reflexes of Clinical Interest

Exteroceptive reflexes, such as described in the preceding paragraph, include all those organized around afferent input from skin and subcutaneous tissue. This input is in the form of primary sensations such as touch, pressure, flutter-vibration, pain and temperature. Adequate stimuli may be either mechanical, thermal or, in the laboratory, electrical. When using the latter, however, one must recognize that electrical stimulation of mixed nerve also activates fibres responsible for proprioceptive input. At the outset, it is important to stress that nociceptive reflexes are merely one subgroup of exteroceptive reflexes; many or most of the latter do not require painful stimuli. Whether or not the necessary input for some of them is via unmyelinated fibres is another question, but even then, as PERL has shown [1969], other forms of adequate stimuli than harmful or painful ones give rise to impulses in C-fibers.

Before considering the clinical significance of exteroceptive reflexes and the profound influences exerted upon spinal motoneurone pools by cutaneous input, it is of interest to list the exteroceptive reflexes familiar to the clinician: blink [SHAHANI and YOUNG, 1972, this volume: Blink reflexes], corneal [RUSHWORTH, 1962], gag, palmomental [REIS, 1961], grasp [SHAHANI et al., 1970], abdominal [HAGBARTH and KUGELBERG, 1958; KUGELBERG and HAG-BARTH, 1958], gluteal, cremasteric, plantar [GRIMBY, 1963a,b; 1965a,b; KUGELBERG et al., 1960] and other flexor reflexes and contact placing, striated muscle being the effector for each. The swallowing and anal reflexes utilize both striated and smooth muscles [see PETERSEN, this series, volume 2]. When cutaneous input produces 'gooseflesh' (piloerection), dermographism, scrotal contraction (dartos muscle), or the ciliospinal or Mass reflexes [HEAD and RIDDOCH, 1917] (a condition where spasms of bladder and bowel evacuation, sweating, and hypertension are present in a patient with a high cord transection – much more frequent subclinically than in its full-blown classical form [CORBETT et al., 1971]), autonomically innervated smooth muscles are the effector organs. Cutaneous input also produces complex sexual reflexes in which skeletal muscle (posturing) and smooth muscles (erection, ejaculation) may be active in a coordinated fashion even below a complete spinal transection [TALBOT, 1949]. The latter reflexes, though of obvious interest, have received little mention in the neurophysiological literature.

Earlier Human Studies

From the masterful study of the physiology of the plantar reflex and its abnormalities by KUGELBERG *et al.* [1960], at least 2 very important points emerge. The first is that under normal circumstances, painful stimulation of the under-surface of the great toe results in its dorsiflexion. An abnormal plantar response (a Babinski response) is present when the same movement of great toe follows stimulation of the sole of the foot – that is, a reflex pattern which is normally produced by stimulation of one cutaneous area now is produced, after damage higher in the nervous system, by stimulation else-where. The 'local sign', a very important aspect of exteroceptive reflexes, is different in the latter instance as compared with normal, and this presumably relates to an abnormality in the processing of afferent information once it reaches the cord [SHAHANI and YOUNG, this volume: Flexor spasms]. Secondly, they, and subsequently GRIMBY [1963a, b, 1965a, b], showed that the group of muscles responsible for dorsiflexing the great toe and those for plantar flexion are both more or less simultaneously active following plantar stimulation. Normally, the plantar flexors are activated more strongly so the net force moves the toe down, whereas under abnormal circumstances the reverse is true. Because both groups of muscles are activated together, intermediate situations arise in which activity is approximately equal in the 2 groups. Then the great toe moves very little, or wavers up-down, or *vice versa*. Earlier, when it was assumed that normally only plantar flexors were activated and abnor-mally the opposite, clinicians felt compelled to classify every plantar response as 'up' or 'down'. Though we can now see that such a clear-cut differentiation is not always possible, in former times the senior clinician present often arbitrarily categorized the response – an artificial oversimplification which would, in fact, result in an unreliable bit of data. KUGELBERG's studies de-monstrated the virtue of applying objective, quantitative methods to the study of clinical phenomena. These studies of the plantar response, and more es-pecially KUGELBERG and HAGBARTH's investigations of the abdominal re-flexes [1958], revealed beautifully coordinated segmental reflex mechanisms which, by excitation of agonists and inhibition of antagonists, serve to with-draw the limb or trunk in a most efficient fashion from an offensive extero-ceptive stimulus. HAGBARTH, first in cat [1952], and then in man [HAGBARTH, 1960; HAGBARTH and FINER, 1963] studied the effects of natural cutaneous input on behavior of the spinal motoneurone pool. His work again revealed an elegantly coordinated reflex system of muscular contraction and relaxation which seems to function to remove the limb from a noxious stimulus. His findings broadened the concept of nociceptive reflexes from that of activity restricted entirely to flexor muscles to contraction, in addition, of certain extensors of distal joints if the stimulus was over that muscle. In this way, the reflex behavior appears to be organized around withdrawal of the limb from the stimulus rather than a stereotyped contraction of all flexor muscles, which would, in the case of stimuli over extensor muscles, pull the limb more tightly against the stimulus. Since KUGELBERG reviewed polysynaptic reflexes in 1962, a number of new observations have come to light.

Recent Animal Studies and Possible Clinical Correlations

During the past decade, LUNDBERG [1966] and his associates have, in the animal laboratory, provided new insight into the mechanisms of exteroceptive reflexes. They have found that, in addition to the numerous and well-known intraspinal pathways which are functional in the acutely isolated cat spinal cord (from proprio- and exteroceptive inputs to alpha and gamma motoneurones and to primary afferents themselves), others are present but can be demonstrated only after activation of facilitatory or suppression of inhibitory control [ANDÉN et al., 1964]. For example, short latency pathways from afferents which ordinarily produce contraction of flexor muscles (so-called flexor reflex afferents: FRA) can excite extensors, but these pathways are not demonstrable in the stage shortly following spinal transection. These may be part of the mechanisms responsible for the 'local sign' so prominent in the studies by HAGBARTH (above) whereby extensor muscles contract if the noxious stimulus is over them. DIMITRIJEVIĆ and NATHAN [1968] reported that, after spinal transection, they could not find evidence of the finely tuned cutaneo-muscular organization described by HAGBARTH in man. It is clear that, under those circumstances, coarse and stereotyped flexor reflexes eventually predominate in man, but in our studies [SHAHANI and YOUNG, 1971] of patients with severe or complete spinal lesions, we could, with minimal stimuli, demonstrate the reciprocal relationships and segmental interaction shown by SHERRINGTON [1906a, 1910] and HAGBARTH. Nevertheless, in the isolated cord of man and cat, it is more difficult than normal, soon after spinal shock has receded, to demonstrate intricate mechanisms of coordination. Later, flexor reflex activity becomes even more predominant, and we have then been unable to record an isolated first component of the flexor reflex – something present normally and, even in the isolated human cord, immediately after recovery from spinal shock. This evolution of change in reflex patterns which goes on for some time after spinal transection suggests that spinal shock may, in part at least, represent an active process [like the inhibition of reflexes during REM sleep HISHIKAWA et al., 1965; HODES and DEMENT, 1964; SAUERLAND et al., 1967] related to transmitter release from transected descending fibers rather than something entirely passive, such as the cessation of excitatory impulses of supraspinal origin. Further efforts are being directed toward pharmacological and physiological exploration of these phenomena. It is interesting, in the same context, that a GABA-congener markedly reduces the 'spontaneous' and induced flexor activity in the chronically isolated human spinal cord but does not effect the reappearance of such normal flexor activity as the first component, mentioned above [SHAHANI and YOUNG, this volume, page 734].

LUNDBERG and colleagues have also shown that, in the acutely transected cat spinal cord, it is possible, by systemic administration of DOPA, to depress the short latency pathways mentioned above, simultaneously opening long latency reflex pathways mainly from FRA [ANDÉN et al., 1964, 1966]. The DOPA is thought to activate a descending monoaminergic system responsible for these effects under normal conditions. After the administration of DOPA, they have recorded unusually sustained flexor motoneurone reflex activity which has a long latency and is not accompanied by inhibition of antagonist motoneurones. Under these circumstances, it is also possible to demonstrate presynaptic inhibition of I_A terminals produced by FRA input; that is, exteroceptive control of simple, often monosynaptic, proprioceptive circuits. DELWAIDE [1971, and this volume] and SHAHANI and YOUNG [this volume: Flexor spasms] have discussed from different points of view possible roles played by abnormalities of presynaptic or other prolonged spinal inhibition in the production of abnormal clinical signs, such as (1) hyperactive myotatic reflexes and their failure to be diminished by tonic vibratory input, or (2) labile flexor reflexes and flexor spasms. Very likely, ultimate appreciation of the pathophysiology of Babinski's response will evolve from understanding of gating mechanisms controlling the inflow of sensation through the posterior gray horns of the cord into local interneuronal pathways [MELZACK and WALL, 1965].

Mechanisms of After-Discharge

After-discharge, defined by SHERRINGTON as the persistence of efferent reflex activity long after the cessation of the responsible stimulus, is characteristic of exteroceptive, as opposed to proprioceptive, reflexes, and this phenomenon has been amply demonstrated clinically and, in recent times, physiologically in studies of the isolated human cord as well [DIMITRIJEVIĆ and NATHAN, 1968; PEDERSEN, 1954; SHAHANI and YOUNG, 1971]. SHERRINGTON himself [1921] considered it to reflect repetitive discharge of the 'reflex centers', and LUNDBERG [1966] and colleagues have recorded prolonged discharges of this sort in spinal interneurones. FORBES [1922], in a 'provisional hypothesis', suggested that 'delay paths' or lengthy circuits of interneurones in chains might account for the time delays required to explain long latency or prolonged reflexes. This concept appealed to clinicians and physiologists alike and has long been part of the neurological folklore. 'Long loop reflexes' [ECCLES, 1966; SHIMAMURA et al., 1964], in which the chain of neurones travels rostral to the spinal cord, through cerebellum, cerebrum, or brain stem and then back to cord, do represent examples of neuronal chains where, for example, [SHIMAMURA, this volume] in oligosynaptic pathways the anatomical extent of the chain can be shown. In more multisynaptic networks, however, it becomes impossible, with present techniques, to trace the actual circuitry. LUNDBERG's work showing interactions of long and short latency pathways affords an alternative mechanism for explaining delayed reflexes other than the postulation of multiple synapses. We have demonstrated [SHAHANI and YOUNG, 1971] a constant repeatable brief reflex with latency, measured from a single small stimulus to the skin of the foot to the discharge in ipsilateral tibialis anterior, of nearly 400 msec in a patient with a complete mid-thoracic spinal transection. The single afferent volley in myelinated fibres [MOUNTCASTLE and POWELL, 1959] sets up activity in the cord which leads to activation of flexor motoneurones only after a precise delay of hundreds of milliseconds. With synaptic delays of the order of 0.5 msec, a chain of hundreds of neurones (but the same number each time) would be required to account for the delay if one were to explain it in these terms. Perhaps the correct explanation may be shown to be in repetitive activity in relatively few neurones in shorter circuits. This mechanism would account for the time lags seen in reflex after-discharge as well as in the brief reflexes with very long latencies described above.

In another attempt to explain after-discharge. TUREEN [1941] postulated that afferent activity in large, group A fibers gave rise to the initial part of a flexor reflex, whereas the after-discharge was related to input in small, δ fibres. LLOYD [1943] thought group II input accounted for the initial and group III for the latter part of a flexor reflex. Both had described what we now call 2 separate components of the flexor reflex, and BROOKS and FUORTES [1952] reclassified these components – the first as a reflex twitch and the second as the after-discharge – in terms slightly different from SHERRINGTON's original definition. Many others have reported recording 2 components from blink [KUGELBERG, 1952; RUSHWORTH, 1962; SHAHANI and YOUNG, 1972] and flexor reflexes [SHAHANI, 1968; SHAHANI and YOUNG, 1971], and we consider the presence of 2 components to be another characteristic feature of extero-ceptive reflexes (though 2 components can also be seen under certain unusual circumstances in proprioceptive [MELVILL-JONES and WATT, 1971a] and tele-ceptive [SHAHANI and YOUNG, this volume, page 644]). The second com-ponent is associated with the movement produced by the reflex; the signifi-cance of the first component is less clear. As outlined elsewhere [SHAHANI and YOUNG, 1971], we believe that the 2 components are not simply a reflection of groups of faster and slower conducting afferents being differentially activated; for example, a stereotyped, repeatable stimulus can under different conditions be made to produce both components or either in isolation. KUGELBERG [1948] suggested, following procainization of peripheral nerve, that A-δ and C-fiber activity gave rise to the flexor reflex. Both SHAHANI's studies [1970] of the timing of normal flexor reflexes and our studies [SHAHANI and YOUNG, 1971] of patients with Friedreich's ataxia [where late in the course of the disease the largest myelinated afferents in cutaneous nerves are missing, whereas the smaller myelinated and C-fibers are intact: DYCK et al., 1968; DYCK and LAMBERT, 1968; HUGHES et al., 1968; MCLEOD, 1969] point to small myelin-ated afferents, conducting at roughly 30–40 m/sec as those responsible for human flexor reflexes. However, subsequent studies of a patient with primary amyloidosis [where C-fibers and small myelinated fibers are severely depleted, leaving the larger myelinated fibres intact; DYCK and LAMBERT, 1969], in whom we found a good first component but no second one, suggest that the matter of exactly which human FRA are sufficient to produce both components of the human flexor reflex is more complicated than it first appeared [see HUGON, this volume, page 713].

Parkinson's Disease

LUNDBERG's work with the effect of DOPA on spinal reflexes led DEL-WAIDE, SCHWAB and YOUNG [1971] to study exteroceptive reflexes in parkin-sonians before and during treatment with L-DOPA. Several obvious clinical abnormalities can be seen in patients with Parkinson's disease; among these are such different and mutually independent abnormalities as rigidity, resting tremor, action tremor and akinesia, so that there are at least several pathophysiologies of Parkinson's disease, some more important in any given patient than another. Even so, it has been difficult to define a physiological abnormality underlying any of these aspects of the syndrome. DENNY-BROWN [1962] felt that rigidity was associated with an unusual stretch reflex in which certain units were activated by passive elongation of muscle only to be silenced, perhaps by heightened Golgi tendon organ inhibition, and simultaneously replaced by other active units in rotation. However, LANDAU et al. [1966] demonstrated subsequently that much, at least, of the 'rotation of units' seen on EMG is an artifact of electrode movement. The absence of changes clinically in tendon jerks or plantar reflexes in Parkinson's disease is well-known, and no significant or consistent abnormalities have been demonstrated in (1) the silent periods produced either by release reflexes [ANGEL et al., 1966] or by stimulation of a mixed nerve in these patients [HOFMANN, 1962], or (2) their H-reflex excitability curves [DIAMANTOPOULOS and OLSEN, 1965; IOKU et al., 1965; TAKAMORI, 1967; YAP, 1967; ZANDER-OLSEN and DIAMANTOPOULOS, 1967].

Most assume, because of the tendency toward a flexed posture in this disease, that an abnormality of flexor reflexes exists though one does not find a Babinski response, flexor spasms or unusually brisk withdrawal reflexes in these patients. In our studies of exteroceptive reflexes in parkinsonians [DELWAIDE et al., 1971; SHAHANI and YOUNG, 1971], 4 abnormalities are noted, though it remains to be seen which, if any, will be found consistently in such a patient population. First, the threshold is quite reduced for the first component of the flexor reflex, and it is preserved during sleep – findings unique in parkinsonism in our experience. Second, the normal reciprocal firing relationships in antagonistic leg motoneurone pools are disturbed and a great deal of 'co-contraction' is seen; that is to say, stimuli which ordinarily produce polysynaptic reflex activity in physiological extensors, now produce concomitant activity in flexors. Third, cutaneous silent periods [SHAHANI and YOUNG, this volume, page 589] are abnormally brief, or vitually absent, in this condition, whereas others report that proprioceptive silent periods are either prolonged or normal in duration. Fourth, habituation, another characteristic of exteroceptive reflexes, is less than normally evident in the blink reflex *and* the flexor reflex of the lower extremity in Parkinson's disease. Furthermore, these abnormalities are largely, though not completely, reversible by the administration of L-DOPA in those parkinsonians who respond clinically to this form of therapy. Whether any of the L-DOPA effect is exerted at the spinal level remains to be seen; it may be effective by virtue of actions upon the nigro-striatal or other cerebral system.

Habituation of Exteroceptive Reflexes

Habituation, defined as a progressive reduction in reflex response to repeated stimulation [HARRIS, 1943; THOMPSON and SPENCER, 1966], has long been known to be a characteristic feature of the normal blink reflex [MYERSON, 1944; RUSHWORTH, 1962] (particularly, but not exclusively, the second reflex component which is responsible for the visible blink, [SHAHANI and YOUNG, 1972]) and recently has been seen with other exteroceptive reflexes as well [DELWAIDE *et al.*, 1971; DIMITRIJEVIĆ and NATHAN, 1970; HAGBARTH and KUGELBERG, 1958; SHAHANI and YOUNG, 1971]. This reduction in response can normally be overcome or reversed in several ways.

Dishabituation occurs if (1) the stimulus is stopped for some time and then restarted, (2) the character of the stimulus is changed, or (3) the patient's expectation in regard to the stimulus is changed. For example, dishabituation will suddenly occur in the presence of repeated identical stimulation if the patient believes that the next stimulus will be painful. It has been assumed that dishabituation is specific in some way for the stimulus, but our observations suggest that there are more interesting, widespread mechanisms involved. For example, the blink to repeated identical stimulation becomes dishabituated if the patient is more apprehensive or anxious in general, even if this is without specific relation to the stimulus. On the other hand, under conditions of minimal apprehension or anxiety, when the patient is permitted to tap himself on the glabella or to press the button triggering the electrical stimulus, one sees only a first component [SHAHANI and YOUNG, 1972]; that is, the second component appears to be maximally habituated and is not seen even with the first stimulus. The status of the patient as regards apprehension or anxiety can, therefore, be seen to exert a profound effect on the reflex performance even if stimulating and recording conditions are otherwise invariable. This applies to blink and flexor reflexes as well as to natural phenomena, such as the response termed 'tickle', produced by a moving touch of modest intensity over the ribs or other 'ticklish' areas; one cannot tickle himself in this manner. In addition to interesting speculations about brain function, this observation leads to considerations of very real experimental significance; it would appear that one may not generalize from the results of reflex studies performed on oneself, a close associate, or sophisticated subject, in which circumstances the effects of apprehension or anxiety would tend to be minimized. Exteroceptive reflex studies should, therefore, be carried out as broad surveys of a sizeable population of subjects, normal or otherwise, who are naïve as far as the experimental situation is concerned.

When considering other natural states in which apprehension or anxiety are at a minimum level, one might imagine that light sleep is the paradigm. However, SHAHANI [1968] has shown that, during non-REM sleep, it is the first component which is absent rather than the second, whereas the opposite is true when the subjects are awake, relaxed and habituated to the stimulus. Clearly, the 2 components of these exteroceptive reflexes differ in their supra-spinal control, and careful analyses of their behavior might illuminate, or at least aid in, the quantification of such psychic phenomena as concentration, apprehension, or relaxation. At the same time, fluctuations in these and other variables are associated with moment-to-moment variations in the extero-ceptive reflexes recorded; this variability must always be borne in mind in evaluation of these studies. In addition to (1) these background fluctuations, and (2) excitability cycles of various reflexes, as found in the animal laboratory, measured in tens or, at most, hundreds of milliseconds, there is evidence in the human cord of (3) very long-lived effects that outlast stimuli by 45–60 sec or more [SHAHANI and YOUNG, 1971]. These relatively long-lasting effects are present even in the isolated cord where, because it represents the simplest human reflex system, many of the most clear-cut studies have been done. How does the artifact of spinal transection affect exteroceptive reflexes – how do they differ then from normal ?

Exteroceptive Reflexes
in Patients with Upper-Motoneurone Lesions

Studies of flexor reflexes below spinal lesions reveal several abnormalities. Obviously the excitability of the clinical reflex (that is, the second component) then becomes markedly increased and its duration prolonged [RIDDOCH, 1917]. In addition, this reflex activity, which is ordinarily almost restricted to flexors, may then be present in all groups of muscles. Simultaneously, the excitability of the first component is decreased, its threshold elevated, and it becomes difficult to produce in isolation [SHAHANI and YOUNG, 1971]. As noted above, normal reciprocal activity at the spinal segmental level appears basically to be preserved though not easy to demonstrate because it is greatly overshadowed by the extreme lability of flexor reflexes in general. DIMITRI-JEVIĆ and NATHAN [1968] report that this apparent loss of segmental coordi-nation after spinal cord transection can be overcome in part by rather non-specific deafferentation. They think that much of the dysfunction of extero-ceptive reflexes below a spinal lesion relates to excessive input into cord [SHA-

HANI and YOUNG, this volume, page 734]. If this is reduced by section of peripheral nerves, roots, or phenol injections, a more normal spinal organization can be demonstrated almost regardless of which nerves are cut. This confirms clinical experience that ankle clonus could be reduced or abolished following obturator neurectomy for adduction contractures, an observation difficult to explain on purely anatomical grounds.

When one turns to patients with rather restricted cerebral lesions for comparison, the story becomes more complex because different lesions in different sites, including variable amounts of specific cortical and subcortical structures, produce very different clinical syndromes, all of which tend to be grouped together, even together with spinal cord disease, under the general heading of 'spasticity'. In patients with discrete cerebral lesions, extensor reflexes tend to be released; the threshold of the first component of the flexor reflex is also increased, but this decrease in excitability may not be due to the same process as occurs with spinal cord disease. The threshold for the second component decreases somewhat, but this is partially masked by the increase in extensor reflexes. Reciprocal inhibition and other reciprocal relationships at the spinal segmental level are better preserved or more easily demonstrable in these patients than in those with spinal lesions. It is interesting, moreover, that the flexor reflex excitability curve in the presence of certain cerebral lesions shows an absence of delayed inhibition that is to date unique in our experience [SHAHANI and YOUNG, 1971].

Discussion and Conclusion

In summary, lesions at various levels of the neuraxis produce abnormalities of function which are, upon close inspection, very different from one another not only as to the clinically localizable level of the lesion, but also physiologically and pharmacologically – the latter difference determining the types of treatment most likely to be effective in their amelioration. To call them all 'spasticity' is to shed little light upon their various pathophysiologies or therapeutic needs; it is better to describe the abnormality one is discussing, such as flexor spasms, clonus, dystonic or hemiplegic posturing, etc. MAGLA-DERY, TEASDALL, PARK and LANGUTH [1952] described differences in the H-reflex excitability curves when patients with spinal lesions were contrasted with those who had supraspinal lesions. Others have shown similar differences in exteroceptive reflexes in the presence of these two types of upper motoneurone lesion [SHAHANI and YOUNG, 1971]. Both these and the clear-cut clinical differences between the 2 groups of patients support LANDAU's contention [1969] that conditions with hyperexcitability of proprioceptive reflexes and those with hyperexcitable exteroceptive reflexes not be grouped together under the all-inclusive term 'spasticity'. In fact, many recent publications refer to 'spasticity' but actually concern themselves with hyperactive exteroceptive reflexes [DIMITRIJEVIĆ and NATHAN, 1968, 1970, 1971; JONES et al., 1970; RUSHWORTH, 1966] or confuse myotatic with flexor reflexes [SORIANO and HERMAN, 1971]. Surely, in 1972, our knowledge of these phenomena, though extremely preliminary, is nonetheless sufficient to justify more precise terminology. Even the simple dichotomy between overactive stretch reflexes and flexor reflexes [LANDAU, 1969] gives more information about the underlying observations than use of a term like 'spasticity' which tells little more about the actual situation than the term 'weakness' when it is used to describe the clinical counterpart of dysfunction anywhere from muscle, neuromuscular junction and nerve throughout all levels of the CNS, including such entities as depression and neurasthenia. The term 'spasticity' catches the eye of the reader but must be used in a carefully restricted fashion if it is to have fundamental meaning.

Previously, most discussions of the physiology, and certainly the pathophysiology, of movement revolved around proprioceptive mechanisms. These have proven to be too simple to account for everyday observations, and preliminary studies of exteroceptive reflexes reveal that they too are abnormal, perhaps characteristically so, in various disease states. The facial musculature, capable of the most highly refined and delicate shadings of movement and posture, is dependent only upon exteroceptive input, no muscle spindles or tendon organs being present there [BOWDEN and MAHRAN, 1956; BRUESCH, 1944; COOPER, 1966; KADANOFF, 1956].

If proprioceptive input is not necessary for normal facial motility and deafferented limbs can be used with fair skill, physiologists might be encouraged to turn their attention away from proprioceptive mechanisms as the fundamental building blocks of movement. The cerebellum, for example, is apparently much more concerned with exteroceptive than proprioceptive input [ECCLES, this volume; ECCLES et al., 1971]. Clearly, useful and interesting data can come from studies of exteroceptive reflexes [HUGON, 1967, and this volume, page 713] which, after all, probably make a larger contribution to ordinary everyday motility than the extent to which they have been studied would suggest. Furthermore, these exteroceptive reflexes are, in many ways, more readily studied in human subjects than proprioceptive ones. The stimuli required are simpler and less complicated than, for example, stimulation of the

muscle spindle, and are largely painless. A very active interest in the neuro-
pharmacology of these reflexes is now emanating from laboratories such as
LUNDBERG's, and this is especially exciting for a clinical neurophysiologist
because many of these pharmacological agents are available for use in human
studies. Fascination with proprioceptive reflexes, and particularly with the role
of the muscle spindle in human movement disorders, is understandable because, after all, this has been the prime focus of much of neurophysiology for
the past few decades. However, it is clear that the techniques used in the animal
laboratory for study of muscle afferents, recording from dorsal root filaments,
stimulation by stretch of single muscles without joint movement, or activation
of fibres in ventral root, are difficult to apply to human studies, and it remains
to be seen [VALLBO; HAGBARTH and VALLBO, this volume] if such techniques
can, in the future, be used clinically with the same precision as in the cat or
monkey. It would appear that the extent to which monosynaptic reflexes, proprioceptive in type, have been studied bears little relation to the importance
they play in nature [MELVILL-JONES and WATT, 1971 a]. Furthermore, studies
of the silent period or the effects of vibration on human limbs are characteristically discussed in terms of the proprioceptive mechanisms involved. Our
studies of the cutaneous silent period [DELWAIDE et al., 1971; SHAHANI and
YOUNG, this volume, page 589] and observations that vibration, when applied to a tendon, also stimulates much of the skin overlying joints, and so
forth, lead one to the conclusion that exteroceptive reflex behavior probably
plays a larger role in these so-called proprioceptive experimental situations
than is usually realized. Comprehensive physiological explanations for various
abnormalities (and normal behavior) in the motor system must remain
woefully inadequate and, therefore, highly suspect, if these explanations are
drawn largely from theorization about proprioceptive phenomena since these
not only are incomplete in themselves but also totally ignore all non-proprioceptive factors. Despite the seductive requests for easy explanations and our
desire to fulfill them, we would be wise to note that animal neurophysiology is
clearly of crucial importance, but extrapolation of these findings to clinical
situations, though earnestly requested by physicians and sanctioned by authority over the years, has resulted in suprisingly little understanding of the real
physiological basis of human motility and its disorders. In fact, many of these
facile 'explanations' may have directed attention (1) away from the more
difficult and challenging task of making controlled physiological observations
in man, and (2) toward reliance upon the great mass of interesting laboratory
neurophysiology which, apart from speculations, has little of substance to
offer the clinician. More careful studies of human physiology are in order with

the investigators, clinical and otherwise, keeping in m.   ̇n-
tained in the following words of CLAUDE BERNARD: '*Recuː*
s'astreindre à les interpréter qu'ensuite est la condition indispensu.
à la vérité.'

Author's address: Dr. R.R. YOUNG, Department of Neurology, Massachusetts General Hospital, *Boston, MA 02114* (USA)

New Developments in Electromyography and Clinical Neurophysiology,
edited by J. E. Desmedt, vol. 3, pp. 713–729 (Karger, Basel 1973)

Exteroceptive Reflexes to Stimulation of the Sural Nerve in Normal Man

M. HUGON

Department of comparative Neurophysiology, University of Provence, Marseilles

The sural nerve is composed of myelinated fibres of large diameter, A-α and A-β [or group II of LLOYD, 1943], and of small diameter A-δ (or group III of LLOYD). It also contains unmyelinated C fibres (or group IV). This composition has been defined for the cat and applies also to man [O'SULLIVAN and SWALLOW, 1968].

Electrical stimulation of the sural nerve with increasing intensities excites first the group II afferents which evoke tactile sensations, and then the group III and, lastly, the unmyelinated afferents, both of which elicit painful sensations [COLLINS, NULSEN and RANDT, 1960]. These observations have been confirmed by the verbal reports of normal human subjects in whom the sural nerve was stimulated through surface electrodes on the skin [HUGON, 1967].

Stimulation of the sural nerve in man was found to elicit exteroceptive reflexes involving polysynaptic pathways and I could differentiate reflexes with quite distinct features according to whether the sural nerve stimulus evoked a tactile sensation or a painful sensation [HUGON, 1967]. The significance of the corresponding spinal polysynaptic reflexes is, of course, different. A painful stimulus to the nerve will elicit a flexor reflex withdrawal whereas a tactile stimulus would produce a reflex which can be considered as related to the exteroceptive control of the postural positioning of the foot. Interactions of these exteroceptive afferent inputs with the proprioceptive input from muscles and with the motor systems descending from supraspinal levels have opened new vistas and methodologies for the analysis of the functional organization of spinal motor mechanisms in man.

Material and Methods

The experiments reported in this paper were performed on 21 normal subjects, 16 men and 5 women, whose ages ranged from 18 to 55 years. The subject sat comfortably in a chair with easy supports for arms, feet and head. He was isolated by a curtain and had to remain completely relaxed during the session which lasted 2–4 h. The nerve was stimulated with square electrical pulses of 1-msec duration. The pulses were delivered singly or in trains of 4–10 shocks at 300/sec. The stimuli were applied at intervals of 5 sec or more to the lateral sural nerve by skin electrodes placed below the external malleolus, on the same side as recording. With correct positioning the subject experienced a tactile sensation which was *projected* along the outer side of the foot and small toe. For each subject, unit stimulation (S) was taken as that necessary to produce threshold tactile perception, thus involving group II fibres. The skin is also stimulated and produces a *local* tactile sensation (at 0.7 S) or a painful sensation (at 3 S); these direct skin stimulation effects combine with the afferents from the sural nerve trunk.

For recording, surface electrodes were placed on the skin over the muscles investigated: biceps femoris capitis brevis (SBi); biceps femoris capitis longus (LBi); gastrocnemius medialis (GM) or lateralis (GL); soleus (Sol); quadriceps femoris (Q); tibialis anterior (TA). Standard differential recording was used and the EMG responses were generally integrated and sometimes averaged.

Results

Nociceptive Reflex of the Short Femoral Biceps

A sural nerve stimulus, single pulse or a brief train of pulses, producing pain elicits a polyphasic reflex which involves first the short biceps and then spreads to other flexor muscles such as LBi and TA, then to the adductors when the stimulus is increased. The experiments were limited to the threshold case where the reflexes involve only the SBi (fig. 1). The sensation associated with the stimulus (brief, acute and localized pain) remains supportable during a long experimental session.

The latency of SBi reflexes varies between 120 and 85 msec. The duration can be very short (a few msec), or longer (up to 100 msec) (fig. 2A).

Several arguments suggest that these SBi nociceptive reflexes are mediated by group III afferents of the lateral saphenous nerve and involve only a spinal pathway, namely the wave form of the compound action potential of the sural nerve recorded and averaged during such stimulations, the conduction velocity of the afferent volley and the latency of the reflex responses (fig. 3) [HUGON, 1967]. This nociceptive reflex was therefore called RA III since it involves group III afferents. Unmyelinated afferents conducting at about 1 m/sec

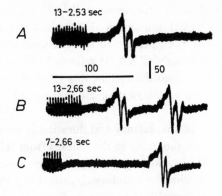

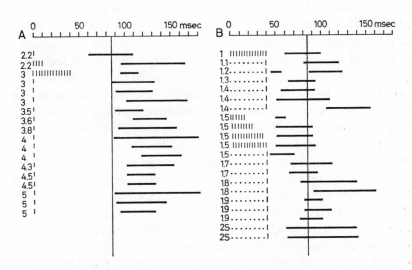

Fig. 1. Reflexes of the short biceps (SBi) muscle elicited by electrical stimulation of the sural nerve at the malleolus externus. *A* RA II tactile reflex elicited by a train of stimuli which do not elicit pain. *B* combined RA II and III reflexes elicited by a train of painful stimuli. *C* RA III nociceptive reflex elicited alone by a shorter train of painful shocks. Notice the similarity of wave forms of the two reflexes. Calibrations: 100 msec and 50 µV.

Fig. 2. Comparison in different subjects of latencies and durations of RA III nociceptive reflexes (A) and RA II tactile reflexes (B) of the SBi muscle. Abscissa: time from the start of the stimulus which is single in most instances of *A*, and repetitive in *B* (see time of occurrence of the stimuli indicated by small dots). The figures before each line indicate intensities relative to tactile threshold. The horizontal black lines indicate the time of occurrence of the reflex EMG bursts in each subject.

are too slow to account for the reflex latency. Furthermore, RA III has been recorded in a man with a complete surgical transection of the spinal cord at the level of Th 12. This transection had been performed for treatment after a gunshot wound and the experimental study of RA III reflexes in SBi was done 8 months after surgery. This leaves no doubt that this reflex can be organized in the spinal cord even after withdrawal of all supraspinal influences.

The polyphasic RA III reflex of variable latency and duration is interpreted in man as a polysynaptic reflex comparable to the reflexes from skin described in animals [LLOYD, 1943]. It is evoked by nociceptive stimulation and can be viewed as a reflex of protection by withdrawal, resembling the triple flexion seen in animals [SHERRINGTON, 1906] and in man [KUGELBERG, EKLUND and GRIMBY, 1960; HAGBARTH, 1960].

Tactile Reflex of the Short Femoral Biceps

A train of 6–10 weak electric pulses producing a tactile sensation in the lateral sural nerve territory elicits a polyphasic reflex of shorter latency in the SBi (fig. 1). This reflex to tactile stimulation appears first, with the lowest threshold, in the SBi. The latency estimated between the beginning of the pulse train and the reflex EMG response varies between 40 and 60 msec (fig. 2B). For a minimal stimulus train which would elicit the SBi reflex every two trials, the last pulse of the train precedes the reflex EMG response by at least 35 msec. A similar study as was performed for the RA III nociceptive reflex indicates that this tactile reflex involving the same SBi muscle must be a spinal reflex triggered by group II afferents from the skin. Evidence for this is provided by the configuration of the sural nerve potential associated with such stimuli, the velocity of the sensory and motor reflex pathways (fig. 3) and by the presence of the reflex in the spinal subject studied. This tactile reflex of the SBi was therefore called RA II [HUGON, 1967]. The central summation time for the reflex is roughly 10–30 msec. Repetitive electric stimulation with mild (tactile) intensities are best used to elicit this reflex. Figure 2B shows the usual fluctuations of latency and duration of the reflex. RA II being a phasic reflex of variable latency and duration may be interpreted as a polysynaptic reflex homologous to the 'flexion reflex' which LLOYD [1943] attributed to the activity of group II afferents.

The discovery of this reflex elicited by tactile afferents rested upon a systematic search for EMG responses among the muscles of the lower limb

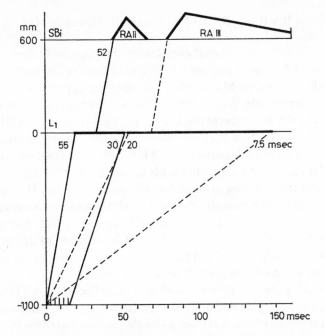

Fig.3. Analysis of latencies of RA II and III SBi reflexes. Abscissa: time in msec. Ordinate, pathway travelled from the external malleolus (−1100 mm) to the spinal cord (0) and then to the SBi muscle (about 600 mm). In the lower section, the afferent volleys of conduction velocity between 55 and 7.5 m/sec are shown. The 2 thick horizontal lines indicate the times of arrival at the cord of afferent volleys elicited by a train of small stimuli (tactile sensation) and by a single strong painful shock, respectively. After a tentative spinal cord delay, the motor volleys travelling to the SBi muscle are shown, as well as the approximate EMG discharges of the RA II and III reflexes.

[HUGON, 1967]. Having found that the SBi disclosed consistently the best responses for liminal stimulation, I noticed that the SBi in man is embryologically homologous to the tenuissimus muscle of the cat. Now, HUNT [1951] showed that activation of hair receptors in the sural territory in cats elicited a polysynaptic reflex response in the tenuissimus. The latter response is probably homologous to the RA II reflex found in man. WALSHE had already shown [1914] that the plantar arch and the lateral side of the foot are skin areas from which a reflex can be elicited which involves at threshold the knee flexor muscles, the effective stimulus being moving pressure on the skin. All these responses are also non-nociceptive and related to a tactile input.

Comparison of RA II and RA III Reflexes of SBi Muscle in Man

On the effector side, these two reflexes can involve apparently the same motor units in the SBi. This is suggested by the wave form of the responses when liminal reflexes are evoked and when the recording is rather selective (fig. 1). On the afferent side the difference between either types of stimuli evoking RA II and RA III, respectively, is quite clear subjectively. RA III is evoked by stimuli producing a pain sensation and since stronger electrical pulses activate the smaller nociceptive axons, it is not surprising that the critical parameter for recruiting a larger RA III is the intensity of the stimulus. On the other hand the sensation produced by stimuli evoking RA II is of a tactile nature; therefore the intensity of the electrical stimulus should remain low enough not to involve the smaller nociceptive axons. In fact, a single stimulus does not succeed in eliciting RA II and it is necessary to deliver a brief train of mild stimuli. The critical factor here is the number of stimuli, the strength of which should remain below the pain threshold.

When comparing the latency and duration of the reflex EMG discharge in a number of subjects (fig. 2) it appears that the latency is shorter and more variable for RA II (B) than for RA III. On the whole the duration of the RA II EMG response appears shorter. There are many differences among subjects, especially for the tactile RA II response which will not easily be observed unless the subject is relaxed and familiar with the experimental conditions. The second session on a given subject often gives better results than the first for RA II. Another point which must be stressed is that these reflexes should not be elicited at intervals below 5 sec, otherwise they would be depressed. I have also emphasized this point which is all too frequently neglected in the chapter on H-reflex methodology [HUGON, this volume, page 277].

The RA II has been considered above as a placing reaction while the RA III was interpreted as a nociceptive flexion reflex. This point deserves discussion as one would not perhaps see clearly how the same muscle SBi would subserve such diverse functions. When the lower limb is extended as in the upright position, the SBi muscle functions as a flexor muscle. When the knee is flexed at 90°, contraction of SBi produces external rotation of the leg and foot. Thus a tactile stimulation on the outer border of the foot will produce an RA II which would flex the limb if it were extended, and which would rotate the limb outward if it were flexed. In the latter case, the contact of the external aspect of the foot with the stimulus would be increased, much as one would expect from a 'placing' reaction of tactile, non-nociceptive, origin. This raises interesting questions about the contribution of the tactile input to the

functional organization of postural reactions. Thus the biomechanical significance of the RA II reflex pathway would depend on the initial position of the
limb, and probably also on the background of supraspinal influences which
would set the stage of spinal motor systems either toward positioning the
foot or toward withdrawal [DENNY-BROWN, 1966].

When now considering the RA III nociceptive reflex elicited by stronger
stimuli, one must concede that, if the knee is flexed, the SBi muscle contraction would also rotate the foot towards the stimulus, which would appear odd.
However, such a strong stimulus will elicit reflex responses which will not be
confined to the SBi and will involve many other flexor muscles whereby effective withdrawal of the foot is achieved. For nociceptive stimuli, the above
considerations on the SBi placing effect on the foot no longer have much bio-
mechanical significance and the total reflex reaction is just withdrawal.

Interactions between Tactile and Nociceptive Input

Current studies initiated by MELZACK and WALL [1965] have emphasized
the far-reaching significance of interactions between tactile and nociceptive
afferents supposed to be organized in the dorsal horn of the spinal cord, and
possibly at higher levels as well, for the perception of pain. It appears obvious
that our identification of tactile and nociceptive reflexes involving the same
muscle SBi should provide outstanding opportunities for the detailed analysis
of similat interactions in man. Whereas it would appear difficult to study
dorsal horn mechanisms in man as has been done for the cat, it is quite feasible
to approach the same kind of problem using as a test the changes in reflex
muscle responses to the two types of afferent stimulations, respectively. This
approach indeed proved feasible and fruitful and the results so far obtained
are thought to extend to a significant degree the MELZACK-WALL approach
to the human subject.

Effect of Nociceptive Stimulation

When an electrical stimulus 3 times threshold eliciting a pain sensation is
repeated every 2 sec, the RA III reflex in SBi is progressively increased in
amplitude and the reflex EMG discharge is more and more prolonged (fig. 4).
At the same time the pain which is elicited by the successive stimuli increases
(HUGON, BATHIEN and DUCOURTHIAL, 1965; HUGON, 1967). A similar effect
can be obtained by double stimulation with intense shocks (fig. 5C). This
kind of facilitation is prolonged and can be regarded as a 'sensitization'

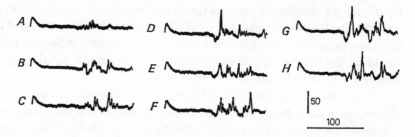

Fig. 4. Facilitation of the RA III nociceptive reflex discharge by repeating the stimulus. An electrical stimulus of 3.5 times the tactile threshold is delivered to the lateral saphenous nerve every 2 sec. The records represent the SBi discharge elicited respectively by the first stimulus (A), the third (B), the fifth (C), the seventh (D), the ninth (E), the eleventh (F), the thirteenth (G) and the fifteenth (H) of the same series. Notice the increase in amplitude and duration of the discharge. Calibrations: 100 msec and 50 µV.

as commented upon by PROSSER and HUNTER [1936] and SPENCER, THOMPSON and NEILSON [1966]. The question whether this involves a purely spinal mechanism has been answered by repeating the experiment in the patient with a spinal transection at thoracic level. This is in line with observations on other reflexes in spinal man [TUREEN, 1941; DIMITRIJEVIC and NATHAN, 1970]. While the latter authors describe the sensitization as transient and easily habituated, I found the sensitization to be a rather peristent phenomenon during repetition as far as the RA III nociceptive reflex of SBi is concerned.

An interesting point is that the SBi muscle can also be tested by tendon percussion which elicits in this muscle a monosynaptic reflex [HUGON, 1967]. It thus becomes possible to test the SBi motoneurones by this proprioceptiv input in the course of the repeated nociceptive stimulation which elicits RA III reflexes of the same muscle. These experiments have shown that the proprioceptive reflex of SBi is not facilitated at the time the RA III reflex of the same muscle in the same subject shows this sensitization. Thus the SBi motoneurones do not appear to have an increased excitability and one must conclude that the sensitization of the RA III nociceptive pathway through repetition involves a mechanism located uphill to the SBi motoneurones, possibly in the primary afferent terminals or in the involved interneurons.

Effect of Preceding Tactile Stimulation
When the sural nerve is stimulated by shocks below twice tactile threshold intensity which do not evoke a pain sensation, the subsequent RA II and

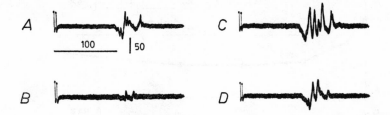

Fig.5. Modification of the RA III nociceptive reflex of the SBi muscle by a single electrical stimulus delivered 50 msec before. *A* and *D* are control RA III reflexes elicited by 2 stimuli of 3.5 times the threshold intensity for a tactile sensation. This pair of stimuli produces a pain sensation. *B* The same nociceptive stimulation is preceded by a single stimulus of threshold intensity to the lateral saphenous nerve; this tactile stimulation does not elicit an RA II reflex but depresses the subsequent RA III nociceptive reflex. *C* The same pair of stimuli is now preceded by a single shock of 3.0 times threshold intensity and this preceding painful stimulation facilitates the RA III. Calibrations: 100 msec and 50 µV.

RA III SBi reflexes are depressed. Figure 5B shows this effect for a single stimulus preceding the nociceptive group III afferent volley. This depression is rather prolonged and lasts for several hundred msec. It is maximal for intervals of about 100–300 msec between the tactile stimulus and the nociceptive stimulus eliciting the testing RA III reflex discharge (fig. 6A). A mild electrical stimulation of the peroneal nerve with a train of 3 electrical pulses of intensity equal to tactile threshold also produces a depression of the RA II discharge in the SBi muscle, with a slightly shorter time course in this example (fig. 6B). A mechanical stimulation of the outer aspect of the ipsilateral foot (tactile sensation) also produces a depression of the RA III nociceptive reflex, as well as of the RA II reflex (fig. 6C).

Effect of the Interval in Sequential SBi Reflex Stimulations

The above observations can explain the difficulties encountered for eliciting the RA II tactile reflex when the intervals between stimuli are below about 5 sec. The depression following each tactile stimulus reduces the efficiency of the subsequent volleys and one may have to augment the intervals between stimuli up to 10 sec to observe a stable RA II response in the SBi muscle. On the other hand the RA III nociceptive reflex is more readily observed because the painful stimuli used leave behind a facilitation so that even with intervals of a few seconds the RA III response will be maintained, or indeed progressively potentiated.

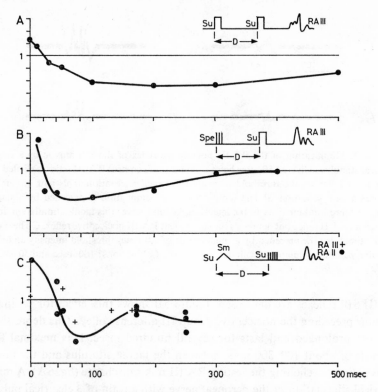

Fig.6. Long-lasting depression produced by a preceding tactile stimulus. Each inset defines the experimental protocol which involves first a tactile stimulation followed at various intervals (abscissa) by a testing stimulus evoking either an RA III nociceptive reflex in SBi muscle (A, B, C: crosses) or an RA II tactile reflex in the same muscle (C: dots). Each point in the diagrams represents the mean of 10 trials at that interval, for *B* and *C*. In *A* the pooled results obtained in 11 normal adult subjects are assembled and each point represents the mean of 220 trials. *A* The first stimulus is delivered to the sural nerve. *B* The first stimulus is delivered to the peroneal nerve. *C* The conditioning stimulus is a mechanical pressure applied to the outer aspect of the ipsilateral foot (tactile sensation).

Perceptual Interactions of Skin Stimuli

The decrease of the amplitude of the RA II EMG response in SBi by a preceding tactile stimulation points to a depressive interaction involving the polysynaptic pathway from the skin. It is interesting that a similar interaction can be observed when the perceptual threshold is tested (fig. 7). The testing stimulus is a single electrical pulse on the lateral saphenous nerve at the malleolus, and its intensity is adjusted so that it is just perceived by the subject.

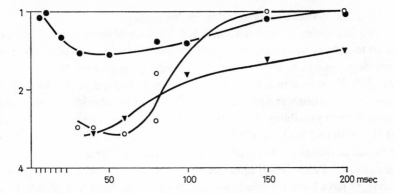

Fig. 7. Depression of tactile sensation by a preceding tactile stimulation. Abscissa: interval between the conditioning train to lateral saphenous nerve (triangles and circles) or to the thigh (dots) and the testing stimulus to the sural nerve. Ordinate, ratio of the threshold intensity of the testing stimulus preceded by the conditioning train with respect to the threshold for the testing stimulus applied alone. The time of occurrence of the conditioning train is indicated to the left on the abscissa.

This stimulus is preceded at various intervals by a brief train of mild stimuli to the same sural nerve or on the thigh. The changes in intensity of the testing stimulus which are necessary for it to be perceived are plotted on the ordinate. A clear increase in threshold (or decrease in perception) is observed. The threshold is more than doubled for homonymous conditioning and it increases by about 50% for heteronymous conditioning. The effect dissipates for intervals of 100–200 msec. This evidence should, of course, be strengthened by further experiments but it suggests that a similar depressive interaction obtains between sequential group II tactile inputs as far as corticipetal volleys mediating tactile sensation are concerned. It is indeed tempting to consider that the neural mechanism involved influences simultaneously the neural pathways corresponding to the RA II SBi reflex and the ascending dorsal column volley, respectively, at a common neural link.

On the other hand when the testing stimulus is of such a strength that it elicits pain, the preceding tactile stimulation reduces the intensity of the perceived pain. This observation confirms those of MELZACK and WALL [1965] and of WALL and SWEET [1967]. I have also confirmed that sequential nociceptive stimulation exacerbates the pain sensation elicited by each successive stimulus [MELZACK and WALL, 1965], which again provides a similar pattern of interaction to the one found for the RA II and RA III exteroceptive reflexes of the SBi muscle.

Possible Mechanisms Involved in the Interaction

The facilitation of the nociceptive RA III reflex of the SBi muscle may be related to a decrease in the level of presynaptic inhibition and of primary afferent depolarization, according to the model proposed by MELZACK and WALL [1965], or else to a facilitation of neurons located in the nociceptive pathway. WHITEHORN and BURGESS [1970] have presented evidence apparently against the first possibility. WALL [1967], POMERANZ, WALL and WEBER [1968] and HILLMAN and WALL [1969] have shown that neurons of Rexed layer V, in the dorsal horn, are easily excited by high-threshold afferents in the cat, and WAGMAN and PRICE [1969] have found this also in the monkey. Repetitive discharges have been recorded in neurons of Rexed layer V and in cells of layer I as well [CHRISTENSEN and PERL, 1970].

While layer V neurons become increasingly active and facilitated when the cord receives afferent impulses in smaller afferents, it has been found that inhibitory mechanisms become more active when the afferent volleys contain a high proportion of impulses in the larger diameter, lower threshold afferents [HILLMAN and WALL, 1969]. These postsynaptic effects in laminae IV and V neurons [cf. HONGO, JANKOWSKA and LUNDBERG, 1965] and the presynaptic inhibition on primary afferents [cf. ECCLES, 1964] might both be involved and it is difficult to evaluate their relative importance in the interaction phenomenon recorded both for perceived skin sensations [MELZACK and WALL, 1965] and for the RA II and III reflexes of the SBi muscle [HUGON, 1967]. My own observations on the spinal cat indicate that mechanical stimulation of the hairs depresses spinal polysynaptic reflexes and decreases the dorsal root potentials and dorsal root reflexes. Such tactile stimulation also decreases the response of second-order neurons of the dorsal horn which are activated by group III afferents (fig. 8) [HUGON, 1967, 1969]. These observations would perhaps emphasize the possible contribution of presynaptic inhibitory mechanisms in the depression of nociceptive pathways by preceding tactile stimulation.

Effects of Afferents from Skin on Proprioceptive Reflexes

A low intensity stimulus to the sural nerve, thus eliciting a tactile sensation, facilitates in a non-reciprocal way the proprioceptive reflexes of the ipsilateral limb (fig. 9 A–C). The diagrams present pooled data obtained in several different subjects (number indicated in the upper right corner) and each point represents the mean of 10 trials in each of these subjects. When there are 10 subjects the mean is thus based on 100 trials. The facilitation is as marked for the Achilles T-reflex as for the H-reflex of the soleus (fig. 9 A). It is also very

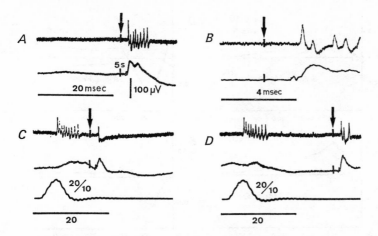

Fig. 8. Depression of a second-order neurone response to a nociceptive stimulus by a preceding tactile input in the spinal cat. The top record in each frame represents the discharge of a second-order neurone in the dorsal horn recorded by an extracellular microelectrode. The second trace represents the dorsal cord potential. At each arrow an electrical stimulus 5 times threshold is delivered to the medial saphenous nerve. Calibrations in msec and in μV. *A* Control response. *B* Same response on a fast time base, indicating that the latency of the first spike is short enough for it to be monosynaptic. *C, D* The same stimulus is preceded by mechanical stimulation (see bottom trace indicating movement produced) of the hairs of the skin innervated by the sural nerve. The response to the shock is depressed more for a shorter interval (C).

clear for the monosynaptic reflex elicited in the quadriceps muscle by percussion of the patellar tendon (fig. 9 B) and for the monosynaptic reflex of the SBi elicited by percussion of its own tendon (fig. 9 C). The facilitation lasts more than 150 msec [HUGON and BATHIEN, 1967].

When a stronger stimulus is delivered to the sural nerve so as to produce pain, it produces a depression of the proprioceptive reflexes in the extensor muscles, soleus (fig. 9 D) and quadriceps (fig. 9 E). The same effect is recorded whether the soleus motoneurones are tested by Achilles T-reflex or by H-reflex. The effect on the monosynaptic reflex of the flexor SBi muscle is generally facilitatory (fig. 9 F, bigger dots; 9 subjects), but can sometimes be depressive (smaller dots, 2 subjects).

Exteroceptive Silent Periods

Silent periods in the voluntary background EMG can be elicited by a stimulus to the motor nerve or by unloading the contracting muscle [see

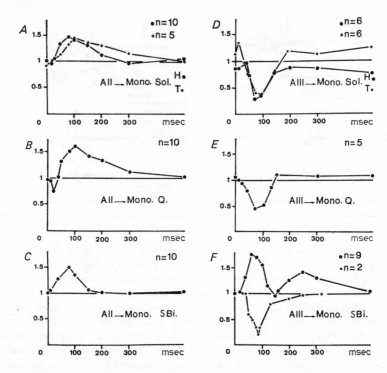

Fig. 9. Effect of skin afferents on proprioceptive reflexes. The conditioning stimulus to the lateral saphenous nerve is a group II afferent volley eliciting a tactile sensation in *A*, *B* and *C*. The stimulus is stronger and elicits a pain sensation for *D*, *E* and *F*. The responses tested are the Achilles T-reflex (smaller dots) and the H-reflex of the soleus muscle (A, D), the patellar reflex of the quadriceps (B, E) and the monosynaptic tendon reflex of SBi muscle (C, F). The abscissa represents the intervals between the two stimuli, corrected for afferent conduction up to spinal cord entry. The number of subjects tested is indicated in the upper right corner of each diagram. Each point is the mean of 10 trials in each of these subjects.

ANGEL; STRUPPLER *et al.*, this volume]. It is of great potential interest that afferent stimulation confined to the skin also elicits silent periods in a contracting muscle [LIBERSON, 1963, 1971; HUGON, 1967; SHAHANI and YOUNG, this volume]. A detailed study has been performed of the effect of sural nerve stimulation on the voluntary EMG of various muscles of the ipsilateral limb. The subject is asked to contract a given muscle in a moderate, stable manner during the test. Figures 10 and 11 illustrate several findings. When comparing the effect of a group II tactile input with a stronger stimulation eliciting pain, the exteroceptive silent periods present different features. Their latencies cor-

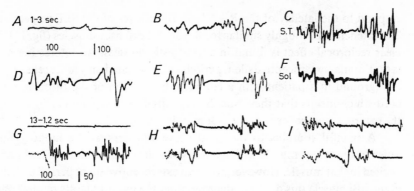

Fig. 10. Exteroceptive silent periods, in muscle under slight voluntary contraction and recorded with a concentric electrode. *A–F* Effect of a strong electrical stimulus 3 times threshold eliciting pain. *G–I* Effect of a train of 13 electrical pulses 1.2 times threshold eliciting a tactile sensation. *A, B* Recording from SBi muscle. *C* Semi-tendinosus. *D* Rectus femoris muscle. *E* Tibialis anterior muscle. *F* Soleus muscle. *G, H, I* Tibialis anterior (upper trace) and SBi muscle. Calibrations: 100 msec and 50 μV. In *G*, the SBi is flexing the knee and the tibialis is relaxed. In *H, I*, both muscles are activated in flexion and external rotation of foot.

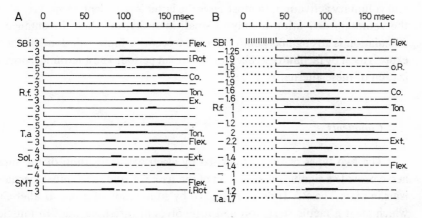

Fig. 11. Modifications of voluntary EMG in various limb muscles by stimulation of the sural nerve with strong painful shocks (A) or with mild electrical pulses eliciting a tactile sensation (B). Each line represents a different observation. The slight EMG background is indicated by a thin continuous line. The thick lines indicate an increase in EMG activity and the interrupted line represents a depression of the EMG activity. The muscles recorded from are indicated on the left: short biceps (SBi), rectus femoris (Rf), tibialis anterior (Ta) soleus (Sol), semi-tendinosus (SMT). The type of tonic activity of each muscle is indicated on the right: flexion (Flex), rotation, internal (iRot) or external (oR), contraction of the muscle and of its antagonists (Co), slight non-deliberate tonic EMG activity (Ton), extension (Ex).

respond to the latency of the RA II and III reflexes of the SBi, respectively. The latencies are roughly similar for flexor and extensor muscles (fig. 11). No clear reciprocal effect is found in antagonistic muscles. An interesting point is that similar exteroceptive silent periods can be elicited on a tonic reflex EMG background in a patient with a complete transection of the thoracic spinal cord which means that they must be organized in the spinal cord itself, like the exteroceptive reflexes of the SBi muscle.

A painful stimulus to the sural nerve will depress the monosynaptic reflex in the soleus muscle with a latency which is similar to the silent period elicited in that muscle. However, when an exteroceptive silent period is elicited in the SBi muscle this does not depress either the monosynaptic tendon reflex of that muscle or the RA III reflex. Thus, the inhibition of voluntary EMG activity does not seem to involve the reflex discharges in the SBi motoneurone pool, as if the silent period was organized at the level of interneurons associated with the voluntary commands [HUGON, 1967]. Finally, when voluntary activation is stronger, exteroceptive silent periods and reflexes are no longer visible while the monosynaptic reflexes of the same muscles are facilitated. There would seem to be a sort of competition between exteroceptive inputs and voluntary activation. In most cases the latter is prevalent. However, when the voluntary activation is only of small intensity, or when the cutaneous stimulation is strong and painful, the exteroceptive input will dominate and express itself by facilitatory effects or by a silent period. The anatomical organization of these alternative effects is not yet clear.

Comments

The two types of reflexes elicited by electrical stimulation of the sural nerve provide a significant extension of the scope of studies of spinal mechanisms in man. The RA II reflex is evoked by stimulation of group II afferents which elicit a tactile sensation and it involves at threshold the SBi muscle. These tactile reflexes can be large and consistently present in some subjects while they are difficult to demonstrate in others. Several technical points should be kept in mind, chiefly the depressive interaction between sequential tactile RA II reflexes. Unless evoked at intervals of 5 or 10 sec these reflexes will appear labile. On the other hand, the nociceptive RA III reflexes of the SBi muscle are evoked by group III afferents which evoke pain. The latter input appears to facilitate the polysynaptic pathway so that RA III reflexes can easily be demonstrated during sequential activation. For quantification of these reflex EMG discharges it is necessary to use an integrator and to average about 10 responses.

This paper also describes a series of interactions between the two types of afferents from the skin, as disclosed by the recording of the SBi reflexes. These results are in line with MELZACK and WALL's [1965] views of the inter-action between larger and smaller diameter afferents in the cat dorsal horn [HILLMAN and WALL, 1969] and indeed provide definite evidence for the oper-ation of similar mechanisms in man [HUGON, 1967]. Although the nature of these mechanisms is not yet entirely clear, the study of RA II and III SBi re-flexes opens up a promising line of quantitative investigations of events as-sociated with exteroceptive sensations. The exteroceptive input also influ-ences the proprioceptive reflexes of the SBi muscle and of other flexor or ex-tensor muscles of the ipsilateral limb, and it induces so-called exteroceptive silent periods in the voluntary EMG.

Acknowledgement

I wish to thank Prof. J. E. DESMEDT for his suggestions during the preparation of this paper.

Author's address: Prof. MAURICE HUGON, Département de Neurophysiologie com-parée Université de Provence, *F-13 Marseille* (France)

New Developments in Electromyography and Clinical Neurophysiology,
edited by J. E. Desmedt, vol. 3, pp. 730–733 (Karger, Basel 1973)

Electromyographic Analysis of Human Flexion Reflex Components[1]

J. Faganel

Institute of Clinical Neurophysiology, The University Hospitals of Ljubljana,
Ljubljana

The flexion reflex is a polysynaptic defence response to noxious stimulation of the skin. The pattern of the muscles activated in the reflex is in accordance with its biological function, which is withdrawal of the stimulated limb away from the offending stimulus [Kugelberg, 1962; Grimby, 1963a; Hagbarth and Finer, 1963]. The flexion reflex is usually described as an EMG discharge consisting of several components, which are separated by periods of more or less pronounced EMG silence [Pedersen, 1954; Hagbarth, 1960; Kugelberg, Eklund and Grimby, 1960; Dimitrijević and Nathan, 1968]. According to some authors [Lloyd, 1943; Kugelberg, 1948; Lloyd, 1960] this pattern is due to the activation of groups of afferent fibres with different conduction velocities. Lloyd [1943] has shown that it is possible to evoke a 2-component flexion reflex in the spinal cat; the first component has a lower threshold and is mediated by group II fibres. The second component appears after an interval of silence and is only evoked by strong stimuli. It was presumed to be due to the activation of high-threshold slow conducting fibres. In a group of patients with various spastic conditions Kugelberg [1948] studied the role of group A and C fibres in the flexion reflex; he concluded that fibres of group C mediated the latest components. The cyclical pattern of the flexion reflex has also been ascribed to central mechanisms [Pedersen, 1954; Lloyd, 1960; Grimby, 1963b; Hagbarth and Finer, 1963]. Sherrington [1906] used the term 'after-discharge' to describe the later components which he thought to be due to delay in conduction through the central part of the reflex arc. Pedersen [1954] supposed that the flexion reflex results from

1   This work was supported by the Boris Kidrič Fund of Slovenia (grant No. 306/20:71) and by the US Department of Health, Education and Welfare, Social Rehabilitation Service (grant No. 19-P-58397-F-01).

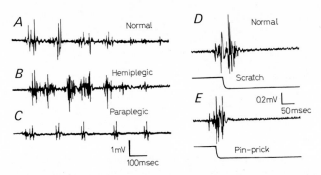

Fig.1. A–C. Electrically elicited flexion reflex in a normal subject (*A*), in a hemiplegic patient (*B*) and in a paraplegic patient (*C*). *D* and *E* Mechanically elicited flexion reflex in a normal subject. The 2nd and 4th trace show the duration of scratch and pinprick, respectively.

bombardment of the ventral horn cells from different internuncial neuron pathways. GRIMBY [1963 b] also believed that later components were more probably due to a longer central reflex time than to slower afferent impulses, but whether this response is of spinal or of cerebral origin cannot be decided judging by the latency values; nor can it be ruled out that it is of a different origin in different cases.

The purpose of this study was to see whether the cyclical pattern of the flexion reflex is due to the type of the stimulus used, to temporal dispersion of the afferent volley as a consequence of the wide range of afferent conduction velocities, or whether it is due to some events in the central part of the reflex arc.

A train of electrical pulses lasting 20 msec of suitable amplitude applied to the skin of the excavated part of the foot with the Copland-Davies electrodes evokes a reflex response in m. tibialis anterior at a latency between 50 and 80 msec. Normal subjects perceive such a stimulus as a sharp burning pain. The reflex response detected by surface electrodes from m. tibialis anterior may consist of a single EMG burst; usually, however, the first burst is followed by one or several later bursts separated by periods of more or less complete EMG silence. Such responses were observed and studied in normal subjects, in patients with capsular hemiplegia and in patients with spastic paraplegia due to complete traumatic division of the spinal cord (fig. 1A,C).

To elucidate the peripheral mechanisms which might contribute to the cyclical pattern we also used mechanical stimulation, pricking with a pin and scratching the sole with a sharp point. The duration of the two stimuli was 70–150 and 150–200 msec, respectively, and the resulting afferent volleys must

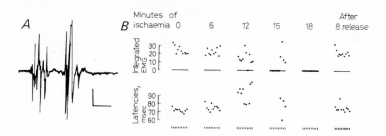

Fig. 2. A Electrically elicited flexion reflex with two components. *B* Magnitudes and latencies of the flexion reflex before, during and after transitory ischaemia. Each dot represents latency or magnitude of one reflex response. Stimuli were delivered in series of 10 at a rate of 1/5 sec, separated by resting periods of 3 min.

have been much more dispersed in time than that of the electrical stimulus, which was only 20 msec long. The mechanical stimuli produced a weaker, but much longer sensation than the electrical stimulus (fig. 1 D, E).

The responses to pin-prick and scratch usually consisted of a single burst, and only in a few cases was there a second burst of EMG activity. We also introduced transitory ischaemia in the stimulated limb with a cuff in order to produce successive differential blocking of afferent fibres of different diameters.

Figure 2, *A* shows a flexion reflex with two EMG bursts, which was analyzed before, during and after ischaemia. The initial mean latency was about 70 msec. In the 12th minute of ischaemia the responses became more variable in size and some responses failed. In the 15th minute 6 of 10 stimuli failed to produce a response, but the mean latency was still almost the same as in the beginning (75 msec). The responses gradually decreased in size but the EMG bursts remained equally well synchronized and their number did not change. In the 18th minute of ischaemia all responses failed; there was no recovery even when stimulus strength was raised by a 100%, though the subject still perceived the stimulus as a sharp pain. The cuff was released after 30 min; when the reflex responses reappeared they had similar latencies to before ischaemia (fig. 2 B).

This kind of change was obtained in all cases. So we could not essentially modify the pattern of the flexion reflex—neither by changing the type and length of the stimulus nor by selective blocking of afferent fibres of larger diameters. Neither in normal subjects nor in hemiplegic and paraplegic patients was it possible to evoke a reflex discharge at a latency long enough to correspond to afferent conduction along unmyelinated group C fibres.

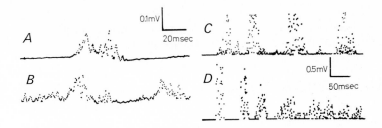

Fig.3. A and *B* Effect of voluntary contraction on the flexion reflex in m. tibialis anterior of a normal subject. Each trace shows the results of summation of 10 rectified responses. *A* Muscle relaxed. *B* During voluntary contraction. Notice shortening of the latency of the first component, and appearance of a later component. *C* Summated rectified flexion reflex in a hemiplegic patient. *D* Same in a paraplegic patient.

The alternative possibility, i. e. that the periodicity in the flexion reflex discharge is due to central mechanisms, was studied by supraspinal conditioning of the central part of the reflex arc. This was done by voluntary plantar and dorsal flexion of the stimulated foot. Plantar flexion characteristically lengthened the latency of the reflex response, while dorsal flexion shortened the latency. Besides, dorsal flexion of the foot resulted in appearance of a later component in the reflex response, which was previously not present (fig. 3A, B).

The cyclical activity, which is manifested in the flexion reflex in the form of repetitive EMG bursts seems, therefore, to be due more to central mechanisms than to the temporal pattern of the afferent volley. This assumption is also supported by the changes of the flexion reflex in capsular hemiplegia and spinal paraplegia. The increased motoneurone excitability in these pathological conditions resulted in a prolongation of the flexion reflex which here has quite a number of later components even when elicited with very weak stimuli just above the threshold. The EMG bursts in figure 3C, D were summated from several responses; they appeared at a frequency of 8–10/sec. To sum up, the noxious stimulus applied to the plantar surface evokes a reflex response in m. tibialis anterior with cyclical bursts of EMG activity both in normal subjects and in patients with hemiplegia and paraplegia. We could not confirm any correlation between this cyclical activity and the pattern of afferent volley. Increase in excitability of the central part of the reflex arc was found to facilitate and prolong the cyclical activity.

Author's address: Dr. JANEZ FAGANEL, Klinični center, Halzapflova ulica, *61000 Ljubljana 5* (Yugoslavia)

New Developments in Electromyography and Clinical Neurophysiology,
edited by J. E. Desmedt, vol. 3, pp. 734–743 (Karger, Basel 1973)

Human Flexor Spasms

B. T. SHAHANI and R. R. YOUNG

Department of Neurology, Harvard Medical School, and Massachusetts General
Hospital, Boston, Mass.

The symptomatology associated with partial and total transection of the
spinal cord has for many years been of interest to both physicians and physiologists. As long ago as the latter half of the last century, it was recognized by
clinical neurologists [ROMBERG, 1853; GOWERS, 1886; CHARCOT, 1894] that
after complete transection of the cord in man, a brief period of total areflexia
is followed by a chronic state of heightened reflex activity in the isolated cord.
This generally accepted view was questioned by BASTIAN in 1890. On the basis
of clinical and pathological evidence in man, he concluded that complete
transection causes complete paralysis and permanent abolition of all reflexes
below the lesion, the only reflex response ever obtainable being occasional
slight movement of toes when the soles are strongly tickled. BASTIAN's law
was an accepted doctrine until HEAD and RIDDOCH [1917; RIDDOCH, 1917]
demonstrated that if the patients with completely divided spinal cords were
kept alive with good nursing care, the flexor reflexes returned in the paralyzed
limbs after a period of 2–3 weeks. Although they found the reflex response to
be feeble and gentle at first, it soon gained strength and was sometimes associated with premature bladder evacuation and excessive sweating ('mass reflex').
For many years after publication of their classic studies, it was assumed that
men with total transection of the cord exhibited only flexor reflexes or spasms
and that the occurrence of extensor reflexes was a sign of an incomplete lesion
of the spinal cord [FULTON, 1949]. This myth was dispelled by KUHN and
MACHT [1949] who studied 27 patients in whom verified complete transection
of the cord had occurred two or more years previously. They were able to demonstrate extensor reflex activity in the paralyzed limbs of 22 of these patients.

It has become clear that the duration of human spinal shock varies inversely with general health. Following a severe cord lesion, patients with

multiple other injuries, infections (especially of lung, bladder or skin) or poor nutrition may remain in the state described by Bastian for a very long time whether or not the transection is complete. Under better circumstances reflexes begin to reappear in a few days or weeks. The Babinski response and other flexor reflexes are seen first and later, if recovery continues, extensor reflexes appear. The opposite sequence is seen during regression of these reflexes if infection, fever or systemic illness supervenes at any time thereafter.

Unfortunately, as they become more easily elicitable, certain of these reflexes, especially the flexor ones, become troublesome to the patient. For example, the involuntary, predominantly flexor, movements seen in the legs of patients with spinal cord disease are termed, in the absence of known provocative stimuli, *flexor spasms*. They are often associated with a painful sensation, but even otherwise the movement itself and any bowel and bladder contraction may be quite distressing. Though they have attracted much clinical attention, there has been little study of their pathophysiology. Most clinicians consider flexor spasms to be merely heightened and labile flexor reflexes, whereas Rushworth [1966], for instance, felt that 'flexor spasms are not just heightened flexor reflexes, for their latency bears no relation to the true spinal flexor reflex'. Flexor reflexes *have* been investigated by a number of people during the past 50 years, and it is therefore of interest to study flexor spasms in the same way.

This study and another [Shahani and Young, 1971], investigating the behavior of flexor reflexes in normal subjects and patients with lesions at various levels of the neuraxis, were undertaken to compare flexor spasms with normal and abnormal human flexor reflexes and to study the effect on these of a new pharmacological agent which had proved useful clinically in the treatment of painful flexor spasms [Jones, Burke, Marosszeky and Gillies, 1970].

Methods

Parameters of testing and methods for the recording of both spontaneous flexor spasms and evoked flexor reflexes have been described elsewhere [Shahani and Young, 1971]. In brief, surface EMG recordings are made with electrodes over tibialis anterior and gastrocnemius. Flexor responses are evoked by 20 msec long supramaximal trains of 0.1 msec square waves at 500/sec. These are led through intradermal needles inserted into the medial surface of the sole of the foot. Five adult patients with extensive spinal cord lesions as a result of multiple sclerosis or trauma had severe, painful, spontaneous flexor spasms. They were studied before, during and after the administration of Ciba Ba-34647

(Lioresal; β 4-chlorophenyl γ amino butyric acid), 10 mg by mouth, 3 or 4 times daily for 2 weeks or, acutely, 10–15 mg intravenously. Care was always taken to monitor spontaneous flexor spasms preceding a test stimulus because of the long-term effect of such spontaneous activity on reflex excitability.

Results

Flexor spasms, when considered from almost any viewpoint, are extremely variable. The degree to which they are painful or otherwise unpleasant varies, in part at least, with the nature and amount of afferent information able to ascend past the spinal lesion(s). Though flexion of hip, knee and ankle characteristically occur together, in this study we have restricted ourselves to recording primarily from tibialis anterior. This muscle was chosen for the study of flexor spasms because of our extensive experience with its behavior in studies of normal and abnormal flexor reflexes [SHAHANI and YOUNG, 1971]. We did not carry out detailed sampling of various lower limb muscles to determine the patterns and timing of their responses in flexor spasms; nevertheless, our preliminary observations did lead us to believe that the behavior of tibialis anterior is quite representative of that of the other muscles in a flexor spasm. However, even with this restricted view, the frequency of recorded flexor spasms ranged from one every 2–3 sec to one every 2, 3 h or more. Within the recording range of our electrodes, the size of these spasms varied from a single unit, firing 1–10 times in a second or two, to many units firing together to produce a complete interference pattern and, of course, the movement varied concomitantly.

Figure 1 A shows a 'mini-flexor spasm' during which a single unit is seen to be active. In figure 1 B, the same unit fires at a slightly higher frequency (an average of 9/sec *versus* 8/sec for the first 6 responses) before it is joined by other units in a larger flexor spasm. When the intensity of a spasm was rather slight, it was often possible to see acceleration of the firing rate of an initial unit followed by recruitment of other units when the first reached rates of 10–12/sec. In mild spasms, this presumed increase in 'central excitatory state' builds up over several hundred msec, and even in larger spasms (fig. 1 C), it may take almost as long to reach a maximal level.

Though the exact source of neural activity responsible for flexor spasms is unknown, it presumably comes from spinal afferent input since dorsal rhizotomy can, in the isolated cord, abolish flexor spasms. Furthermore, they probably originate largely in the leg itself. We have observed that the frequency of spontaneous flexor spasms is much higher during periods when the

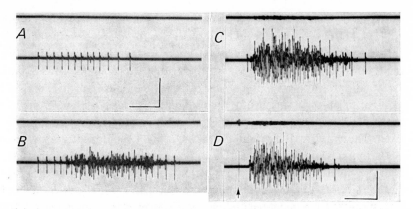

Fig. 1. *A, B* Two spontaneous flexor spasms recorded from tibialis anterior (lower tracing of each pair) and gastrocnemius (upper tracing) of a patient with a rostral spinal cord lesion. Note the single unit activity in *A* during a very small flexor spasm and in *B* the same unit joined subsequently by others in a larger spasm. *C, D* Comparison of a spontaneous flexor spasm *(C)* with an electrically evoked flexor reflex *(D)* in a patient with a rostral spinal cord lesion. Upper tracing from gastrocnemius and lower from tibialis anterior; onset of stimulus at the arrow. Note the similarity in duration and configuration with the characteristic crescendo buildup in each. Calibrations are 500 msec and 500 μV.

limb is the site of a pressure ulcer, a local cellulitis, or other presumed source of noxious afferent stimuli, irrespective of the patient's awareness of pain. It also proved very easy to increase the frequency of flexor spasms without changing their patterns or other characteristics, by light rubbing, scratching, or tickling of the foot, regardless of whether the patient could feel it.

Flexor spasms are indistinguishable clinically (with regard to the movement produced, its duration, the pattern of muscles seen to be involved, and the presence, absence or degree of associated pain) from flexor reflexes produced in the same patient by stronger mechanical (pinch, touch, tickle) or electrocutaneous stimulation. Furthermore, if an electrically produced flexor reflex is interposed when flexor spasms are occurring at relatively regular intervals, the cycle is preserved by the omission of what one would have expected to be the next flexor spasm. Similarly (fig. 1 C, D), neither the EMG amplitude, duration and shape of these flexor spasms or electrically evoked reflexes in these patients, nor the patterns of cutaneously produced changes in these reflex patterns [SHAHANI and YOUNG, 1971] are in any way different from those found in electrically provoked flexor reflexes in other patients with cord disease. The latencies of the electrically evoked reflexes are also similar in all these patients, and neither these mechanically produced reflexes nor the spontaneous spasms consist of the two separate components seen in normal subjects. The flexor reflexes evoked in patients with spontaneous flexor spasms differ then in no clearly apparent way from those found in other patients with spinal cord disease. In all these patients, a single stimulus may suffice (unlike the situation with Sherrington's 'scratch reflex'), or a train of stimuli may be required to produce a reflex, depending on the threshold in general as well as the phase of the spontaneous flexor spasm cycle in which the stimulus is given. Though formal excitability curves, using a spontaneous flexor spasm as the initial response, were not carried out, it was apparent in our study of both the spinal man and the other patients with spontaneous flexor spasms that the size of a flexor reflex produced by electrical stimuli varied markedly from moment to moment depending upon how recently a spontaneous flexor spasm had occurred. That is, flexor spasms appear to 'exhaust' the central flexor reflex mechanisms just like a preceding flexor reflex does. This encouraged us to study excitability curves over periods of seconds to minutes in these patients as well as in normal controls [SHAHANI and YOUNG, 1971]. Though we did not measure stimulus current, the thresholds for electrically induced flexor reflexes in patients with spontaneous flexor spasms appeared little different from those in the other patients with spinal cord disease who did not have clinically significant flexor spasms.

The γ amino butyric acid (GABA) derivative proved clinically to be very effective in the 5 patients treated for severe flexor spasms. Though it did not change voluntary power, 'spasticity', sphincter control or general mobility, it markedly reduced the frequency and amplitude of flexor spasms and thereby alleviated much of the patients' discomfort. These patients were studied before, during and after chronic administration of the drug in an effort to see which parameters of the electrically evoked flexor reflexes might be amenable to pharmacological alteration. In addition to longer intervals between spontaneous flexor spasms, the apparent threshold, both electrical and mechanical, for production of flexor reflexes appeared to rise markedly during the time the patient was receiving Ciba Ba-34647. This effect was produced within the first 24 h when the agent was given by mouth, and the effect outlasted the receipt of the drug by 1–2 days. When it was given intravenously, equally dramatic effects were produced but only 2–3 h after the injection. The minimal flexor reflex latency obtainable after supramaximal stimulation of the foot increased from the pretreatment range of 60–80 to 90–130 msec, or sometimes even greater. Figure 2 illustrates this change in latency (as well as amplitude). In figure 2 E, the effects of this compound are illustrated for one patient; note the extremely unusual increase in reflex latency with increasing strength of

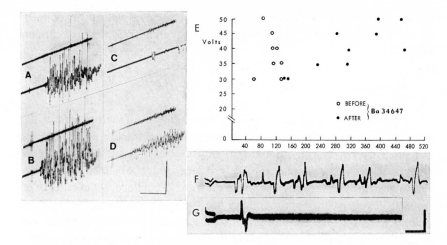

Fig.2. A–D Comparison of electrically evoked flexor reflexes in a patient with a rostral spinal cord lesion before (*A* and *B*) and during (*C* and *D*) the receipt of Lioresal (Ciba). Stimuli for *A* and *C* are single square waves and for *B* and *D* trains of square waves. All stimuli were adjusted to be supramaximal and the stimulus artifacts can be seen. Gastrocnemius is recorded above and tibialis anterior below in each pair. The minimal reflex latency is shorter with trains of stimuli in each instance. Note that the reflex latency is prolonged in *C* and *D* and the response is briefer and of considerably lower amplitude. Calibrations are 200 msec and 500 μV. *E* Relationship between the latency in msec of the flexor reflex response (abscissa) and the strength of electrical stimulus provoking it (ordinate) in a patient with spinal cord lesions and spontaneous flexor spasms before and after receipt of Lioresal. Note the prolongation of latencies in the latter instance. The points plotted represent latencies recorded as the voltage of the stimulus train was gradually increased and then gradually decreased. *F, G* Electrically evoked flexor reflexes in tibialis anterior of a patient with a rostral spinal cord lesion and spontaneous flexor spasms. *F* before and *G* during receipt of Lioresal. The stimulus artifacts visible on the left represent supramaximal trains in both instances. Note that in *G* the reflex is much diminished and its latency somewhat prolonged. Furthermore, it appears as an isolated 'first component' (see text) which is most unusual in the presence of spinal lesions. Calibrations are 50 msec and 500 μV.

stimulation. This was seen in several patients while they were receiving Ciba Ba-34647 and at no other time. The first component of the flexor reflex, which is characteristically very difficult or impossible to isolate in patients with spinal cord disease, did not become more apparent in 4 of the 5 patients during the receipt of Ciba Ba-34647. The one exception to this is illustrated in figure 2F, G, which show the flexor reflex respectively before and during treatment. In this one instance, the electrically evoked flexor reflex became almost normal in appearance. However, the excitability curves were not changed in any patient, nor was the pattern of movement associated with a flexor reflex or flexor spasm, once it had been produced, different from the pretreatment state.

Discussion

Flexor spasms are extremely variable, ranging from the brief, repetitive firing of a single motor unit without any visible contraction in the limb to sustained firing of many motor units in many muscles with massive movements. Because the adequate stimulus for flexor spasms cannot be defined, it is impossible to measure their latency or compare it with that of flexor reflexes. In other respects flexor spasms appear to have the same spinal organization as flexor reflexes. Noxious or painful stimuli, such as trains of electrical pulses, can evoke flexor reflexes in normal subjects and in patients with spinal cord disease. Though there are clear reflex differences between those two groups, in the patient group we cannot differentiate on reflex grounds between those with and those without spontaneous flexor spasms. In spinal cord disease, however, brisk flexor reflexes can also be produced by more natural non-noxious stimuli which do not commonly produce flexor movements in normal subjects. Why some patients with cord disease develop flexor spasms and others do not is far from clear. Those with flexor spasms appear to have even lower thresholds for the production of flexor reflexes by non-noxious stimuli than do those without spasms, but we have no quantitative data in that regard. At any rate, flexor spasms appear to consist of the overly vigorous reflex response of the isolated cord to certain segmental inputs. It seems unlikely that new or different flexor reflex circuits would arise after more rostral lesions, and though transection of the cord may result more caudally in sprouting from existing cutaneous afferent fibers to centers responsible for flexor reflexes [MCCOUCH, AUSTIN, LIU and LIU, 1958], this is unlikely to be acutely reversed by Ba-34647. Furthermore, cutaneous input of sufficient magnitude can produce flexor reflexes in perfectly normal people. Electrical stimulation of the skin of the lower extremity has long been used to do just that, and our earlier report dealt with this in some detail. Other types of painful stimuli can also produce flexor withdrawal movements in normal subjects; when the adequate stimulus is apparent, as when one touches a hot surface or steps on a thorn, the movements are termed flexor reflexes, whereas if the stimulus is less obvious, they are called flexor spasms. We have seen typical flexor spasms in patients without any disorder of the nervous system patients, for example, who have paroxysms of painful input on the basis of spontaneous fractures in association with osteomalacia or osteoporosis. These patients appear to differ from the spinal patients only in the magnitude of noxious input necessary to evoke the spasm. Both central and peripheral nervous pathways for the production of flexor reflexes in response to painful input are, therefore, present normally. In patients with flexor spasms, light touch or other non-noxious cutaneous stimuli are also effective in the production of a spasm. Such stimuli are only capable of eliciting flexor reflexes in normal subjects under rather unusual circumstances. For example, it is possible in normal subjects to produce flexor withdrawal movements in tibialis anterior with latencies much shorter than the 'reaction time' by a light touch to the sole of the foot when the subject concentrates very hard to feel the stimulus which he thinks may be painful and to withdraw his foot immediately. The same short latency flexor response occurs in certain very apprehensive subjects, and the systems responsible for these responses may be the same ones responsible for startle in normal persons and certain types of myoclonus under pathological circumstances. The fact that these flexor responses *can* be elicited in normal subjects at all demonstrates that mechanisms are present to produce flexor spasms. In patients with lesions rostral in the cord the flexor spasms seem to arise because normal spontaneous inputs are now more effective. The pathophysiology of flexor spasm, therefore, would include dysfunction of mechanisms to limit the input of normal everyday flexor reflex afferent activity into pathways which then lead on irrevocably to flexor spasms and the accompanying pain. Spinal gating mechanisms involving presynaptic inhibition [ECCLES, 1964] may affect the activity in pathways responsible for the sensation of pain [MELZACK and WALL, 1965]. The same arguments pertain to activity in pathways responsible for flexor spasms, and CONRADI [1969] has electron-microscopic evidence for axo-axonic synapses, presumably mediating presynaptic inhibition, in the vicinity of the flexor motoneuron pool in the lateral portion of the lumbar ventral horn in the cat. Pain is a frequent accompaniment of flexor spasms and is generally ascribed to the muscular contraction and movement associated with the spasm. It seems equally likely that when the normal spinal segmental afferent bombardment is allowed to overflow past the now non-functional gate, it might reach supraspinal structures and elicit pain. Pain and spasm would both, therefore, be manifestations of deficient inhibition in the same gating system rather than one causing the other. We have seen patients with spinal lesions in whom episodic pain was more troublesome than actual muscular spasms; in one such patient Ba-34647 was successfully used.

Ba-34647 clearly is effective in reducing the frequency or probability of occurrence of flexor spasms, painful or otherwise. Furthermore, it makes flexor reflexes more difficult to produce, diminishes their amplitude, duration and painful quality and prolongs their minimal latency to values much longer than usually found in patients with spinal lesions (more like those in patients with cerebral lesions). Since all of these effects have been recorded below a complete transection, Ba-34647 exerts its primary effect at the spinal level apparently to raise the threshold for flexor reflex activity or, put another way, to reduce the facility with which much ordinary afferent input produces flexor reflexes. Whether it does this by enhancing inhibitory gating function remains to be proven. How far one may carry the analogy between the short and longer flexor reflex latencies before and after Ba-34647 and the short and long latency flexor reflex pathways described by LUNDBERG [1966] and colleagues in cats also remains to be seen. The dorsal horns where the postulated presynaptic inhibition and other gating mechanisms related to pain and flexor spasms would be expected to occur, are in the cat spinal cord the areas of highest GABA concentration [GRAHAM, SHANK, WERMAN and APRISON, 1967]. GABA is likely to function as the inhibitory transmitter in the pathway mediating presynaptic inhibition [LEVY, REPKIN and ANDERSON, 1971]. CURTIS and colleagues [CURTIS, DUGGAN, FELIX and JOHNSTON, 1971] consider GABA to be the inhibitory transmitter active on dorsal horn interneurons and elsewhere in the cord [cf. CONRADI, 1969; above] mediating what they term 'prolonged inhibition', a concept encompassing presynaptic as well as remote inhibition, either or both of which we presume to be deficient in patients with flexor spasms.

These experimental findings, our results using Ba-34647 and our theory of the pathophysiology of flexor spasms combine to make an interesting story. We have no knowledge, however, of the mechanisms by which Ba-34647 is metabolically active. DELWAIDE [1971, and this volume] has produced compelling arguments that the failure of muscle vibration to inhibit monosynaptic reflexes in spastic patients is due to a lack of presynaptic inhibition and suggests that the clinical hyperreflexia characteristically seen as increased tendon jerks and stretch reflexes in such patients also results from deficient presynaptic inhibition, but further studies are needed to support that contention. If subconvulsive doses of bicuculline (a GABA antagonist) provoked flexor spasms or labile flexor reflexes in otherwise normal animals, the hypothesis would be strengthened. In any case, it is important to consider briefly the impact of present knowledge on the management of patients with flexor spasms. Ba-34647 should be very useful when it becomes widely available, but meanwhile, if the primary disorder in flexor spasms is the marked increase of afferent input past the initial few synapses into the spinal cord, it would seem wise to decrease as much of the afferent activity to the cord as possible. That is, to avoid pressure sores and ulcers, to avoid distention and/or infections of bladder, and to use peripheral nerve blocks or neurectomies when indicated. It may also be possible, by the use of modern techniques for maintenance of patients in baths of various kinds, to reduce input from the periphery related to gravitational pressure and surface contact.

Summary

As far as can be determined, flexor spasms have the same clinical and physiological properties as do evoked flexor reflexes in the same patients. Flexor spasms, therefore, may be considered labile or heightened flexor reflexes, the stimuli for which are not apparent but probably relate to rather ordinary background afferent input. They occur characteristically following more rostral spinal cord lesions but have also been seen under certain unusual circumstances in subjects without lesions of the nervous system. Lioresal (Ciba Ba-34647) is very effective in lessening the frequency and severity of flexor spasms. It also results in flexor reflexes becoming less easily elicitable. We suggest that the pathophysiology of flexor spasms must include an abnormality of the normal segmental afferent gating function, perhaps related to deficient presynaptic inhibition. The painful quality of many flexor spasms may be the subjective result of the same excessive egress into pathways within the cord of such relatively normal afferent activity as that responsible for the motor component of the spasm; that is, both pain and muscular spasm would be caused by excessive CNS input rather than the efferent discharge itself being responsible of the pain.

Acknowledgement

This work was begun under the aegis of the Association for the Aid of Crippled Children, New York and with the aid of the National Institute of Neurological Diseases and Blindness, Special Fellowship 2F11NB1539-02. It was completed with the aid of a grant from Mrs. W.B. Lloyd, and the support of the Ciba Pharmaceutical Company, and Dr. A.S.C. Ling who provided us with Ba-34647.

Author's address: Dr. R.R. Young, Massachusetts General Hospital, *Boston, MA 02114* (USA)

New Developments in Electromyography and Clinical Neurophysiology,
edited by J. E. Desmedt, vol. 3, pp. 744–750 (Karger, Basel 1973)

Withdrawal Reflexes[1]

M. R. DIMITRIJEVIĆ

Institute of Clinical Neurophysiology, Ljubljana

Sudden, unexpected exposure of any external surface of the human body
to an injurious stimulus results in a withdrawal reflex. The categories under
which the various withdrawal reflexes are known are numerous, and generally
descriptive. Some refer to the site of stimulation: plantar, palmar, abdominal
skin and corneal reflexes; others to the pattern of the reflex movement:
flexion, extension and blink reflexes; others to the anatomical structures and
reflex arcs involved: cutaneo-muscular, cutaneo-spinal, spino-bulbo-spinal
and polysynaptic reflexes; others to the receptors stimulated: nociceptive and
exteroceptive reflexes. Common to all these reflexes is their biological significance: each reflex response is appropriately oriented to the stimulus so as to
remove the body surface from the offending agent.

The various methods of recording the electrical manifestations of active
somatic muscles give different views of what is happening in the CNS. Figure 1
shows recordings, with 4 different types of electrodes, of the activity of tibialis
anterior following stimulation of the ipsilateral plantar surface of the foot.
In *A* skin electrodes, 3 cm apart, sample a large region of muscle and show
many poorly defined spikes. In *B*, a concentric needle electrode shows less
activity with some better defined spikes. In *C*, a bipolar needle electrode shows
less activity, as it records from a smaller amount of muscle, with clearly
defined spikes of different amplitudes. In *D*, a semimicro-electrode [EKSTEDT,
HÄGGQVIST and STÅLBERG, 1969] records essentially from one muscle fiber;
the activity of one or two other muscle fibers can be seen also as very small

1   This work was supported by the Boris Kidrič Fund of Slovenia (grant No. 306/20:70)
and by the US Department of Health, Education and Welfare, Social and Rehabilitation
Service (grant No. 19-P-58397-F-01).

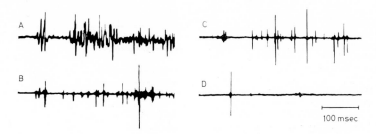

Fig.1. Flexor response recorded from tibialis anterior. *A* Skin electrodes; *B* concentric needle electrode; *C* bipolar needle electrode; *D* semimicro-electrode.

spikes. These various recording techniques thus allow one to record activity of a single muscle fiber, a single motor unit, many local motor units, or a large region of muscle. They have all shown that withdrawal reflexes have longer and more variable latencies than proprioceptive reflexes.

EMG has contributed to the study of functional organization in the central pathways of human withdrawal reflexes. The work of HAGBARTH [1960], KUGELBERG [1948], PEDERSén [1954] and TEASDALL and MAGLADERY [1959] has shown that the plantar flexion reflex and the abdominal skin reflex are typical skin reflexes. In 1952, HAGBARTH described the system of cutaneomuscular reflexes organized at spinal level in the cat. Further investigations of this system of co-ordination were published by ELDRED and HAGBARTH [1954], HAGBARTH and KUGELBERG [1958], KUGELBERG and HAGBARTH [1958], EKLUND, GRIMBY and KUGELBERG [1959], KUGELBERG, EKLUND and GRIMBY [1960], MEGIRIAN [1962], GRIMBY [1963a,b], HAGBARTH and FINER [1963], and CLARKE [1966]. In this system, the pattern of responding in the trunk, abdominal wall and lower limb muscles is determined by the skin region stimulated; each muscle is excited or facilitated by stimuli applied to the skin covering it and inhibited by stimuli applied elsewhere.

KUGELBERG [1962] emphasized that nociceptive reflexes are organized not only on an anatomical basis. He demonstrated that the normal plantar reflex has a discriminative capacity as regards strength, modality and site of the stimulus. He suggested that the increase of the flexion reflex in patients with suprasegmental lesions indicates that this reflex is normally subjected to a strong cerebral inhibition. The Babinski sign apparently results from a shift in balance in favour of reflex dorsiflexion, due to impaired cerebral control. GRIMBY [1963a,b, 1965] has studied the supraspinal control of the plantar reflex and the integration of its flexor and extensor components, in normal subjects and in patients with cortical, capsular and spinal lesions, with

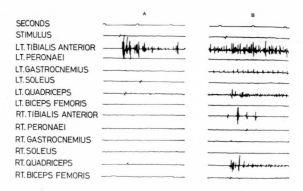

SECONDS
STIMULUS
LT. TIBIALIS ANTERIOR
LT. PERONAEI
LT. GASTROCNEMIUS
LT. SOLEUS
LT. QUADRICEPS
LT. BICEPS FEMORIS
RT. TIBIALIS ANTERIOR
RT. PERONAEI
RT. GASTROCNEMIUS
RT. SOLEUS
RT. QUADRICEPS
RT. BICEPS FEMORIS

Fig. 2. EMG of lower limbs. *A* Hemiplegic (left) limb; *B* unaffected (right) limb.

present or absent Babinski sign. On the basis of his results in normal subjects, he assumed the spinal reflex center to be controlled by two descending pathways: one regulating the relationship between extensor and flexor activity and the other determining the amount of contrast between the plantar and hallux pattern. A pathological process, he suggested, may involve one or other of the pathways independently.

In patients with the spinal cord totally or partially divided [Dimitrijević and Nathan, 1967a, b, 1968] there is such excessive activity of the motoneurons that it overrides reciprocal inhibition and the delicate organization of cutaneo-muscular reflexes of Hagbarth. Even weak stimuli activate flexor and extensor muscles simultaneously both in withdrawal reflexes and in stretch reflexes. The withdrawal response to noxious stimuli is abnormally prolonged and the response may continue as a continuous discharge or as a cyclical response.

The amount of motor unit activity depends on intensity of stimulation, the presence or absence of motor unit activity at the time of stimulation, and the position or the phase of movement of the limb if it is moving at the time of stimulation. Reduction of the input via the posterior roots to the chronically isolated spinal cord reduces muscle activity, flexor and extensor spasms, and the excessive reflex responses to all forms of stimulation, tactile, noxious, and proprioceptive.

In hemiplegic patients with long-standing capsular lesions the threshold of the flexion reflex is lowered on the hemiplegic side; and there is also an irradiation of activity to the muscles of that side when the unaffected side is stimulated. In figure 2A, the plantar surface of the hemiplegic side was

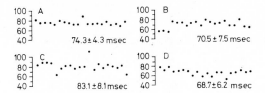

Fig.3. Latencies of the flexion reflex. *A* Normal subject; *B* paraplegic patient; *C, D* hemiplegic patient, on the affected (*C*) and nonaffected (*D*) sides. The mean latencies ± SD are indicated.

stimulated. The response was largely confined to the ipsilateral tibialis anterior, with only slight activity in the quadriceps. In figure 2*B*, the plantar surface of the unaffected side was stimulated. There was a large response of the ipsilateral muscles and a much larger response of the hemiplegic muscles to this stimulus on the contralateral limb.

The low threshold and the widespread irradiation of activity from the unaffected to the hemiplegic side may result from insufficiency of supra-segmental inhibitory control. On the other hand, the latency of the first EMG burst in the tibialis anterior was significantly longer on the affected side. This difference was not due to temperature differences between the two limbs and conduction velocity of the two peroneal nerves was the same (± 1 m/sec).

Figure 3 shows the latency in msec of the response of the tibialis anterior to a series of plantar stimulations. The latencies are all within a certain narrow range in the normal, the paraplegic, and the unaffected limbs of the hemi-plegic; they are longer in the hemiplegic limb. We suggest that this difference in latency is due to a lack of suprasegmental facilitation thus implying that a capsular lesion may involve both facilitatory and inhibitory descending path-ways controlling the flexion reflex.

By contrast, when the cord is totally severed from the brain the latency of the flexion reflex is the same as in the normal; thus the spinal interneuronal system can substitute for the absent supraspinal facilitation but not for spinal inhibitory control, since the reflex threshold is abnormally low, there is irradiation to distant motoneurons and the response lasts an excessively long time.

Repetitive stimulation at the same site will evoke repetitive withdrawal reflexes which decline progressively and eventually disappear. This habitu-ation can be interpreted as adaptive, the decrease and final cessation of response occurring after the nervous system has recognized the irrelevance of the stimulation. Habituation occurs in withdrawal reflexes of spinal and mid-

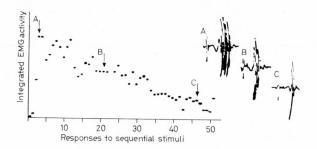

Fig.4. Histogram of sequential EMG responses of tibialis anterior to repetitive plantar stimulation in a patient with the spinal cord divided; integrated EMG on ordinate. *A, B, C* samples of EMG responses in m. tibialis anterior recorded at the times indicated.

brain type in normal subjects. It also occurs in man with the spinal cord completely divided [DIMITRIJEVIĆ and NATHAN, 1970]. It had generally been assumed that this habituation required involvement of the cerebral hemispheres, but it can be observed in spinal patients.

In figure 4, a mild noxious electric stimulus was applied to the plantar surface in a patient with the spinal cord completely divided. The response shows several phases that we have called, respectively, build-up, fluctuation, diminution and cessation.

In this example, the build-up phase occurred for the first 4 stimuli and the phase of fluctuation led into that of diminution. This experiment was not continued until the response ceased. The habituated response in spinal man can easily be dishabituated by introducing an extra stimulus, close to or far away from the site of original stimulation [DIMITRIJEVIĆ and NATHAN, 1971].

An example of such dishabituation is shown in figure 5. Flexion reflexes were elicited with stimuli applied to the plantar surface in a paraplegic patient. In this experiment, by the 25th plantar stimulus the phase of diminution passes into that of fluctuation. At the 50th stimulus a simultaneous stimulus is added to the posterior surface of the thigh. The same motor units are immediately reactivated. Both stimuli are then given repetitively. After 25 combined stimuli have been given, the response has become less than it was before stimulation at the second site began.

Recently, we have found that habituation can be prevented or delayed by presenting the stimulus in an irregular pattern [DIMITRIJEVIĆ, FAGANEL, GREGORIČ, NATHAN and TRONTELJ, 1972]. Habituation does not occur if stimulation is made random either in time, stimulus intervals or intensity. In figure 6*A*, the rapid habituation of the response to regular repetitive stimu-

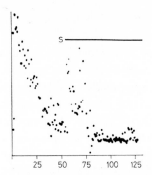

Fig.5. Histogram of sequential EMG responses of tibialis anterior. Abscissa: rank of stimuli to plantar surface. Horizontal line marked S, additional stimuli to posterior surface of thigh.

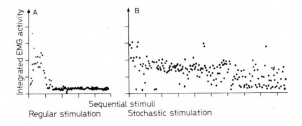

Fig.6. Histogram of sequential EMG response of tibialis anterior. *A* Regular stimulation; *B* stimulation at random intensities.

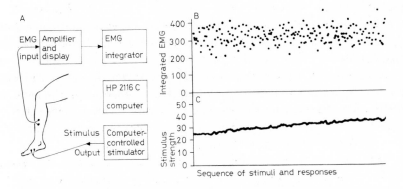

Fig.7. A Block diagram of feedback control of stimulation. *B* Sequential EMG responses of tibialis anterior of a paraplegic patient. *C* Feedback-controlled stimulus intensities.

lation is seen; in figure 6 B there is minimal and very late diminution in the response.

The changing response of the spinal cord to noxious stimulation was examined in another way. An attempt was made to remove habituation by changing the stimulus intensity by means of a computer-controlled feedback loop. Figure 7 A shows the experimental design. The amount of EMG response to each plantar stimulus was determined (fig. 7 B). Decreases in the mean quantity of response were compensated by increase in stimulus intensity, and *vice versa* (fig. 7 C). It can be seen that the variation of the responses was greatly reduced by this method. Habituation could be suppressed very effectively, although the average stimulus strength was even less than in the series with stochastic and regular stimulation. Such studies on qualitative and quantitative aspects of habituation and dishabituation of withdrawal reflexes in healthy and spinal man provide an opportunity to estimate behavioral capacities of certain isolated parts of the nervous system, as well as to understand the pathophysiology of motor organization in different clinical conditions.

Summary

Different methods of recording muscle activity electromyographically have been considered. They permit the examination of single muscle fibers, single motor units, many local motor units, and a large region of muscle.

The contribution of the Swedish workers to the study of withdrawal reflexes is briefly reviewed.

A study of the flexion reflex in hemiplegia is reported. One finding is that the latency of the reflex is longer on the hemiplegic side, although this is a spinal reflex.

The work of DIMITRIJEVIĆ and NATHAN on withdrawal and proprioceptive reflexes and on habituation of the flexion reflex in spinal man is also briefly summarized. Factors influencing habituation of the reflex are discussed.

Author's address: Dr. M. R. DIMITRIJEVIĆ, Inštitut za klinično nevrofiziologijo, Holzapflova 1, *61000 Ljubljana* (Yugoslavia)

New Developments in Electromyography and Clinical Neurophysiology,
edited by J. E. Desmedt, vol. 3, pp. 751–760 (Karger, Basel 1973)

Pathological Interoceptive Responses in Respiratory Muscles and the Mechanism of Hiccup

J. Newsom Davis

Department of Neurophysiology, Institute of Neurology, National Hospital, Queen Square, London

Disorder of the neural mechanisms controlling the respiratory muscles has received less attention from neurologists than has that affecting the limbs. Although characterized as *respiratory*, these muscles (diaphragm, intercostal and abdominal muscles) have other important functions besides respiration. They are concerned, for instance, in viscerally evoked (interoceptive) responses like coughing and vomiting. Here I want to illustrate how the neural mechanism controlling these responses can be disturbed by considering as an example the nature of hiccup, a disorder usually classed as an interoceptive response, which not only seems to lack any obvious function but which one must regard as a potentially pathological process since in some cases it may be intractably and exhaustingly recurrent. There are thus good clinical reasons for wishing to know more about its genesis.

Most of those who have written about hiccup have concerned themselves with the factors which precipitate or cure it. These clinical observations have in fact provided some clues to its nature, but analysis of the properties of hiccup itself has proved to be more rewarding [Newsom Davis, 1970]. Two things are usually said about the physiological nature of hiccup: first, that it is a respiratory phenomenon, being served by the 'respiratory centres' in the brain stem, and second, that it is a reflex action [Kuntz, 1953]. As we shall see, there are grounds for doubting both these assertions.

Precipitating Factors

Everyone probably experiences at some time short bouts of hiccup, usually following food. More prolonged attacks of hiccup occur in association with a large number of disorders. The most comprehensive clinical review of the aetiology of hiccup is that by SAMUELS [1952] who classified the 'organic' causes on an anatomical basis into disorders involving the abdomen, thorax, neck and central nervous system (CNS). Amongst the psychogenic disorders believed to cause hiccup were mental shock, cardiospasm, malingering, publicity seeking and prolonged nervous strain. SAMUELS' list of more than 40 diverse disorders, which is not even complete, does not at first seem to be very helpful. Nevertheless, some tentative conclusions can be drawn. If hiccup is a reflex, then the receptive field from which the reflex can be evoked would seem to be extraordinarily large. SALEM, BARAKA, RATTENBORG and HOLADAY [1967], for example, have suggested that it should include vagal afferents, phrenic afferents, and the sympathetic chain from the 6th to the 12th thoracic segments. This at once raises doubts about the reflex nature of hiccup, for it is characteristic of reflexes that the response is specific both for the type of stimulus and for the structure from which it can be evoked.

If we consider more closely the CNS disorders which cause chronic hiccup, it appears that many of them directly involve medullary structures. Most neurologists can probably recall cases of low brain stem infarction, for example, in which hiccup has occurred in the acute phase of the illness along with vertigo, dysphagia, dysarthria or vomiting. A patient in whom the initial manifestation of his medullary glioma was hiccup, borborygmi and watery diarrhoea provided a particularly striking example of the association of hiccup and medullary disease. These symptoms were soon followed by dysphagia, vagal sensory loss and disturbance in spontaneous breathing. At autopsy a few months later, the tumour was found to be centred around the vagal nuclei (fig. 1). These clinical observations would seem to point to the importance of vagal or paravagal structures in the genesis of hiccup, the initial association of symptoms in this patient suggesting that visceral efferent fibres in the vagus may have a specific role. To learn more about the nature of the disorder, however, we need to examine the characteristics of hiccup and some of the factors which influence it.

But there is one other feature about the occurrence of hiccup which we ought to consider in relation to its function. In health, hiccup is known to be most frequent in infancy and early childhood, becoming less frequent as age advances. It is less well-known that hiccup occurs in the fetus during the later months of gestation; it can be distinguished from other fetal movements by its abrupt nature and regular recurrence every 2 or 3 sec. This was recognized in the 18th century by STORCH [1750, quoted by PEIPER 1958], and is probably much commoner than its rarely reported occurrence suggests. This raises the

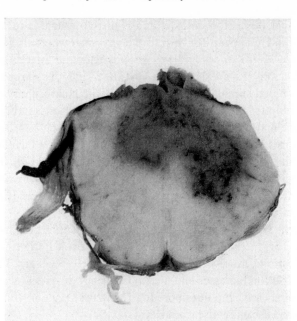

Fig. 1. Transverse section through the caudal part of medulla oblongata of a 46-year-old man in whom the initial manifestation of a brain-stem glioma were hiccup, borborygmi and diarrhoea, followed by difficulty in swallowing, signs of vagal deafferentation and disturbance of spontaneous breathing. Macroscopically, the tumour was centred at the region containing the dorsal motor nuclei of the vagus nerves, and extended on one side to involve the region of nucleus ambiguus.

possibility either that hiccup may have some as yet undetermined function in the fetus or that it is a vestige of an act which has significance in lower animals.

Inhibitory Factors

Several techniques can unquestionably end a bout of hiccups, although in chronic hiccup it is either unresponsive or quickly recurs. These remedies provide some information about the nature of hiccup. The classic cure by PLATO is given in the symposium [416 B.C.]. ARISTOPHANES develops hiccup during an after-dinner discussion on love, and is therefore unable to make his own contribution. He appeals for help to the physician who tells him to make himself sneeze several times and this happily ends the attack. More recently, SALEM, BARAKA, RATTENBORG and HOLADAY [1967] have reported a similar observation that stimulation of the pharynx with a naso-pharyngeal tube immediately inhibited hiccup in all but one of their 85 cases. Stimulation of the pharynx alone may also be effective, and manoeuvres such as drinking out of a cup backwards or swallowing iced water probably achieve their effect in this way. Coughing can also inhibit hiccup [NEWSOM DAVIS, 1970]. Thus stimulation in a widespread area of the upper respiratory tract involving mucosal afferents travelling in the trigeminal, glosso-pharyngeal and vagal nerves can inhibit hiccup.

A quite different type of stimulus which may stop hiccup is breathing a CO_2 rich mixture. This method, recommended by HAMILTON BAILEY [1943] can lead to complications, however, if used indiscriminately [SAMUELS, 1952]. As PCO_2 rises, ventilation increases and hiccup frequency progressively falls although the size of each hiccup does not alter [NEWSOM DAVIS, 1970]. The implications of these effects are discussed below.

Finally, it is interesting to note that events causing emotion or excitement, including sexual intercourse, may also terminate hiccup.

Properties of Hiccup

The principle characteristics of hiccup have been established by an electrophysiological analysis in which the EMG activity in the diaphragm and intercostal muscles was recorded from indwelling wire electrodes with simultaneous recording of respiratory airflow and lung volume [NEWSOM DAVIS, 1970]. Each hiccup consists of one or several discrete bursts of activity occurring not only in the diaphragm but also in inspiratory intercostal muscles, the activity having a very similar time course in the two groups of muscles. Expiratory intercostal muscles are inhibited coincidentally with the inspiratory muscle discharge. The discrete bursts within the hiccup discharge occur at a rate of 4–5 c/sec, the total duration of hiccup discharge usually being 500–750 msec. At about 35 msec after the onset of the inspiratory discharge, laryngeal closure occurs and persists until the inspiratory discharge has virtually ceased. A consequence of this is that although the level of inspiratory muscle activation in hiccup far exceeds that occurring in quiet breathing, its ventilatory effects remain trivial. The intense inspiratory muscle discharge taking place against a closed airway must be one of the chief reasons why hiccup is uncomfortable. Unlike most involuntary movements, hiccup continues during sleep.

The frequency of hiccup varies considerably among individuals, values from 15 to 60/min having been reported [BROOKS, 1931; ANGLE, 1932; SAMUELS, 1952; NEWSOM DAVIS, 1970]. Hiccup frequency within the individual may be remarkably constant, although plotting the instantaneous frequency of hiccup (i.e. the reciprocal of the inter-hiccup interval) reveals slow, recurring fluctuations. The amplitude of hiccup, using as the index the rectified diaphragmatic EMG integrated for a 1-sec period, shows considerable spontaneous variation. There appears to be no correlation of amplitude and instantaneous frequency, but instantaneous frequency is negatively correlated with the amplitude of the subsequent hiccup, i.e. when hiccup amplitude is large, the occurrence of the subsequent hiccup is delayed [NEWSOM DAVIS, 1970].

Hiccup has a tendency to occur during inspiration rather than expiration. The probability density of hiccup in relation to the fraction of the respiratory cycle in which it occurs is such that the peak incidence of hiccup is at mid- or end-inspiration, sometimes with a subsidiary peak at end-expiration after airflow has ceased [NEWSOM DAVIS, 1970].

Relationship of Hiccup to Breathing

The case against hiccup being respiratory in nature has been argued in detail elsewhere [NEWSOM DAVIS, 1970]. The evidence is based first on the effects of changes in PCO_2 on hiccup. Increasing PCO_2 causes a decrease in hiccup frequency without any obvious change in its amplitude, while reducing PCO_2 by hyperventilation has no clear effect on hiccup frequency but causes a highly significant increase in its amplitude. Neither of these effects would be consistent with the view that the neurons constituting the 'respiratory centre' are directly concerned in the genesis of hiccup. The second line of evidence comes from the relationship between hiccup amplitude and the fraction of the respiratory cycle in which hiccup occurs. The mean amplitude is greatest in mid-inspiration, being significantly smaller than this at end-inspiration ($p < 0.05$) and in early expiration ($p < 0.001$) but not significantly different at end-expiration. This again is not the relationship that would be predicted if hiccup were generated by the neurons constituting the inspiratory centre, for their excitability is maximal at end-inspiration rather than mid-inspiration and is least towards the end of expiration. Thus, although there is a respiratory modulation of hiccup amplitude, it is out of phase with the excitability changes of the inspiratory centre. The phase relationship is in fact such that hiccup amplitude is greatest when inspiratory airflow is maximal (i.e. when the rate of change of lung volume is maximal in an inspiratory direction), and least when expiratory airflow is maximal. As pointed out earlier, the probability density of hiccup has a similar phase relationship.

If excitability changes in the inspiratory centre are not the immediate cause of the respiratory modulation of hiccup, what other sources exist? Perhaps the most likely source is pulmonary vagal afferents. This is supported by observations made when the frequency of breathing is controlled voluntarily by instructing the subject to time his inspiration with a regularly recurring auditory cue. Fixed frequency breathing of this type leads to hyperventilation and a decrease in PCO_2, removing the stimulus for spontaneous breathing and allowing ventilation to be determined by voluntary action. Under these circumstances, the tendency for hiccup to occur at the point of maximal inspiratory airflow is again maintained, the frequency of hiccup becoming locked to that of breathing [NEWSOM DAVIS, 1970]. That this timing relationship of hiccup with inspiratory airflow persists during both spontaneous and voluntary breathing indicates that it is a consequence of the ventilatory movement itself, rather than of its source.

One way in which vagal pulmonary afferents might exert this influence on hiccup is through their presynaptic inhibition of the largest afferent fibres of the superior laryngeal nerve, which would include those serving mucosal touch receptors [RUDOMIN, 1967]. These laryngeal afferents can inhibit hiccup so that one might predict that their dysinhibition by this means would favour the occurence of hiccup. The time course of presynaptic inhibition exerted by pulmonary afferents during passive lung inflation is in fact in parallel with the rate of change of lung volume, and could thus account for the changes in hiccup amplitude and probability density, which have a similar phase relationship with inspiratory airflow.

Category of Movement

Is hiccup a reflex action, as commonly supposed? Generally accepted criteria of a reflex action are that it is unlearned, uniform, adjustive or protective in character, and that it requires excitation of peripheral receptors, components of the input being identifiable in the pattern of the output [BIZZI and EVARTS, 1971]. The first two of these conditions are met by hiccup, and to the third no answer can yet be given. With regard to the last, we cannot

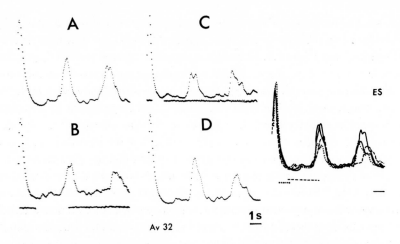

Fig. 2. The averaged records (32 sweeps) of the integrated EMG recorded from the right diaphragm with surface electrodes over the lower chest wall in a patient with chronic hiccup. The sweep duration was 10.24 sec so that a sequence of 3 hiccups could be recorded, the inter-hiccup interval being about 4 sec; the averaging computer (*Biomac* 1000) was triggered by the first hiccup of the sequence. *A* and *D* controls; *B* and *C* nasal mucosa stimulation for 3 sec and 1 sec respectively indicated by break in bar line; stimulus intensity 8 V, width 0.2 msec, frequency 300/sec. Note decrease in amplitude of averaged hiccup discharge in *B* and *C* compared to controls taken before and after the test runs. In *B*, stimulation also appears to have delayed and reduced the amplitude of the 3rd hiccup in the sequence. On the right of the figure, the traces have been graphically superimposed and aligned with respect to the 2nd hiccup.

yet define an adequate stimulus for the response nor even the receptive field from which hiccup can be evoked, although the variety of clinical conditions which seem to evoke hiccup suggest that this field is large. One can, however, investigate the mechanism concerned in generating hiccup without knowledge of the exitatory stimulus by examining the effect of an inhibitory stimulus, such as stimulation of the nasal mucosa which is known to be effective in stopping hiccup. The results of such a study show that events preceding an individual hiccup by several seconds are concerned in determining its size.

The technique used in this study will be briefly described. The rectified and integrated hiccup discharge in the diaphragm recorded with surface electrodes over the lower chest wall was used as the index of hiccup amplitude. A trigger circuit was arranged so that the integrated EMG of the hiccup triggered the sweep of an averaging computer (*Biomac* 1000), the sweep duration of 5 or 10 sec being chosen so as to encompass the next one or two hiccups. The frequency of hiccup is sufficiently constant for an average hiccup discharge to be built up in this way. Figure 2A shows the averaged records (32 sweeps, sweep duration 10 sec) for a sequence of three hiccups; the decrement in size of the averaged records is a consequence of the slight variation in inter-hiccup interval. The trigger circuit also initiated the timing cycle for the stimulus. A train of shocks at 300/sec could be delivered to the nasal mucosa for a pre-determined period through electrodes inserted 5 cm up each nostril. The stimulus intensity was kept well below the pain threshold, and gave rise to a tingling sensation comparable to that preceding a sneeze.

Low intensity stimulation for a 3-second period immediately preceding the hiccup (fig. 2B) causes a reduction in the amplitude of the averaged hiccup discharge without any clear change in its latency when compared to the control records taken before and after the test run (fig. 2A and D). But stimulation at the same intensity lasting for only 1 sec and ending 2 sec before the onset of the next hiccup causes an even more obvious decrease in amplitude of the averaged discharge again without changing its latency. To the right of the figure, the records have been graphically superimposed and aligned with respect to the second hiccup of the sequence. The decrease in the amplitude of the averaged discharge does not seem to be due to temporal dispersion in view of the good fit. Similar results have been obtained in two other subjects; in these subjects the stimulus not only decreased the amplitude of the discharge but also increased its latency.

Experiments in which the individual runs were simultaneously averaged and photographically superimposed confirmed that the decrease in size of the averaged hiccup discharge was due to a decrease in size of the individual hiccup amplitude, rather than to temporal dispersion, as shown in figure 3. This figure also illustrates that a stimulus applied close to the onset of a hiccup has little, if any, effect on its amplitude in contrast to its effects when applied earlier, as shown in trace D. The onset of the stimulus here was timed to occur approximately at the same time as the hiccup. The slight variation in the inter-hiccup interval results in some taking origin just before the stimulus and others just after. The averaged discharge shows no obvious decrease in amplitude which is confirmed in the photographically superimposed traces.

It thus looks as if the size of an individual hiccup is determined by events occurring for up to several seconds before the hiccup is 'due', but that once the hiccup is initiated and for a brief period beforehand such events no longer influence it. The results also indicate that the frequency and amplitude of hiccup are controlled independently, a conclusion that has already been reached from the lack of correlation between amplitude and frequency, and

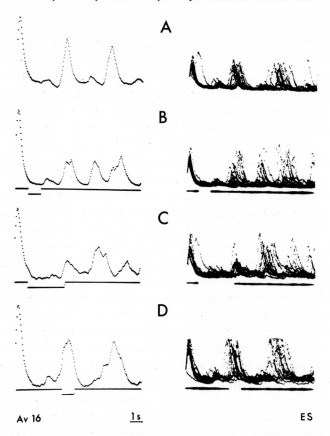

Fig. 3. The left-hand traces show the averaged records (16 sweeps, sweep duration 10.24 sec) of the integrated diaphragmatic EMG obtained during hiccup as in figure 2. In the right-hand traces, the individual runs constituting the averaged responses have been photographically superimposed. *A* control; *B*, *C* and *D* nasal mucosa stimulation as in figure 2. Note in *B* the reduction in amplitude of some of the individual hiccups and change in timing of the next hiccup, an effect which was enhanced when the stimulus duration was increased from 1 to 3 sec. In *D*, a stimulus applied so that some hiccups had their onset during the period of stimulation had little effect on hiccup amplitude.

by the differential effects of changes in PCO_2. In other words, factors which influence the probability of a hiccup occurring do not necessarily cause any accompanying changes in size, while factors which change its size can do so without altering its frequency. These observations imply that the neural mechanism concerned in initiating hiccup is independent of that which executes it.

What, then, can be said about the mechanism concerned in initiating hiccup? If hiccup were a reflex, we would need to assume that its frequency characteristics are determined by a suitably timed phasic discharge in the putative afferent system which evokes the reflex, a possibility that seems unlikely. The observed frequency characteristics of hiccup would be more readily accounted for if it were assumed: (1) that the initiation of a hiccup was a 'threshold' event, hiccup occurring when excitability changes in a hypothetical triggering mechanism reached a critical level, and (2) that the occurrence of hiccup leads to a temporary increase in its threshold, a fact which is suggested by the negative correlation between hiccup amplitude and the instantaneous frequency of the subsequent hiccup and by the upper limits of hiccup frequency. Excitation within the triggering mechanism would be increased by gastro-intestinal afferents in particular, and decreased by a raised PCO_2 Chronic hiccup would be explained by a lowering of the triggering threshold.

With regard to the mechanism executing hiccup, we have already seen that the pattern of hiccup is in fact quite stereotyped consisting of inspiratory muscle activation, expiratory muscle inhibition and laryngeal closure occurring very soon after the onset of the inspiratory discharge. The size of the discharge in inspiratory muscles and the degree of inhibition of expiratory muscles can be influenced by factors acting at spinal level and centrally, but these do not alter the main features of the phenomenon.

One may conclude, therefore, that the features of hiccup are not those expected of a reflex action. Hiccup is best described as a centrally patterned movement whose initiation is determined by excitatory changes in some central neural mechanism to which several different groups of afferents may contribute and which, once triggered, runs its course, its features being determined primarily by the existing arrangement of the neural connections rather than by components of the stimulus as in a reflex action.

Author's address: Dr. J. Newsom Davis, Department of Neurophysiology, Institute of Neurology, National Hospital, Queen Square, *London, WC 1 N3 BG* (England)

New Developments in Electromyography and Clinical Neurophysiology,
edited by J. E. Desmedt, vol. 3, pp. 761–766 (Karger, Basel 1973)

Neural Mechanisms of the Startle Reflex in Cerebral Palsy, with Special Reference to its Relationship with Spino-Bulbo-Spinal Reflexes

M. Shimamura

Laboratory of Neurophysiology, Tokyo Metropolitan Institute for Neurosciences,
Tokyo

Introduction

The startle reflex, a general jerky muscular contraction, is one of the abnormal movements in children with cerebral palsy, which is elicited by a small stimulus, e.g. acoustic click or light touch to the body.

In this study attempts have been made to analyze the neural mechanisms of the startle reflex in cases of cerebral palsy, using motion analysis and EMG observations.

Methods

Five children with cerebral palsy were investigated. To elicit the startle reflex an auditory or tactile stimulus was used which also triggered the beam of a Braun CRO simultaneously. The general jerky movements were recorded on 16-mm film, and by EMG from the various muscles in the limbs.

Results

A. Neurological Observations

The children had severe physical and mental disorders. They could not stand or sit up, and were usually confined to bed. They showed athetosis, spastic palsy, muscle atrophy, ankylosis and greatly reduced voluntary movements. They were inarticulate and lacking in emotional expression. They took a liquid diet only. Neurological examinations showed that tendon reflexes, surface reflexes, etc., were not generally enhanced or decreased. The EEG was abnormal in all cases, with spike potentials, seizure discharges, etc. However, they did not show any EEG findings in common nor any seizure discharges simultaneous with the startle reflexes.

B. Movements of the Startle Reflex

When a click stimulus (castanett) was applied near the ear, generalized rapid movements were obtained involving blinking, arm flexion and leg flexion, as shown in figure 1. These generalized flexions were slightly different depending on the type of case (ankylosis, muscle palsy or atrophy).

Similar generalized flexions were elicited by a tactile stimulus to the legs and trunk. With repetitive stimulation generalized movements were not evoked by every stimulus, except the blink reflex. Similar movements were elicited during sleep. Under light barbiturate anesthesia they were not obtained to any stimulation.

C. EMG Observations of the Startle Reflex

EMG recordings were made from various muscles using bipolar surface electrodes: orbicularis oculi, pectoralis major, biceps brachii, triceps brachii, flexor carpi radialis, extensor longus, quadriceps femoris, hamstring, tibialis anterior, gastrocnemius, etc. With a click stimulus, EMGs were obtained from the flexor muscles as shown in figure 2, but were not found in the extensor muscles. Latencies measured from the stimulus to the begining of the EMG were reduced in m. orbicularis oculi and increased in m. tibialis anterior. There were obvious increases of latencies as recording points descended from the head. In figure 2 averages of 30 evoked EMG responses in B demonstrate the latency pattern mentioned above.

Similar latency patterns were evoked by tactile stimulation to the hand. However, the latency was about 25 msec longer in each muscle than that to click stimulation (table I).

When tactile stimulus was applied to the trunk, a similar latency pattern was obtained in the evoked EMGs but latencies were about 15 msec longer than those to click stimulation.

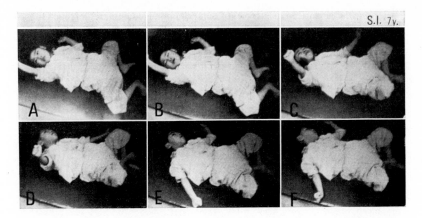

Fig. 1. Photographs showing generalized movements obtained by click stimulation in cerebral palsy child (7 years old). *A* Control. *B–F* After click, photographs taken on 64 frames/sec film; *B* shows blinking, *C–F* show movements of arms, neck and legs.

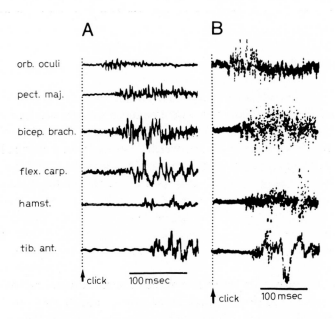

Fig. 2. Child 8 years. *A* Latencies of the EMG in various muscles. A click stimulation was applied at the starting point of each record. Latency of EMG was shortest in the orbicularis oculi and longest in the tibialis anterior. *B* Averaged latencies of EMG in the same muscles as in *A*. (Average of 30 responses.)

Table I. Comparison of latencies of the EMG obtained by click and touch to the hand. There is about 25 msec difference between click and tactile stimuli in all muscles for all subjects

Muscles	Name									
	T.K. 8y♂		S.I. 7y♂		M.K. 4y♂		S.K. 6y♂		M.A. 6y♀	
	C	T	C	T	C	T	C	T	C	T
orb. oculi	25.9	50.5	29.5	42.9	23.2	41.2	25.3	45.3	23.9	49.8
pect. maj	36.2	63.3	40.2	66.2	35.3	59.1	35.5	62.0	35.2	64.2
bicep. brach.	42.3	70.0	45.5	71.2	40.5	70.1	41.2	72.2	41.5	71.5
flex. carp.	50.3	79.5	57.2	80.3	49.2	75.2	49.9	78.3	50.9	76.5
hamst.	54.1	83.1	56.6	85.2	52.1	84.0	56.1	80.2	55.2	78.2
tib. ant.	70.5	98.5	80.3	105.5	69.8	95.5	70.2	99.2	70.9	109.2

Time in msec, C = click, T = touch stimulus.

D. The SBS Reflex in Cerebral Palsy

Electrical pulse stimulation was applied to the tibial nerve at the popliteal fossa and generalized jerky movements were obtained similar to those with click stimulation. EMG recorded from m. tibialis anterior showed longer latency reflex responses of about 60 msec which compared with the H and M reflex responses. This may be the spino-bulbo-spinal (SBS) reflex. This latency was slightly shorter (about 40 msec) than that of the click-induced startle reflex (fig. 3).

Discussion

The click-induced startle reflex is a rather complicated phenomenon [cf. the Moro extension reflex, WIESER, DOMANOWSKY and HEINEN, 1957]. However, when EMGs were recorded from various muscles, they showed a simple latency pattern. The latency of EMG of the startle reflex was shorter proximally, i.e. m. orbicularis oculi, and longer distally, i.e. m. tibialis anterior. There were obvious increases of latency as recording points became further

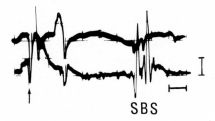

SBS

Fig. 3. The SBS reflex in cerebral palsy. An electrical stimulus was applied to the tibial nerve at the popliteal fossa. EMG recorded with surface electrodes from the tibialis anterior (below) and gastrocnemius (above). Longer latency responses (SBS) at about 60 msec and of high amplitude showed in tibialis anterior. Time scale 10 msec; vertical scale, 0.2 mV.

from the brain. Similar latency patterns were obtained to tactile and electrical stimuli, but latencies were different on each stimulation. This may be an important part of the explanation of neural mechanisms of the startle reflex. That is to say, latencies of the EMG in the same muscle were different depending on the part of the body stimulated. These latency differences may be related to the distance of the ascending pathway from the stimulating point. These latency patterns may be explained by the mechanisms of SBS reflesex [SHIMAMURA and LIVINGSTON, 1963]. It may be that the cerebral palsy children showed SBS reflex with lower threshold and higher amplitude than normal [SHIMAMURA, MORI, MATSUSHIMA and FUJIMORI, 1964]. Chloralose anesthetized cats showed jerky muscular contraction generally associated with the SBS reflex [SHIMAMURA and YAMAUCHI, 1967]. This phenomenon is very similar to the startle reflex in the present observation.

For the above-mentioned reasons, we used the term *startle reflex* rather than *startle reaction*. The startle reaction involves the Moro's reflex of infants and the startle reaction elicited by a pistol-shot in the normal human [BROWN, MERYMAN and MARZOCCO, 1956; DUENSING, 1952; HUNT and LANDIS, 1936; SZABO, 1967]. These startle reactions may be related to cortical mechanisms, but this is still unclear.

It is known that man shows the SBS reflex, but it is difficult to obtain in the normal subject [SHIMAMURA, MORI, MATSUSHIMA and FUJIMORI, 1964]. Why does the SBS reflex occur in the cerebral palsy children? This may be related to the brain damage, disinhibition and/or enhancement, similar to decerebration or decortication, but at present detailed mechanisms have not been identified.

Summary

The startle reflex, a generalized jerky flexion, was elicited by click and tactile stimuli in cerebral palsy children. The neural mechanisms were analyzed by the use of movie film and EMG observations.

The latency of the EMG was different in muscles in the legs and arms. The earliest appeared in the orbicularis oculi muscle. This was followed by the pectoralis major, biceps brachii, semitendinosus and tibialis anterior muscles. These similar latency patterns of EMG were obtained by tactile stimulation to the trunk and foot. However, latencies were different at the stimulated part. The latency was about 25 msec longer for tactile stimulation to the foot than for the click-induced EMG, in all muscles.

These observations were analogous to the SBS reflex in the cat. Therefore, this suggests that the mechanisms of the startle reflex in cerebral palsy patients may be the same as those underlying the SBS reflexes.

Author's address: Dr. M. SHIMAMURA, Laboratory of Neurophysiology, Tokyo Metropolitan Institute for Neurosciences, 2–6 Musashidai, *Fuchu-city*, *Tokyo* (Japan)

New Developments in Electromyography and Clinical Neurophysiology,
edited by J. E. Desmedt, vol. 3, pp. 767–772 (Karger, Basel 1973)

Averaged Muscle-Responses to Repetitive Sensory Stimuli

K. Meier-Ewert, C. Schmidt, G. Nordmann, U. Hümme and J. Dahm

Neurologische Klinik der Technischen Universität, München

Electronic averaging instruments are extensively used for investigating cortical evoked potentials (centripetal averaging), and also muscle reflex potentials (centrifugal averaging) after sensory stimulation. In the case of scalp-recorded responses to sound it was shown that early components actually involved a vestibular mediated reflex response of neck muscles [Bickford, Jacobson and Cody, 1964]. These and similar myogenic artifacts represent previously unrecognized types of reflex responses to sensory stimuli [Bickford, 1964]. To separate them from the classical reflexes they have been called microreflexes [Bickford, 1964]. With respect to the somatosensory modality the scalp-recorded cerebral evoked potentials are not contaminated by such myogenic artifacts in the relaxed subject [Halliday, 1967; Desmedt, 1971] but, as shown by Cracco and Bickford [1968], muscle responses can be extracted by averaging the interference EMG pattern from muscles in voluntary contraction. In view of their small voltage (5–40 μV) such microreflexes evoked by sensory stimulation are not seen in the interference EMG record and 50 or more responses must be averaged to disclose the masked reflex transient.

Somatomotor Response Experiments

Electrical stimuli of 0.2-msec duration, intensity 3–4 times sensory threshold and frequency 1–7/sec elicited a somatomotor response in various muscles of 80% of 130 normal subjects. This response usually consisted of a triphasic potential of 30–40 msec duration and amplitudes of 15–40 μV and could be evoked from cutaneous afferents at the finger tips, the soles of the

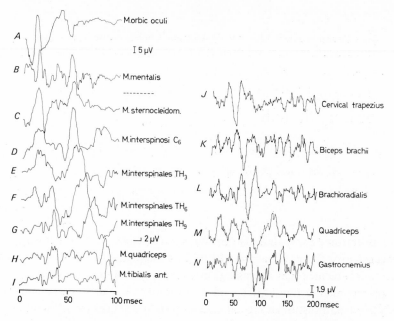

Fig. 1. Somatomotor and blink reflexes in a normal subject. *A* Two components of the blink reflex elicited in the right m. orbicularis oculi by single electric shocks to the right supraorbital nerve (in *A* and *B:* needle electrodes, stimulus duration 0.2 msec, stimulus intensity slightly over sensory threshold, frequency 1/sec, N = 100). *B* Two components of somatomotor responses from the partly contracted m. mentalis after stimulation of the right supraorbital nerve. *C–I* Somatomotor responses from interspinous and limb muscles following 400 single electric stimuli to the skin of the forehead (rectangular electrodes 3 × 5 cm, stimulus intensity 3–4 times sensory threshold, stimulus duration 0.2 msec, frequency 1/sec). Records *A–I* are from the same normal subject. The blink reflex from the relaxed m. orbicularis oculi as well as the somatomotor responses from the innervated facial, interspinous and limb muscles exhibit two components which show progressively greater delay with increasing distance from the cephalic region. *J–N* Somatomotor responses in a normal subject recorded from neck-, arm- and leg muscles of the right side of the body following 800 single electric stimuli to the left median nerve. Bipolar recordings with surface EMG electrodes.

feet and various other stimulation sites (base of the second and third finger, forehead, perioral regions). Two components of the response separated by 30–40 msec were observed frequently following median nerve stimulation and nearly always after cutaneous stimulation of the forehead (fig. 1). Occasionally, an additional smaller early component was observed. Similar results as far as mixed nerve stimulation is concerned were reported by PODIVINSKY [1971].

The earliest reflex response always occurred in the cephalic and cervical muscles, regardless of the site of stimulation, and showed a progressively greater delay with increasing distance of the sampled muscles from the brain stem (fig. 1 J–N). This was the case even with plantar stimulation [MEIER-EWERT, DAHM and NIEDERMEIER, 1971]. The somatomotor response to plantar stimulation thus represents a spino-bulbo-spinal reflex in man similar to that described by SHIMAMURA, MORI, MATSUSHIMA and FUJIMORI [1964]. Our results in man show that the conduction speed of the plantar evoked somatomotor response is faster in the afferent part of the spino-bulbo-spinal reflex arc than in the efferent one. Similar findings have been reported for the spino-bulbo-spinal reflex in animals by SHIMAMURA. In man the earliest cephalic response after plantar stimulation had a latency of 28 msec for the m. temporalis (body height 1.70 m) and in other patients 30 msec for the EEG response in the centro-parietal region [DAWSON, 1947, reported latencies between 30 and 35 msec for the EEG response following tapping of a patellar tendon (body height not given)]. If a conduction speed of 55 m/sec is assumed for the n. ischiadicus of 1 m length, 10 msec are left for the final 70 cm to the trigeminal motoneurons. Considering synaptic delay and outflow time, this would suggest an afferent spinal conduction speed higher than 70 m/sec. The efferent spinal conduction speed (if estimated from sampling of the mono-segmentally innervated interspinous muscles) would be usually about 20 m/sec (fig. 2). The length of the efferent motor α neurons to the dorsal interspinous muscles being approximately the same from cervical to lumbar regions, it is evident that the velocity of 20 m/sec reflects conduction in descending spinal pathways.

Following electrical stimulation of the skin of the forehead, somatomotor responses may be recorded from interspinous and limb muscles (fig. 1 A–I). These responses exhibit two components that can be demonstrated in cervical, interspinous and even in m. tibialis anterior (fig. 1). The efferent conduction velocity of these reflexes was about 20 m/sec in one subject (fig. 1) and between 40 and 50 m/sec in another.

A comparison between the latencies of the normal blink reflex components and those of the forehead-elicited somatomotor response components is given in figure 1. The data suggest that both phenomena have the same origin. Apparently, the electrical stimulus to the n. supraorbitalis (needle electrodes) or to the skin of the forehead (rectangular electrodes 5–3 cm) provokes two consecutive descending waves of increased motoneuron excitability which can be indentified by microreflex-recording in muscles as far apart as orbicularis oculi and tibialis anterior (fig. 1 A, I). The reflex potential

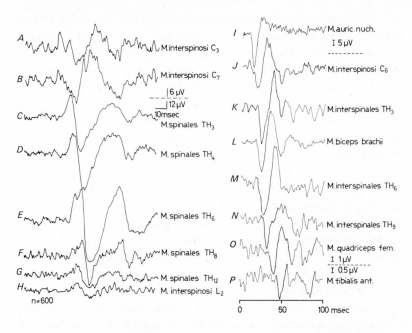

Fig. 2. A–H, somatomotor responses in a normal subject recorded from interspinous muscles following 400 single electric stimuli to the base of the 2nd and 3rd finger. From the distance of the recording sites C_3-Th_8 and the different latencies of these muscles a cranio-caudal conduction speed of about 20 m/sec is estimated. Bipolar recordings with needle electrodes (interelectrode distance 2 cm). *I–P* Sonomotor responses in a normal subject recorded from interspinous and limb muscles following 100, 400 or 1,000 single acoustic stimuli to both ears. The responses show a cranio-caudal conduction speed of about 20 m/sec, as do the somatomotor reflex potentials. Stimulus duration 0.1 sec, 750/sec, rate 1/sec.

can be seen in m. orbicularis oculi in either a relaxed or a contracted state of the muscle, but in all other facial, cervical, interspinous and limb muscles a background of voluntary innervation is required. From the latter, by means of electronic averaging of 50–100 epochs, the microreflex can be demonstrated. The fact that the first, but also the second blink reflex – and somatomotor response component could be shown in many muscles from orbicularis oculi down to tibialis anterior is of special interest, as it confirms the findings of Shahani and Young [1968], and suggests that the first component is of cutaneous, rather than proprioceptive, origin [see Young and Shahani, Penders and Delwaide, this volume, page 641].

PODIVINSKY [1971] described in patients with torticollis significantly higher amplitudes of somatomotor responses from the affected sternocleidomastoid and showed that the reflex in these patients could be modified or abolished by cutaneous stimulation. One cannot profitably speculate on the site of interaction or occlusion of these microreflexes until more is known of their pathways.

Sonomotor Responses

BICKFORD [1964] showed that the latencies of the sonomotor responses are shorter than those of the photo- and of the somatomotor responses. We recorded a postauricular response following acoustic stimuli (750/sec, 0.1-sec duration, frequency 1/sec) with a latency of 12 msec (fig. 2 I–P).

The samples from the partially contracted interspinous and limb muscles exhibited responses which also showed progressively greater delay with increasing distance from the cephalic and cervical region (fig. 2). The latencies we found are in good accord with those reported by BICKFORD, JACOBSON and CODY [1964]. They also compare with those of pathological startle syndromes described by DUENSING [1952], by SUHREN, BRUYN and TUYNMAN [1966] and by GASTAUT and VILLENEUVE [1967].

Though exhibiting shorter latencies, the sonomotor response usually shows the same cranio-caudal conduction speed of about 20 m/sec shown for the photo- and somatomotor responses. If, during a sequence of acoustic stimuli, sensory stimulation was constantly applied to the skin of the face in normal subjects no alteration of the sonomotor response in the sternocleidomastoid and other muscles could be detected. Thus interaction could not be demonstrated by PODIVINSKY's method [1971] between cutaneous and acoustic stimuli in normal subjects.

Summary

Electrically and acoustically evoked microreflexes were recorded in 130 normal subjects. Electric stimuli were applied to mixed nerves, cutaneous afferents at the finger tips, the skin of the forehead and the sole of the foot. From results with plantar stimulation it has been suggested that the somatomotor responses are mediated by a spino-bulbo-spinal reflex arc having an afferent conduction velocity of about 80 m/sec, while the efferent conduction speed is, in most subjects, around 20 m/sec.

Electric stimuli to the skin of the forehead evoked two reflex components in the voluntarily activated muscles of the face, trunk and limbs. The latencies of both components appeared comparable to those of the blink reflex components. We have, therefore, assumed that the blink reflex and the somatomotor responses both have their origin in two consecutive descending waves of increased motoneuron excitability which are evoked by the electrical stimulus to the skin of the forehead. Our findings confirm recent evidence that the first blink reflex component is not a myotatic reflex of proprioceptive origin, but rather an exteroceptive reflex.

Author's address: Priv.-Doz. Dr. K. MEIER-EWERT, Neurologische Klinik der Technischen Universität, Möhlstrasse 28, *D-8000 München 80* (FRG)

Methodology of the Triceps Surae Proprioceptive Reflexes

New Developments in Electromyography and Clinical Neurophysiology,
edited by J. E. Desmedt, vol. 3, pp. 773–780 (Karger, Basel 1973)

A Discussion of the Methodology of the Triceps Surae T- and H-Reflexes

Studies of human reflexes have developed significantly in the last few years and the pioneer studies of the previous decade have expanded into a considerable body of detailed information. Human reflexes attracted much interest throughout the Brussels International EMG Congress in September, 1971, and this segment of clinical neurophysiology will no doubt assume increasing significance both in research and in clinical diagnosis in the near future.

The human reflex studies seem to have reached a stage of development similar to that of Electroencephalography about 20 years ago when standardization of, for example, electrode positions on the scalp (the International Ten-Twenty System) proved so useful. Many methods are used in human reflex studies and some of the more sophisticated ones may be considered more relevant to specific research problems than to clinical diagnosis. The standardization of methods and relevant parameters will require a long-term effort on the part of many experts and it could well be assigned as a task to an *ad hoc* international commission. On the other hand, there is an urgent need to propose a number of practical suggestions about the procedures used in T- and H-reflex studies of triceps surae. These abbreviations describe the phasic stretch reflexes elicited either by mechanical percussion of the tendon (T-reflex) or by electrical stimulation of the I_A afferents from the spindles (Hoffmann or H-reflex).

Informal discussions at the Brussels Congress led to the suggestion that some standardization could be achieved for the most established methods in spite of some differences of opinion about a number of points. Techniques cannot be prescribed but it may prove useful to call attention to some details of procedures which should anyway be made explicit when results are published. It was also pointed out that such a preliminary report would provide guidance to those entering the field and would prepare for the work of a subsequent standardization commission.

A preliminary draft of the present paper was prepared by M. HUGON, P. DELWAIDE, E. PIERROT-DESEILLIGNY and J. E. DESMEDT and circulated for additional suggestions among interested investigators whose names appear at the end of the present report[1].

Proprioceptive Reflexes of Triceps Surae

Position of the Subject

Studies are currently performed either in the *prone position* with the subject lying on a couch with the front of the body turned toward the supporting surface, or in the *supine position* with the subject lying face upward on the back or, better, sitting in an easy chair. It is not yet known for certain whether results on H- and T-reflexes obtained with either positions can be compared in detail.

The *prone position* with a foot suspended over the end of the bed may appear easier to use and offers quick access to the popliteal fossa for electrical stimulation, to the calf muscles for recording and to the Achilles tendon. When the subject is on a simple couch it is not possible to maintain the knee and ankle joints. However, special accessories can be used to maintain these joints at standard angles [see GASSEL, this volume, page 345]. For experiments lasting for 1 h or more, the prone position is tedious to maintain, especially for fat people or for older patients with respiratory problems. The subject tends to become restless in this position which is not a normal one. It may be difficult to study a large group of patients with this method and be reasonably sure that consistent postural conditions have been achieved.

The *sitting position* is more comfortable for the subject, including older people with respiratory problems, and it permits long experiments without inconvenience. However, it requires a specially equipped chair with devices for holding the leg in specified positions (fig. 1).

---

1   As far as possible, the many suggestions received have been included in this report. Different investigators expressed divergent preferences only for a few points and the corresponding statements have been re-written accordingly by the Editor. It can fairly be said that the 20 persons consulted considered the final report useful and quite acceptable.

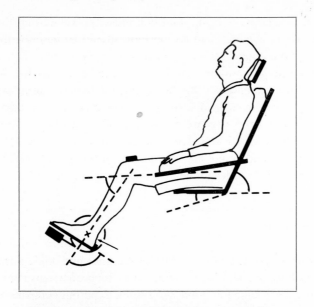

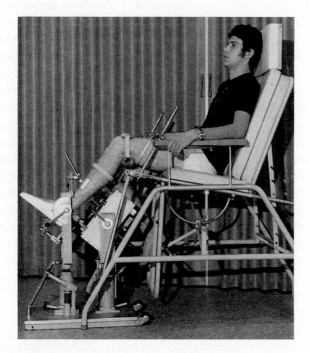

Fig. 1. Example of devices for holding the leg with the subject in sitting position.

No matter which method is being used one must appreciate that this can influence the results obtained and the position should be described in detail in the publication. A general requirement with both methods is that the patient should be relaxed and passive throughout the tests and that the limb positions should be maintained by the equipment rather than by the subject himself. It is recommended that the subject be given time to adapt himself to the apparatus before recording starts. The state of relaxation should be monitored by having EMG electrodes on both calf and pre-tibial muscles. Finally there should be no phasic head movements nor alterations in head and neck position in order to avoid vestibular influences.

Position of the Limb

Many studies have shown that the angles of the knee and ankle joints influence the excitability of the monosynaptic reflex loop of triceps surae. These angles should be specified in published results to permit comparison of the data recorded by different investigators. It is difficult to recommend specific positions, which may be decided upon according to the type of study performed and to special features of the patient (for example: tendon retractions, spasticity). It is useful to have the knee maintained by soft cushions on both sides and semi-flexed about 120° which relaxes the gastrocnemius muscles inserted on the femur. Stretch of the gastrocnemius may have an inhibitory effect on the soleus monosynaptic reflex. The angle of the ankle joint (usually 100–120°) determines the extension of the soleus which is inserted on the upper part of the tibia.

One should also specify whether the soleus muscle contracts under isometric or isotonic conditions. With isotonic conditions the movement of the foot will extend the antagonistic pre-tibial muscles, thus possibly eliciting afferent volleys from these muscles. On the other hand, with isometric conditions the contraction of soleus can elicit other proprioceptive effects, namely of the I_B afferents. A mechanogram of the reflex response is easily recorded with strain gauges arranged in the foot-plate.

Recording Electrodes

Surface electrodes are preferred to concentric needle electrodes because they pick up more consistently the contributions of a large number of motor units and, thus, provide a global reproducible estimate of the response of the muscle. Plate electrodes or silver discs filled with electrode jelly can be fixed to the prepared skin and the impedance between electrodes should be maintained low during the session. Stable recording conditions can also be achieved by inserting thin stainless steel needles through the skin *over* the muscle (not *into* the muscle). Flexible double-wire electrodes inserted into the muscle can be used for special purposes, e. g. when recording from single motor units during movement.

With a relaxed subject tested under standard conditions the H-response practically involves only the soleus muscle whereas the direct M-response involves both soleus and gastrocnemius [see HUGON, this volume, page 277]. The recording electrodes should be placed so as to pick up selectively the activity of the soleus muscle, thereby permitting meaningful comparison between the M- and H-responses recorded in the experiment. A convenient way to achieve this end is by placing the bipolar surface electrodes 3 cm apart along the longitudinal axis of the calf, the proximal (active) electrode of the pair being 2 cm distal to the insertion of the gastrocnemius muscle on the Achilles tendon [HUGON, this volume] (fig. 2). Other possible locations are on the lateral side of the calf where the soleus muscle is not covered by the gastrocnemius tendon. When studying the T-reflex, one may prefer more proximal locations in order to minimize the mechanical artefact associated with the tendon tap.

The amplitude of the response in millivolts is only meaningful if belly-tendon recording is used with the reference electrode on a relatively silent area at a distance of the active electrode. With the generally used bipolar recording from electrodes only a few centimeters apart, the size of responses is better expressed relative to the maximum M-response recorded through the same electrodes from the muscle. In view of the known variability of reflex responses it is suggested that the mean of 10 consecutive reflex responses evoked at intervals of at least 10 sec be calculated.

The Stimulus Eliciting the T-Reflex

It must be stressed that a conventional reflex hammer fitted with a switch to trigger the oscilloscope sweep is only suitable to demonstrate the presence of a reflex and to estimate its latency. For quantitative studies of the T-reflex 2 methods can be used: (1) a constant tendon tap produced by a solenoid plunger or a compressed air device fixed on a vibration-free

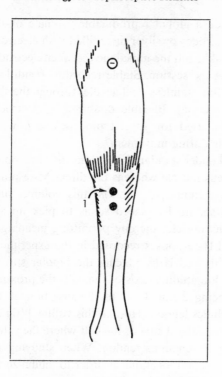

Fig.2. Position of the surface recording electrodes over the soleus with the active electrode (1) located 2 cm below the insertion of the gastrocnemius on the Achilles tendon in Hugon's method.

stand supporting the foot in a chosen position. The mechanical stimulus is then delivered with specified intensity to a fixed site on the tendon, and (2) a reflex hammer equipped with a momentum measuring device whereby the intensity of the mechanical stimulus can be compared to the reflex responses. The amplitude and rate of stretch of the muscle depend both on the mechanical features of the stimulus (site of impact, angle of impact, force delivered...) and on the compliance of the muscle tissue.

The Stimulus Eliciting the H-Reflex

While soleus H-reflexes can be elicited by even careless popliteal stimulations, it is important to standardize the stimulating conditions to

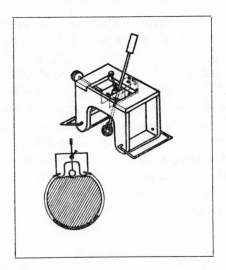

Fig. 3. The SIMON [1962] electrode for stimulating the tibial nerve through the skin. Such an electrode in place is photographed in figure 1. The ball joint holds a rod terminated by a small handle and a ball electrode, and it can be blocked in position after adjustment. The ball joint is held in a plexiglass frame which is placed in the popliteal fossa and maintained by bands of tissue around the leg as shown in figure 1 B on a smaller scale. The anode plate can be maintained on the skin above the patella by the same band.

obtain reproducible H-responses which are as large as possible with respect to the M-response picked up by the same recording electrodes. Monopolar stimulation with respect to a reference electrode placed on the skin above the patella is frequently used. The stimulating cathode should maintain a stable position with respect to the popliteal nerve throughout the experiment. When using electrodes on the skin the impedance between them should be reasonably low and stable to minimize variations in the electric field produced by the stimulus. Measurement of the stimulating current by a current probe can be recommended. A device such as the SIMON [1962] electrode is convenient when the subject is sitting in a chair and the popliteal fossa thus less accessible because it allows easy adjustment of electrode position as well as fixation (fig. 3). In experiments where posture and active muscle contractions are experimental variables one may consider using flexible wire leads for nerve stimulation in order to minimize inadvertent electrode movements.

The position of the cathode should optimize activation of I_A afferents with respect to α motor axons. Electric pulses of long duration, 1 msec, favour activation at threshold of the I_A afferents which have longer utilization times (lower accommodation) than α axons [see e.g. VEALE, REES and MARK, this volume, page 526].

When evaluating inhibitory and facilitatory effects on the H-reflex, it may be useful to choose a stimulus intensity such that the test H-reflex is about half the amplitude of the maximum H-reflex recorded under identical conditions in the muscle, thus permitting both increases or decreases of the test response. On the other hand, using such submaximal H-reflexes can increase the variability and it has been argued that the 'maximum' H-reflexes elicited by the test stimulus are not necessarily maximum reflexes and that they can be facilitated. No matter which condition is chosen this should be specified in published reports.

Other Parameters

Cumulative depression of the H-reflexes evoked sequentially will disturb the tests unless intervals of at least 10 sec separate the stimuli. This critical parameter failed to be correctly appreciated in some of the previous studies.

When estimating the latency of responses, the temperature along the nerves should have a value of 35–37 °C. This can be measured by a needle thermistor or, more conveniently, by checking that the skin temperature above the popliteal nerve is at least 34 °C. Since it is difficult to keep a patient warm in a cool laboratory it is advisable to do the testing in a room at 26–28 °C.

The methods considered are mild and require no sedation; the room should be reasonably quiet, and the subject should remain awake throughout the session.

C.H.M.BRUNIA (Tilburg), D.BURKE (Sydney), P.J.DELWAIDE (Liège), J.E.DESMEDT (Brussels), P.DIETRICHSON (Oslo), M.M.GASSEL (San Francisco), K.E.HAGBARTH (Uppsala), R.HERMAN (Philadelphia), M.HUGON (Marseilles), F.ISCH (Strasbourg), J.W.LANCE (Sydney), W.M.LANDAU (St-Louis), R.F.MARK (Melbourne), R.F.MAYER (Baltimore), E.PIERROT-DESEILLIGNY (Paris), P.PINELLI (Rome), G.RUSHWORTH (Oxford), E.SCHENCK (Freiburg i.Br.), A.STRUPPLER (Munich), M.WIESENDANGER (Zurich) and R.R.YOUNG (Boston)

Bibliography of Human Reflexes

ADAL, M.N.: The fine structure of the sensory region of cat muscle spindles. J.Ultrastruct. Res. 26: 332–354 (1969).

ADAL, M.N. and BARKER, D.: Motor supply to hindlimb muscles of the cat and rabbit. J.Anat., Lond. 99: 918–919 (1965).

ADAL, M.N. and BARKER, D.: Intramuscular branching of fusimotor fibres. J.Physiol., Lond. 177: 288–299 (1965).

ADOLPH, A.R.: Feedback in physiological systems: an application of feedback analysis and stochastic models to neurophysiology. Bull. Math. Biophys. 21: 195–215 (1959).

AGARWAL, G.C. and GOTTLIEB. G.L.: Analysis of step tracking in normal human subjects. IEEE Trans. Man-Machine Systems 10: 132–137 (1969).

AGARWAL, G.C. and GOTTLIEB, G.L.: Sampling in the human motor control system. IEEE Trans. Automatic Control 16: 180–183 (1971).

AKIMOTO, H. and SAITO, Y.: Synchronizing influences and their interactions on cortical and thalamic neurons; in TOKIZANE and SCHADE Correlative neurosciences. Progr. Brain Res., vol. 21A, pp. 323–351 (Elsevier, Amsterdam 1966).

ALBE FESSARD, D. et LIEBESKIND, J.: Origine des messages somatosensitifs activant les cellules du cortex moteur chez le singe. Exp. Brain Res. 1: 127–146 (1966).

ALBRECHT, M.H. and FERNSTRÖM, R.C.: A modified Nauta-Gygax method for human brain and spinal cord. Stain Technol. 34: 91–94 (1959).

ALNAES, E.: Static and dynamic properties of Golgi tendon organs in the anterior tibial and soleus muscles of the cat. Acta physiol. scand. 70: 176–187 (1967).

ALNAES, E.; JANSEN, J.K.S., and RUDJORD, T.: Fusimotor activity in the spinal cat. Acta physiol. scand. 63: 197–212 (1965).

ALSTON, W.; ANGEL, R.W.; FINK, F.S., and HOFMANN, W.W.: Motor activity following the silent period in human muscle. J. Physiol., Lond. 190: 189–202 (1967).

ALTMAN, J. and CARPENTER, M.B.: Fiber projections of the superior colliculus in the cat. J. comp. Neurol. 116: 157–177 (1961).

ALTMANN, H.; BRUGGENCATE, G. TEN, and SONNHOF, U.: Differential strength of action of glycine and GABA in hypoglossus nucleus. Pflügers Arch. ges. Physiol. 331: 90–94 (1972).

AMASSIAN, V.E. and BERLIN, L.: Early cortical projection of group I afferents in the forelimb muscle nerves of the cat. J. Physiol., Lond. *143:* 61P (1958).

AMASSIAN, V.E.; ROSENBLUM, M., and WEINER, H.: Role of thalamic nuclei ventralis-lateralis and its cerebellar input in contact placing. Fed. Proc. *29:* 792 (1970).

AMASSIAN, V.E.; WEINER, H., and ROSENBLUM, M.: Neural systems subserving the tactile placing reaction: a model for the study of higher level control of movement. Brain Res. *40:* 171–178 (1972).

AMINOFF, M.J. and SEARS, T.A.: Spinal integration of segmental cortical and breathing inputs to thoracic respiratory motoneurons. J. Physiol., Lond. *215:* 557–575 (1971).

ANASTASIJEVIĆ, R.; ANOJČIĆ, M.; TODOROVIĆ, B., and VUČO, J.: Effect of fusimotor stimulation on the reflex response of spinal alpha motoneurons to sinusoidal stretching of the muscle. Exp. Neurol. *25:* 559–570 (1969).

ANDEN, N.E.; JUKES, M.G.M.; LUNDBERG, A., and VYKLICKY, L.: A new spinal flexor reflex. Nature, Lond. *202:* 1344–1345 (1964).

ANDEN, N.E.; JUKES, M.G.M.; LUNDBERG, A., and VYKLICKY, L.: The effect of DOPA on the spinal cord. I. Influence on transmission from primary afferents. Acta physiol. scand. *67:* 373–386 (1966).

ANDERSEN, P.; ECCLES, J.C., and SEARS, T.A.: Cortically evoked depolarization of primary afferent fibres in the spinal cord. J. Physiol., Lond. *172:* 63–77 (1964).

ANDERSEN, P. and SEARS, T.A.: Medullary activation of intercostal fusimotor and alpha motoneurones. J. Physiol., Lond. *209:* 739–755 (1970).

ANDERSEN, P. and SEARS, T.A.: The mechanical properties and innervation of fast and slow motor units in the intercostal muscles of the cat. J. Physiol., Lond. *173:* 114–129 (1964).

ANDREWS, C.; KNOWLES, L., and HANCOCK, J.: Control of the tonic vibration reflex by the brainstem reticular formation in the cat (submitted for publication, 1972).

ANDREWS, C.; KNOWLES, L., and LANCE, J.W.: The effect of stimulation of the motor cortex and basal ganglia on the tonic vibration reflex of the cat (submitted for publication, 1972).

ANGAUT, P.: The ascending projections of the nucleus interpositus posterior of the cat cerebellum: an experimental anatomical study using silver impregnation methods. Brain Res. *24:* 377–394 (1970).

ANGAUT, P. and BOWSHER, D.: Ascending projections of the medial cerebellar (fastigial) nucleus: an experimental study in the cat. Brain Res. *24:* 49–68 (1970).

ANGEL, R.W.: Unloading reflex in patients with hemiparesis. Neurology, Minneap. *18:* 497–503 (1968).

ANGEL, R.W.; AGUILAR, J., and HOFMANN, W.W.: Action tremor and thalamotomy. Electroenceph. clin. Neurophysiol *26:* 80–85 (1969).

ANGEL, R.W. and ALSTON, W.: Spindle afferent conduction velocity. Neurology, Minneap. *14:* 647–676 (1964).

ANGEL, R.W.; EPPLER, W., and IANNONE, A.: Silent period produced by unloading of muscle during voluntary contraction. J. Physiol., Lond. *180:* 864–870 (1965).

ANGEL, R.W. and HOFMANN, W.W.: The H-reflex in normal, spastic and rigid subjects. Arch. Neurol., Chicago *8:* 591–596 (1963).

ANGEL, R.W.; HOFMANN, W.W., and EPPLER, W.: Silent period in patients with parkisonian rigidity. Neurology, Minneap. *16:* 529–532 (1966).

ANGLE, L. W.: Bilateral phrenicectomy in the treatment of persistent hiccoughs. Sth. med. J. *25:* 1012–1013 (1932).

AOYAMA, M.; HONGO, T.; KUDO, N., and TANAKA, R.: Convergent effects from bilateral vestibulospinal tracts on spinal interneurons. Brain Res. *35:* 250–253 (1971).

APPELBERG, B.; BESSOU, P., and LAPORTE, Y.: Effects of dynamic and static fusimotor gamma fibres on the responses of primary and secondary endings belonging to the same spindle. J. Physiol., Lond. *177:* 20P–30P (1965).

APPELBERG, B.; BESSOU, P., and LAPORTE, Y.: Action of static and dynamic fusimotor fibres on secondary endings of cat's spindles. J. Physiol., Lond. *185:* 160–171 (1966).

APPELBERG, B. and EMONET-DÉNAND, F.: Central control of static and dynamic sensitivities of muscle spindle primary endings. Acta physiol. scand. *63:* 487–494 (1965).

APPELBERG, B. and EMONET-DÉNAND, F.: Motor units of the first superficial lumbrical muscle of the cat. J. Neurophysiol. *30:* 154–160 (1967).

APTER, J. T.: Eye movements following strychninization of the superior colliculus of cat. J. Neurophysiol. *9:* 73–86 (1946).

ARIËNS KAPPERS, C. U.; HUBER, G. C., and CROSBY, E. C.: The comparative anatomy of the nervous system of vertebrates, including man (MacMillan, New York 1936).

ASANUMA, H. and ROSÉN, I.: Topographical organization of cortical efferent zones projecting to distal forelimb muscles in the monkey. Exp. Brain Res. *14:* 243–273 (1972).

ASANUMA, H.; STONEY, S. D., and THOMPSON, W. D.: Characteristics of cervical interneurons which mediate motor outflow to distal forelimb muscles in cats. Brain Res. *27:* 79–96 (1971).

ASHBY, P.; ANDREWS, C.; KNOWLES, L., and LANCE, J. W.: Pyramidal and extrapyramidal control of tonic mechanisms in the cat. Brain *95:* 21–30 (1972).

ASHBY, P. and BURKE, D.: Stretch reflexes in the upper limb of spastic man. J. Neurol. Neurosurg. Psychiat. *34:* 765–771 (1971).

ASHBY, P.; BURKE, D.; RAO, S., and JONES, R. F.: The assessment of cyclobenzaprine in the treatment of spasticity. J. Neurol. Neurosurg. Psych. *35:* 599–605 (1972).

ASHWORTH, B.; GRIMBY, L., and KUGELBERG, E.: Comparison of voluntary and reflex activation of motor units. Functional organization of motor neurones. J. Neurol. Neurosurg. Psychiat. *30:* 91–98 (1967).

BABINSKI, J.: Du phénomène des orteils et de sa valeur sémiologique. Semaine méd. *18:* 321–322 (1898).

BACH-Y-RITA, P. and ITO, F.: *In vivo* studies on fast and slow muscle fibers in cat extraocular muscles. J. gen. Physiol. *49:* 1177–1198 (1966).

BAILEY, H.: Persistent hiccup. Practitioner *150:* 173–177 (1943).

BARKER, D.: The structure and distribution of muscle receptors; in BARKER Symposium Muscle Receptors. pp. 227–240 (Hong Kong University Press, Hong Kong 1962).

BARKER, D. and CHIN, N. K.: The number and distribution of muscle spindles in certain muscles of the cat. J. Anat., Lond. *94:* 473–486 (1960).

BARKER, D. and COPE, M.: The innervation of individual muscle fibres; in BARKER Symposium Muscle Receptors, pp. 263–269 (Hong Kong University Press, Hong Kong 1962).

BARKER, D.; EMONET-DÉNAND, F.; LAPORTE, Y.; PROSKE, U., et STACEY, M.: Identification des terminaisons motrices des fibres fusimotrices statiques chez le chat. C. R. Acad. Sci. *271:* 1203–1206 (1970).

BARKER, D. and IP, M.C.: A study of single and tandem type of muscle-spindle in the cat. Proc. roy. Soc. B *154:* 377–397 (1961).

BARKER, D. and IP, M.C.: The motor innervation of cat and rabbit muscle spindles. J. Physiol., Lond. *177:* 27P–28P (1964).

BARKER, D.; STACEY, M.J., and ADAL, M.N.: Fusimotor innervation in the cat. Philos. Trans. B *258:* 315–346 (1970).

BARNARD, R.J.; EDGERTON, V.R.; FURUKAWA, T., and PETER, J.B.: Histochemical, biochemical and contractile properties of red, white and intermediate fibers. Amer. J. Physiol. *220:* 410–414 (1971).

BARNES, C.D. and POMPEIANO, O.: Inhibition of monosynaptic extensor reflex attributable to presynaptic depolarization of the group Ia afferent fibres produced by vibration of flexor muscle. Arch. ital. Biol. *108:* 233–258 (1970a).

BARNES, C.D. and POMPEIANO, O.: Presynaptic and postsynaptic effects in the monosynaptic reflex pathway to extensor mononeurons following vibration of synergic muscles. Arch. ital. Biol. *108:* 259–294 (1970b).

BARNES, C.D. and POMPEIANO, O.: Dissociation of presynaptic and postsynaptic effects produced in the lumbar cord by vestibular volleys. Arch. ital. Biol. *108:* 295–324 (1970c).

BARRETT, J.N. and CRILL, W.E.: Specific membrane resistivity of dye-injected cat motoneurons. Brain Res. *28:* 556–561 (1971).

BARRIOS, P.; HAASE, J., und HEINRICH, W.: Fusimotorische Alphareflexe an prätibialen Flexorenspindeln der Katze. Pflügers Arch. ges. Physiol. *296:* 49–69 (1967).

BASMAJIAN, J.V.: Control and training of individual motor units. Science *141:* 440–441 (1963).

BASTIAN, H.C.: On the symptomatology of total transverse lesions of the spinal cord; with special reference to the condition of the various reflexes. Med.-chir. Trans. *73:* 151–217 (1890).

BATES, J.A.V.: Some characteristics of a human operator. J. Inst. elect. Engrs, Lond. *94:* 298–304 (1947).

BATHIEN, N.: Réflexes spinaux chez l'homme et niveaux d'attention. Electroenceph. clin. Neurophysiol. *30:* 32–37 (1971).

BATHIEN, N. and BOURDARIAS, H.: Lower limb cutaneous reflexes in hemiplegia. Brain *95:* 447–456 (1972).

BATHIEN, N. et HUGELIN, A.: Réflexes monosynaptiques et polysynaptiques de l'homme au cours de l'attention. Electroenceph. clin. Neurophysiol. *26:* 604–612 (1969).

BATHIEN, N. et HUGON, M.: Etude chez l'homme de la dépression d'un réflexe monosynaptique par stimulation d'un nerf cutané J. Physiol., Paris *56:* 285–286 (1964).

BAUTISTA, N.S. and MATZKE, H.A.: A degeneration study of the course and extent of the pyramidal tract of the opossum. J. comp. Neurol. *124:* 367–376 (1965).

BECK, C.H. and CHAMBERS, W.W.: Speed, accuracy and strength of forelimb movement after unilateral pyramidotomy in rhesus monkeys. J. comp. physiol. Psychol. *70:* 1–22 (1970).

BEHSE, F. and BUCHTHAL, F.: Normal sensory conduction in the nerves of the leg in man. J. Neurol. Neurosurg. Psychiat. *34:* 404–414 (1971).

BENDER, L.F.; MAYNARD, F.M., and HASTINGS, S.V.: The blink reflex as a diagnostic procedure. Arch. phys. Med. Rehabil. *50:* 27–31 (1969).

BENOIST, J. M.; BESSON, J. M.; CONSEILLER, C., and LE BARS, D.: Action of bicuculline on presynaptic inhibition of various origins in the cat's spinal cord. Brain Res. *43:* 672–676 (1972).

BENSON, A. J.: Effect of labyrinthine stimulation on reflex and postural activity in gastrocnemius-soleus muscle group in man. J. Physiol., Lond. *146:* 37P–38P (1959).

BERGMAN, P. S.; HIRSCHBERG, G. G., and NATHANSON, M.: Measurement of quadriceps reflex in spastic paralysis. Neurology, Minneap. *5:* 542–549 (1955).

BERNHARD, C. G. and BOHM, E.: Cortical representation and functional significance of the cortico-motoneuronal system. Arch. Neurol. Psychiat., Chicago *72:* 473–502 (1954).

BERNHARD, C. G. and REXED, B.: The localization of the premotor interneurons discharging through the peroneal nerve. J. Neurophysiol. *8:* 387–392 (1945).

BESSOU, P.; EMONET-DÉNAND, F. et LAPORTE, Y.: Relation entre la vitesse de conduction des fibres nerveuses et le temps de contraction de leurs unités motrices. C. R. Acad. Sci. *256:* 5625–5627 (1963).

BESSOU, P.; EMONET-DÉNAND, F., and LAPORTE, Y.: Motor fibres innervating extrafusal and intrafusal muscle fibres in the cat. J. Physiol., Lond. *180:* 649–672 (1965).

BESSOU, P. and LAPORTE, Y.: Responses from primary and secondary endings of the same neuromuscular spindle of the tenuissimus muscle of the cat; in BARKER Symposium Muscle Receptors, pp. 105–119 (Honk Kong University Press, Hong Kong 1962).

BESSOU, P. and LAPORTE, Y.: Observations on static fusimotor fibers; in GRANIT Nobel Symposium 1, Muscular afferents and motor control, pp. 81–91 (Almqvist & Wiksell, Stockholm 1966).

BESSOU, P.; LAPORTE, Y. et PAGÈS, B.: Similitude des effets (statiques ou dynamiques) exercés par des fibres fusimotrices uniques sur les terminaisons primaires de plusieurs fuseaux chez le chat. J. Physiol., Paris *58:* 31–39 (1966).

BESSOU, P.; LAPORTE, Y., and PAGÈS, B.: Frequencygrams of spindle primary endings elicited by stimulation of static and dynamic fusimotor fibres. J. Physiol., Lond. *196:* 47–63 (1968).

BESSOU, P. and PAGÈS, B.: Intracellular recording from spindle muscle fibres of potentials elicited by static fusimotor axons in the cat. Life Sci. *8:* 417–419 (1969).

BIANCONI, R.; GRANIT, R. and REIS, D. J.: The effects of extensor muscle spindles and tendon organs on homonymous motoneurones in relation to gamma-bias and curarization. Acta physiol. scand. *61:* 331–347 (1964).

BIANCONI, R.; GRANIT, R., and REIS, D. J.: The effects of flexor muscle spindles and tendon organs on homonymous motoneurones in relation to γ-bias and curarization. Acta physiol. scand. *61:* 348–356 (1964).

BIANCONI, R. and MEULEN, J. P. VAN DER: The response to vibration of the end-organs of mammalian muscle spindles. J. Neurophysiol. *26:* 177–190 (1963).

BICKFORD, R. G.: Fast motor-systems in man. Trans. amer. neurol. Ass. *89:* 56–58 (1964).

BICKFORD, R. G.: Microreflexes. Electroenceph. clin. Neurophysiol. Suppl. *31* (in the press) (1972).

BICKFORD, R. G.; JACOBSON, J. L., and CODY, D. TH.: Nature of average evoked potentials to sound and other stimuli in man. Ann. N.Y. Acad. Sci. *112:* 204–218 (1964).

BILODEAU, I.: Information feedback; in BILODEAU Acquisition of skill (Academic Press, New York 1966).

Bischof, W.: Die longitudinale Myelotomie. Zbl. Neurochir. *11:* 79–88 (1951).

Bishop, A.: Use of the hand in lower primates, in Buettner-Janusch Evolutionary and genetic biology of primates, pp. 133–225 (Academic Press, London 1964).

Bisti, S.; Iosif, G.; Marchesi, G. F., and Strata, P.: Pharmacological properties on inhibitions in the cerebellar cortex. Exp. Brain Res. *14:* 24–37 (1971).

Bizzi, E. and Evarts, E. V.: Central control of movement. III. Translational mechanisms between input and output. Neurosc. Res. Progr. Bull. *9:* 31–59 (1971).

Bizzi, E.; Kalil, R. E., and Morasso, P.: Two modes of active eye-head coordination in monkeys. Brain Res. *40:* 45–48 (1972).

Block, A. M.: Expériences sur les sensations musculaires. Rev. Sci. *45:* 294–301 (1890).

Blom. S.; Hagbarth, K-E., and Skoglund, S.: Post-tetanic potentiation of H reflexes in human infants. Exp. Neurol. *9:* 198–211 (1964).

Bodian, D: Development of fine structure of spinal cord in monkey fetuses. I. The motoneuron neuropil at the time of onset of reflex activity. Bull. Johns Hopk. Hosp. *119:* 129–149 (1966).

Borenstein. S.: Réponses électriques globales à la stimulation du nerf moteur chez le nouveau-né normal. C. R. Soc. Biol. *162:* 2334–2336 (1968).

Bornstein, A. und Saenger, A.: Untersuchungen über den Tremor und andere pathologische Bewegungsformen mittelst des Saitengalvanometers. Dtsch. Z. Nervenheilk. *52:* 1–27 (1914).

Bosemark, B.: Some aspects of the crossed extensor reflex in relation to motoneurons supplying fast and slow contracting muscles; in Granit Nobel Symposium 1, Muscular afferents and motor control, pp. 261–268 (Almqvist & Wiksell, Stockholm 1966).

Bossom, J. and Ommaya, A. K.: Visuo-motor adaptation (to prismatic transformation of the retinal image) in monkeys with bilateral dorsal rhizotomy. Brain *91:* 161–172 (1968).

Botterell, E. H. and Fulton, J. F.: Functional localization in the cerebellum of primates. I. Unilateral section of the peduncles J. comp. Neurol. *69:* 31–46 (1938).

Botterell, E. H. and Fulton, J. F.: Functional localization in the cerebellum of primates. III. Lesions of hemispheres (neocerebellum). J. comp. Neurol. *69:* 63–87 (1938).

Bouhuys, A.; Proctor, D. F., and Mead, J.: Kinetic aspects of singing. J. appl. Physiol. *21:* 483–496 (1966).

Bowden, R. E. M. and Mahran, Z. Y.: The functional significance of the pattern of innervation of the muscle quadratus labii superioris of the rabbit, cat and rat. J. Anat., Lond. *90:* 217–227 (1956).

Bowditch, H. P. and Southard, W. F.: A comparison of sight and touch. J. Physiol., Lond. *3:* 232–245 (1880).

Boyd, I. A.: The structure and innervation of the nuclear bag muscle fibre system and the nuclear chain muscle fibre system in mammalian muscle spindles. Philos. Trans. B *245:* 81–136 (1962).

Boyd, I. A.: The behaviour of isolated mammalian muscle spindles with intact innervation. J. Physiol., Lond. *186:* 109P–110P (1966).

Boyd, I. A.: Specific fusimotor control of nuclear bag and nuclear chain fibres in cat spindles. J. Physiol., Lond. *214:* 30P–31P (1971).

Boyd, I. A. and Davey, M. R.: The distribution of two types of small motor nerve fibre to different muscles in the hind limb of the cat; in Granit Nobel Symposium 1, Afferent and motor control, pp. 59–68 (Almqvist & Wiksell, Stockholm 1966).

BOYD, I.A. and DAVEY, M.R.: Composition of pheripheral nerves (Livingston, London 1968).

BREININ, G.M. and MOLDAVER, J.: Electromyography of the human extraocular muscles. Arch. Ophthal., Chicago 54: 200–210 (1955).

BREMER, F.: Le cervelet; dans ROGER et BINET Traité de physiologie normale et patholo-gique, vol. 10, pp. 39–134 (Masson, Paris 1935).

BRINKMAN, J. and KUYPERS, H.G.: Splitbrain monkeys: cerebral control of ipsilateral and contralateral arm, hand and finger movements. Science: 176: 536–539 (1972).

BRINKMAN, J.; KUYPERS, H.G.J.M., and LAWRENCE, D.G.: Ipsilateral and contralateral eye-hand control in split-brain rhesus monkeys. Brain Res. 24: 559 (1970).

BROCK, L.G.; COOMBS, J.S., and ECCLES, J.C.: The recording of potentials from moto-neurones with an intracellular electrode. J.Physiol., Lond. 117: 431–460 (1952).

BROCK, L.G.; ECCLES, J.C., and RALL, W.: Experimental investigations on the afferent fibres in muscle nerve. Proc.roy. Soc.B 138: 453–475 (1951).

BRODAL, A.: Spasticity – anatomical aspects. Acta neurol.scand. 38: suppl. 3: pp.9–40 (1962).

BRODAL, A: Anatomical studies of cerebellar fibre connections with special reference to problems of functional localization; in FOX and SNIDER The cerebellum. Prog.Brain Res. vol. 25, 135–173 (Elsevier, Amsterdam 1967).

BRODAL, A.: Neurological anatomy in relation to clinical medicine (Oxford University Press, New York 1969).

BRODAL, A. and POMPEIANO, O.: The origin of ascending fibers of the medial longitudinal fasciculus from the vestibular nuclei. An experimental study in the cat. Acta morph. neerl.scand. 1: 306–328 (1957).

BRODAL, A.; POMPEIANO, O., and WALBERG, F.: The vestibular nuclei and their connections. Anatomy and functional correlations. The Henderson Trust Lectures (Oliver & Boyd, Edinburgh 1962).

BROMAN, T.: Electro-mecanographic registration of passive movements in normal and pathological subjects. Acta psychiat. scand., suppl. 53: 1–63 (1949).

BROOKHART, J.M.: A study of cortico-spinal activation of motor neurones. Res. Publ. Ass. nerv. ment. Dis. 30: 157–173 (1952).

BROOKS, C.McC. and FUORTES M.G.F.: Electrical correlates of the spinal flexor reflex. Brain 75: 91–95 (1952).

BROOKS, V.B.; CURTIS, D.R., and ECCLES, J.C.: The action of tetanus toxin on the inhibition of motoneurons. J. Physiol., Lond. 135: 655–672 (1957).

BROOKS, V.B. and WILSON, V.J.: Localization of stretch reflexes by recurrent inhibition. Science 127: 472–473 (1958).

BROOKS, V.B. and WILSON, V.J.: Recurrent inhibition in the cat's spinal cord. J.Physiol., Lond. 146: 380–391 (1959).

BROOKS, W.D.W.: Zoster, hiccup and varicella. Brit.med.J. ii: 298–299 (1931).

BROWN, J.S.; MERYMAN, J.W., and MARZOCCO, F.N.: Sound-induced startle response as a function of time since shock. J.comp.physiol.Psychol. 49: 190–194 (1956).

BROWN, K.; LEE, J., and RING, P.A.: The sensation of passive movement at the metatarso-phalangeal joint of the great toe in man. J.Physiol., Lond. 126: 448–458 (1954).

BROWN, M.C.; CROWE, A., and MATTHEWS, P.B.C.: Observations on the fusimotor fibres of the tibialis posterior muscle of the cat. J.Physiol., Lond. 177: 140–159 (1965).

BROWN, M. C.; ENGBERG, I., and MATTHEWS, P. B. C.: The relative sensitivity to vibration of muscle receptors of the cat. J. Physiol., Lond. *192:* 773–800 (1967).

BROWN, M. C.; GOODWIN, G. M., and MATTHEWS, P. B. C.: After effects of fusimotor stimulation on the response of muscle spindle primary afferent endings. J. Physiol., Lond. *205:* 677–694 (1969).

BROWN, M. C.; LAWRENCE, D. G., and MATTHEWS, P. B. C.: Reflex inhibition by Ia afferent input of spontaneously discharging motoneurones in the decerebrate cat. J. Physiol., Lond. *198:* 5P–7P (1968).

BROWN, M. C. and MATTHEWS, P. B. C.: On the subdivision of the efferent fibres to muscle spindles into static and dynamic fusimotor fibres; in ANDREW Control and innervation of skeletal muscle, pp. 17–31 (Thomson, Dundee 1966).

BROWN, T. G.: The intrinsic factors in the act of progression in the mammal. Proc. roy. Soc. B *84:* 308–319 (1911).

BROWN, W. J. and FANG, H. C. H.: Spastic hemiplegia in man associated with unilateral infarct of the corticospinal tract at the ponto-medullary junction. Trans. amer. neurol. Ass.: 22–26 (1956).

BRUESCH, S. R.: The distribution of myelinated afferent fibres in the branches of the cat's facial nerve. J. comp. Neurol. *81:* 169–191 (1944).

BRUGGENCATE, G. TEN and ENGBERG, I.: Iontophoretic studies in deiters' nucleus of the inhibitory actions of GABA and related amino acids and the interactions of strychnine and picrotoxin. Brain Res. *25:* 431–448 (1971).

BRUGGENCATE, G. TEN and SONNHOF, U.: Effects of glycine and GABA, and blocking actions of strychnine and picrotoxin in the hypoglossus nucleus. Pflügers Arch. ges. Physiol. *334:* 240–252 (1972).

BRUNIA, C. H. M.: Reflex amplitudes and increasing mental load. Electroenceph. clin. Neurophysiol. *29:* 320 (1970).

BRUNIA, C. H. M.: Alertheid en de veranderingen van de Achillespees – en Hoffmann reflex (de Grazendonk, Breda 1970).

BRUNIA, C. H. M.: The influence of a task on the Achilles tendon and Hoffmann reflex. Physiol. Behav. *6:* 367–373 (1971).

BRUNIA, C. H. M. en DIESVELDT, H.: Onderdrukking van de sinusaritmie tijdens een taak. T. soc. Geneesk. *49:* 130–132 (1971).

BUCHTHAL, F. and SCHMALBRUCH, H.: Contraction times of twitches evoked by H-reflexes. Acta physiol. scand. *80:* 378–382 (1970).

BUCHWALD, J. S.; STANDISH, M.; ELDRED, E., and HALAS, E. S.: Contribution of muscle spindle circuits to learning as suggested by training under Flaxedil. Electroenceph. clin. Neurophysiol. *16:* 582–594 (1964).

BUCY, P. C.: Is there a pyramidal tract? Brain *80:* 376–392 (1957).

BUCY, P. C. and KEPLINGER, J. E.: Section of the cerebral peduncles. Arch. Neurol., Chicago *5:* 132–139 (1961).

BUCY, P. C.; KEPLINGER, J. C., and SIGNERA, E. B.: Destruction of the pyramidal tract in man. J. Neurosurg. *21:* 385–398 (1964).

BUCY, P. C.; LADPLI, R., and EHRLICH, A.: Destruction of the pyramidal tract in the monkey. J. Neurosurg. *25:* 1–23 (1966).

BUIST, W. G.: Klinische EMG onderzoekmethoden bij enkele supranucleaire syndromen (van Gorcum, Assen 1970).

BULLER, A.J.: The ankle-jerk in early hemiplegia. Lancet *ii:* 1262–1263 (1957).

BULLER, A.J. and DORNHORST, A.C.: The reinforcement of tendon-reflexes. Lancet *273:* 1260–1262 (1957).

BÜRGI, S. und MONNIER, M.: Motorische Erscheinungen bei Reizung und Ausschaltung der Substantia reticularis pontis. Helv. physiol. pharmacol. Acta *1:* 489–510 (1943).

BURKE, D.; ANDREWS, C., and ASHBY, P.: Autogenic effects of static muscle stretch in spastic man. Arch. Neurol. Chicago *25:* 367–372 (1971).

BURKE, D.; ANDREWS, C.J., and GILLIES, J.D.: The reflex response to sinusoidal stretching in spastic man. Brain *94:* 455–470 (1971).

BURKE, D.; ANDREWS, C.J., and KNOWLES, L.: The action of a GABA derivative in spasticity. J. neurol. Sci. *14:* 199–208 (1971).

BURKE, D.; ANDREWS, C.J., and LANCE, J.W.: Tonic vibration reflex in spasticity, Parkinson's disease, and normal subjects. J. Neurol. Neurosurg. Psychiat. *35:* 477–486 (1972).

BURKE, D. and ASHBY, P.: Are spinal 'presynaptic' inhibitory mechanisms suppressed in spasticity? J. neurol. Sci. *15:* 321–326 (1972).

BURKE, D.; GILLIES, J.D., and LANCE, J.W.: The quadriceps stretch reflex in human spasticity. J. Neurol. Neurosurg. Psychiat. *33:* 216–223 (1970).

BURKE, D.; GILLIES, J.D., and LANCE, J.W.: Hamstrings stretch reflex in human spasticity. J. Neurol. Neurosurg. Psychiat. *34:* 231–235 (1971).

BURKE, D.; KNOWLES, L.; ANDREWS, C., and ASHBY, P.: Spasticity, decerebrate rigidity and the clasp-knife phenomenon: an experimental study in the cat. Brain *95:* 31–48 (1972).

BURKE, R.E.: Motor unit types of cat triceps surae muscle. J. Physiol., Lond. *193:* 141–160 (1967a).

BURKE, R.E.: Composite nature of monosynaptic excitatory postsynaptic potential. J. Neurophysiol. *30:* 1114–1137 (1967b).

BURKE, R.E.: Group Ia synaptic input to fast and slow twitch motor units of cat triceps surale. J. Physiol., Lond. *196:* 605–630 (1968a).

BURKE, R.E.: Firing patterns of gastrocnemius motor units in the decerebrate cat. J. Physiol., Lond. *196:* 631–654 (1968b).

BURKE, R.E.: Control systems operating on spinal reflex mechanisms. Neurosci. Res. Progr. Bull. *9:* 60–85 (1971).

BURKE, R.E.; FEDINA, L., and LUNDBERG, A.: Spatial synaptic distribution of recurrent and group Ia inhibitory systems in cat spinal motoneurones. J. Physiol., Lond. *214:* 305–326 (1971).

BURKE, R.E.; JANKOWSKA, E., and BRUGGENCATE, G. TEN: A comparison of peripheral and rubrospinal synaptic input to slow and fast twitch motor units of triceps surae. J. Physiol., Lond. *207:* 709–732 (1970).

BURKE, R.E.; LEVINE, D.N.; ZAJAC, F.E. III; TSAIRIS, P., and ENGEL, W.K.: Mammalian motor units: physiological-histochemical correlation in three types in cat gastrocnemius. Science *174:* 709–712 (1971).

BURKE, R.E. and NELSON, P.G.: Accomodation to current ramps in motoneurons of fast and slow twitch motor units. Int. J. Neurosci. *1:* 347–356 (1971).

BURKE, R.E.; RUDOMIN, P., and ZAJAC, F.E.: Catch property in single mammalian motor units. Science *168:* 122–124 (1970).

BURKE, R. E. and BRUGGENCATE, G. TEN: Electrotonic characteristics of alpha motoneurones of varying size. J. Physiol., Lond. *212:* 1–20 (1971).

BUSCH, H. F. M.: Anatomical aspects of the anterior and lateral funiculi at the spinobulbar junction; in ECCLES and SCHADÉ Organization of the spinal cord. Progr. Brain Res. vol. 11, (p. 285 Elsevier, Amsterdam 1964).

BUSKIRK E. R. and KOMI, P. V.: Reproducibility of electromyographic measurements with inserted wire electrodes and surface electrodes. Acta physiol. scand. *79:* 29A (1970).

BUXTON, D. F. and GOODMAN, D. C.: Motor function and the corticospinal tracts in the dog and racoon. J. comp. Neurol. *129:* 341–360 (1967).

BYNKE, O.: Facial reflexes and their clinical uses. Lancet *i:* 137–138 (1971).

CALVIN, W. H. and STEVENS, C. F.: Synaptic noise and other sources of randomness in motoneurone interspike intervals. J. Neurophysiol. *31:* 574–587 (1968).

CAMPBELL, C. P. G.; YASHON, D., and JANE, J. A.: The origin, course and terminations of corticospinal fibers in the slow loris, *Nycticebus coucang* (Boddaert). J. comp. Neurol. *127:* 101–112 (1966).

CANNON, W.; MAGOUN, H. W., and WINDLE, W.: Paralysis with hypotonicity and hyper-reflexia subsequent to section of the basis peduncli in monkeys. J. Neurophysiol. *7:* 425–437 (1944).

CARMICHAEL, L.: The onset and early development of behavior; in Manuel of child psychology (Wiley, New York 1954).

CARPENTER, D.; ENGBERG, I.; FUNKENSTEIN, H., and LUNDBERG, A.: Decerebrate control of reflexes to primary afferents. Acta physiol. scand. *59:* 242–437 (1964).

CARPENTER, M. B.; HARBISON, J. W., and PETER, P.: Accessory oculomotor nuclei in the monkey: projections and effects of discrete lesions. J. comp. Neurol. *140:* 131–154 (1970).

CARPENTER, M. B. and PINES, J.: Rubrobulbar tract: anatomical relationships, course and termination in the rhesus monkey. Anat. Rec. *128:* 171–185 (1956).

CARREA, R. M. E. and METTLER, F. A.: Physiologic consequences following extensive removals of the cerebellar cortex and deep cerebellar nuclei and effect of secondary cerebral ablations in the primate. J. comp. Neurol. *87:* 169–288 (1947).

CHAMBERS, W. W. and LIU, C. N.: Corticospinal tract of the cat. An attempt to correlate the pattern of degeneration with deficits in reflex activity following neocortical lesions, J. comp. Neurol. *108:* 23–55 (1957).

CHAMBERS, W. W. and LIU, C. N.: An experimental study of the corticospinal system in the monkey *(Macaca mulatta)*, J. comp. Neurol. *123:* 257–284 (1965).

CHAMBERS, W. W. and SPRAGUE, J. M.: Functional localization in the cerebellum. I. Organization in longitudinal cortico-nuclear zones and their contributions to the control of posture extra-pyramidal and pyramidal. J. comp. Neurol. *103:* 105–129 (1955a).

CHAMBERS, W. W. and SPRAGUE, J. M.: Functional localization in the cerebellum. II. Somatotopic organization in cortex and nuclei. Arch. Neurol. Psychiat., Chicago *74:* 653–680 (1955b).

CHANG, H.-T.: The repetitive discharges of corticothalamic reverberating circuit. J. Neurophysiol. *13:* 235–257 (1950).

CHARCOT, J.-M.: Leçons sur les maladies du système nerveux. Bourneville, vol. 2 (Progrès Médical, Paris 1894).

CHASE, R. A.; CULLEN, J. K.; SULLIVAN, S. A., and OMMAYA, A. K.: Modification of intention tremor in man. Nature, Lond. *206:* 485–487 (1965).

CHIN, N. K.; COPE, M., and PANG, M.: Number and distribution of spindle capsules in seven hind limb muscles of the cat; in BARKER Symposium Muscle Receptors, 241–248 (Hong Kong University Press, Hong Kong 1962).

CHRISTENSEN, B. N. and PERL, E. R.: Spinal neurons specifically excited by noxious or thermal stimuli. Marginal zone of dorsal horn. J. Neurophysiol. *33:* 293–307 (1970).

CLAMANN, H. P.: Statistical analysis of motor unit firing pattern in human skeletal muscle. Biophys. J. *9:* 1233–1251 (1969).

CLARK, M. S. G. and RAND, M. J.: A pharmacological effect of tobacco smoke. Nature, Lond. *201:* 507–508 (1964).

CLARK, M. S. G.; RAND, M. J., and VANOV, S.: Comparison of pharmacological activity of nicotine and related alkaloids occuring in cigarette smoke. Arch. int. Pharmacol. *156:* 363–379 (1965).

CLARKE, A. M.: Effect of stimulation of certain skin areas on extensor motoneurons in the plastic reaction of a stretch reflex in normal human subjects. Electroenceph. clin. Neurophysiol. *21:* 185–193 (1966).

CLOSE, R.: Properties of motor units in fast and slow skeletal muscles of the rat. J. Physiol., Lond. *193:* 45–55 (1967).

CLOUGH, J. F. M.; KERNELL, D., and PHILLIPS, C. G.: The distribution of monosynaptic excitation from the pyramidal tract and from primary spindle afferents to motoneurones of the baboon's hand and forearm. J. Physiol., Lond. *198:* 145–166 (1968).

CLOUGH, J. F. M.; PHILLIPS, C. G., and SHERIDAN, J. D.: The short-latency projection from the baboon's motor cortex to fusimotor neurones of the forearm and hand. J. Physiol., Lond. *216:* 257–279 (1971).

COGAN, D. G.: Neurology of the ocular muscles; 2nd ed., pp. 129 (Thomas, Springfield 1969).

COHEN, D.; CHAMBERS, W. W., and SPRAGUE, J. M.: Experimental study of the efferent projections from the cerebellar nuclei to the brain stem of the cat. J. comp. Neurol. *109:* 233–259 (1958).

COHEN, L. A.: Localization of stretch reflex. J. Neurophysiol. *16:* 274–285 (1953).

COLLIER, J. and BUZZARD, F.: Descending mesencephalic tracts in cat, monkey and man; Monakow's bundle, etc. – Brain *24:* 177–221 (1901).

COLLINS, W. F.; NULSEN, F. E., and RANDT, C. T.: Relation of peripheral nerve size and sensation in man. Arch. Neurol., Chicago *3:* 381–385 (1960).

COMBS, C. M.: Course of fibres of brachium conjunctivum revealed by evoked potential method. Exp. Neurol. *1:* 13–27 (1959).

CONRADI, S.: Ultrastructure of dorsal root boutons on lumbosacral motoneurons of the adult cat, as revealed by dorsal root section. Acta physiol. scand. *78:* suppl. 332: 85–115 (1969).

CONRADI, S. and SKOGLUND, S.: On motoneurones synaptology in kittens. An electron microscopic study of the structure and location of neuronal and glial elements on cat lumbosacral motoneurons in the normal state and after dorsal root section. Acta physiol. scand., suppl. *333* (1969).

COOK, C. D.; MEAD, J., and ORZALESI, M. M.: Static volume-pressure characteristics of the respiratory system during maximal efforts. J. appl. Physiol. *19:* 1016–1022 (1964).

COOKS, W. A., jr.; NEILSON, D. R. jr., and BROOKHART, J. M.: Primary afferent depolarization and monosynaptic reflex depression following succinylcholine administration. J. Neurophysiol. *23:* 290–311 (1965).

COOPER, I. S.: A cerebellar mechanism in resting tremor. Neurobiology *16:* 1003–1015 (1966).

COOPER, I. S.: Involuntary movement disorders (Harper & Row, New York 1969).

COOPER, S.: Muscle spindles in the intrinsic muscles of the human tongue. J. Physiol., Lond. *122:* 193–202 (1953).

COOPER, S.: The secondary endings of muscle spindles. J. Physiol., Lond. *149:* 27P–28P (1959).

COOPER, S.: The responses of the primary and secondary endings of muscle spindles with intact motor innervation during applied stretch. Quart. J. exp. Physiol. *46:* 389–398 (1961).

COOPER, S.: Muscle spindles and motor units; in ANDREW Control and innervation of skeletal muscle, p. 15 (University St. Andrews Press, Aberdeen 1966).

COOPER, S. and DANIEL, P. M.: Muscle spindles in human extrinsic eye muscles. Brain *72:* 1–24 (1949).

COOPER, S. and DANIEL, P. M.: Human muscle spindles. J. Physiol., Lond. *133:* 1P (1956).

COOPER, S. and DANIEL, P. M.: Muscle spindles in man; their morphology in the lumbrical and the deep muscles of the neck. Brain *86:* 563–586 (1963).

COPLAND, J. G. and DAVIES, C. T. M.: A simple clinical skin electrode. Lancet *i: 416* (1964).

COQUERY, J. M.: Les variations spontanées du réflexe de Hoffmann; thèse Marseille (1972).

COQUERY, J. M. et COULMANCE, M.: Variations d'amplitude des réflexes monosynaptiques avant un mouvement volontaire. Physiol. Behav. *6:* 65–71 (1971).

COQUERY, J. M.; MARK, R. F. et PAILLARD, J.: Les fluctuations spontanées du réflexe de Hoffmann à différents niveaux de la courbe de recrutement. Electroenceph. clin. Neurophysiol., suppl. *22:* 90–92 (1962).

CORAZZA, R.; FADIGA, E., and PARMEGGIANI, P. L.: Patterns of pyramidal activation of cat's motoneurons. Arch. ital. Biol. *101:* 337–364 (1963).

CORBETT, J. L., FRANKEL, H. L., and HARRIS, P. J.: Cardiovascular changes associated with skeletal muscle spasm in tetraplegic man. J. Physiol., Lond. *215:* 381–393 (1971).

CORBETT, J. L.; KERR, J. H.; PRYS-ROBERTS, C.; CRAMPTON SMITH, A., and SPALDING, J. M. K.: Cardiovascular disturbances in severe tetanus due to overactivity of the sympathetic nervous system. Anaesthesia *24:* 198–212 (1969).

CORDA, M.; EKLUND, C., and EULER, C. VON: External intercostal and phrenic alpha motor responses to changes in respiratory load. Acta physiol. scand. *63:* 391–400 (1965).

CORDA, M.; EULER, C. VON, and LENNESTRAND, G.: Reflex and cerebellar influences on alpha and on rhythmic and tonic gamma activity in the intercostal muscle. J. Physiol., Lond. *184:* 898–923 (1966).

CORRIE, W. S. and HARDIN, W. B., jr.: Post-tetanic potentiation of H reflex in normal man. Arch. Neurol., Chicago *11:* 317–323 (1964).

CORVAJA, N.; MARINOZZI, V., and POMPEIANO, O.: Muscle spindles in the lumbrical muscle of the adult cat. Electron microscopic observations and functional considerations. Arch. ital. Biol. *107:* 365–543 (1969).

CORVAJA, N. and POMPEIANO, O.: The differentiation of two types of intrafusal fibres in rabbit muscle spindles. Pflügers Arch. ges. Physiol. *317:* 187–197 (1970).

COURVILLE, J.: Somatotopical organization of the projection from the nucleus interpositus anterior of the cerebellum to the red nucleus. An experimental study in the cat with silver impregnation methods. Exp. Brain Res. *2:* 191–215 (1966).

CRACCO, R. Q. and BICKFORD, R. G.: Somatomotor and somatosensory evoked responses: median nerve stimulation in man. Arch. Neurol., Chicago *18:* 52–68 (1968).

CREED, R. S.; DENNY-BROWN, D.; ECCLES, J. C.; LIDDELL, E. G. T., and SHERRINGTON, D. S.: Reflex activity of the spinal cord (Oxford University Press, London 1932).

CRITCHLOW, V. and EULER, C. VON: Rhythmic control of intercostal muscle spindles. Experientia *18:* 426 (1962).

CRITCHLOW, V. and EULER, C. VON: Intercostal muscle spindle activity and its γ-motor control. J. Physiol., Lond. *168:* 820–847 (1963).

CROWE, A. and MATTHEWS, P. B. C.: The effects of stimulation of static and dynamic fusimotor fibres on the response to stretching of the primary endings of muscle spindles. J. Physiol., Lond. *174:* 109–131 (1964a).

CROWE, A. and MATTHEWS, P. B. C.: Further studies of static and dynamic fusimotor fibres. J. Physiol., Lond. *174:* 132–151 (1964b).

CURTIS, D. R.; DUGGAN, D. W.; FELIX, D., and JOHNSTON, G. A. R.: Bicuculline, an antagonist of GABA and synaptic inhibition in the spinal cord of the cat. Brain Res. *32:* 69–96 (1971).

CURTIS, D. R. and ECCLES, J. C.: Synaptic action during and after repetitive stimulation. J. Physiol., Lond. *150:* 374–398 (1960).

CURTIS, D. R.; ECCLES, J. C., and ECCLES, R. M.: Pharmacological studies on spinal reflexes. J. Physiol., Lond. *136:* 420–434 (1957).

CURTIS, D. R. and RYALL, R. W.: The synaptic excitation of Renshaw cells. Exp. Brain Res. *2:* 81–96 (1966).

DASGUPTA, A.: Behaviour of single motor units in human skeletal muscle, Ph. D. thesis, Edinburgh (1962).

DA SILVA, K. M. C.: Neuromuscular activity and respiratory dynamics in the cat; Ph. D. thesis, London (1971).

DAVEY, M. R. and TAYLOR, A.: The activity of jaw muscle spindles recorded together with active jaw movement in the cat recovering from anaesthesia. J. Physiol., Lond. *190:* 8P (1967).

DAVIDOFF, R. A.: Gamma-Aminobutyric acid antagonism and presynaptic inhibition in the frog spinal cord. Science *175:* 331–333 (1972).

DAVIS, W. J.: Functional significance of motoneuron size and soma position in swimmeret system of the lobster. J. Neurophysiol. *34:* 274–288 (1971).

DAVISON, C.: Syndrome of the anterior spinal artery of the medulla oblongata. Arch. Neurol. Psychiat., Chicago *37:* 91–107 (1937).

Dawson, G. D.: The relative excitability and conduction velocity of sensory and motor nerve fibers in man. J. Physiol., Lond. *131:* 436–451 (1956).

DAWSON, G. D. and MERTON, P. A.: Recurrent discharges from motoneurones. 2nd Int. Congr. Physiol. Sci., Bruxelles, abstr., pp. 221–222 (1956).

DECANDIA, M.; PROVINI, L., and TÁBOŘÍKOVÁ, H.: Mechanism of the reflex discharge depression in the spinal motoneurone during repetitive orthodromic stimulation. Brain Res. *4:* 284–291 (1967a).

DECANDIA, M.; PROVINI, L., and TÁBOŘÍKOVÁ, H.: Presynaptic inhibition of the mono-synaptic reflex following the stimulation of nerves to extensor muscles of the ankle. Exp. Brain Res. *4:* 34–42 (1967b).

DEKABAN, A.: Neurology in infancy (Williams & Wilkins, Baltimore 1959).

DELONG, M.R.: Central patterning of movement. Neurosci. Res. Progr. Bull. *9:* 10–30 (1971).

DELONG, M.R.: Activity of basal ganglia neurons during movement. Brain Res. *40:* 127–135 (1972).

DELWAIDE, P.J.: Approche de la physiopathologie de la spasticité: Réflexe de Hoffmann et vibrations appliquées sur le tendon d'Achille. Rev. neurol. *121:* 72–74 (1969).

DELWAIDE, P.J.: Etude expérimentale de l'hyperréflexie tendineuse en clinique neurolo-gique (Arscia, Bruxelles 1971).

DELWAIDE, P.J. et BONNET, M.: Nouveau mécanisme à considérer dans l'inhibition du réflexe de Hoffmann. J. Physiol., Paris *61:* 119 (1969).

DELWAIDE, P.J.; HUGON, M., and THIRIET, E.: Slow stretching of human triceps surae by a pneumatic device. Electromyography *4:* 399–405 (1970).

DELWAIDE, P.J.; SCHWAB, R.S., and YOUNG, R.R.: Polysynaptic spinal reflexes in Parkin-son's diseases: effects of L-dopa treatment. Neurology, Minneap. *21:* 407 (1971).

DENNY-BROWN, D.: On inhibition as a reflex accompaniment of the tendon jerk and other forms of active muscular response. Proc. roy. Soc. B *103:* 321–326 (1928).

DENNY-BROWN, D.: On the nature of postural reflexes. Proc. roy. Soc. B *104:* 252–301 (1929).

DENNY-BROWN, D.: The histological features of striped muscle in relation to its functional activity. Proc. roy. Soc. B *104:* 371–411 (1929).

DENNY-BROWN, D.: Disintegration of motor function resulting from cerebral lesions. J. nerv. ment. Dis. *112:* 1–45 (1950).

DENNY-BROWN, D.: The basal ganglia and their relation to disorders of movement (Oxford University Press, London 1962).

DENNY-BROWN, D.: The extrapyramidal system and postural mechanisms. Clin. Pharmacol. Ther. *5:* 812–827 (1964).

DENNY-BROWN, D.: The cerebral control of movement (Liverpool University Press, Liver-pool 1966).

DENNY-BROWN, D. and GILMAN, S.: Depression of gamma innervation by cerebellectomy. Trans. amer. neurol. Ass. *90:* 96–101 (1965).

DENNY-BROWN, D. and PENNYBACKER, J.B.: Fibrillation and fasciculation in voluntary muscle. Brain *61:* 311–334 (1938).

DESCHUYTERE, J. and ROSSELLE, N.: Electromyographic and neurophysiological investiga-tion in root compression syndromes in man. Electromyography *10:* 339–340 (1970).

DESMEDT, J.E.: Méthodes d'étude de la fonction neuromusculaire chez l'homme. Myo-gramme isométrique, électromyogramme d'excitation et topographie de l'innervation terminale. Acta neurol. belg. *58:* 977–1017 (1958).

DESMEDT, J.E.: Somatosensory cerebral evoked potentials in man; in RÉMOND Handbook of EEG and clinical neurophysiology, vol. 9, pp. 56–82 (Elsevier, Amsterdam 1971).

DESMEDT, J.E. and BORENSTEIN, S.: The testing of the neuromuscular transmission; in VINKEN and BRUYN Handbook of neurology, vol. 7, pp. 104–115 (North Holland, Amsterdam 1970).

DESMEDT, J.E. et DELWAIDE, P.J.: Physiopathologie des temblements et organisation fonctionnelle de la motricité. Acta neurol. belg. *67:* 239–297 (1967).

DEVANANDAN, M.S.; ECCLES, R.M., and YOKOTA, T.: Depolarization of afferent terminals evoked by muscle stretch. J. Physiol., Lond. *179:* 417–429 (1965a).

DEVANANDAN, M.S.; ECCLES, R.M., and YOKOTA, T.: Muscle stretch and presynaptic inhibition of the group Ia pathway to motoneurones. J. Physiol., Lond. *179:* 430–441 (1965b).

DEVANANDAN, M.S.; ECCLES, R.M., and WESTERMAN, R.A.: Single motor units of mammalian muscle. J. Physiol., Lond. *178:* 359–367 (1965c).

DIAMANTOPOULOS, E. and GASSEL, M.M.: Electrically induced monosynaptic reflexes in man. J. Neurol. Neurosurg. Psychiat. *28:* 496–502 (1965).

DIAMANTOPOULOS, E. and ZANDER OLSEN, P.: Motoneurone excitability in normal subjects and patients with spasticity, rigidity and cerebellar signs. Electroenceph. clin. Neurophysiol. *18:* 207–208 (1965).

DIAMANTOPOULOS, E. and ZANDER OLSEN, P.: Excitability of motor neurones in spinal shock in man. J. Neurol. Neurosurg. Psychiat. *30:* 427–431 (1967).

DIETRICHSON, P.: Phasic ankle reflex in spasticity and Parkinsonian rigidity. The possible role of the fusimotor system. Acta neurol. scand. *47:* 22–51 (1971a).

DIETRICHSON, P.: Tonic ankle reflex in Parkinsonian rigidity and in spasticity. The possible role of the fusimotor system. Acta neurol. scand. *47:* 163–182 (1971b).

DIETRICHSON, P.: The silent period in spastic, rigid and normal subjects during isotonic and isometric muscle contractions. Acta neurol. scand. *47:* 183–193 (1971c).

DIETRICHSON, P.: The role of the fusimotor system in spasticity and Parkinsonian rigidity Universitetsforlaget, Oslo 1971d).

DIETRICHSON, P. and SORBYE, R.: Clinical method for electrical and mechanical recording of the mechanically and electrically elicited ankle reflex. Acta neurol. scand. *47:* 1–21 (1971).

DIMITRIJEVIC, M.R.; FAGANEL, J.; GREGORIC, M.; NATHAN, P.W., and TRONTELJ, J.K.: Habituation of regular and stochastic stimulation. J. Neurol. Neurosurg. Psychiat. *35:* 234–242 (1972).

DIMITRIJEVIC, M.R. and NATHAN, P.W.: Studies of spasticity in man. 1. Some features of spasticity. Brain *90:* 1–30 (1967a).

DIMITRIJEVIC, M.R. and NATHAN, P.W.: Studies of spasticity in man. 2. Analysis of stretch reflexes in spasticity. Brain *90:* 333–358 (1967b).

DIMITRIJEVIC, M.R. and NATHAN, P.W.: Studies of spasticity in man. 3. Analysis of reflex activity evoked by noxious cutaneous stimulation. Brain *91:* 349–368 (1968).

DIMITRIJEVIC, M.R. and NATHAN, P.W.: Studies of spasticity in man. 4. Changes in flexion reflex with repetitive cutaneous stimulation in spinal man. Brain *93:* 743–768 (1970).

DIMITRIJEVIC, M.R. and NATHAN, P.W.: Studies of spasticity in man. 5. Dishabituation of the flexion reflex in spinal man. Brain *94:* 77–90 (1971).

DOMINIK, F. and WIESENDANGER, M.: Pyramidal and non-pyramidal motor cortical effects on distal forelimb muscles of monkeys. Exp. Brain Res. *12:* 81–91 (1971).

DOW, R.S. and MORUZZI, G.: The physiology and pathology of the cerebellum (University of Minnesota Press, Minneapolis 1958).

DRAPER, M.H.; LADEFOGED, P., and WHITTERIDGE, D.: Expiratory pressures and air flow during speech. Brit. med. J. *i:* 1837–1843 (1960).

DUENSING, F.: Schreckreflex und Schreckreaction als hirnorganische Zeichen. Arch. Psychiat. Nervenkr. *188:* 162–192 (1952).

◑ DURKOVIC, R.G.; PIWONKA, R.W., and PRESTON, J.B.: Cortical modulation of spindle afferent discharge patterns in the pyramidal cat. Brain Res. *40:* 179–186 (1972).

DYCK, P.J.; GUTRECHT, J.A.; BASTRON, J.A.; KARNES, W.E., and DALE, A.J.D.: Histologic and teased-fiber measurements of sural nerve in disorders of lower motor and primary sensory neurons. Mayo Clin. Proc. *43:* 81–123 (1968).

DYCK, P.J. and LAMBERT, E.H.: Lower motor and primary sensory neuron disease with peroneal muscular atrophy. II. Neurologic, genetic, and electrophysiologic findings in various neuronal degenerations. Arch. Neurol., Chicago *18:* 619–625 (1968).

DYCK, P.J. and LAMBERT, E.H.: Dissociated sensation in amyloidosis. Arch. Neurol., Chicago *20:* 490–507 (1969).

ECCLES, J.C.: The central action of antidromic impulses in motor nerve fibres. Pflügers Arch. ges. Physiol. *260:* 385–415 (1955).

ECCLES, J.C.: Presynaptic inhibition in the spinal cord; in ECCLES and SCHADÉ Physiology of spinal neurons. Progr. Brain Res., vol. 12, pp. 65–91 (Elsevier, Amsterdam 1964).

ECCLES, J.C.: The physiology of synapses (Springer, Berlin 1964).

ECCLES, J.C.: Long-loop reflexes from the spinal cord to the brain stem and cerebellum. Atti Accad. med. lom. *21:* 1–19 (1966).

ECCLES, J.C.: Circuits in the cerebellar control of movement. Proc. nat. Acad. Sci., Wash. *58:* 336–343 (1967).

ECCLES, J.C.: The dynamic loop hypothesis of movement control; in LEIBOVIC Information processing in the nervous system, pp. 245–269 (Springer, New York 1969).

ECCLES, J.C.; ECCLES, R.M.; IGGO, A., and ITO, M.: Distribution of recurrent inhibition among motoneurones. J. Physiol., Lond. *159:* 479–499 (1961).

ECCLES, J.C.; ECCLES, R.M.; IGGO, A., and LUNDBERG, A.: Electrophysiological studies on gamma motoneurones. Acta physiol. scand *50:* 32–40 (1960).

ECCLES, J.C.; ECCLES, R.M., and LUNDBERG, A.: The convergence of monosynaptic excitatory afferents on to many different species of alpha motoneurones. J. Physiol., Lond. *137:* 22–50 (1957a).

ECCLES, J.C.; ECCLES, R.M., and LUNDBERG, A.: Duration of after-hyperpolarization of motoneurones supplying fast and slow muscles. Nature, Lond. *179:* 866–868 (1957b).

ECCLES, J.C.; ECCLES, R.M., and LUNDBERG, A.: Synaptic actions on motoneurones in relation to the two components of the group I muscle afferent volley. J. Physiol., Lond. *136:* 527–546 (1957c).

ECCLES, J.C.; ECCLES, R.M., and LUNDBERG, A.: Synaptic actions on motoneurones caused by impulses in Golgi tendon organ afferents. J. Physiol., Lond. *138:* 227–252 (1957d).

ECCLES, J.C.; ECCLES, R.M., and LUNDBERG, A.: The action potentials of the alpha motoneurones supplying fast and slow muscles. J. Physiol., Lond. *142:*275–291 (1958).

ECCLES, J.C.; ECCLES, R.M., and MAGNI, F.: Central inhibitory action attributable to presynaptic depolarization produced by muscle afferent volleys. J. Physiol., Lond. *159:* 147–166 (1961).

ECCLES, J.C.; FABER, D.S.; MURPHY, J.T.; SABAH, N.H., and TÁBOŘÍKOVÁ, H.: Afferent volleys in limb nerves influencing impulse discharges in cerebellar cortex. II. In Purkyne cells. Exp. Brain Res. *13:* 36–53 (1971a).

ECCLES, J.C.; FABER, D.S.; MURPHY, J.T.; SABAH, N.H., and TÁBOŘÍKOVÁ, H.: Investigations on integration of mossy fiber inputs to Purkyne cells in the anterior lobe. Exp. Brain Res. *13:* 54–77 (1971b).

ECCLES, J.C.; FATT, P., and KOKETSU, K.: Cholinergic and inhibitory synapses in a pathway from motor-axon collaterals to motoneurones. J. Physiol., Lond. *126:* 524–562 (1954).

ECCLES, J.C.; FATT, P., and LANDGREN, S.: Central pathway for direct inhibitory action of impulses in largest afferent nerve fibres to muscle. J. Neurophysiol. *19:* 75–98 (1956).

ECCLES, J.C. and HOFF, H.E.: Rhythmic discharge of motoneurones. Proc. roy. Soc. B *110:* 483–514 (1932).

ECCLES, J.C.; ITO, M., and SZENTAGOTHAI, J.: The cerebellum as a neuronal machine (Springer, Berlin 1967).

ECCLES, J.C.; MAGNI, F., and WILLIS, W.D.: Depolarization of central terminals of group I afferent fibres from muscle. J. Physiol., Lond. *160:* 282–297 (1962).

ECCLES, J.C.; SABAH, N.H.; SCHMIDT, R.F., and TÁBOŘÍKOVÁ, H.: Cerebellar Purkyne cell responses to cutaneous mechanoreceptors. Brain Res. *30:* 419–424 (1971).

ECCLES, J.C.; SABAH, N.H., and TÁBOŘÍKOVÁ, H.: Responses in neurones of the fastigial nucleus by cutaneous mechanoreceptors. Brain Res. *35:* 523–527 (1972).

ECCLES, J.C. and SHERRINGTON, C.S.: Numbers and contraction values of individual motor units examined in some muscles of the limb. Proc. roy. Soc. B *13:* 326–357 (1930).

ECCLES, R.M. and LUNDBERG, A.: Integrative patterns of Ia synaptic actions on motoneurones of hip and knee muscles J. Physiol., Lond. *144:* 271–298 (1958a).

ECCLES, R.M. and LUNDBERG, A.: The synaptic linkage of 'direct' inhibition. Acta physiol. scand. *43:* 204–215 (1958b).

ECCLES, R.M. and LUNDBERG, A.: Synaptic actions in motoneurones by afferents which may evoke the flexion reflex. Arch. ital. Biol. *97:* 199–221 (1959a).

ECCLES, R.M. and LUNDBERG, A.: Supraspinal control of interneurones mediating spinal reflexes. J. Physiol., Lond. *147:* 565–584 (1959b).

ECCLES, R.M.; PHILLIPS, C.G., and WU, C.-P.: Motor innervation, motor unit organization and afferent innervation of m. extensor digitorum communis of the baboon's forearm. J. Physiol., Lond. *198:* 179–192 (1968).

ECCLES, R.M.; SEARS, T.A., and SHEALEY, C.N.: Intracellular recording from respiratory motoneurones of the thoracic spinal cord. Nature, Lond. *193:* 844–846 (1962).

ECCLES, R.M. and WILLIS, R.D.: The effect of repetitive stimulation upon monosynaptic transmission in kittens. J. Physiol., Lond. *176:* 311–321 (1965).

ECHLIN, F. and FESSARD, A.: Synchronized impulse discharges from receptors in the deep tissues in response to a vibrating stimulus. J. Physiol., Lond. *93:* 312–334 (1938).

EDSTRÖM, L. and KUGELBERG, E.: Histochemical composition, distribution of fibres and fatiguability of single motor units. J. Neurol. Neurosurg. Psychiat. *31:* 424–433 (1968).

EFRON, R.: The conditioned reflex: a meaningless concept. Perspect. Biol. Med. *11:* 488–514 (1966).

EGGER, M.D. and WALL, P.D.: The plantar cushion reflex circuit: an oligosynaptic reflex. J. Physiol., Lond. *216:* 483–502 (1971).

EKBOM, K.A.; JERNELIUS, B., and KUGELBERG, E.: Notions on variations in muscle stretch reflexes in relation to tremor in Parkinsonism. Acta med. scand. *66:* 301–304 (1952).

EKHOLM, J.: Postnatal changes in cutaneous reflexes and in the discharge pattern of cutaneous and articular sense organs. A morphological and physiological study in the cat. Acta physiol. scand., suppl. *297:* 1–124 (1967).

EKLUND, G.: Influence of muscle vibration on balance in man. Acta Soc. Med. Uppsala *74:* 113–117 (1969).

EKLUND, G.: On muscle vibration in man; an amplitude-dependent inhibition, inversely related to muscle length. Acta physiol. scand. *83:* 425–426 (1971).

EKLUND, G.: Some physical properties of muscle vibrators used to elicit tonic proprioceptive reflexes in man. Acta Soc. Med. Uppsala *76:* 271–280 (1971).

EKLUND, G.: Position sense and state of contraction. J. Neurol. Neurosurg. Psychiat. *35:* 606–611 (1972).

EKLUND, G.; GRIMBY, L., and KUGELBERG, E.: Nociceptive reflexes of the human foot. The plantar responses. Acta physiol. scand. *47:* 297–298 (1959).

EKLUND, G. and HAGBARTH, K. E.: Motor effects of vibratory muscle stimuli in man. Electroenceph. clin. Neurophysiol. *19:* 619P (1965).

EKLUND, G. and HAGBARTH, K. E.: Normal variability of tonic vibration reflexes in man. Exp. Neurol. *16:* 80–92 (1966).

EKLUND, G. and LÖFSTEDT, L.: Bio-mechanical analysis of balance. Bio-med. Engineer. *5:* 333–337 (1970).

EKLUND, G. and STEEN, M.: Muscle vibration therapy in children with cerebral palsy. Scand. J. Rehab. Med. *1:* 35–37 (1969).

EKLUND, G.; EULER, C. VON, and RUTKOWSKI, S.: Intercostal γ-motor activity. Acta. physiol. scand. *57:* 481–482 (1963).

EKLUND, G.; EULER, C. VON, and RUTKOWSKI, S.: Spontaneous and reflex activity of intercostal gamma motoneurones. J. Physiol., Lond. *171:* 139–163 (1964).

EKSTEDT, J.: Human single muscle fiber action potentials. Acta physiol. scand. *61:* suppl. 226: pp. 1–96 (1964).

EKSTEDT, J.; HÄGGQVIST, P., and STALBERG, E.: The construction of needle multi-electrodes for single fibre electromyography. Electroenceph. clin. Neurophysiol. *27:* 540–543 (1969).

EKSTEDT, J. and STALBERG, E.: A method of recording extracellular action potentials of single muscle fibres and measuring their propagation velocity in voluntary activated human muscles. Bull. Amer. Ass. EMG Electrodiagn. *10:* 16 (1963).

ELDRED, E.; BRIDGMAN, C. F.; SWETT, J. E., and ELDRED, B.: Quantitative comparisons of muscle receptors of the cat's medial gastrocnemius, soleus and extensor digitorum brevis muscles, in BARKER Symposium Muscle Receptors, pp. 207–213 (Hong Kong University Press, Hong Kong 1972).

ELDRED, E.; GRANIT, R., and MERTON, P. A.: Supraspinal control of the muscle spindles and its significance J. Physiol., Lond. *122:* 498–523 (1953).

ELDRED, E. and HAGBARTH, K. E.: Facilitation and inhibition of gamma efferents by stimulation of certain skin areas. J. Neurophysiol. *17:* 59–65 (1954).

ELDRED, E.; LINDSLEY, D. F., and BUCHWALD, J. S.: The effects of cooling on mammalian muscle spindles. Exp. Neurol. *2:* 144–157 (1960).

ELDRED, E.; YELLIN, H.; GADBOIS, L., and SWEENEY, S.: Bibliography on muscle receptors; their morphology, pathology and physiology. Exp. Neurol., suppl. *3* (1967).

ELLAWAY, P.H.: Antidromic inhibition of fusimotor neurones. J.Physiol., Lond. *198:* 39P–40P (1968).

EMONET-DÉNAND, F.; JANKOWSKA, E., and LAPORTE, Y.: Skeleto-fusimotor fibres in the rabbit. J. Physiol., Lond. *210:* 669–680 (1970).

ENGBERG, I.: Reflexes to foot muscles in the cat. Acta physiol. scand. *62:* suppl. 235 (1964).

ENGBERG, I. and LUNDBERG, A.: An electromyographic analysis of muscular activity in the hinlimb of the cat during unrestrained locomotion. Acta physiol. scand. *75:* 614–630 (1969).

ENGBERG, I.; LUNDBERG, A., and RYALL, R.W.: Reticulospinal inhibition of transmission in reflex pathways. J. Physiol., Lond. *194:* 201–223 (1968a).

ENGBERG, I., LUNDBERG, A., and RYALL, R.W.: Reticulospinal inhibition of interneurones. J. Physiol., Lond. *194:* 225–236 (1968b).

ERB, W.: Über Sehnenreflexe bei Gesunden und bei Rückenmarkskranken. Arch. Psychiat. Nervenkr. *5:* 792–802 (1875).

ERBEL, F.: Zur Entstehung des Entlastungsreflexes; Diss. München (1969).

ERBEL, F. und STRUPPLER, A.: Weitere Untersuchungen zum Entlastungsreflex am Menschen. Pflügers Arch. ges. Physiol. *316:* 59 (1970).

EULER, C. VON: The control of respiratory movement; in HOWELL and CAMPBELL Breathlessness, pp. 19–32 (Blackwell, Oxford 1966).

EULER, C. VON: Proprioceptive control in respiration; in GRANIT Nobel Symposium 1, Muscular afferents and motor control (Almqvist & Wiksell, Stockholm 1966).

EULER, C. VON: Fusimotor activity in spindle control of natural movements with special reference to respiration; in ANDERSEN and JANSEN Excitatory synaptic mechanisms (Universitetsforlaget, Oslo 1970).

EULER, C. VON and FRITTS, H.W.: Quantitative aspects on respiratory reflexes from lungs and chest wall. Acta physiol. scand. *57:* 284–300 (1963).

EULER, C. VON and PERETTI, C.: Dynamic and static contributions to the rhythmic γ-activation of primary and secondary spindle endings in external intercostal muscle. J. Physiol., Lond. *187:* 501–516 (1966).

EVARTS, E.V.: Pyramidal tract activity associated with a conditioned hand movement in the monkey. J. Neurophysiol. *29:* 1011–1027 (1966).

EVARTS, E.V.: Representation of movements and muscles by pyramidal tract neurons of the precentral motor cortex; in YAHR and PURPURA Neurophysiological basis of normal and abnormal motor activities (Raven Press, New York 1967).

EVARTS, E.V.: Relation of pyramidal tract activity to force exerted during voluntary movement J. Neurophysiol. *31:* 14–27 (1968).

EVARTS, E.V.; BIZZI, E.; BURKE, R.E.; DELONG, M., and THACH, W.T.: Central control of movement. Neurosci. Res. Progr. Bull. *9:* 1–170 (1971).

EVARTS, E.V. and THACH, W.T.: Motor mechanisms of the CNS, cerebrocerebellar relations Annu. Rev. Physiol. *31:* 451–498 (1969).

FABER, D.S.; ISHIKAWA, K., and ROWE, M.J.: The responses of cerebellar Purkyně cells to muscle vibration. Brain Res. *26:* 184–187 (1971).

FELIX, D. and WIESENDANGER, M.: Pyramidal and non-pyramidal motor cortical effects on distal forelimb muscles of monkeys. Exp. Brain Res. *12:* 81–91 (1971).

FERRARI, E. and MESSINA, C.: Blink reflex during sleep and wakefulness in man. Electroenceph. clin. Neurophysiol. *32:* 55–62 (1972).

FETZ, E. E.: Pyramidal tract effects on interneurons in the cat lumbar dorsal horn. J. Neurophysiol. *31:* 69–80 (1968).

FETZ, E. E. and FINOCCHIO, D. V.: Operant conditioning of isolated activity in specific muscles and precentral cells. Brain Res. *40:* 19–23 (1972).

FIDONE, S. J. and PRESTON, J. B.: Patterns of motor cortex control of flexor and extensor cat fusimotor neurons. J. Neurophysiol. *32:* 103–115 (1969).

FINK, R. P. and HEIMER, L.: Two methods for selective silver impregnation of degenerating axons and their synaptic endings in the central nervous system. Brain Res. *4:* 369–374 (1967).

FISHER, C. M. and CURRY, H. B.: Pure motor hemiplegia of vascular origin. Arch. Neurol., Chicago *13:* 30–44 (1965).

FOERSTER, O.: On the indications and results of the excision of posterior spinal nerve roots in men. Surg. Gynec. Obstet. *16:* 463–474 (1913).

FOERSTER, O.: Zur Analyse und Pathophysiologie der striären Bewegungsstörungen. Z. ges. Neurol. Psychiat. *73:* 1–169 (1921).

FOERSTER, O.: Schlaffe und spastische Lähmung; in BERGMAN, EMBDEN und ELLINGER Des Normalen und Pathologischen Physiologie (Springer, Berlin 1927).

FOERSTER, O. und ALTENBURGER, H.: Beiträge zur Physiologie und Pathophysiologie der Sehnen- und Knochenphänomene und der Dehnungsreflexe. III. Die Sehnen- und Knochenphänomene beim Pyramidenbahnsyndrom. Z. ges. Neurol. Psychiat. *147:* 779–790 (1933).

FOIX, CH.: L'automatisme médullaire. Questions neurologiques d'actualité, p. 389 (Masson, Paris 1922).

FOIX, CH. et THÉVÉNARD, A.: Le réflexe de posture. Rev. neurol. *30:* 449–468 (1924).

FOLEY, J.: The stiffness of spastic muscle. J. Neurol. Neurosurg. Psychiat. *24:* 125–131 (1961).

FORBES, A.: The interpretation of spinal reflexes in terms of present knowledge of nerve conduction. Physiol. Rev. *2:* 361–414 (1922).

FRANK, K. and FUORTES, M. G.: Presynaptic and postsynaptic inhibition of monosynaptic reflexes. Fed. Proc. *16:* 39–40 (1957).

FREEDMAN, S. J.: On the mechanisms of perceptual compensation; in FREEDMAN The neuropsychology of spatially oriented behavior (Dorsey Press, Illinois 1968).

FREEDMAN, S. J. and WILSON, L.: Compensation auditory rearrangement following exposure to auditory-tactile discordance. Percept. mot. Skills *25:* 861–866 (1967).

FREEDMAN, W.: Systems analysis of the myotatic reflex in normal, hemiplegic and paraplegic man; Ph. D. thesis Philadelphia (1971).

FREEMAN, M. A. R. and WYKE, B.: Articular contributions to limb muscle reflexes. The effects of partial neurectomy of the knee joint on postural reflexes. Brit. J. Surg. *53:* 61–69 (1966).

FRENCH, J. H.; CLARK, D. B.; BUTLER, H. G., and TEASDALL, R. D.: Phenylketonuria: Some observations on reflex activity. J. Pediat. *58:* 17–22 (1961).

FREUND, N. J. und WITA, C. W.: Computeranalyse des Intervallmusters einzelner motorischer Einheiten bei Gesunden und Patienten mit supraspinalen motorischen Störungen. Arch. Psychiat. Nervenkr. *214:* 56–71 (1971).

FREYSCHUSS, U. and KNUTSSON, E.: Discharge patterns in motor nerve fibres during voluntary effort in man. Acta physiol. scand. *83:* 278–279 (1971).

FROMM, C. und HASSE, J.: Positionsempfindlichkeit und fusimotorische Aktivierung prätibialer Muskelspindelendigungen vor und nach Deafferentierung. Pflügers Arch. ges. Physiol. *321:* 242–252 (1970).

FUJIMORI, B.; KATO, M.; MATSUSHIMA, S.; MORI, S., and SHIMAMURA, M.: Studies on the mechanism of spasticity following spinal hemisection in cat; in GRANIT Nobel Symposium 1, Muscular afferents and motor control (Almqvist & Wiksell, Stockholm 1966).

FULTON, J.F.: Physiology of the nervous system, p. 150 (Oxford University Press, New York 1949).

FULTON, J.F. and DOW, R.S.: The cerebellum: a summary of functional localization. Yale J. Biol. Med. *10:* 89–119 (1937).

FULTON, J.F. and LIDDELL, E.G.T.: Electrical responses of extensor muscles during postural (myotatic) contraction. Proc. roy. Soc. B *98:* 577–589 (1925).

FULTON, J.F.; LIDDELL, E.G.T., and RIOCH, D.M.: The influence of unilateral destruction of the vestibular nuclei upon posture and the knee-jerk. Brain *53:* 327–343 (1931).

GAIL, P. DE; LANCE, J.W., and NEILSON, P.D.: Differential effects on tonic and phasic reflex mechanisms produced by vibration of muscles in man. J. Neurol. Neurosurg. Psychiat. *29:* 1–11 (1966).

GANDIGLIO, G. and FRA, L.: Further observations on facial reflexes. J. neurol. Sci. *5:* 273–285 (1967).

GARCÍA-AUSTT, E.; BOGACZ, J., and VANZULLI, A.: Effects of attention and inattention upon visual evoked response. Electroenceph. clin. Neurophysiol. *17:* 136–143 (1964).

GARCIA-MULLIN, R. and MAYER, R.F.: Studies of motoneuron excitability in hemiplegics. Excerpta med., Int. Cong. Ser. *193:* 220 (1969).

GARCIA-MULLIN, R. and MAYER, R.F.: Acute and chronic hemiplegia. I. Associated motor unit activity. Electromyography *12:* 5–18 (1972a).

GARCIA-MULLIN, R. and MAYER, R.F.: Acute and chronic hemiplegia. II. H-reflex studies. Brain *95:* 559–572 (1972b).

GARTEN, S.: Zur Kenntnis des zeitlichen Ablaufes der Lidschläge. Pflügers Arch. ges. Physiol. *71:* 477–491 (1898).

GASSEL, M.M.: A study of femoral nerve conduction time. Arch. Neurol., Chicago *9:* 607–615 (1963).

GASSEL, M.M.: A test of nerve conduction to muscles of the shoulder girth as an aid in the diagnosis of proximal neurogenic and muscle disease. J. Neurol. Neurosurg. Psychiat. *27:* 200–205 (1964).

GASSEL, M.M.: Monosynaptic reflexes (H-reflex) and motoneurone excitability in man. Develop. Med. Child Neurol. *11:* 193–197 (1969).

GASSEL, M.M.: The role of skin areas adjacent to extensor muscles in motoneurone excitability; evidence bearing on the physiology of Babinski's response. J. Neurol. Neurosurg. Psychiat. *33:* 121–126 (1970).

GASSEL, M.M.: Critical review of evidence concerning long-loop reflexes excited by muscle afferents in man. J. Neurol. Neurosurg. Psychiat. *33:* 358–362 (1970).

GASSEL, M.M. and DIAMANTOPOULOS, E.: The Jendrassik maneuver. I. The pattern of reinforcement of monosynaptic reflexes in normal subjects and patients with spasticity or rigidity. Neurology, Minneap. *14:* 555–560 (1964a).

GASSEL, M. M. and DIAMANTOPOULOS, E.: The Jendrassik maneuver. II. An analysis of the mechanism. Neurology, Minneap. *14:* 640–642 (1964b).

GASSEL, M. M. and DIAMANTOPOULOS, E.: The effect of procaïne nerve block on neuromuscular reflex regulation in man. (An appraisal of the role of the fusimotor system.) Brain *87:* 729–742 (1964c).

GASSEL, M. M. and DIAMANTOPOULOS, E.: Nerve potential recordings during electrically and mechanically evoked monosynaptic reflexes in man. Nature, Lond. *208:* 1004–1005 (1965).

GASSEL, M. M. and DIAMANTOPOULOS, E.: Mechanically and electrically elicited monosynaptic reflexes in man. J. appl. Physiol. *21:* 1053–1058 (1966).

GASSEL, M. M.; MARCHIAFAVA, P. L., and POMPEIANO, O.: Tonic and phasic inhibition of spinal reflexes during deep desynchronized sleep in unrestrained cats. Arch. ital. Biol. *102:* 471–499 (1964).

GASSEL, M. M.; MARCHIAFAVA, P. L., and POMPEIANO, O.: An analysis of the supraspinal influences acting on motoneurons during sleep in the unrestrained cat. Arch. ital. Biol. *103:* 25–44 (1965).

GASSEL, M. M. and OTT, K. H.: Motoneuron excitability in man: a novel method of evaluation by modulation of tonic muscle activity. Electroenceph. clin. Neurophysiol. *29:* 190–195 (1970).

GASSEL, M. M. and OTT, K. M.: Local sign and late effects on motoneuron excitability of cutaneous stimulation in man. Brain *93:* 95–106 (1970).

GASSEL, M. M. and POMPEIANO, O.: Fusimotor function during sleep in unrestrained cats. An account of the modulation of the mechanically and electrically evoked monosynaptic reflexes. Arch. ital. Biol. *103:* 347–368 (1965).

GASSEL, M. M. and TROJABORG, W.: Clinical and electrophysiological study of the pattern of conduction times in the distribution of the sciatic nerve. J. Neurol. Neurosurg. Psychiat. *27:* 351–357 (1964).

GASSEL, M. M. and WIESENDANGER, M.: Recurrent and reflex discharges in plantar muscles of the cat. Acta physiol. scand. *65:* 138–142 (1965).

GASTAUT, H. and VILLENEUVE, A.: The startle disease or hyperexplexia. J. neurol. Sci. *53:* 523–542 (1967).

GAUTHIER, G. F.: On the relationship of ultrastructural and cytochemical features to color in mammalian skeletal muscle. Z. Zellforsch. *95:* 462–482 (1969).

GELFAN, S. and CARTER, S.: Muscle sense in man. Exp. Neurol. *18:* 469–473 (1967).

GELFAN, S.; KAO, G., and LING, H.: The dendritic tree of spinal neurons in dogs with experimental hindlimb rigidity. J. comp. Neurol. *146:* 143–174 (1972).

GELFAN, S. and RAPISARDA, A. F.: Synaptic density on spinal neurons of normal dogs and dogs with experimental hind-limb rigidity. J. comp. Neurol. *123:* 73–95 (1964).

GERNANDT, B. E. and THULIN, C.-A.: Reciprocal effects upon spinal motoneurons from stimulation of bulbar reticular formation. J. Neurophysiol. *18:* 113–129 (1955).

GESCHWIND, N.: The clinical syndromes of the cortical connections. Mod. Trends Neurol., vol. 5, pp. 29–40 (Butterworth, London 1970).

GHEZ, C. and PISA, M.: Inhibition of afferent transmission in cuneate nucleus during voluntary movement in the cat. Brain Res. *40:* 145–151 (1972).

GILLIES, J. D.; BURKE, D., and LANCE, J. W.: The tonic vibration reflex in the cat. J. Neurophysiol. *34:* 252–262 (1971a).

GILLIES, J.D.; BURKE, D., and LANCE, J.W.: Supraspinal control of tonic vibration reflex. J. Neurophysiol. *34:* 302–309 (1971b).

GILLIES, D.; LANCE, J.W.; NEILSON, P.D., and TASSINARI, C.A.: Presynaptic inhibition of the monosynaptic reflex by vibration. J. Physiol., Lond. *205:* 329–339 (1969).

GILLIES, J.D.; LANCE, J.W., and TASSINARI, C.A.: The mechanism of the suppression of the monosynaptic reflex by vibration. Proc. austr. Ass. Neurol. *7:* 97–102 (1970).

GILMAN, S.: The mechanism of cerebellar hypotonia. An experimental study in the monkey. Brain *92:* 621–638 (1969).

GILMAN, S. and EBEL, H.C.: Fusimotor neuron responses to natural stimuli as a function of prestimulus fusimotor activity in decerebellate cats. Brain Res. *21:* 367–384 (1970).

GILMAN, S. and McDONALD, W.I.: Cerebellar facilitation of muscle spindle activity. J. Neurophysiol. *30:* 1494–1512 (1967a).

GILMAN, S. and McDONALD, W.I.: Relation of afferent fiber conduction velocity to reactivity of muscle spindle receptors after cerebellectomy. J. Neurophysiol. *30:* 1513–1522 (1967b).

GILMAN, S. and MARCO, L.A.: Effects of medullary pyramidotomy in the monkey. I. Clinical and electromyographic abnormalities. Brain *94:* 495–514 (1971).

GILMAN, S.; MARCO, L.A., and EBEL, H.C.: Effects of medullary pyramidotomy in the monkey. II. Abnormalities of spindle afferent responses. Brain *94:* 515–530 (1971).

GILMAN, S.; MARCO, L.A., and LIEBERMAN, J.S.: Experimental hypertonia in the monkey: interruption of pyramidal or pyramidal-extrapyramidal cortical projections. Trans. amer. neurol. Ass. *96:* 162–168 (1971).

GINZEL, K.H.: A restatement of criteria for evaluation of centrally acting muscle relaxants with special reference to orphenadrine. Symp. CNS Drugs, pp. 70–83 (1966).

GINZEL, K.H.: The blockade of reticular and spinal facilitation of motor function by orphenadrine. J. Pharmacol. exp. Ther. *154:* 128–141 (1966).

GINZEL, K.H.: The action of nicotine and smoking on reflex pathways. Austr. J. Pharm. *48:* suppl. 52: pp. 30–33 (1967).

GIOVANELLI BARILARI, M. and KUYPERS, H.G.J.M.: Propriospinal fibers interconnecting the spinal enlargements in the cat. Brain Res. *14:* 321–330 (1969).

GLASER, E.M. and GRIFFIN, J.P.: Influence of cerebral cortex on habituation. J. Physiol., Lond. *160:* 429–445 (1962).

GLASER, G.H. and HIGGINS, D.C.: Motor stability, stretch responses and the cerebellum, in GRANIT Nobel Symposium 1, Muscular afferents and motor control, pp. 121–138 (Almqvist & Wiksell, Stockholm 1966).

GODWIN-AUSTEN, R.B.: The mechanoreceptors of the costo-vertebral joints. J. Physiol., Lond. *202:* 737–753 (1969).

GOGAN, P.: The startle and orienting reactions in man. A study of their characteristics and habituation. Brain Res. *18:* 117–135 (1970).

GOLDBERGER, M.E.: The extrapyramidal systems of the spinal cord. II. Results of combined pyramidal and extrapyramidal lesions in the macaque. J. comp. Neurol. *135:* 1–26 (1969).

GOLDSCHEIDER, A.: Gesammelte Abhandlungen. Physiologie des Muskelsinnes (Barth, Leipzig 1898).

GOODMAN, D.C.; JARRARD, L.E., and NELSON, J.F.: Corticospinal pathways and their sites of termination in the albino rat. Anat. Rec. *154:* 462 (1966).

GOODWIN, G. M.; MCCLOSKEY, D. I., and MATTHEWS, P. B. C.: The persistence of appreciable kinesthesia after paralysing joints afferents but preserving muscle afferents. Brain Res. *37:* 326–329 (1972).

GOODWIN, G. M.; MCCLOSKEY, D. J., and MITCHELL, J. H.: Cardiovascular and respiratory responses to changes in central command during isometric exercise at constant muscle tension. J. Physiol., Lond. *219:* 43–44 (1971).

GOODWIN, G. M. and MATTHEWS, P. B. C.: Effects of fusimotor stimulation on the sensitivity of muscle spindle endings to small-amplitude sinusoidal stretching. J. Physiol., Lond. *221:* 8P–9P (1972).

GORDON, G.: Recherches sur la fonction des noyaux somesthésiques primaires chez le chat. Actualités neurophysiol. *8:* 89–110 (1968).

GORDON, G. and MILLER, R.: Identification of cortical cells projecting to the dorsal column nuclei of the cat. Quart. J. exp. Physiol. *54:* 85–98 (1969).

GORDON, G. and PHILLIPS, C. G.: Slow and rapid components in a flexor muscle. J. exp. Physiol. *38:* 35–45 (1953).

GOTTLIEB, G. L. and AGARWAL, G. C.: Dynamic relationship between isometric tension and the electromyogram in man. J. appl. Physiol. *30:* 345–351 (1971a).

GOTTLIEB, G. L. and AGARWAL, G. C.: The effects of initial conditions on the Hoffmann reflex. J. Neurol. Neurosurg. Psychiat. *34:* 226–230 (1971b).

GOTTLIEB, G. L. and AGARWAL, G. C.: The role of the myotatic reflex in the voluntary control of movements. Brain Res. *40:* 139–144 (1972).

GOTTLIEB, G. L.; AGARWAL, G. C., and STARK, L.: Interaction between voluntary and postural mechanisms of the human motor system. J. Neurophysiol. *33:* 365–381 (1970).

GOWERS, W. R.: A manual of diseases of the nervous system (Churchill, London 1886).

GRAHAM, L. T.; SHANK, R. P.; WERMAN, R., and APRISON, M. H.: Distribution of some synaptic transmitter suspects in cat spinal cord: glutaminic acid, aspartic acid, gamma aminobutyric acid, glycine and glutamine. J. Neurochem. *14:* 465–472 (1967).

GRANIT, R.: Reflex self-regulation of the muscle contraction and autogenic inhibition. J. Neurophysiol. *13:* 351–372 (1950).

GRANIT, R.: Receptors and sensory perception (Yale University Press, New Haven 1955).

GRANIT, R.: Neuromuscular interaction in postural tone of the cat's isometric soleus muscle. J. Physiol., Lond. *143:* 387–402 (1958).

GRANIT, R.: Muscular afferents and motor control Nobel Symposium 1 (Almqvist & Wiksell, Stockholm 1966).

GRANIT, R.: The functional role of the muscle spindle's primary end organs. Proc. roy. Soc. Med. *61:* 69–78 (1968).

GRANIT, R.: The basis of motor control (Academic Press, London 1970).

GRANIT, R.; HAASE, J., and RUTLEDGE, L. T.: Recurrent inhibition in relation to frequency of firing and limitation of discharge rate of extensor motoneurones. J. Physiol., Lond. *154:* 308–328 (1960).

GRANIT, R. and HENATSCH, H.-D.: Gamma control of dynamic properties of muscle spindles. J. Neurophysiol. *19:* 356–366 (1956).

GRANIT, R.; HENATSCH, H.-D., and STEG, G.: Tonic and phasic ventral horn cells differentiated by post-tetanic potentiation in cat extensors. Acta physiol. scand. *37:* 114–126 (1956).

GRANIT, R.; HOLMGREN, B., and MERTON, P.A.: Two routes for excitation of muscle and their subservience to the cerebellum. J. Physiol., Lond. *130:* 213–224 (1955).

GRANIT, R. and JOB, C.: Electromyographic and monosynaptic definition of reflex excitability during muscle stretch. J. Neurophysiol. *15:* 409–420 (1952).

GRANIT, R. and KAADA, B.R.: Influence of stimulation of central nervous structures on muscle spindles in cat. Acta physiol. scand. *27:* 130–160 (1952).

GRANIT, R.; KERNELL, D., and SHORTESS, G.K.: Quantitative aspects of repetitive firing of mammalian motoneurones, caused by injected currents. J. Physiol., Lond. *168:* 911–931 (1963).

GRANIT, R.; KELLERTH, J.-O., and SZUMSKI, A.J.: Intracellular recording from extensor motoneurons acrivated across the gamma loop. J. Neurophysiol. *29:* 530–544 (1966).

GRANIT, R.; PASCOE, J.E., and STEG, G.: The behaviour of tonic and gamma-motoneurones during stimulation of recurrent collaterals. J. Physiol., Lond. *138:* 381–400 (1957).

GRANIT, R.; PHILLIPS, C.G.; SKOGLUND, S., and STEG, G.: Differentiation of tonic from phasic ventral horn cells by stretch, pinna and crossed extensor reflexes. J. Neurophysiol. *20:* 470–481 (1957).

GRAY, H.: Gray's anatomy, ed. by JOHNSTON and WHILLIS (Longmans & Green, London 1954).

GREEN, D.G. and KELLERTH, J.-O.: Intracellular autogenetic and synergistic effects of muscular contraction on flexor motoneurones. J. Physiol., Lond. *193:* 73–94 (1967).

GREGOR, A.: Über die Vertheilung der Muskelspindeln in der Muskulatur des menschlichen Fötus. Arch. Anat. Physiol. 112–196 (1904).

GRIFFIN, J.P. and PEARSON, J.A.: The effect of lesions of the frontal areas of the cerebral cortex on habituation of the flexor withdrawal response in the rat. Brain Res. *8:* 177–184 (1968).

GRIGG, P. and PRESTON, J.B.: Baboon flexor and extensor fusimotor neurons and their modulation by motor cortex. J. Neurophysiol. *34:* 428–436 (1971).

GRILLNER, S.: Supraspinal and segmental control of static and dynamic gamma-motoneurones in the cat. Acta physiol. scand. suppl. *327:* 1–34 (1969).

GRILLNER, S.: Is the tonic stretch reflex dependent upon group II excitation? Acta physiol. scand. *78:* 431–432 (1970).

GRILLNER, S.; HONGO, T., and LUND, S.: The vestibulospinal tract. Effects on alpha motoneurons in the lumbosacral spinal cord in the cat. Exp. Brain Res. *10:* 94–120 (1970).

GRILLNER, S. and UDO, M.: Is the tonic stretch reflex dependent on suppression of autogenic inhibitory reflexes? Acta physiol. scand. *79:* 13A–14A (1970).

GRIMBY, L.: Normal plantar response: integration of flexor and extensor reflex components. J. Neurol. Neurosurg. Psychiat. *26:* 39–50 (1963a).

GRIMBY, L.: Pathological plantar response: disturbances of the normal integration of flexor and extensor reflex components. J. Neurol. Neurosurg. Psychiat. *26:* 314–321 (1963b).

GRIMBY, L.: Pathological plantar response. J. Neurol. Neurosurg. Psychiat. *28:* 469–481 (1965).

GRIMBY, L. and HANNERZ, J.: Recruitment order of motor units on voluntary contraction: changes induced by proprioceptive afferent activity. J. Neurol. Neurosurg. Psychiat. *31:* 565–573 (1968).

GRIMBY, L. and HANNERZ, J.: Differences in recruitment order of motor units in phasic and tonic reflex in spinal man. J. Neurol. Neurosurg. Psychiat. *33:* 562–570 (1970).

GRIMBY, L. and HANNERZ, J.: Disturbances in the voluntary recruitment order of motor units in Parkinsonian bradykinesia. J. Neurol. Neurosurg. Psychiat. (1972).

GRÜSSER, O. J. und THIELE, B.: Reaktionen primärer und sekundärer Muskelspindelafferenzen auf sinusförmige mechanische Reizung. Pflügers Arch. ges. Physiol. *300:* 161–184 (1968).

GRÜTZNER, P.: Zur Anatomie und Physiologie der quergestreiften Muskeln. Rec. zool. Suisse *1:* 665–684 (1884).

GURFINKEL, V. S.; KOTS, Y. M., and SHIK, M. L.: Regulation of human posture (Nauka, Moscow 1965) (in Russian).

GURFINKEL, V. S. and PAL'TSEY, YA. I.: Effect of the state of the segmental apparatus of the spinal cord on the execution of a simple motor reaction. Biofizika *10:* 855–860 (1965) English translation by Pergamon Press).

GUTRECHT, J. A. and DYCK, P. J.: Quantitative teased-fiber and histologic studies of human sural nerve during postnatal development. J. comp. Neurol. *138:* 117–130 (1970).

HAARTSEN, A. B. and VERHAART, W. J. C.: Cortical projections to brain stem and spinal cord in the goat by way of the pyramidal tract and the bundle of Bagley. J. comp. Neurol. *129:* 189–202 (1967).

HAASE, J. and MEULEN, J. P. VAN DER: Effects of supraspinal stimulation on Renshaw cells belonging to extensor motoneurones J. Neurophysiol. *24:* 510–520 (1961).

HAASE, J. und VOGEL, B.: Die reflektorische Aktivierung prätibialer Muskelspindeln durch Spindelafferenzen. Pflügers Arch. ges. Physiol. *311:* 168–178 (1969).

HAGBARTH, K.-E.: Excitatory and inhibitory skin areas for flexor and extensor motoneurones. Acta physiol. scand. *26:* suppl. 94: 1–58 (1952).

HAGBARTH, K. E.: Spinal withdrawal reflex in the human lower limbs. J. Neurol. Neurosurg. Psychiat. *23:* 222–227 (1960).

HAGBARTH, K.-E.: Post tetanic potentiation of myotatic reflexes in man. J. Neurol. Neurosurg. Psychiat. *25:* 1–10 (1962).

HAGBARTH, K.-E.: Electromyographic studies of stretch reflex in man. Electroenceph. clin. Neurophysiol., suppl. *25:* 74–79 (1967).

HAGBARTH, K.-E. and EKLUND, G.: Motor effects of vibratory stimuli in man: in GRANIT Nobel Symposium 1, Muscular afferents and motor control., pp. 177–186 (Almqvist & Wiksell, Stockholm 1966a).

HAGBARTH, K.-E. and EKLUND, G.: Tonic vibration reflexes (TVR) in spasticity. Brain Res. *2:* 201–203 (1966b).

HAGBARTH, K.-E. and EKLUND, G.: The effects of muscle vibration in spasticity, rigidity and cerebellar disorders. J. Neurol. Neurosurg. Psychiat. *31:* 207–213 (1968).

HAGBARTH, K.-E. and EKLUND, G.: The muscle vibrator – a useful tool in neurological therapeutic work. Scand. J. Rehab. Med. *1:* 26–34 (1969).

HAGBARTH, K.-E. and FINER, B. L.: The plasticity of human withdrawal reflexes to noxious skin stimuli in lower limbs; in MORUZZI, FESSARD and JASPER Brain mechanisms. Progr. Brain Res. vol. 19, pp. 65–81 (Elsevier, Amsterdam 1963).

HAGBARTH, K.-E.; HONGELL, A., and WALLIN, G.: Parkinson's disease: afferent muscle nerve activity in rigid patients Acta Soc. Med. Uppsala *75:* 70–76 (1970).

HAGBARTH, K.-E.; HONGELL, A., and WALLIN, B. G.: The effect of gamma fibre block on afferent muscle nerve activity during voluntary contractions. Acta physiol. scand. *79:* 27–28A (1970).

HAGBARTH, K.-E. and KUGELBERG, E.: Plasticity of the human abdominal skin reflexes. Brain *81:* 305–318 (1958).

HAGBARTH, K.-E. and VALLBO, Å. B.: Afferent response to mechanical stimulation of muscle receptors in man. Acta Soc. Med. Uppsala. *72:* 102–104 (1967a).

HAGBARTH, K.-E. and VALLBO, Å. B.: Mechanoreceptor activity recorded percutaneously with semi-microelectrodes in human peripheral nerves. Acta physiol. scand. *69:* 121–122 (1967b).

HAGBARTH, K.-E. and VALLBO, Å. B.: Discharge characteristics of human muscle afferents during muscle stretch and contraction. Exp. Neurol. *22:* 674–694 (1968).

HAGBARTH, K.-E. and VALLBO, Å. B.: Single unit recording from muscle nerves in human subjects. Acta physiol. scand. *76:* 321–334 (1969).

HAMMOND, P. H.: Involuntary activity in biceps following the sudden application of velocity to the abducted forearm. J. Physiol., Lond. *127:* 23–25 (1954).

HAMMOND, P. H.: The influence of prior instruction to the subject on an apparently neuro-muscular response. J. Physiol., Lond. *132:* 17P–18P (1956).

HAMMOND, P. H.: An experimental study of servo action in human muscular control. Proc. 3rd Int. Conf. med. Electron., London 1960, pp. 190–199.

HAMMOND, P. H.; MERTON, P. A., and SUTTON, G. G.: Nervous gradation of muscular contraction. Brit. med. Bull. *12:* 214–218 (1956).

HAMORI, J. and SZENTAGOTHAI, J.: The 'crossing over' synapse. An electron microscope study of the molecular layer in the cerebellar cortex. Acta Biol. hung. *15:* 95–117 (1964).

HANNERZ, J.: Differential voluntary activation of high and low frequency motor units. Experientia (1972).

HANNERZ, J. and GRIMBY, L.: Phasic and tonic recruitment order of motor units in normal man. J. Neurol. Neurosurg. Psychiat. (1972).

HANSEN, K. und HOFFMANN, P.: Weitere Untersuchungen über die Bedeutung der Eigenreflexe für unsere Bewegungen. I. Anspannungs- und Entspannungsreflexe. Z. Biol. *75:* 293–304 (1922).

HANSEN, K. und RECH, W.: Beziehungen des Kleinhirns zu den Eigenreflexen. Dtsch. Z. Nervenheilk. *87:* 207–222 (1925).

HARRIS, C. S.: Adaptation to displaced vision: visual, motor or proprioceptive change? Science *140:* 812–813 (1963).

HARRIS, F.; JABBUR, S. J.; MORSE, R. W., and TOWE, A. L.: Influence of the cerebral cortex on the cuneate nucleus of the monkey. Nature, Lond. *208:* 1215–1216 (1965).

HARRIS, J. P.: Habituatory response decrement in the intact organism. Psychol. Bull. *40:* 385–422 (1943).

HARVEY, R. J. and MATTHEWS, P. B. C.: The response of de-efferented muscle spindle endings in the cat's soleus to slow extension of the muscle. J. Physiol., Lond. *157:* 370–392 (1961).

HASSLER, R.: Über die Kleinhirnprojektionen zum Mittelhirn und Thalamus beim Menschen. Dtsch. Z. Nervenheilk. *163:* 629–671 (1950).

HASSLER, R.: Die extrapyramidalen Rindensysteme und die zentrale Regelung der Motorik. Dtsch. Z. Nervenheilk. *175:* 233–258 (1956).

HASSLER, R.: Thalamo-corticale Systeme der Körperhaltung und Augenbewegungen; in TOWER and SCHADÉ Structure and function of the cerebral cortex, pp. 124–130 (Elsevier, Amsterdam 1960).

HASSLER, R.: Thalamic regulation of muscle tone and the speed of movements; in PURPURA and YAHR The thalamus (Columbia University Press, New York 1966).

HASSLER, R. und HESS, W. R.: Experimentelle und anatomische Befunde über die Drehbewegungen und ihre nervösen Apparate. Arch. Psychiat. Z. Neurol. *192:* 488–526 (1954).

HASSLER, R.; MUNDINGER, F., and RIECHERT, T.: Pathophysiology of tremor at rest derived from the correlation of anatomical and clinical data. Confin. neurol. *32:* 79–87 (1970).

HAY, J.: Red and pale muscle. Liverpool med.-chir. J. *21:* 431–451 (1901).

HEAD, H. and RIDDOCH, G.: The automatic bladder, excessive sweating, and some other conditions in gross injuries of the spinal cord. Brain *40:* 188–263 (1917).

HEDBERG, Å.; OLDBERG, B., and TOVE, P. A.: EMG-controlled muscle vibrators to aid mobility in spastic paresis. 7th Int. Conf. Med. Biol. Engng. p. 197 (1967).

HELD, R.: Exposure – history as a factor in maintaining stability of perception and coordination. J. nerv. ment. Dis. *132:* 26–32 (1961).

HELD, R. and FREEDMAN, S.: Plasticity in human sensorimotor control. Science *142:* 455–462 (1963).

HELMHOLTZ, H. L. F. VON: Handbuch der physiologischen Optik, vol. 3 (Voss, Leipzig 1867).

HENATSCH, H. D. und INGVAR, D. H.: Chlorpromazin und Spastizität. Arch. Psychiat. Z. Neurol. *195:* 77–93 (1956).

HENATSCH, H.-D.; SCHULTE, F. J. und BUSCH, G.: Wandelbarkeit des tonisch-phasischen Reaktionstyps einzelner Extensor-Motoneurone bei Variation ihres Antriebes. Pflügers Arch. ges. Physiol. *270:* 161–173 (1959).

HENNEMAN, E.: Relation between size of neurons and their susceptibility to discharge. Science *126:* 1345–1347 (1957).

HENNEMAN, E. and OLSON, C. B.: Relations between structure and function in the design of skeletal muscles. J. Neurophysiol. *28:* 581–598 (1965).

HENNEMAN, E.; SOMJEN, G., and CARPENTER, D. O.: Functional significance of cell size in spinal motoneurones. J. Neurophysiol. *28:* 560–580 (1965a).

HENNEMAN, E.; SOMJEN, G., and CARPENTER, D. O.: Excitability and inhibitability of motoneurons of different sizes. J. Neurophysiol. *28:* 599–620 (1965b).

HEPP-REYMOND, M.-C. and WIESENDANGER, M.: Unilateral pyramidotomy in monkeys: effect on force and speed of a conditioned precision grip. Brain Res. *36:* 117–131 (1972).

HERMAN, R.: Reflex activity in spinal cord lesions. Amer. J. phys. Med. *43:* 252–259 (1964).

HERMAN, R.: Silent period during isometric and isotonic contraction. Arch. phys. Med. *50:* 642–646 (1969).

HERMAN, R.: The myotatic reflex. Clinico-physiological aspects of spasticity and contracture. Brain *93:* 273–312 (1970).

HERMAN, R. and BRAGIN, S: Function of the gastrocnemius and soleus muscle. A preliminary study in the normal human subject. J. amer. phys. ther. Ass. *47:* 105–113 (1967).

HERMAN, R. and MAYER, N. H.: The silent period and control of isometric contraction of the triceps surae muscle. Electromyography *9:* 79–84 (1969).

HERMAN, R. and MECOMBER, S. A.: Vibration-elicited reflexes in normal and spastic muscle in man. Amer. J. phys. Med. *50:* 169–183 (1971).

HERMAN, R. and SCHAUMBERG, H.: Dynamic and static sensitivity of the stretch reflex in spastic cerebral palsy. Develop. Med. Child Neurol. *9:* 487–492 (1967).

HERMAN, R.; SCHAUMBERG, H., and REINER, S.: A rotational joint apparatus. A device for study of tension-lenght relations of human muscle. Med. Res. Engng. *6:* 18–20 (1967).

HERRICK, C.J.: Origin and evolution of the cerebellum. Arch. Neurol. Psychiat. Chicago *11:* 621–652 (1924).

HESS, A. and PILAR, G.: Slow fibers in the extraocular muscles of the cat. J. Physiol., Lond. *169:* 780–798 (1963).

HESS, W.R.: Hypothalamus und Thalamus, Experimental. Dokumente (Thieme, Stuttgart 1968).

HESS, W.R.; BÜRGI, S. und BUCHER, V.: Motorische Funktion des Tektal- und Tegmental-gebietes. Mschr. Psychiat. Neurol. *112:* 1–52 (1946).

HILL, D.K.: Tension due to interaction between the sliding filaments in resting striated muscle. The effect of stimulation. J. Physiol., Lond. *199:* 637–684 (1968).

HILLMAN, P. and WALL, P.D.: Inhibitory and excitatory factors influencing the receptive fields of lamina 5 spinal cord cells. Exp. Brain Res. *9:* 284–306 (1969).

HINSEY, J.C.; RANSON, S.W., and DIXON, H.H.: Responses elicited by stimulation of the mesencephalic tegmentum in the cat. Arch. Neurol. Psychiat., Chicago *24:* 966–977 (1930).

HISHIKAWA, Y.; SUMITSUJI, N.; MATSUMOTO, K., and KANEKO, Z.: H-reflex and EMG of the mental and hyoid muscles during sleep, with special reference to narcolepsy. Electroenceph. clin. Neurophysiol. *18:* 487–492 (1965).

HODES, R. and DEMENT, W.C.: Depression of electrically induced reflexes (H-reflexes) in man during low voltage EEG sleep. Electroenceph. clin. Neurophysiol. *17:* 617–629 (1964).

HODES, R. and GRIBETZ, I.: H-reflexes in normal human infants: depression of these electrically induced reflexes (EIR's) in sleep. Proc. Soc. exp. Biol. Med. *110:* 577–580 (1962).

HODES, R.; GRIBETZ, I., and HODES, H.L.: Abnormal occurrence of the ulnar nerve-hypothenar muscle H-reflex in Sydenham's chorea. Pediatrics *30:* 49–56 (1962).

HOEFER, P.F.A.: Innervation and 'tonus' of striated muscle in man. Arch. Neurol. Psychiat., Chicago *46:* 947–972 (1941).

HOEFER, P.F.A.: Electromyographic study of the motor system in man. Mschr. Psychiat. Neurol. *117:* 241–256 (1949).

HOEFER, P.F.A.: Physiological mechanisms in spasticity. Brit. J. phys. Med. *15:* 88–90 (1952).

HOEFER, P.F.A. and PUTNAM, T.J.: Action potentials of muscles in 'spastic' conditions. Arch. Neurol. Psychiat., Chicago *43:* 1–22 (1940a).

HOEFER, P.F.A. and PUTNAM, T.J.: Action potentials in muscles in rigidity and tremor. Arch. Neurol. Psychiat., Chicago *43:* 704–725 (1940b).

HOFFMANN, P.: Über die Beziehungen der Sehnenreflexe zur willkürlichen Bewegung und zum Tonus. Z. Biol. *68:* 351–370 (1918).

HOFFMANN, P.: Demonstration eines Hemmungsreflexes im menschlichen Rückenmark. Z. Biol. *70:* 515–524 (1919).

HOFFMANN, P.: Untersuchungen über die Eigenreflexe (Sehnenreflexe) menschlicher Muskeln (Springer, Berlin 1922).

HOFFMANN, P.: Untersuchungen über die refraktäre Periode des menschlichen Rückenmarkes. Z. Biol. *81:* 37–48 (1924).

HOFFMANN, P.: Die physiologischen Eigenschaften der Eigenreflexe. Ergebn. Physiol. *36:* 15–108 (1934).

HOFFMANN, P.: Die Aufklärung der Wirkung des Jendrassikschen Handgriffs durch die Arbeiten von Sommer und Kuffler. Dtsch. Z. Nervenheilk. *166:* 60–64 (1951).

HOFFMANN, P. und KELLER, C.J.: Über gleichzeitige willkürliche und künstliche Reizung von Nerven. Z. Biol. *87:* 527–536 (1928).

HOFFMANN, P.; SCHENCK, E. und TÖNNIES, J.F.: Über den Beugereflex des normalen Menschen. Pflügers Arch. ges. Physiol. *250:* 724–732 (1948).

HOFFMANN, P. und TÖNNIES, J.F.: Nachweis des völlig konstanten Vorkommens des Zungen-Kiefer-Reflexes beim Menschen. Pflügers Arch. ges. Physiol. *250:* 103–108 (1948).

HOFMANN, W.W.: Observations on peripheral servomechanisms in Parkinsonian rigidity. J. Neurol. Neurosurg. Psychiat. *25:* 203–207 (1962).

HOFMANN, W.W. and ANGEL, R.W.: Transient responses of the normal and diseased motor system to sudden load changes. Neurology, Minneap. *17:* 952–960 (1967).

HOLMES, G.: The croonian lectures on the clinical symptoms of cerebellar diseases and their interpretation. Lancet *100 (1):* 1177–1182, 1231–1237; *100 (2):* 59–65, 111–115 (1922).

HOLMES, G.: The cerebellum of man. Brain *62:* 21–30 (1939).

HOLMQVIST, B. and LUNDBERG, A.: On the organization of the supraspinal inhibitory control of interneurones of various spinal reflex arcs. Arch. ital. Biol. *97:* 340–356 (1959).

HOLMQVIST, B. and LUNDBERG, A.: Differential supraspinal control of synaptic actions evoked by volleys in the flexion reflex afferents in alpha motoneurones. Acta physiol. scand. *54:* suppl. 186: 1–51 (1961).

HOLST, E. VON: Relations between the central nervous system and the peripheral organs. Brit. J. anim. Behav. *2:* 89–94 (1954).

HOMMA, S.: Phasic stretch of muscle and afferent impulse transmission in tonic and phasic motoneurones. Jap. J. Physiol. *13:* 351–365 (1963).

HOMMA, S.; ISHIKAWA, K., and STUART, D.G.: Motoneuron responses to linearly rising muscle stretch. Amer. J. phys. Med. *49:* 290–306 (1970).

HOMMA, S.; ISHIKAWA, K., and WATANABE, S.: Optimal frequency of muscle vibration for motoneuron firing. J. Chiba Med. Soc. *43:* 190–196 (1967).

HOMMA, S.; KANDA, K., and WATANABE, S.: Monosynaptic coding of group Ia afferent discharges during vibratory stimulation of muscle. Jap. J. Physiol. *21:* 405–417 (1971a).

HOMMA, S.; KANDA, K., and WATANABE, S.: Tonic vibration reflex in human and monkey subjects. Jap. J. Physiol. *21:* 419–430 (1971b).

HOMMA, S. and KANO, M.: Electrical properties of the tonic reflex arc in the human proprioceptive reflex; in BARKER Symposium Muscle Receptors, pp. 167–174 (Hong Kong University Press, Hong Kong 1962).

HOMMA, S.; KOBAYASHI, H., and WATANABE, S.: Vibratory stimulation of muscles and stretch reflex. Jap. J. Physiol. *20:* 309–319 (1970).

HOMMA, S. and ZETANI, T.: Unitary EMG spikes evoked reflexly during vibratory stimulation of muscles. 9th Int. Conf. med. biol. Eng., Melbourne 1971.

HONGO, T.; JANKOWSKA, E., and LUNDBERG, A.: The rubrospinal tract. I. Effects on alpha-motoneurones innervating hindlimb muscles in cats. Exp. Brain Res. *7:* 344–364 (1969).

HONGO, T.; JANKOWSKA, E., and LUNDBERG, A.: The rubrospinal tract. II. Facilitation of interneuronal transmission in reflex paths to motoneurones. Exp. Brain Res. *7:* 365–391 (1969).

HONGO, T.; KUBOTA, K., and SHIMAZU, H.: EEG spindle and depression of gamma motor activity. J. Neurophysiol. *26:* 568–580 (1963).

HONGO, T.; KUDO, N., and TANAKA, R.: Effects from the vestibulospinal tract on the contralateral hindlimb motoneurones in the cat. Brain Res. *31:* 220–223 (1971).

HOOKER, D.: The prenatal origin of behavior. 18th Porter Lecture (University of Kansas Press, Lawrence 1952).

HOOKER, D.: Early human fetal behavior, with a preliminary note on double simultaneous fetal stimulation. Res. Publ. Ass. nerv. ment. Dis. *33:* 98–113 (1954).

HOPF, H.C.; HANDWERKER, H.; HAUSMANNS, J. und POLZIEN, F.: Die rasche Willkürbewegung des Menschen. Dtsch. Z. Nervenheilk. *191:* 186–209 (1967).

HORI, H.: Evoked EMG by the repetitive stimulation. Shinkei Kenkyu no Shinpo *3:* 413–434 (1959) (in Japanese).

HORI, H.: Tonic vibration reflex. Rinsho Nooha (J. clin. Electromyography) *13:* 439–474 (1971) (in Japanese).

HORVATH, F.E.; ATKIN, A.; KOZLOVSKAYA, I.; FULLER, D.R.G., and BROOKS, V.B.: Effects of cooling the dentate nucleus on alternating bar-pressing performance in monkey. Intern. J. Neurol. *7:* 252–270 (1970).

HOSOKAWA, H.: Proprioceptive innervation of striated muscles in the territory of cranial nerves. Texas Rep. Biol. Med. *19:* 405–464 (1961).

HOUK, J. and HENNEMAN, E.: Responses of Golgi tendon organs to active contractions of the soleus muscle of the cat. J. Neurophysiol. *30:* 466–481 (1967).

HOUK, J.C.; SINGER, J.J., and GOLDMAN, M.R.: An evaluation of length and force feedback to soleus muscles of decerebrate cats. J. Neurophysiol. *33:* 784–811 (1970).

HOWARD, I.P. and TEMPLETON, W.B.: Human spatial orientation (Wiley, London 1966).

HUDSON, R.D. and DOMINO, E.F.: Effects of chlorpromazine on some motor reflexes. Int. J. Pharmacol. *2:* 143–162 (1963).

HUFFMAN, R.D. and McFADIN, L.S.: Effects of bicuculline on central inhibition. Neuropharmacol. *11:* 789–799 (1972).

HUFSCHMIDT, H.-J.: Wird durch Muskelvibration eine Eigenreflexreihe erzeugt? Pflügers Arch. ges. Physiol. *267:* 508–516 (1958).

HUFSCHMIDT, H.-J.: Über den Spannungsreflex beim Menschen. Z. Biol. *111:* 75–80 (1959).

HUFSCHMIDT, H.-J.: Questions on tremor and ataxia. Electroenceph. clin. Neurophysiol., suppl. *22:* 114–119 (1962).

HUFSCHMIDT, H.-J.: The demonstration of autogenic inhibition and its significance in human voluntary movement; in GRANIT Nobel Symposium 1, Muscular afferents and motor control, pp. 269–274 (Almqvist & Wiksell, Stockholm 1966).

HUFSCHMIDT, H.-J. and HUFSCHMIDT, T.: Antagonist inhibition as the earliest sign of a sensory motor reaction. Nature, Lond. *174:* 607 (1954).

HUFSCHMIDT, H.-J. und SCHWIND, F.: Die Willkürkontraktion des Spastikers und ihre pharmakologische Beeinflussung. Dtsch. Z. Nervenheilk. *181:* 517–531 (1960).

HUGHES, J.T.; BROWNELL, B., and HEWER, R.L.: The peripheral sensory pathway in Friedreich's ataxia. Brain *91:* 803–818 (1968).

HUGON, M.: Potentiel d'action developpé par activation synchrone du muscle soléaire chez l'homme. Electroenceph. clin. Neurophysiol., suppl. *22:* 176–179 (1962).

HUGON, M.: Réflexes polysynaptiques cutanés et commande volontaire; thèse Paris (1967).

HUGON, M.: Réflexes polysynaptiques et réflexe monosynaptique évoqués dans le muscle biceps femoris capitis brevis chez l'homme normal. Rev. neurol. *120:* 492–494 (1969).

HUGON, M.: Inhibition présynaptique évoquée par stimulation naturelle chez le chat. Arch. int. Physiol. *77:* 130–132 (1969).

HUGON, M. et BATHIEN, N.: Influence de la stimulation du nerf sural sur divers réflexes monosynaptiques de l'homme. J. Physiol., Paris *59:* 244 (1967).

HUGON, M.; BATHIEN, N. et DUCOURTHIAL, F.: Le réflexe saphéno-bicipital chez l'homme, étudié par la technique de double stimulation. J. Physiol., Paris *57:* 631–632 (1965).

HUGON, M.; CHOUTEAU, J.; ROLL, J.P.; BONNET, M., and IMBERT, G.: A study of spinal reflexes in baboons during saturation exposures to 16 and 21 ATA of $N_2 + O_2$ (in preparation) (1972).

HULTBORN, H.; JANKOWSKA, E., and LINDSTROM, S.: Recurrent inhibition from motor axon collaterals of transmission in the Ia inhibitory pathway to motoneurones. J. Physiol., Lond. *215:* 591–612 (1971a).

HULTBORN, H.; JANKOWSKA, E., and LINDSTROM, S.: Recurrent inhibition of interneurones monosynaptically activated from group Ia afferents. J. Physiol., Lond. *215:* 613–636 (1971b).

HULTBORN, H.; JANKOWSKA, E., and LINDSTROM, S.: Relative contribution from different nerves to recurrent depression of Ia IPSP's in motoneurones. J. Physiol., Lond. *215:* 637–664 (1971c).

HUMPHREY, D.R.: Relating motor cortex spike trains to measures of motor performance. Brain Res. *40:* 7–18 (1972).

HUNT, C.C.: The reflex activity of mammalian small nerve fibers. J. Physiol., Lond. *115:* 456–469 (1951).

HUNT, W.A. and LANDIS, C.: Studies of the startle pattern. Amer. J. Psychol. *2:* 201–205 (1936).

HUNT, C.C. and PAINTAL, A.S.: Spinal reflex regulation of fusimotor neurones. J. Physiol., Lond. *143:* 195–212 (1958).

IMAI, Y. and KUSAMA, T.: Distribution of the dorsal root fibers in the cat. An experimental study with the Nauta method. Brain Res. *13:* 338–359 (1969).

INGRAM, W.R.; RANSON, S.W.; HANNETT, F.I.; ZEISS, F.R., and TERWILLIGER, E.H.: Results of stimulation of the tegmentum with the Horsley-Clarke stereotaxic apparatus. Arch. Neurol. Psychiat., Chicago *28:* 513–541 (1932).

INMAN, B.T.; RALSTON, H.J.; SAUNDERS, J.; FEINSTEIN, B., and WRIGHT, E.W.: Relation of human electromyogram to muscular tension. Electroenceph. clin. Neurophysiol. *4:* 187–194 (1952).

IOKU, M.; RIBERA, V.A.; COOPER, I.S., and MATSUOKA, S.: Parkinsonism, electromyographic studies of monosynaptic reflex. Science *150:* 1472–1475 (1965).

ISCH, F.; ISCH-TREUSSARD, C. et JESEL, M.: La stimulation répétitive de la réponse H dans le clonus et le tremblement parkinsonien. Electroenceph. clin. Neurophysiol., suppl. *22:* 124–125 (1962).

ITO, M.: Neural design of the cerebellar motor control system. Brain Res. *40:* 81–84 (1972).

ITO, M.; UDO, M., and MANO, N.: Long inhibitory and excitatory pathways onto cat reticular and Deiter's neurons and their relevance to reticulofugal axons. J. Neurophysiol. *33:* 210–226 (1970).

JACK, J.J.B.; MILLER, S.; PORTER, R., and REDMAN, S.J.: The distribution of group Ia synapses on lumbosacral spinal motoneurones in the cat; in ANDERSEN and JANSEN Excitatory synaptic mechanisms, pp. 199–205 (Universitetsforlaget, Oslo 1970).

JACK, J.J.B.; MILLER, S.; PORTER, R., and REDMAN, S.J.: The time course of minimal excitatory post-synaptic potentials evoked in spinal motoneurones by group Ia afferent fibres. J. Physiol., Lond. 215: 353–380 (1971).

JACKSON, J.H.: On epilepsies and the after effects of epileptic discharges. Tood and Robertson's hypothesis; in HOLMES and WALSH Selected writings of John Hughling Jackson, vol. 1 (1876).

JACOBSEN, C.F.: Influence of motor and premotor area lesions upon the retention of skilled movements in monkeys and chimpanzees. Proc. Ass. Res. nerv. ment. Dis. 13: 225–247 (1932).

JANE, J.A.; CAMPBELL, C.B.G., and YASHON, D.: Pyramidal tract. A comparison of two prosimian primates. Science 147: 153–155 (1965).

JANE, J.A.; CAMPBELL, C.B.G., and YASHON, D.: The origin of the corticospinal tract of the tree shrew (Tupaia glis) with observations on its brain stem and spinal termination. Brain, Behav. Evol. 2: 160–182 (1969).

JANKOWSKA, E. and LINDSTRÖM, S.: Morphology of interneurones mediating I a reciprocal inhibition of motoneurones in the spinal cord of the cat. J. Physiol. Lond. 226: 805–823 (1972).

JANSEN, J. K. S.: Spasticity – functional aspects. Acta neurol. scand. 38: suppl. 3: 41–51 (1962).

JANSEN, J.K.S.: On fusimotor reflex activity; in GRANIT Nobel Symposium 1, Muscular afferents and motor control, pp. 91–105 (Almqvist & Wiksell, Stockholm 1966).

JANSEN, J. and BRODAL, A.: Aspects of cerebellar anatomy (Tanum, Oslo 1954).

JANSEN, J.K.S. and MATTHEWS, P.B.C.: The central control of the dynamic response of muscle spindle receptors. J. Physiol., Lond. 161: 357–378 (1962).

JANSEN, J.K.S.; NICOLAYSEN, K., and RUDJORD, T.: Discharge pattern of neurones of the dorsal spinocerebellar tract activated by static extension of primary endings of muscle spindles. J. Neurophysiol. 29: 1061–1086 (1966).

JANSEN, J.K.S. and RACK, P.M.: The reflex response to sinusoidal stretching of soleus in the decerebrate cat. J. Physiol., Lond. 183: 15–36 (1966).

JANSEN, J.K.S. and RUDJORD, T.: On the silent period and Golgi tendon organs of the soleus muscle of the cat. Acta physiol. scand. 62: 364–379 (1964).

JANSEN, J.K.S. and RUDJORD, T.: Dorsal spinocerebellar tract: response pattern of nerve fibers to muscle stretch. Science 149: 1109–1111 (1965).

JANSEN, J.K.S. and RUDJORD, T.: Fusimotor activity in a flexor muscle of the decerebrate cat. Acta physiol. scand. 63: 236–246 (1965).

JOHNSTON, R.M.; BISHOP, B., and COFFREY, G.M.: Mechanical vibration of skeletal muscles. Phys. Ther. Rev. 50: 499–505 (1970).

JONES, R.F.; BURKE, D.; MAROSSZEKY, J.E., and GILLIES, J.D.: A new agent for the control of spasticity. J. Neurol. Neurosurg. Psychiat. 33: 464–468 (1970).

JOYCE, G.C.; RACK, P.M.H., and WESTBURY, D.R.: The mechanical properties of cat soleus muscle during controlled lengthening and shortening movements. J. Physiol., Lond. 204: 461–474 (1969).

KADANOFF, D.: Die sensiblen Nervenendigungen in der mimischen Muskulatur des Menschen. Z. mikr.-anat. Forsch. *62:* 1–15 (1956).

KATO, H.; TAKAMURA, H., and FUJIMORI, B.: Studies on effects of pyramid stimulation upon flexor and extensor motoneurones and gamma motoneurones. Jap. J. Physiol. *14:* 34–44 (1964).

KEHOE, J.: Ionic mechanisms of a two-component cholinergic inhibition in aplysia neurones. J. Physiol., Lond. *225:* 85–114 (1972).

KELLER, E. L. and ROBINSON, N. D. A.: Absence of stretch reflex in extraocular muscles of the monkey. J. Neurophysiol. *34:* 908–919 (1971).

KERNELL, D.: High-frequency repetitive firing of cat lumbosacral motoneurones stimulated by long-lasting injected currents. Acta physiol. scand. *65:* 74–86 (1965a).

KERNELL, D.: The limits of firing frequency in cat lumbosacral motoneurones possessing different time course of afterhyperpolarization. Acta physiol. scand. *65:* 87–100 (1965b).

KERNELL, D.: Input resistance, electrical excitability, and size of ventral horn cells in the cat spinal cord. Science *152:* 1637–1640 (1966a).

KERNELL, D.: The repetitive discharge of motoneurones; in GRANIT Nobel Symposium 1, Muscular afferents and motor control, pp. 351–362 (Almqvist & Wiksell, Stockholm 1966b).

KIMURA, J.: Alteration of the orbicularis oculi reflex by pontine lesions: study in multiple sclerosis. Arch. Neurol., Chicago *22:* 156–161 (1970).

KIMURA, J.: An evaluation of the facial nerve and the trigeminal nerve in polyneuropathy: electrodiagnostic study in Charcot-Marie-Tooth disease, Guillain-Barré syndrome and diabetic neuropathy. Neurology, Minneap. *21:* 745–752 (1971a).

KIMURA, J.: Electrodiagnostic study of brainstem strokes: alteration of the orbicularis oculi reflex in 39 vascular, 6 anoxic and 2 traumatic lesions. Stroke *2:* 576–580 (1971b).

KIMURA, J. and HARADA, O.: Excitability of the orbicularis oculi reflex in all night sleep – its suppression in non REM and recovery in REM sleep. Electroenceph. clin. Neurophysiol. *33:* 369–377 (1972).

KIMURA, J. and LYON, L. W.: The orbicularis oculi reflex in the Wallenberg syndrome: alteration of the late reflex by lesions of the spinal tract and nucleus of the trigeminal nerve. J. Neurol. Neurosurg. Psychiat. *35:* 228–233 (1972a).

KIMURA, J. and LYON, L. W.: Alteration of orbicularis oculi reflex by posterior fossa tumors. J. Neurosurg. (1972b).

KIMURA, J.; POWERS, J. M., and ALLEN, M. W. VAN: Reflex response of the orbicularis oculi muscle to supraorbital nerve stimulation: study in normal subjects and in peripheral facial paresis. Arch. Neurol., Chicago *21:* 193–199 (1969).

KIMURA, J.; RODNITZKY, R., and ALLEN, M. W. VAN: Electrodiagnostic study of trigeminal nerve: orbicularis oculi reflex and masseter reflex in trigeminal neuralgia, paratrigeminal syndrome and other lesions of the trigeminal nerve. Neurology, Minneap. *20:* 574–583 (1970).

KINGSBURY, B. F.: The law of cephalocaudal differential growth in its application to the nervous system. J. comp. Neurol. *56:* 431–464 (1932).

KNAPP, H. D.; TAUB, E., and BERMAN, A. J.: Movements in monkeys with deafferented forelimbs. Exp. Neurol. *7:* 305–315 (1963).

KNUTSSON, E.: On effects of local cooling upon motor functions in spastic paresis. Progr. Phys. Ther. *1:* 124–131 (1970).

KNUTSSON, E.; LINDBLOM, U., and ODEEN, I.: Reflex facilitation by muscle vibration in the treatment of spastic hemiparesis. Scand. J. Rehab. Med. *2:* 110–116 (1970).

KOEZE, T.H.; PHILLIPS, C.G., and SHERIDAN, J.D.: Thresholds of cortical activation of muscle spindles and alpha-motoneurones of the baboon's hand. J. Physiol., Lond. *195:* 419–449 (1968).

KONORSKI, J.: Integration activity of the brain (University of Chicago Press, Chicago 1967).

KORNMÜLLER, A.E.: Eine experimentelle Anästhesie der äusseren Augenmuskeln am Menschen und ihre Auswirkungen. J. psychol. Neurol., Lpz. *41:* 354–366 (1931).

KOSTYUK, P.G. and PILYASKY, A.L.: A possible direct interneuronal pathway from rubrospinal tract to motoneurones. Brain Res. *14:* 526–528 (1969).

KOSTYUK, P.G.; VASILENKO, D.A., and LANG, E.: Propriospinal pathways in the dorsolateral funiculus and their effects on lumbosacral motoneuronal pools. Brain Res. *28:* 233–249 (1971).

KOTS, Y.M.: Supraspinal control of the segmental centres of muscle antagonists in man. I. Reflex excitability of the motor neurones of muscle antagonists in the period of organization of voluntary movement. Biofizika *14:* 167–172 (1969) (translation by Pergamon Press).

KOTS, Y.M.: Supraspinal control of the segmental centres of muscle antagonists in man. II. Reflex excitability of the motor neurones of muscle antagonists on organization of sequential activity. Biofizika *14:* 1087–1094 (1969) (translation by Pergamon Press).

KOTS, Y.M. et KRINSKI, A.: Réflexe monosynaptique H chez l'homme dans les muscles soléaires et gastrocnémien médian au repos. J. Physiol. URSS *53 (7):* 784 (1967).

KOTS, Y.M. and MART'YANOV, V.A.: Use of the technique of monosynaptic H-reflex to record the effects of electric stimulation of the vestibular apparatus in man. Kosm. Biol. Med. *1:* 81–85 (1967) (in Russian).

KOTS, Y.M. and MART'YANOV, V.A.: Cutting out the vestibulospinal influences in the periodic organization of voluntary movement. Biofizika *13:* 818–826 (1968) (translation by Pergamon Press).

KRASSOIEVITCH, M.; DELWAIDE, P.J., and HUGON, M.: Modifications du réflexe de Hoffmann par l'étirement: contrôle du muscle soléaire. Electroenceph. Clin. Neurophysiol. *32:* 465–470 (1972).

KREMER, M.: Sitting, standing and walking. Brit. med. J. *1:* 63–68 (1958).

KRÜGER, L. and PORTER, P.: A behavioural study of the functions of the Rolandic cortex in the monkey. J. comp. Neurol. *109:* 439–469 (1958).

KUFFLER, S.W. and HUNT, C.G.: The mammalian small-nerve fibres; a system for efferent nervous regulation of muscle spindle discharge. Res. Publ. Ass. nerv. ment. Dis. *30:* 24–47 (1952).

KUFFLER, S.W.; HUNT, C.C., and QUILLIAM, J.P.: Function of medullated small-nerve fibres in mammalian ventral roots: efferent muscle spindle innervation. J. Neurophysiol. *14:* 29–54 (1951).

KUGELBERG, E.: Demonstration of A and C fibre components in the Babinski plantar response and the pathological flexion reflex. Brain *71:* 304–319 (1948).

KUGELBERG, E.: Facial reflexes. Brain *75:* 385–396 (1952).

KUGELBERG, E.: Polysynaptic reflexes of clinical importance. Electroenceph. clin. Neurophysiol., suppl. *22:* 103–111 (1962).

KUGELBERG, E. and EDSTRÖM, L.: Differential histochemical effects of muscle contractions on phosphorylase and glycogen in various types of fibres: relation to fatigue. J. Neurol. Neurosurg. Psychiat. *31:* 415–423 (1968).

KUGELBERG, E.; EKLUND, K., and GRIMBY, L.: An electromyographic study of the nociceptive reflexes of the lower limb. Mechanism of the plantar responses. Brain *83:* 394–410 (1960).

KUGELBERG, E. and HAGBARTH, K.-E.: Spinal mechanism of the abdominal and erector spinae skin reflexes. Brain *81:* 390–404 (1958).

KUHN, R. A. and MACHT, M. B.: Some manifestations of reflex activity in spinal man with particular reference to the occurrence of extensor spasm. Bull. Johns Hopk. Hosp. *84:* 43–75 (1949).

KUNO, M.: Excitability following antidromic activation in spinal motoneurones supplying red muscles. J. Physiol., Lond. *149:* 374–393 (1959).

KUNO, M.: Quantal components of excitatory synaptic potentials in spinal motoneurones. J. Physiol., Lond. *175:* 81–99 (1964).

KUNO, M. and MIYAHARA, J. T.: Factors responsible for multiple discharge of neurons in Clarke's column. J. Neurophysiol. *31:* 624–638 (1968).

KUNO, M. and MIYAHARA, J. T.: Analysis of synaptic efficacy in spinal motoneurones from 'quantum' aspects. J. Physiol., Lond. *201:* 479–493 (1969).

KUNO, M. and PERL, J. T.: Alteration of spinal reflexes by interaction with suprasegmental and dorsal root activity. J. Physiol., Lond. *151:* 103–122 (1960).

KUNTZ, A.: The autonomic nervous system; 4th ed. (Lea & Febiger, Philadelphia 1953).

KUYPERS, H. G. J. M.: An anatomical analysis of cortico-bulbar connexions to the pons and the lower brain stem in the cat. J. Anat., Lond. *92:* 198–218 (1958a).

KUYPERS, H. G. J. M.: Some projections from the pericentral cortex to the pons and lower brain stem in monkey and chimpanzee. J. comp. Neurol. *110:* 221–225 (1958b).

KUYPERS, H. G. J. M.: Corticobulbar connexions to the pons and lower brain stem in man. An anatomical study. Brain *81:* 364–388 (1958c).

KUYPERS, H. G. J. M.: Central cortical projections to motor and somatosensory cell groups. An experimental study in the rhesus monkey. Brain *83:* 161–184 (1960).

KUYPERS, H. G. J. M.: Corticospinal connections: postnatal development in the rhesus monkey. Science *138:* 678–680 (1962).

KUYPERS, H. G. J. M.: The organization of the 'motor system'. Int. J. Neurol. *4:* 78–91 (1963).

KUYPERS, H. G. J. M.: The descending pathways to the spinal cord, their anatomy and function; in ECCLES and SCHADÉ Organization of the spinal cord. Progr. Brain Res., vol. 11, pp. 178–200 (Elsevier, Amsterdam 1964).

KUYPERS, H. G. J. M.: Cortical projections to the red nucleus in the monkey and chimpanzee; in PURPURA and YAHR The thalamus, pp. 122–126 (Columbia University Press, New York 1966).

KUYPERS, H. G. J. M. and BRINKMAN, J.: Precentral projections to different parts of the spinal intermediate zone in the rhesus monkey. Brain Res. *24:* 29–48 (1970).

KUYPERS, H. G. J. M.; FLEMING, W. R., and FARINHOLT, J. W.: Subcortical projections in the rhesus monkey. J. comp. Neurol. *118:* 107–131 (1962).

KUYPERS, H. G. J. M. and LAWRENCE, D. G.: Cortical projections to the red nucleus and the brain stem in the rhesus monkey. Brain Res. *4:* 151–188 (1967).

LADPLI, R. and BRODAL, A.: Experimental studies of commissural and reticular formation projections from the vestibular nuclei in the cat. Brain Res. *8:* 65–96 (1968).

LAMBERT, E. H.: Defects of neuromuscular transmission in syndromes other than myasthenia gravis. Ann. N. Y. Acad. Sci. *135:* 367–384 (1966).

LANCE, J. W.: The mechanism of reflex irradiation. Proc. austr. Ass. Neurol. *3:* 77–82 (1965).

LANCE, J. W.: A physiological approach to clinical neurology (Butterworths, London 1970).

LANCE, J. W. and GAIL, P. DE: Spread of phasic muscle reflexes in normal and spastic subjects. J. Neurol. Neurosurg. Psychiat. *28:* 328–334 (1965).

LANCE, J. W.; GAIL, P. DE, and NEILSON, P. D.: Tonic and phasic spinal cord mechanisms in man. J. Neurol. Neurosurg. Psychiat. *29:* 535–544 (1966).

LANCE, J. W.; NEILSON, P. D., and TASSINARI, C. A.: Suppression of the H-reflex by peripheral vibration. Proc. austr. Ass. Neurol. *5:* 45–49 (1968).

LANCE, J. W.; SCHWAB, R. S., and PETERSON, E. A.: Action tremor and the cogwheel phenomenon in Parkinson's disease. Brain *86:* 95–110 (1963).

LANDAU, W. M.: Spasticity and rigidity; in PLUM Adv. Neurol., pp. 1–32 (Davis, Philadelphia 1969).

LANDAU, W. M. and CLARE, M. H.: Fusimotor function. IV. Reinforcement of the H reflex in normal subjects. Arch. Neurol., Chicago *10:* 117–122 (1964a).

LANDAU, W. M. and CLARE, M. H.: Fusimotor function. VI. H-reflexes, tendon jerk and reinforcement in hemiplegia. Arch. Neurol., Chicago *10:* 128–134 (1964b).

LANDAU, W. M.; STRUPPLER, A., and MEHLS, O.: A comparative electromyographic study of the reactions to passive movement in Parkinsonian and normal subjects. Neurology, Minneap. *16:* 34–48 (1966).

LANDAU, W. M.; WEAVER, R. A., and HORNBEIN, T. F.: Fusimotor nerve function in man: differential block studies in normal subjects and in spasticity and rigidity. Arch. Neurol., Chicago *3:* 10–23 (1960).

LANDGREN, S.; PHILLIPS, C. G., and PORTER, R.: Minimal synaptic actions of pyramidal impulses on some alpha motoneurones of the baboon's hand and forearm. J. Physiol., Lond. *161:* 91–111 (1962).

LANDGREN, S. and SILVENIUS, H.: Projection to cerebral cortex of group I muscle afferents from the cat's hind-limb. J. Physiol., Lond. *200:* 353–372 (1969).

LANDIS, C. and HUNT, W. A.: The startle pattern (Farrar & Rinehart, New York 1939).

LANG, A. H. and VALLBO, Å. B.: Motoneuron activation by low intensity tetanic stimulation of muscle afferents in man. Exp. Neurol. *18:* 383–391 (1967).

LANGUTH, H. W.; TEASDALL, R. D., and MAGLADERY, J. W.: Motoneurone excitability following afferent nerve volleys in patients with rostrally adjacent spinal cord damage. Bull. Johns Hopk. Hosp. *91:* 257–266 (1952).

LANGWORTHY, O. R.: Development of behavior patterns and myelination of the nervous system in the human fetus and infant. Contrib. Embryol. Carneg. Inst. *24:* 1–57 (1933).

LAPORTE, Y. et BESSOU, P.: Modifications d'excitabilité de motoneurones homonymes provoqués par l'activation physiologique de fibres afférentes d'origine musculaire du groupe II. J. Physiol., Paris *51:* 897–908 (1959).

LAPORTE, Y. and LLOYD, D. P. C.: Nature and significance of the reflex connections established by large afferent fibers of muscular origin. Amer. J. Physiol. *169:* 609–621 (1952).

LARSSON, L.E.: Correlation between the psychological significance of stimuli and the magnitudes of the startle blink and evoked EEG potentials in man. Acta physiol. scand. *48:* 276–294 (1960).

LARSSON, L.E.: The relation of the nonspecific EEG-response to different phases of the orienting reflex. Mechanisms of orienting reaction in man. Trans. Int. Coll., Bratislava 1966 (Slovak Academy of Sciences, 1967).

LARUELLE, L.: Contribution à l'étude du névraxe végétatif. C. R. Ass. Anat. *31:* 210–229 (1936).

LARUELLE, L.: La structure de la moelle épinière en coupes longitudinales. Rev. neurol. *44:* 695–725 (1937).

LARUELLE, L. et REUMONT, M.: La colonne de Stilling-Clarke (colonne dorsale spino-cérébelleuse). Anatomie microscopique. Signification fonctionnelle générale. C.R. Ass. Anat. *33:* 267–286 (1938).

LASHLEY, K.S.: The accuracy of movement in the absence of excitation from the moving organ. Amer. J. Physiol. *43:* 169–194 (1917).

LASSEK, A.M.: Inactivation of voluntary motor function following rhizotomy. J. Neuropath. exp. Neurol. *12:* 83–87 (1953).

LASSEK, A.M.: The pyramidal tract. Its status in medicine (Thomas, Springfield 1954).

LAURSEN, A.M.: Static and phasic muscle activity of monkeys with pyramidal lesions. *40:* 125–126 (1972).

LAURSEN, A.M. and WIESENDANGER, M.: Pyramidal effect on alpha and gamma motoneurones. Acta physiol. scand. *67:* 165–172 (1966).

LAWRENCE, D.G. and HOPKINS, D.A.: Bilateral pyramidal lesions in infant rhesus monkeys. Brain Res. *24:* 543–544 (1970).

LAWRENCE, D.G. and HOPKINS, D.A.: Developmental aspects of pyramidal motor control in the rhesus monkey. Brain Res. *40:* 117–118 (1972).

LAWRENCE, D.G. and KUYPERS, H.G.J.M.: Pyramidal and nonpyramidal pathways in monkeys, anatomical and functional correlations. Science *148:* 973–975 (1965).

LAWRENCE, D.G. and KUYPERS, H.G.J.M.: The functional organization of the motor system in the monkey. I. The effects of bilateral pyramidal lesions. Brain *91:* 1–14 (1968a).

LAWRENCE, D.G. and KUYPERS, H.G.J.M.: The functional organization of the motor system in the monkey. II. The effect of lesions of the descending brain stem pathways. Brain *91:* 15–36 (1968b).

LEE, M.A.M. and KLEITMAN, N.: Studies on the physiology of sleep. II. Attempt to demonstrate functional changes in the nervous system during experimental insomnia. Amer. J. Physiol. *67:* 141–152 (1923).

LEFEBVRE, J. et SCHERRER, J.: Renseignements fournis par l'EMG dans l'étude des mouvements anormaux. Rev. Neurol. *86:* 584–595 (1952).

LEGROS CLARK, W.E.: The antecedents of man (University Press, Edinburgh 1962).

LEKSELL, L.: The action potential and excitatory effects of the small ventral root fibres to skeletal muscle. Acta physiol. scand. *10:* suppl. 31: 1–84 (1945).

LENNERSTRAND, G. and THODEN, U.: Dynamic analysis of muscle spindle endings in the cat using length changes of different length-time relations. Acta physiol. scand. *73:* 234–250 (1968a).

LENNERSTRAND, G. and THODEN, U.: Muscle spindle responses to concomitant variations in length and fusimotor activation. Acta physiol. scand. *74:* 153–165 (1968b).

LEVY, R. A.; REPKIN, A. H., and ANDERSON, E. G.: The effect of bicuculline on primary afferent terminal excitability. Brain Res. *32:* 261–265 (1971).

LEWIS, R. and BRINDLEY, G. S.: The extrapyramidal cortical motor map. Brain *88:* 397–406 (1965).

LIBERSON, W. T.: Electromyographic studies of the spinal cord inhibition in man. Electroenceph. clin. Neurophysiol. *13:* 312P (1961).

LIBERSON, W. T.: Monosynaptic reflexes and clinical significance. Electroenceph. clin. Neurophysiol., suppl. *22:* 79–89 (1962).

LIBERSON, W. T.: Sensory conduction velocities in normal individuals and in patients with peripheral neuropathies. Arch. phys. Med. *44:* 313–320 (1963).

LIBERSON, W. T.: 'Silent periods' and wide spread inhibition in man. Abstr. 4th Int. Congr. EMG, Brussels 1971, p. 91.

LIBERSON, W. T.; DONDEY, M., and ASA, M. M.: Brief repeated isometric maximal exercises. Amer. J. phys. Med. *41:* 3–14 (1962).

LIBERSON, W. T.; ZALIS, A.; GRATZER, M., and GRABINSKI, B.: Comparison of conduction velocities determined in motor and sensory fibers by different methods. Electroenceph. clin. Neurophysiol. *17:* 98–99 (1964).

LIBERSON, W. T.; ZALIS, A.; GRATZER, M., and GRABINSKI, B.: Comparison of conduction velocities determined in motor and sensory fibers by different methods. Arch. phys. Med. Rehabil. *47:* 17–22 (1966).

LIDDELL, E. G. T. and SHERRINGTON, C. S.: Reflexes in response to stretch (myotatic reflexes). Proc. roy. Soc. B *96:* 212–242 (1924).

LIDDELL, E. G. T. and SHERRINGTON, C. S.: Further observations on myotatic reflexes. Proc. roy. Soc. B *97:* 267–283 (1925).

LILLY, J. C.: Correlations between neurophysiological activity in the cortex and short-term behaviour in the monkey; in HARLOW and WOOLSEY Biological and biochemical bases of behaviour, pp. 83–100 (University of Wisconsin Press, Madison 1958).

LINDQUIST, C. and MARTENSSON, A.: Mechanisms involved in the cat's blink reflex. Acta physiol. scand. *80:* 149–159 (1970).

LINDSLEY, D. B.: Electromyographic studies of neuromuscular disorders. Arch. Neurol. Psychiat., Chicago *36:* 128–157 (1936).

LIPPOLD, O. C. J.; REDFEARN, J. W. T., and VUCO, J.: The rhythmical activity of groups of motor units in voluntary contraction of muscle. J. Physiol., Lond. *137:* 473–487 (1957).

LIPPOLD, O. C. J.; REDFEARN, J. W. T., and VUCO, J.: The effect of sinusoidal stretching upon the activity of stretch receptors in voluntary muscle and their reflex responses. J. Physiol., Lond. *144:* 373–386 (1958).

LLOYD, A. J. and CALDWELL, L. S.: Accuracy of active and passive positioning of the leg on the basis of kinesthetic cues. J. comp. physiol. Psychol. *60:* 102–106 (1965).

LLOYD, D. P. C.: Activity in neurons of the bulbospinal correlation system. J. Neurophysiol. *4:* 116–134 (1941a).

LLOYD, D. P. C.: The spinal mechanism of the pyramidal system in cats. J. Neurophysiol. *4:* 525–546 (1941b).

LLOYD, D. P. C.: Mediation of descending long spinal reflex activity. J. Neurophysiol. *5:* 435–458 (1942).

LLOYD, D. P. C.: Neural patterns controlling transmission of ipsilateral hind limb reflexes in cats. J. Neurophysiol. *6:* 293–315 (1943a).

LLOYD, D. P. C.: Conduction and synaptic transmission of the reflex response to stretch in spinal cats. J. Neurophysiol. *6:* 317–326 (1943b).

LLOYD, D. P. C.: Facilitation and inhibition of spinal motoneurones. J. Neurophysiol. *9:* 421–438 (1946).

LLOYD, D. P. C.: Spinal mechanisms involved in somatic activities; in FIELD Handbook of physiology, section I. Neurophysiology, vol. 2, pp. 929–949 (1960).

LLOYD, D. P. C. and CHANG, H. T.: Afferent fibers in muscle nerves. J. Neurophysiol. *11:* 199–207 (1948).

LLOYD, D. P. C. and McINTYRE, A. K.: Dorsal column conduction of group I muscle afferent impulses and their relay through Clarke's column. J. Neurophysiol. *13:* 39–54 (1950).

LOEB, J.: Untersuchungen über die Orientierung im Fühlraum der Hand und im Blickraum. Pflügers Arch. ges. Physiol. *46:* 1–146 (1890).

LOMBARD, W. P.: On the nature of the knee-jerk. J. Physiol., Lond. *10:* 122–148 (1889).

LORENTE DE Nó, R.: Vestibulo-ocular reflex arc. Arch. Neurol. Psychiat., Chicago *30:* 245–291 (1933).

LORENTE de Nó, R.: A study of nerve physiology. Studies from Rockefeller Institute for Medical Research 131,132 (The Rockefeller Institute for Medical Research, 1947).

LUCIANI, L.: Il cervelleto: nuovi studi di fisiologia normale et patologica (Le Monnier, Firenze 1891).

LUND, S.; LUNDBERG, A., and VYKLICKY, L.: Inhibition action from the flexor reflex afferents on transmission to Ia afferents. Acta physiol. scand. *64:* 345–355 (1965).

LUND, S. and POMPEIANO, O.: Monosynaptic excitation of alpha-motor neurons from supraspinal structures in the cat. Acta physiol. scand. *73:* 1–21 (1968).

LUNDBERG, A.: Integrative significance of patterns of connexions made by muscle afferents in the spinal cord. Symp. 21st Congr. int. Ciencias Fisol., Buenos Aires 1959, pp. 100–105.

LUNDBERG, A.: Integration in the reflex pathway; in GRANIT Nobel Symposium 1, Muscular afferents and motor control, pp. 275–305 (Almqvist & Wiksell, Stockholm 1966).

LUNDBERG, A.: The supraspinal control of transmission in spinal reflex pathways. Electroenceph. clin. Neurophysiol. *25:* 35–46 (1967).

LUNDBERG, A.: Reflex control of stepping. Proc. norw. Acad. Sci. (Universitetsforlaget, Oslo 1969a).

LUNDBERG, A.: Convergence of excitability and inhibitory action on interneurones in the spinal cord; in BRAZIER The interneuron, pp. 231–265 (University of California Press, Los Angeles 1969b).

LUNDBERG, A.; NORSELL, U., and VOORHOEVE, P.: Pyramidal effects on lumbosacral interneurones activated by somatic afferents. Acta physiol. scand. *56:* 220–229 (1962).

LUNDBERG, A. and VOORHOEVE, P.: Effects from the pyramidal tract on spinal reflex arcs. Acta physiol. scand. *56:* 201–219 (1962).

LUX, H. D.: Eigenschaften eines Neuron-Modells mit Dendriten begrenzter Länge. Pflügers Arch. ges. Physiol. *297:* 238–255 (1967).

LUX, H. D.: Ammonium and chloride extrusion: hyperpolarizing synaptic inhibition in spinal motoneurons. Science *173:* 555–557 (1971).

LUX, H. D.; SCHUBERT, P., and KREUTZBERG, G. W.: Direct matching of morphological and electrophysiological data in cat spinal motoneurones; in ANDERSEN and JANSEN Excitatory synaptic mechanisms, pp. 189–198 (Universitetsforlaget, Oslo 1970).

LYON, L.W. and VAN ALLEN, M.W.: Alteration of the orbicularis oculi reflex by acoustic neuroma. Arch. Otolaryng. *95:* 100–103 (1972a).

LYON, L.W. and VAN ALLEN, M.W.: Orbicularis oculi reflex in internuclear ophthalmoplegia and pseudo-internuclear ophthalmoplegia. Arch. Ophthal. *87:* 148–154 (1972b).

LYON, L.W.; KIMURA, J., and MCCORMICK, W.F.: Orbicularis oculi reflex in coma: Clinical electrophysiological, pathological correlations. J. Neurol. Neurosurg. Psychiat. *35:* 582–588 (1972).

MABUCHI, M.: Rotatory head responses evoked by stimulating and destroying the interstitial nucleus and surrounding region. Exp. Neurol. *27:* 175–193 (1970).

MACCARTHY, C.S. and KIEFER, E.J.: Thoracic, lumbar and sacral spinal cordectomy: preliminary report. Staff Meet. Mayo Clin. *24:* 108–115 (1949).

MACCHI, G.: Organizzazione morfologica della connessioni thalamocorticali. Atti. Soc. ital. anat., 18. Convegno sociale, Arch. ital. Biol., suppl. *66:* 25–124 (1958).

MCCOMAS, A.J. and THOMAS, H.C.: Fast and slow twitch muscles in man. J. Neurol. Sci. *7:* 301–307 (1968).

MCCOUCH, G.P.; AUSTIN, G.M.; LIU, C.N., and LIU, C.Y.: Sprouting as a cause of spasticity. J. Neurophysiol. *21:* 205–216 (1958).

MCCOUCH, G.P.; LIU, C.N., and CHAMBERS, W.W.: Descending tracts and spinal shock in the monkey *(Macaca mulatta)*. Brain *89:* 359–376 (1966).

MCCULLOCH, W.S.; GRAF, C., and MAGOUN, H.W.: A cortico-bulbo-reticular pathway from areas 4S. J. Neurophysiol. *9:* 127–132 (1946).

MCINTYRE, A.K. and PROSKE, U.: Reflex potency of cutaneous afferent fibres. Austr. J. exp. Biol. med. Sci. *49:* 19 (1968).

MCKEAN, T.A.; POPPELE, R.E.; ROSENTHAL, N.P., and TERZUOLO, C.A.: The biologically relevant parameter in nerve impulse trains. Kybernetik *6:* 168–170 (1970).

MACLEAN, J.B. and LEFFMAN, H.: Supraspinal control of Renshaw cells. Exp. Neurol. *18:* 94–104 (1967).

MCLEOD, J.G.: Electrophysiological and histological studies in patients with Friedreich's ataxia. Electroenceph. clin. Neurophysiol. *27:* 723P (1969).

MCLEOD, J.G. and MEULEN, J.P. VAN DER: Effect of cerebellar ablation on the H reflex in the cat. Arch. Neurol., Chicago *16:* 421–432 (1967).

MCLEOD, J.G. and WRAY, H.S.: An experimental study of the F wave in the baboon. J. Neurol. Neurosurg. Psychiat. *29:* 196–200 (1966).

MCMASTERS, R.E.; WEISS, A.H., and CARPENTER, M.B.: Vestibular projections to the nuclei of the extraocular muscles. Degeneration resulting form discrete partial lesions of the vestibular nuclei in the monkey. Amer. J. Anat. *118:* 163–194 (1966).

MCPHEDRAN, A.M.; WUERKER, R.B., and HENNEMAN, E.: Properties of motor units in a homogeneous red muscle (soleus) of the cat. J. Neurophysiol. *28:* 71–84 (1965).

MAGENDIE, F.: Leçons sur les fonctions et les maladies du système nerveux (Ebrard, Paris 1839).

MAGLADERY, J.W.: Some observations on spinal reflexes in man. Pflügers Arch. ges. Physiol. *261:* 302–321 (1955).

MAGLADERY, J.W.: Central facilitating and inhibiting mechanisms in the control of muscle tone. Clin. Pharmacol. Ther. *5:* 805–811 (1964).

MAGLADERY, J.W. and McDOUGAL, D.B., jr.: Electrophysiological studies of nerve and reflex activity in normal man. I. Identification of certain reflexes in the electromyogram and the conduction velocity of peripheral nerve fibers. Johns Hopk. Hosp. Bull. *86:* 265–290 (1950).

MAGLADERY, J.W.; McDOUGAL, D.B., and STOLL, J.: Electrophysiological studies of nerve and reflex activity in normal man. II. Effects of peripheral ischemia. Johns Hopk. Hosp. Bull. *86:* 291–312 (1950a).

MAGLADERY, J.W.; McDOUGAL, D.B., and STOLL, J.: Electrophysiological studies of nerve and reflex activity in normal man. III. The post-ischemic state. Johns Hopk. Hosp. Bull. *86:* 313–340 (1950b).

MAGLADERY, J.W.; PARK, A.M.; PORTER, W.E., and TEASDALL, R.D.: Spinal reflex patterns in man. Res. Publ. Ass. nerv. ment. Dis. *30:* 118–151 (1952).

MAGLADERY, J.; PORTER, W.E.; PARK, A.M., and TEASDALL, R.D.: Electrophysiological studies of nerve and reflex activity in normal man. IV. The two neurone reflex and identification of certain action potentials from spinal roots and cord. Johns Hopk. Hosp. Bull. *88:* 499–519 (1951).

MAGLADERY, J.W. and TEASDALL, R.D.: Stretch reflexes in patients with spinal cord lesions. Johns Hopkins Hosp. Bull. *103:* 236–241 (1958).

MAGLADERY, J.W.; TEASDALL, R.D.; PARK, A.M., and LANGUTH, H.W.: Electrophysiological studies of reflex activity in patients with lesions of the nervous system. I. A comparison of spinal motoneurone excitability following afferent nerve volleys in normal persons and patients with upper motor neurone lesions. Johns Hopk. Hosp. Bull. *91:* 219–244 (1952).

MAGLADERY, J.W.; TEASDALL, R.D.; PARK, A.M., and PORTER, W.E.: Electrophysiological studies of nerve and reflex activity in normal man. V. Excitation and inhibition of two-neurone reflexes by afferent impulses in the same nerve trunk. Johns Hopk. Hosp. Bull. *88:* 520–537 (1951).

MAGOUN, H.W. and RHINES, R.: Spasticity: the stretch reflex and extrapyramidal systems (Thomas, Springfield 1947).

MANNEN, H.: Noyau fermé et noyau ouvert. Contribution à l'étude cytoarchitectonique du tronc cérébral envisagée du point de vue du mode d'arborisation dentrique. Arch. ital. Biol. *98:* 330–350 (1960).

MARCHIAVAFA, P.L. and POMPEIANO, O.: Pyramidal influence on spinal cord during desynchronized sleep. Arch. ital. Biol. *102:* 500–529 (1964).

MARIE, P. et FOIX, C.: Les réflexes d'automatisme médullaire et le phénomène des raccourcisseurs. Leur valeur sémiologique. Leur signification physiologique. Rev. neurol. *10:* 657–676 (1912).

MARINESCO, G.: Recherches sur les localisations motrices spinales. Sem. méd., Paris *29:* 225–231 (1904).

MARK, R.F.: Analyse électrophysiologique des réflexes d'étirement du muscle triceps sural chez l'homme; thèse Marseille (1962).

MARK, R.F.; COQUERY, J.M., and PAILLARD, J.: Autogenic reflex effects of slow or steady stretch of the calf muscles in man. Exp. Brain Res. *6:* 130–145 (1968).

MARR, D.: A theory of cerebellar cortex. J. Physiol., Lond. *202:* 437–470 (1969).

MARSDEN, C.D.; MEADOWS, J.C., and HODGSON, H.J.: Observations on the reflex response to muscle vibration and its voluntary control. Brain *92:* 829–846 (1969).

MARSDEN, C.D.; MERTON, P.A., and MORTON, H.B.: Servo action and stretch reflex in human muscle and its apparent dependence on peripheral sensation. J. Physiol., Lond. *216:* 21P–22P (1971).

MARSDEN, C.D.; MERTON, P.A., and MORTON, H.B.: Changes in loop gain with force in the human muscle servo. J. Physiol., Lond. *222:* 32P–34P (1972).

MARSHALL, C.: The functions of the pyramidal tracts. Quart. Rev. Biol. *11:* 35–56 (1936).

MARTIN, G.F. and DOM, R.: Rubrobulbar projections of the opossum *(Didelphis virginiana).* J. comp. Neurol. *139:* 199–214 (1970).

MASPES, P.E. and PAGNI, C.A.: Surgical treatment of dystonia and choreo-athetosis in infantile cerebral palsy by pedunculotomy. J. Neurosurg. *21:* 1076–1086 (1964).

MASSION, J. and CROIZE, B.: Tonic facilitatory action of the cerebellum on the ventrolateral nucleus; in FIELDS and WILLIS The cerebellum in health and disease, pp. 332–338 (Green, St. Louis 1970).

MASSION, J.; MEULDERS, M. et COLLE, J.: Fonctions posturales des muscles respiratoires. Arch. int. Physiol. *58:* 314–326 (1960).

MASSION, J. and PADEL-RISPAL, L.: Spatial organization of the cerebello-thalamo-cortical pathway. Brain Res. *40:* 61–66 (1972).

MATSUDA, T.: Studies of the method of neuromuscular facilitation by H-reflex, especially on vibratory stimulation. J. Chiba med. Soc. *44:* 759–778 (1969).

MATSUSHITA, M.: Some aspects of the interneuronal connections in the cat's spinal gray matter. J. comp. Neurol. *136:* 57–80 (1969).

MATSUSHITA, M.: The axonal pathways of spinal neurones in the cat. J. comp. Neurol. *138:* 391–418 (1970).

MATTHEWS, B.H.C.: Nerve endings in mammalian muscle. J. Physiol., Lond. *78:* 1–53 (1933).

MATTHEWS, P.B.C.: Experimental observations on hypertonus. 2. Hypertonus and the gamma motoneurones. Cerebr. Palsy Bull. *1:* 2–3 (1959).

MATTHEWS, P.B.C.: The dependence of tension upon extension in the stretch reflex of the soleus muscle of the decerebrate cat. J. Physiol., Lond. *147:* 521–546 (1959).

MATTHEWS, P.B.C.: The differentiation of two types of fusimotor fibres by their effects on the dynamic response of muscle spindle primary endings. Quart. J. exp. Physiol. *47:* 324–333 (1962).

MATTHEWS, P.B.C.: The response of de-efferented muscle spindle receptors to stretching at different velocities. J. Physiol., Lond. *168:* 660–678 (1963).

MATTHEWS, P.B.C.: Muscle spindles and their motor control. Physiol. Rev. *44:* 219–288 (1964).

MATTHEWS, P.B.C.: The reflex excitation of the soleus muscle of the decerebrate cat caused by vibration applied to its tendon. J. Physiol., Lond. *184:* 450–472 (1966).

MATTHEWS, P.B.C.: Evidence that the secondary as well as the primary endings of the muscle spindles may be responsible for the tonic stretch reflex of the decerebrate cat. J. Physiol., Lond. *204:* 365–393 (1969).

MATTHEWS, P.B.C.: The origin and functional significance of the stretch reflex; in ANDERSEN and JANSEN Excitatory synaptic mechanism, pp. 301–305 (Universitetsforlaget, Oslo 1970).

MATTHEWS, P.B.C.: Recent advances in the understanding of the muscle spindle; in GILLILAND and FRANCIS The scientific basis of medicine, pp. 99–128 (Athlone Press, London 1971).

MATTHEWS, P.B.C.: Mammalian muscle receptors and their central actions (Arnold, London 1972).

MATTHEWS, P.B.C. and RUSHWORTH, G.: The selective effect of procaine on the stretch reflex and tendon jerk of soleus muscle when applied to its nerve. J.Physiol., Lond. *135:* 245–262 (1957).

MATTHEWS, P.B. and RUSHWORTH, G.: The discharge from muscle spindles as an indicator of γ efferent paralysis by procaïne. J. Physiol., Lond. *140:* 421–426 (1958).

MATTHEWS, P.B.C.: Vibration and the stretch reflex; in DE REUCK and KNIGHT Myotatic, kinesthetic and vestibular mechanisms (Churchill, London 1967).

MATTHEWS, P.B.C. and STEIN, R.B.: The sensitivity of muscle spindle afferents to small sinusoidal changes of length. J. Physiol., Lond. *200:* 723–743 (1969a).

MATTHEWS, P.B.C. and STEIN, R.B.: The regularity of primary and secondary muscle spindle afferent discharges. J. Physiol., Lond. *202:* 59–82 (1969b).

MATTHEWS, W.B.: The action of chlorproethazine on spasticity. Brain *88:* 1057–1064 (1965).

MAYER, R.F.: Nerve conduction studies in man. Neurology (Minneap.) *13:* 1021–1030 (1963).

MAYER, R.F. and FELDMAN, R.G.: Observation on the nature of the F wave in man. Neurology, Minneap. *17:* 147–156 (1967).

MAYER, R.F. and MAWDSLEY, C.: Studies in man and cat of the significance of the H wave. J. Neurol. Neurosurg. Psychiat. *28:* 201–211 (1965).

MAYER, R.F. and MOSSER, R.S.: Vestibular effects on monosynaptic (H) reflexes. Trans. amer. neurol. Ass. *92:* 269–270 (1967).

MAYER, R.F. and MOSSER, R.S.: Maturation of H-reflexes in infants. Neurology, Minneap. *19:* 319 (1969a).

MAYER, R.F. and MOSSER, R.S.: Excitability of motoneurons in infants. Neurology, Minneap. *19:* 932–945 (1969b).

MEGIRIAN, D.: Bilateral facilitatory and inhibitory skin areas of spinal interneurones of cat. J. Neurophysiol. *25:* 127–137 (1962).

MEIER-EWERT, K.; DAHM, J. und NIEDERMEIER, E.: Optisch und elektrisch ausgelöste Mikroreflexe des Menschen. Z. Neurol. *199:* 167–182 (1971).

MEIER-EWERT, K.; HÜMME, U., and DAHM, J.: New evidence favouring long loop reflexes in man. Arch. Psychiat. Nervenkr. *215:* 121–128 (1972).

MELLSTRÖM, A. and SKOGLUND, S.: Quantitative morphological changes in some spinal cord segments during postnatal development. A study in the cat. Acta physiol. scand., suppl. *331:* 1–84 (1969).

MELVILL-JONES, G. and WATT, D.G.D.: Observations on the control of stepping and hopping movements in man. J. Physiol., Lond. *219:* 709–727 (1971a).

MELVILL-JONES, G. and WATT, D.G.D.: Muscular control of landing from unexpected falls in man. J. Physiol., Lond. *219:* 729–737 (1971b).

MELZACK, R. and WALL, P.D.: Pain mechanisms, a new theory. Science *150:* 971–979 (1965).

MENDELL, L.: Properties and distribution of peripherally evoked presynaptic hyperpolarization in cat lumbar spinal cord. J. Physiol. Lond. *226:* 769–792 (1972).

MENDELL, L.M. and HENNEMAN, E.: Terminals of single Ia fibers: distribution within a pool of 300 homonymous motor neurons. Science *160:* 96–98 (1968).

MENDELL, L. M. and HENNEMAN, E.: Terminals of single Ia fibers: location density and distribution within a pool of 300 homonymous motoneurones. J. Neurophysiol. *34:* 171–187 (1971).

MERRILL, E. G. and WALL, P. D.: Factors forming the edge of a receptive field: the presence of relatively ineffective afferent terminals. J. Physiol., Lond. *226:* 825–846 (1972).

MERTON, P. A.: The silent period in a muscle of the human hand. J. Physiol., Lond. *114:* 183–198 (1951).

MERTON, P. A.: Speculations on the servo-control of movement; in MALCOLM, GRAY and WOLSTENHOLME The spinal cord, Ciba Foundation Symposium, London, pp. 247–260 (Churchill, London 1953).

MERTON, P.: Human position sense and sense of effort. Symp. Soc. exp. Biol. Med. *18:* 387–400 (1964).

MESSINA, C.: L'abitudine dei riflessi trigemino-facciali in parkinsoniani sottoposti a trattamento con L-dopa. Riv. Neurol. *40:* 327–336 (1970).

METTLER, F. A.: Physiological consequences and anatomic degenerations following lesions of the primate brain stem: plantar and patellar reflexes. J. comp. Neurol. *80:* 69–148 (1944).

MEYER, J. C. and HERNDON, R. M.: Bilateral infarctions of the pyramidal tracts in man. Neurology, Minneap. *12:* 637–642 (1962).

MIGLIETTA, O.: Electromyographic characteristics of clonus and influence of cold. Arch. phys. Med. Rehabil. *45:* 508–512 (1964).

MILLER, S.: Excitatory and inhibitory propriospinal pathways from lumbo-sacral to cervical segments in the cat. Acta physiol. scand. *80:* 25A (1970).

MINKOWSKI, A.: Regional development of the brain in early life (Blackwell, Oxford 1967).

MISHELEVICH, D. J.: Repetitive firing to current in cat motoneurons as a function of muscle unit twitch type. Exp. Neurol. *25:* 401–409 (1969).

MOLDAVER, J.: Facial nerve physiology and electrical analysis. Trans. amer. Acad. Ophthal. Otolaryng. *68:* 1045–1059 (1964).

MONAKOW, C. VON: Die Gehirnpathologie; pp. 722–723 (Hölder, Wien 1897).

MONNIER, M.: Les formations réticulées tegmentales. Equilibration des postures du regard, de la tête et du tronc. Rev. neurol. *78:* 422–452 (1946).

MONTANDON, P. and MONNIER, M.: Correlation of the diencephalic nystagmogenic area with the bulbovestibular nystagmogenic area. Brain *87:* 673–690 (1964).

MORRISON, A. R. and POMPEIANO, O.: An analysis of the supraspinal influences acting on motoneurons during sleep in the unrestrained cat. Responses of the alpha motoneurons to direct electrical stimulation during sleep. Arch. ital. Biol. *103:* 497–516 (1965a).

MORRISON, A. R. and POMPEIANO, O.: Pyramidal discharge from somatosensory cortex and cortical control of primary afferents during sleep. Arch. ital. Biol. *103:* 538–568 (1965b).

MORTIMER, E. M. and AKERT, K.: Cortical control and representation of fusimotor neurons. Amer. J. phys. Med. *40:* 228–247 (1961).

MOTT, F. W. and SHERRINGTON, C. S.: Experiments on the influence of sensory nerves upon movement and nutrition of the limbs. Proc. roy. Soc. B *57:* 481–488 (1895).

MOUNTCASTLE, V. B.; POGGIO, G. F., and WERNER, G.: The relation of thalamic cell response to peripheral stimuli varied over an intensive continuum. J. Neurophysiol. *26:* 807–834 (1963).

MOUNTCASTLE, V.B. and POWELL, T.P.S.: Central nervous mechanisms subserving position sense and kinesthesia. Johns Hopk. Hosp. Bull. *105:* 173–200 (1959a).

MOUNTCASTLE, V.B. and POWELL, T.P.S.: Neural mechanisms subserving cutaneous sensibility with special reference to the role of afferent inhibition in sensory perception and discrimination. Johns Hopk. Hosp. Bull. *105:* 201–232 (1959b).

MUNRO, D.: Anterior rhizotomy for spastic paraplegia. New Engl.J.Med. *233:* 453–461 (1945).

MYERS, R.E.; SPERRY, R.W., and McCURDY, N.N.: Neural mechanisms in visual guidance of limb movement. Arch. Neurol., Chicago *7:* 195–202 (1962).

MYERSON, A.: Tap and thrust responses in Parkinson's disease. Arch. Neurol. Psychiat., Chicago *51:* 480 (1944).

NAMEROW, N.S.: Somatosensory evoked responses in multiple sclerosis patients with varying sensory loss. Neurology, Minneap. *18:* 1197–1204 (1968).

NAMEROW, N.S. and ETEMADI, A.: The orbicularis oculi reflex in multiple sclerosis. Neurology, Minneap. *20:* 1200–1203 (1970).

NAMEROW, N.S. and THOMPSON, L.R.: Plaques, symptoms and the remitting course in multiple sclerosis. Neurology, Minneap. *19:* 765–774 (1969).

NAPIER, J.R.: The prehensile movements of the human hand. J.Bone Jt Surg. *38B:* 902 (1956).

NASHNER, L.M.: Sensory feedback in human posture control; Ph.D. thesis, Massachusetts Institute of Technology, Cambridge (1970).

NATHAN, P.W. and SEARS, T.A.: Effects of posterior root section on the activity of some muscles in man. J. Neurol. Neurosurg. Psychiat. *23:* 10–22 (1960).

NATHAN, P.W. and SMITH, M.C.: Long descending tracts in man. I. Review of present knowledge. Brain *78:* 248–303 (1955).

NAUTA, W.J.H. and GYGAX, P.A.: Silver impregnation of degenerating axons in the central nervous system. A modified technic. Stain Technol. *29:* 91–94 (1954).

NAUTA, W.J.H. and KUYPERS, H.G.J.M.: Some ascending pathways in the brain stem reticular formation; in JASPER, PROCTOR, KNIGHTON, NOSHAY and COSTELLO Reticular formation of the brain, pp. 3–30 (Little Brown, Boston 1958).

NAVAS, F. and STARK, L.: Sampling of intermittency in hand control system dynamics. Biophys. J. *8:* 252–302 (1968).

NELSON, P.G. and LUX, H.D.: Some electrical measurements of motoneuron parameters. Biophys. J. *10:* 55–73 (1970).

NEWSOM DAVIS, J.: An experimental study of hiccup. Brain *93:* 851–872 (1970).

NEWSOM DAVIS, J. and SEARS, T.A.: The effect of sudden alterations in load on human intercostal muscle during voluntary activation. J. Physiol., Lond. *190:* 36P–38P (1967).

NEWSOM DAVIS, J. and SEARS, T.A.: The proprioceptive reflex control of the intercostal muscles during their voluntary activation. J. Physiol., Lond. *209:* 711–738 (1970).

NEWSOM DAVIS, J.; SEARS, T.A.; STAGG, D., and TAYLOR, A.: A quantitative method for determining the effect of increased airway resistance on the electrical activity of human respiratory muscles. J. Physiol., Lond. *178:* 33P–34P (1965).

NEWSOM DAVIS, J.; SEARS, T.A.; STAGG, D., and TAYLOR, A.: The effect of airway obstruction on the electrical activity of intercostal muscles in man. J. Physiol., Lond. *185:* 19P (1966).

NGAI, S. H.; TSENG, D. T. C., and WANG, S. C.: Effect of diazepam and other central nervous system depressants on spinal reflexes in cats: a study of site of action. J. Pharmacol. exp. Ther. *153:* 344–351 (1966).

NYBERG-HANSEN, R.: Sites and mode of termination of reticulospinal fibres in the cat. An experimental study with silver impregnation methods. J. comp. Neurol. *124:* 71–100 (1965).

NYBERG-HANSEN, R.: Functional organization of descending supraspinal fiber systems to the spinal cord. Anatomical observations and physiological correlations. Ergebn. Anat. EntwGesch. *39:* 6–48 (1966).

NYBERG-HANSEN, R.: Do cat spinal motoneurones receive direct supraspinal fibre connections? A supplementary silver study. Arch. ital. Biol. *107:* 67–78 (1969).

NYSTROM, B. and SKOGLUND, S.: Calibre spectra of spinal nerves and roots in newborn man. Acta morph. neerl. scand. *6:* 115–127 (1966).

OERTEL, G.: Zur Zyto- und Myeloarchitektonik des Rhombencephalon des Rhesus-Affen. J. Hirnforsch. *11:* 377–405 (1969).

OKA, M.; TOKUNAGA, T.; MURAO, T.; YOKOI, H.; OKUMURA, T.; HIRATA, T.; MIYASHITA, Y., and YOSHITATSU, S.: Trigemino-facial reflex, its evoked electromyographic study on several neurologic disorders. Med. J. Osaka Univ. *9:* 389–396 (1958).

OKADA, Y.: Fundamental researches on repetitive evoked EMG. Rhythmic fluctuations in amplitude of H- and M-waves evoked by repetitive stimulation. J. physiol. Soc. Japan *24:* 37–47 (1962).

OLSEN, P. Z. and DIAMANTOPOULOS, E.: Excitability of spinal motor neurones in normal subjects and patients with spasticity, parkinsonian rigidity and cerebellar hypotonia. J. Neurol. Neurosurg. Psychiat. *30:* 325–331 (1967).

OLSON, C. B.; CARPENTER, D. O., and HENNEMAN, E.: Orderly recruitment of muscle action potentials. Arch. Neurol., Chicago *19:* 591–597 (1968).

OLSON, C. B. and SWETT, C. P.: A functional and histochemical characterization of motor units in a heterogeneous muscle (flexor digitorum longus) of the cat. J. comp. Neurol. *128:* 475–498 (1966).

OLSON, C. B. and SWETT, C. P.: Effect of prior activity on properties of different types of motor units. J. Neurophysiol. *34:* 1–16 (1971).

OLSZEWSKI, J. and BAXTER, D.: Cytoarchitecture of the human brain stem, p. 199 (Karger, Basel 1954).

ORIOLI, F. L. and METTLER, F. A.: Rubrospinal tract in *Macaca mulatta.* J. comp. Neurol. *106:* 299–318 (1956).

ORIOLI, F. L. and METTLER, F. A.: Effect of rubrospinal tract section on ataxia. J. comp. Neurol. *107:* 305–313 (1957).

ORLOVSKY, G. N.: The effect of different descending systems on flexor and extensor activity during locomotion. Brain Res. *40:* 359–371 (1972).

ORLOVSKY, G. N.: Activity of vestibulospinal neurons during locomotion. Brain Res. *46:* 85–98 (1972).

ORLOVSKY, G. N.: Activity of rubrospinal neurons during locomotion. Brain Res. *46:* 99–112 (1972).

OSCARSSON, O.: Functional organization of the spino- and cuneocerebellar tracts. Physiol. Rev. *45:* 495–522 (1965).

OSCARSSON, O. and ROSEN, I.: Projection to cerebral cortex of large muscle spindle afferent in forelimb nerves of the cat. J. Physiol., Lond. *169:* 924–945 (1963).

O'SULLIVAN, D.J. and SWALLOW, M.: The fibre size and content of the radial and sural nerves. J. Neurophysiol. Neurosurg. Psychiat. *31:* 464–470 (1968).

OTT, K.H. and GASSEL, M.M.: Methods of tendon jerk reinforcement; the role of muscle activity in reflex excitability. J. Neurol. Neurosurg. Psychiat. *32:* 541–547 (1969).

PADYKULA, H.A. and GAUTHIER, G.F.: Morphological and cytochemical characteristics of fiber types in normal mammalian skeletal muscle; in MILHORAT Exploratory concepts in muscular dystrophy and related disorders, pp. 117–128 (Excerpta Medica Foundation, Amsterdam 1967).

PAILLARD, J.: Réflexes et régulations d'origine proprioceptive chez l'homme. Etude neurophysiologique et neuropsychologique (Arnette, Paris 1955).

PAILLARD, J.: Rapports entre les durées de la période de silence et du myogramme dans le triceps sural chez l'homme. J. Physiol., Paris *47:* 259–262 (1955).

PAILLARD, J.: Functional organization of afferent innervation of muscle studied in man by monosynaptic testing. Amer. J. phys. Med. *38:* 239–247 (1959).

PAILLARD, J.: The patterning of skilled movements; in MAGOUN and HALL Handbook of physiology, neurophysiology *3:* 1679–1708 (American Physiological Society, Washington 1960).

PAILLARD, J. and BROUCHON, M.: Active and passive movements in calibration of position sense; in FREEDMAN The neuropsychology of spatially oriented behavior (Dorsey Press, Illinois 1968).

PARTRIDGE, L.D.: Modification of neuronal output signals by muscles: frequency response study. J. appl. Physiol. *20:* 150–156 (1965).

PARTRIDGE, L.D.: Intrinsic feedback factors producing inertial compensation in muscle. Biophys. J. *7:* 853–863 (1967).

PARTRIDGE, L.D. and GLASER, G.H.: Adaptation in regulation of movement and posture. A study of stretch responses in spastic animals. J. Neurophysiol. *23:* 257–268 (1960).

PATTON, H.D.: Reflex regulation of movement and posture; in RUCH, PATTON, WOODBURY and TOWE Neurophysiology, pp. 181–206 (Saunders, Philadelphia 1965).

PEARCE, G.: Tecto-reticular fibres; in JASPER, PROCTOR, KNIGHTON, NOSHAY and COSTELLO Reticular formation of the brain, pp. 65–68 (Little, Brown, Boston 1958).

PEARCE, J.; HOSAN, A., and GALLAGHER, J.C.: Primitive reflex activity in primary and symptomatic Parkinsonism. J. Neurol. Neurosurg. Psychiat. *31:* 501–508 (1968).

PEDERSEN, E.: Studies on the central pathway of the flexion reflex in man and animal. Acta psychiat. scand., suppl. *88:* 1–81 (1954).

PEIPER, A.: Chronik der Kinderheilkunde; 3. Aufl., p. 458 (Thieme, Leipzig 1958).

PENDERS, C.A.: Contribution à l'étude électrophysiologique des réflexes faciaux chez l'homme. Unpublished memoir, Concours des Bourses de Voyage, Gouvernement Belge (1969).

PENDERS, C.A. et DELWAIDE, P.J.: Etude chez l'homme du cycle d'excitabilité de la réponse précoce du réflexe de clignement. C. R. Soc. Biol. *163:* 228–232 (1969).

PENDERS, C.A. et DELWAIDE, P.J.: Le réflexe de clignement chez l'homme. Particularités électrophysiologiques de la réponse précoce. Arch. int. Physiol. Biochem. *77:* 351–354 (1969).

PENDERS, C. A. and DELWAIDE, P. J.: Blink reflex studies in parkinsonism before and during therapy. J. Neurol. Neurosurg. Psychiat. *34:* 674–678 (1972).

PENFIELD, W.: Discussion of Dr. Phillips' paper; in ECCLES Brain and conscious experience, p. 412 (Springer, New York 1964).

PERL, E. R.: Somatic sensation: transfer and processing of information. 1. Peripheral receptors. Electroenceph. clin. Neurophysiol. *27:* 650–651 (1969).

PETAJAN, J. H. and PHILIP, B. A.: Frequency control of motorunit action potentials. Electroenceph. clin. Neurophysiol. *27:* 66–72 (1969).

PETRAS, J. M.: Cortical, tectal and tegmental fiber connections in the spinal cord of the cat. Brain Res. *6:* 275–324 (1967).

PETRAS, J. M. and LEHMAN, R. A. W.: Corticospinal fibers in the racoon. Brain Res. *3:* 195–197 (1966).

PHILLIPS, C. G.: Changing concepts of the precentral motor area; in ECCLES Brain and conscious experience, pp. 389–421 (Springer, New York 1966).

PHILLIPS, C. G.: Motor apparatus of the baboon's hand. Proc. roy. Soc. B *173:* 141–174 (1969).

PHILLIPS, C. G.: See discussion in EVARTS, BIZZI, BURKE, DELONG and THACH Central control of movement. Neurosc. Res. Progr. Bull. *9:* 135–138 (1971).

PHILLIPS, C. G.: Evolution of the corticospinal tract in primates with special reference to the hand; in BIEGERT and LEUTENEGGER Neurobiology, immunology, cytology; Proc. 3rd Int. Congr. Primatol., Zürich. vol. 2, pp. 2–23 (Karger, Basel 1971).

PHILLIPS, C. G. and PORTER, R.: The pyramidal projection to motoneurones of some muscle groups of the baboon's forelimb; in ECCLES and SCHADÉ Physiology of spinal neurons. Progr. Brain Res., 12: 222–242 (Elsevier, Amsterdam 1964).

PHILLIPS, C. G.; POWELL, T. P. S., and WIESENDANGER, M.: Projection from low-threshold muscle afferents of hand forearm to area 3a of baboon's cortex. J. Physiol., Lond. *217:* 419–446 (1971).

PIERROT-DESEILLIGNY, E.: Etude par le réflexe monosynaptique (réflexe H) de certains mécanismes impliqués dans la régulation tonique posturale normale et pathologique; thèse méd. Paris (1966).

PIERROT-DESEILLIGNY, E.; LACERT, P. et CATHALA, H.-P.: Amplitude et variabilité des réflexes monosynaptiques avant un mouvement volontaire. Physiol. Behav. *7:* 495–508 (1971).

PLATO: The symposium; transl. by W. HAMILTON, pp. 53, 58 (Penguin Books, London 1951).

PODIVINSKY, F.: Somatomotor response in torticollis. Confin. neurol. *33:* 231–243 (1971).

POMERANZ, B.; WALL, P. D., and WEBER, W. V.: Cord cells responding to fine myelinated afferents from viscera, muscle and skin. J. Physiol., Lond. *199:* 511–532 (1968).

POMPEIANO, O.: Analisi degli effetti della stimolazione elettrica del nucleo rosso nel gatto decerebrato. R. C. Accad. naz. Lincei Cl. Sci., Ser. VIII *22:* 100–103 (1957).

POMPEIANO, O.: Muscular afferents and motor control during sleep; in GRANIT Nobel Symposium 1, Muscular afferents and motor control, pp. 415–436 (Almqvist & Wiksell, Stockholm 1966).

POMPEIANO, O.: Functional organization of the cerebellar projections to the spinal cord; in FOX and SNIDER The cerebellum, Progr. Brain Res. *25:* 282–321 (Elsevier, Amsterdam 1967a).

POMPEIANO, O. and BRODAL, A.: Experimental demonstration of a somatotopical origin of rubrospinal fibers in the cat. J. comp. Neurol. *108:* 225–252 (1957).

POMPEIANO, O.: Sensory inhibition during motor activity in sleep; in PURPURA and YAHR Neurophysiological basis of normal and abnormal motor activities, pp. 323–373 (Raven Press, New York 1967b).

POMPEIANO, O. and WALBERG, F.: Descending connections to the vestibular nuclei. An experimental study in the cat. J. comp. Neurol. *108:* 465–503 (1957).

POPPELE, R. E. and BOWMAN, R. J.: Quantitative description of linear behavior of mammalian muscle spindles. J. Neurophysiol. *33:* 59–72 (1970).

POPPELE, R. E. and TERZUOLO, C. A.: Myotatic reflex: its input-output relation. Science *159:* 743–745 (1968).

PORTER, R.: Synaptic potentials in hypoglossal motoneurones. J. Physiol., Lond. *180:* 209–224 (1965).

PORTER, R.: Early facilitation at cortico-motoneuronal synapses. J. Physiol., Lond. *207:* 733–745 (1970).

PORTER, R.: Relationship of the discharges of cortical neurones to movement in free-to-move monkeys. Brain Res. *40:* 39–43 (1972).

PORTER, R. and HORSE, J.: Time course of minimal corticomotoneuronal excitatory post-synaptic potentials in lumbar motoneurons of the monkey. J. Neurophysiol. *32:* 443–451 (1969).

POURPRE, M. H.: Traitement neuro-chirurgical des contractures chez les paraplégiques post-traumatiques. Neuro-chir., Paris *6:* 229–236 (1960).

PRECHTL, H. F. R.; VLACH, V.; LENARD, H. G., and GRANT, D. K.: Exteroceptive and tendon reflexes in various behavioral states in newborn infants. Biol. Neonat. *11:* 159–175 (1967).

PRESTON, J. B.; SHENDE, M. C., and UEMARA, K.: The motor cortex. Pyramidal system: patterns of facilition and inhibition on motoneurons innervating limb musculature of cat and baboon and their possible adaptive significance; in YAHR and PURPURA Neurophysiological basis of normal and abnormal motor activities, pp. 61–72 (Raven Press, New York 1967).

PRESTON, J. B. and WHITLOCK, D. G.: Precentral facilitation and inhibition of spinal motoneurones. J. Neurophysiol. *23:* 154–170 (1960).

PRESTON, J. B. and WHITLOCK, D. G.: Intracellular potentials recorded from motoneurones following precentral gyrus stimulation in primate. J. Neurophysiol. *24:* 91–100 (1961).

PRESTON, J. B. and WHITLOCK, D. G.: A comparison of motor cortex effects on slow and fast muscle innervations in the monkey. Exp. Neurol. *7:* 327–341 (1963).

PRITCHARD, E. A. B.: The EMG of voluntary movements in man. Brain *53:* 344–375 (1930).

PROSKE, U. and LEWIS, D. M.: The effects of muscle stretch and vibration on fusimotor activity in the lightly anaesthetised cat. Brain Res. *46:* 55–59 (1972).

PROSSER, C. L. and HUNTER, W. S.: The extinction of the startle response and spinal reflexes in the white rat. Amer. J. Physiol. *117:* 609–618 (1936).

PROVINI, L.; REDMAN, S., and STRATA, P.: Somatotopic organization of mossy and climbing fibres to the anterior lobe of cerebellum activated by the sensorimotor cortex. Brain Res. *6:* 378–381 (1967).

PROVINS, K. A.: The effect of peripheral nerve block on the appreciation and execution of finger movements. J. Physiol., Lond. *143:* 55–67 (1958).

PRYS-ROBERTS, C.; CORBETT, J. L.; KERR, J. H.; CRAMPTON-SMITH, A., and SPALDING, J. M. K.: Treatment of sympathetic overactivity in tetanus. Lancet *i:* 542–545 (1969).

PURPURA, D. P. and SCHADÉ, J. P.: Growth and maturation of the brain. Progr. Brain Res. *4:* 1–289 (Elsevier, Amsterdam 1964).

PUTNAM, T. J.: Treatment of unilateral paralysis agitans by section of the lateral pyramidal tract. Arch. Neurol. Psychiat., Chicago 44: 950–976 (1940).

RACK, P. M. H.: The behaviour of a mammalian muscle during sinusoidal stretching. J. Physiol., Lond. *183:* 1–14 (1966).

RADEMAKER, G. G. J.: Das Stehen (Springer, Berlin 1931).

RADEMAKER, G. G. J.: On the lengthening and shortening reactions and their occurrence in man. Brain *70:* 109–126 (1947).

RALL, W.: Branching dendritic trees and motoneuron membrane resistivity. Exp. Neurol. *1:* 491–527 (1959).

RALL, W.: Theory of physiological properties of dendrites. Ann. N. Y. Acad. Sci. *96:* 1071–1092 (1962).

RALL, W.: Theoretical significance of dendritic trees for neuronal input-output relations; in REISS Neural theory and modeling, pp. 73–97 (Stanford University Press, Stanford 1964).

RALL, W.: Distinguishing theoretical synaptic potentials computed for different soma-dendritic distributions of synaptic input. J. Neurophysiol. *30:* 1138–1168 (1967).

RALL, W.: Cable properties of dendrites and effects of synaptic location; in ANDERSEN and JANSEN Excitatory synaptic mechanisms, pp. 175–187 (Universitetsforlaget, Oslo 1970).

RALL, W.; BURKE, R. E.; SMITH, T. G.; NELSON, P. G., and FRANK, K.: Dendritic location of synapses and possible mechanisms for the monosynaptic EPSP in motoneurons. J. Neurophysiol. *30:* 1169–1193 (1967).

RAMÓN-MOLINER, E.: An attempt at classifying nerve cells on the basis of their dendritic patterns. J. comp. Neurol. *119:* 211–227 (1962).

RAMON Y CAJAL, S.: Histologie du système nerveux de l'homme et des vertébrés, vol. 1 (Maloine, Paris 1909).

RAMON Y CAJAL, S.: Histologie du système nerveux de l'homme et des vertébrés, vol. 2 (Maloine, Paris 1911).

RANVIER, L.: De quelques faits relatifs à l'histologie et à la physiologie des muscles striés. Arch. physiol. norm. path. *6:* 1–15 (1874).

RASMUSSEN, A. T.: Secondary vestibular tracts in the cat. J. comp. Neurol. *54:* 143–171 (1932).

REED, A. F.: The nuclear masses in the cervical spinal cord of *Macaca mulatta.* J. comp. Neurol. *72:* 187–206 (1940).

REIS, D. J.: The palmomental reflex. Arch. Neurol., Chicago *4:* 486–495 (1961).

RENOU, G.: Electromyographie unitaire du tremblement parkinsonien; thèse méd. Paris (1971).

RENOU, G.; RONDOT, P., et METRAL, S.: Analyse de décharges itératives d'une même unité motrice dans les bouffées de tremblement. Rev. neurol. *122:* 420–423 (1970).

RENSHAW, B.: Influence of discharge of motoneurones upon excitation of neighbouring motoneurones. J. Neurophysiol. *4:* 167–183 (1941).

RENSHAW, B.: Central effects of centripetal impulses in axons of spinal ventral roots. J. Neurophysiol. *9:* 190–204 (1946).

REQUIN, J.: Some data on neurophysiological processes involved in the preparatory motor activity to reaction time performance. Acta psychol., Amst. *30:* 358–367 (1969).

REQUIN, J. and PAILLARD, J.: Depression of spinal monosynaptic reflexes as a specific aspect of preparatory motor set in visual reaction time. Communic. Symp. Visual information processing and control of motor activity (Sofia 1969).

RÉTHELYI, M.: The Golgi architecture of Clarke's column. Acta morph. Acad. Sci. hung. *16:* 311–330 (1968).

RÉTHELYI, M.: Ultrastructural synaptology of Clarke's column. Exp. Brain Res. *11:* 159–174 (1970).

RÉTHELYI, M.: Cell and neuropil architecture of the intermediolateral (sympathetic) nucleus of cat spinal cord. Brain Res. *46:* 203–213 (1972).

RÉTHELYI, M. and SZENTÁGOTHAI, J.: The large synaptic complexes of the substantia gelatinosa. Exp. Brain Res. *7:* 258–274 (1969).

RÉTHELYI, M. and SZENTÁGOTHAI, J.: Distribution and connections of afferent fibers in the spinal cord; in IGGO Handbook of sensory physiology, vol. 2 (Springer, Berlin, in press 1972).

REXED, B.: Contribution to the knowledge of the postnatal development of the peripheral nervous system in man. Acta psychiat. scand., suppl. *33:* 1–206 (1944).

REXED, B.: The cytoarchitectonic organization of the spinal cord in the cat. J. comp. Neurol. *96:* 415–495 (1952).

REXED, B.: A cytoarchitectonic atlas of the spinal cord in the cat. J. comp. Neurol. *100:* 297–379 (1954).

REXED, B. and THERMAN, P.O.: Calibre spectra of motor and sensory fibres to flexor and extensor muscles. J. Neurophysiol. *11:* 133–139 (1948).

RIDDOCH, G.: The reflex functions of the completely divided spinal cord in man, compared with those associated with less severe lesions. Brain *40:* 264–402 (1917).

RINVIK, E. and WALBERG, F.: Demonstration of a somatotopically arranged cortico-rubral projection in the cat. J. comp. Neurol. *121:* 393–407 (1963).

RISPAL-Padel, L.; LATREILLE, J. et VANUXEM, P.: Répartition sur le cortex moteur des projections des différents noyaux cérébelleux chez le chat C.R. Acad. Sci. *272:* 451–454 (1971).

RISPAL-PADEL, L. and MASSION, J.: Relations between the ventrolateral nucleus and the motor cortex in the cat. Exp. Brain Res. *10:* 331–339 (1970).

RISTOW, W. und STRUPPLER, A.: Labyrinthausschaltung und tonische Innervation. Hals-Nas.-Ohrenarzt *189:* 291–295 (1964).

ROBERTS, T.D.M.: Neurophysiology of postural mechanisms (Butterworth, London 1967).

ROMANES, G.J.: The motor cell columns of the lumbosacral spinal cord of the cat. J. comp. Neurol. *94:* 313–364 (1951).

ROMANUL, F.C.A.: Enzymes in muscle. I. Histochemical studies of individual skeletal muscle fibers of the rat. Arch. Neurol., Chicago *11:* 355–368 (1964).

ROMBERG, M.H.: A manual of the nervous diseases of man (New Sydenham Society, London 1853).

RONDOT, P.: Les contractures. Etude clinique et physio-pathologique. Encéphale *3–4:* 242–285, 287–332 (1968).

RONDOT, P.: Etude clinique et physiopathologique des contractures. Rev. neurol. *118:* 321–342 (1968).

RONDOT, P. et RENOU, G.: Bases physiologiques de la régulation motrice extrapyramidale. Mécanismes périphériques. Réunion neurologique franco-allemande, Munich 1970.

RONDOT, P.; DALLOZ. J.-C. et TARDIEU, G.: Mesure de la force des réactions musculaires à l'étirement passif au cours des raideurs pathologiques par lésions cérébrales. Rev. franç. Et. clin. biol. *3:* 585–592 (1958).

RONDOT, P. et SCHERRER, J.: Contraction réflexe provoquée par le raccourcissement passif du muscle dans l'athéthose et les dystonies d'attitude. Rev. neurol. *114:* 329–337 (1966).

ROSÉN, I. and ASANUMA, H.: Peripheral afferent inputs to the forelimb area of monkey motor cortex: input-output relations. Exp. Brain Res. *14:* 257–273 (1972).

ROSENBERG, M. E.: Synaptic connexions of alpha extensor motoneurones with ipsilateral and contralateral cutaneous afferents. J. Physiol., Lond. *207:* 231–255 (1970).

ROSENTHAL, N. P.; MCKEAN, T. A.; ROBERTS, W. J., and TERZUOLO, C. A.: Frequency analysis of stretch reflex and its main subsystems in triceps surae muscles of the cat. J. Neurophysiol. *33:* 713–749 (1970).

ROSSI, G. and CORTESINA, G.: Morphological study of the laryngeal muscles in man. Insertions and courses of the muscle fibres, motor end-plates and proprioceptors. Acta oto-laryng., Stockh. *59:* 575–592 (1965).

ROTHBALLER, B. A.: Experience with longitudinal frontal cordotomy for the relief of mass spasms. Proc. 17th V.A. Spinal Cord Injury Conf., pp. 82–87 (New York 1969).

ROTHMANN, M.: Über die physiologische Wertung der corticospinalen (Pyramiden-)Bahn. Zugleich ein Beitrag zur Frage der elektrischen Reizbarkeit und Funktion der Extremitätenregion der Grosshirnrinde. Arch. Anat. Physiol. (physiol. Abt.) *31:* 217–275 (1907).

RUDOLPH, G.: Spiral nerve-endings (proprioceptors) in the human vocal muscle. Nature, Lond. *190:* 726–727 (1961).

RUDOMIN, P.: Presynaptic inhibition induced by vagal afferent volleys. J. Neurophysiol. *30:* 964–981 (1967).

RUDOMIN, P. and DUTTON, H.: Effects of conditioning afferent volleys on variability of monosynaptic response of extensor motoneurones. J. Neurophysiol. *32:* 140–157 (1969).

RUDOMIN, P. and DUTTON, H.: Effects of muscle and cutaneous afferent nerve volleys on excitability fluctuations of Ia terminals. J. Neurophysiol. *32:* 158–169 (1969).

RUFFINI, A.: On the minute anatomy of the neuromuscular spindles of the cat, and on their physiological significance. J. Physiol., Lond. *23:* 190–208 (1898).

RUSHWORTH, G.: Spasticity and rigidity: an experimental study and review. J. Neurol. Neurosurg. Psychiat. *23:* 99–118 (1960).

RUSHWORTH, G.: Observations on blink reflexes. J. Neurol. Neurosurg. Psychiat. *25:* 93–108 (1962).

RUSHWORTH, G.: Some aspects of the pathophysiology of spaticity and rigidity. Clin. Pharmacol. Ther. *5:* 828–836 (1964a).

RUSHWORTH, G.: The pathophysiology of spasticity. Proc. roy. Soc. B *57:* 715–720 (1964b).

RUSHWORTH, G.: Some studies on the pathophysiology of spasticity. Paraplegia *4:* 130–141 (1966a).

RUSHWORTH, G.: Some functional properties of deep facial afferents; in ANDREW Control and innervation of skeletal muscle (Livingstone, Edinburgh 1966b).

RUSHWORTH, G.: Diagnostic value of the electromyohraphic study of reflex activity in man. Electroenceph. clin. Neurophysiol. *25:* 65–71 (1967).

RUSHWORTH, G.: Neurophysiologie clinique de quelques réflexes dans le domaine du nerf facial. Electromyography *8:* 349–366 (1968).

RUSHWORTH, G. and YOUNG, R. R.: The effect of vibration on tonic and phasic reflexes in man. J. Physiol., Lond. *185:* 63–64 (1966).

RUSTIONI, A.; KUYPERS, H. G. J. M., and HOLSTEGE, G.: Propriospinal projections from the ventral and lateral funiculi to the motoneurons in the lumbosacral cord of the cat. Brain Res. *35:* 255–275 (1971).

RYALL, R. W.: Renshaw cell mediated inhibition of Renshaw cells: patterns of excitation and inhibition from impulses in motor axon collaterals. J. Neurophysiol. *33:* 257–270 (1970).

RYALL, R. W. and PIERCEY, M. F.: Excitation and inhibition of Renshaw cells by impulses in peripheral afferent nerve fibres. J. Neurophysiol. *34:* 242–251 (1971).

RYALL, R. W.; PIERCEY, M. F., and POLOSA, C.: Intersegmental and intrasegmental distribution of mutual inhibition of Renshaw cells. J. Neurophysiol. *34:* 700–707 (1971).

RYALL, R. W.; PIERCEY, M. F.; POLOSA, C., and GOLDFARB, J.: Excitation of RENSHAW cells in relation to orthodromic and antidromic excitation of motoneurones. J. Neurophysiol. *35:* 137–148 (1972).

SALEM, M. R.; BARAKA, A.; RATTENBORG, C. C., and HOLADAY, D. A.: Treatment of hiccups by pharyngeal stimulation in anesthetized and conscious subjects. J. amer. med. Ass. *202:* 32–36 (1967).

SAMOJLOFF, A. und KISSELEFF, M.: Die Verkürzungs- und Verlängerungsreaktion des Knieextensors der decerabrierten Katze. Pflügers Arch. ges. Physiol. *218:* 267–284 (1927).

SAMUELS, L.: Hiccup: a ten year review of anatomy, etiology and treatment. Canad. med. Ass. J. *67:* 315–322 (1952).

SANDLER, S. G.; TOBIN, W., and HENDERSON, E. S.: Vincristine-induced neuropathy. Neurology, Minneap. *19:* 367–374 (1969).

SASAKI, K.; NAMIKAWA, A., and HASHIRAMOTO, S.: The effect of midbrain stimulation upon alpha motoneurones in lumbar spinal cord of the cat. Jap. J. Physiol. *10:* 303–316 (1960).

SASAKI, K. and OTANI, T.: Accomodation in spinal motoneurones of the cat. Jap. J. Physiol. *11:* 443–456 (1961).

SASAKI, K. and TANAKA, T.: Phasic and tonic innervation of spinal motoneurones from upper brain stem centers. Jap. J. Physiol. *14:* 56–66 (1964).

SASAKI, K.; TANATA, T., and MORI, K.: Effects of stimulation of pontine and bulbar reticular formation upon spinal motoneurons of the cat. Jap. J. Physiol. *12:* 45–62 (1962).

SAUERLAND, E. K.; NAKAMURA, Y., and CLEMENTE, C. D.: The role of the lower brain stem in cortically induced inhibition of somatic reflexes in the cat. Brain Res. *6:* 164–180 (1967).

SCHAEFER, E. A.: Textbook of physiology. The cerebral cortex, pp. 697–782 (Pentland, London 1900).

SCHALTENBRAND, G. und HUFSCHMIDT, H. J.: Myographische Analyse des Parkinsonsyndroms 1er Congr. Int. Sci. Neurol. Brussels, vol. 1, pp. 94–99 (1957).

SCHEIBEL, A. B.: Axonal afferent patterns in the bulbar reticular formation. Anat. Rec. *121:* 361 (1955).

SCHEIBEL, M. E. and SCHEIBEL, A. B.: Structural substrates for integrative patterns in the brain stem reticular core; in JASPER, PROCTOR, KNIGHTON, NOSHAY and COSTELLO Reticular formation of the brain, pp. 31–55 (Little, Brown, Boston 1958).

SCHEIBEL, M. E. and SCHEIBEL, A. B.: Spinal motoneurones, interneurones and Renshaw cells. A Golgi study. Arch. ital. Biol. *104:* 328–353 (1966a).

SCHEIBEL, M.E. and SCHEIBEL, A.B.: Terminal axonal patterns in cat spinal cord. I. The lateral corticospinal tract. Brain Res. *2:* 333–350 (1966b).

SCHEIBEL, M.E. and SCHEIBEL, A.B.: Terminal axonal patterns in cat spinal cord. II. The dorsal horn. Brain Res. *9:* 32–58 (1968).

SCHEIBEL, M.E. and SCHEIBEL, A.B.: Terminal patterns in cat spinal cord. III. Primary afferent collaterals. Brain Res. *13:* 417–443 (1969).

SCHEIBEL, M.E. and SCHEIBEL, A.B.: Developmental relationship between spinal moto-neuron dendrite bundles and patterned activity in the hind limb of cats. Exp. Neurol. *29:* 328–335 (1970).

SCHEIBEL, M.E. and SCHEIBEL, A.B.: Inhibition and the Renshaw cells. A structural critique. Brain, Behav. Evol. *4:* 53–93 (1971a).

SCHEIBEL, M.E. and SCHEIBEL, A.B.: Developmental relationship between spinal moto-neuron dentrite bundles and patterned activity in the forelimb of cats. Exp. Neurol. *30:* 367–373 (1971b).

SCHENCK, E. und KOEHLER, B.: Über eine reflektorische Hemmung in der Willkürinnerva-tion von Beugemuskeln beim Menschen. Pflügers. Arch. ges. Physiol. *251:* 404–412 (1949).

SCHIMERT, J.: Die Endigungsweise des Tractus vestibulospinalis. Z. Anat. EntwGesch. *108:* 761–767 (1938).

SCHIMERT, J.: Das Verhalten der Hinterwurzelkollateralen im Rückenmark. Z. Anat. EntwGesch. *109:* 665–687 (1939).

SCHLEGEL, H.-J. und SONNTAG, K.-H.: Aktivitätssteigerungen in primären Flexor-Muskel-spindelafferenzen der Katze bei tetanischer Reizung von Streckreflexafferenzen. Pflügers Arch. ges. Physiol. *311:* 159–167 (1969).

SCHLEGEL, H.-J. und SONNTAG, K.-H.: Reflektorische Aktivierung prätibialer Fusimoto-neurone der Katze durch Reizung niedrigschwelliger antagonistischer Muskelafferen-zen. Pflügers Arch. ges. Physiol. *319:* 200–204 (1970).

SCHOEN, J.H.R.: Comparative aspects of the descending fibre systems in the spinal cord; in ECCLES and SCHADÉ Organization of the spinal cord. Progr. Brain Res. *11:* 203–222 (Elsevier, Amsterdam 1964).

SCHOEN, J.H.R.: Het aspect van de rubrospinale baan bij de mens. Ned. T. Geneesk. *113:* 680 (1969a).

SCHOEN, J.H.R.: The corticofugal projection on the brain stem and spinal cord in man. Psychiat. Neurol. Neurochir. *72:* 121–128 (1969b).

SCHRIVER, J.E. and NOBACK, C.R.: Cortical projections to the lower brain stem and spinal cord in the tree shrew *(Tupaia glis)*. J. comp. Neurol. *130:* 25–54 (1967).

SCHULLER, A.: Experimentelle Pyramidendurchschneidung beim Hunde und beim Affen. Wien klin. Wschr. *19:* 57–62 (1906).

SCHULTE, F.J.; LINKE, I.; MICHAELIS, R., and NOLTE, R.: Electromyographic evaluation of the Moro-reflex in preterm, term and small-for-date newborn infants. Develop. Psychol. *1:* 41–47 (1968).

SCHULTE, F.J.; MICHAELIS, R.; LINKE, I., and NOLTE, R.: Motor nerve conduction velocity in term, preterm and small-for-date newborn infants. Pediatrics *42:* 17–26 (1968).

SCHULTZ, A.H.: Some factors influencing the social life of primates in general and of early man in particular, in WASHBURN Social life of early man, pp. 58–90 (Aldine, Chicago 1961).

SCHWEITZER, A. and WRIGHT, S.: Action of nicotine on the spinal cord. J. Physiol., Lond. *94:* 136–147 (1938).

SCHWERIN, O.: Untersuchungen über den Entlastungsreflex des Menschen. Dtsch. Z. Nervenheilk. *140:* 240–244 (1936).

SEARS, T. A.: Electrical activity in expiratory muscles of the cat during inflation of the chest. J. Physiol., Lond. *142:* 35P (1958).

SEARS, T. A.: The activity of the small motor fibre system innervating respiratory muscles of the cat. Austr. J. Sci. *25:* 102 (1962).

SEARS, T. A.: Activity of fusimotor fibres innervating muscle spindles in the intercostal muscles of the cat. Nature, Lond. *197:* 1013–1014 (1963a).

SEARS, T. A.: Investigations on respiratory motoneurones; Ph. D. thesis, Australian National University (1963).

SEARS, T. A.: Investigations on respiratory motoneurones of thoracic spinal cord; in ECCLES and SCHADÉ Physiology of spinal neurons. Progr. Brain Res. *12:* 259–273 (Amsterdam, Elsevier 1964a).

SEARS, T. A.: The fibre calibre spectra of sensory and motor fibres in the intercostal of the cat. J. Physiol., Lond. *172:* 150–161 (1964b).

SEARS, T. A.: Efferent discharges in alpha and fusimotor fibres of intercostal nerves of the cat. J. Physiol., Lond. *174:* 295–315 (1964c).

SEARS, T. A.: Some properties and reflex connexions of respiratory motoneurones of the cat's thoracic spinal cord. J. Physiol., Lond. *175:* 386–403 (1964d).

SEARS, T. A.: The slow potentials of thoracic respiratory motoneurones and their relation to breathing. J. Physiol., Lond. *175:* 404–424 (1964e).

SEARS, T. A.: Pathways of supraspinal origin regulating the activity of respiratory motoneurones; in GRANIT Nobel Symposium 1, Muscular afferents and motor control, pp. 187–196 (Almqvist & Wiksell, Stockholm 1966).

SEARS, T. A.: Breathing, a sensori-motor act; in GILLILAND and FRANCIS Scientific basis of medicine annuals reviews, pp. 129–147 (Athlone Press, London 1971).

SEARS, T. A. and NEWSOM DAVIS, J.: The control of respiratory muscles during voluntary breathing. Conf. Sound Production in Man. Ann. N. Y. Acad. Sci. *155:* 183–190 (1968).

SEVERIN, F. V.; ORLOVSKII, G. N., and SHIK, M. L.: Work of the muscle receptors during controlled locomotion. Biophysics *12:* 575–586 (1967).

SEVERIN, F. V.; ORLOVSKII, G. N., and SHIK, M. L.: Reciprocal influences on work of single motoneurones during controlled locomotion. Bull. exp. Biol. Med. *66:* 713–716 (1968).

SHAHANI, B. T.: Effects of sleep on human reflexes with a double component. J. Neurol. Neurosurg. Psychiat. *31:* 574–579 (1968).

SHAHANI, B. T.: Flexor reflex afferent nerve fibres in man. J. Neurol. Neurosurg. Psychiat. *33:* 786–791 (1970a).

SHAHANI, B. T.: The human blink reflex. J. Neurol. Neurosurg. Psychiat. *33:* 792–800 (1970b).

SHAHANI, B. T.; BURROWS, P., and WHITTY, C. W. M.: The grasp reflex and perseveration. Brain *93:* 181–192 (1970).

SHAHANI, B. T. and YOUNG, R. R.: A note on blink reflexes. J. Physiol., Lond. *198:* 103P–104P (1968).

SHAHANI, B. T. and YOUNG, R. R.: Human flexor reflexes. J. Neurol. Neurosurg. Psychiat. *34:* 616–627 (1971).

SHAHANI, B.T. and YOUNG, R.R.: Human orbicularis oculi reflexes. Neurology, Minneap. 22: 149–154 (1972).

SHAHANI, B.T. and YOUNG, R.R.: The cutaneous nature of the first component of the monkey's blink reflex. Neurology 22: 438 (1972).

SHAPOVALOV, A.I.: Excitation and inhibition of spinal neurons during supraspinal stimulation; in GRANIT Nobel Symposium 1, Muscular afferents and motor control, pp. 331–348 (Almqvist & Wiksell, Stockholm 1966).

SHAPOVALOV, A.I.: Post-tetanic potentiation of monosynaptic and disynaptic actions from supraspinal structures on lumbar motoneurons. J. Neurophysiol. 32: 948–959 (1969).

SHAPOVALOV, A.I.: Extrapyramidal monosynaptic and disynaptic control of mammalian alpha-motoneurons. Brain Res. 40: 105–116 (1972).

SHAPOVALOV, A.I.; GRANTYN, A.A., and KURCHAVYI, G.G.: Short-latency reticulospinal synaptic projections to alpha-motoneurons. Bull. exp. Biol. Med. USSR 64: 3–15 (1967).

SHAPOVALOV, A.I. and GUREVITCH, N.R.: Monosynaptic and disynaptic reticulospinal actions on lumbar motoneurons of the rat. Brain Res. 21: 249–263 (1970).

SHAPOVALOV, A.I.; KARAMYAN, O.A.; KURCHAVYI, G.G., and REPINA, Z.A.: Synaptic actions evoked from the red nucleus on the spinal alpha-motoneurons in the rhesus monkey. Brain Res. 32: 325–348 (1971).

SHAPOVALOV, A.I.; KARAMJAN, O.A.; TAMAROVA, Z.A., and KURCHAVYI, G.G.: Cerebello-rubrospinal effects of hindlimb motoneurons in the monkey. Brain Res. 47: 49–59 (1972).

SHAPOVALOV, A.I.; KURCHAVYI, G.G.; KARAMYAN, O.A., and REPINA, Z.A.: Extrapyramidal pathways with monosynaptic effects upon primate α-motoneurons. Experientia 27: 522–524 (1971).

SHAPOVALOV, A.I.; KURCHAVYI, G.G., and STROGANOVA, M.P.: Synaptic mechanisms of vestibulospinal influences on alpha-motoneurones. Sechenov J. Physiol. USSR 52: 1401–1409 (1966).

SHAPOVALOV, A.I.; TAMAROVA, Z.A.; KARAMYAN, O.A., and KURCHAVYI, G.G.: Reticulospinal and vestibulospinal synaptic actions on lumbar motoneurones of the monkey. Neurofisiologia USSR 3: 408–417 (1971).

SHARRARD, W.J.W.: The distribution of the permanent paralysis in the lower limb in poliomyelitis. J. Bone Jt Surg. 37: 540–558 (1955).

SHERRINGTON, C.S.: Decerebrate rigidity, and reflex coordination of movements. J. Physiol., Lond. 22: 319–322 (1898).

SHERRINGTON, C.S.: The muscular sense, in SCHAEFER Textbook of physiology, vol. 2, pp. 1002–1025 (Pentland, London 1900).

SHERRINGTON, C.S.: The integrative action of the nervous system (Scribners, New York 1906).

SHERRINGTON, C.S.: On plastic tonus and proprioceptive reflexes. Quart. J. exp. Physiol. 2: 109–156 (1909).

SHERRINGTON, C.S.: Flexion-reflex of the limb, crossed extension-reflex and reflex stepping and standing. J. Physiol., Lond. 40: 28–121 (1910).

SHERRINGTON, C.S.: Break-shock reflexes and 'supramaximal' contraction – response of mammalian nerve-muscle to single shock stimuli. Proc. roy. Soc. B 92: 245–258 (1921).

SHERRINGTON, C.S.: Remarks on some aspects of reflex inhibition. Proc. roy. Soc. B 97: 519–545 (1925).

SHIK, M.L.; ORLOVSKII, G.N., and SEVERIN, F.V.: Locomotion of the mesencephalic cat elicited by stimulation of the pyramids. Biofyzika *13:* 127–135 (1968).

SHIMAMURA, M. and LIVINGSTON, R.B.: Longitudinal conduction systems serving spinal and brainstem coordination. J. Neurophysiol. *26:* 258–272 (1963).

SHIMAMURA, M.; MORI, S.; MATSUSHIMA, S., and FUJIMORI, B.: On the spino-bulbo-spinal reflex in dogs, monkeys and man. Jap. J. Physiol. *14:* 411–421 (1964).

SHIMAMURA, M. and YAMAUCHI, T.: Neural mechanisms of the chloralose jerk with special reference to its relationship with the spino-bulbo-spinal reflex. Jap. J. Physiol. *17:* 738–745 (1967).

SHIMAZU, H.; HONGO, T., and KUBOTA, K.: Two types of central influences on gamma motor system. J. Neurophysiol. *25:* 309–323 (1962).

SHIMAZU, H.; HONGO, T.; KUBOTA, K., and NARABAYASHI, H.: Rigidity and spasticity in man. Electromyographic analysis with reference to the role of the globus pallidus. Arch. Neurol., Chicago *6:* 10–17 (1962).

SHIMIZU, A.; YAMADA, Y.; YAMAMOTO, J.; FUJIKI, A., and KANEKO, Z.: Pathways of descending influences on H-reflex during sleep. Electroenceph. clin. Neurophysiol. *20:* 337–347 (1966).

SHULEIKINA, K.V. and GLAKOVICH, N.G.: Synaptic endings of reticular fibers on motor neurons of the human embryonic spinal cord. Bull. exp. Biol. Med. USSR *59:* 106–111 (1965).

SIE PEK GIOK: Localization of fibre systems within the white matter of the medulla oblongata and the cervical cord in man; thesis, Leiden (1956).

SIGWALD, J. and RAVERDY, P.: Muscle tone; in VINKEN and BRUYN Handbook of clinical neurology, *1:* 257–276 (North Holland, Amsterdam 1969).

SILVETTE, H.; HOFF, E.C.; LARSON, P.S., and HAAG, H.B.: The actions of nicotine on central nervous system functions. Pharmacol. Rev. *14:* 137–173 (1962).

SIMON, J.N.: Dispositif de contention des électrodes de stimulation pour l'étude du réflexe de Hoffmann chez l'homme. Electroenceph. clin. Neurophysiol., suppl. *22:* 174–176 (1962).

SIMONS, D.G. and BINGEL, A.G.A.: Quantitative comparison of passive motion and tendon reflex responses in biceps and triceps brachii muscles in hemiplegic or hemiparetic man. Stroke *2:* 58–66 (1971).

SKOGLUND, S.: Anatomical and physiological studies of knee joint innervation in the cat. Acta physiol. scand. *36:* suppl. 124: 3–101 (1956).

SKOGLUND, S.: The activity of muscle receptors in the kitten. Acta physiol. scand. *50:* 203–221 (1960a).

SKOGLUND, S.: Central connections and functions of muscle nerves in the kitten. Acta physiol. scand. *50:* 222–237 (1960b).

SKOGLUND, S.: The reaction of tetanic stimulation of the two neuron arcs in the kitten. Acta physiol. scand. *50:* 238–253 (1960c).

SKOGLUND, S.: Muscle afferents and motor control in the kitten; in GRANIT Nobel Symposium 1, Muscle afferents and motor control, pp. 245–259 (Almqvist & Wiksell, Stockholm 1966).

SKOGLUND, S.: Growth and differentiation with special emphasis on the central nervous system. Ann. Rev. Physiol. *31:* 19–42 (1969).

SLINGER, R.T. and HORSLEY, V.: Upon orientation of points in space by the muscular arthroidal and tactile senses of the upper limbs in normal individuals. Brain *29:* 1–27 (1906).

SMITH, A. M.: Effects of rubral area lesions and stimulation on conditioned forelimb flexion responses in the cat. Brain Res. *24:* 549 (1970).

SMITH, O. C.: Action potentials from single motor units in voluntary contraction. Amer. J. Physiol. *108:* 629–638 (1934).

SMITH, R. S.: Properties of intrafusal muscle fibres; in GRANIT Nobel Symposium 1, Muscular afferents and motor control (Almqvist & Wiksell, Stockholm 1966).

SOMJEN, G.; CARPENTER, D. O., and HENNEMAN, E.: Responses of motoneurons of different sizes to graded stimulation of supraspinal centers of the brain. J. Neurophysiol. *28:* 958–965 (1965).

SOMMER, J.: Der Entlastungsreflex des menschlichen Muskels. Dtsch. Z. Nervenheilk. *150:* 83–92 (1939).

SOMMER, J.: Periphere Bahnung von Muskeleigenreflexen als Wesen des Jendrassikschen Phänomens. Dtsch. Z. Nervenheilk. *150:* 249–262 (1940).

SORBYE, R.: Reflex hammer for registering and analysis of reflexes in animals. Nord. Med. *75:* 548 (1966).

SORIANO, D. and HERMAN, R.: Radiofrequency cordotomy for the relief of spasticity in decerebrate cats. J. Neurol. Neurosurg. Psychiat. *34:* 628–636 (1971).

SPENCER, W. R.; THOMPSON, R. F., and NEILSON, D. R.: Response decrement of the flexion reflex in the acute spinal cat and transient restoration by strong stimuli. J. Neurophysiol. *29:* 221–239 (1966).

SPERRY, R. W.: Cerebral regulation of motor coordination in monkeys following multiple transection of sensorimotor cortex. J. Neurophysiol. *10:* 275–294 (1947).

SPERRY, R. W.: Neural basis of spontaneous optokinetic response produced by visual inversion. J. comp. physiol. Psychol. *43:* 482–489 (1950).

SPERRY, R. W.: A modified concept of consciousness. Psychol. Rev. *76:* 532–536 (1969).

SPERRY, R. W.; GAZZANIGA, M. S., and BOGEN, J. E.: Interhemispheric relationships; the neocortical commissures; syndromes of hemisphere disconnection; in VINKEN and DE BRUYN Handbook of clinical neurology *4:* 273–290 (North-Holland, Amsterdam. 1968).

SPRAGUE, J. M.: A study of motor cell localization in the spinal cord of the rhes snmonkey. Amer. J. Anat. *82:* 1–26 (1948).

SPRAGUE, J. M. and CHAMBERS, W. W.: Control of posture by reticular formation and cerebellum in the intact, anaesthetized and unanaesthetized and in the decerebrate cat. Amer. J. Physiol. *176:* 52–64 (1954).

STALBERG, E.: Propagation velocity in human muscle fibers *in situ.* Acta physiol. scand. *70:* suppl. 287: 1–112 (1941).

STALBERG, E.; EKSTEDT, J., and BROMAN, A.: The electromyographic jitter in normal human muscles. Electroenceph. clin. Neurophysiol. *31:* 429–438 (1971).

STALBERG, E. and TRONTELJ, J. V.: Demonstration of axon reflexes in human motor nerve fibres. J. Neurol. Neurosurg. Psychiat. *33:* 571–579 (1970).

STEG, G.: The function of muscle spindles in spasticity and rigidity. Acta neurol. scand. *39:* 53–59 (1962).

STEG, G.: Efferent muscle innervation and rigidity. Acta physiol. scand. *61:* suppl. 225: 1–53 (1964).

STEIN, R. B. and MATTHEWS, P. B. C.: Difference in variability of discharge frequency between primary and secondary muscle spindle afferent endings of the cat. Nature, Lond. *208:* 1217–1218 (1965).

STEINBRECHER, W.: Die Wirkung von Muskelrelaxation bei zerebraler und spinaler Spastik. Dtsch. med. Wschr. *84:* 2295–2298 (1959).

STEINBRECHER, W.: Elektromyographie in Klinik und Praxis (Thieme, Stuttgart 1965).

STELTER, W. J.; SPAAN, G. und KLUSSMAN, F. W.: Der Einfluss der spinalen und peripheren Temperatur auf die Reflexspannung «roter» und «blasser» Muskeln. Pflügers Arch. ges. Physiol. *312:* 1–17 (1969).

STERLING, P. and KUYPERS, H.G.J.M.: Anatomical organization of the brachial spinal cord of the cat. I. The distribution of dorsal root fibers. Brain Res. *4:* 1–15 (1967a).

STERLING, P. and KUYPERS, H.G.J.M.: Anatomical organization of the brachial spinal cord of the cat. II. The motoneuron plexus. Brain Res. *4:* 16–32 (1967b).

STERLING, P. and KUYPERS, H.G.J.M.: Anatomical organization of the brachial spinal cord of the cat. III. The propriospinal connections. Brain Res. *7:* 419–443 (1968).

STERN, K.: Note on nucleus ruber magnocellularis and its efferent pathway in man. Brain *61:* 284–289 (1938).

STEVENS, S. S.: On the psychophysical law. Psychol. Rev. *64:* 153–181 (1957).

STIMSON, C. W.; KHEDER, N.; HICKS, R. G., and ORLANDO, R.: Nerve conduction velocity and H-reflex studies in two groups of severely retarded children. Arch. phys. Med. Rehabil. *50:* 626–631 (1969).

STRUPPLER, A. und BOBBELSTEIN, H.: Elektromyographische Untersuchung des Glabella-reflexes bei verschiedenen Störungen. Nervenarzt *34:* 347–352 (1963).

STRUPPLER, A.; LANDAU, W. M. und MEHLS, H.: Analyse des Entlastungsreflexes (ER) am Menschen. Pflügers Arch. ges. Physiol. *279:* 18–19 (1964).

STRUPPLER, A.; LANDAU, W. M. und MEHLS, H. O.: Analyse des Entlastungsreflexes am Menschen. Pflügers Arch. ges. Physiol. *313:* 155–167 (1969).

STRUPPLER, A. und SCHENCK, E.: Der sogenannte Entlastungsreflex bei zerebellaren und anderen Ataxien. Fortschr. Neurol. Psychiat. *26:* 421–429 (1958).

STRUPPLER, A. und STRUPPLER, E.: Neurologische Untersuchungen an deafferentierten Muskeln des Menschen. Z. Biol. *11:* 438–448 (1960).

STRUPPLER, A.; STRUPPLER, E., and ADAMS, R. D.: Local tetanus in man. Arch. Neurol., Chicago *8:* 162–178 (1963).

STUART, D. G.; GOSLOW, G. E.; MOSHER, C. G., and REINKING, R. M.: Stretch responsiveness of Golgi tendon organs. Exp. Brain Res. *10:* 463–476 (1970).

STUART, D. G.; MOSHER, C. G., and GERLACH, R. L.: Properties and central connections of Golgi tendon organs with special reference to locomotion; in BANKER, PRZYBYLSKI, VAN DER MEULEN and VICTOR Research in muscle development and the muscle spindle, part II, pp. 437–467 (Excerpta Medica, Amsterdam 1972).

STUART, D. G.; MOSHER, C. G.; GERLACH, R. L., and REINKING, R. M.: Selective activation of Ia afferents by transient muscle stretch. Exp. Brain Res. *10:* 477–487 (1970).

STUART, D. G.; OTT, K.; ISHIKAWA, K., and ELDRED, E.: Muscle receptor responses to sinusoidal stretch. Exp. Neurol. *13:* 82–95 (1965).

SUHREN, O.; BRUYN, G. W., and TUYMAN, J. A.: Hyperexplexia and hereditary startle syndrome. J. neurol. Sci. *3:* 577–605 (1966).

SUMI, T.: Functional differentiation of hypoglossal neurones in cats. Jap. J. Physiol. *19:* 68–79 (1969).

SUZUKI, S.: Muscle contraction during vibration. Tairyoku Kagaku *10:* 106 (1961) (in Japanese).

SWETT, J.E. and ELDRED, E.: Comparisons in structure of stretch receptors in medial gastrocnemius and soleus muscles of the cat. Anat. Rec. *137:* 461–473 (1960).

SZABO, I.: Magnitude of sound-induced startle response as a function of primary hunger drive. Acta physiol. Acad. Sci. hung. *32:* 241–252 (1967).

SZENTÁGOTHAI, J.: Die zentrale Innervation der Augenbewegungen. Arch. Psychiat. Nervenkr. *116:* 721–760 (1943).

SZENTÁGOTHAI, J.: Anatomical aspects of inhibitory pathways and synapses; in FLOREY Nervous inhibitions, pp. 32–46 (Pergamon, Oxford 1961a).

SZENTÁGOTHAI, J.: Somatotopic arrangement of synapses of primary sensory neurons in Clarke's colum. Acta morph. Acad. Sci. hung. *10:* 307–311 (1961b).

SZENTÁGOTHAI, J.: Propriospinal pathways and their synapses, in ECCLES and SCHADÉ Organization of the spinal cord. Progr. Brain Res., *11:* 155–177 (Elsevier, Amsterdam 1964a).

SZENTÁGOTHAI, J.: Neuronal and synaptic arrangement in the substantia gelatinosa Rolandi. J. comp. Neurol. *122:* 219–240 (1964b).

SZENTÁGOTHAI, J.: The anatomy of complex integrative units in the nervous system; in LISSAK Results in neuroanatomy, neurohistology, neuromorphology and neurophysiology, pp. 9–45 (Akademiai Kiadó, Budapest 1967a).

SZENTÁGOTHAI, J.: Synaptic architecture of the spinal motoneuron pool. Electroenceph. clin. Neurophysiol., suppl. *25:* 4–19 (1967b).

SZENTÁGOTHAI, J. and ALBERT, A.: The synaptology of Clark's column. Acta morph. Acad. Sci. hung. *5:* 43–51 (1955).

SZENTÁGOTHAI, J. and KISS, T.: Projections of dermatomes on the substantia gelatinosa. Arch. Neurol. Psychiat., Chicago *62:* 734–744 (1949).

SZENTÁGOTHAI, J. und SCHIMERT, J.: Die Endigungsweise der absteigenden Rückenmarksbahnen. Z. Anat. EntwGesch. *111:* 322–330 (1941).

TABARY, J.C.; TARDIEU, C.; TARDIEU, G. et CHANTRAINE, A.: Etude cinématographique et EMG du maintien postural avec changement de charge. J. Physiol., Paris *57:* 799–810 (1965).

TÁBOŘÍKOVÁ, H.; DECANDIA, M., and PROVINI, L.: Evidence that muscle stretch evokes long-loop reflexes from higher centres. Brain Res. *2:* 192–194 (1966).

TÁBOŘÍKOVÁ, H. and SAX, D.S.: Motoneurone pool and the H-reflex. J. Neurol. Neurosurg. Psychiat. *31:* 354–361 (1968).

TÁBOŘÍKOVÁ, H. and SAX, D.S.: Conditioning of H-reflex by a preceeding sub-threshold H-reflex stimulus. Brain *92:* 203–212 (1969).

TAKAMORI, M.H.: Reflex study in upper motoneuron disease. Neurology, Minneap. *17:* 32–40 (1967).

TALBOT, H.S.: A report on sexual function in paraplegics. J. Urol. *61:* 265–270 (1949).

TARDIEU, C.; TABARY, J.C. et TARDIEU, G.: Etude mécanique et électromyographique des réponses à différentes perturbations du maintien postural. J. Physiol., Paris *60:* 243–260 (1968).

TAUB, E. and BERMAN, A.J.: Movement and learning in the absence of sensory feedback; in FREEDMAN The neuropsychology of spatially oriented behavior, pp. 173–192 (Dorsey Press, Illinois 1968).

TAYLOR, A.: The contribution of the intercostal muscles to the effort of respiration in man. J. Physiol., Lond. *151:* 390–402 (1960).

TEASDALL, R.D.; LANGUTH, H.W., and MAGLADERY, J.W.: Electrophysiological studies of reflex activity in patients with lesions of the nervous system. IV. A note on the tendon jerk. Johns Hopk. Hosp. Bull. *91:* 267–275 (1952).

TEASDALL, R.D. and MAGLADERY, J.W.: Superficial abdominal reflexes in man. Arch. Neurol. Psychiat., Chicago *81:* 28 (1959).

TEASDALL, R.D.; MAGLADERY, J.W., and RAMEY, E.: Changes in reflex patterns following spinal cord hemisections in cats. Johns Hopk. Hosp. Bull. *103:* 223–236 (1958).

TEASDALL, R.D.; PARK, A.M.; LANGUTH, H.W., and MAGLADERY, J.W.: Disclosure of normally suppressed monosynaptic reflex discharge of spinal motoneurones by lesions of lower brain stem and spinal cord. Johns Hopk.Hosp.Bull. *91:* 245–256 (1952).

TEBECIS, A.K. and DiMARIA, A.: Strychnine-sensitive inhibition in the medullary reticular formation: evidence for glycine as an inhibitory transmitter. Brain Res. *40:* 373–383 (1972).

TESTA, C.: Functional implication of the morphology of spinal ventral horn neurones of the cat. J. comp. Neurol. *123:* 425–444 (1964).

THACH, W.T.: Discharge of Purkinje and cerebellar nuclear neurons during rapidly alternating arm movements in the monkey. J. Neurophysiol. *31:* 785–797 (1968).

THACH, W.T.: Discharge of cerebellar neurons related to two maintained postures and two prompt movements. I. Nuclear cell output. J. Neurophysiol. *33:* 527–536 (1970a).

THACH, W.T.: Discharge of cerebellar neurons related to two maintained postures and two prompt movements. II. Purkinje cell output and input. J.Neurophysiol. *33:* 537–547 (1970b).

THACH, W.T.: Cerebellar output: properties, synthesis and uses. Brain Res. *40:* 89–97 (1972).

THIÉBAUT, F. et ISCH, F.: Etude clinique, cinématographique et bio-électrique de quatre formes d'athétose. Rev. neurol. *87:* 26–40 (1952).

THODEN, U.; MAGHERINI, P.C., and POMPEIANO, O.: Evidence that presynaptic inhibition may decrease the autogenetic excitation caused by vibration of extensor muscles. Arch. ital. Biol. *110:* 90–116 (1972).

THOMAS, J.E.: Muscle tone, spasticity and rigidity. J.nerv.ment.Dis. *132:* 505–514 (1961).

THOMAS, D.M.; KAUFMAN, R.P.; SPRAGUE, J.M., and CHAMBERS, W.W.: Experimental studies of the vermal cerebellar projections in the brain stem of the cat (fastigiobulbar tract). J. Anat., London *90:* 371–385 (1956).

THOMAS, J.E. and LAMBERT, E.H.: Ulnar nerve conduction velocity and H-reflex in infants and children. J. appl. Physiol. *15:* 1–9 (1960).

THOMPSON, R.F. and SPENCER, W.A.: Habituation: a model phenomenon for the study of neuronal substrates of behaviour. Psychol. Rev. *73:* 16–43 (1966).

THORNE, J.: Central responses to electrical activation of the peripheral nerves supplying the intrinsic hand muscles. J. Neurol. Neurosurg. Psychiat. *28:* 482–485 (1965).

THULIN, C.A.: Effects of electrical stimulation of the red nucleus on the alpha motor system. Exp. Neurol. *7:* 464–484 (1963).

TOKIZANE, T. and SHIMAZU, H.: Functional differentiation of human skeletal muscle. Corticalization of movement (University of Tokyo Press, Tokyo 1964).

TOKUNAGA, A.; OKA, M.; MURAO, T.; YOKOI, H.; OKUMURA, T.; HIRATA, T.; MIGASHITA, Y., and YOSHITATSU, S.: An experimental study on facial reflex by evoked electromyography. Med. J. Osaka Univ. *9:* 397–411 (1958).

Torvik, A.: Afferent connections to the sensory trigeminal nuclei, the nucleus of the solitary tract and adjacent structures. J. comp. Neurol. *106:* 51–141 (1956).

Torvik, A. and Brodal, A.: The origin of reticulospinal fibers in the cat. An experimental study. Anat. Rec. *128:* 113–137 (1957).

Towe, A. L. and Jabbur, S. J.: Cortical inhibition of neurons in dorsal columni nuclei of cat. J. Neurophysiol. *24:* 488–498 (1961).

Tower, S. S.: Pyramidal lesion in the monkey. Brain *63:* 36–90 (1940).

Tower, S. S.: The pyramidal tract; in Bucy The precentral motor cortex, pp. 149–172 (University of Illinois Press, Urbana 1949).

Trontelj, J.: H-reflex of single motoneurones in man. Nature, Lond. *220:* 1043–1044 (1968).

Tureen, L. L.: Form of the reflex response in relation to the pattern of afferent stimulation. Proc. Soc. exp. Biol. Med. *K6:* 543–550 (1941).

Tuttle, W. W.: The effect of sleep upon the patellar tendon reflex. Amer. J. Physiol. *68:* 345–348 (1924).

Twitchell, T. E.: Sensory factors in purposive movement. J. Neurophysiol. *17:* 239–252 (1954).

Uno, M.; Yoshida, M., and Hirota, J.: The mode of cerebello-thalamic relay transmission investigated with intracellular recording from cells of the ventrolateral nucleus of cat's thalamus. Exp. Brain Res. *10:* 121–139 (1970).

Ushiyama, J.; Koizumi, K., and Broks, C. McC: Accomodative reactions of neuronal elements in the spinal cord. J. Neurophysiol. *29:* 1028–1045 (1966).

Uttley, A. M.: The probability of neural connexions. Proc. roy. Soc. B *144:* 229–240 (1955).

Vallbo, Å. B.: Slowly adapting muscle receptors in man. Acta physiol. scand. *78:* 315–333 (1970a).

Vallbo, Å. B.: Discharge patterns in human muscle spindle afferents during isometric voluntary contraction. Acta physiol. scand. *80:* 552–566 (1970b).

Vallbo, Å. B.: Muscle spindle response at the onset of isometric voluntary contraction in man. Time difference between fusimotor and skeletomotor effects. J. Physiol., Lond. *218:* 405–431 (1971).

Vallbo, Å. B. and Hagbarth, K. E.: Impulses recorded with micro-electrodes in human muscle nerves during stimulation of mechanoreceptors and voluntary contractions. Electroenceph. clin. Neurophysiol. *23:* 392 (1967).

Vallbo, Å. B. and Hagbarth, K. E.: Activity from skin mechanoreceptors recorded percutaneously in awake human subjects. Exp. Neurol. *21:* 270–289 (1968).

Van der Meulen, J. P. and Gilman, S.: Recovery of muscle spindle activity in cats after cerebellar ablation. J. Neurophysiol. *28:* 943–957 (1965).

Vasilenko, D. A.; Zadorozhny, A. G., and Kostyuk, P. G.: Synaptic processes in the spinal neurons, monosynaptically activated by the pyramidal tract. Bull. exp. Biol. Med. USSR *64:* 20–25 (1967) (in Russian).

Veale, J. L.; Mark, R. F., and Rees, S. M.: Differential stimulation of the human ulnar nerve. Proc. austr. Phys. Pharm. Soc. *2:* 85 (1971).

Vedel, J. P. et Mouillac-Baudevin, J.: Contrôle de l'activité des fibres fusimotrices dynamiques et statiques par la formation réticulée mésencéphalique chez le chat. Exp. Brain Res. *9:* 307–324 (1969a).

VEDEL, J.P. et MOUILLAC-BAUDEVIN, J.: Etude fonctionnelle du contrôle de l'activité des fibres fusimotrices dynamiques et statiques par les formations réticulées mésencéphalique, pontique et bulbaire chez le chat. Exp. Brain Res. 9: 325–345 (1969b).

VEDEL, J.P. et MOUILLAC-BAUDEVIN, J.: Contrôle pyramidal de l'activité des fibres fusimotrices dynamiques et statiques chez le chat. Exp. Brain Res. 10: 39–63 (1970).

VISSER, S.L.: Influence of L-DOPA on the EEG and EMG of Parkinson patients. Electroenceph. clin. Neurophysiol. 31: 107P–108P (1971).

VISSER, S.L. and POSTMA, J.U.: Influence of L-Dopa on the EEG and EMG in Parkinson patients. Psychiat. Neurol. Neurochir. 74: 315–321 (1971).

VIZOSO, A.D. and YOUNG, J.Z.: Internodal length and fibre diameter in developing and regenerating nerves. J. Anat., Lond. 82: 110–124 (1948).

VOOGD, J.: The cerebellum of the cat, structure and fibre connexions, p.215 (Van Gorcum, Assen 1964).

WAGMAN, I.H.: Eye movements induced by electrical stimulation of cerebrum in monkeys and their relationship to bodily movements; in BENDER The oculomotor system, pp. 18–39 (Harper & Row, New York 1964).

WAGMAN, I.H.; PIERCE, D.S., and BURGER, R.E.: Proprioceptive influence in volitional control of individual motor units. Nature, Lond. 207: 957–958 (1965).

WAGMAN, I.H. and PRICE, D.D.: Responses of dorsal horn of Macaca mulatta to cutaneous and sural nerve A and C fiber stimuli. J. Neurophysiol. 32: 803–817 (1969).

WALBERG, F.: Corticofugal fibers to the nuclei of the dorsal columns. An experimental study in the cat. Brain 80: 273–287 (1957).

WALBERG, F. and JANSEN, J.: Cerebellar corticovestibular fibers in the cat. Exp. Neurol. 3: 32–52 (1961).

WALBERG, F.; POMPEIANO, O.; BRODAL, A., and JANSEN, J.: The fastigiovestibular projection in the cat. An experimental study with silver impregnation methods. J. comp. Neurol. 118: 49–75 (1962).

WALBERG, F.; POMPEIANO, O.; WESTRUM, L.E., and HANGLIE-HANSSEN, E.: Fastigioreticular fibers in the cat. An experimental study with silver methods. J. comp. Neurol. 119: 187–199 (1962).

WALKER, A.E.: Oscillographic study of cerebello-cerebral relationships. J. Neurophysiol. 1: 16–23 (1938).

WALKER, A.E.: The primate thalamus, p.305 (University of Chicago Press, Chicago 1938).

WALKER, A.E.: Cerebral pedunculotomy for the relief of involuntary movements: hemiballismus. Acta psychiat. scand. 24: 723–726 (1949).

WALKER, A.E. and RICHTER, H.: Section of the cerebral peduncle in the monkey. Arch. Neurol., Chicago 14: 231–240 (1966).

WALKER, L.B., jr.: Diameter spectrum of intrafusal muscle fibers in muscle spindles of the dog. Anat. Rec. 130: 385 (1958).

WALL, P.D.: The laminar organization of the dorsal horn and effects of descending impulses. J. Physiol., Lond. 188: 403–423 (1967).

WALL, P.D. and SWEET, W.H.: Temporary abolition of pain in man. Science 155: 108–109 (1967).

WALSHE, F.M.R.: On the genesis and physiological significance of spasticity and other disorders of motor innervation with a consideration of the functional relationships of the pyramidal tract. Brain 42: 1–28 (1919).

WALSHE, F. M. R.: The physiological significance of the reflex phenomenon in spastic paralysis of lower limbs. Brain *37:* 269–336 (1914).

WALSHE, F. M. R.: Oliver-Sharpey Lectures on the physiological analysis of some clinically observed disorders of movement. Lancet *I:* 963–968 (1929).

WARTENBERG, R.: Studies in reflexes. History, physiology synthesis and nomenclature: study 1. Arch. Neurol. Psychiat., Chicago *51:* 113–133 (1944).

WARTENBERG, R.: The examination of reflexes: a simplification (Year Book, Chicago 1945).

WEAVER, R. A.; LANDAU, W. M., and HIGGINS, J. F.: Fusimotor function. 2. Evidence of fusimotor depression in human spinal shock. Arch. Neurol., Chicago *9:* 127–132 (1963).

WEBSTER, D. D.: The dynamic quantitation of spasticity with automated integrals of passive motion resistance. Clin. Pharmacol. Ther. *5:* 900–908 (1964).

WEDDELL, G.; FEINSTEIN, B., and PATTLE, R. E.: The electrical activity of voluntary muscle in man under normal and pathological conditions. Brain *67:* 178–257 (1944).

WEINMANN, H. M.; MEITINGER, CH. und VLACH, V.: Polygraphische Untersuchungen exterozeptiver Reflexe bei Frühgeborenen. Z. Kinderheilk. *107:* 74–86 (1969).

WEISS, P.: Experimental analysis of coordination by the disarrangement of central-peripheral relations. Symp. Soc. Exp. Biol. *4:* 92–111 (1950).

WEIZSÄCKER, VON: Über Willkürbewegungen und Reflexe bei Erkrankungen des Zentralnervensystems. Dtsch. Z. Nervenheilk. *70:* 115–130 (1921).

WERNER, G. and MOUNTCASTLE, V. B.: The variability of central neural activity in a sensory system, and its implications for the central reflections of sensory events. J. Neurophysiol. *26:* 958–977 (1963).

WERTHEIM SALOMONSON, J. K. A.: Verkürzungsreflexe. Neurolog. Centralblatt *33:* 1180–1188 (1914).

WERTHEIM SALOMONSON, J. K. A.: Tonus and the reflexes. Brain *43:* 369–389 (1920).

WESTBURY, D. R.: A comparison of stretch and vibration responses at the motoneurone. J. Physiol., Lond. *213:* 25P–26P (1971).

WESTPHAL, C.: Über einige Bewegungserscheinungen an gelähmten Gliedern. Arch. Psychiat. Nervenkr. *5:* 803–834 (1875).

WESTPHAL, C.: Unterschenkelphänomen und Nervendehnung. Arch. Psychiat. Nervenkr. *7:* 666–670 (1877).

WESTPHAL, C.: Über eine Art paradoxer Muskelcontraction. Arch. Psychiat. Nervenkr. *10:* 243–248 (1880).

WIESENDANGER, M.: Rigidity produced by deafferentiation. Acta physiol. Scand. *62:* 160–168 (1964).

WIESENDANGER, M.: The pyramidal tract. Recent investigations on its morphology and function. Ergebn. Physiol. *61:* 73–136 (1969).

WIESENDANGER, M.: Effects of electrical stimulation of peripheral nerves to the hand and forearm on pyramidal tract neurones of the baboon. Brain Res. *40:* 193–197 (1972).

WIESENDANGER, M. und TARNECKI, R.: Die Rolle des pyramidalen Septums bei der sensomotorischen Integration. Bull. schweiz. Akad. med. Wiss. *22:* 306–328 (1966).

WIESER, S.; DOMANOWSKY, K. und HEINEN, G.: Morosche Reflex- und Schreckreaktion beim Säugling. Arch. Kinderheilk. *155:* 17–23 (1957).

WILLIS, W. D. and WILLIS, J. C.: Properties of interneurones in the ventral spinal cord. Arch. ital. Biol. *104:* 354–386 (1966).

WILSON, V. J.: Recurrent facilitation of spinal reflexes. J. gen. Physiol. *42:* 703–713 (1959).

WILSON, V.J.: Reflex transmission in the kitten. J. Neurophysiol. *25:* 263–275 (1962).

WILSON, V.J.: Regulation and function of Renshaw cells discharge; in GRANIT Nobel Symposium 1, Muscular afferents and motor control pp. 317–329 (Almqvist & Wiksell, Stockholm 1966).

WILSON, V.J. and BURGESS, P.R.: Disinhibition in the cat spinal cord. J. Neurophysiol. *25:* 392–404 (1962).

WILSON, V.J.; TALBOT, W.H., and DIEKE, F.P.J.: Distribution of recurrent facilitation and inhibition in cat spinal cord. J. Neurophysiol. *23:* 144–153 (1960).

WILSON, V.J.; TALBOT, W.H., and KABO, M.: Inhibition convergence upon Renshaw cells. J. Neurophysiol. *27:* 1063–1079 (1964).

WILSON, V.J. and YOSHIDA, M.: Vestibulospinal and reticulospinal effects on hindlimb, forelimb and neck alpha motoneurons of the cat. Proc. nat. Acad. Sci., Wash. *60:* 836–840 (1968).

WILSON, V.J. and YOSHIDA, M.: Monosynaptic inhibition of neck motoneurons by the medial vestibular nucleus. Exp. Brain Res. *9:* 365–380 (1969a).

WILSON, V.J. and YOSHIDA, M.: Comparison of effects of stimulation of Deiter's nucleus and medial longitudinal fasciculus on neck, forelimb and hindlimb motoneurons. J. Neurophysiol. *32:* 743–758 (1969b).

WINDLE, W.F.: Physiology of the fetus: origin and extent of function of prenatal life (Saunders, Philadelphia 1940).

WOOLSEY, C.N.; GORSKA, T.; WETZEL, A.; ERICKSON, T.C.; EARLS, F.J., and ALLMAN, J.M.: Complete unilateral section of the pyramidal tract at the medullary level in *Macaca mulatta*. Brain Res. *40:* 119–123 (1972).

WOOLSEY, C.N.; SETTLAGE, P.H.; MEYER, D.R.; SPENCER, W.; HAMUY, T.P., and TRAVIS, A.M.: Patterns of localization in precentral and 'supplementary' motor areas and their relation to the concept of a premotor area. Res. Publ. Ass. nerv. ment. Dis. *30:* 238–264 (1950).

WUERKER, R.B.; McPHEDRAN, A.M., and HENNEMAN, E.: Properties of motor units in a heterogeneous pale muscle (m. gastrocnemius) of the cat. J. Neurophysiol. *28:* 85–99 (1965).

YAMANAKA, T.: Effect of high frequency vibration on muscle spindle in the human body. J. Chiba med. Soc. *40:* 338–346 (1964).

YAP, C.-B.: Spinal segmental and long-loop reflexes on spinal motoneurone excitability in spasticity and rigidity. Brain *90:* 887–896 (1967).

YELLIN, H. and GUTH, L.: The histochemical classification of muscle fibers. Exp. Neurol. *26:* 424–432 (1970).

YOKOTA, T. and VOORHOEVE, P.E.: Pyramidal control of fusimotor neurons supplying tensor muscles in the cat's forelimb. Exp. Brain Res. *9:* 96–115 (1969).

YOSHIDA, M.; YAJIMA, K., and UNO, M.: Different activation of the two types of the pyramidal tract neurons through the cerebello-thalamo-cortical pathway. Experientia *22:* 331–332 (1966).

ZANDER OLSEN, P. and DIAMANTOPOULOS, E.: Excitability of spinal motor neurones in normal subjects and patients with spasticity, Parkinsonian rigidity, and cerebellar hypotonia. J. Neurol. Neurosurg. Psychiat. *30:* 325–331 (1967).

ZBINDEN, G. and RANDALL, L.O.: Pharmacology of benzodiazepines. Adv. Pharmacol. *5:* 213–291 (1967).

Appendix

New Developments in Electromyography and Clinical Neurophysiology,
edited by J. E. Desmedt, vol. 3, pp. 847–852 (Karger, Basel 1973)

The Brussels International Congress of Electromyography

J. E. DESMEDT

Brain Research Unit, University of Brussels, Brussels

The Fourth International Congress of Electromyography was held in Brussels, Belgium, September 12–15, 1971. The congress placed under the High Protection of H. M. BAUDOUIN, King of the Belgians was sponsored by the International Federation of Societies for Electroencephalography and Clinical Neurophysiology (IFSECN). The International Society for Electromyographic Kinesiology (ISEK) collaborated by holding a Regional Meeting which was integrated in the same program and the recently established European Alliance of Muscular Dystrophy Associations (EAMDA) organized a one-day session devoted to current social problems of patients with neuromuscular diseases.

The Brussels EMG Congress was the fourth of a series initiated with the First International EMG Congress held in Pavia (1961) and continued with the Second (Copenhagen 1963) and Third (Glasgow 1967). The current trend of these meetings has been to promote critical and detailed discussions of recent EMG methodologies and concepts. Their function can be viewed as complementary to the EMG participations in the bigger congresses of neurology, physical medicine, …in which EMG results are presented to fellow scientists and colleagues sothat the diagnostic or other contributions of EMG become more visible from the outside.

The special impact of the Brussels congress was the result of EMG being now at a critical stage of its growth and expansion. It is also related to a careful design of the scientific sessions and to the enthusiastic participation of so many experts in both clinical neurophysiology and basic neurosciences. The active part played by eminent scientists like Sir JOHN ECCLES (Buffalo, N. Y.), ANDREW F. HUXLEY (London), BERNHARD FRANKENHAEUSER (Stockholm), CHARLES PHILLIPS (Oxford), SHIRLEY H. BRYANT (Cincinnati, Ohio), PETER

B. C. Matthews (Oxford), Robert E. Burke (Bethesda, Md.) and A. Paintal (Delhi) (this list includes only persons with no personal experience with human patients) contributed to reach wider multidisciplinary perspectives. EMG data have now reached adequate consistency and precision sothat such strong interactions with the basic neurosciences become feasible on the large scale of an international congress.

Each participant in the Brussels congress came back home with the idea that EMG had achieved a significant step forward, that new methods were available both for research and for diagnostic practice, and that the current problems had been put into new perspectives. Many new data had indeed been acquired since 1969 and were discussed at the congress for the first time. For example, the plenary session on New Concepts of the Motor Unit marked an important date in the development of basic concepts of the discipline. This symposium was attended by over 750 persons. The symposia on Computer EMG Analysis and on Human Reflexes also attracted unsuspectedly large audiences and an explosive development can probably be anticipated in these areas. Paul Hoffmann studied the H-reflex before Hans Berger discovered the Electroencephalogram and it is now time for the human reflex studies to establish their territories among the clinical procedures derived from neurophysiology.

Table I. International congresses of electromyography: quantitative data

	Copenhagen 1963	Glasgow 1967	Brussels 1971
Active members	201	230	572
Papers (by topic)			
Motor control and human reflexes	3	20	70
Clinical EMG and nerve stimulation, Motor units, etc.	27	27	57
Nerve conduction and Neuropathies	16	19	48
EMG kinesiology	0	4	30
Computer EMG analysis	0	4	7
Total	46	74	212
Rejected papers (read by title)	20	20	0

CE DOCUMENT A ETE DÉLIVRÉ AU DOCTEUR

pour avoir participé au cours d'électromyographie organisé dans le cadre du 4ᵉ Congrès International d'EMG, tenu à Bruxelles du 12 au 16 septembre 1971.

Bruxelles, 16 septembre 1971.

J. DEBECKER
Secrétaire

J. E. DESMEDT
Président

FACULTÉ:

Dr J. Bergmans (Leuven)	Dr J. Ekstedt (Uppsala)
Dr S. Borenstein (Bruxelles)	Dr K. E. Hagbarth (Uppsala)
Dr P. Brink Henriksen (Copenhagen)	Dr M. Hugon (Marseille)
Dr A. Chantraine (Liège)	Dr E. H. Lambert (Rochester, USA)
Dr J. Debecker (Bruxelles)	Dr C. Penders (Liège)
Dr L. Delhez (Liège)	Dr N. Rosselle (Leuven)
Dr P. Delwaide (Liège)	Dr E. Stalberg (Uppsala)
Dr J. Deschuytere (Leuven)	Dr G. Wallin (Uppsala)
Dr J. E. Desmedt	Dr R. Willison (London)

The present volumes have drawn freely on the results, expertise and enthusiasm displayed at the Brussels congress and they should no doubt further accelerate the progress of EMG by rapidly disseminating an up-dated wealth of information.

Quantitative data presented in table I disclose interesting trends in the last three international EMG congresses. The number of active members more than doubled at Brussels and the number of scientific papers also increased considerably. The policy followed in the Brussels congress has been *not* to reject any submitted voluntary paper. In spite of this policy the general level of the free communications was very good, indeed much better than it was ten years ago, which bears testimony to the coming of age of EMG and to the remarkable improvement in methodologies and in electronic equipment.

On the day following the congress (September 16) a Practical EMG Course was organized to demonstrate on volunteer subjects and patients with neuromuscular diseases the new procedures discussed during the meeting. The number of registrants had to be limited to 150 and 12 demonstrations were organized in separate rooms. An international faculty was appointed for the course and a diploma featuring the celebrated Brussels anatomist VESALIUS (fig. 1) was issued later to the participants. Among the highlights of the course were: microneurography in normal man, computer EMG analysis, microphysiology of the motor unit and the jitter phenomenon, diagnostic EMG on patients with sacral root lesions and sphincter disorders, proprioceptive and blink reflexes in man, sensory nerve potentials, selective stimulation of single motor units in intact man, new diagnostic procedures for myasthenia gravis, lumber reflexes in root compression syndromes, exteroceptive reflexes elicited in man by sural nerve stimulation and clinical neurophysiology of the phrenic and intercostal nerves. These demonstrations were attended throughout the day and followed by a round-table discussion of current diagnostic problems.

Officers of the Congress

Executive Committee

President: Prof. J. E. Desmedt, Neurophysiology and Clinical Neurophysiology, University of Brussels
Vice-Presidents: Prof. X. Aubert, Physiology, University of Louvain
    Prof. H. Claessens, Orthopaedics and Physical Medicine, University of Ghent
    Prof. E. Colinet, Physical Medicine, University of Brussels
    Prof. W. Esser, Physical Medicine, University of Liège
    Prof. N. Rosselle, Electromyography and Physical Medicine, University of Louvain
    Prof. H. van Cauwenberge, Medicine, University of Liège
    Prof. H. Vander Eecken, Neurology, University of Ghent
Secretary: Dr. K. Hainaut, Neurophysiology, University of Brussels
Treasurer: Prof. L. Franken, Neurology, University of Brussels
Local Representative of ISEK: Prof. J. Debecker, Neurophysiology and clinical neurophysiology, University of Brussels

Program Committee

Dr. J. Bergmans (Louvain)
Dr. A. Bonne (Louvain)
Dr. A. Chantraine (Liège)
Dr. C. Claes (Antwerp)
Prof. J. Debecker (Brussels)
Dr. L. Delhez (Liège)
Dr. P. J. Delwaide (Liège)
Dr. J. Deschuytere (Schelle)
Prof. J. E. Desmedt (Brussels)
Dr. J. Dumoulin (Charleroi)
Dr. M. Ectors (Brussels)

Dr. M. Faes (Louvain)
Dr. J. M. Gillis (Louvain)
Prof. J. Gijbels (Louvain)
Dr. K. Hainaut (Brussels)
Dr. J. Mortier (Ghent)
Dr. P. Noël (Brussels)
Dr. C. Penders (Liège)
Prof. N. Rosselle (Louvain)
Dr. A. Stevens (Louvain)
Dr. L. Vanden Bulcke (Bruges)

Sponsored by the

International Federation of Societies for Electroencephalography and Clinical Neurophysiology (IFSECN)
President: C. Ajmone-Marsan (Bethesda, Md.)
Past-President: A. Remond (Paris)
Secretary: R. J. Ellingson (Omaha, Nebr.)
Treasurer: R. Hess (Zürich)
Editor of Electroenceph. clin. Neurophysiol.: W. A. Cobb (London)
Member at large: D. H. Ingvar (Lund)
Chairman of EMG Commission: F. Buchthal (Copenhagen)
EMG Commission: F. Buchthal, J. E. Desmedt and F. Isch

Host to the Congress

Belgian Society of Electromyography
(Member of IFSECN in Belgium)

Honorary Members:
Prof. Ph. Bauwens (Ostend)   Prof. A. F. Huxley (London)
Prof. F. Bremer (Brussels)   Prof. F. Isch (Strasbourg)
Prof. Sir John Eccles (Buffalo, N.Y.)   Prof. E. H. Lambert (Rochester, Minn.)
Prof. W. Esser (Liège)   Prof. W. T. Liberson (Miami, Fla.)
Prof. R. W. Gilliatt (London)   Prof. T. Tokizane (Tokyo)
Prof. K. E. Hagbarth (Uppsala)   Prof. P. van Gehuchten (Louvain)

Committee:
President: Prof. E. Colinet (Brussels)
Secretary: Prof. J. E. Desmedt (Brussels)
Publications: Prof. N. Rosselle (Louvain)
Members: Dr. F. Bostem (Liège)
        Prof. H. Claessens (Ghent)
        Prof. L. Franken (Brussels)

Six symposia were organized during the congress:

1. New Concepts of the Motor Unit:
   Chairmen: E. H. Lambert (Rochester, Minn.) and T. Tokizane (Tokyo)
   Secretary: S. Borenstein (Brussels)
2. Computer EMG Analysis:
   Chairman: R. Willison (London)
   Secretary: J. Bergmans (Louvain)
3. Intracellular Electromyography:
   Chairman: A. F. Huxley (London)
   Secretary: J. M. Gillis (Louvain)

4. Pathological Conduction in Nerve Fibres:
   Chairman: R.W. Gilliatt (London)
   Secretary: P. Noel (Brussels)
5. Electromyography in Biomechanical Studies (ISEK Regional Meeting):
   Chairmen: B. Jonsson (Göteborg) and S. Bouisset (Lille)
   Secretary: H. Faes (Louvain)
6. Human Reflexes:
   Chairmen: Sir John Eccles (Buffalo, N.Y.) and K.E. Hagbarth (Uppsala)
   Secretary: J. Debecker (Brussels)

Sessions of voluntary communications:[1]

Electromyography and motor units:
   Chairmen: B. Drechsler (Würzburg), W.K. Engel (Bethesda, Md.), F. Isch (Strasbourg), J. Lefèbvre (Paris) and J. Lerique (Paris)
   Secretaries: J. Deschuytere (Louvain), M. Ectors (Brussels)
Nerve conduction:
   Chairmen: F. Buchthal (Copenhagen), D. Grob (New York, N.Y.), I. Hausmanowa-Petrusewicz (Warsaw), H.E. Kaeser (Basel), J. Simpson (Glasgow), and D. Van der Most van Spyk (Utrecht)
   Secretaries: C. Claes (Antwerp), G. Mortier (Ghent), L. Vanden Bulcke (Bruges)
Electromyographic Kinesiology (ISEK Regional Meeting):
   Chairmen: J.V. Basmajian (Atlanta, Ga.), J. Joseph (London), A. Franklin (Toronto)
   Secretary: A. Stevens (Louvain)
Human Reflexes:
   Chairmen: P. Delwaide (Liège), M.M. Gassel (San Francisco), M. Hugon (Marseille), J.A. Lenman (Dundee), G. Rushworth (Oxford), R.R. Young (Boston, Mass.)
   Secretaries: P. Delwaide (Liège), M. Ectors (Brussels), C. Penders (Liège)
Microphysiology of muscle:
   Chairman: S.H. Bryant (Cincinnati, Ohio)
   Secretary: J.M. Gillis (Louvain)

1   The volume of abstracts of communications presented at the Brussels congress (given to each registrant) can still be purchased (300 B.F.) by writing to Mme Lemaire, Brain Research Unit, 115, boulevard de Waterloo, B-1000 Brussels.

Authors' Index

The figure before the slant line indicates the volume number, the figure after it the page number in the respective volume.

Subject Index

The figures in parentheses refer to the page numbers of the relevant volume, the number of which is stated before the parentheses.